# The Handbook of Displacement

Peter Adey • Janet C. Bowstead
Katherine Brickell • Vandana Desai
Mike Dolton • Alasdair Pinkerton
Ayesha Siddiqi
Editors

# The Handbook of Displacement

palgrave
macmillan

*Editors*
Peter Adey
Department of Geography
Royal Holloway, University of London
Egham, UK

Katherine Brickell
Department of Geography
Royal Holloway, University of London
Egham, UK

Mike Dolton
Department of Geography
Royal Holloway, University of London
Egham, UK

Ayesha Siddiqi
Department of Geography
University of Cambridge
Cambridge, UK

Janet C. Bowstead
Department of Geography
Royal Holloway, University of London
Egham, UK

Vandana Desai
Department of Geography
Royal Holloway, University of London
Egham, UK

Alasdair Pinkerton
Department of Geography
Royal Holloway, University of London
Egham, UK

ISBN 978-3-030-47177-4     ISBN 978-3-030-47178-1  (eBook)
https://doi.org/10.1007/978-3-030-47178-1

© The Editor(s) (if applicable) and The Author(s) 2020
This work is subject to copyright. All rights are solely and exclusively licensed by the Publisher, whether the whole or part of the material is concerned, specifically the rights of translation, reprinting, reuse of illustrations, recitation, broadcasting, reproduction on microfilms or in any other physical way, and transmission or information storage and retrieval, electronic adaptation, computer software, or by similar or dissimilar methodology now known or hereafter developed.
The use of general descriptive names, registered names, trademarks, service marks, etc. in this publication does not imply, even in the absence of a specific statement, that such names are exempt from the relevant protective laws and regulations and therefore free for general use.
The publisher, the authors and the editors are safe to assume that the advice and information in this book are believed to be true and accurate at the date of publication. Neither the publisher nor the authors or the editors give a warranty, expressed or implied, with respect to the material contained herein or for any errors or omissions that may have been made. The publisher remains neutral with regard to jurisdictional claims in published maps and institutional affiliations.

Cover illustration: Mary Turner / Contributor/getty images

This Palgrave Macmillan imprint is published by the registered company Springer Nature Switzerland AG.
The registered company address is: Gewerbestrasse 11, 6330 Cham, Switzerland

# Acknowledgements

We are grateful to the Department of Geography at Royal Holloway, University of London, for the financial support provided through the Geopolitics, Development, Security and Justice Research Group for us to go away together to plan the handbook and to assist us with extra administrative and copyediting capacity needed in the final stages of submission. Thank you to Carolyn Dodds for the dedicated work on the handbook in the final stages.

# Contents

1 Introduction to Displacement Studies: Knowledges, Concepts, Practices ... 1
*Peter Adey, Janet C. Bowstead, Katherine Brickell, Vandana Desai, Mike Dolton, Alasdair Pinkerton, and Ayesha Siddiqi*

**Part I   Section One: Conceptualising Displacement** ... 39

2 Mobilities and Displacement ... 41
*Mimi Sheller*

3 Political Ecologies of Displacement ... 55
*Rebecca Elmhirst*

4 Displacement Economies: A Relational Approach to Displacement ... 67
*Amanda Hammar*

5 The Slow and the Fast Violence of Displacement ... 79
*James A. Tyner*

6 Assembling Climate Change-Related Displacement ... 89
*Leonie Tuitjer*

7 Affect and Displacement ... 99
*Mark Griffiths*

| | | |
|---|---|---|
| 8 | **Protection of Displaced Persons and the Rights-Based Approach**<br>*Rónán McDermott, Pat Gibbons, and Sinéad McGrath* | 109 |
| 9 | **Queering Displacement/The Displacement of Queers**<br>*Scott McKinnon* | 121 |
| 10 | **Gendered and Feminist Approaches to Displacement**<br>*Katherine Brickell and Jessie Speer* | 131 |
| 11 | **'Race,' Ethnicity, and Forced Displacement**<br>*Luisa F. Freier, Matthew D. Bird, and Soledad Castillo Jara* | 143 |
| 12 | **Conceptualising Postcolonial Displacement Beyond Aid and Protection**<br>*Jose Jowel Canuday* | 157 |

| | | |
|---|---|---|
| Part II | **Section Two: Technologies of Displacement** | 171 |
| 13 | **Intervention: Displacement Aesthetics**<br>*Kaya Barry and Peter Adey* | 173 |
| 14 | **The Artwashing of Gentrification and Social Cleansing**<br>*Stephen Pritchard* | 179 |
| 15 | **Taking the Weather with You: Remittances, Translocality, and the Climate Migrant Within**<br>*Laurie Parsons* | 199 |
| 16 | **Barbed Displacement: Walls to the Disciplined Migrant**<br>*Bénédicte Michalon* | 217 |
| 17 | **Technologies of Deportation**<br>*William Walters* | 237 |
| 18 | **Street Technologies of Displacement: Disposable Bodies, Dispossessed Space**<br>*Elijah Adiv Edelman* | 255 |

| 19 | Olympic Favela Evictions in Rio de Janeiro: The Consolidation of a Neoliberal Displacement Regime<br>James Freeman | 271 |

| Part III | Section Three: Journeys of Displacement | 287 |

| 20 | Intervention: Women's Narratives from Refugee Camps in the Kurdistan Region of Iraq<br>Nazand Begikhani | 289 |
| 21 | Constraints and Transgressions in Journeys of Displacement<br>Joris Schapendonk and Milena Belloni | 297 |
| 22 | Migrants' Displacements at the Internal Frontiers of Europe<br>Martina Tazzioli | 313 |
| 23 | Carceral Journeys<br>Nick Gill and Oriane Simon | 329 |
| 24 | Precarious Migrations and Maritime Displacement<br>Vicki Squire and Maurice Stierl | 345 |
| 25 | Maintaining Health on the Move: Access and Availability for Displaced People<br>Jennifer Cole | 363 |

| Part IV | Section Four: Traces of Displacement | 379 |

| 26 | Intervention: Disasters and Displacement: When There Is No Time to Stop<br>Ayesha Siddiqi | 381 |
| 27 | Antipodean Architectures of Displacement<br>Anoma Pieris | 383 |
| 28 | Spiritual Geographies of Displacement and Resilience<br>Julia Christensen and Veronica Madsen | 399 |

29  Mapping Trajectories of Displacement 413
    Nishat Awan

30  Uncovering Internally Displaced People in the Global North
    Through Administrative Data: Case Studies of Residential
    Displacement in the UK 431
    Janet C. Bowstead, Stuart Hodkinson, and Andy Turner

Part V  Section Five: Governing Displacement 451

31  Intervention: Forensic Oceanography—Tracing Violence
    Within and Against the Mediterranean Frontier's Aesthetic
    Regime 453
    Charles Heller and Lorenzo Pezzani

32  Governing the Displaced: Contradictory Constellations of
    Actors, Ideas, and Strategies 459
    Lama Tawakkol, Ali Bhagat, and Sarah E. Sharma

33  Bureaucracies of Displacement: From Immigrants' Social and
    Physical Exclusion to Their Judicial Removal 475
    Cecilia Menjívar and Andrea Gómez Cervantes

34  Police, Bailiffs, and Hired Hands: Researching the
    Distribution and Dissolution of Eviction Enforcement 493
    Alexander G. Baker

35  Governing the Unwanted: Measuring European Migration
    Enforcement at Street Level 507
    Lisa Marie Borrelli

36  A Forced Displacement and Atrocity Crime Nexus:
    Displacement as Transfer, Annihilation, and Homogenisation 521
    Andrew R. Basso

## Part VI  Section Six: More-Than-Human Displacements — 539

**37** Intervention: Flower Power—Khmer Women's Protests Against Displacement in Cambodia and the United States — 541
*Katherine Brickell*

**38** Animals, People, and Places in Displacement — 549
*Benjamin Thomas White*

**39** Energy on the Move: Displaced Objects in Knowledge and Practice — 569
*Jamie Cross, Craig Martin, and G. Arno Verhoeven*

**40** Smartphones: Digital Infrastructures of the Displaced — 583
*Koen Leurs and Jeffrey Patterson*

**41** Family Photographs in Displacement — 599
*Penelope Pitt*

**42** Displaced Home-Objects in Homing Experiences — 613
*Mastoureh Fathi*

**43** The Role of Design in Displacement: Moving Beyond Quick-Fix Solutions in Rebuilding Housing After Disaster — 629
*Esther Charlesworth and John Fien*

## Part VII  Section Seven: Representing Displacement — 651

**44** Intervention: Activism, Research and Film-Making—Fighting for the Right to Housing in Bucharest, Romania — 653
*Michele Lancione*

**45** How Not to Eat Human Stories: Ruts, Complicities, and Methods in Visual Representations of Refugees — 659
*Dominika Blachnicka-Ciacek*

**46** Displacements of Experience: The Case of Immersion and Virtual Reality — 673
*Emma Bond*

| 47 | Displacement in Contemporary Art<br>*John Potts* | 687 |
|---|---|---|
| 48 | Reclaiming Safe Spaces: Arts-Based Research, Advocacy, and Social Justice<br>*Nelli Stavropoulou* | 701 |

| Part VIII | Section Eight: Resisting Displacement | 719 |
|---|---|---|
| 49 | Intervention: An Interview with Anna Minton<br>*Vandana Desai* | 721 |
| 50 | 'Housing is a Human Right. Here to Stay, Here to Fight': Resisting Housing Displacement Through Gendered, Legal, and Tenured Activism<br>*Mel Nowicki* | 725 |
| 51 | Contesting Displacement Through Radical Emplacement and Occupations in Austerity Europe<br>*Mara Ferreri* | 739 |
| 52 | Legal Geographies of Resistance to Gentrification and Displacement: Lessons from the Aylesbury Estate in London<br>*Loretta Lees and Phil Hubbard* | 753 |
| 53 | Local Faith Communities and Responses to Displacement<br>*Susanna Trotta and Olivia Wilkinson* | 771 |
| 54 | Hosting the Displaced: From Sanctuary Cities to Hospitable Homes<br>*Jonathan Darling* | 785 |
| 55 | Food and the Politics of Refuge: The Transformative Power of Asylum Seeker and Refugee Food Initiatives<br>*Fiona Murphy* | 799 |

| Index | 813 |
|---|---|

# Notes on Contributors

**Peter Adey** is Professor in Human Geography at Royal Holloway, University of London and works at the intersections of space, security, and mobility. He is former chair of the Social and Cultural Geography research group of the Royal Geographical Society with the Institute of British Geographers, has published widely in academic journals and edited collections, and is co-editor of the journal *Mobilities*. Among other volumes he is author of *Mobility* (2009, 2017 2nd edition); *Aerial Life: spaces, mobilities, affects* (2010); co-editor of the *Handbook of Mobilities* (2014), and co-editor of the Routledge Changing Mobilities book series with Monika Büscher. He is finishing the book *The Way We Evacuate* (with Duke University Press).

**Nishat Awan** is Senior Lecturer in Visual Cultures at Goldsmiths, University of London, based at the Centre for Research Architecture. An architect by training, Nishat's work focuses on the intersection of geopolitics and space, including questions related to Diasporas, migration, and border regimes. She is interested in modes of visual and spatial representation and ethical forms of engagements with places at a distance.

**Alexander G. Baker** received a PhD from Newcastle University in 2017 on Eviction Enforcement in the United Kingdom, and his articles work on technology, eviction, and housing. Alex is a Leverhulme Early Career Fellow in Urban Studies and Planning at the University of Sheffield.

**Kaya Barry** is an artist and geographer working in the areas of mobilities, migration, tourism, material cultures, and arts research. She is a postdoctoral research fellow at Griffith University, Australia, exploring how migration experiences are conditioned through materiality, everyday routines, and visual aesthetics. Her creative arts practice informs both the methods and focus of research, with a keen interest in how mobilities and geographical research intersect with creative, participatory, and community-oriented forms of engagement. Recent publications include the monograph: *Creative Measures of the Anthropocene: Art, Mobilities and Participatory Geographies* (Barry & Keane, 2019, Palgrave Macmillan).

**Andrew R. Basso** is Social Sciences and Humanities Research Council Postdoctoral Fellow at Western University, Canada. His research focuses on atrocities and political violence, human rights, transitional justice, and forced displacement. He is researching historical and contemporary cases of Displacement Atrocities as well as Domicide in various spaces and times.

**Nazand Begikhani** is Honorary Senior Research Fellow at the University of Bristol, is the Vincent Wright chair and visiting professor, 2019–2020, at Paris's Grande Ecole Sciences Po. Begikhani is a leading researcher on gender-based violence (GBV), having conducted research on many aspects of violence and gender relations, including honour-based violence and honour killings in the United Kingdom and in Iraq: She has particularly focused on women and war, examining rape and sexual violence during conflict in Iraq and Syria. She recently finished a two-year research project on gender-based violence and displacement, which was funded by the UK Economic and Social Research Council (ESRC). She has worked as an expert advisor with a range of national and international organisations and government departments, including the United Nation's Assistance Mission to Iraqi (UNAMI), UN Women, the UK Metropolitan Police, the Swedish Ministry of Integration, and Amnesty International. She has addressed social policy in post-conflict Iraq and advises the Kurdistan Region's President on Higher Education and Gender. In addition, she is the principal editor/publisher of a Specialised Imprint (Collection) at the L'Harmattan, entitled *Peuples cultures et littératures de l'Orient*. Nazand is also an internationally known poet, having won several poetry prizes, including France's Simone Landrey's Feminine Poetry Prize in 2012.

**Milena Belloni** is FWO postdoctoral researcher at the University of Antwerp (Belgium). Her research concerns refugees' migration, integration pathways, smuggling, and ethnographic methods. She is the author of the book *The Big Gamble* (2019), a book on Eritrean migration. Many of her articles have appeared in international journals, such as *JRS, Global Networks,* and the *IJCS*.

**Ali Bhagat** is at the Department of Politics, University of Manchester, Manchester, UK. He is finishing his dissertation titled 'Forced Migration in Racial Capitalism: Urban Refugee Survival in Paris and Nairobi' and holds the IDRC doctoral research award. Most recently, he received the Robert and Jessie Cox award at the International Studies Association (ISA) for best graduate student paper of critical inquiry in international relations. His articles, concerning his wider research interests in forced displacement, refugees, and racial capitalism, have appeared in the journals Geoforum and Sexualities.

**Matthew D. Bird** is a professor at the Universidad del Pacífico Graduate School of Business in Lima, Peru, where he also consults for private and non-profit clients in the region helping them address organisational change and social impact challenges. His research seeks to design and evaluate innovative interventions that better harvest local solutions to solve common social challenges. Areas of focus include the leveraging of digital technologies and behavioural insights in the areas of financial inclusion, SMEs, early childhood development, and post-conflict reconstruction. He received his PhD in Human Development from the University of Chicago, where he studied cultural influences on economic decisions, especially in relationship to trust, an interest developed after working as a management consultant in Barcelona, Spain. Prior to joining Universidad del Pacífico, he served as research director for the Advanced Leadership Initiative at Harvard University and as a research associate at Harvard Business School. His research has been funded by grants from the Ford Foundation, National Science Foundation, Fulbright-Hays, the National Institute of Mental Health, Citi Foundation, and Innovations for Poverty Action. He holds a BA in History from Yale University. He is working as primary investigator for projects in the United States, Colombia, and Peru.

**Dominika Blachnicka-Ciacek** is a visual sociologist (PhD 2016, Goldsmiths, University of London) with an interest in researching relationships between people, spaces, and memory in migration and forced migration contexts, with a regional focus on migrants and refugees in the Middle East and Europe. She is an assistant professor at the SWPS University in Warsaw.

**Emma Bond** is Reader in Modern Languages at the University of St Andrews. She works on the transnational circulation of people, texts, and cultural artefacts, and on contemporary migration. Her publications include *Writing Migration through the Body* (2018) and the co-edited volume *Destination Italy: Representing Migration in Contemporary Media and Narrative* (2015).

**Lisa Marie Borrelli** is post-doctoral researcher at the HES-SO Valais–Wallis, Switzerland. She finalised her PhD in 2018 at the Institute of Sociology, University of Bern, Switzerland, conducting ethnographic fieldwork with government agencies dealing with irregular migration in Italy, Switzerland, Germany, Sweden, Lithuania, and Latvia.

**Janet C. Bowstead** is a researcher at Royal Holloway, University of London, with a professional background in frontline, policy, and coordination work on violence against women. Her research is interdisciplinary in nature, across geography, social policy, and sociology; integrating quantitative, spatial, qualitative, and creative methods. www.womensjourneyscapes.net.

**Katherine Brickell** is Professor of Human Geography at Royal Holloway, University of London (RHUL), UK. At RHUL, she is director of the Geopolitics, Development, Security and Justice Research Group. She is editor of the journal *Gender, Place and Culture* and is former chair of the RGS-IBG Gender and Feminist Geographies Research Group. Katherine's feminist-oriented research cross-cuts social, political, development, and legal geography, with a long-standing focus on the domestic sphere as a precarious space of contemporary everyday life. Katherine has published multiple co-edited collections which speak to her connected interests in home (un)making, displacement, and (im)mobilities, including *Translocal Geographies* (2011 with Ayona Datta) and *Geographies of Forced Eviction* (2017 with Melissa Fernández Arrigoitia and Alex Vasudevan). Her most recent writing aims to reaffirm and reprioritise the home as a political entity which is foundational to the concerns of human geography. Katherine's monograph *Home SOS: Gender, Violence and Survival in Crisis Ordinary Cambodia* (Wiley RGS-IBG Book Series, 2020) brings domestic violence and forced eviction into twin view to pursue this.

**Jose Jowel Canuday** works on the visual anthropology of displacement in conflict and disaster risk contexts, following closely acts of human creativity and innovations, grounded cosmopolitanism, and interfaith relations. He is associate professor at the Ateneo de Manila University and editor of *Social Transformations: Journal of the Global South*.

**Esther Charlesworth** is a professor in the School of Architecture and Design at RMIT University, Melbourne. She is the academic director of the RMIT 'Master of Disaster, Design and Development' degree (MoDDD) and also the founding director of Architects without Frontiers (AWF). Since 2002, AWF has undertaken over forty health, education, and social infrastructure projects in twelve countries for vulnerable communities and has been described by

ABC radio broadcaster Phillip Adams as 'destined to develop into one of the greater forces of good on this battered planet.' Esther has published seven books on the theme of social justice and architecture, including *Humanitarian Architecture* (2014) and *Sustainable Housing Reconstruction* (2015).

**Julia Christensen** is an Associate Professor in Geography and Canada Research Chair in Northern Governance and Public Policy at Memorial University in St. John's, Canada. She has published extensively on housing, health, and home in the circumpolar North and is the author of *No Home in a Homeland: Indigenous Peoples and Homelessness in the Canadian North* (2017).

**Jennifer Cole** is a research fellow in the Geography Department of Royal Holloway and is the Northern Europe Hub Coordinator of the Planetary Health Alliance. She studied biological anthropology at Cambridge University and has held positions as Senior Research Fellow in Resilience and Emergency Management at the Royal United Services Institute think tank and as public health policy advisor to the Rockefeller Foundation Economic Council on Planetary Health at Oxford University.

**Jamie Cross** is Senior Lecturer in Social Anthropology at the University of Edinburgh. His research interests are situated in the 'anthropology of development,' including examination of infrastructures, low carbon energy futures, corporations and social enterprises, work, labour, and global supply chains.

**Jonathan Darling** is Assistant Professor in Human Geography at Durham University. His research focuses on the spatial politics of asylum, sanctuary, and the urban dynamics of forced migration. He is the co-editor of *Encountering the City* (2016) and *Sanctuary Cities and Urban Struggles* (2019).

**Vandana Desai** is Senior Lecturer in Development Geography at Royal Holloway, University of London. Her research is on issues of community participation, grassroots non-governmental organisation, gender, informal housing, and currently on widowhood and ageing, equality and diversity, and relationship between NGOs and their corporate partners. Her research has been funded by British Academy, British Council, Economic and Social Research Council, Department for International Development, Development Studies Association, and the Swedish Research Council. She is one of the editors of three editions (2002, 2008, 2014) of the *Companion to Development Studies* and of *Doing Development Research (2006)*. She is a member of the advisory editorial board of the journal *Progress in Development Studies* and has supervised twelve PhDs to completion.

**Mike Dolton** is Senior Lecturer in Human Geography in the Department of Geography, Royal Holloway. His background is in urban studies and sustainability with research interests in the politics of urban regeneration, governance, democracy, and urban sustainability.

**Elijah Adiv Edelman** is an assistant professor in the Department of Anthropology at Rhode Island College in Providence, Rhode Island. His work has appeared in *GLQ*, *Journal of Homosexuality*, *Porn Studies*, and *Social Text*, as well as in the collections *Queer Necropolitics* (2014), *Queer Excursions* (2014), and *Out in Public* (2009).

**Rebecca Elmhirst** is a human geographer based at the University of Brighton in the United Kingdom. Her research interests lie in feminist political ecology and the politics of nature, resource governance, displacement, and migration, with an empirical focus on Southeast Asia. She is editor of the journal *Gender, Technology and Development*.

**Mastoureh Fathi** is a Marie Skłodowska-Curie fellow at Institute of Social Science Research in twenty-first century at University College Cork. Her research revolves around everyday experiences of migration, intersectionality, gender, and class in migration processes, identity, home-making, and belonging in diaspora and the importance of objects in displacement.

**Mara Ferreri** is VC Fellow in Human Geography at Northumbria University, prior to which she held a Marie Curie Fellowship at the Universitat Autónoma de Barcelona, Spain. Her research interests include temporary urbanism, precarity, and emerging housing practices. She is a founding editor of the *Radical Housing Journal*.

**John Fien** is Professor of Practice in the interdisciplinary field of Disaster, Design and Development at RMIT University, Melbourne. His role focuses on developing national and global partnerships and fostering an applied research culture in the Humanitarian Architecture Research Bureau [HARB] at RMIT. John has been a professor at Griffith and Swinburne Universities and has led development projects and evaluations in Kenya, South Africa, Thailand, Vietnam, the Philippines, and Fiji.

**James Freeman** is a lecturer in the Department of Geography, Planning and Environment at Concordia University in Montreal. He received his PhD in Geography from the University of California at Berkeley and researches the social, cultural, and economic geography of the Latin American city.

**Luisa F. Freier** is Assistant Professor of Political Sciences at the Universidad del Pacífico (Lima, Peru). She is a leading scholar of Latin American migration and refugee policies, south-south migration, and the Venezuelan displacement crisis. She has published widely in both academic and media outlets, and her interdisciplinary and multi-method publications have appeared in leading migration journals such as *International Migration Review* and *Migration Studies*. She has been cited on the Venezuelan displacement crisis in international media, including *BBC, CBC, El Comercio, El País, La Presse, Liberation, Newsdeeply, PBS*, and *The Economist*, and has provided advice to various international institutions and organisations such as the International Organization for Migration (IOM), the European Union (EU), and the Inter-American Development Bank (IDB). She holds a PhD in Political Science from the London School of Economics and Political Science (LSE), an MA in Latin American and Caribbean Studies from the University of Wisconsin, Madison, and a bachelor's degree in Economics from Universität zu Köln. Her research has been funded by the Fulbright program, the Naumann Foundation for Freedom, and the IDB.

**Pat Gibbons, PhD** is director of UCD's Centre for Humanitarian Action (CHA). He served as president of Network on Humanitarian Action (NOHA) from 2007 to 2013 and chaired the inter-sectoral group that shaped Ireland's contribution to the World Humanitarian Summit. He is principal investigator of a number of Horizon 2020-funded programmes relating to resilience. Other research interests include the humanitarian principles, localising humanitarian response, and gender-based violence.

**Nick Gill** is Professor of Human Geography at the University of Exeter, UK. His work focuses on justice and injustice, especially in the context of border control, mobility and its confiscation, incarceration, and the law. His research examines access to justice in European asylum claim determination.

**Andrea Gómez Cervantes** is a University of California President's Postdoctoral Fellow in the Department of Chicana/o & Central American Studies at UCLA and will begin her appointment as assistant professor in the Department of Sociology at Wake Forest University in the Fall of 2020. Gómez Cervantes received her PhD and MA in sociology at the University of Kansas and BS from Grand Valley State University. Her research investigates immigration policy and ethno-racial divides among Latin American immigrants living in the United States. Her most recent publication, '"Looking Mexican": Indigenous and non-Indigenous Latina/o Immigrants and the Racialization of Illegality in the Midwest,' appeared in the journal *Social Problems* (2019).

**Mark Griffiths** is a NUAcT research fellow at Newcastle University. He is a geographer with a focus on the embodied aspects of the occupation of Palestine and the ethics of geographical research. His work has been published in *Antipode, Political Geography, Transactions of the Institute of British Geographers, Gender, Place & Culture,* and *Area.*

**Amanda Hammar** directs the Centre of African Studies, University of Copenhagen, and is president (2019–2023) of the European African Studies Association (AEGIS). She has worked on displacement in southern Africa for several decades. Her publications include *Displacement Economies in Africa* (2014). Current research focuses on urban displacement and resettlement, property and personhood, and certifications of citizenship in Africa.

**Charles Heller** is a researcher and filmmaker whose work has a long-standing focus on the politics of migration. In 2015, he completed a PhD in Research Architecture at Goldsmiths, University of London. He is based in Geneva, where he is a research associate at Graduate Institute's Centre on Conflict, Development, and Peacebuilding. Working together since 2011, Heller and Lorenzo Pezzani co-directed the Forensic Oceanography project that critically investigates the militarised border regime and the politics of migration in the Mediterranean Sea. Their collaborative work has been used as evidence in courts of law, published across different media and academic outlets, as well as exhibited and screened internationally.

**Stuart Hodkinson** is Associate Professor of Critical Urban Geography at the University of Leeds. His research focuses on the social harms of housing privatisation under neoliberalism. His most recent book is *Safe as Houses: Private Greed, Political Negligence and Housing Policy After Grenfell* (2019).

**Phil Hubbard** is Professor of Urban Studies at King's College, London. He has written widely cited work on urban sociology, urban geography, and social geographies. He is also a leading figure in the study of sexuality and space and retail gentrification.

**Soledad Castillo Jara** is a research assistant at Universidad del Pacífico and teaching assistant of an Introductory Course in International Relations at Pontifical Catholic University of Peru (PUCP). She holds a BA in Political Science from the Pontifical Catholic University of Peru (PUCP). Soledad has co-published various blog posts and academic articles on the Venezuelan displacement crisis.

**Michele Lancione** is an urban ethnographer and activist interested in issues of marginality, diversity, and radical politics. His most recent writing has focused on homelessness, racialised displacement, and underground life in Bucharest, Romania. He is one of founders and editors of the *Radical Housing Journal*, editor of *City*, and the corresponding editor for Europe at *IJURR*. He is a Senior Research Fellow based at the *Urban Institute*, University of Sheffield (UK).

**Loretta Lees** (FAcSS, FRSA, FHEA) is Professor of Human Geography at the University of Leicester. She is an urban geographer and scholar-activist who is internationally known for her research on gentrification/urban regeneration, global urbanism, urban policy, urban public space, architecture, and urban social theory.

**Koen Leurs** is Assistant Professor in Gender and Postcolonial Studies, Graduate Gender Programme, Department of Media and Culture Studies, Utrecht University, Utrecht, The Netherlands. He chairs the European Communication Research and Education (ECREA) Diaspora, Migration, and the Media section. Recently, he co-edited the *Sage Handbook of Media and Migration* and journal special issues on 'Connected migrants' for *Popular Communication* and Forced migration and digital connectivity (in)to Europe for *Social Media + Society*. He is now working on a book titled *Digital Migration* (Sage Publications).

**Veronica Madsen** is a graduate student in Geography at Memorial University in St. John's, Canada, where she studies northern and Indigenous housing and home. She is very active in her role as student support assistant at Memorial University's Aboriginal Resource Office.

**Craig Martin** is Reader in Design Cultures at the University of Edinburgh. His research interests involve the social, cultural, and spatial complexity of design and how this is manifest in a range of social practices, involving three substantive themes: informal design, social complexity, and mobilities.

**Rónán McDermott, PhD** is a research coordinator at University College Dublin's Centre for Humanitarian Action where he contributes primarily to the humanitarian action in urban settings research theme. He also teaches in the area of social research methods and legal dimensions of humanitarian action.

**Sinéad McGrath** is a doctoral researcher in the School of Politics and International Relations at University College Dublin. Her research examines the link between gender, displacement, and transitional justice, particularly the role of international organisations. She also teaches in the Department of International Development at Maynooth University.

**Scott McKinnon** is a Vice Chancellor's Postdoctoral Research Fellow in the Australian Centre for Culture, Environment, Society, and Space (ACCESS), University of Wollongong. He is the author of *Gay Men at the Movies: Cinema, Memory and the History of a Gay Male Community* (2016) and co-editor (with Margaret Cook) of the forthcoming *Histories of Disaster in Australia and New Zealand: Historical Approaches to Understanding Catastrophe* (Palgrave Macmillan, 2020).

**Cecilia Menjívar** is Dorothy L. Meier Social Equities Chair and Professor of Sociology at the University of California—Los Angeles. Her research focuses on manifestations of state power in everyday life—through immigration laws that shape immigrants' experiences and through institutional systems that ignore or foment gender-based violence against women. Empirically, she focuses on Central American immigrants in the United States and on women in Central America. Most recently, she co-edited *The Oxford Handbook of Migration* Crises (2019) and co-authored, 'The Impact of Adjacent Laws on Implementing Violence Against Women Laws: Legal Violence in the Lives of Costa Rican Women' (*Law & Social Inquiry*, 2020).

**Bénédicte Michalon** is a Senior Research Fellow at the French CNRS, Bordeaux. She works on detention and house arrest for foreigners and asylum seekers housing in rural areas, with fieldwork in Romania and France. She coordinated the interdisciplinary team *TerrFerme* into spatial readings of confinement (http://terrferme.hypotheses.org/). She is a member of the Migreurop network (http://www.migreurop.org).

**Fiona Murphy** is a Research Fellow in the Senator George J. Mitchell Institute for Global Peace, Security, and Justice, Queen's University Belfast. She specialises in Indigenous politics and movements, refugees and mobility studies, and sustainability in Australia, France, Turkey, and Ireland.

**Mel Nowicki** is Lecturer in Urban Geography at Oxford Brookes University. Her research focuses on urban housing exclusion and critical geographies of home. She has worked predominately in the United Kingdom and Ireland on a range of issues, including the criminalisation of squatting and experiences of family homelessness.

**Laurie Parsons** is British Academy Postdoctoral Research Fellow at Royal Holloway, University of London. His work investigates the impact of mass labour migration and climate change on socio-economic inequalities, working in particular towards more politically grounded perspectives on the nexus of environment and mobility. It combines a variety of approaches—including visual and statistical analysis of social networks and qualitative methods—to discern how norms and social structures both reflect and mediate these new conditions. He has conducted large-scale projects examining Cambodia's uneven economic development for Transparency International, Plan International, Save the Children, CARE International, ActionAid, the IDRC, and the Royal University of Phnom Penh, among others.

**Jeffrey Patterson** graduated with honours from the master program Youth, Education, and Society in the Department of Social and Behavioral Sciences at Utrecht University, the Netherlands. Some of his recent publications are in high-standing international academic journals, such as *Media & Communication and European Journal of Cultural Studies*. He assisted a research project investigating the digital practices of young adult forced and voluntary migrants, which is funded by the Netherlands Organization for Scientific Research. He is enrolled in the International Postgraduate program at the Netherlands Research School of Gender Studies run by Utrecht University. He is seeking PhD opportunities investigating the digital intimacies of young adult expatriates.

**Lorenzo Pezzani** is an architect and researcher. In 2015, he completed a PhD in Research Architecture at Goldsmiths, University of London, where he is lecturer and leads the MA studio in Forensic Architecture. His work deals with the spatial politics and visual cultures of migration, with a particular focus on the geography of the ocean. Working together since 2011, Pezzani and Charles Heller co-direct the Forensic Oceanography project that critically investigates the militarised border regime and the politics of migration in the Mediterranean Sea. Their collaborative work has been used as evidence in courts of law, published across different media and academic outlets, as well as exhibited and screened internationally.

**Anoma Pieris** is a professor at the Faculty of Architecture Building and Planning, The University of Melbourne. She has published widely on nationalism, citizenship, and sovereignty. Recent publications include *Hidden Hands and Divided Landscapes: a penal history of Singapore's plural society* (2009); *Architecture and Nationalism in Sri Lanka: The trouser under the cloth* (2012); *Assembling the Centre: Architecture for Indigenous Cultures. Australia and Beyond* (2015), co-authored with Janet McGaw, *Sovereignty, Space and Civil War in Sri Lanka* (2018) and the anthology, *Architecture on the Borderline* (2019).

**Alasdair Pinkerton** is Reader (Associate Professor) in Geopolitics in the Department of Geography at Royal Holloway, University of London. Alasdair's research examines the interaction of popular culture and foreign/

security policy and practices, with a particular focus on the United Kingdom, Commonwealth, and British Overseas Territories. Since 2015, in conjunction with the Royal Geographical Society, Land Rover, and Google Arts and Culture, Alasdair has co-led a large, international project using some of the latest Virtual Reality technologies to explore and narrate the histories, cultures, and politics of 'no-man's land.' Alasdair publishes widely within and beyond the academy. His first book, *Radio: making waves in sound* is a groundbreaking political/cultural history of the world's original electronic means of communication and published by Reaktion in collaboration with The Science Museum in London.

**Penelope Pitt** is advocate and research coordinator at Deakin University Student Association, Melbourne, Australia. Penelope researches and writes on topics including migration, temporality, visual methods and ethics, and new material feminism. Her recent publications include articles in journals such as *Gender, Place and Culture*; *Teaching in Higher Education*; and *Australian Educational Researcher*.

**John Potts** is Professor of Media at Macquarie University, Sydney. His articles on media and technology studies, intellectual history, and art history have appeared in many publications. He is the author of the books *The New Time and Space*, *Ideas in Time*, *A History of Charisma*, and *Culture and Technology* (with Andrew Murphie). He is editor of the books *The Future of Writing* and *After the Event: New Perspectives on Art History* (with Charles Merewether).

**Stephen Pritchard** is an independent academic, researcher, art historian, critical theorist, activist, writer, curator, filmmaker, and community artist. His interdisciplinary approach to research is grounded in post-critical ethnography, radical art history, Frankfurt School Critical Theory, and Critical Urban Theory. Pritchard has contributed significantly to the understanding of artwashing and how art is employed in relation to urban environments and placemaking. He presents papers internationally, lectures widely, and his writing has been widely published. He is a cultural commentator. Pritchard is co-organiser of the Movement for Cultural Democracy and co-founder of Art Not Arms, Artists Against Social Cleansing, and the Socially Engaged and Participatory Arts Network. He is also working as a critical friend for Super Slow Way—a Creative People and Places project—and as lead artist on a project in 'Category D' pit villages in north-east England for Northern Heartlands Great Place project.

**Joris Schapendonk** is Assistant Professor in Human Geography and an active member of Nijmegen Centre for Border Research (NCBR) of Radboud

University. His research concentrates on im/mobility trajectories in the context of African migration and the role of migration industry actors in shaping mobility processes. The book *Finding Ways Through Eurospace* is his latest publication (2020).

**Sarah E. Sharma** is a doctoral candidate and Vanier Canada Scholar in the Department of Politics at Queen's University. She is also an IDRC Doctoral Research Awardee and a visiting fellow at the International Centre for Climate Change and Development (ICCCAD) at the Independent University, Bangladesh (IUB). Her dissertation examines the political economy of urbanisation and climate change in Dhaka, Bangladesh, and Amsterdam, Netherlands.

**Mimi Sheller, PhD** is Professor of Sociology and founding director of the Center for Mobilities Research and Policy at Drexel University in Philadelphia. She is founding co-editor of the journal *Mobilities*, associate editor of *Transfers*, and past president of the International Association for the History of Transport, Traffic, and Mobility. She has helped to establish the interdisciplinary field of mobilities research. She is author or co-editor of twelve books, including most recently *Mobility Justice: The Politics of Movement in an Age of Extremes* (2018).

**Ayesha Siddiqi** is University Lecturer in Human Geography at the University of Cambridge. She is a development and post-colonial geographer. Her core research interests are around hazard-based disasters and their intersection with politics, security, and development in the Global South. In particular, her research explores questions of political space in the aftermath of disasters and uses a social contract framework to shed new light on the way disasters are lived and experienced on the margins of the post colony. Prior to joining Cambridge, Ayesha was based at the Department of Geography at Royal Holloway, University of London. She has an interdisciplinary background and has done considerable work on the interface of academia and policy, most recently for the United Kingdom's Houses of Parliament.

**Oriane Simon** did her PhD at UNSW in Canberra researching extraordinary rendition—the practice of abduction, transfer, detention, interrogation, and torture of suspected terrorists. She uses a Deleuzian–Spinozist micropolitical perspective to unpack the lived experiences of victims and perpetrators. Oriane works as consultant in information security.

**Jessie Speer** is a British Academy Postdoctoral Fellow in the School of Geography at Queen Mary University of London. Her work examines housing displacement and homelessness in the United States and United Kingdom, bringing together analyses of intimate domestic politics, urban political economies, and legal geographies of displacement.

**Vicki Squire** is Reader in International Security at University of Warwick, the United Kingdom. Her research is situated at the intersections of border, migration, security, and citizenship studies, to focus on struggles over migration involving governing authorities, civil society groups as well as people on the move themselves.

**Nelli Stavropoulou** is a final year PhD student based at Durham University, exploring the role of participatory arts research methods as a vehicle for self-expression for individuals from a refugee/asylum seeker background. Nelli is a trustee of Durham City of Sanctuary and is involved in different creative projects with local community groups seeking asylum in the North East of England. She is also the curator of the *Moving Worlds* special film programme for Refugee Week produced by Counterpoints Arts.

**Maurice Stierl** is a Leverhulme Research Fellow at the University of Warwick. Between 2015 and 2017, he was an assistant professor at the University of California, Davis. His research focuses on migration struggles in contemporary Europe and is situated in international relations, international political sociology, and migration, citizenship & border studies.

**Lama Tawakkol** is a doctoral candidate at Queen's University. She is interested in the lived experiences of people under capitalism, particularly vulnerable populations along the axes of class, gender, and race, in grave social inequalities under the current political-economic order, and in theorizing alternatives to it. Her PhD dissertation research focuses on the intersection(s) between capitalism, informality, and displacement at the urban level, with a focus on informal settlement residents and refugees, particularly in the context of the Middle East.

**Martina Tazzioli** is Lecturer in Politics and Technology at Goldsmiths, University of London. She is the author of *The Making of Migration. The Biopolitics of Mobility at Europe's Borders* (2019), *Spaces of Governmentality, Autonomous Migration and the Arab Uprisings* (2015), co-author with Glenda Garelli of *Tunisia As a Revolutionised Space of Migration* (2017), co-editor of *Foucault and the History of Our Present* (2015), and *Foucault and the Making of Subjects* (2016). Her new book project is entitled *Border Abolitionism: Migration Containment and the Genealogies of Rescue and Struggles*. She is on the editorial board of *Radical Philosophy*.

**Susanna Trotta** is a research associate for the Joint Learning Initiative on Faith and Local Communities. She holds an MSc in global migration from

University College London and has previously worked as a social worker with asylum seekers in Italy.

**Leonie Tuitjer** is a postdoc at Leibniz University Hanover, Germany. She obtained her PhD from Durham University (UK). Her dissertation investigated the relation between climate change and displacement in Bangkok, Thailand. Her research interests are global environmental change, urban studies, infrastructures, and migration studies.

**Andy Turner** is a researcher specialising in computational geography and is based in the School of Geography at the University of Leeds.

**James A. Tyner** is Professor of Geography at Kent State University and Fellow of the American Association of Geographers. He is the author of eighteen books, including *War, Violence, and Population: Making the Body Count*, which received the AAG Meridian Book Award for Outstanding Scholarly Work in Geography. His other honours include the AAG Glenda Laws Award, which recognizes outstanding contributions to geographic research on social issues.

**G. Arno Verhoeven** is Senior Lecturer in Design, and programme director of Design for Change [MA] at the University of Edinburgh. His interdisciplinary research is situated in critical design methods and practices facilitating change through communicating complexity, fostering collective decision-making, and exploring alternative futures through constructivist approaches to design.

**William Walters** is Professor of Political Sociology at Carleton University, Canada. His main research interests are secrecy and security, borders and migration, and mobility and politics.

**Benjamin Thomas White** teaches history at the University of Glasgow, Scotland, where he is also a member of the Glasgow Refugee, Asylum, and Migration Network. A Middle East historian by background, he now teaches refugee history more broadly and is researching the global history of the refugee camp.

**Olivia Wilkinson** is the director of research at the Joint Learning Initiative on Faith and Local Communities. Her research focuses on secular and religious influences in humanitarian action, and her monograph on this topic is due out in 2020.

# List of Figures

| | | |
|---|---|---|
| Fig. 1.1 | *Shoes: Discarded and Donated*. Croatia-Slovenia border crossing, September 2015. (Source: Into No Man's Land/Elliot Graves, courtesy of Alasdair Pinkerton) | 3 |
| Fig. 1.2 | Handbook planning tablecloth, July 2017. (Source: Katherine Brickell) | 12 |
| Fig. 1.3 | Workshop participants explaining their journeys at the exhibition, Johannesburg Holocaust and Genocide Centre, June 2019. (Source: Ayesha Siddiqi) | 20 |
| Fig. 7.1 | A Palestinian flag flies at Um Al-Khair, a community in the Occupied Palestinian West Bank threated with displacement by house demolition, 2018 (Source: Mark Griffiths) | 104 |
| Fig. 8.1 | The different societal levels that influence protection (Action Aid 2010) | 118 |
| Fig. 13.1 | *Inflation (night)* (Kaya Barry 2019) | 175 |
| Fig. 13.2 | *Inflation (life vest)* (Kaya Barry 2016) | 175 |
| Fig. 13.3 | *Emergency exit (water)* (Kaya Barry 2019) | 176 |
| Fig. 14.1 | Montage of demolition of Robin Hood Gardens estate and Blackwall Reach promotional video. (Film still from *A Cacophony of Crows* by S. Pritchard 2018d) | 180 |
| Fig. 14.2 | Montage of demolition of Robin Hood Gardens estate and Blackwall Reach promotional video. (Film still from *A Cacophony of Crows* by S. Pritchard, 2018d) | 182 |
| Fig. 14.3 | People of Southwark mural at The Artworks Elephant box park, Southwark, London. (Photograph by S. Pritchard 2016) | 183 |
| Fig. 14.4 | One of the walkways at Robin Hood Gardens before demolition. (Photograph by S. Pritchard 2016) | 187 |
| Fig. 14.5 | View of Robin Hood Gardens with some units boarded up. (Photograph by S. Pritchard 2016) | 189 |

# List of Figures

| | | |
|---|---|---|
| Fig. 14.6 | London Olympics mural in playground at Robin Hood Gardens with the Blackwall Reach development on other side of road. (Photograph by S. Pritchard 2016) | 191 |
| Fig. 14.7 | Montage of section retrieved from Robin Hood Gardens estate installed at V&A's *Robin Hood Gardens: A Ruin in Reverse* exhibition at the 2018 Venice Architecture Biennale and footage from V&A's Biennale promotional video. (Film still from *A Cacophony of Crows* by S. Pritchard 2018d) | 192 |
| Fig. 14.8 | Robin Hood Gardens awaiting demolition. (Photograph by S. Pritchard 2017b) | 194 |
| Fig. 18.1 | *Jan's map* (Example 1): Participant in 2010 Trans DC Panel, consent for reproduction provided by producer | 263 |
| Fig. 18.2 | *Brett's map* (Example 2): Participant in 2010 Trans DC Panel, consent for reproduction provided by producer | 264 |
| Fig. 18.3 | *Cameron's map* (Example 3): Participant in 2010 Trans DC Panel, consent for reproduction provided by producer | 265 |
| Fig. 20.1 | Untitled, painting by a Yazidi ISIS survivor, Sharya Refugee Camp, Kurdistan | 289 |
| Fig. 20.2 | Untitled, by Sohaila, Yazidi ISIS survivor, Sharya Refugee Camp, Kurdistan | 293 |
| Fig. 29.1 | 'Migrant Narratives of Citizenship' exhibition (Awan 2016); interview with Afghan woman in Odessa, Ukraine. (Image: Cressida Kocienski) | 418 |
| Fig. 31.1 | Reconnaissance picture of the 'left-to-die boat' taken by a French patrol aircraft on 27 March 2011. (Credit: Council of Europe) | 455 |
| Fig. 31.2 | Analysis of the 29 March 2011 Envisat satellite image showing the modelled position of the 'left-to-die boat' (yellow diagonal hatch) and the nearby presence of several military vessels who did not intervene to rescue the migrants. (Credit: Forensic Oceanography and SITU Research, *Report on the Left-to-Die Boat Case*) | 456 |
| Fig. 37.1 | Activist offering a lotus flower across the barricade line, Phnom Penh, Cambodia, 2012 (© Erika Piñeros) | 543 |
| Fig. 37.2 | Free the 15 'Stop the Violence' T-shirt, Phnom Penh, Cambodia, 2012 (© Erika Piñeros) | 544 |
| Fig. 37.3 | Protest outside Sacramento Capitol building, 2019. (Photo courtesy of Asian Law Caucus) | 545 |
| Fig. 37.4 | Pink paper carnation, Sacramento, 2019. (Photo courtesy of Kevin Lo) | 546 |
| Fig. 38.1 | Derek Robertson, *The Desert Is Full of Promises* (by permission of the artist) | 562 |
| Fig. 38.2 | Kamruzzaman Shadhin, from *Elephant in the Room* (by permission of the artist) | 563 |
| Fig. 39.1 | Illustration of double-bag bellows employed by Tuareg blacksmith apprentice (Illustration by Ann-Kathrin Müller 2018) | 572 |

## List of Figures

| | | |
|---|---|---|
| Fig. 39.2 | Illustration of goatskin water carrier (Illustration by Ann-Kathrin Müller 2018) | 577 |
| Fig. 43.1 | Cyclone Aila washed away the southern-most island of Bangladesh and people became refugees. This picture showed how the cyclone smashed people's houses there, 2009—description by the photographer Mayeenul Islam. (https://commons.wikimedia.org/wiki/File:Cyclone_Aila_Climate_Change_Nijhum_Dwip_2009_Dec_Bangladesh.jpg—used under Creative Commons Attribution-Share Alike 3.0 Unported License https://creativecommons.org/licenses/by-sa/3.0/legalcode) | 630 |
| Fig. 43.2 | KPMG Housing Village, Seenigama, 2013. (Photo Credit: Foundation of Goodness, 2019) | 638 |
| Fig. 43.3 | Rebuilt houses in Villa Rosa. (Photo credit: Koolhaus and Urhahn, 2016. https://favelapainting.com/HAITI-PAINTING-2015-2016) | 644 |
| Fig. 44.1 | (Source: *All pictures in the text are from the author*) | 654 |
| Fig. 44.2 | (Source: *All pictures in the text are from the author*) | 655 |
| Fig. 44.3 | (Source: *All pictures in the text are from the author*) | 656 |
| Fig. 44.4 | (Source: *All pictures in the text are from the author*) | 657 |
| Fig. 48.1 | One of 48 panels painted as a result of interviews and workshops with the Dreamers Group of young accompanied refugees and asylum seekers in Loughborough. *I Had a Dream* by Paul Gent, Sense of Belonging—AHRC-funded project led by Maggie O'Neill and Phil Hubbard. (Courtesy of Prof. Maggie O'Neill and Paul Gent) | 708 |
| Fig. 48.2 | One of 48 panels painted as a result of interviews and workshops with the Dreamers Group of young accompanied refugees and asylum seekers in Loughborough. Paul Gent, Sense of Belonging—AHRC-funded project led by Maggie O'Neill and Phil Hubbard. (Courtesy of Prof. Maggie O'Neill and Paul Gent) | 709 |
| Fig. 48.3 | *Missed Call.* Photograph by Amine Oulmakki. Arts for Advocacy 2017. (Courtesy of Prof. Laura Jeffery, Dr. Mariangela Palladino, and Dr. Sébastien Bachelet) | 710 |
| Fig. 48.4 | Screenshot of *Scattered People* mini-documentary. (Available on YouTube http://www.youtube.com/watch?v=GV_US5RyIgM. Courtesy of Assoc. Prof. Caroline Lenette) | 712 |
| Fig. 48.5 | *My Country is Unforgettable* Mohamad El Hamood. Dispersed Belongings 2018. (Courtesy of Dr. Caitlin Nunn and Mohamad El Hamood) | 713 |
| Fig. 51.1 | Graffiti: 'Occupation against speculation,' Barcelona, March 2019. (Source: Mara Ferreri) | 747 |
| Fig. 52.1 | 'Arry's bar: Site of first Aylesbury Estate Public Inquiry with heavy security. (Photo courtesy of 35% campaign) | 758 |

# 1

# Introduction to Displacement Studies: Knowledges, Concepts, Practices

Peter Adey, Janet C. Bowstead, Katherine Brickell, Vandana Desai, Mike Dolton, Alasdair Pinkerton, and Ayesha Siddiqi

## Introduction

Eight or nine pairs of shoes are lined up in a row outside a faded grey tent on the bare concrete floor of a disused warehouse in Idomeni, a small village in Greece, near the border with Macedonia. The shoes belong to a family of ten, their youngest twin toddlers. The picture on the front cover of this handbook, by documentary photographer Mary Turner, was taken before the informal camp—which is located on and around the train tracks—was later cleared by Greek riot police following the closure of the Macedonian border in 2015. The same thing happened again in 2016 leading to a second eviction. The migrants were seeking to enter central Europe through the so-called Balkan route, some after being bused by private companies from Athens after arriving in Greece from Syria, Iraq, and Afghanistan, but also Iran, and North and Central Africa. Many had blocked the railway line in protest of the closure by using rocks, tents, and their bodies. Police encouraged the refugees to board

---

P. Adey (✉) • J. C. Bowstead • K. Brickell • V. Desai • M. Dolton • A. Pinkerton
Department of Geography, Royal Holloway, University of London, Egham, UK
e-mail: Peter.Adey@rhul.ac.uk; Janet.Bowstead@rhul.ac.uk; katherine.brickell@rhul.ac.uk; V.Desai@rhul.ac.uk; M.Dolton@rhul.ac.uk; A.D.Pinkerton@rhul.ac.uk

A. Siddiqi
Department of Geography, University of Cambridge, Cambridge, UK
e-mail: as3017@hermes.cam.ac.uk

© The Author(s) 2020
P. Adey et al. (eds.), *The Handbook of Displacement*,
https://doi.org/10.1007/978-3-030-47178-1_1

buses that would take them to permanent camps away from the border in Greece.

We were drawn to this image, and indeed many other images of shoes, in the course of putting together this handbook and inspired by our own research encounters with displacement. For co-editor Alasdair Pinkerton, Mary Turner's photographs resonated closely with his own experience of the Croatian-Slovenian border in 2015 where, only a few hours after the closure of the Slovenian border, thousands of displaced migrants became trapped in a politico-legal "no-man's land" alongside the formal international frontier (see Leshem and Pinkerton 2016). As the neighbouring sovereign states refused and/or withdrew the mobility of and care for the bodies of the displaced, so alternative economies of care materialised. NGOs and humanitarian organisations rapidly filled the void left by the Slovenian and Croatian governments, supplying food, water, tents, clothes, and, crucially, shoes. Far from being just a random jumble of footwear, the discarded and donated shoes (Fig. 1.1) can be understood as elements within a much larger and more complex network of geopolitical, legal, judicial, and humanitarian forces, pushing and pulling whole communities into forced movements and equally forced periods of immobility. In opening up these spatial and temporal dimensions through these photographs, the shoes reveal that displacement is never a singular event. It is a grounded process bound up in inequities of power and injustice.

As co-editor Ayesha Siddiqi's intervention also demonstrates later in the volume, in a photograph of just visible shoes buried beneath a mudslide, shoes are powerful objects which haunt. They are reminders of not only the intense violences and traumas of displacement but the intensities and intimacies of displacement amidst apparent global processes. Even without the person who once wore them, or who might wear them still, shoes communicate how displacement is an utterly intimate act. Shoes invite us in, to some degree, to the personal and embodied lives of those forced to move, sometimes across vast distances (Arizpe 2019). Shoes walk us backwards and forwards. As much as those outside the tent on the front cover of the handbook may serve as indicators of the stoppages, pauses, immobilisations, and wearings out that displacement inevitably involves, their shoes also point forward at possibilities, about displacements and/or rebuildings in the making.

As of 2018 there were approximately 70.8 million displaced persons in the world, with an average rate of 37,000 people being newly displaced every day (UNHCR 2018). That is 37,000 pairs of shoes, and lives, displaced every day. This record high will only be further exacerbated by the threats of political instability, climate change, and a range of other factors which will further

**Fig. 1.1** *Shoes: Discarded and Donated.* Croatia-Slovenia border crossing, September 2015. (Source: Into No Man's Land/Elliot Graves, courtesy of Alasdair Pinkerton)

complicate a world which is losing sight of human rights and global governance mechanisms. And yet, the kinds of figures that demonstrate the vast scale of human displacement, if taken on their own, tend to render displacement into a kind of fact of global-scale flows. These are well wielded, quite rightly, by some critical academics, NGOs, human rights activists, development agencies, and global institutions as signals of global inequality, as outcomes of conflict or climatic instability, as symptoms of states acting inadequately, incompetently, or even with zealous intent. Such statistics are also politically mobilised to persuade populations of the permeability of borders that are under threat by displacement. The European border agency Frontex has consistently mobilised its map of 'invasion arrows' (van Houtum and Bueno Lacy 2019), for example, and they have also been harnessed and manipulated in such a way by others. This has served a geographical imagination of mobile risks, which often transform displaced peoples from conflict, economic uncertainty, and climatic instability into an invading vector from an external and volatile outside. And to such an extent, that even those deserving of care are perceived as clear and present threats to ways of life.

In lots of ways, these debates sit right at the core of why displacement is such a crucial subject for study and discussion. But to equate displacement with popular portrayals of migrants and refugees is far too narrow and loses sight of a broader picture of displacement. They say little to how displacement, as the shoe shows, is walked, is lived, and felt. If the shoe challenges the kind of global projections and senses of scale at play in the above, it helps point to a much more complex picture wherein the refugees at the Greek-Macedonian border find themselves in a stop-start interrupted chain of mobility and immobility which they cannot control. Displacement, for many, might mean actually long periods of temporary immobility and precarity in a camp, or in a long-term period of circulatory displacement within a state's migration governance system (Tazzioli and Garelli 2018). More still, it reinforces a sense that displacement is something that is, for some, only 'out there.' It is for others to do, to wear, to endure. Perhaps shoes, as those most ordinary of objects, remind us that displacement is far more universal than this.

Outside of the lens of the nation-state or the global imaginaries of planetary flows of people and labour, there is a multiplicity of forms of displacement. Just as there have been a variety of different approaches, traditions, subdisciplines, and disciplinary approaches to those different kinds of displacement, from housing eviction to displacement from gender-based violence, from displacement in and from emergencies and disasters to international migration, *The Handbook of Displacement* offers understandings of the fragility of an unstable world and these multiple threats to human security and well-being. As such, the collection aims to offer dynamic and original analyses of displacement and how it affects individuals, states, and systems during times of war, atrocity, and peace.

The *Handbook of Displacement* traces knowledges, concepts, and practices in the interdisciplinary study of displacement, and as such provides a crucial redefinition of what, who, and where counts as displaced. The handbook aims, therefore, to not only expand the conceptualisation of the framework of displacement but also provide a foundational text for facilitating future discussion and research that will follow. In this sense, the handbook is necessarily broad and interdisciplinary, drawing from a variety of traditions and perspectives, from urban studies to the study of internal displacement, from migration studies to security. Numerous conceptual lineages and energies therefore run through this book, from postcolonial to feminist theory, from notions of gentrification to human rights. In the aggregation of different cuts and approaches to displacement, we see the volume not as an attempt to build a kind of global theory of displacement. We recognise the potential dangers of doing so as other handbooks and collections naturally grapple with some sense of conceptual dilution. The *Handbook of Gentrification* (Lees and Phillips 2018), for example, which includes a lot of different kinds of displacement throughout, worries about superficiality. Its editors are equally concerned by being crushed by the weight of theoretical baggage if easy equivalences are too readily drawn between one form of displacement and another. We take inspiration from the editors' approach as one of 'mid-range theory,' in the sense that we may work with displacement, like gentrification, as an umbrella concept, able to 'consolidate and articulate empirical regularities that would otherwise appear disarticulated' (Lees and Phillips 2018, pp. 6–7). Our approach is to afford by juxtaposition the dissonances and consonances to emerge between the chapters, within the sections and between them. We are, however, driven in our selection of contributions and topics, as well as the organisation of the book, by a number of cardinal reference points that orientate our approach.

To anticipate this a little further, the handbook is positioned in several ways as a product of a team of editors from a geography department with diverse interests spanning Global North and Global South. Our own research has also contributed to debates within fields beyond or that intersect human geography, such as international development, disaster and hazards research, urban sustainability, gender and development, the politics of emergency, and security studies, to name but a few. While we do not necessarily see the handbook as a product of our disciplinary inclinations, we do want to suggest four possible articulations of displacement that a spatial or geographical perspective on displacement might afford.

## Scale: Displacements Beyond the Nation-State

Firstly, the orthodoxy of existing overview texts on displacement is to focus on mass migration and refugees across transnational boundaries, and these are typically authored from within disciplines pertinent to migration studies, namely, sociology, human geography, and anthropology. *The Handbook of Displacement* includes, but also widens, this purview. It encompasses other spatial scales of analysis (beyond national boundaries) and lived experience so that internal displacements and their everyday (re)occurrences are not overlooked. This approach mirrors recognition of 'the ever-expanding variety of mobilities and displacements' (Peteet 2007, p. 643) which can be witnessed today. This might mean a focus upon displacement from the home, for example. Displacement within a city. Displacement from a whole region. And also, for too many, their combination. Such a shift of view to a kaleidoscope of scales moves in tune within a wider trend in scholarship on displacement which substantively deals with, but also shows a direction of travel beyond, the study of the refugee crisis. As Horn and Parekh (2018, p. 503) identify in their introductory editorial to a special issue of *Signs: Journal of Women in Culture and Society* on the topic, 'Displacement is not the same as the global refugee crisis, though that is likely to come to mind for many readers.' Thinking creatively and differently about scale can mean to move beyond the usual frames of analysis for a topic like displacement and outside, as well as within, the traditional limits of nation-state boundaries and classical geopolitical logics. By reframing displacement, the handbook brings together the existing literature on displacement with the contemporary concerns of why people move.

This might mean more than widening our perspective to include other kinds of (im)mobility within or as displacement which would simply expand displacement to a wider set of socially constructed scales, but explored independently of one another, or to prioritise one set of scales over another, perhaps along degrees of magnitude or size. Instead we mean something more akin to concepts of intimacy which Pain and Staeheli (2014) consider in relation to notions of violence. They suggest that intimacy helps reorder violence away from hierarchical spatial scales. Their approach to intimacy might be particularly helpful in destabilising displacement from its apparent scalar coordinates, and to a rather more flexible and messy understanding of relations 'stretching from proximate to distant,' from the body to the globe. Such an approach would help recognise or expose the 'already-thereness' (Pain and Staeheli 2014, p. 345) of the intimate, bodily, local, and global within different processes and experiences of displacement.

## Complicating Terms of Reference Regarding 'Displaced Populations'

Secondly, and related to this former point, the handbook expands the 'terms of reference' concerning the subjects of displacement. It includes 'displaced' people who are less obviously identifiable and a clearly circumscribed or categorised group. Furthermore, it explores and calls into question the plenitude of categories and nomenclatures that go along with various forms of displacement as not only a process but a designation of subject or population that also connotes various kinds of rights or services under law (domestic and international), possible sympathy or enmity from authorities and communities, and, moreover, the subjectification of individuals and communities to bureaucracies, technologies, systems, and procedures which sustain their treatment in displacement.

The majority of scholarship on displacement focuses on contexts of war and conflict, environmental risk, and climate refugees, yet there are other processes of displacement which warrant a place in the orbit of displacement studies. A central aim of the handbook is to bring greater scholarly view to displacements which are 'hidden in plain sight.' For example, just as war is marked by the huge displacement of people from their homes, domestic violence victims are also displaced en masse during 'peacetime' (Pain 2015). Landau (2018) and others show in the *Oxford Handbook of Refugee and Forced Migration Studies* (Fiddian-Qasmiyeh et al. 2018) how people are displaced without having crossed an internationally recognised state border. Many displacements struggle to be made visible. This may be a product of the inevitable scaling that society and academia uses to make sense of particular kinds of displacement and mobility. As Landau (2018, p. 140) argues in his examination of urban refugees, there are compelling reasons for 'sustained attention to both cities and the people seeking protection within them. However, such inquiries' potential will only be achieved through a substantial redefinition of the modes though which we "see" and understand displacement and humanitarian intervention.'

Formal recognition of who is, and is not, 'displaced' matters. For example, it is unusual for anyone in the Global North to claim the status of an 'internally displaced person' (IDP), yet there are many who feel they are. In the context of Hurricane Katrina, lawyers tried to claim IDP status but without success (Kromm and Sturgis 2008). The handbook therefore counterpoises the officialdom of displacement with the realities that many people experience. In other words, it questions in different legal, regulatory, human rights arenas and jurisdictions who *counts* and *for whom* they count? And what implications does this have for different people? Also at stake here is precisely what a subject, individual, or population is assumed to be?

Scaled geopolitical imaginaries of displacement have a tendency to render the complex causes of displacement into simplistic categories of force, perhaps from vast and monstrous non-human global forces: hurricanes and typhoons, earthquake, landslides, and the political equivalent of them. Yet displacement is much more complex than this and requires fine-grained analyses and understandings of agency amongst those people who have experienced displacement. Moreover, just as the way displacement is governed, regulated, surveilled, and administered requires research, the study of police officers, bailiffs, policy officers, border guards, traffickers, and any number of actors within agencies grappling amidst institutional and technological assemblages, policy instruments, laws, and more, are also important. So often the state, for example, is an 'absent presence' in academic research, and direct contact is limited, either by lack of access and/or because academics exclude state actors from their analyses. As Mountz (2004, p. 324) encountered in her research on Canada's response to human smuggling, it is the case that 'state practices are often concealed in praxis,' 'decision-making processes are obscured,' and academics 'reify this disembodiment when they marginalize [these] people from their analyses.' As such, *The Handbook of Displacement* is attentive not just to those who are displaced but also to those who are involved in the business of, or activisms against, displacement.

## Stuck in Displacement

Thirdly, displacement suggests mobility, but there are also periods and spaces of enforced stillness that are not adequately reflected in the displacement literature (although see Bissell and Fuller 2011). Literature from within the putative 'mobility studies' is as much concerned with (im)mobility as it is mobility, where the relations between movement and stasis are integral. As the late John Urry suggested, we need to ask 'Who is moving? Who is moving whom? Who has to move? Who can stay put?' (Urry cited in Adey and Bissell 2010, p. 7). Doreen Massey's (1993) notion of a power-geometry, a way of understanding how societal structures and social differences place everyone in highly uneven relations and capacities to move or stay, is still utterly central. The handbook therefore includes perspectives on displacement being developed by scholars on in situ displacement (Pitkanen 2017), limbo (Brun and Fábos 2015), waiting (Hyndman and Giles 2011), protracted uncertainty (Ehrkamp 2017; El-Shaarawi 2015), and involuntary immobility (Horst and Ibrahim Nur 2016; Siu 2007). Various power geometries, processes, and experiences of suspension and of being 'stuck' in place or circulation will be woven through the handbook.

This attention to the temporalities of displacement also extends to the multiple or overlapping displacements that people can experience across the lifecourse which warrant greater acknowledgement to counteract the focus on singular moments of rupture. For example, while there may be an initial dislocation and trauma, survivors sometimes must deal with long-term or gradual harms associated with protracted circumstances of displacement (Tanyag 2018) that can be conceptualised as slow violence (see Tyner 2020). Journeys of displacement involve myriad adjustments to unfamiliar ecologies and challenge survival 'in a truncated, severed present, torn by involuntary displacement from the numinous fabric that had woven extended meaning from time-in-place' (Nixon 2011, p. 162).

As such, this collection is about displacement as process as well as event. Displacement is often framed in the guise of spatial exclusions or 'expulsions' (Sassen 2014) from home and living space as part of a wider diagnosis of unstable and disconcerting times (Brickell 2020). Yet, as Brickell (2020, p. 6) argues, displacement is commonly experienced as an 'ongoing loss that a language of expulsion risks eliding.' 'Less sensational, yet nevertheless devastating' dislocations from home (Vaz-Jones 2018, p. 711) are important and too often neglected. They can also be multiple rather than singular, and multiple in the sense of persisting over peoples' lives in a slow kind of wounding dislocation or in a manner of a stretched-out duress. Jasbir K. Puar (2017) might call this a kind of 'debility' through displacement, well beyond any kind of severing punctual moment (see also Hyndman and Giles 2016).

## Place and Displacement

Fourthly, as a cohort of geographers, we are well positioned to re-evoke and explore the 'place' in displacement. This connects to the significance of place in 'dis*place*ment' (Davidson 2009, p. 226, emphasis in original). The significance of placemaking, unmaking, and remaking will run across the handbook as will their non-linear fashioning in peoples' lives. It is becoming increasingly critical that investigations of displacement explore the struggle for place that is at play.

In addition to examining peoples' 'right to place' and related ideas of (non) belonging, the technologies and materialities used to displace people from their homes and communities will be a forefront focus of the handbook. While displacement challenges the displaced 'to take their proper place [of non-being] instead of taking place' (Butler and Athanasiou 2013, p. 20), 'emplacement' (Roy 2017; Çaglar and Glick Schiller 2018) is one means of contestation. To see place as a kind of right is an interesting pairing to consider with a right to move or leave. Displacement might be effectively contested by a 'right to dwell' (Davidson 2009, p. 232) in place. Such a right to dwell emphasises 'a right to inhabit the abstract space comprising "home" in a wider sense' than just its physical infrastructure (Hubbard and Lees 2018, p. 18; Baeten and Listerborn 2015). Indeed, Vukov's (2015) recent advocacy of a mobility justice (see also Sheller 2018) identifies not only 'just forms of movement' but also 'dwelling,' both a call for a 'right to immobility' and an end to imposed forms of immobility. Reclaiming can also be a powerful process of finding place or what Ahmed et al. (2003) once called 'homing.' 'Homing' for these authors could depend 'on the reclaiming and reprocessing of habits, objects, names, and histories that have been up-rooted-in migration, displacement or colonization' (p. 9).

To see displacement through place and emplacement also powerfully focuses displacement outside of the abstract lenses of flow and trajectory, which turn mobility and displacement into a kind of technical vector or trace of movement. Some traditions within geography have also sought to add motion, mobility, and displacement into interiorised, static, and even regressive notions of place and home (Verstraete and Cresswell 2005). Doreen Massey's concept of a 'progressive sense of place' attempted precisely to understand the exterior and interior relations, the local and global movements, and trajectories of throwntogetherness (Massey 2005), which constitute even the most and seemingly bounded of places. Perhaps one of the other faces to displacement that we regularly see is through the interiorised senses of place that worried Massey and which have been so irritated in recent years. For those confronted with the displaced in the form of the figure of the migrant or the homeless, and subjected to the undersides of a globalism and neoliberal politics which have hollowed out their employment prospects, or undermined wage growth, and reduced the quality and extent of public services, displacement is reflected in new kinds of *re-placing*. Displacement seems to have easily disturbed an anxiety of being replaced by an unwelcome other/outsider. Ensuing toxicity of debate has, as a result, 'intensified over the last few years, with the politics of fear and division increasingly framing discussions' (IOM 2020, p. 161).

There is, of course, much danger in reducing boundedness to some kind of social construction; displacement might force a realisation of the insistence of place even in the experiences of those most and continually displaced from their homes. The anthropologist Annika Lems (2016) has sought to reassert and problematise place within studies of global displacement. She explores the story of a Somali refugee Halima, who, having fled Mogadishu to an indeterminate and precarious status in the UAE, and under the threat of deportation, arrived with her family at Melbourne airport in order to accept settled status in Australia. In Halima's context, the presence and absence of place through displacement was in play. 'At the same time as the place made itself felt, it also created an utter sense of displacement within her. Taking in all the strange and unknown features of this new place made her realise how much she had become part and parcel of another place, a place she had left behind' (Lems 2016, pp. 316–317). For Lems, this story of arrival elicits a reciprocity of emplacement and displacement, 'it forces us to look into the inescapable *presence* of places and into the ways they continue to shape us existentially—even in the face of violent disruption and displacement' (2016, p. 317, emphasis in original).

## Organisation of the Handbook

We organised the handbook, and our discussion in the reminder of this introduction, into eight major sections: (1) *Conceptualising Displacement*; (2) *Technologies of Displacement*; (3) *Journeys of Displacement*; (4) *Traces of Displacement*; (5) *Governing Displacement*; (6) *More-than-Human Displacements*; (7) *Representing Displacement*; and (8) *Resisting Displacement*. The first section covering key theories should not be taken as an exhaustive or definitive set of ideas, but rather a suite of approaches rooted in different conceptual lineages which reflect the breadth of recent and emerging work in interdisciplinary displacement studies. They include mobilities, political ecology, political economy, slow and fast violences, assemblage thinking, affect, rights-based approaches, queer, gender and feminist, race and ethnicity, and postcolonial perspectives. Together they help tease out and make sense of displacement as an event and process linked to highly contingent and often contested rights to place and belonging. We hope the seven thematic sections that follow from it offer the reader the background knowledge and tools to understand how displacement becomes known, is conceptualised, and is practised in different ways. Given the expanding study of displacement, the handbook aims to be a forward-thinking text and provides discussions of key contemporary areas of debate and controversy across these sections. With regard to digital infrastructures of displacement, for example, the handbook includes critical explication and reflection on surveillance technologies for monitoring and tracking displaced populations, to immersive virtual reality kits used to understand and temporarily 'inhabit' the bodies of the displaced.

Each thematic section begins with what we have called an 'intervention.' These try to ensure that actions to creatively and/or strategically intervene in displacement events and processes are not lost in the volume. These interventions explore myriad meanings and manifestations of displacement and its contestation from the perspective of displaced people, artists, writers, activists, scholar-activists, and scholars involved in practice-oriented research. As the interventions will richly demonstrate, these 'categories' are neither mutually exclusive nor preclude collaboration across. Their conception and naming as 'interventions' also arose from our desire for the handbook not to be exclusively beholden to academic prose, but to be inviting of other ways of communicating knowledge of, and on, displacement.

The concepts, themes, and interventions which comprise the collection as a whole also look to push beyond disciplinary boundaries. As the biographies testify, academic contributors to the book come from a wide range of backgrounds, namely, architecture and design, visual arts, media studies, geography, politics and international relations, sociology, international development, history, modern languages, business, and area studies. While all editors are themselves geographers as previously noted, the handbook is innovative in its reframing of displacement beyond the usual purview of human geography and international development. It brings together cutting-edge work from broader interdisciplinary realms and is attentive to the study of displacement as being radically reshaped by newer work in the humanities, including fine art and literature (see, as examples, Espiritu and Duong 2018; Stonebridge 2018; Qasmiyeh 2016).

In order to understand and make sense of the expanding landscape of displacement research, we came around a table (Fig. 1.2) for three days to talk, share, and plan. As members of the Geopolitics, Development, Security and Justice (GDSJ) Research Group located in the Department of Geography at Royal Holloway, University of London, we had become increasingly aware that our individual research endeavours, albeit quite distinct, all had displacement in common. We were motivated by there being productive value in thinking more about their connective tissue, not just for the sake of group

**Fig. 1.2** Handbook planning tablecloth, July 2017. (Source: Katherine Brickell)

cohesion, or carving out space and time for collegiality and scholarship in the academy, but because the insights we could collectively harness could elucidate and make connections between key issues in displacement studies. We have done so not just as editors but also as (co-)authors to interventions and thematic chapters in the volume. Several of our departmental colleagues (Laurie Parsons and Jennifer Cole) also contributed to this endeavour through their own chapters.

In some ways, the tablecloth became a different kind of map for us to consider displacement with. It was less a reductive representation (as we discussed above), but a means to organise the relations between disparate and discrete concepts of displacement and stories of displaced peoples. Where the folds in the cloth look like vertiginous boundaries between these parts and types of displacement, reasserting perhaps the disciplinary and conceptual demarcations which could keep different types and apprehensions of displacement apart, we like to think of this eventual manuscript as a refolded or scrumpled up version of that surface. In two dimensions, displacement appears legible and orderable, whereas in the volume of this book, the boundaries become rearticulated as relations and connections, and entities become proximate in ways they weren't before. The potential and multiple valences of displacement might become visible.

While this RHUL orientation is a distinctive feature of the book, we were conscious that research on displacement is obviously a much wider endeavour. Contributors identify from Europe (United Kingdom, Italy, Switzerland, Ireland, Denmark, the Netherlands, France, and Germany), North America (United States and Canada), Australia, South America (Peru), and South East Asia (the Philippines). These geographical designations render clear how the production of scholarly knowledge published in books and journal articles on displacement is authored by mainly Global North-located academic voices and how far there is to go in addressing this centrism. This is all the more stark given the global nature of displacement and its significance in and between Global South countries featured in the handbook, from Haiti to Mexico, Brazil, Kenya, Burkina Faso, Iran, Iraq, Sri Lanka, and Cambodia, and in respect to refugee populations from a diverse set of countries entering Europe and North America. In the rest of the introduction that follows, we provide an overview of each section and a flavour of what is to come across the seven organisational themes.

# Technologies of Displacement

'Measure and control—the legal and the technical' (Elden 2010, p. 799). These are perhaps some of the key ways we might consider technologies of displacement, or rather displacement *as* technology. What does this mean? It could mean drawing on the long body of scholarship which has elaborated Foucault's various writings on political technologies exercised through biopolitics and the securing, management, and regulation of the social body through regimes of power (Foucault 1997). As Elden (2010) suggests, apparatuses of power summon technologies that measure and control, that may move through legal and juridical terrains and more technical sets of machines. And so, the technologies of displacement may be much more than actual technologies, but the *techniques* of the technical: the practices of human and non-human, administrative and bureaucratic systems, underpinned by particular political rationalities and beliefs. Elden gives us some sense of the broad array of ways in which territory might be considered a 'political technology,' suggesting that it may include:

> legal systems and arguments; political debates, theories, concepts, and practices; colonization and military excursions; works of literature and dictionaries; historical studies, myths, and—the technical in the narrower sense—geometrical instruments, statistical handbooks, maps, land-surveying instruments, and population controls. (2013, p. 17)

The entries in this section and scholars elsewhere too (see, e.g. Lea 2015) recognise that the above might present an overly 'technical' reading of what constitute technologies of displacement. As Foucault himself suggested, the technologies of life are also 'arts of governing.' Taking this in an admittedly literal sense, the section begins with a co-authored intervention by artist-researcher Barry, and Adey, which shows how technologies of displacement are inherently aesthetic and aestheticisable. The intervention explores Barry's subversive reworkings of aircraft 'safety cards,' which are found in the pockets of seats during air travel. While safety cards are a form of aesthetic technology used to manage (the potential for) emergency amongst the relatively rich and privileged, Barry's artworks speak to representations of precarious displacements that have become much more visible in the current visual regimes of the so-called migrant crisis, at least in the case of African and Middle Eastern migration to Europe. Just as some sites of protest, such as the 2016 'die-ins' for refugees in London's St. Pancras Eurostar station, have become possible sites of rupture (see Verifeye 2016 for a video), the reimagined safety cards are too. They can present moments of public awareness of distant and once invisible displacements, but also antagonisms between different kinds of displacement politics.

In the following thematic chapter by Pritchard on 'The Artwashing of Gentrification and Social Cleansing,' aestheticiation is also placed centre stage. His chapter shows how aestheticiation has become a common technique in processes of gentrification and one which tends to rely upon certain orderings of visual and sensory experiences and feelings of places which are privileged and highly exclusive. To evidence this, Pritchard explores the actions of the V&A Museum in London, and its attempts to 'preserve' a utopian project of urban renewal. Even while driven by some positive intentions, the museumification of once-lived-in council homes becomes a continuation of the dispossession and displacement caused by gentrification practices premised on real estate speculation, aestheticised assumptions around working-class families and their housing, and inadequate processes of community and resident consultation. His examination of the artwashing of the demolition of the Robin Hood Gardens estate, and wider gentrification of east London—through the V&A's purchase of a chunk of the apartment buildings—also leads us to the need for a diverse emphasis on technologies of displacement. The chapters by Barry and Adey and then Pritchard demand and practise a way of wrenching displacement technologies from out of the purview of the technical, or artwashed, from the hidden, silenced, even glossy aesthetic registers of displacement.

If the first two chapters in the section demonstrate the need for a focus on technologies of displacement as exerted through the visual and material, then the third, by Parsons on 'Remittances, Translocality, and the Climate Migrant Within,' asks the handbook reader to additionally consider the significance of policy and media discourse in the construction of populations currently or likely to be displaced by climate change in the future. Here, the 'epistemic object' of displacement technology is terminology: the 'climate migrant' or 'refugee,' who has been treated as 'exceptional,' through a variety of visual representations, words and utterances, discourses, and policies which 'other' them. Michalon's chapter 'Barbed Displacement' focuses too on governmental power within the framework of migration policy, but hones in on the carceral parameters of detention through a explicitly Foucauldian lens. The chapter acts as a powerful synthesis of contemporary governmental control over migration and works with the figuration of barbed wire as a literal and metaphorical device to understand the various walls that channel or keep in place displaced migrant bodies.

The section also moves through a range of different scales of displacement technologies in relation to deportation. As Walters makes clear, for example, in his chapter 'Technologies of Deportation,' deportation could at once be understood in terms of the technologies that constitute its variety of forms—an aircraft, a detection regime, places of detention, and paperwork—just as deportation itself, in different contexts, could be understood as a kind of technology, made up of a 'contradictory field of practices, programmes, strategies, and struggles.' For Walters, the task is to dismantle or unpack the black box of the machinery of deportation.

While the chapters by Parsons, Michalon, and Walters focus on national and international policymaking in relation to displaced populations, the fifth chapter in the section, by Edelman on 'Street Technologies of Displacement,' turns to city-scale technologies of management. In the context of Washington, DC, Edelman examines transgender experiences and responses to Prostitution Free Zones (PFZs). These zones are based on relatively strict and normative demarcations and definitions of space, which bodies may inhabit it, and what behaviours they may perform. Edelman shows how the apparent regulation of sex work actually becomes a way in which to profile and displace different categories of race, class, and sexuality, especially those who appear to transgress the heteronormative and neoliberal norms within the city. The PFZs exist as lines on a map, but they are translated and performed by a whole apparatus of law, police, practices and procedures, bodies, and architectures. For this reason, the transgender community were some of the strongest activists working to repeal the PFZs.

As Freeman shows in the next, and final, chapter in the section, 'Olympic Favela Evictions in Rio de Janeiro,' the quality of such 'practices, programmes, strategies, and struggles' is inherently based on a history of neoliberal policies and therefore a much wider global pattern and regime of displacement. And yet, across these different categorisations or types of displacement technologies/displacement as technology, between and within each contribution, we also find that displacement technologies are not consistent or necessarily moving in the same direction or towards the same goal.

## Journeys of Displacement

This section focuses the displacement lens on the actual journeys themselves. Drawing conceptual and empirical insights from research into relocation and migration across land and sea, it moves with the people concerned through time and space. Key themes include the complexity of journey trajectories, the implications of borders and boundary-crossing, state-driven displacements which keep people on the move, and the disorientation of journeys without end.

The section starts with the written and painted story given by an anonymous Iraqi woman who fled ISIS (provided by Begikhani). In the story 'Our Last Day in the Orchard,' she recalls their journey, from their family orchard to a life lived in the refugee camps of the Iraqi Kurdistan Region (IKR). Further reflecting on the unpredictable dynamics of journeying, the first thematic chapter in the section by Schapendonk and Belloni, 'Constraints and Transgressions in Journeys of Displacement,' highlights the ways that the journeys themselves become sources of threat, risk, and trauma. The conventional binary of force or agency is expanded into agency *and* constraints, choice *and* force within fragmented journeys full of geographical detours, unexpected transit situations, and successive moments of displacement and replacement. They disrupt conventional notions of displaced people as simply victims, including through their consideration of the gendered dynamics of displacement journeys.

Tazzioli's chapter, 'Migrants' Displacements at the Internal Frontiers of Europe,' focuses on the role of state policies in not just triggering displacement but in keeping people on the move. Within Europe, she discusses the ways in which state authorities chase people away from internal borders, disrupting their journeys and forcing them to restart their journeys multiple times. She focuses on the costs to migrants in terms of the expense of travel, but also in terms of the exhaustion of repetition of the same routes time and again. These repetitive journey attempts steal migrants' time and lives, delaying their futures and even their possibility to think about their futures as they remain stuck in an endless journeying.

A similar focus on how state power both causes and maintains displacement is discussed in Gill and Simon's chapter on 'Carceral Journeys.' Moving people through a complex infrastructure of institutions can make individuals invisible to family, friends, and supporters and even somewhat to themselves as their ways of thinking and sense of identity are disrupted. Using examples from research into 'extraordinary rendition,' they highlight how these displacements can be multinational and the importance of focusing on the journeys as well as the institutions implicated in the networks used. The profitability of operating these displacement networks is highlighted as both cause and consequence of their increase across the globe. The journeys incorporate the unpredictability of both forced mobility and forced inertia for the people being displaced but can also take their toll on the staff of both the means of transport and the carceral institutions.

The interaction between humanitarian and security concerns on displacement routes is discussed by Squire and Stierl in their chapter 'Precarious Migrations and Maritime Displacement.' Increased security at land borders has redirected many displacement routes to the sea, with the distinctive precarities of the means of transport on water. They highlight the complex legal and regulatory frameworks, the technologies of surveillance, and the range of vessels involved in maritime journeys. The deep cultural and historical understandings and symbolic significance of the sea, with a particular focus on the Mediterranean, have fed into the current military-humanitarian search and rescue operations which attempt to manage desperate and dangerous journeys on the borders of Europe.

The humanitarian responses to displacement are highlighted next by Cole in her chapter 'Maintaining Health on the Move.' Geographical journeys, both within states and across borders, affect individuals' existing health conditions and access to treatment, often causing or spreading health problems. Complex and multistage journeys make consistency of healthcare difficult, which is compounded by lack of clarity over legal status and rights to health. Attempts to provide continuities of care, such as during pregnancy, can be confounded by administrative actions of state and humanitarian providers which force further displacement. In the same way as borders and conflicts block the movement of people, they can block the movement of medicines and healthcare workers.

Overall, the section explores how a focus on the journeys of displacement unlocks insights into the causes and consequences of such journeys, and their complex trajectories over time and space. Drawing on a rich conceptual language of journeying, and grounding it in the realities of borders, institutions, networks, and transport, the chapters start to piece together the often fragmentary understandings of the routes of people on the move. Shining an empirical and conceptual light on the gulf between here and there brings to the forefront the gendered and racialised processes that trap people in ongoing displacement and uncompleted journeys. Furthering insights from the first thematic section on 'Technologies of Displacement,' this second thematic section highlights the significance of state and multinational actions and inactions for intentionally ensuring disorientation in the present, and the putting of futures permanently on hold.

## Traces of Displacement

In this section chapters explore 'Traces of Displacement' in two respects. The first explores how current-day displacements can be identified and traced. Chapters blur the approaches of how the displaced are monitored, tracked, and traced through a variety of visual and close-sensing techniques, with research methodologies available to academic researchers. The second traces histories and legacies of displacement in built and spiritual landscapes from former times and/or former places. The chapters expose the diverse temporalities of displacement and point to some of the methodological challenges and opportunities for studies of displacement.

These insights are woven through the chapters presented and were compelled by the editors' experiences in their own research practices. The opening intervention, 'When There Is No Time to Stop,' for example, is authored by Ayesha Siddiqi in connection to her work on disasters and displacement. In 2019, Siddiqi worked with refugees in South Africa from the Democratic Republic of Congo (DRC) on an exhibition that shared their experiences of displacement. These citizens of the DRC, displaced primarily by the violence of war, participated in a workshop at the Johannesburg Holocaust and Genocide Centre to co-create their stories and produce materials for the exhibit. A key feature at the exhibition was a 5 foot by 6 foot map of Africa (Fig. 1.3) that used coloured string to trace the journeys of these participants from DRC to Johannesburg in South Africa. The map exhibit looked deceptively simple, but populating it with each individual journey was an immensely challenging exercise and epitomises the ethical and material difficulty of 'tracing displacement.'

The most evident challenge, repeated in a number of participant narratives, was that these journeys were not planned; those who fled left in urgency. They neither knew where exactly they were going, what route they would take, or when they would arrive. Participants often crossed Lake Tanganyika by boat, hitched rides with truck drivers, crossed Zimbabwe and Zambia, and eventually arrived in South Africa. For many this was the extent of the details they could, or indeed wanted to, recall of their journey, obscuring names of specific towns and places, perhaps even consciously, to let others follow in their paths.

While those fleeing wanted to remain invisible to border regimes during their displacement journey, once in South Africa, they desired nothing more than to be visible to the state, going to great lengths to 'get papers' that would enable them to work legally in Johannesburg. It is therefore not surprising that migrants use both visibility and invisibility as part of their tactics and

**Fig. 1.3** Workshop participants explaining their journeys at the exhibition, Johannesburg Holocaust and Genocide Centre, June 2019. (Source: Ayesha Siddiqi)

strategies through difficult journeys and later settlement in host communities (Polzer and Hammond 2008). Displaced peoples, refugees, and asylum seekers navigate between invisibility and visibility not just as helpless victims but using their own discernibility as active agents. Visibility can prove dangerous if it results in those considered 'illegal' to be subject to control by state, UNHCR, border regimes, or indeed any Foucauldian disciplinary power. In a different context, visibility is a much-coveted asset, helping to access economic opportunities or social and legal protection for the displaced.

The two contributions by Pieris ('Antipodean Architectures of Displacement') and Christensen and Madsen ('Spiritual Geographies of Displacement and Resilience') work to make displacement visible in the context of settler colonies. Pieris's piece explores narratives of displacement threaded through Australia's built environment. This includes carceral facilities where Aboriginal populations and Anglophile convicts were forcibly displaced and were integral to the state building project. The disciplined geometric structures tamed the vast and open spaces of the continent while also re-educating 'errant subjects.' Displacement, and incarceration of the displaced, has been integral in creating normative citizenship in Australia. The chapter therefore makes visible the architectures of displacement in the colonial project. Christensen and Madsen make another form of displacement visible in their piece, while Indigenous populations in settler colonies have not been physically displaced, and they have experienced deep and violent spiritual displacement. Their work demonstrates that 'Indigenous ontology *is* spirituality' and in dispossessing Indigenous peoples of the land they have been connected to for generations, capitalism and settler colonialism have been part of their ongoing experience of 'rootedlessness.'

The contributions in this section demonstrate not only that recording and documenting displacement is an immensely political exercise that always serves particular interests but also that its hiding and concealment is too. As Awan argues in her chapter, 'Mapping Trajectories of Displacement,' methods employed to tag and trace individuals cannot be 'either apolitical or neutral' in the context of heavily surveilled border regimes. Research using tracking techniques such as GPS for location positioning of refugees is reproducing the oppressive structures of these regimes for discipline and control. Tracking individuals on the move in this way strips away their agency to be invisible, leaving 'deterritorialised and desubjectified lines' feigning neutrality and masking the politics of such a project. At the same time, we are also provided lucid examples of research that works with migrants using a more practice-based approach, protecting the identities of vulnerable migrants, yet using their narratives to question dominant power structures. She labels this research 'unapologetically political' and in doing so is clear about where her scholarship on mapping displacement sits and whose interests it is choosing to serve.

While making vulnerable migrants visible to oppressive surveillance regimes reveals a strong leaning towards elite politics and hegemonic power structures, keeping someone invisible also involves power. Making the presence of the invisible felt and their voices heard brings to the fore research that is challenging hegemonic power structures. Bowstead, Turner, and Hodkinson's contribution, 'Uncovering Internally Displaced People in the Global North through Administrative Data,' speaks truth to this power by revealing internal displacement in the United Kingdom using administrative data. Their research on social policy, specifically housing benefit reform in Leeds and support for survivors of domestic violence, enables the scale of residential displacement to be newly identified. While the UK government is hesitant to visibilise the residentially displaced, as that would make the state liable 'to address the material needs of the displacees,' these authors illustrate that thousands more people have been residentially displaced as a result of the housing benefit reform or domestic violence. Their methodological approach uses state-generated data to trace the movement of the very people the state would rather remain invisible.

## Governing Displacement

The governing of the displaced, the policing of the unwanted, and the disciplining of migrants are the key themes in the 'Governing Displacement' section which flows logically from the former on the politics of presence and visibility. The chapters in this section explore the evolution of and (often violent) practices of governing displaced populations from spatial and temporal perspectives and their implications. They interrogate historical and contemporary neoliberal global patterns of governance while drawing on the specificities of localised governance regimes. The intersectionalities of ethnicity, gender, religion, and class are shown to have profound effects on the life chances of displaced groups as they are subjected to targeted governance practices from state to street level which decide who can enter, who can stay, who cannot, and ultimately who can live or die. As such, the chapters build on those in the 'Technologies of Displacement' section which are attuned to a critical geography of vulnerability which maps the spatial linkages between the hurt and the hurter (Philo 2005)—between the vulnerable and the governing 'bodies' of displacement.

In the opening intervention on 'Forensic Oceanography,' Heller and Pezzani explore their scholar-activism work in piecing together photographs and satellite images to corroborate survivor testimonies of deliberate non-assistance as migrants are left to perish as they try to cross the Mediterranean Sea. With a wider overview remit, in the first thematic chapter, 'Governing the Displaced,' Tawakkol, Bhagat, and Sharma explore the complex and uneven governance of the displaced by a range of state, non-state, and private sector actors. Governance is framed here within the contexts of neoliberalism, austerity, and the contradictions of how official policies differ from actual practices. The chapter outlines how the categorisation of migrants, across and within borders, can have profound impacts on shaping their life chances including citizenship rights, self-determination, access to adequate housing, employment, finance, and mobility. Surrounding these life chances are the pressures of austerity and the increasing commodification of displaced populations. They note, for example, how people living in situations of displacement are often expected to house themselves whilst constrained within often long-established refugee camps or competing with pre-existing marginalised populations who are also struggling to find homes. The displaced are thus constructed as the 'other' through state-led discourses and policies as the marginalised and the displaced struggle over limited resources.

The uneven and unequal nature of governing, or policing, of displaced groups across time and space is further developed by Menjívar and Gómez Cervantes who use what they describe as institutionally racist US immigration policies towards Latino and Muslim migrants to illustrate the wider issues of how immigration bureaucracy can target specific immigrant groupings. In their chapter, 'Bureaucracies of Displacement,' the violence perpetrated by the state under the premise of legal mechanisms such as Executive Orders is explored. Their discussion hones in on the social and mental impacts of the techniques employed, currently amplified by Trump's regime, such as the deliberate separation of families, detention of mothers and their children, and consequent trauma experienced by the detained and the deported.

Baker focuses in on the actual 'doing' of displacement via 'Police, Bailiffs, and Hired Hands.' The chapter explores the often postcolonial specificities of eviction practices and situates them within the wider context of eviction as a planetary process of accumulation by dispossession within global private property markets. Baker thinks through the genealogy of systems and practices of eviction which evolve and react to new techniques of resistance and contestation. Eviction is, importantly, discussed as the duration in which eviction does not begin and end with the violence of actual eviction but includes deliberate physical and emotional policing and coercion of the targets of eviction leading up to their displacement, as well as post-eviction issues such as instability and trauma. Baker concludes with a thoughtful discussion of the ethical positionality of academics who research displacement given that research outputs have been used by policing agencies to refine their techniques of eviction. A more interventionist mode of research is advocated with a call for academics to take a lead in the pursuit of abolition of displacement policies and practices outlined here.

The actual 'doing' of elements of displacement such as categorisation, detention, movement, and removal of migrants is examined by Borrelli through the Foucauldian lens of biopower. In her chapter, 'Governing the Unwanted,' Borrelli outlines how the technologies and practices of displacement deliberately discipline the migrant though practices such as incarceration, constant movement between places, deciding where migrants can/cannot go, and deciding who is a migrant regardless of how long they have lived in their (or their parents') adopted country. Local enforcement is key with street-level bureaucrats having the power to decide on the forms of disciplining and potentially who has the right to live or die (as the intervention at the start of the section rendered stark).

The governance of the right to life is taken to its logical conclusion by Basso in the final chapter in this section entitled 'A Forced Displacement and

Atrocity Crime Nexus.' Basso outlines a historical account of the relationship between displacement, atrocity, and genocide. The removal and transportation of displaced people are seen as key in genocidal processes, often in the pursuit of ethnic cleansing and identity-based homogenisation. They are moved out of, or driven from, their communities to killing sites using methods such as starvation, forced marches, forced labour, and executions to annihilate targeted populations. Displacement, in this context, is therefore both event and process with the event of forced displacement being part of wider processes such as ethnic cleansing and annihilation. Forced displacement is likely to grow as climate change creates the conditions for ever greater displacement and political violence within the context of existing social and economic relations such as local political economies, entitlements to resources, and ethnic and religious tensions. Climate change also provides an additional discursive tool for existing state policies towards migrant populations cast as security risk, threat, or 'other' (e.g. see Hartmann 2010).

The governance of the displaced is thus a key element of the displacement 'story,' and this is reflected in the links which can read across the thematic sections so far. The identification and selection of the displaced, the legal frameworks developed over time and space to facilitate the violence of displacement, the agency of state and non-state actors deployed to physically enforce (or prevent) displacement, and the possibilities to resist are all situated within the ever-evolving governance frameworks and consequent power relations set out here.

## More-than-Human Displacements

Experiences of displacement have literal and emotional dimensions which are mediated through materiality—the belongings and kit of everyday life which are destroyed, left behind, carried en route, provided in situ, and/or (re)found upon arrival (and not necessarily in that linear order). In this section of the book, the chapters focus on survival, endurance, resilience, hope, and love mediated through human, but also more-than-human materialities of displacement. Our use of the term 'more-than-human' takes its cue from geographical and wider social science writing on the 'livingness of the world' (Whatmore 2006, p. 602) and the co-production of 'humans, animals, plants and other actors and intermediaries' (Bear and Eden 2008, p. 448).

The need for this open approach is perhaps most clearly articulated at the start of section six through the flower-focused intervention and animal-centred chapter which follows it. More specifically, Brickell's 'Flower Power' intervention on the use of lotuses and carnations to resist displacement emphasises their shared capacities to peacefully demonstrate against forced eviction in, and deportation to, Cambodia (see also Sect. 8 on 'Resisting Displacement'). White's chapter, on 'Animals, People, and Places in Displacement,' focuses, meanwhile, on how livestock, working animals, pets, and wild animals influence the lives of displaced people living in purpose-built camps because of war. Written from a historical and contemporary perspective, incorporating artwork on the subject, and providing insights that have implications for the study of other displacement contexts, White contends that 'our understanding of human experiences of displacement deepens when we recognise the roles that animals play in them.' To demonstrate this, the chapter deals with myriad and at times vexed issues pertaining to the representation and treatment of displaced peoples as animals; animals as displaced people's livelihoods; butchery, farming, and yarn production in and outside of camps; and the sense of place, comfort, and identity that animals can afford displaced people. Together the intervention and this chapter foreground human-nature relationships and show the importance of interactions between humans and more-than-humans in the context of displacement.

A second theme that runs through the section is technology and its role and significance for displaced people. In their chapter, 'Energy on the Move,' Cross, Martin, and Verhoeven focus on 'energy objects' that meet refugees' needs for heat and water in the Goudobou camp in Burkina Faso. The authors emphasise the material indispensability of local technologies to everyday life. By exploring the use of bellows and a goatskin water carrier, they reveal how such Indigenous technologies have an important place in everyday practice. While UNHCR provide blowtorches and plastic water containers for the equivalent functions of producing heat and storing water, these are seen, by some, as inferior. They preclude, too, the intergenerational transfer of knowledge and skills associated with more traditional and familiar technologies. The authors warn, therefore, that Indigenous technological heritages have been unduly sidelined in the humanitarian community. Leurs and Patterson focus next on 'Smartphones: Digital Infrastructures of the Displaced' and their diverse 'material, portable, embodied, and affective' roles in supporting displaced populations. Drawing on conceptual ideas of infrastructure, they argue that matter is not passive, but rather is processual and agentic. In this regard, their chapter speaks to the more-than-human orientation of the section and White's chapter specifically. As Whatmore (2006, p. 604) writes, both 'animals and technological devices have variously been used as "agents provocateurs" in tackling the question of difference and rigorously working it through the specific materialities and multiplicities of subjectivity and agency.'

Smartphones, Leurs and Patterson argue, are critical infrastructures of survival and surveillance, transnational communication and emotional management, and mediating technologies of digital self-representation.

The everyday stuff of belonging is a third theme that connects the chapters. In the aforementioned chapters, White details the affection shown to animals to engender comfort, Cross et al. impress the significance of Indigenous technologies for placemaking and identity, and Leurs and Patterson talk similarly to the affective work of the smartphone. The ensuing chapters by Pitt on 'Family Photographs in Displacement,' and Fathi on 'Displaced Home-Objects in Homing Experiences,' also link strongly to this theme. These chapters are trained on 'home' as a material and/or metaphorical site where 'found' family photography (Pitt) and 'home-objects' (Fathi) have an important role for people who are currently, or at one time have been, displaced. Pitt explores different genres of photographs produced in displacement, including 'celebration selfies' today and formal portraiture in the past, to prove, share, and enhance family belonging and togetherness under challenging circumstances. Her chapter speaks to the experiences of the family whose shoes feature on the front cover of the handbook. The photographer of the book cover image remembers the father of the family showing her a whole set of printed photographs he had carried with him: photographs of family both with him and those left behind. For Fathi too, the traumas of leaving home are tempered by household objects which hold cultural meaning in her chapter. Here, food is also identified as critical to 'the construction of national identity and belonging' with the sending and receiving of food commonplace (see also Murphy later in this volume).

Pulling through the material and infrastructural focus of chapters that have preceding it, the final entry in the section considers the role of design in displacement and 'Moving beyond Quick-Fix Solutions in Rebuilding Housing after Disaster.' Charlesworth and Fien, both scholars of architecture and urban design, argue that 'facilitating the provision of culturally appropriate and technically efficient housing is one of the most important tasks for those working with displaced people.' Through case study examples they explore what kinds of shelter best meet the needs of displaced groups and question universalist approaches to resettlement and rebuilding which neglect the knowledges and aspirations of those most affected. Akin to the points raised by Cross et al., they contend that the use of local technologies, materials, and labour is too often dismissed.

## Representing Displacement

> The photo is haunting, a vivid reminder of the danger many face when they try to cross into the United States. It shows the human toll of a crisis at the border that's often debated with abstract statistics and detached policy arguments.

So begins a CNN news report, from 26 June 2019, reflecting upon the representational power of a photograph taken on the southern banks of the Rio Grande river—the fluid boundary between Mexico and the United States (Gallón et al. 2019). The image in question, taken by the Mexican photojournalist Julia Le Duc and published initially in *La Jornada* newspaper, shows the deceased bodies of Óscar Alberto Martínez Ramírez and his 23-month-old daughter, Valeria, floating face down in the murky and littered water of the Rio Grande. Valeria's body is tucked inside Óscar's T-shirt, her right arm still apparently clinging to her father's neck. From the pages of *La Jornada*, Le Duc's photograph travelled around the world, propelled by social media and large multinational news organisations, drawing international reaction to the unfolding humanitarian crisis. For some, the parallels between the lives of Óscar and Valeria Ramírez and that of Alan Kurdi—the Syrian-Kurdish toddler whose body was found (and photographed) on a Turkish beach in 2015—were striking: displaced bodies bound together by water, infancy, powerlessness, victimhood, and, in death, their mutual documentation by photojournalists. Sky News reproduced the picture under the headline 'Why Shocking Photos of Drowned Migrants Must Be Seen' (Walker 2019), while others, including the New Jersey senator and presidential hopeful Cory Booker, urged followers not to 'look away' (Booker 2019; also cited in Rahim 2019). For Nadine Batchelor-Hunt (2019), writing in the *Huffington Post*, these two ends of the visual economy of displacement are bound up in the media's (and, by extension, the public's) 'dehumanising, macabre voyeurism' of displaced bodies that is all-too-often characterised by momentary, fleeting, outrage at the death of un-named or even misnamed victims—transient, ephemeral interest in transient, ephemeral lives (Anderson 2015).

The chapters within this section examine different aesthetic forms, both established and experimental, precisely in order to disrupt commonly held understandings of displacement and the dominant narratives, expectations, and power dynamics that have shaped the ways in which displacement is seen, heard, and experienced and displaced people are all-too-often constituted for public 'consumption.' The chapters are interdisciplinary in their scholarly origins, international in their coverage, and work across multiple representational modes—from news media, to documentary film, immersive and virtual reality experiences, through to the work of contemporary artists.

The visual sociologist Dominika Blachnicka-Ciacek discusses 'How Not to Eat Human Stories' through a close examination of the production of photographic and cinematic images of the displaced, and the ways in which representations of refugees are constructed, canonised, and appropriated. By focussing on two films that have been produced in response to the recent 'migration crisis' in Europe, Blachnicka-Ciacek explores the power relations involved in the production and consumption of images of displacement, and the problematic nature of the 'gaze' and spectatorship.

The contested and increasingly circular relationship between the 'producer' and 'consumer' is further explored in Emma's Bond's chapter, 'The Case of Immersion and Virtual Reality,' which maps out the blurring of agency, empathy, and representation enabled by 'immersive' techniques and technologies that promise to communicate the refugee experience to their audiences in new and innovative ways. Drawing on affect, haptic theory, and the work of Jacques Rancière on the image and the spectator, Bond focuses analysis on three recent 'immersive' representations while also considering oppositional perspectives that push back against the notion of the 'empathetic audience' and the ability of aesthetics to contribute real social and political change. Ultimately, it weighs up the potential risks and benefits in claiming to represent the experience of displacement through media that blur the lines between activism and entertainment and seeks to determine how far experience itself can be 'displaced' through such immersive and virtual representations.

The entanglement of displacement and the contemporary art movement is the focus of the chapter by John Potts, who considers the role of displacement as a theme and artistic provocation by drawing on work from around the world, including South Africa, China, and Australia. His writing on 'Displacement in Contemporary Art' reveals the power and politics of displacement: the bureaucracies, institutions, and bodies that assemble the experience of being displaced and maintained in a state of displacement. In working through multiple examples, Potts's chapter also considers the entanglements of the globalist art industry with the concept of displacement through an attentiveness towards the displacement of ideas and images that is inherent to the processes of globalisation that enable artists to function as 'semionauts' (navigating the virtual seas of global images), and which—through war, conflict, and environmental change—is entrenching displacement as an increasingly widespread human condition.

Indeed, we are increasingly bombarded in our news channels by different images and particular narratives of refugees and asylum seekers (e.g. the European refugee crisis) representing displacement and anti-immigrant rhetoric. Overall the media plays a dominant role in constructing and informing our understanding, perceptions, and opinions. Stavropoulou in this section explores the transformative role of creative participatory art-based methodologies as a means of exposing personal experiences and providing self-expression for refugee/asylum seekers to challenge, resist, and reimagine their experiences of displacement. Through local participatory engagement, dominant narratives in the media can be resisted, giving the displaced a voice and creating representative spaces for co-creating knowledge production of the displacement process. Her chapter, 'Reclaiming Safe Spaces,' shows how these can accomplish social justice and raise awareness through advocacy. Artistic and creative practices have become an increasingly important set of resources and methods used for public engagement that involve participants and respondents in imagining, making, and telling as a set of interrelated practices. These can be both educational in the broadest sense and mutually transformative for the participants, practitioners, and researchers involved. Creative practice can encourage conversation around issues that might otherwise be difficult to articulate and provide avenues for discussions, debates, and broader understanding of complex information. From this chapter, it is possible to see how such artistic and creative practice might facilitate a more inclusive politics with capacity to bring in more diverse voices in decision-making.

Throughout many of the chapters in this section is a critical attentiveness to techniques, technologies, and methodologies of and for representing displacement (as a process) and those who are displaced—often in ways that seek to blur the boundaries between 'producer' and 'subject.' In Michele Lancione's important intervention, 'Activism, Research, and Film-Making,' this blurring is explored in the context of the author's own activist-academic film-making with the displaced Roma community in Bucharest, using participatory techniques that actively engage the Roma community in representing their politico-legal challenges with local, national, and supranational powers. Lancione's intervention which begins the section and Stavropoulou's which ends it link to the focus of the final section on resisting displacement.

## Resisting Displacement

As the editors of *Geographies of Forced Eviction* (Brickell et al. 2017, p. 14) set out, 'the injustices involved in the expulsions of people, economies and lifeworlds have enabled subjective and collective identifications that have, in turn, produced a rich historical geography of resistance.' In this section, the chapters have been designed to explore a range of ways in which different forms of displacement are resisted, subverted, unsettled, or reversed. The invention interview with award-winning writer, journalist, and academic Anna Minton, in combination with the chapters that follow, speaks to the complex repertoire of resistance that brings bodies, spaces, and livelihoods into new forms of solidarity and relation in the contestation of displacement.

There are three chapters in this section that discuss creative grassroots resistance and activism in recent years against the neoliberal privatisation of social housing and construction of homeownership, rising rent arrears, and insecurity of tenure. Precarious relations with housing markets arise due to lack of affordable housing and the resultant increase in housing displacement across the globe. Nowicki's chapter, '"Housing is a Human Right. Here to Stay, Here to Fight",' draws on the range of methods utilised by women activists to resist and subvert housing injustice in a range of different cultural contexts in the Global North as well as in the Global South. It highlights the multifaceted nature of resistance, the role of women in resistance, and the use of social media to access and cultivate means of legal knowledge exchange, co-constructing bottom-up legal resistance, in the context of exclusionary and coercive state policies which can also be a marginalising force in housing activism. The chapter makes a strong argument on how women's access to domestic life is eroded by displacement and can become oppressive, highlighting moral claims to a legal home.

Meanwhile, the second chapter by Ferreri, entitled 'Contesting Displacement through Radical Emplacement and Occupations in Austerity Europe,' offers a critical synthesis of literature on different forms of resistance to new processes of gentrification, urban regeneration, and speculative real estate development that cause residential displacement. It mainly focuses on practices of mobilised spatial occupation of public spaces and vacant buildings as key to resistance to displacement. It provides an overview of debates and examples surrounding practices of occupation in both northern and southern European cities in response to post-2008 financial recession, particularly with austerity-fuelled dispossession.

Thirdly, Lees and Hubbard's chapter, 'Legal Geographies of Resistance to Gentrification and Displacement,' highlights the importance of law (here public inquiries) in resisting displacement of existing communities in the midst of council estate redevelopment. Legal challenges are made against the state institutions and developers by ordinary people which also raise public awareness and campaigns against gentrification-induced displacement. The case study of Aylesbury Estate in Southwark (London) highlights the increasing importance and role of the Equality Act 2010 Public Sector Equality Duty, given the ruling that children, the elderly, and black ethnic minority residents are disproportionately affected by the compulsory purchase orders issued by councils and that they would have a negative impact on their ability to retain their cultural ties.

The focus on contesting the breakup of communities through legal means is followed by Trotta and Wilkinson's chapter, 'Local Faith Communities and Responses to Displacement,' which reviews extensive literature and draws on a variety of examples from the Global North and Global South to trace the complex role played by local humanitarian faith community organisations in providing diverse forms of support and solidarity to people at various stages of the process of displacement. Local faith communities are often the first responders to communities affected by conflict and displacement, but also operate along migration routes, providing food, shelter, and other material needs in addition to providing spiritual and pastoral support. Increasingly, international agencies and international nongovernmental organisations are engaging local faith-based organisations to support refugees. However, the reality of engaging with and working through local faith-based communities is much more nuanced and complex. Their chapter engages, for example, with debates pertaining to the inclusionary and exclusionary nature of local faith responses to displacement and examines the complex position of local faith communities within the context of the so-called 'localisation of aid' agenda. This chapter highlights the way in which local faith communities implement humanitarian principles and provide more effective and appropriate assistance than the humanitarian system itself, in spite of its standards and procedures.

Providing accommodation, support, and services to asylum seekers and refugees (particularly undocumented migrants) in the Global North is further explored in Darling's chapter, 'Hosting the Displaced,' which focuses on the work of sanctuary movements which promote the values of urban hospitality. These also emphasise the contribution to the social and cultural life of their 'host' communities. This chapter brings out the selective ways in which displaced peoples are managed as well as supported by state-led resettlement programmes and community initiatives. It highlights the unanticipated effects on the displaced, the established communities, and political activists. It further explores the political and ethical tensions that evolve in their efforts to accommodate asylum seekers and refugees. These are interesting debates in the context of integration, cohesion, and citizenship, but also underline the fragmented nature of support and services available to displaced people.

On this front, research on asylum seeker and refugee food initiatives, social enterprises, and international movements are the mainstay focus of Murphy's chapter on 'Food and the Politics of Refuge' (which flows from Darling's writing). Murphy effectively demonstrates how these food projects, and the foodscape they have tried to generate, have become a relational tool of solidarity and action. The 'vibrant' materialisms of food (see Bennett 2010; Goodman 2015) in situations of displacement should, therefore, not be discounted. In the Irish context, she finds that their emergence was largely in response to, and in turn further problematised, the failings of the system of 'direct provision' of asylum seeker accommodation owned or managed by contractors for the Irish state.

Across the section, the focus of resistance has shifted from claiming the right to 'stay put' to well-being. The section offers valuable insights into emerging experiences of collective empowerment and intersectional solidarity and makes visible the vulnerability, insecurity, and dispossession which displaced individuals, families, and other groups navigate. In doing so, it strategically makes visible material and political causes of displacement, understanding wider socio-symbolic processes and impacts connecting the local to the global social justice movement. The reader will find that, in striving for social justice, the section interrogates a potential range of methods in resisting displacement.

## Conclusion

In the chapters that follow on from this introduction, the contributions both deepen and widen understanding of the interdisciplinary study of displacement in terms of knowledges, concepts, and practices. Its thematic focus on the technologies, journeys, traces, governance, more-than-human, representation, and resisting of displacement brings into convened view the richness and dynamism of current attention, both inside and outside of academia, to displacement in its myriad forms, scales, and places. The 50-plus chapters show how the handbook as a collective of writing, and the study of displacement more broadly, requires us to think of, and respond to, displacement as so much more than simply people being forced to leave their place of residence.

Displacement 'transpires in kaleidoscopic forms,' and losses are 'corporeal, cultural, haunting, and real' (Anti-Eviction Mapping Project 2019; see also Maharawal and McElroy 2018), just as responses to it are lived as complex, politically charged, and intimately felt in the short and long term. Many of the contributions demonstrate how displaced groups are positioned and treated as 'other,' as a threat used by populist, far right governments and movements who have re-emerged in mainstream politics over the past decade. The management of displaced populations has also become, in many countries, commodified and privatised such that there is the contracting out of the policing of evictions, the profit-making 'warehousing' of refugees, and the exploitation of vulnerable groups through labour violations.

Such negative and contemptuously skewed discussions of movement, and the disaster capitalism that has emerged to profit from record numbers of displaced persons, should not, however, mean that we lose sight of human endeavours to improve lives. Returning to the footwear theme we began this introduction with, another set of images we considered for the front cover of the handbook were the shoes donated, in a brief moment of sympathy and solidarity, to the 'caravan' of migrants who had made their way from Central America through the Guatemalan-Mexican border to the United States in 2018. While such acts are unable to contest the structural and direct violences which have led to fleeing Central America, they do speak to the political significance of kindness in how humanity thinks, speaks, and, most importantly, acts towards those encountering displacement. This handbook aims to provide the knowledge and tools to understand how displacement is lived, governed, and mediated in different ways and to usefully accompany future discussions on how to find and forge solidarity, care, and home in an era of displacement.

# References

Adey, P., & Bissell, D. (2010). Mobilities, meetings, and ruptures: An interview with John Urry. *Environment and Planning D: Society and Space, 28*(1), 1–16.

Ahmed, S., Castañeda, C., Fortier, A. M., & Sheller, M. (2003). *Uproot-ings/regroundings: Questions of home and migration*. Oxford: Berg.

Anderson, S. (2015). Fugitive borders. *Fabrications, 25*(3), 344–375.

Anti-Eviction Mapping Project. (2019). Retrieved from https://www.antievictionmap.com.

Arizpe, E. (2019). Migrant shoes and forced walking in children's literature about refugees: Material testimony and embodied simulation. *Migration Studies*. Advance online publication. https://doi.org/10.1093/migration/mnz047.

Baeten, G., & Listerborn, C. (2015). Renewing urban renewal in Landskrona, Sweden: Pursuing displacement through housing policies. *Geografiska Annaler: Series B, Human Geography, 97*(3), 249–261.

Batchelor-Hunt, N. (2019, June 28). Photos of dead migrants Óscar and Valeria Ramírez typify how the media dehumanises people of colour's bodies. *Huffington Post*. Retrieved from https://www.huffingtonpost.co.uk/entry/oscar-valeria-ramirez_uk_5d1544cfe4b03d611638f18f.

Bear, C., & Eden, S. (2008). Making space for fish: The regional, network and fluid spaces of fisheries certification. *Social & Cultural Geography, 9*(5), 487–504.

Bennett, J. (2010). *Vibrant matter: A political ecology of things*. London: Duke University Press.

Bissell, B., & Fuller, G. (Eds.). (2011). *Stillness in a mobile world*. London: Routledge.

Booker, C. (2019, June 25). We should not look away [Twitter post]. *Twitter*. Retrieved from https://twitter.com/corybooker/status/1143642829765435393.

Brickell, K. (2020). *Home SOS: Gender, violence and survival in crisis ordinary Cambodia*. Oxford: Wiley RGS-IGS Series.

Brickell, K., Fernández Arrigoitia, M., & Vasudevan, A. (2017). Geographies of forced eviction: Dispossession, violence, resistance. In K. Brickell, M. Fernández Arrigoitia, & A. Vasudevan (Eds.), *Geographies of forced eviction: Dispossession, violence, resistance* (pp. 1–23). Basingstoke: Palgrave Macmillan.

Brun, C., & Fábos, A. (2015). Making homes in limbo? A conceptual framework. *Refuge, 31*(1), 5–17.

Butler, J., & Athanasiou, A. (2013). *Dispossession: The performance in the political*. Cambridge: Polity Press.

Çaglar, A., & Glick Schiller, N. (2018). *Migrants and city-making: Dispossession, displacement, and urban regeneration*. Durham, NC: Duke University Press.

Davidson, M. (2009). Displacement, space and dwelling: Placing gentrification debate. *Ethics, Place and Environment: A Journal of Philosophy and Geography, 12*(2), 219–234.

Ehrkamp, P. (2017). Geographies of migration I: Refugees. *Progress in Human Geography, 41*(6), 813–822.

Elden, S. (2010). Land, terrain, territory. *Progress in Human Geography, 34*(6), 799–817.

Elden, S. (2013). *The birth of territory*. Chicago, IL: University of Chicago Press.

El-Shaarawi, N. (2015). Living an uncertain future: Temporality, uncertainty, and well-being among Iraqi refugees in Egypt. *Social Analysis, 59*, 38–56.

Espiritu, Y. E., & Duong, L. (2018). Feminist refugee epistemology: Reading displacement in Vietnamese and Syrian refugee art. *Signs: Journal of Women in Culture and Society, 43*(3), 587–615.

Fiddian-Qasmiyeh, E., Long, K., Sigona, N., & Loescher, G. (Eds.). (2018). *Oxford handbook of refugee and forced migration studies*. Oxford: Oxford University Press.

Foucault, M. (1997). The birth of biopolitics. In P. Rabinow (Ed.), *Michel Foucault: Ethics, subjectivity, and truth* (pp. 73–80). New York, NY: New Press.

Gallón, N., Melgar, A., & Almasy, S. (2019, 26 June). A shocking image of a drowned man and his daughter underscores the crisis at the US-Mexico border. *CNN*. Retrieved from https://edition.cnn.com/2019/06/25/americas/mexico-photo-of-father-and-daughter-dead-in-rio-grande/index.html.

Goodman, M. K. (2015). Food geographies I: Relational foodscapes and the busyness of being more-than-food. *Progress in Human Geography, 40*(2), 257–266.

Hartmann, B. (2010). Rethinking climate refugees and climate conflict: Rhetoric, reality and the politics of policy discourse. *Journal of International Development, 22*(2), 233–246.

Horn, D. M., & Parekh, S. (2018). Introduction to 'displacement'. *Signs: Journal of Women in Culture and Society, 43*(3), 503–514.

Horst, C., & Nur, A. I. (2016). Governing mobility through humanitarianism in Somalia: Compromising protection for the sake of return. *Development and Change, 47*(3), 542–562.

Hubbard, P., & Lees, L. (2018). The right to community? *City, 22*(1), 8–25.

Hyndman, J., & Giles, W. (2011). Waiting for what? The feminization of asylum in protracted situations. *Gender, Place & Culture, 18*(3), 361–379.

Hyndman, J., & Giles, W. (2016). *Refugees in extended exile: Living on the edge.* London: Routledge.

International Organization for Migration. (2020). *World migration report 2020.* Geneva: IOM. Retrieved from https://publications.iom.int/system/files/pdf/wmr_2020.pdf.

Kromm, C., & Sturgis, S. (2008). *Hurricane Katrina and the guiding principles on internal displacement.* Institute for Southern Studies and Southern Exposure. Retrieved from https://www.brookings.edu/wp-content/uploads/2012/04/0114_ISSKatrina.pdf.

Landau, L. B. (2018). Urban refugees and IDPs. In E. Fiddian-Qasmiyeh, K. Long, N. Sigona, & G. Loescher (Eds.), *Oxford handbook of refugee and forced migration studies* (pp. 139–150). Oxford: Oxford University Press.

Lea, T. (2015). What has water got to do with it? Indigenous public housing and Australian settler-colonial relations. *Settler Colonial Studies, 5*(4), 375–386.

Lees, L., & Phillips, M. (Eds.). (2018). *Handbook of gentrification studies.* Cheltenham: Edward Elgar.

Lems, A. (2016). Placing displacement: Place-making in a world of movement. *Ethnos, 81*(2), 315–337.

Leshem, N., & Pinkerton, A. (2016). Re-inhabiting no-man's land: Genealogies, political life and critical agendas. *Transactions of the Institute of British Geography, 41*(1), 41–53.

Maharawal, M. M., & McElroy, E. (2018). The anti-eviction mapping project: Counter mapping and oral history toward Bay Area Housing Justice. *Annals of the American Association of Geographers, 108*(2), 380–389.

Massey, D. (1993). Power-geometry and a progressive sense of place. In J. Bird, B. Curtis, T. Putnam, G. Robertson, & L. Tickner (Eds.), *Mapping the futures: Local cultures, globalchange* (pp. 59–69). London: Routledge.

Massey, D. (2005). *For space*. London: Sage.

Mountz, A. (2004). Embodying the nation-state: Canada's response to human smuggling. *Political Geography, 23*(3), 323–345.

Nixon, R. (2011). *Slow violence and the environmentalism of the poor*. Cambridge, MA: Harvard University Press.

Pain, R. (2015). Intimate war. *Political Geography, 44*, 64–73.

Pain, R., & Staeheli, L. A. (2014). Introduction: Intimacy-geopolitics and violence. *Area, 46*(4), 344–347.

Peteet, J. (2007). Problematizing a Palestinian diaspora. *International Journal of Middle East Studies, 39*(4), 627–646.

Philo, C. (2005). The geographies that wound. *Population, Space and Place, 11*, 441–454.

Pitkanen, L. (2017). The state comes home: Radiation and in-situ dispossession in Canada. *Political Geography, 61*, 99–109.

Polzer, T., & Hammond, L. (2008). Invisible displacement. *Journal of Refugee Studies, 21*(4), 41–431.

Puar, J. K. (2017). *The right to maim: Debility, capacity, disability*. Durham, NC: Duke University Press.

Qasmiyeh, Y. M. (2016). At the feast of asylum. *GeoHumanities, 2*(1), 248–253.

Rahim, Z. (2019, June 26). 'Trump is responsible': Democrats, celebrities and activists react to photo of drowned father and toddler. *The Independent*. Retrieved from https://www.independent.co.uk/news/world/americas/us-politics/us-mexico-border-deaths-rio-grande-drowned-man-girl-migrants-reaction-a8975916.html.

Roy, A. (2017). Dis/possessive collectivism: Property and personhood at city's end. *Geoforum, 80*, A1–A11.

Sassen, S. (2014). *Expulsions: Brutality and complexity in the global economy*. Cambridge, MA: Harvard University Press.

Sheller, M. (2018). *Mobility justice: The politics of movement in an age of extremes*. London: Verso.

Siu, H. F. (2007). Grounding displacement: Uncivil urban spaces in postreform South China. *American Ethnologist, 34*(2), 329–350.

Stonebridge, L. (2018). *Placeless people: Writings, rights, and refugees*. Oxford: Oxford University Press.

Tanyag, M. (2018). Resilience, female altruism, and bodily autonomy: Disaster-induced displacement in post-Haiyan Philippines. *Signs: Journal of Women in Culture and Society, 43*(3), 563–585.

Tazzioli, M., & Garelli, G. (2018). Containment beyond detention: The hotspot system and disrupted migration movements across Europe. *Environment and Planning D: Society and Space*. Advance online publication. https://doi.org/10.1177/0263775818759335.

Tyner, J. (2020). The Fast and Slow Violence of Displacement. In Adey et al. (Eds.), *The Handbook of Displacement*. Palgrave MacMillan: London.

UNHCR. (2018). *Global Trends Forced Displacement in 2018*. Geneva: UNHCR. Retrieved from https://www.unhcr.org/globaltrends2018/.

van Houtum, H., & Bueno Lacy, R. (2019). The migration map trap. On the invasion arrows in the cartography of migration. *Mobilities*. Advance online publication. https://doi.org/10.1080/17450101.2019.1676031.

Vaz-Jones, L. (2018). Struggles over land, livelihood, and future possibilities: Reframing displacement through feminist political ecology. *Signs: Journal of Women in Culture and Society, 43*(3), 711–735.

Verifeye. (2016, January 16). Protesters hold 'die-in' for refugees at Eurostar terminal in London's St Pancras station [Video]. *The Guardian*. Retrieved from https://www.theguardian.com/uk-news/video/2016/jan/16/protesters-hold-die-in-for-refugees-at-eurostar-terminal-in-londons-st-pancras-station-video.

Verstraete, G., & Cresswell, T. (Eds.). (2005). *Mobilising place, placing mobility: The politics of representation in a globalized word*. Amsterdam: Rodopi.

Vukov, T. (2015). Strange moves: Speculations and propositions on mobility justice. In L. Montegary & M. A. White (Eds.), *Mobile desires: The politics and erotics of mobility justice* (pp. 108–121). London: Palgrave Pivot.

Walker, A. (2019, June 29). Why shocking photos of drowned migrants must be seen—Warning: Distressing images. *Sky News*. Retrieved from https://news.sky.com/story/why-shocking-photos-of-drowned-migrants-must-be-seen-warning-distressing-images-11749589.

Whatmore, S. (2006). Materialist returns: Practising cultural geography in and for a more-than-human world. *Cultural Geographies, 13*(4), 600–609.

# Part I

## Section One: Conceptualising Displacement

# 2

# Mobilities and Displacement

Mimi Sheller

## Introduction

Recognising the complex and intimate relation between multiple (im)mobilities and varied (dis)placements was central to the emergence of transnational feminist/postcolonial theory in the 1990s and became one of the instigations for the emergence of critical mobilities theory at the turn of the millennium. In the face of prevailing discourses about globalisation and free trade, the 'new mobilities paradigm' sought to refocus attention on the uneven and unequal access of different people to 'embodied and material practices of movement, digital and communicative mobilities, the infrastructures and systems of governance that enable or disable movement, and the representations, ideologies and meanings attached to both movement and stillness' (Sheller 2014, p. 789; Sheller and Urry 2006; Urry 2007). In emphasising the multi-scalar spatial formations and power relations of (im)mobilities and their infrastructural and political moorings (Hannam et al. 2006; Cresswell 2006, 2010), this new field of interdisciplinary mobilities research extended transnational feminist theories of space, place, and travel as power relations, and built on earlier postcolonial approaches to diaspora, de-territorialisation, and displacement, which in the 1990s had deconstructed bounded static categories of race, nation, ethnicity, and state borders (e.g. Kaplan and Grewal 1994; Basch et al. 1994; Appadurai 1996; Clifford 1997).

---

M. Sheller (✉)
Center for Mobilities Research and Policy, Drexel University, Philadelphia, PA, USA
e-mail: mimi.sheller@drexel.edu

In her influential book *Questions of Travel*, feminist theorist Caren Kaplan suggested that 'the emergence of terms of travel and displacement (as well as their oppositional counterparts, home and location) must be linked to the history of the production of colonial discourses' (1996, p. 2). Yet Kaplan's influential analysis was not simply about a critique of displacement but a far more subtle unravelling of the relation between concepts such as mobility/travel versus settlement/home. Displacement, she argued, is a multivalent concept that was central not only to thinking about the postcolonial condition but also the postmodern condition: 'The oscillation and tension between the liberating promise of mobility and the security of fixed location is one of modernity's most enduring and complex oppositional binaries' (Kaplan 2003, p. 212).

Surprisingly perhaps, given the current negative connotations of displacement, this oscillation refers to the ways in which displacement and uprooting, for a time in the late twentieth century, came to be associated not simply with negative loss of place but also with a liberatory potential as found in 'the bittersweet pathos of displacement's enabling powers for the artist, the energizing jolt that influxes of new populations brings to economies and cultures of the metropole, and the emergence of postcolonial aesthetics and cultural practices' (Kaplan 2003, p. 212). Such ideas are found in the Frankfurt School, in avant-garde movements, and the work of Raymond Williams, but also later became associated with postcolonial theories of the vibrancy of diaspora cultures, creolisation, and hybridity as models of travel and translation (Clifford 1997). Displacement could liberate us from fixed identities. For anthropologist James C. Scott, moreover, escaping the state and its fixed ethnic identities is also part of the 'art of not being governed,' in which subaltern people deliberately remained stateless, engaged in mobile agricultural practices, and dispersed across physically challenging terrains (Scott 2010).

With the emergence of critical mobility studies, however, there was a wave of attention on the 'contingent relations between movements' within a 'relational politics of (im)mobilities' (Adey 2006). This approach suggested that the fundamental relationality of (im)mobilities was co-produced through relations of power. The making of place involves legal, social, and material infrastructures of displacement, as well as the exercise of gendered, sexual, racial, and ethno-national power through displacement. Thinking through the dialectics of mobilities and moorings drew attention to experiences of, for example, forced migration (Gill et al. 2011; Mookherjee 2011), involuntary immobility (Lubkemann 2008), or detention within various kinds of carceral spaces (Mountz et al. 2013). It also brought out more subtle experiences of daily mobility, accessibility, and social exclusion (Ureta 2008) or the myth of nomadism within the social construction of 'gypsies/travellers' (Drakakis-Smith 2007). Finally, it also elicited theoretical reflections on 'stillness in a mobile world' as a site of potentiality, suspension, flow, and possible reanimation and turbulence (Bissell and Fuller 2011).

In returning to Kaplan's cultural analysis of displacement, therefore, I begin this essay from a complex point of view that emphasises the entanglement of concepts such as (im)mobilities, (dis)placement, and uprootings/regroundings. These terms are not binary oppositions but are the kernels of oscillating and flickering conceptual assemblages that form influential vortices of conjoined meaning and complex institutional practices. Processes of displacement can range from eviction, homelessness, anti-loitering laws, sexual harassment, and criminalisation of the illicit use of place (e.g. squatting, sidewalk-vending, pan-handling) to state-scale processes of colonial appropriation, land grabs, enslavement, deportation, detention, and mass incarceration.

In thinking about this core relationality of (im)mobilities and (dis)placement, one of the key collections on relational (im)mobilities that influenced my own thinking was the volume *Uprootings/Regroundings: Questions of Home and Migration*, which I co-edited with Sara Ahmed, Claudia Castañeda, and Anne-Marie Fortier when we were all at Lancaster University, and I was involved in co-founding the Centre for Mobilities Research there. In that book we suggest that staying put and displacing others are deeply simultaneous relations of power: 'How can movement or staying put be a form of privilege that "extends" the reach of some bodies, for example when the movement of some takes place through "fixing the bodies of others" (Ahmed 2000), or when staying put takes place through displacing others?' (Ahmed et al. 2003, p. 6). Sometimes powers of 'extension' for some groups are closely tied with fixing others in place, and vice versa. The concept of displacement was also central to our thinking in the book, in which we wrote:

> Home and migration cannot be adequately theorized outside of these spatialized relations of power. Mobility can be foisted upon bodies through homelessness, exile and forced migration just as the purported comforts of the familial 'home' may be sites of alienation and violence (for women, children, queers). The founding of homelands and places of belonging can entail the displacement of others from their homes. It can also involve the spoliation of the homes of those who nevertheless remain 'in place', as is so evident in the migration of European settlers that has historically entailed the desecration of indigenous peoples' homelands. (Ahmed et al. 2003, p. 6)

Thus, from the beginning of the field we were attempting to articulate a complex instability within key terms; to think through conditions of coloniality, indigeneity, and queer embodiment; and to suggest that settling often involves displacing others, that homes may be places of violence that can be alienating, and that place and home may be displaced even for those who remain in place.

Place, home, roots, and settlement, in other words, share a political genealogy with displacement, uprooting, mobility, and migration, and each term is dialectically (and painfully) entangled with the other, not only at the scale of nations but deeply within the body. How can we make sense of these reverberating cycles of multiple compounded (dis)placements? And how can we do justice to the complex dialectic of (im)mobilities and place? Mobilities research has sought to think through the relational 'power-geometries' (Massey 1993) that at once enable and coerce people to live 'mobile lives' (Elliott and Urry 2010), to be 'on the move' (Cresswell 2006), and to seek 'mobility justice' (Sheller 2018). In the following sections, I will analyse these relational concepts through a discussion of, first, bordering practices in relation to migration; second, urban gentrification in relation to homelessness; and, third, the pursuit of more just mobilities through commoning mobility.

## (Dis)placement and Bordering

Control over (im)mobility and (dis)placement are forms of power that have deep historical roots in Western liberal modes of state formation, governance, and territoriality. In her study of the liberal governances of mobility, Hagar Kotef traces the history of liberalism as a regime of movement in which the 'free movement of some [people] limits, hides, even denies the existence of others' (Kotef 2015, p. 54). This nuanced view of freedom of movement poses some interesting conundrums that are suggestive of the dialectical linkage of mobility and displacement. 'Control over movement,' Kotef argues, 'was always central to the ways in which subject-positions are formed and by which different regimes establish and shape their particular political orders' (Kotef 2015, p. 37). The liberal subject and 'free' citizen forms through rights to (well-regulated) movement, protected by the territorial state; yet this simultaneously produces an 'other' who potentially suffers not only exclusion, enclosure, incarceration, and violence but also coerced mobility via displacement or alienation through the destruction of place. Gannit Ankori, for example, interrogates the 'dis-orienting' relation between displaced bodies and embodied displacements through a close reading of the artwork of Mona Hatoum and Khalil Rabah, describing the 'central dialectic that defines contemporary Palestinian identity, [as] the oscillation between rootedness and displacement' that is almost carved into the flesh and the soil (Ankori 2003, p. 61).

Kotef describes a split in early sources of liberal theory beginning with Hobbes and Locke, who each in slightly different ways differentiate the mobility of the citizen and its others: '(I) the citizen (often a racialized, classed, ethnically marked, and gendered entity more than a juridical one), as a figure of "good," "purposive," even "rational," and often "progressive" mobility that should be maximized; and (II) other(ed) groups, whose patterns of movement are both marked and produced as a disruption, a danger, a delinquency' (Kotef 2015, p. 63). It is such groups who suffer most from displacement, being subjected to the force and control of others in determining their movements and their possibilities for settling or dwelling in particular places. This divide remains a split between populations today: 'between those whose movement is a manifestation of liberty, and should therefore be maximized, and those whose freedom is a problem, and should therefore be tightly regulated.' Kotef develops this argument through an analysis of Israeli checkpoints and control over Palestinian mobility. But she also suggests more widely, 'In our global and local travel, in patterns of migration and border crossing, in deployments of checkpoints—be it in poor neighborhoods or occupied lands—we can still witness this split' (Kotef 2015, p. 100).

It is precisely this split that we see coming to the fore in current policies over borders and security in Europe, the United States, and other white settler nations such as Australia but also internally in US American controversies over policing, racial profiling, and mass incarceration of people of colour. There are an estimated sixty-five million people who are today forcibly displaced from their homes, including those internally displaced, and twenty-five million who are living outside their home country, said to be more than at any time since World War II. Approximately twenty-eight million children were driven from their homes in 2017 alone due to conflict, according to the United Nations Children's Fund. The International Organization for Migration has calculated that between 2006 and 2015 more than 40,000 people died trying to cross borders around the world (IOM 2018). Political geographers of borders attribute these deaths to the consequences 'of states expanding the reach of their security and detention practices to capture, intercede, or make intentionally perilous the movement of people' (Jones et al. 2017, p. 1). Such efforts include the growth in construction of border walls, with almost seventy such around the world now, as well as a 'much wider set of state practices to control movement such as deployments of more border guards, seaborne patrols, and investments in new technologies to monitor more comprehensively events within state space, at the edges of their territories, and beyond' (Jones et al. 2017, p. 1). Border walls, in other words, are a kind of displacement of displacement, shifting the onus of responsibility from the mover to the moved.

Expulsions are not accidental, they are made, argues Saskia Sassen: 'The instruments for this making range from elementary policies to complex institutions, systems, and techniques that require knowledge and intricate organizational formats … Our advanced political economies have created a world where complexity too often tends to produce elementary brutalities' (Sassen 2014, p. 2). This reminds us that the kinds of migrations that the Trump administration in the United States is currently trying to prevent are driven in large part by US American military domination and economic policies (including globalisation, neoliberalism, and the North American Free Trade Agreement) that undermined rural agricultural economies in Central America and Mexico, displacing people from the land, and strengthened the narco-economies of international drug cartels, bringing violent gangs into many small towns across Central America, as well as drug-trade-associated violence into US cities. Thousands of rural Central Americans displaced from their homes and subjected to violence in cities controlled by gangs travelled across Mexico, seeking shelter in the United States, while the Trump administration demonised the 'caravan' and militarised the border with more than 5000 troops in the fear-filled run-up to the 2018 US elections and the subsequent standoff with Congress over funding his border wall. This too was a manoeuvre to displace political attention from the causes of displacement to the protection of a supposed homeland against dangerous alien invaders.

We can situate these arguments in relation to a wider turn in border studies that has found many theorists rethinking the relation between territory and borders, mobility, and containment, or 'bordering and ordering' (Popescu 2011). In contrast to theories that distinguish between fixed territorial logics of state power versus the mobility of capital accumulation, relational theorists emphasise that it is through acts of bordering and debordering that a territory is at once produced, stabilised, and sometimes de-territorialised. Borders are not simply edges, limits, or barriers for controlling mobility in and out of adjacent territories, nor does territory pre-exist the border, but instead the relation between the two terms can be understood as 'bordering practices' (Sassen 2005). The 'state border is not simply a border-line,' Sassen writes. 'It is a mix of regimes with variable contents and geographic and institutional locations,' including different flows of capital, information, professionals, undocumented migrants, smuggled goods, and so on (Sassen 2013, p. 30). And thus, borders produce various kinds of placements and displacements.

Bordering practices, in other words, are performed through multiple relations of (im)mobilities and (dis)placements. If this is true at the national scale, it is also true at the urban scale. In the following section, I turn to the spatial formations of (im)mobility and (dis)placement within American cities, especially in regard to deportation regimes, mass incarceration, and homelessness as racialised practices of exclusion and mobility injustice.

## Internal Displacements: Eviction, Mass Incarceration, and Homelessness

The capacity to move and the coercive power of stopped movement, detention, or imprisonment are fundamental dimensions of human rights and justice. But just as crucial is the right to remain in place, to dwell, and to self-determine one's place of belonging. While much of the literature on displacement focuses on cross-border migration, refugees, and exile, it is equally important to show how the governance of (im)mobilities and (dis)placement also occurs within the nation-state through practices such as eviction, mass incarceration and homelessness. A growing body of work on homelessness has drawn on concepts of displacement and expulsion to show how urban policies around austerity, infrastructure, policing, real estate development, and restructuring all serve to displace people (e.g. Peck 2012; Peters and Robillard 2009) and to remake urban place—including in the design practice of 'placemaking'—in ways that are exclusionary (Stehlin 2019).

Existing social movements have begun to make scale-jumping connections from the migrant justice movement to the racial justice movement, from refugee protection to Black Lives Matter, and from anti-border-wall protests to the right to the city. Mass incarceration, in particular, has been analytically traced to histories of slavery, Jim Crow segregation, white privilege, and racial oppression in the United States (Gilmore 2007; Alexander 2012). These forms of forcible displacement and mobility control of racialised minorities blur with institutions for migrant detention and deportation enforced against 'brown' people. As Ashley Dawson notes, 'The United States deports nearly 400,000 people annually. The US prison-industrial complex keeps over 2.3 million people in cages; in any particular day, 19,000 people are in federal prison for criminal convictions of violating immigration laws, and an additional 33,000 are civilly detained by the Immigration and Customs Enforcement (ICE) agency' (Dawson 2017, p. 197). This is a massive 'detention and deportation regime,' exercised in the name of national sovereignty, 'but its practices extend far beyond the borders of the nation-state' (Dawson 2017, p. 197). Its practices also extend far *within* the borders of the nation-state.

That is to say, this massive (im)mobility regime involves not only interception of people at sea before they reach the state border but also extends to every street, every public space, and every workplace and public institution, where ICE raids are frequently taking place. Hence undocumented migrants, but also their relatives and employers, are subjected to hiding, avoiding, and being displaced from public space and from social protection. Citizenship is placed under threat, and 'sanctuary cities' that seek to protect migrants are threatened with federal penalties. The federal government seeks to spark what it calls 'self-deportation' by effectively violating human rights and undermining civil rights, and exercising these powers in the name of a white ethnonationalist majority.

Violence at home, and inside cities, also denies people a place in the world, leading to their alienation, their othering, and ultimately their displacement. Detention and deportation are forcible displacements that continue to be exercised against racial and ethnic minorities in white settler societies. As Julia Sudbury argues in the introduction to *Global Lockdown*, a book written by activists working at the forefront of the prison abolition movement, 'The global prison is a local manifestation of transnational flows of people, products, capital, and ideas … [So that] both the fabric of the prison and the people caged within it are shaped by global factors, from free trade agreements and neoliberal restructuring to multinational corporate expansion' (Sudbury 2005, p. xii). But international deportation is also related to more local forms of eviction and the causes of homelessness, which disproportionately affect women of colour and families with children (Desmond 2016). The same racialised minority groups subjected to mass incarceration are also denied public housing, while their families are caught up in cycles of eviction, homelessness, and lack of shelter, leading to a kind of existential lack of place.

Processes of uneven development that promote infrastructural improvements for those who already have high "motility" often feed into gentrification, high housing costs, and eviction of the mobility poor (Stehlin 2019). So many urban populations today experience housing insecurity and place instability, with moments of anchorage or place attachment being more like a temporary mooring than a permanent place in the world. How can we make more just mobilities and more just places under such conditions? Perhaps home needs to be a place that is not exclusive and private but is in some ways shared.

## In Pursuit of Mobility Justice Through Commoning Mobilities

As described above, mobilities theory enables us to think about (dis)placement in new ways: it challenges dualistic thinking, works across scales, and entails complex relationality. In this concluding section, I examine how this unlocks new political visions for more just mobilities. Mobility justice has emerged as a key concept within recent work that seeks to mobilise resistance against the powerful processes of displacement (Cook and Butz 2018; Sheller 2018). How does the problem of mobility justice relate to the ethics of placement? And can the political struggles over occupation, commons, and collective place-making offer alternatives to liberal narratives of freedom of movement that mask underlying processes of displacement?

Mobilities theorist Tamara Vukov has written about the political implications of the struggle for mobility justice, and offers some ideas towards a vision of mobility justice, including, first, 'the building of a world in which safe, accessible, and just forms of movement and dwelling are open and available to all' (Vukov 2015, p. 120). She incorporates not only migrant justice but also concerns related to dwelling, such as equitable policies around homelessness, and rights to occupy and remain in place, such as equitable policies around residency and citizenship. Second, Vukov calls for 'an end to the many macro and micro forms of *forced* mobility and displacement (from colonial and war-based displacements to deportation and evictions due to gentrification)' (Vukov 2015, p. 120). Here, too, we see a combination of scales, a sensitivity to colonial histories but also to contemporary urban issues such as gentrification, eviction, and homelessness. And finally, she also calls for 'The dismantling of imposed forms of immobility, including detention, incarceration, the legacy of colonial confinement (such as reservations) and separation walls and barriers' (Vukov 2015, p. 120).

As I have argued in my book *Mobility Justice* (Sheller 2018), these kinds of everyday displacements especially affect Black, Indigenous, and people of colour, as well as queer people. The critical mobilities perspective allows us to make these scale-jumping connections across areas that are usually studied separately. As mobility justice organisation The Untokening notes: 'When people live at the intersection of multiple vectors of oppression, unfettered access to mobility and public space are not guaranteed. Racism, sexism, classism, ableism, xenophobia, homophobia, and constraints imposed upon gender-non-conforming folks can make the public space hostile to many. Bodies encounter different risks and have different needs' (The Untokening

2018, pp. 11–12). Displacements, therefore, range from urban-scale processes of eviction, homelessness, anti-loitering laws, sexual harassment, and criminalisation of the illicit use of place (e.g. squatting, sidewalk-vending, panhandling) to state-scale processes of colonial appropriation, land grabs, enslavement, deportation, detention, and mass incarceration. There is an uneven social distribution of exposure to such displacements, and a range of ways in which some groups benefit from displacing others.

By applying the lens of mobility justice to a consideration of multiple forms of (dis)placement, this interrogation into the dialectics of (im)mobilities suggests the possibility for an ethics of placement. Given the instabilities of displacement, is it possible to imagine political struggles over occupation that might reclaim place without imposing settler-colonial, racial-capitalist, or privileged extensions of mobile subjectivities that displace others? I would argue that underlying the multiple political dimensions of complex (dis)placements there is the need for a new sense of the mobile commons as an alternative to displacement premised on both a rethinking of private property as an exclusive right to a place and of the individualised right to mobility (see Sheller 2018). Commoning mobility has emerged out of mobilities research as a new conceptualisation of more ethical mobilities (Nikolaeva et al. 2019) and even of ways of assembling more 'moral mobilities' (Scott 2019).

Rather than counter-posing mobility against place/roots/home, which falls into the trap of liberal citizen-subjects as the only 'good' mobile subjects, we should try to imagine the fragile coalescence of belonging and mobility, openness towards others and self-protection, place, and movement. We might think of this as a kind of collective 'placing' that maintains an ethics of mindful mobility through shared spaces of entangled being. Mobile placing might imply being mindfully inclusive not only towards other humans but also towards the more-than-human occupants of a particular place and those who pass through it. By limiting the harms of various kinds of mobility that violently take place, we can reclaim a place for new kinds of collectives not premised on the liberal mobile subject and its displaced others. By 'commoning' mobility, we can perhaps keep places open to the participation of others in dwelling together and moving together, rather than using the territorial state for displacing the other through violent forms of uprooting. Let us oscillate in mobile places and displace ourselves to make room for others.

# References

Adey, P. (2006). If mobility is everything then it is nothing: Towards a relational politics of (im)mobilities. *Mobilities, 1*(1), 75–94.

Ahmed, S. (2000). *Strange encounters: Embodied others in post-coloniality.* London and New York: Routledge.

Ahmed, S., Castañeda, C., Fortier, A. M., & Sheller, M. (Eds.). (2003). *Uprootings/regroundings: Questions of home and migration.* Oxford: Berg.

Alexander, M. (2012). *The new Jim Crow: Mass incarceration in the age of color blindness.* New York: New Press.

Ankori, G. (2003). 'Dis-orientalisms': Displaced bodies/embodied displacements in contemporary Palestinian art. In S. Ahmed, C. Castañeda, A. M. Fortier, & M. Sheller (Eds.), *Uprootings/regroundings: Questions of home and migration* (pp. 59–90). Oxford: Berg.

Appadurai, A. (1996). *Modernity at large: Cultural dimensions of globalization.* Minneapolis: University of Minnesota Press.

Basch, N., Glick Schiller, N., & Szanton Blanc, C. (1994). *Nations unbound: Transnational projects, postcolonial predicaments, and deterritorialized nation-states.* Amsterdam: Gordon and Breach.

Bissell, D., & Fuller, G. (Eds.). (2011). *Stillness in a mobile world.* London and New York: Routledge.

Clifford, J. (1997). *Routes: Travel and translation in the late twentieth century.* Cambridge, MA: Harvard University Press.

Cook, N., & Butz, D. (Eds.). (2018). *Mobilities, mobility justice and social justice.* London and New York: Routledge.

Cresswell, T. (2006). *On the move: Mobility in the modern Western world.* London: Routledge.

Cresswell, T. (2010). Towards a politics of mobility. *Environment and Planning D: Society and Space, 28*(1), 17–31.

Dawson, A. (2017). *Extreme cities: The peril and promise of urban life in the age of climate change.* London and New York: Verso.

Desmond, M. (2016). *Evicted: Poverty and profit in the American city.* New York: Penguin Random House.

Drakakis-Smith, A. (2007). Nomadism a moving myth? Policies of exclusion and the gypsy/traveller response. *Mobilities, 2*(3), 463–487. https://doi.org/10.1080/17450100701597467.

Elliott, A., & Urry, J. (2010). *Mobile lives.* New York and London: Routledge.

Gill, N., Caletrío, J., & Mason, V. (2011). Introduction: Mobilities and forced migration. *Mobilities, 6*(3), 301–316.

Gilmore, R. (2007) *Golden Gulag: Prisons, Surplus, Crisis, and Opposition in Globalizing California.* Berkeley: University of California Press.

Hannam, K., Sheller, M., & Urry, J. (2006). Mobilities, immobilities and moorings. *Mobilities, 1*(1), 1–22.

IOM. (2018). *IOM's missing migrants project*. Joint initiative of the International Organization for Migration (IOM) and the Global Migration Data Analysis Centre (GMDAC). Retrieved from http://missingmigrants.iom.int

Jones, R., Johnson, C., Brown, W., Popescu, G., Pallister-Wilkins, P., Mountz, A., & Glibert, E. (2017). Corridors, camps, and spaces of confinement. *Political Geography, 59*, 1–10.

Kaplan, C. (1996). *Questions of travel: Postmodern discourses of displacement*. Durham, NC: Duke University Press.

Kaplan, C. (2003). Transporting the subject: Technologies of mobility and location in an era of globalization. In S. Ahmed, C. Castañeda, A. M. Fortier, & M. Sheller (Eds.), *Uprootings/regroundings: Questions of home and migration* (pp. 207–224). Oxford: Berg.

Kaplan, C., & Grewal, I. (Eds.). (1994). *Scattered hegemonies: Postmodernity and transnational feminist practices*. Minneapolis, MN: University of Minnesota Press.

Kotef, H. (2015). *Movement and the ordering of freedom: On liberal governances of mobility*. Durham and London: Duke University Press.

Lubkemann, S. (2008). Involuntary immobility: On a theoretical invisibility in forced migration studies. *Journal of Refugee Studies, 21*(4), 454–475.

Massey, D. (1993). Power-geometry and a progressive sense of place. In J. Bird, B. Curtis, T. Putnam, & G. Robertson (Eds.), *Mapping the futures: Local cultures, global change* (pp. 59–69). London and New York: Routledge.

Mookherjee, N. (2011). Mobilising images: Encounters of 'forced' migrants and the Bangladesh war of 1971. *Mobilities, 6*(3), 399–414.

Mountz, A., Coddington, K., Catania, T., & Loyd, J. (2013). Conceptualizing detention: Mobility, containment, bordering, and exclusion. *Progress in Human Geography, 37*(4), 522–541.

Nikolaeva, A., Adey, P., Cresswell, T., Lee, J. Y., Novoa, A., & Temenos, C. (2019). Commoning mobility: Towards a new politics of mobility transitions. *Transactions of the Institute of British Geographers, 44*(2), 346–360. https://doi.org/10.1111/tran.12287.

Peck, J. (2012). Austerity urbanism. *City, 16*(6), 626–655.

Peters, E., & Robillard, V. (2009). 'Everything you want is there': The place of the reserve in First Nations' homeless mobility. *Urban Geography, 30*(6), 652–680.

Popescu, G. (2011). Controlling mobility. In G. Popescu (Ed.), *Bordering and ordering the twenty-first century: Understanding borders* (pp. 91–120). Lanham: Rowman and Littlefield.

Sassen, S. (2005). When national territory is home to the global: Old borders to novel borderings. *New Political Economy, 10*(4), 523–541.

Sassen, S. (2013). When territory deborders territoriality. *Territory, Politics, Governance, 1*(1), 21–45.

Sassen, S. (2014). *Expulsions: Brutality and complexity in the global economy*. Cambridge, MA: Belknap Press.

Scott, J. (2010). *The art of not being governed: An anarchist history of upland Southeast Asia*. New Haven, CT: Yale University Press.

Scott, N. (2019). *Assembling moral mobilities*. Lincoln, NE: University of Nebraska Press.

Sheller, M. (2014). The new mobilities paradigm for a live sociology. *Current Sociology Review, 62*(6), 789–811.

Sheller, M. (2018). *Mobility justice: The politics of movement in an age of extremes*. London: Verso.

Sheller, M., & Urry, J. (2006). The new mobilities paradigm. *Environment and Planning A, 38*(2), 207–226.

Stehlin, J.G. (2019). *Cyclescapes of the Uneven City: Bicycle Infrastructure and Uneven Development*. Minneapolis: University of Minnesota Press.

Sudbury, J. (Ed.). (2005). *Global lockdown: Race, gender, and the prison-industrial complex*. Abingdon and New York: Routledge.

The Untokening. (2018, January). *Untokening mobility: Beyond pavement, paint and place* (A. Lugo, N. Doerner, D. Lee, S. McCullough, S. Sulaiman, & C. Szczepanski, Eds.). Retrieved from https://www.untokening.org/updates/2018/1/27/untokening-mobility-beyond-pavement-paint-and-place

Ureta, S. (2008). To move or not to move? Social exclusion, accessibility and daily mobility among the low-income population in Santiago, Chile. *Mobilities, 3*(2), 269–289.

Urry, J. (2007). *Mobilities*. London: Polity.

Vukov, T. (2015). Strange moves: Speculations and propositions on mobility justice. In L. Montegary & M. A. White (Eds.), *Mobile desires: The politics and erotics of mobility justice*. London: Palgrave Macmillan.

# 3

# Political Ecologies of Displacement

Rebecca Elmhirst

## Introduction

Political ecology is a loosely defined interdisciplinary field of inquiry and practice that foregrounds questions of power in society–nature relations (Robbins 2011; Tetreault 2017). Political ecologists share a normative commitment to social justice and ethical regard for human and more-than-human natures (Collard et al. 2015), and an attentiveness to the situated politics of knowledge (Rocheleau 2016). The field has grown from a combination of Anglophone political ecology that explores the impacts of power on socio-environmental relations (Perreault et al. 2015) and a political ecology of the South constituted as a dialogue of knowledges created through alliance with ecological justice resistance movements (Martinez-Alier 2014; Leff 2015). In attending to questions of power, social justice, and the environment, political ecologists have drawn variously, sometimes simultaneously, on Marxist political economy, Foucaultian concepts of governance and biopolitics, feminist theories of embodiment and social reproduction, and post-human/more-than-human ontologies (Collard et al. 2015). Amidst a range of conceptual approaches, methodologies, and political strategies, political ecology contributes multiscalar, historicized, and richly contextual accounts that explore the

R. Elmhirst (✉)
University of Brighton, Brighton, UK
e-mail: R.J.Elmhirst@brighton.ac.uk

power relations between society and nature embedded in social interests, institutions, and situated knowledges (Perreault et al. 2015).

Environmental displacement—the dislocation of people from the land they inhabit—features as a core area of concern within political ecology's 'common space for reflection and analysis' (Tetreault 2017, p. 3). Some of the earliest iterations of political ecology in the 1980s were devoted to understanding the political economic underpinnings of environmental shocks (disasters) and stresses (soil erosion and other forms of environmental degradation) (Blaikie and Brookfield 1987). Contemporary political ecology analyses focus on the socio-material dynamics of displacement associated with climate change (and interventions directed towards adaptation and mitigation), resource extraction, and biodiversity conservation (Fairhead et al. 2012; Lunstrum et al. 2016; Huber et al. 2017). Many displaced populations are already vulnerable, lacking secure access to land or political rights (Lunstrum et al. 2016). Concepts such as dispossession and dislocation are deployed to show that displacement does not necessarily mean relocation (Vandergeest 2003; Nixon 2011) but can also mean forms of dispossession that reduce or remove the possibility of meaningful livelihoods and cultural practices. Moreover, feminist political ecologists are extending the agenda to consider incremental and quotidian displacements (Mollett 2015; Doshi 2019).

This chapter charts the ways political ecology approaches have been applied in analysing and challenging displacement across a range of sociopolitical and ecological contexts. It begins by considering the contribution made by materialist political ecology in challenging positivist and behavioural science-inspired analyses of displacement, hazard, and risk. New questions and themes emerge from later analyses that have drawn on poststructuralist concepts to examine displacements associated with the governance of commodified nature through technical fixes that rework society–nature relationships, in practices such as infrastructure development, conservation, and environmental restoration. The chapter concludes by considering the knowledge politics that make displacement appear as a 'natural' outcome, and which underlie the messy and embodied politics of resistance that challenge its inevitability.

## Materialist Political Ecologies of Degradation, Disaster, and Displacement

In the early 1980s, displacement emerged as a key theme within political ecology as the field directed the conceptual tools of Marxist analysis towards the analysis of the structural causes of environmental crisis. Specifically, 'apolitical' positivist and behavioural approaches to environmental hazards and population displacement were subjected to a historical materialist critique. This sought to reveal how environments (and hazards) were socially constructed and political in nature, and thus 'displacement' was ultimately rooted in capitalist surplus extraction, class inequality, and spatial injustice (Blaikie and Brookfield 1987).

Much of the work in this vein was framed around 'ecological marginalization', locating the roots of displacement within the capitalist system and class structures, as states and elite actors imposed unsustainable extractive regimes of accumulation and social stress on small-scale land managers and peasant farmers (Blaikie and Brookfield 1987; Robbins 2011). Watts's work on famine and 'slow onset' displacement in Northern Nigeria is an early exemplar of this type of analysis in political ecology (Watts 1983). A similar 'chain of explanation' was adopted by researchers seeking to take the 'naturalness' out of 'natural disaster' (O'Keefe et al. 1976). Here, a materialist political ecology was applied to show how disasters, and associated displacement, far from being entirely 'natural', were effectively rooted in global economic processes, urbanization, resource dependency, and environmental degradation writ large within a capitalist world economy (Blaikie et al. 1994).

Later studies follow a similar train in showing how displacement is enabled and accentuated when disasters become opportunities for capital accumulation for the powerful. Collins, for example, traces the production of risk related to urban wildfires in Arizona's White Mountains (Collins 2008). His study shows how marginalization is also accompanied by facilitation: described as a process that enables powerful groups to exploit hazardous places as environmental opportunities for private gain. This chimes with what Klein (2007) has referred to as 'disaster capitalism', a concept used to describe how disaster reconstruction (including displacement and resettlement) opens space for neoliberal capitalist economic policies that redistribute wealth and exacerbate socioeconomic divisions, as exemplified in New Orleans, following Hurricane Katrina (Adams et al. 2009). A raft of political ecology studies demonstrate how displacement becomes less 'contestable' during moments of collective trauma, as exemplified in Honduras following Hurricane Mitch (Timms

2011), and in both Sri Lanka (Gunewardena 2008) and Thailand (Grundy-Warr and Rigg 2016) following the Indian Ocean tsunami of 2004.

Similar lines of political ecology argument have challenged high-profile and influential studies on climate change and mass migration that undertheorize the relationship between society and nature and work with a depoliticized notion of the environment (Oliver-Smith 2012; Morrissey 2012). Contributions from political ecology have sought to interrogate the environmental politics underpinning translocality (Greiner and Sakdapolrak 2016) and to explore the ways displacement is embedded in complex socio-natural relationships between flooding, capitalist economic development, and dispossession amidst 'already mobile populations' (Middleton et al. 2017).

## Displacement and the Governance of Commodified Nature

The studies of displacement outlined above have moved the terrain of political ecology beyond analysis of the politics of environmental degradation, hazards, and risk, and towards the study of the governance of commodified nature through technical fixes that rework society–nature relationships, in practices such as conservation, infrastructure development, and environmental restoration. A strong thread within political ecology draws on the concept of territorialization (Roth 2008) to demonstrate the links between the securing of state power and the control of 'unruly' populations through a strategy described by Robbins as 'conservation as control' (Robbins 2011). Displacements include the removal and resettlement of those found to be 'out of place' in colonial and postcolonial orderings of territory, as lands represented as 'Edenic wilderness' became spaces for colonial agriculture, recreational hunting by white elites, or nature reserves (Neumann 1996; Goldman 2003).

A political ecology analysis has been adopted to address contemporary conservation-related displacements (Rangarajan and Shahabuddin 2006; Adams and Hutton 2007), particularly following the rapid expansion of protected areas in the 1990s. Reviewing studies of eviction for conservation, Brockington and Igoe note that evictions are led by powerful and well-funded international conservation organizations animated by Western ideals of people-less wilderness landscapes, but eviction itself is undertaken by developmental states. They also note that the act of eviction alone was 'but one part of a whole series of marginalisations, inconveniences and impoverishments' brought upon those being displaced (Brockington and Igoe 2006, p. 469).

Ethnographic and discourse-based research by political ecologists shows this pattern being repeated across Asia and Africa (Roth 2008), as the classification of lands leads to some users being discursively represented as 'forest squatters' (Elmhirst 2012) or poachers (Neumann 1996). Displacement associated with state-led strategies of sedentarization of swidden cultivators in Laos and the Philippines (Lestrilin 2011; Dressler and McDermott 2010) or the ecological resettlement of pastoralists (Zhang 2018) is made possible through contested discursive framings of 'underutilized' or 'unproductive' lands that provide a development-related moral justification (Harms and Baird 2014).

Whilst such studies focus primarily on the relationship between state power, environmental governance, and displacement, more recent work draws on an analysis of neoliberal natures (Bigger and Dempsey 2018) to attend to the role of transnational capital and the financialization of the environment as a vehicle for exclusion and dispossession. Here, climate change (and its mitigation), financial crisis, extractivism, and the global commodities boom provide a context for forms of displacement that accompany the establishment and growth of environmental markets (Sullivan 2013) within an 'economy of repair' (Fairhead et al. 2012, p. 242). Effectively, this involves nature being valued for its marketability: as carbon, minerals, and aesthetic values are sold off in turn (Fletcher 2010).

Financial and physical infrastructures for climate change mitigation are increasingly associated with displacement. For example, REDD+ is an initiative of the UN designed to place economic value on the carbon storage services that forests provide as a mechanism for reducing emissions from deforestation and thereby contributing to climate change mitigation. Political ecologists have pointed to how such schemes can effectively enclose the land and forest resources of marginalized communities, rendering their traditional livelihoods unviable (Milne and Adams 2012; Osborne 2015). Others point to the physical infrastructures of climate change adaptation in low-lying megacities impacted by climate-related sea level rises. Justification for displacement is provided by capital-fixing technical measures such as canal improvements, raised roads, flood walls, and slum clearance (Bose 2016; Ranganathan 2015; Vaz-Jones 2018; Doshi 2019), furthering the precarity of low-income and politically marginalized urban populations. A twenty-first-century version of enclosure for conservation has seen an alignment between international conservation organizations, tourist operators, and the state, as efforts to promote ecotourism schemes have also been associated with displacement of people and livelihoods in various contexts, including Colombia (Ojeda 2012) and the marine parks of Malaysia (Hill 2017).

Displacement is associated with the reworking of land as an investable resource by powerful transnational actors, in response to rising demands for food, water, energy, and other commodities, which has raised their relative value in commodity markets (Kay 2017). Political ecology analyses underscore many studies of boom-crop-related land grabbing (Peluso and Lund 2011), whilst feminist political ecology analyses have documented small-scale incremental displacements as Indigenous communities are unfavourably drawn into oil palm value chains (Elmhirst et al. 2017).

Other kinds of commodity boom relate to mineral commodities. Research in political ecology has documented the dramatic environmental and social implications of mineral extraction, and the displacements that accompany this. Analysis has centred on impacts and responses of the 'new extractivism', where marginalized Indigenous communities have been displaced by extractive industries, just as governments were justifying resource extraction as a revenue source for national development (Gudynas 2010; Bebbington 2012; Acosta 2013; Perreault 2013).

## To Conclude: Knowledge Politics and Embodied Resistance

Political ecology is, in many respects, an environmental activist science and practice, with a normative commitment to challenging unjust relationships within and between humans and more-than-human nature. One strand of this involves challenging expert knowledges that work in the service of capital, by examining the knowledge politics through which scientific expertise is constructed in line with developmentalist ideas about growth and the redistribution of risk (Huber et al. 2017). There is a growing alignment of university-based political ecology research with environmental justice organizations to challenge displacement, enabling international networking, advocacy, and communication on ecological struggles worldwide (ICTA-UAB n.d.). This kind of engagement has involved attending to what kinds of knowledge become dominant and with what effects. For example, political ecologists have shown how challenges to the financialization of nature (and the displacements this entails) are being made via other languages (and knowledges) of (non-economic) valuation, including Indigenous concepts of well-being or 'being with nature' (Martinez-Alier 2014; see also Radcliffe 2014). Decolonial ontologies of this kind have the effect of contesting the scientific inevitability of neoliberal technocratic development and of displacement itself. Finally, political ecology debates around nature, development, and displacement have been a cauldron not only for critiquing (and decolonizing) environmental knowledges that justify displacement but also for setting out repertoires of protest over maldevelopment and ecological injustices. Political ecology research has focused on the production of political subjectivities in the context of exclusion and displacement (Nightingale 2019), and explored a range of resistances, from violent repertoires of protest (Jenkins 2015) to forms of quiet encroachment to reclaim space through occupation (Lawhon et al. 2014; Vaz-Jones 2018). Multiscaled analyses navigate between global and intimate scales as political ecologists demonstrate how embodied emotions of anger and sorrow sustain Indigenous resistance to displacement (González-Hidalgo and Zografos 2017; Doshi 2019).

In documenting the contribution of political ecology to the analysis of displacement, this chapter has highlighted the breadth and depth of this field, where a combination of scholarly and activist work across a range of sociopolitical and ecological contexts has done much to reveal and challenge the workings of power. Its shifting focus reflects changes in the nature of global environmental issues more broadly but also a growing recognition of the situated politics of knowledge that fundamentally challenge the inevitability of development and displacement in its current form. Whilst the chapter has only touched on a sample of the work on displacement being undertaken through a political ecology framing, the rich illuminations yielded within this literature are testament to the potency of political ecology as a field of research, advocacy, and resistance at a point where there is urgent need for transformative change.

## References

Acosta, A. (2013). Extractivism and neoextractivism: Two sides of the same curse. In M. Lang & D. Mokrani (Eds.), *Beyond development alternative visions from Latin America* (pp. 61–86). Amsterdam: Transnational Institute and Quito, Ecuador: Rosa Luxemburg Foundation.

Adams, V., Van Hattum, T., & English, D. (2009). Chronic disaster syndrome: Displacement, disaster capitalism, and the eviction of the poor from New Orleans. *American Ethnologist, 36*(4), 615–636.

Adams, W. M., & Hutton, J. (2007). People, parks and poverty: Political ecology and biodiversity conservation. *Conservation and Society, 5*, 147–183.

Bebbington, A. (2012). Underground political ecologies: The second annual lecture of the Cultural and Political Ecology Specialty Group of the Association of American Geographers. *Geoforum, 43*(6), 1152–1162.

Bigger, P., & Dempsey, J. (2018). The ins and outs of neoliberal natures. *Environment and Planning E: Nature and Space, 1*(1–2), 25–43.

Blaikie, P., & Brookfield, H. (1987). *Land degradation and society*. London: Longman Development Series.

Blaikie, P., Cannon, T., Davis, I., & Wisner, B. (1994). *At risk: Natural hazards, people's vulnerability and disasters*. London: Routledge.

Bose, P. S. (2016). Vulnerabilities and displacements: Adaptation and mitigation to climate change as a new development mantra. *Area, 48*(2), 168–175.

Brockington, D., & Igoe, J. (2006). Eviction for conservation: A global overview. *Conservation and Society, 4*(3), 424–470.

Collard, R. C., Dempsey, J., & Sundberg, J. (2015). A manifesto for abundant futures. *Annals of the Association of American Geographers, 105*(2), 322–330.

Collins, T. W. (2008). The political ecology of hazard vulnerability: Marginalization, facilitation and the production of differential risk to urban wildfires in Arizona's White Mountains. *Journal of Political Ecology, 15*(1), 21–43.

Doshi, S. (2019). Greening displacements, displacing green: Environmental subjectivity, slum clearance, and the embodied political ecologies of dispossession in Mumbai. *International Journal of Urban and Regional Research, 43*(1), 112–132.

Dressler, W. H., & McDermott, M. H. (2010). Indigenous peoples and migrants: Social categories, rights, and policies for protected areas in the Philippine Uplands. *Journal of Sustainable Forestry, 29*(2–4), 328–361.

Elmhirst, R. (2012). Displacement, resettlement, and multi-local livelihoods: Positioning migrant legitimacy in Lampung, Indonesia. *Critical Asian Studies, 44*(1), 131–152.

Elmhirst, R., Siscawati, M., Sijapati Basnett, B., & Ekowati, D. (2017). Gender and generation in engagements with oil palm in East Kalimantan, Indonesia: Insights from feminist political ecology. *The Journal of Peasant Studies, 44*(6), 1135–1157.

Fairhead, J., Leach, M., & Scoones, I. (2012). Green grabbing: A new appropriation of nature? *Journal of Peasant Studies, 39*(2), 237–261.

Fletcher, R. (2010). Neoliberal environmentality: Towards a poststructuralist political ecology of the conservation debate. *Conservation and Society, 8*(3), 171.

Goldman, M. (2003). Partitioned nature, privileged knowledge: Community-based conservation in Tanzania. *Development and Change, 34*(3), 833–862.

González-Hidalgo, M., & Zografos, C. (2017). How sovereignty claims and 'negative' emotions influence the process of subject-making: Evidence from a case of conflict over tree plantations from Southern Chile. *Geoforum, 78*, 61–73.

Greiner, C., & Sakdapolrak, P. (2016). Migration, environment and inequality: Perspectives of a political ecology of translocal relations. In J. Schade, T. Faist, & R. McLeman (Eds.), *Environmental migration and social inequality* (pp. 151–163). Dordrecht, Switzerland: Springer International Publishing.

Grundy-Warr, C., & Rigg, J. (2016). The reconfiguration of political, economic and cultural landscapes in post-tsunami, Thailand. In P. Daley & R. M. Freener (Eds.), *Rebuilding Asia following natural disasters: Approaches to reconstruction in the Asia-Pacific* (pp. 210–235). Cambridge: Cambridge University Press.

Gudynas, E. (2010). The new extractivism of the 21st century: Ten urgent theses about extractivism in relation to current South American progressivism. *Americas Program Report, 21*, 1–14.

Gunewardena, N. (2008). Peddling paradise, rebuilding Serendib: The 100-meter refugees versus the tourism industry in post-tsunami Sri Lanka. In N. Gunewardena & M. Schuller (Eds.), *Capitalizing on catastrophe: Neoliberal strategies in disaster reconstruction* (pp. 69–92). Lanham, MD: Altamira Press.

Harms, E., & Baird, I. G. (2014). Wastelands, degraded lands and forests, and the class(ification) struggle: Three critical perspectives from mainland Southeast Asia. *Singapore Journal of Tropical Geography, 35*(3), 289–294.

Hill, A. (2017). Blue grabbing: Reviewing marine conservation in Redang Island marine park, Malaysia. *Geoforum, 79*, 97–100.

Huber, A., Gorostiza, S., Kotsila, P., Beltrán, M. J., & Armiero, M. (2017). Beyond "socially constructed" disasters: Re-politicizing the debate on large dams through a political ecology of risk. *Capitalism Nature Socialism, 28*(3), 48–68.

ICTA-UAB (Institute of Environmental Science and Technology from the Universitat Autònoma de Barcelona). (n.d.). *Environmental justice Atlas*. Retrieved June 9, 2019, from https://www.ejatlas.org

Jenkins, K. (2015). Unearthing women's anti-mining activism in the Andes: Pachamama and the 'Mad Old Women'. *Antipode, 47*(2), 442–460.

Kay, C. (2017). Contemporary dynamics of agrarian change. In H. Veltmeyer & P. Bowles (Eds.), *The essential guide to critical development studies* (pp. 291–300). London: Routledge.

Klein, N. (2007). *The shock doctrine: The rise of disaster capitalism*. London: Penguin Books.

Lawhon, M., Ernstson, H., & Silver, J. (2014). Provincializing urban political ecology: Towards a situated UPE through African urbanism. *Antipode, 46*(2), 497–516.

Leff, E. (2015). The power-full distribution of knowledge in political ecology. In T. Perreault, G. Bridge, & J. McCarthy (Eds.), *The Routledge handbook of political ecology* (pp. 64–75). London and New York: Routledge.

Lestrilin, G. (2011). Rethinking state–ethnic minority relations in Laos: Internal resettlement, land reform and counter-territorialization. *Political Geography, 30*(6), 311–319.

Lunstrum, E., Bose, P., & Zalik, A. (2016). Environmental displacement: The common ground of climate change, extraction and conservation. *Area, 48*(2), 130–133.

Martinez-Alier, J. (2014). The environmentalism of the poor. *Geoforum, 54*, 239–241.

Middleton, C., Elmhirst, R., & Chantavanich, S. (Eds.). (2017). *Living with floods in a mobile Southeast Asia: A political ecology of vulnerability, migration and environmental change*. London: Routledge.

Milne, S., & Adams, B. (2012). Market masquerades: Uncovering the politics of community-level payments for environmental services in Cambodia. *Development and Change, 43*(1), 133–158.

Mollett, S. (2015). 'Displaced futures': Indigeneity, land struggle, and mothering in Honduras. *Politics, Groups and Identities, 3*(4), 678–683.

Morrissey, J. (2012). Rethinking the 'debate on environmental refugees': From 'maximilists and minimalists' to 'proponents and critics'. *Journal of Political Ecology, 19*(1), 36–49.

Neumann, R. P. (1996). Dukes, earls, and ersatz Edens: Aristocratic nature preservationists in colonial Africa. *Environment and Planning D: Society and Space, 14*(1), 79–98.

Nightingale, A. J. (2019). Commoning for inclusion? Political communities, commons, exclusion, property and socio-natural becomings. *International Journal of the Commons, 13*(1), 16–35.

Nixon, R. (2011). *Slow Violence and the Environmentalism of the Poor*. Boston: Harvard University Press.

O'Keefe, P., Westgate, K., & Wisner, B. (1976). Taking the naturalness out of natural disasters. *Nature, 260*, 566–567.

Ojeda, D. (2012). Green pretexts: Ecotourism, neoliberal conservation and land grabbing in Tayrona National Natural Park, Colombia. *Journal of Peasant Studies, 39*(2), 357–375.

Oliver-Smith, A. (2012). Debating environmental migration: Society, nature and population displacement in climate change. *Journal of International Development, 24*(8), 1058–1070.

Osborne, T. (2015). Tradeoffs in carbon commodification: A political ecology of common property forest governance. *Geoforum, 67*, 64–77.

Peluso, N. L., & Lund, C. (2011). New frontiers of land control: Introduction. *Journal of Peasant Studies, 38*(4), 667–681.

Perreault, T. (2013). Dispossession by accumulation? Mining, water and the nature of enclosure on the Bolivian Altiplano. *Antipode, 45*(5), 1050–1069.

Perreault, T., Bridge, G., & McCarthy, J. (2015). *The Routledge handbook of political ecology*. Oxford: Routledge.

Radcliffe, S. A. (2014). Gendered frontiers of land control: Indigenous territory, women and contests over land in Ecuador. *Gender, Place & Culture, 21*(7), 854–871.

Ranganathan, M. (2015). Storm drains as assemblages: The political ecology of flood risk in post-colonial Bangalore. *Antipode, 47*(5), 1300–1320.

Rangarajan, M., & Shahabuddin, G. (2006). Displacement and relocation from protected areas: Towards a biological and historical synthesis. *Conservation and Society, 4*, 359–378.

Robbins, P. (2011). *Political ecology: A critical introduction* (2nd ed.). Chichester: John Wiley & Sons.

Rocheleau, D. (2016). Rooted networks, webs of relation, and the power of situated science: Bringing the models back down to earth in Zambrana. In W. Harcourt

(Ed.), *The Palgrave handbook of gender and development* (pp. 213–231). London: Palgrave Macmillan.

Roth, R. J. (2008). 'Fixing' the forest: The spatiality of conservation conflict in Thailand. *Annals of the Association of American Geographers, 98*(2), 373–391.

Sullivan, S. (2013). Banking nature? The spectacular financialisation of environmental conservation. *Antipode, 45*(1), 198–217.

Tetreault, D. (2017). Three forms of political ecology. *Ethics and the Environment, 22*(2), 1–23.

Timms, B. F. (2011). The (mis) use of disaster as opportunity: Coerced relocation from Celaque National Park, Honduras. *Antipode, 43*(4), 1357–1379.

Vandergeest, P. (2003). Land to some tillers: Development-induced displacement in Laos. *International Social Science Journal, 55*(175), 47–56.

Vaz-Jones, L. (2018). Struggles over land, livelihood and future possibilities: Reframing displacement through feminist political ecology. *Signs: Journal of Women in Culture and Society, 43*(3), 711–735.

Watts, M. J. (1983). *Silent violence: Food, famine, and peasantry in northern Nigeria.* Berkeley, CA: University of California Press..

Zhang, Q. (2018). Managing sandstorms through resettling pastoralists in China: How multiple forms of power govern the environment at/across scales. *Journal of Political Ecology, 25*(1), 364–380.

# 4

# Displacement Economies: A Relational Approach to Displacement

Amanda Hammar

## Introduction

Displacement—here conceived in terms of interlinked ideas, acts, experiences, and effects of *enforced* spatial, social, symbolic, and material disruption—is a constant of human history. Always paradoxical, it entails, simultaneously, movement and confinement, rupture and continuity, unpredictability and order, loss and opportunity, and vulnerability and capability. As an analytical concept, it can generate insights into an array of complex empirical realities, varied by spatial and historical context. Yet it is also both a 'buzzword' of sorts (Cornwall 2007) and a technocratic term whose naturalised, common-sense meanings—often associated with war, mass exodus, and passive victims—problematically reproduce simplifications that mask a great deal.[1] It is a term that is applied and contested by a wide spectrum of differentiated actors who, directly or indirectly, are affected by, generate, or in some way govern different conditions, regimes, relations, and lived experiences of multilayered disruption (Hammar 2015; McDowell and Morrell 2010; Bakewell 2008; Colson 2007 [2004]).

---

A. Hammar (✉)
Centre of African Studies, University of Copenhagen, København, Denmark
e-mail: aha@teol.ku.dk

## The Dominance of Operational Approaches

Despite displacement's conceptual and contextual variability, scholarly, policy, popular, and practitioner orientations largely remain lodged within global(ised) humanitarian discursive frameworks. These originated within the historical and ideological conjunctures of mid-twentieth century postwar Europe and America, expanding with independence-driven changes on the African continent after 1960. In more recent times, particularly Euro-American projections and populist fears of mass invasion by 'hoards of refugees' from troubled and troubling elsewheres have intensified. Both national and multilateral policies and funding systems are aimed increasingly at border control and keeping unwanted migrants at a safe distance, partly by attempting to 'improve' original displacement 'sending' environments. Displacement associated with dispossession and forced resettlement, often linked to forms of authoritarian rule—be this royal, religious, colonial, or postcolonial—and/or private capital, is addressed extensively within such scholarly fields as agrarian or urban studies.[2] This aspect of displacement is less commonly included within humanitarian discourses and emergency protection regimes. These have tended to rest on rather de-historicised, simplified, and operational tropes of conflict, mass flight, and emergency, and of victimhood, care, and protection.

That being said, global humanitarian policies and programmes have become somewhat more dynamic in recent times. Partly, this has been in response to new empirical and political challenges—of magnitude and protraction, for example[3]—and in recognition of greater variation and complexity in displacement contexts (McDowell and Morrell 2010; Omata 2017; Betts et al. 2017). Even so, the dominant 'forced displacement' paradigm remains firmly grounded in a mix of legalistic and highly technocratic, solution-driven forms of problematisation and intervention (Bacchi 2012; Hammar 2014b). This works mostly within the formalised conventions and protocols of refugeehood and internal displacement and its associated population categories (Zetter 2007), albeit more recently these have expanded to include hosts (Rodgers 2008) as well as non-camp and urban populations (Sanyal 2012).

There is no disputing the profound need for emergency support and protection in conditions of both extreme and chronic crisis linked to forced dislocation or enclosure. The argument made here, however, is that there is far more to the causes, conditions, and consequences of displacement than is revealed through approaches concerned with 'what needs to be done to assist, protect, manage, integrate, repatriate or resettle' those counted within specific country 'caseloads' (Hammar 2014b, p. 11).

## Displacement Economies: An Alternative Relational Approach

The challenge of thinking about displacement beyond official, operationally driven perspectives requires, among other things, a broader and more inclusive *relational* approach. Key to this is an understanding of displacement as multidimensional, with respect to the simultaneity of and interplay between its multiple spatial, historical, temporal, social, political, cultural, and economic elements. The approach recognises both material and symbolic causes and forms of enforcement and their effects, thus embracing a combination of political economy and cultural politics lenses in tracing the lived implications of difference in displacement contexts. It requires a longer historical and relational view of any given displacement context or 'event': not seeing it as a singular or discrete phenomenon but rather as part of a more complex, layered continuum of inclusions and exclusions, of possibilities and impossibilities. These may be related to struggles over 'the nation' and the boundaries of citizenship and belonging, or over specific material or symbolic resources. They might reflect older or new structural patterns of power, authority, differentiation, vulnerability, and violence, as well as old or emerging forms of resistance.

Empirically, displacement might manifest as enforced physical dislocation at different scales within or outside the territorial boundaries of a given state, be this conflict-generated, politically motivated, or as development-induced forced removals or resettlement.[4] It may also be expressed as forced immobility, including involuntary encampment or abandonment (Lubkemann 2008; Magaramombe 2010; Turner 2010). In such settings, loss of the basic certainties of everyday life constitutes the most widely recognised effect: shelter, food, water, health care, livelihoods, education, and so on. However, displacement is also encountered in non-physical or symbolic forms of disruption and loss. This could be the disappearance of formal jobs or steady incomes and the status attached to them, the undoing or reconstitution of familiar family patterns including gender or generational hierarchies, or the denial of basic citizenship rights and the decline in human dignity. Yet in almost all displacement circumstances, unexpected openings and gains emerge alongside closures and losses, either sharpening or reshaping older social and economic hierarchies.

For example, as shortages in essentials intensify through displacement, business opportunities arise for those with relative access to resources; or younger and potentially more adaptable men and women might replace older men as primary breadwinners in the household. Notably, a spectrum of new forms of individual and collective agency emerges even in the face of extreme forms of dissolution.

Displacement-related actors include not only those who have been forcibly dislocated, confined, or left behind. They also include those that host or in some cases refuse to host displacees whether at local or national level, or among individuals, communities, or institutions. It includes those that precipitate or perpetuate particular conditions of displacement, who might overtly or covertly benefit from such conditions, for example, different kinds of state authorities, political parties, private companies, the formal security sector, or informal militia or rebel movements, or even environmental non-governmental organisations. At the same time, new and old entrepreneurs, gatekeepers, or brokers may arise among the displaced themselves, creating or reinforcing historical forms of differentiation and exclusion (Omata 2017). Finally, one needs to take account of the material or symbolic benefits—and not just the costs—accruing to those managing or governing the displaced, be these different national state agencies, multilateral or local humanitarian agencies, or religious organisations.

What these considerations point to more broadly—not least conceptually—is a necessary focus not only on what and who generates displacement and why but on *what displacement itself produces*. This question is at the core of the *displacement economies* approach. Crucial to note is that this approach to displacement does not merely refer to specific types of economy, even while, as already indicated, it necessarily includes relevant economic and political economy dimensions of displacement.[5] This contrasts with what Betts et al. (2017) refer to as 'refugee economies,' which focuses on 'the economic lives of refugees,' in so far as refugees 'occupy a particular legal status and position, which in turn places [them] … in a distinctive institutional context' (p. 40). A focus specifically on economic aspects within the refugee-oriented literature is still relatively recent. The scholarship that has begun to fill a recognised gap in this regard (Jacobsen and Landau 2003) is well summarised by Betts et al. (2017). They highlight important insights emerging, for example, in relation to refugee livelihoods, resource access and allocation, remittances, self-reliance, effects on host economies and on local labour markets, the straddling of formal and informal sectors and national and transborder economies, and the economic costs of refugee management itself. Taking a more multilayered, development-focused rather than humanitarian approach, and addressing displacement contexts other than merely refugee camps, they nonetheless remain within a broadly solution-oriented perspective. The analytical limits of this in terms of capturing a wider spectrum of displacement causes, conditions, actors, experiences, and consequences, and some of its problematic political, policy, and practical implications, have been addressed elsewhere (Hammar 2014b, 2015; Hammar et al. 2010; Hammar and Rodgers 2008).

## Empirical Origins of an Approach

The contours of the displacement economies approach outlined so far emerged not as an abstract formulation but out of empirically grounded realities whose complexities (Hammar et al. 2010) demanded a conceptual and analytical language that did not yet exist (Hammar 2014b). Specifically, this was the context of Zimbabwe's multilayered and deeply displacement-inflected political and economic crisis, which began in the early 2000s and has continued as a chronic condition, albeit with varying periods of greater or lesser intensity. From the outset, the crisis was marked by a combination of mostly party-state-generated physical and symbolic displacements, partly supported and implemented by nationalist war veterans, land-short peasants, and ruling party loyalists. These occurred at larger and smaller scales and within both rural and urban settings. In rural areas, this entailed extensive white farmer and black farmworker evictions (in the thousands and hundreds of thousands respectively) in the first half of the 2000s. The reallocation of occupied and appropriated land to a combination of black smallholders and larger-scale farmers dramatically reshaped the agrarian landscape in both economic and social terms. In urban areas, in mid-2005, the party-state undertook mass evictions of poorer township dwellers who were seen by the ruling party (Zanu PF) as supporting the political opposition Movement for Democratic Change (MDC). In addition, in both rural and urban areas, extreme forms of political violence were perpetrated against individual opposition members, including beatings, torture, and disappearances.

Zanu PF's legitimations for violent acts of both direct and indirect dispossession, dislocation, and relocation, imposed upon carefully targeted populations, drew on a selective mix of discourses. With respect to extensive land invasions of large-scale, white-owned farms, radical nationalist and nativist discourses were applied with respect to reversing unequal, colonial-era racialised land ownership patterns. Alongside this, anti-imperialist rhetoric and claims of sovereignty, especially aimed at Britain, were used to attack and delegitimise the opposition's more liberal and human rights-based critiques of violence and displacement, framing them as attempts to 'reverse the revolution' (Phimister and Raftopoulos 2004). At the same time, new legislation and policies were introduced that retroactively legalised the land appropriations (Matondi 2012). With respect to the mass urban evictions across the entire country, a combination of legalistic and technocratic discourses were used to redefine certain categories of urban dwellers and urban structures, both residential and commercial, as 'illegal,' and to criminalise certain urban practices (Potts 2006). This was used to justify the mass destruction of houses and small businesses and the forced removal and/or relocation of three-quarters of a million urbanites, negatively affecting an estimated 2.4 million people overall. The dual aim of this was to both punish and fragment urban opposition voters. Unsurprisingly, to facilitate these and other partisan actions against selected citizens, government agencies became highly politicised and in some cases militarised through appointing former security sector personnel to key positions. The general effect was a de-professionalisation of large sections of the bureaucracy.

In the course of the mass physical displacements and political violence across the country, in addition to the direct human suffering it caused to millions of Zimbabweans, the economy as a whole went into radical decline. Production in commercial agriculture, in particular, dropped substantially with critical losses in foreign currency earnings that had severe consequences for the once-burgeoning manufacturing sector, which systematically started to close down. Formal unemployment soared to over 90 per cent, generating widespread informalisation of the economy in all spheres. Ironically, 'black market' trade in foreign currency itself became a new growth sector, favouring a small business, political, and military elite with privileged access to such currency while also stimulating lower-level street trade. By 2007–2008, hyperinflation was in the trillions of per cent, so that even for those few who remained in formal employment, in either the private or public sectors, the value of wages dropped to meaningless levels.

For both middle-class and working-class urban families, the former possibilities for social mobility closed down entirely (Hammar 2014c). At the same time, basic survival needs prompted profound changes in livelihood strategies that altered both gender and generational positions and relations. In place of the formally employed senior male breadwinner as the more traditional core of family income, women of various ages started to undertake informal cross-border trading, while young men engaged in a range of ad hoc *kukiya-kiya* activities as and where opportunities arose (Jones 2010). At the same time, the combination of direct physical dislocation, economic devastation, and political persecution drove a mass exodus of Zimbabweans across borders, both legally and illegally, especially, although not only, to South Africa. This created an extensive global Zimbabwean diaspora that barely existed prior to the 2000s, one extremely varied in class, race, gender, and generational terms (McGregor and Primorac 2010). This has become key to supporting many of those remaining at home through remittances, although the regularity and distribution of such funds is highly uneven. An additional effect of the past two decades of displacement, disruption, and decline has been the generation of a substantial body of both scholarship and creative writing, as well as theatre, art, and music, that attempts to expose, comment on, and counteract that which sustains the crisis.

The overall conditions, forms, experiences, and effects of displacement observed from the early 2000s onwards in relation to the Zimbabwe crisis were key to an acute consciousness of the paradoxes of displacement, generating in turn the search for a multi-lensed approach through which to make sense of this. Core to this consciousness was evidence of the range of 'new physical, social, economic and political spaces, relations, systems and practices that displacement itself was *producing*' (Hammar 2014b, p. 3).[6]

## Asking What Displacement Produces

The question of what displacement itself produces, which together with a relational perspective, is core to the displacement economies approach, is both conceptually grounded and empirically generative. As has been demonstrated, it emerged out of an analytical lacuna, especially in trying to confront the paradoxes of displacement. Conceptually, it is crucial for expanding ways of thinking about displacement beyond operational interpretations and interventions. At the same time, it speaks to the methodological imperatives that drive empirical investigations into diverse displacement contexts, anywhere and at any time. In this sense, in order to unearth, analyse, and learn from displacement's multilayered, lived complexities and implications, the core question of what displacement produces can be extended into the following range of questions:

> How are physical spaces altered: how are they damaged or reconstructed, differently occupied or abandoned, controlled and (re)allocated, used or repurposed? How do notions and experiences of time change in contexts of crisis and displacement, and in so doing, affect economic, social, and political practices? What happens to forms and patterns of ownership, access and control of

property and people? What changes are generated in the value and commodification of different objects, spaces, natural resources, bodies, or money itself, and how does this shape related markets and modes of exchange? What shifts occur in forms and conditions of labour, in the means and modes of production, in patterns of accumulation and distribution? What happens to both official and alternative economies (formal and informal, licit and illicit) and their ways of articulating with each other, and with what implications for those differentially positioned in relation to either or both? How are dynamics of authority altered, and political and administrative relations, spaces, systems, and practices revised, consolidated, or diminished? How are definitions, practices, and experiences of citizenship and personhood altered, and forms of agency reshaped? In what ways are social identities and social relations reconstituted, and how do forms and sites of social reproduction change?

In relation to all these questions, one needs to ask additionally, on what and whom do these changes have differential effects?

## A Southern-Driven Approach with General Relevance

The context within which the displacement economies approach originally emerged was specific in both a national and regional sense (Hammar and Rodgers 2008; Hammar et al. 2010). However, it generated a set of conceptual and methodological orientations and questions that are applicable well beyond the specific 'case' of Zimbabwe and southern Africa. Indeed, elements of the approach were both applied and further developed within scholarship covering a range of cases across the African continent (Hammar 2014a). This included both more classic conflict-related refugee and IDP contexts, as well as other kinds of displacement conditions. Yet beyond Africa, or even the Global South, the multi-dimensionality of displacement in any context calls for a relational approach that makes visible and comprehensible the interconnectedness of precipitating factors, diverse actors, and paradoxical, productive effects. A narrower operational approach, despite—or because of—its best intentions to protect and care within the limits of prescribed legal frames and categories, is in danger of failing to recognise or address the more complex and contradictory dimensions of displacement and, in some cases, may even reinforce its most damaging patterns and effects.

## Notes

1. See Polzer and Hammond (2008) and Lubkemann (2008) among others, for critiques of problematic invisibilities within more classic operational approaches. McDowell and Morrell (2010) make a strong argument for attention to a greater diversity of displacement conditions and experiences, proposing greater analytical convergence between the more commonly recognised 'conflict-generated' displacement and what they argue is misleadingly termed 'non-conflict displacement.' In the latter case, similar effects are generated. However, legal protection and humanitarian responses generally are only provided to vulnerable populations in the former.
2. One focus of this is the scholarship on 'development-induced displacement.' See, for example, De Wet (2006) and Vandergeest et al. (2007). But see Hammar (2008) for a discussion specifically on the relationship between displacement and state making.
3. See, for example, the Comprehensive Refugee Response Framework (CRRF) turn in global refugee policy (UNHCR 2016).
4. Environmental causes—which are never purely 'natural'—are excluded here. The main focus is on more overtly anthropocentric forms of enforced displacement.
5. Although not the original conceptual intention, the term 'displacement economies' has come to be applied by some more narrowly, to describe certain kinds of economies produced by displacement, much as others speak of 'war economies' or 'shadow economies' (e.g. see Pugh and Cooper 2004).
6. Such paradoxical dynamics of displacement were observed at a much smaller scale in Zimbabwe's agrarian margins during doctoral research in the late 1990s (see Hammar 2001) but were not conceptualised in these terms at that time.

## References

Bacchi, C. (2012). Why study problematizations? Making politics visible. *Open Journal of Political Science, 2*(1), 1–8.

Bakewell, O. (2008). Research beyond categories: The importance of policy irrelevant research into forced migration. *Journal of Refugee Studies, 21*(4), 432–453.

Betts, A., Bloom, L., Kaplan, J., & Omata, N. (2017). *Refugee economies: Forced displacement and development.* Oxford, UK: Oxford University Press.

Colson, E. (2007 [2004]). Displacement. In D. Nugent & J. Vincent (Eds.), *A companion to the anthropology of politics* (pp. 107–120). Malden, MA; Oxford, UK; and Carlton, Australia: Blackwell.

Cornwall, A. (2007). Buzzwords and fuzzwords: Deconstructing development discourse. *Development in Practice, 17*(4–5), 471–484.

De Wet, C. (Ed.). (2006). *Development-induced displacement: Problems, programs and people*. New York, NY: Berghahn.

Hammar, A. (2001). 'Day of Burning': Eviction and reinvention in the margins of northwest Zimbabwe. *Journal of Agrarian Change, 1*(4), 550–574.

Hammar, A. (2008). In the name of sovereignty: Displacement and state making in post independence Zimbabwe. *Journal of Contemporary African Studies, 26*(4), 417–434.

Hammar, A. (Ed.). (2014a). *Displacement economies in Africa: Paradoxes of crisis and creativity*. London, UK and Uppsala, Sweden: Zed Books and Nordiska Afrikainstitutet.

Hammar, A. (2014b). Introduction: Displacement economies in Africa. In A. Hammar (Ed.), *Displacement economies in Africa: Paradoxes of crisis and creativity* (pp. 3–32). London and Uppsala, Sweden: Zed Books and Nordiska Afrikainstitutet.

Hammar, A. (2014c). The paradoxes of class: Crisis, displacement and repositioning in post-2000 Zimbabwe. In A. Hammar (Ed.), *Displacement economies in Africa: Paradoxes of crisis and creativity* (pp. 79–104). London, UK and Uppsala, Sweden: Zed Books and Nordiska Afrikainstitutet.

Hammar, A. (2015). The concept and paradoxes of displacement. In K. Havnevik, T. Oestigaard, & E. Tobisson (Eds.), *Framing African development—challenging concepts* (pp. 111–134). Leiden, Netherlands and Boston, MA: Brill.

Hammar, A., McGregor, J., & Landau, L. (2010). Introduction: Displacing Zimbabwe: Crisis and construction in Southern Africa. *Journal of Southern African Studies., 36*(2), 263–283.

Hammar, A., & Rodgers, G. (2008). Introduction: Notes on political economies of displacement in Southern Africa. *Journal of Contemporary African Studies, 26*(4), 355–370.

Jacobsen, K., & Landau, L. (2003). The dual imperative in refugee research: Some methodological and ethical considerations in social science research on forced migration. *Disasters, 27*(3), 185–206.

Jones, J. (2010). 'Nothing is straight in Zimbabwe': The rise of the *kukiya-kiya* economy 2000–2008. *Journal of Southern African Studies, 36*(2), 285–299.

Lubkemann, S. C. (2008). Involuntary immobility: On a theoretical invisibility in forced migration studies. *Journal of Refugee Studies, 21*(4), 454–475.

Magaramombe, G. (2010). Agrarian displacements, replacements and resettlement: 'Displaced in place' farm workers in Mazowe District. *Journal of Southern African Studies, 36*(2), 361–375.

Matondi, P. B. (2012). *Zimbabwe's fast track land reform*. London, UK and Uppsala, Sweden: Zed Books and Nordiska Afrikainstitutet.

McDowell, C., & Morrell, G. (2010). *Displacement beyond conflict: Challenges for the 21st century*. New York, NY: Berghahn.

McGregor, J., & Primorac, R. (Eds.). (2010). *Zimbabwe's new diaspora. Displacement and the cultural politics of survival*. New York, NY and Oxford, UK: Berghahn.

Omata, N. (2017). *The myth of self-reliance. Economic lives inside a Liberian refugee camp*. New York, NY and Oxford, UK: Berghahn.

Phimister, I., & Raftopoulos, B. (2004). Mugabe, Mbeki & the politics of anti-imperialism. *Review of African Political Economy, 31*(101), 385–400.

Polzer, T., & Hammond, L. (2008). Invisible displacement (editorial introduction). *Journal of Refugee Studies, 21*(4), 417–431.

Potts, D. (2006). 'Restoring order'? Operation Murambatsvina and the urban crisis in Zimbabwe. *Journal of Southern African Studies, 32*(2), 273–291.

Pugh, M., & Cooper, N. (2004). *War economies in a regional perspective. Challenges of transformation.* Boulder, CO and London, UK: Lynne Rienner.

Rodgers, G. (2008). Everyday life and the political economy of displacement on the Mozambique–South Africa border. *Journal of Contemporary African Studies, 26*(4), 385–399.

Sanyal, R. (2012). Refugees and the city: An urban discussion. *Geography Compass, 6*(11), 633–644.

Turner, S. (2010). *Politics of innocence: Hutu identity, conflict, and camp life*. New York, NY: Berghahn.

UNHCR. (2016). *Applying comprehensive responses (CRRF) in Africa*. Retrieved from https://www.unhcr.org/publications/operations/5a8fcfff4/applyingcomprehensive-responses-crrf-africa.html

Vandergeest, P., Idahosa, P., & Bose, P. S. (Eds.). (2007). *Development's displacements. Ecologies, economies, and cultures at risk.* Vancouver and Toronto, Canada: UBC Press.

Zetter, R. (2007). More labels, fewer refugees: Remaking the refugee label in an era of globalization. *Journal of Refugee Studies, 20*(2), 172–192.

# 5

# The Slow and the Fast Violence of Displacement

James A. Tyner

## Introduction

We readily understand displacement as inherently spatial. Being confronted with injurious or life-threatening conditions, for example, impels individuals to flee. Similarly, practices of enclosure or eviction force people from their homes, leading to dispossession and displacement. So framed, much scholarship tends also to conceptualize displacement as an abrupt, singular, or catastrophic moment of loss imposed from above and contested from below (Vaz-Jones 2018, p. 711). That is, displacement marks the moment in which people flee some spectacular instance of violence, a violent event that compels people to move. Such a narrow definition of displacement, however, threatens to minimize if not obfuscate less sensational, yet nevertheless devastating and traumatic, forms that more gradual processes of displacement can take (Vaz-Jones 2018, p. 711). In other words, scholars of displacement must remain sensitive to the temporality of displacement as much as the spatiality of displacement.

This is not to say that studies of displacement are atemporal. Indeed, considerable work addresses the travails of waiting *following* displacement: of displaced persons detained in 'refugee' camps or detention centres, of terrorist

---

J. A. Tyner (✉)
Kent State University, Kent, OH, USA
e-mail: jtyner@kent.edu

© The Author(s) 2020
P. Adey et al. (eds.), *The Handbook of Displacement*,
https://doi.org/10.1007/978-3-030-47178-1_5

suspects held in extraterritorial camps, of prison populations, and of men, women, and children waiting in halfway houses, domestic violence shelters, or homeless shelters. In this chapter, I consider the durability of displacement associated with slow violence, of the conditions in place prior to displacement. Here, we understand the ebbs and flows of violence and of the temporal liminality that accompanies violence. More specifically, I reposition displacement as a process that entails both periods of intense activity and prolonged waiting. However, I argue also that displacement as process needs to be expanded, in that displacement itself may manifest as an act of resistance in response to slow violence.

## Time and Violence

Simon Springer's suggestion that violence 'sits in places' is apropos of both the spatiality and temporality of violent displacement. For Springer, this means firstly that the way in which we perceive the manifestation of violence is localized and embodied (see also Tyner 2012, 2016, 2019). This should not translate, however, into an understanding of violence as material expression of an isolated 'event' or localized 'thing.' Rather, we can more appropriately understand violence as an unfolding process, arising from the broader geographical phenomena and temporal patterns of the social world (Springer 2011, p. 91). In other words, to conceive of violence—and, by extension, of the violence of displacement—as process instead of event highlights the *durability* of displacement.

To recalibrate our understanding of violent displacement not as a distinct event but instead as a process is to engage with the concept of *slow violence* (Christian et al. 2016; Davies 2018; Anderson et al. 2019). For Rob Nixon (2011, p. 2), slow violence refers to 'a violence that occurs gradually and out of sight, a violence of delayed destruction that is dispersed across time and space, an attritional violence that is typically not viewed as violence at all.' Especially germane to the temporality of displacement, slow violence constitutes 'a violence that is neither spectacular nor instantaneous, but rather incremental and accretive' (Nixon 2011, p. 2). As Thom Davies (2018, p. 1539) explains, the concept has value in uncovering the slow and hidden brutality of certain spaces, that slow violence is a form of harm that is neither instantaneous nor overtly dramatic, yet has damaging consequences. Slow violence, conceptually, reorients attention away from the immediacy of almighty events and towards the oft-hidden structural inequalities that manifest in a perceived need to flee. From this vantage point, we may understand *the process of displacement* exhibiting a range of temporalities. Certainly, violent ruptures such as hurricanes and conflict are causes of displacement; but so too are conditions that unfold gradually, perhaps inexorably, over months, years, and even generations. By way of illustration, consider the slow violence associated with the abandoned Giant Mine in Yellowknife, Northwest Territories, Canada. For fifty years, arsenic polluted the surrounding lakes, rivers, and streams, poisoning local communities (Sandlos and Keeling 2016). As Nixon (2011, p. 3) argues, 'We need to account for how the temporal dispersion of slow violence affects the way we perceive and respond to a variety of social afflictions—from domestic abuse to posttraumatic stress and, in particular, environmental calamities.'

To this end, we need to understand that displacements can and do arise, for example, from spectacular forms of violence but also from the harmful effects of climate change, environmental degradation, intimate partner abuse, and other forms of violence that unfold slowly and often imperceptibly. As Emma Laurie and Ian Shaw (2018, p. 10) write, violence 'burns in the background of daily life.' In short, we need to consider not only the waiting that happens after displacement but also the waiting that occurs prior to displacement.

## Waiting and Displacement

Superficially, the concept of waiting seems so banal as to not merit attention. For waiting is an ever-present facet of life. We wait at banks, grocery stores, and post offices. We wait at traffic stops and ticket counters. And beyond these *places* of waiting, we wait for things to happen: we wait for job interviews and phone calls and late submissions of manuscripts. As Elizabeth Olson (2015, p. 517) writes, so ubiquitous is waiting that it seems hardly of ethical or moral consequence. However, Olson (2015, p. 517) correctly surmises that the geographies of waiting may 'produce and maintain potentially abusive and harmful arrangements of power and inequality.' She affirms, 'Lines, lists, rooms and rosters dictate an order of being received, and in doing so they may also influence the dignity and safety of those who are required to wait for jobs, housing, asylum or security from intimate terrorism' (Olson 2015, p. 517).

The manifold ways in which human beings in their lifeworlds think and feel about (and act on) time have been the subject of much scholarly work in the social sciences (Auyero 2011, p. 6). In recent years, geographers and other scholars have reconfigured waiting as a power-laden concept. If someone is subjected to waiting, he or she is delayed, or prevented, from engaging in some other activity. The physical act of waiting for visa processing, for example, is time spent *not* moving. For asylum seekers or refugees, bureaucratic delays augment an already fraught and traumatic experience. Accordingly, the subjective effect of waiting secures domination, as undocumented persons find themselves 'waiting hopefully and then frustratedly for others to make decisions, and in effect surrendering to the authority of others' (Auyero 2012, p. 123; see also Olson 2015, p. 522). All too often, 'to be the person that waits … means to be powerless and in a state of uncertainty' (Mueller-Hirth 2018, p. 103). Consequently, the 'fact that waiting mirrors unequal power relations has negative impacts on people's emotional and material well-being and can exacerbate social inequalities' (Mueller-Hirth 2017, p. 202).

Crucially, this condition of powerlessness exacerbates prejudices and other dehumanizing practices. As Barry Schwartz (1974, p. 856) writes, to be kept waiting, 'especially to be kept waiting an unusually long time, is to be the subject of an assertion that one's own time (and therefore, one's social worth) is less valuable than the time and worth of the one who imposes the wait.' Under these conditions, men, women, and children waiting to flee famine, conflict, or eviction endure violence in situ. Thus, rather than simply conceptualize moments of inactivity as waiting, perhaps it is more appropriate to consider various forms of waiting from the standpoint of endurance.

To endure is to withstand some potentially traumatic experience; but it is also potentially to resist. In his study of housing evictions in Saigon, Vietnam, Erik Harms (2013, p. 346) considers the 'complex assortment of temporalities that arise when people are displaced from their land and homes.' With particular emphasis on peoples' lives prior to eviction, Harms (2013, p. 346) provides insight into the ways people in eviction zones cope with and also take advantage of unfamiliar and largely alienating temporal relations marked by uncertainty, ambiguity, and contradiction. Waiting, in other words, marks a period fraught with peril but also a time for action. Crucial to this struggle, of coping with the uncertainties of likely displacement, is the ability to challenge both those conditions that make life precarious and those institutional constraints that impinge on one's ability either to change those conditions or to flee.

Much recent scholarship documents the emotional and psychological pain, suffering, and trauma associated with waiting among displaced persons in post-conflict contexts. Natascha Mueller-Hirth (2017, 2018), for example, examines the experiences and effects of waiting for victim support, recognition, and social change among those affected by apartheid-era human rights violations in South Africa, while Victor Igreja (2012) documents the quotidian experiences of survivors of civil war in Mozambique. In the remainder of this chapter, I direct attention to the waiting that precedes displacement, that is, to questions of 'when' and 'how long' people *endure* before becoming displaced. My concern therefore is not the physical, emotional, and psychological conditions experienced after spectacular moments of displacement, such as the immediacy of natural disasters or armed conflict, but instead the durability of lives in waiting subjected to slow violence. In short, I am concerned with the travails of those people subjected to slow violence before displacement.

## Climate-Induced Displacement

Climate change is a form of slow violence (Nixon 2011; O'Lear 2016). Here, the conditions for sustaining life are not immediately destroyed, but rather slowly degraded, as in the form of deforestation, desertification, and sea-level rise. Some of the most vulnerable regions include areas like low-lying islands and deltas, coastal zones, glacial-fed water systems, and regions subject to persistent drought (Warner 2012, p. 1061). In sub-Saharan Africa, for example, effects of climate change include loss of arable land and lower crop yields, both of which (will) result in malnutrition and famine; and competition and conflict over increasingly scarce resources (notably water) will likewise lead to political instability (Willett 2015). Stated bluntly, the impacts of climate change will most certainly lead to widespread displacement around the world.

It is important to note, however, the effects of climate change and other environmental practices interact in myriad other ways that may induce displacement. As Elizabeth Lunstrum et al. (2016, p. 131) observe, 'If the burning of fossil fuels leads to displacement via climate change, fossil fuel and other forms of extraction themselves lead to more immediate forms of displacement. This includes direct displacement from residences or from agricultural, hunting and fishing lands slated for state-approved extractive projects as well as more indirect forms of dispossession tied to the ecological degradation and toxicity that emerges from the externalities of mining projects.' In other words, when we speak of climate-induced displacement, it is necessary to be cognizant of the multiple and overlapping temporalities of displacement.

Climate-induced environmental degradation will intensify pre-existing vulnerabilities for already struggling communities, especially throughout the Global South (Chu et al. 2016; Willett 2015). However, some discussion has centred also on the prospect that 'the uneven and in some regions potentially dramatic impacts of climate change may catalyze cooperation and transcend enmities' (Tänzler et al. 2010, p. 741). In other words, rather than merely waiting for catastrophic change and massive displacement, key agents are actively working to mitigate risks and to adapt in situ. Such discussions are fraught with difficulties (Warner 2012); nevertheless, scholarship on climate-induced displacement brings focus to the myriad ways in which individuals, institutions, and governments may respond. In rural Kenya, for example, Willett (2015, p. 10) documents how residents engage in adaptive farming activities based on practices that previously helped them under conditions of drought.

## Chronic Trauma and Displacement

Slow violence is not simply the manifestation of structural processes. Indeed, social relations may and often do exhibit a variant of slow violence, manifest, for example, in the form of sustained abuse. Studies of displacement often stress the trauma of leaving one's home. Accordingly, post-displacement waiting entails a longing to return, to re-establish one's life to its pre-violent conditions. And yet for many people—but especially women—the home may itself be where violence resides. Women in the United States, for example, are far more likely to be deliberately injured in their homes than on the streets; in fact, violence in the home accounts for more injuries than car accidents and muggings, combined. Furthermore, domestic homicides account for approximately one-third of all female homicides per year (Hattery 2009).

In this penultimate section, I call attention to the traumatic waiting that takes place within abusive relations. Here, trauma associated with slow violence may include long-term vulnerability to emotional, psychological, and physical abuse (Pain 2019). Crucially, trauma 'can embody multiple temporalities and spatialities as traumatic time and space move both forwards and backwards, occupying here and there, center and margin' (Coddington and Micieli-Voutsinas 2017, p. 55). Especially relevant is Judith Herman's concept of chronic trauma. According to Herman (1992, p. 377), prolonged, repeated trauma may occur when someone is in a state of captivity, unable to flee, and under the control of another. Examples include prisons and concentration camps, but such conditions also exist in abusive relationships. As Rachel Pain (2019) explains, chronic trauma is often associated with forms of gender-based violence, child abuse, and intimate partner abuse that share certain characteristics including repeated exposure to violence in all its myriad forms, and a limited prospect of escape.

Numerous studies document that many women (and men) remain for prolonged periods in abusive relationships (Wuest and Merritt-Gray 1999;

Anderson and Saunders 2003; Bell et al. 2007; Baly 2010; Ben-Porat and Sror-Bondarevsky 2018). Frequently, abused women remain because they are economically dependent on their abusive partner; others remain because they have limited (if any) access to adequate legal assistance (Barnett 2000; Anderson and Saunders 2003; Kim and Gray 2008). When subjected to chronic trauma, displacement becomes an act of resistance, an effort to reassert power within an asymmetric relationship. In other words, with a focus on the temporalities of slow violence, one is better positioned to raise questions of structure and agency, resistance and force. As Andrew Baly (2010, p. 2298) explains, women in abusive relationships often take active steps to ameliorate their oppressive conditions; these coping techniques, however, potentially 'maintain abused women in situations that carry the risk of further psychological and physical harm.' To this end, the physical act of displacement may be read as an act of emancipation, albeit one that may well be fraught with other, unanticipated, challenges. Judith Wuest and Marily Merritt-Gray (1999, p. 118) explain that the 'struggle of breaking free is exhausting and leaves survivors depleted.' For example, 'the requirements for moving forward into the stage of not going back are feeling some measure of safety, wanting to take more control, having a semi-permanent place to live, and having help.' These studies remind us that while displacement may be understood as exhibiting agency in the face of oppression, the mere fact of leaving is not always a panacea (Tamas 2011; Cahill and Pain 2019).

## Conclusions

It is not uncommon to associate a sense of urgency with displacement. When confronted with armed conflict or natural disaster, for example, people are compelled to flee, to seek safety elsewhere. Such urgency directs attention to the immediacy of displacement and, as corollary, to a particular form of spectacular violence, that is, to a violence that is instantaneous in its potentially harmful effects. There are, however, other forms of violence that are slower in effect; these are forms of violence that unfold gradually, indiscernibly, and contribute to forms of slow displacement marked by waiting and endurance. To this end, it is more appropriate to understand 'slow and fast violence as mutually constitutive categories' (Christian and Dowler 2019, p. 1070). In addition, it is necessary to acknowledge that the temporalities of slow and fast violence intersect and overlap across myriad scales, from the body to the global. As Caitlin Cahill and Rachel Pain (2019, p. 1057) suggest, the myriad and overlapping temporalities of displacement and violence reveal sharp, immediate ruptures but also 'continual, incremental discriminatory dispossessions at the intersection of gender, class, race and place.'

# References

Anderson, D., & Saunders, D. G. (2003). Leaving an abusive partner: An empirical review of predictors, the process of leaving, and psychological well-being. *Trauma, Violence & Abuse, 4*(2), 163–191.

Anderson, B., Grove, K., Richards, L., & Kearnes, M. (2019). Slow emergencies: Temporality and the racialized biopolitics of emergency governance. *Progress in Human Geography*. https://doi.org/10.1177/0309132519849263.

Auyero, J. (2011). Patients of the state: An ethnographic account of poor people's waiting. *Latin American Research Review, 46*(1), 5–29.

Auyero, J. (2012). *Patients of the state: The politics of waiting in Argentina*. Durham, NC: Duke University Press.

Baly, A. R. (2010). Leaving abusive relationships: Constructions of self and situation by abused women. *Journal of Interpersonal Violence, 25*(12), 2297–2315.

Barnett, O. W. (2000). Why battered women do not leave, part 1: External inhibiting factors within Society. *Trauma, Violence & Abuse, 1*(4), 343–372.

Bell, M. E., Goodman, L. A., & Dutton, M. A. (2007). The dynamics of staying and leaving: Implications for battered women's emotional well-being and experiences of violence at the end of a year. *Journal of Family Violence, 22*, 413–328.

Ben-Porat, A., & Sror-Bondarevsky, N. (2018). Length of women's stays in domestic violence shelters: Examining the contribution of background variables, level of violence, reasons for entering shelters, and expectations. *Journal of Interpersonal Violence*. https://doi.org/10.1177/0886260518811425.

Cahill, C., & Pain, R. (2019). Representing slow violence and resistance: On hiding and seeing. *ACME: An international Journal for Critical Geographies, 18*(5), 1054–1065.

Christian, J., & Dowler, L. (2019). Slow and fast violence: A feminist critique of binaries. *ACME: An International Journal for Critical Geographies, 18*(5), 1066–1075.

Christian, J., Dowler, L., & Cuomo, D. (2016). Fear, feminist geopolitics and the hot and banal. *Political Geography, 54*, 64–72.

Chu, E., Anguelovski, I., & Carmin, J. (2016). Inclusive approaches to urban climate adaptation planning and implementation in the Global South. *Climate Policy, 16*(3), 372–392.

Coddington, K., & Micieli-Voutsinas, J. (2017). On trauma, geography, and mobility: Towards geographies of trauma. *Emotion, Space and Society, 24*, 52–56.

Davies, T. (2018). Toxic space and time: Slow violence, necropolitics, and petrochemical pollution. *Annals of the American Association of Geographers, 108*(6), 1537–1553.

Harms, E. (2013). Eviction time in the new Saigon: Temporalities of displacement in the rubble of development. *Cultural Anthropology, 28*(2), 344–368.

Hattery, A. (2009). *Intimate partner violence*. Lanham, MD: Rowman & Littlefield.

Herman, J. (1992). Complex PTSD: A syndrome in survivors of prolonged and repeated trauma. *Journal of Traumatic Stress, 5*(3), 377–391.

Igreja, V. (2012). Multiple temporalities in Indigenous justice and healing practices in Mozambique. *The International Journal of Transitional Justice, 6*, 404–422.

Kim, J., & Gray, K. A. (2008). Leave or stay? Battered women's decision after intimate partner violence. *Journal of Interpersonal Violence, 23*(10), 1465–1482.

Laurie, E., & Shaw, I. G. R. (2018). Violent conditions: The injustices of being. *Political Geography, 65*, 8–16.

Lunstrum, E., Bose, P., & Zalik, A. (2016). Environmental displacement: The common ground of climate change, extraction and conservation. *Area, 48*(2), 130–133.

Mueller-Hirth, N. (2017). Temporalities of victimhood: Time in the study of post-conflict societies. *Sociological Forum, 32*(1), 186–206.

Mueller-Hirth, N. (2018). Still waiting: Victim policies, social change and fixed liminality. In N. Mueller-Hirth & S. R. Oyola (Eds.), *Time and temporality in transitional and post-conflict societies* (pp. 102–121). New York: Routledge.

Nixon, R. (2011). *Slow violence and the environmentalism of the poor*. Cambridge, MA: Harvard University Press.

O'Lear, S. (2016). Climate science and slow violence: A view from political geography and STS on mobilizing technoscientific ontologies of climate change. *Political Geography, 52*, 4–13.

Olson, E. (2015). Geography and ethics I: Waiting and urgency. *Progress in Human Geography, 39*(4), 517–526.

Pain, R. (2019). Chronic urban trauma: The slow violence of housing dispossession. *Urban Studies, 56*(2), 385–400.

Sandlos, J., & Keeling, A. (2016). Toxic legacies, slow violence, and environmental injustice at Giant Mine, Northwest Territories. *The Northern Review, 42*, 7–21.

Schwartz, B. (1974). Waiting, exchange, and power: The distribution of time in social systems. *American Journal of Sociology, 79*, 841–870.

Springer, S. (2011). Violence sits in places? Cultural practice, neoliberal rationalism, and virulent imaginative geographies. *Political Geography, 30*, 90–98.

Tamas, S. (2011). *Life after leaving: The remains of spousal abuse*. Walnut Creek, CA: Left Coast Press.

Tänzler, D., Maas, A., & Carius, A. (2010). Climate change adaptation and peace. *Wiley Interdisciplinary Reviews: Climate Change, 1*(5), 741–750.

Tyner, J. A. (2012). *Space, place, and violence: Violence and the embodied geographies of race, sex, and gender*. New York: Routledge.

Tyner, J. A. (2016). *Violence in capitalism: Devaluing life in an age of responsibility*. Lincoln, NE: University of Nebraska Press.

Tyner, J. A. (2019). *Dead labor: Toward a political economy of premature death*. Minneapolis: University of Minnesota Press.

Vaz-Jones, L. (2018). Struggles over land, livelihood, and future possibilities: Reframing displacement through feminist political ecology. *Signs: Journal of Women in Culture and Society, 43*(3), 711–735.

Warner, K. (2012). Human migration and displacement in the context of adaptation to climate change: The Cancun Adaptation Framework and potential for future action. *Environment and Planning C: Government and Policy, 30*, 1061–1077.

Willett, J. L. (2015). The slow violence of climate change in poor rural Kenyan communities: 'Water is Life. Water is Everything'. *Contemporary Rural Social Work, 7*(1), 39–55.

Wuest, J., & Merritt-Gray, M. (1999). Not going back: Sustaining the separation in the process of leaving abusive relationships. *Violence Against Women, 5*(2), 110–133.

# 6

# Assembling Climate Change-Related Displacement

Leonie Tuitjer

## Introduction

Displacement is commonly used within refugee and migration studies to signal a degree of force within the respective relocation processes. Especially within scenarios of conflict and violence, internal displacement is frequently observed and institutionally mediated through the UN Guiding Principles on Internal Displacement. Recently the role of climate change and environmental disasters in such displacements has sparked political debates, media attention, and academic research (Foresight 2011; Ionesco et al. 2017). The Internal Displacement Monitoring Centre (IDMC) 'estimate[s] that more than 19.3 million people were forced to flee their homes by disasters in 100 countries in 2014' (IDMC 2015). In increasing numbers of case studies, so-called climate refugees, climate migrants, or climate change-induced displaced people signal a break between the world as we know it and an uncertain, potentially apocalyptic climate future (Baldwin et al. 2014). Yet, as Morrissey (2012, p. 45) has argued: 'The interaction between environmental and non-environmental factors should be examined in a way that is contextually specific and historically relevant.'

I want to offer an assemblage theory perspective to study such emergent contexts and their affective and material dimensions. It is through the concept of assemblage that I propose a focus on displacement synonymous with both

L. Tuitjer (✉)
Leibniz University Hanover, Hanover, Germany
e-mail: tuitjer@kusogeo.uni-hannover.de

© The Author(s) 2020
P. Adey et al. (eds.), *The Handbook of Displacement*,
https://doi.org/10.1007/978-3-030-47178-1_6

'to move or to shift' and the processes of 'rearrangement' which may reveal a different political dimension within climate displacement. The chapter makes two distinct arguments for an assemblage perspective. First, it argues that such an approach allows us to think of climate displacement as emergent and embedded in local contexts and routine procedures that warrant rearrangements of human and non-humans alike in changing environments. Second, it demonstrates climate change-related displacement involves processes of de- and re-territorialisation that may hold the potential for so-called lines of flight or alternative futures. The chapter draws on ethnographic literature from the Arctic (Cruikshank 2005; Sakakibara 2008; Rattenbury et al. 2009; Bravo 2009; Ignatowski and Rosales 2013) to illustrate these two claims. As this region is frequently portrayed as on the forefront of climatic change within media reports and political debates, it is here where one sees a climate change and displacement assemblage emerging. Importantly, the ongoing rearrangements of human settlements, hunting routes, and animal trails due to changing climatic conditions in the Arctic are deeply embedded within sociopolitical structures and historical power relations, demonstrating the necessity to pay attention to context and historical relevance and further warranting an assemblage approach.

## Assemblages

Assemblage theory has been used to better understand the complexities and ambiguities of urban life (Simone 2010, 2011), social marginality (Lancione 2016), the functioning and failure of infrastructures (Bennett 2005; Ranganathan 2015), human–environment interactions (Braun 2005, 2014; Li 2007, 2014), mobility governance (Salter 2013), and policy processes (Prince 2011) within various social science disciplines. Within these efforts we can find disparate uses of the concept. The following paragraphs try to briefly summarise what an assemblage perspective entails, before relating it to studies of climate change-related displacement.

The concept as it is used here is derived from Deleuze and Guattari's (2013) philosophical work. An assemblage emerges from the alignment of matter and statements. In other words, an assemblage combines both *machinic* (objects, materials, passions) and *enunciative* (statements, sentences, discourses) qualities. Of importance are 'the *arrangement* of these *connections* that gives the concepts their sense' (Phillips 2006, p. 108). McFarlane (2011) analyses human dwellings through an assemblage perspective and teases out the similarities between informal urban life and an assemblage perspective. Dwelling and assemblage both refer to a process or the very labour to carve out a territory and cohere things within such a territory to form a provisional whole. Scrap material, improvised hygienic infrastructures, and a vacant plot of land may form the material components and territorial bases from which slum dwellers may assemble their homes. Angell (2014) highlights how material artefacts, concrete, and clay as well as knowledge about the forces of the moving tectonic plates affect the entire housing assemblage in her analysis of earthquake risks in Istanbul. Within the polar region, Indigenous people carve out their territory despite harsh climatic conditions. Local wisdom sustains such dwelling assemblages.

The materials chosen for human habitation may take on a life of their own, affecting the durability of dwellings and augmenting or diminishing real and perceived risks. Bennett refers to such material capacities as a form of 'thing power' which highlights the agencies of (non-human) others within assemblages. One way of understanding the diverse elements and agencies within an assemblage is to attend to their affective capacities (Coole and Frost 2010). To make an assemble of scrap metal and concrete intelligible as a 'home,' a set of affective and emotional states are enrolled. The smell of the gas cooker or the touch of the cement floor may be involved in forming the dwelling assemblage that turns the improvised machinic alignments into a territory that can be enunciated as 'home.' Small alterations within this material composition can alter the affective dimension of belonging to a particular space, displacing the notion of 'home' and forcing a rearrangement of dwelling space.

Dwelling assemblages are moreover deeply contextual and embedded in wider sociopolitical and economic structures that forestall, for example, other types of dwelling assemblages (e.g. in a more formalised territory). Thus, while materiality is granted a certain degree of agentic capacity and forcefulness, an assemblage is not a level playing field but imbued with competing sets of power and passions (Bignall 2008). Understanding displacement as assemblage thus, first, highlights both the importance of acknowledging the (sometimes only small) socio-material and affective ruptures that lead people 'to move' or 'to shift,' and, second, is attentive to the sociopolitical context in which such shifts are supported or prevented.

While much of assemblage theory research is interested in the ways in which particular elements come to be aligned (as in the dwelling assemblage), an equally strong emphasis is placed on the moments when assemblages fail, when they disintegrate, break apart, and require realignments (e.g. during a blackout (Bennett 2005)). An assemblage is thus a theoretical tool that thinks about both coming together and breaking apart of relations, using the vocabulary of de- and re-territorialisation to make sense of these two processes

(Legg 2009). Crucially, within the moments that an assemblage breaks apart, there is a possibility for radically new relations to emerge (Watt 2016). In particular, Guattari (2011) has thought of such possible new relations as lines of flight in which alternative futures may emerge. A line of flight, however, signals potential for change more than offering a guaranteed positive outcome. Displacement perceived as a process of de- and re-territorialisation thus acknowledges both the act of shift/movement as well as the ongoing process of rearranging oneself within new context. Reinserting oneself in such new context may be painful and difficult, yet it always also entails the potential of a line of flight—a potential for change.

## Thinking About the 'Climate Displacement Assemblage'

How then can we make such a theoretical perspective useful for rethinking climate displacement? I propose turning to ethnographic literature from the Arctic that analyses how Indigenous communities are already affected by climate change and their emergent displacement experiences to demonstrate the usefulness of an assemblage perspective (Cruikshank 2005; Sakakibara 2008; Rattenbury et al. 2009; Bravo 2009; Ignatowski and Rosales 2013).

### Emergent Displacement

Blitz (2011) has argued that climate change can be seen as a form of de- and re-territorialisation of human–environment relations. Climate change signals a moment of physical relations breaking apart and new alignments between them. As Blitz observes, de-territorialisation has gained traction predominantly as a metaphor for unravelling relations, 'yet, there are also situations where re-territorialisation occurs quite literally' (Blitz 2011, p. 433). I propose placing the emerging climate displacement assemblages in such a context of de-territorialisation and re-territorialisation.

Ethnographic literature that deals with Indigenous experiences of climate change in the Arctic reveals how small fishing and reindeer-herding communities are already seeing their livelihoods threatened by the qualitative shifts within the material composition of ice sheets. Such changes are rendering traditional fishing techniques (Ignatowski and Rosales 2013), hunting and travelling routes (Bravo 2009), and territories for settlements and housing constructions (Sakakibara 2008) to be at risk of becoming literally de-territorialised. Within their livelihoods, displacement emerges as an assemblage that involves shifts and rearrangements in various aspects of everyday routines. The altering composition, texture, and volume of ice, as well as their altered temporal occurrences, impact on human and animal livelihoods and trigger the need for mobile rearrangements (see also: Rattenbury et al. 2009). New hunting routes must be tested and old paths through the landscape become increasingly abandoned. Although settlements are not yet abandoned, the displacement of everyday routines, tracks, and interactions already forces rearrangements to happen and threaten particular traditions with being displaced forever. Within these contexts of rearrangement, non-cognitive, affective clues such as sound, touch, and smell are drawn upon (Cruikshank 2005, p. 229).

Importantly, the materiality of the icy Arctic environments is seen as a dynamic, relational unfolding of properties (Bennett 2010a, b; Braun and Whatmore 2010). Rooted in animist belief systems, the research on Alaskan Indigenous communities reveals how they see their surroundings as animated with spirits possessing a type of agency that does not discriminate between the human and non-human world. Sakakibara (2008) writes about the different forms of supernatural beings encountered during fieldwork in Point Hope, northern Alaska, and how seemingly effortlessly the spirits could take the shape of animals, people, or 'nature,' disregarding neat division between such categories held in many strands of Western philosophy. Interestingly, these supernatural forces themselves are fluid and shape-shifting in the same way as the Arctic environment is: 'fieldwork confirmed that the characteristics of the spirit beings have never been static. With their environment becoming more unpredictable, the Inupiat find themselves interacting with new spirit beings in their drowning home' (Sakakibara 2008, p. 460). Using such techniques as oral history or storytelling for sense-making, thus, allows the communities to enunciate their displacement experiences and compare past and present changes. Within such stories the emerging displacement assemblage finds new *enunciative* statements for the *machinic* realignments of the shifting environment.

Moreover, thinking about climate change-related displacement through the material capacity of water to flow, freeze, or evaporate constitutes not only a vital part of human life but connects us—from molecular cell to vast ecosystems—in seemingly endless and multi-scalar ways (Strang 2014; Davies 2014). The emergent climate displacement assemblages are thus interlinked to the shifts within global atmospheric assemblages, local environments, and more-than-human rearrangements. Climate displacement in these remote places is thus a multi-species effort of rearranging life in altering conditions.

Acknowledging the force of materials, objects, and non-human animals signals a deeper awareness that the 'agency of others' (Strang 2014, p. 141) includes not only other people but is networked across a wide range of actants. Ice sheets here are granted an ethical position outside of a capitalistic logic that prescribes seeing them as purely abstract surrounding or as a resource for human exploitation only (Strang 2014). This point resonates with Jane Bennett's (2010a) idea that to dismiss the power of others can potentially be harmful for the entire assemblage of relationships that mark our existence. Perceiving climate change-related displacement as an emerging assemblage thus sensitises us to the role of non-human others within such displacement and offers a post-humanist perspective on displacement in which human and non-humans (materials, vegetation, animals) are displaced alongside each other. Climate change-related displacement is thus a process that involves the rearrangement of collective livelihoods and entire ecological systems rather than a displacement of individual people.

## Lines of Flight

Returning to Blitz's (2011) argument about climate change as a literal form of de-territorialisation, in which I propose to place the emergent displacement assemblage, Blitz highlights the possibility of alternative futures emerging. Within phases of de-territorialisation, potential lines of flight can emerge. They signal the possibility of radical alternative futures. Blitz highlights the (political) necessity of also thinking about how people displaced by climatic change find themselves in a position in which they may become re-territorialised in other territories. Thinking about extreme cases of loss in which, for example, island populations may be forced to resettle, the climate-change-displaced populations may become agents of change. Together they may be able to forge collective futures. Within the displacement assemblage thus not only lies the labour of rearrangement and the suffering of loss but—potentially—a line of flight from which to create alternative futures. Similarly, Clark (2017) highlights the idea that climate change might contribute to new ways of relating to each other within society as much as to earth and climate itself. This is not to downplay the suffering climate change might mean for large parts of humanity—such as flora, fauna, and whole ecosystems—but to show that despite these forces there is the potential for new relations to emerge. Sakakibara's (2008) observation of the spirit world at Point Hope (see section above) seems to confirm just this: territories and spirits are never fully lost, but rather new relations emerge between sea, ice, humans, and animals, forming a line of flight and potential new dwelling assemblage.

## Conclusion

This short contribution has argued that an assemblage perspective is useful to rethink the relation between climate change and displacement for two specific reasons. First, an assemblage perspective is sensitive towards the emergent displacement experiences that involve temporal as well as routine rearrangements of bodies in altering landscapes. Here a focus on everyday practices and emergent change is advanced, by granting agentic capacity to the material alterations within, for example, Alaskan environments. Second, by perceiving environmental change as a physical form of de- and re-territorialisation, a potential line of flight may be forged. Rather than subscribing to apocalyptic visions of the Anthropocene, such a perspective encourages us to think about potential alternatives and moreover signals a need to think about ways of making societies open for welcoming de-territorialised newcomers. Here the climate-displaced person can be perceived of as an agent of change that may potentially contribute to forging alternative—perhaps more hopeful—futures.

## References

Angell, E. (2014). Assembling disaster: Earthquakes and urban politics in Istanbul. *City, 18*(6), 667–678.

Baldwin, A., Methmann, C., & Rothe, D. (2014). Securitizing 'climate refugees': The futurology of climate-induced migration. *Critical Studies on Security, 2*(2), 121–130.

Bennett, J. (2005). The agency of assemblages and the North American blackout. *Public Culture, 17*(3), 445–465.

Bennett, J. (2010a). *Vibrant matter: A political ecology of things*. Durham and London: Duke University Press.

Bennett, J. (2010b). A vitalist stopover on the way to a New Materialism. In D. Coole & S. Frost (Eds.), *New materialisms: Ontology, agency, and politics* (pp. 47–69). London and Durham: Duke University Press.

Bignall, S. (2008). Deleuze and Foucault on desire and power. *Angelaki Journal of the Theoretical Humanities, 13*(1), 127–147.

Blitz, B. (2011). Statelessness and environmental-induced displacement: Future scenarios of deterritorialisation, rescue and recovery examined. *Mobilities, 6*(3), 433–450.

Braun, B. (2005). Environmental issues: Writing a more-than-human urban geography. *Progress in Human Geography, 29*(5), 635–650.

Braun, B. (2014). A new urban dispositif? Governing life in an age of climate change. *Environment and Planning D: Society and Space, 32*(1), 49–64.

Braun, B., & Whatmore, S. J. (Eds.). (2010). *Political matter: Technoscience, democracy and public life*. Minneapolis and London: University of Minnesota Press.

Bravo, M. T. (2009). Voices from the sea ice: The reception of climate impact narratives. *Journal of Historical Geography, 35*, 256–278.

Clark, N. (2017). Strangers on a strange planet: On hospitality and Holocene climate change. In A. Baldwin & G. Bettini (Eds.), *Life adrift: Climate change, migration, critique* (pp. 131–150). London and New York: Rowman & Littlefield International.

Coole, D., & Frost, S. (2010). Introducing the new materialism. In D. Coole & S. Frost (Eds.), *New materialisms: Ontology, agency, and politics* (pp. 1–43). London and Durham: Duke University Press.

Cruikshank, J. (2005). *Do glaciers listen? Local knowledge, colonial encounters & social imagination*. Vancouver: UBC Press.

Davies, M. I. J. (2014). Don't water down your theory. Why we should all embrace materiality but not material determinism. *Archaeological Dialogue, 21*(2), 153–157.

Deleuze, G., & Guattari, F. (2013 [1988]). *A thousand plateaus*. New York: Bloomsbury.

Foresight. (2011). *Migration and global environmental change: Final project report*. London: Government Office for Science. Retrieved from https://sustainabledevelopment.un.org/content/documents/867migrationscience.pdf.

Guattari, F. (2011). *Lines of flight – For another world of possibilities*. London and New York: Bloomsbury.

IDMC. (2015). *Global estimates 2015: People displaced by disasters*. Retrieved from http://www.internal-displacement.org/library/publications/2015/global-estimates-2015-people-displaced-by-disasters/.

Ignatowski, J., & Rosales, J. (2013). Identifying the exposure of two subsistence villages in Alaska to climate change using traditional ecological knowledge. *Climatic Change, 121*(2), 285–299.

Ionesco, D., Mokhnacheva, D., & Gemenne, F. (2017). *The atlas of environmental migration*. Produced for IOM. London and New York: Routledge.

Lancione, M. (Ed.). (2016). *Rethinking life at the margins – The assemblage of contexts, subjects, and politics*. London and New York: Routledge.

Legg, S. (2009). Of scales, networks and assemblages: The League of Nations apparatus and the scalar sovereignty of the Government of India. *Transactions of the Institute of British Geographers, 34*(2), 234–253.

Li, T. M. (2007). Practices of assemblage and community forest management. *Economy and Society, 36*(2), 263–293.

Li, T. M. (2014). What is land? Assembling a resource for global investment. *Transactions of the Institute of British Geographers, 39*, 589–602.

McFarlane, C. (2011). The city as assemblage: Dwelling and urban space. *Environment and Planning D: Society and Space, 29*, 649–672.

Morrissey, J. (2012). Rethinking the 'debate on environmental refugees': From 'maximilists and minimalists' to 'proponents and critics'. *Journal of Political Ecology, 19*, 36–49.

Phillips, J. (2006). Agencement/assemblage. *Theory, Society and Culture, 23*(2–3), 108–109.

Prince, R. (2011). Policy transfer, consultants and the geography of governance. *Progress in Human Geography, 36*(2), 188–203.

Ranganathan, M. (2015). Storm drains as assemblages: The political ecology of flood risk in post-colonial Bangalore. *Antipode, 47*(5), 1300–1320.

Rattenbury, K., Kielland, K., Finstad, G., & Schneider, W. (2009). A reindeer herder's perspective on caribou, weather and socio-economic change on the Seward Peninsula, Alaska. *Polar Research, 28*(1), 71–88.

Sakakibara, C. (2008). 'Our home is drowning': Inupiat storytelling and climate change in Point Hope, Alaska. *Geographical Review, 98*(4), 456–475.

Salter, M. B. (2013). To make move and let stop: Mobility and the assemblage of circulation. *Mobilities, 8*(1), 7–19.

Simone, A. (2010). *City life from Jakarta to Dakar – Movements at the crossroads.* New York and London: Routledge.

Simone, A. (2011). The surfacing of urban life. *City, 15*(3–4), 355–364.

Strang, V. (2014). Fluid consistencies. Material relationality in human engagements with water. *Archeological Dialogues, 21*(2), 133–150.

Watt, P. (2016). A nomadic war machine in the metropolis. *City, 20*(2), 297–320.

# 7

## Affect and Displacement

### Mark Griffiths

This chapter synthesises perspectives on the body and forcible displacement with a specific focus on affect. The commentary is thus situated in the now extensive body of geographical literature that has drawn focus on the pre- and interpersonal affective dimensions of life that constitute subjects, moods, and spaces (prominently: Adey 2009; Anderson 2006; Pile 2010). Such work has heightened geographers' sensitivity to the sometimes imperceptible but nonetheless significant ways that 'the trans-human, the non-cognitive, the inexpressible … underlies and constitutes social life' (Pile 2010, p. 11), and subsequently how affects are therefore manipulated (by powerful elites) and/or irreducibly become 'autonomous' (Massumi 1995). Thus, from such a perspective, subjects and spaces are constituted by the sometimes subtle, sometimes powerful, capacities of the body. This is very much the case in the experience of forcible displacement where—at the most basic conception—strong affects (such as both fear and hope) undoubtedly play parts in the displacement of people from one to another place (Boccagni and Baldassar 2015; Conradson and McKay 2007). We have seen this in the most stark terms in the ongoing refugee crisis, where the threat of harm and the hope of asylum turn people to seek refuge and thus expose the body to the dangers of making passage to safety, negotiating the hopes, fears, and anxieties that push and pull them to act and move (Biehl 2015; Loyd et al. 2018). This is of course no new thing: displacement is always tied to spatial and temporal repertoires of emotions and affects.

---

M. Griffiths (✉)
Newcastle University, Newcastle upon Tyne, UK
e-mail: mark.griffiths@ncl.ac.uk

For the context at hand, affects, along with broader embodiments such as feelings and emotion, are constitutive of 'the refugee condition' (see Ehrkamp 2017), and experiences of displacement—leaving, arriving, and adjusting—are marked by a range of affects from fear, threat, and humiliation to hope, relief, and even joy. This range of embodiments permeates accounts of forcible displacement, most recently in the Syrian and Middle East refugee crises (e.g. Oppedal et al. 2018) but also across broader histories of expulsion. To draw on two prominent examples: Hannah Arendt wrote poignantly of humiliation, degradation, and hopelessness on her arrival in North America (1994, pp. 114–115); and Olaudah Equiano's powerful writing is shot through with feelings of being 'quite oppressed and weighed down by grief after my mother and friends; and my love of liberty' (2005, p. 65). In contemporary contexts, it is well documented that in 'host states' such as Turkey (the world's largest host of refugees), labyrinthine bureaucratic systems and further displacement to peripheral towns and cities induce great anxieties among displaced people such that 'governing through uncertainty' has become a key way of conceptualising the management of the current refugee crisis. Uncertainty in this sense is associated with 'indefinite waiting, imperfect knowledge, and the volatility of legal status' and has become a 'defining element' of seeking asylum in host states such as Turkey (Biehl 2015, p. 58). Research in this area has documented the ways that waiting in uncertainty can 'exacerbate trauma or to cause new trauma as refugees worry about loved ones, their futures or simply—and importantly—about how to have food for the next day' and thus act as 'ongoing stressors [that] cause mental and other health problems' for displaced people (Loyd et al. 2018, p. 382). It is clear, even from a brief review of relevant literatures, that from beginning to (uncertain) end, embodiments are bound up with each stage of forcible displacement.

In this short chapter I do not intend to fit ranging experiences into a neat model of affect and displacement, but I do bring a temporal frame to thinking through affect and forcible displacement. Specifically, the chapter's three sections consider affects that emerge from the different temporalities of displacement, beginning with the often traumatic past, the fearful and humiliating present, and, finally, the threat and hopefulness of displaced people's future. The discussion is grounded with reference to ongoing research with displaced people in the Occupied Palestinian West Bank.

## Becoming Displaced: Traumatic Pasts

Returning momentarily to Arendt, she wrote of the disjoint between living a 'normal' life in the sanctuary of 1940s America surrounded 'by so many stars and famous men' and the anxieties around the feeling that 'if we are saved we feel humiliated, and if we are helped we feel degraded' (1994, p. 114). Arendt roots this anxiety in the fact that she and so many other Jewish refugees had 'become witnesses and victims of worse terrors than death' (1994, p. 112)—'experiences that make their flesh creep' (1994, p. 111). The body—whose 'flesh creeps'—is thus a primary site of recollection, evidencing the powerfully affective dimension of the terrible conditions from which refuge was sought. For Palestinians, the most intense period of forcible displacement came in 1948 at the birth of the Israeli state when 750,000 were driven from villages and towns in a violent military and militia operation (Pappé 2006). Accounts of this displacement, known to Palestinians as the *Nakba* ('the catastrophe'), even seventy years on, retain an affective intensity that is evident in the wealth of testimony collated in both academic and non-academic work. For instance, Sfaa Ghnadre-Naser and Eli Somer's study documents quite extraordinary (yet commonplace) accounts of the Nakba's continuing affective resonance: one Palestinian man, Afefe, recalled 1948 as 'the greatest Nakba of my life [cries and continues talking in a suffocated voice] … and the further it gets, the harder and more painful it becomes in our hearts'; another, Hoda, disclosed that 'we felt humiliation, we felt shame […] And still feel, every time we remember it, we feel the disgust and feel the weakness' (2016, pp. 245–246). What is notable about these accounts is that the body is the locus of relating past experience, and that history has an embodied resonance in the present.

Taking these two short examples, we might make the quite self-evident note that the past moment of displacement retains something of its intensity in the present. In this way, as Giorgio Hadi Curti has written, memory of displacement is not 'simply a matter of episodic compartmentalisation' but an emergent embodiment through 'chains of associations of the (pre- and post-conscious) body' (2008, p. 107). This much is evident where recollection of displacement is embodied in the examples above—the body 'creeps' and cries—and the past is not 'sealed off' but instead resonates through the (displaced) body.

## Being Displaced: Fearful and Humiliating Presents

While the past continues to reverberate, displacement is marked too by the emergence of new affective experiences in the place to which one is displaced. The case of Palestine is particular; there are by now third- and fourth-generation refugees whose protracted exile is maintained by an occupation that explicitly targets the body—specifically its nervous and anxious capacities—to the end of effecting 'political and human asphyxiation' (see Falah 2004, p. 599). We can set this in context in Hebron, a city where two starkly contrasted notions of displacement and return coalesce: for Palestinians, the right of return is recognised by international law but is perpetually suspended and violated through ever-increasing restrictions on movement, while the city's Zionists settlers have exercised a right to return to a land on which they had never before stepped foot and, in doing so, broken Article 49 of the Fourth Geneva Convention.[1] As a consequence, the city centre of Hebron has become a complex of checkpoints, cages, and turnstiles where a notoriously violent Israeli military and settler presence maintains an 'affective sensorium' characterised by fear, threat, and humiliation (see Griffiths 2017). As one Hebron resident made clear in a recent research interview: 'we move with fear, every day is the same and one day we will not be able to be any place in the city centre, it's full of soldiers and settlers and they go everywhere and do what they want.'

While the case of Palestine is particular, there are commonalities with displaced people the world over. For many, the emergence of new affective experiences after displacement occurs primarily in camps, detention centres, and the peripheries of large cities where 'hostile environments' are fomented by state actors and established populations via the means of bureaucratic opaqueness (Joronen 2017), spaces of exception (Pratt 2005), and a rise of populist movements organised around anti-immigration (Coddington 2018). In her study of refugees in Colombia, for example, Pilar Raiño-Alcalá elaborates on an 'economy of affect': 'where powerful, anxious, and sometimes contradictory feelings of uncertainty, anxiety and hope mediate both the process of reconstructing their lives, and their interactions with the forced migration regime and the host society' (2008, p. 16).

From these additional brief examples, we might further note that displaced people's enforced entanglement with 'slow-motion' (Joronen 2017) and purposely opaque and inefficient (Berda 2017) bureaucracies and restrictions hinders and prevents both settling in a new place and returning to the old. In this way, being displaced introduces a repertoire of affects that is both contingent with and in important ways distinct from those that effect displacement in the first place: living with displacement is marked and maintained by an affective experience that subordinates, oppresses, and marginalises displaced people.

## Threatening/Hopeful Futures

The future, the final temporal focus, presents a broad field of affect in the context of displacement. There are many futures that effect/affect the movement of displaced people: the threat of harm often triggers the initial movement away from one's home; the hope of refuge can spur similar movement; and anxieties around bureaucratic applications, rights, and asylum each relate to possible and potential futures whose uncertainty can be profoundly unsettling. Without scope for full discussion of these many ways to consider the future, I close by drawing focus on the key dynamic of uncertainty around how alternatively threatening and hopeful futures connect to debilitating and abilitating affective lives.

Returning to Palestine, my recent collaborative research on the threat of displacement by demolition has explored the ways that uncertain futures can have profound consequences for people's affective lives. In the Bedouin village of Um Al-Khair in the South Hebron Hills, a community that was forcibly displaced from *al Naqab* (the Negev) during the 1967 war, demolition orders have been served on all but two of its residential and communal structures; the population of the village is around ninety. No notice is given of when the demolition will take place, and many structures are allowed to stand for years with an order; others are destroyed quite promptly.[2] This situation has brought the uncertain future of demolition and displacement into the emotional and affective rhythms of life in the village where, as one of the village's spokespeople, Fayez, told us, 'they [the IDF] do whatever they want when they want, we live always in fear for our home ... every day my mother gets up early to check everything, that they are not coming that day, *every morning*.' It is hard to imagine living each day this way, and it is no real wonder that Fayez went on to describe his community's growing anxieties over the prospect that the house that provides shelter will be razed, of how people wake early to scan the surrounding hills for bulldozers, and of the ways that, Fayez described in detail, the continual threat demolition has induced 'stress and anxiety [such that] ... for a long time people have not slept well in the night.' Another spokesperson, Faris, explicated: 'sometimes you hope, but most of

**Fig. 7.1** A Palestinian flag flies at Um Al-Khair, a community in the Occupied Palestinian West Bank threated with displacement by house demolition, 2018 (Source: Mark Griffiths)

the time you wish you were not born, [you] feel powerless, unable to do anything' (see Joronen and Griffiths 2019a) (Fig. 7.1).

Demolition in this case functions not simply through physical loss of property but a persistent threat of a future shaped by loss and displacement. On threat and affect, Brian Massumi has written that it is the very indeterminacy of a prospective future that constitutes threat, where 'uncertainties' disrupt linear time as 'an eventuality that may or may not occur, indifferent to its actual occurrence. The event's consequences precede it' (2010, p. 8). The evidenced stress and anxiety is a case in point: the very threat of losing one's home effects 'demolition before demolition,' transforming it from a place of security and protection to one of disrupted sleep and nervous anticipation (see Shalhoub-Kevorkian 2015, pp. 101–103).

There is little doubt of the deleterious effects of such renderings of the future; the threat of (further) forcible displacement has taken its toll on the emotional and affective well-being of the community of Um Al-Khair. Turning this around, somewhat, as a final focus, it is important to recall, first, the theoretical corrective that affects and bodies are not stable objects of intervention but in fact retain an 'autonomy' or potential to evade efforts to constrain (Massumi 1995) and, second, the ethical imperative to resist positioning Palestinians—and by extension other displaced people—as solely 'victims' of oppression without agency or voice of their own (see Kotef and Amir 2011). To this end, thinking through the affective dimensions of displacement can also draw focus on the vitalising orientations of, for instance, hope, even in the context of severe political oppression. It is significant, therefore, that Fayez in Um Al-Khair continually insisted that the community 'go on ... we have fun, we laugh, *we go on* ... We ask people to help us, to help in resistance against the Israeli state ... I'm a believer ... *we go on*.' And it is important to recognise the agency and voice of people such as Baha, a Palestinian refugee activist based in Beit Sahour whose resoluteness is testament to the persistence of hope: 'the last 5,000 years have witnessed the rising of different powers and empires in Palestine—none of them remains. Most of them left a stone and a bad story behind: what did the crusaders leave? They left a lot of monuments, massacres of people ... a bad story of murder and crime ... all these tyrants and all these oppressors collapsed ... people's hope is motivated by understanding the history of displacement, of human history and realising that *no oppressor lasts forever* here, so we will survive this oppressor' (see Joronen and Griffiths 2019b). For Baha, and also for the communities living with the threat of demolition, there remains a persistence of hope, and that pushes along resistance and activism as a 'dynamic imperative *to action*' (Anderson 2006, p. 744). The corollary is that even in the depths of the severe political oppression that threatens and effects forcible displacement, hope can emerge as an animating affect that enables people to 'go on.'

The examples here are brief and the different contexts of Palestine are specific, but they are also illustrative of the ways that affects emerge from the different temporalities of displacement. The affective dimensions of forcible displacement are broad and profound, more than could be covered in an introductory chapter such as this. No neat ordering of affect and displacement is offered, but the discussion aims to take tentative steps towards a temporal frame that has the potential to attend to the different time-spaces of displacement, from those associated with the past, around nostalgia and trauma, the present to do with living in a new and sometimes hostile environment, and the future connected to extreme uncertainty, anxiety, hope and threat. I hope to have convinced, at the very least, of the importance of focusing on the powerful embodiments that constitute subjects, moods, and spaces that are integral to the different temporalities of forcible displacement.

## Notes

1. Article 49 of the Fourth Geneva Convention states: 'The occupying power shall not deport or transfer parts of its own population into the territories it occupies'
2. This situation is consistent with Israeli tactics across the West Bank where, since 1988, a total of 16,085 demolitions orders have been issued for Palestinian-owned structures. Of these, more than 75 per cent have not been carried out, with only 12 per cent of orders executed between 2014–2016 (OCHA 2017).

## References

Adey, P. (2009). Facing airport security: Affect, biopolitics, and the preemptive securitisation of the mobile body. *Environment and Planning D: Society and Space, 27*(2), 274–295.

Anderson, B. (2006). Becoming and being hopeful: Towards a theory of affect. *Environment and Planning D: Society and Space, 24*(5), 733–752.

Arendt, H. (1994 [1943]). We refugees. In M. Robinson (Ed.), *Altogether elsewhere: Writers on exile* (pp. 10–119). London: Faber and Faber.

Berda, Y. (2017). *Living emergency: Israel's permit regime in the occupied west bank*. Stanford, CA: Stanford University Press.

Biehl, K. (2015). Governing through uncertainty: Experiences of being a refugee in Turkey as a country for temporary asylum. *Social Analysis, 59*(1), 57–75.

Boccagni, P., & Baldassar, L. (2015). Emotions on the move: Mapping the emergent field of emotion and migration. *Emotion, Space and Society, 16*, 73–80.

Coddington, K. (2018). Landscapes of refugee protection. *Transactions of the Institute of British Geographers, 43*(3), 326–340. https://doi.org/10.1111/tran.12224.

Conradson, D., & McKay, D. (2007). Translocal subjectivities: Mobility, connection, emotion. *Mobilities, 2*(2), 167–174.

Curti, G. H. (2008). From a wall of bodies to a body of walls: Politics of affect| Politics of memory| Politics of war. *Emotion, Space and Society, 1*(2), 106–118.

Ehrkamp, P. (2017). Geographies of migration I: Refugees. *Progress in Human Geography, 41*(6), 813–822.

Equiano, O. (2005 [1789]). *The interesting narrative of the life of Olaudah Equiano*. Retrieved from https://www.gutenberg.org/files/15399/15399-h/15399-h.htm

Falah, G. W. (2004). Truth at war and naming the intolerable in Palestine. *Antipode, 36*(4), 596–600.

Ghnadre-Naser, S., & Somer, E. (2016). 'The wound is still open': The Nakba experience among internally displaced Palestinians in Israel. *International Journal of Migration, Health and Social Care, 12*(4), 238–251.

Griffiths, M. (2017). Hope in Hebron: The political affects of activism in a strangle city. *Antipode, 49*(3), 617–635.

Joronen, M. (2017). Spaces of waiting: Politics of precarious recognition in the occupied West Bank. *Environment and Planning D: Society and Space, 35*(6), 994–1011.

Joronen, M., & Griffiths, M. (2019a). The affective politics of precarity: Home demolitions in the occupied West Bank. *Environment and Planning D: Society and Space, 37*(3), 561–576. https://doi.org/10.1177/0263775818824341.

Joronen, M., & Griffiths, M. (2019b). The moment to come: Geographies of hope in the hyperprecarious sites of occupied Palestine. *Geografiska Annaler: Series B, Human Geography, 101*(2), 69–83. https://doi.org/10.1080/04353684.2019.1569473.

Kotef, H., & Amir, M. (2011). Between imaginary lines: Violence and its justifications at the military checkpoints in occupied Palestine. *Theory, Culture & Society, 28*(1), 55–80.

Loyd, J. M., Ehrkamp, P., & Secor, A. (2018). A geopolitics of trauma: Refugee administration and protracted uncertainty in Turkey. *Transactions of the Institute of British Geographers, 43*(3), 377–389.

Massumi, B. (1995). The autonomy of affect. *Cultural critique, 31*, 83–109.

Massumi, B. (2010). The future birth of the affective fact. In M. Greggs & G. J. Seigworth (Eds.), *The affect theory reader*. https://read.dukeupress.edu/books/book/1469/chapter-abstract/170330/The-Future-Birth-of-the-Affective-FactThe?redirectedFrom=fulltext. Retrieved April 20, 2018.

OCHA. (2017). *Israeli demolition orders against Palestinian structures in Area C, 1988–2016*. Retrieved April 20, 2019, from http://data.ochaopt.org/demolitions.aspx

Oppedal, B., Özer, S., & Şirin, S. R. (2018). Traumatic events, social support and depression: Syrian refugee children in Turkish camps. *Vulnerable children and Youth Studies, 13*(1), 46–59.

Pappé, I. (2006). *The ethnic cleansing of Palestine*. Oxford: One World.

Pile, S. (2010). Emotions and affect in recent human geography. *Transactions of the Institute of British Geographers, 35*(1), 5–20.

Pratt, G. (2005). Abandoned women and spaces of the exception. *Antipode, 37*(5), 1052–1078.

Riaño-Alcalá, P. (2008). Journeys and landscapes of forced migration: Memorializing fear among refugees and internally displaced Colombians. *Social Anthropology, 16*(1), 1–18.

Shalhoub-Kevorkian, N. (2015). *Security theology, surveillance and the politics of fear*. Cambridge: Cambridge University Press.

# 8

# Protection of Displaced Persons and the Rights-Based Approach

Rónán McDermott, Pat Gibbons, and Sinéad McGrath

## Introduction

Forced displacement entails a rupture of relationships of protection for those affected, whether those relationships stem from family, community, state, or other societal institutions. The rupture of these relationships threatens the capacity to live with dignity and to identify and implement durable solutions. Actors responsible for protecting the rights of displaced persons seek to provide alternative protection in a bid to save lives, alleviate suffering, and support dignified livelihoods of disaster-affected people. While a variety of protection definitions are used within the aid sector, the most widely agreed definition of protection is the one provided by the UN's Inter-Agency Standing Committee:

> [A]ll activities aimed at obtaining full respect for the rights of the individual in accordance with the letter and the spirit of the relevant bodies of law (i.e. human rights law, humanitarian law and refugee law).[1]

---

R. McDermott (✉) • P. Gibbons
University College Dublin, Dublin, Ireland
e-mail: ronan.mcdermott@ucd.ie; pat.f.gibbons@ucd.ie

S. McGrath
University College Dublin, Dublin, Ireland

Department of International Development, Maynooth University, Maynooth, Ireland
e-mail: sinead.mcgrath@ucdconnect.ie

© The Author(s) 2020
P. Adey et al. (eds.), *The Handbook of Displacement*,
https://doi.org/10.1007/978-3-030-47178-1_8

This chapter firstly provides an overview of the provisions of human rights law, humanitarian law, and refugee law as they relate to the protection of displaced persons. Secondly, while recognising that protection encompasses all activities that aim for full respect for these provisions, it examines ways to operationalise these legal frameworks on the basis of what is commonly referred to as the rights-based approach.

## Protection of Displaced Persons: Overview of the Relevant Legal Frameworks

The bodies of international law that contribute to the current international regime of protection emerged in the aftermath of the Second World War, a war that precipitated displacement on a hitherto unprecedented scale. The resultant 'norm cascade' led to a largely consensual adoption by states of international treaties, declarations, and institutions that sought to remould the relationship between the state and the individual (Finnemore and Sikkink 1998). It did so by binding states into an international legal regime, admittedly built on the sovereignty of states, that nonetheless recognised the inviolability of human dignity. Within this regime, the international community makes a clear commitment to the recognition and upholding of the rights of displaced persons.

### International Refugee Law

The international refugee regime comprises a series of international laws and norms, some of which are legally binding while others are not (Koser 2007). The foundation of international refugee law is Article 14 (1) of the 1948 Universal Declaration of Human Rights, which states that 'everyone has the right to seek and to enjoy in other countries asylum from persecution.' The 1951 Refugee Convention can be considered the backbone of international refugee law, since most regional and national laws are based on its definition of a refugee. The Convention defines a refugee as someone who 'owing to a well-founded fear of being persecuted for reasons of race, religion, nationality, membership of a particular social group or political opinion, is outside the country of his nationality.' Each of the conditions outlined in the definition ought to be fulfilled in order for an individual to meet the refugee definition. A fundamental element of the 1951 Convention is that refugee status can only be considered once a persecuted individual or individuals have crossed a border into a different national jurisdiction and so does not apply to internally displaced persons (IDP). IDPs have, however, been considered under subsequent conventions and guidelines and are protected by international humanitarian law (IHL) and international human rights law (IHRL) as discussed below.

The 1951 Convention establishes the principle of *non-refoulement*—the principle that individuals seeking asylum ought not to be returned to the country from which they have fled while their application is considered—and the rights afforded to those granted refugee status. Although the refugee definition contained in the 1951 Convention remains the dominant definition, modifications have been made by regional human rights treaties in response to instances of displacement not covered by the 1951 Convention, such as events seriously disturbing public order, as addressed in the 1969 OAU Convention in Africa and the 1984 Cartagena Declaration in the Americas.

While basing the existing refugee regime on the 1951 Convention is useful in creating a framework and benchmark for refugee determination across the globe, its focus on individual persecution leaves considerable gaps in terms of those displaced internationally by large-scale violence (in cases of conflict, these will normally be recognised by UNHCR or states as *prima facie* refugees, when individual determination appears impractical). The refugee definition has been interpreted broadly in certain cases such that 'Membership of a Particular Social Group' (PSG) includes gender, sexual orientation and identity, breaching of social norms, and so on. However, the definition does not address other forms of displacement that do not necessarily fall under the title of 'persecution,' such as environmental displacement.

Furthermore, regional bodies and alliances were built on the 1951 Convention to harmonise legal norms for refugee determination, the processing of asylum applications, and reception conditions. Although these vary across regions, the African Union Convention Governing Specific Aspects of Refugee Problems in Africa (1969) influenced the Cartagena Declaration (1984) in Latin America and the establishment of norms at the national level, particularly for victims of events of high-scale societal disturbance and external aggression. Likewise, the Common European Asylum System (CEAS) (2008) is a highly sophisticated move to harmonise European protection measures for those forcibly displaced who have entered the European Union and claimed asylum.

## International Humanitarian Law (IHL)

Armed conflict is a major driver of displacement. IHL (also known as the law of armed conflict or the laws of war) sets minimum standards of humane conduct that must be adhered to in times of armed conflict. It applies in both international armed conflict involving two or more states and non-international armed conflict in which one or more non-state armed groups is party to a conflict. IHL only applies during internal conflicts where there is a degree of organisation of the armed groups involved and a minimum level of intensity of armed confrontations (Akande 2012). Where its provisions are applied, IHL can serve to reduce the likelihood of displacement during wartime as well as provide certain protections to those who do become displaced.

IHL expressly prohibits parties to an armed conflict, whether they are armed forces of states or non-state armed groups, from forcibly displacing civilians. There are two important exceptions to this general rule; civilians may be displaced where the security of such persons so demands and for reasons of military necessity.[2] Once the reasons for the forced displacement no longer exist, displaced persons are accorded a right to voluntarily return to their homes or places of habitual residence (ICRC 2015).[3] This approach to the displacement of persons within international humanitarian law reflects the ethos of this body of law as striking a balance between competing principles of humanity and military necessity during wartime.

IHL also provides that, should persons become displaced, their homes and possessions are protected from direct attack by the parties to an armed conflict so long as they do not directly participate in hostilities and so long as their homes and possessions are not used for military purposes. These protections derive from displaced persons' status as civilians under IHL. Refugees enjoy the same protection within the IHL framework as civilians generally. In addition, special protection of refugees is explicitly accorded under the Fourth Geneva Convention (Article 44) and Additional Protocol I (Article 73).

## International Human Rights Law (IHRL)

IHRL provides normative guidance concerning the relationship between states and individuals within the territory or effective control of states. In contrast to IHL, IHRL applies in all circumstances and not solely during armed conflict. Under IHRL, states are obliged to respect the enjoyment of human rights by all persons in their territory or under their effective control. As such, human rights are accorded not only to citizens of the state concerned. This is significant in the context of displacement. States are also obliged to protect human rights from being violated by other actors and to take positive steps towards their fulfilment. Human rights violations can constitute persecution and can either ignite or fuel conflict, thereby contributing to displacement. As such, adherence to human rights law mitigates the risks posed by these drivers of displacement. Adherence to human rights law also promotes the voluntary return of displaced persons in the aftermath of crisis.

Human rights are also of relevance in the context of displacement arising in the context of disasters. Abebe (2011, p. 102) outlines a threefold relationship between so-called 'natural' and technological disasters and human rights. Firstly, the enjoyment of human rights is heavily impacted upon by the effects of natural and man-made disasters, such as eroding the capacity of public actors to promote the right to life and safeguard the highest attainable standard of health. Secondly, the protection of human rights can be viewed as a key component of rescue, recovery, and reconstruction. Finally, the denial of human rights and their violation contributes to the vulnerabilities/fragilities of disaster-prone individuals and communities. As such, social and political vulnerability and the implications of discrimination, marginalisation, poverty, and other social ills for the enjoyment of human rights can serve to precipitate and exacerbate the impact of natural hazards.

Similar to IHL, IHRL has its origins in the post-World War II period. The Universal Declaration of Human Rights informed the two chief universal human rights treaties: the International Covenant on Civil and Political Rights (1966) and the International Covenant on Economic, Social and Cultural Rights (1966). Each of these treaties addresses a certain category of human rights. Civil and political rights are a category of human rights that protect individuals' freedom from infringement by governments and that ensure individuals' ability to participate in the civil and political life of the state without discrimination or repression. A further category of human rights is economic, social, and cultural rights. The nature of the obligations imposed on states by economic, social, and cultural rights differs to that of civil and political rights. Thus, the drafters of the cornerstone economic social and cultural rights treaty, the 1966 International Covenant on Economic, Social and Cultural Rights (ICESCR), recognises that the full realisation of economic, social, and cultural rights must be achieved progressively and in recognition of states' available resources. Nonetheless, each state is obliged under the Covenant to take steps to the maximum of its available resources to progressively realise all of the rights recognised within the ICESCR.

Human rights may be limited or derogated from in order to address public security, public health, or other public interest concerns. However, such limitations or derogations must be undertaken on a proportionate basis. In times of public emergency (which are likely to generate displacement), states may derogate from important provisions of IHRL, such as liberty and freedom of movement, insofar as it is necessary in order to respond to such emergencies. For example, in June 2016, Ecuador suspended the right to liberty of movement (Article 12 ICCPR) amongst other rights on the basis that the suspensions of the rights in question were necessary to prevent individuals affected by a major earthquake from returning to their homes (Sommario 2018, p. 112). In terms of economic, social, and cultural rights, it has been noted that emergencies can impact on the availability of resources.[4] However, certain immediate obligations are imposed on states. Thus, discrimination in the achievement of economic, social, and cultural rights is prohibited; minimum essential levels of each of the rights must be satisfied; and an obligation exists to provide an effective remedy for violations of economic, social, and cultural rights.

The right to life is the cornerstone human right without which other rights are rendered meaningless. Other civil and political rights of most relevance to displacement include freedom of liberty of movement (which includes a right to remain), and the right to adequate housing (McDermott and Gibbons 2017). Together these rights prohibit unjustified forced evictions. Forced evictions are a breach of the obligations to respect the right to legal security of tenure, defined by the ICESCR in its fourth General Comment as a factor to be considered in assessing the adequacy of housing. Governments must refrain from executing forced evictions and must ensure that the law is enforced against its agents or against third parties such as private companies who undertake them (Barber 2008, p. 460).

It should be noted that the principle of non-refoulement has permeated international human rights law to complement the protection provided by refugee law to those who have been displaced across an international border. Also known as subsidiary protection, in the sense that the protection offered is to be considered subsidiary to refugee law, various provisions of human rights treaties prohibit states from deporting persons within their jurisdiction to states in which they are at substantial risk of being subjected to serious violations of human rights. Thus, depending on the treaties ratified by a state, it is forbidden to refoul a person to a state where they are likely to face the death penalty or torture or cruel, inhuman, and degrading treatment or punishment (Worster 2012).

## Internal Displacement

There is no universal legally binding instrument specifically addressing the plight of internally displaced persons. However, in 1997 the UN Commission on Human Rights adopted a resolution requesting the submission of a report concerning the human rights of internally displaced persons, resulting in the Guiding Principles on Internal Displacement.[5] While the Guiding Principles are not legally binding, they compile numerous rules from existing IHRL and IHL frameworks as well as norms that may be derived from refugee law.[6] Furthermore, the Guiding Principles have enjoyed wide support from the international community, and many states have incorporated them into their domestic legal systems. A key notion affirmed in the Guiding Principles is that states have the primary responsibility to prevent displacement, to protect and assist internally displaced persons under their jurisdiction, and to provide durable solutions to their situation. To fulfil this responsibility, states need to have domestic normative and policy frameworks in place with the necessary implementing structures and processes that facilitate effective response to the specific needs and vulnerabilities of internally displaced persons. At the regional level, the African Union adopted the Kampala Convention in 2009 on internally displaced persons. Inspired by the Guiding Principles, it is the first legally binding regional instrument addressing internal displacement.

A summary of the main bodies of law providing protection to displaced persons is presented in Table 8.1.

## Agency of Displaced Persons and the Rights-Based Approach to the Protection of Such Persons

The lived protection of displaced persons that is envisioned by the relevant bodies of law depends more than ever on an adequate enabling environment for its realisation, given the record number worldwide of persons forcibly displaced.[7] This involves the promotion of effective enforcement mechanisms at the international, regional, and national levels. However, it is also important to create an enabling environment beyond the legal system whereby the rights of displaced persons or those at risk of displacement are promoted. Such an enabling environment extends beyond the strictly legal sphere. It necessitates the accordance to individuals, communities, and organisations

Table 8.1 Summary of protection frameworks for displaced persons

| Body of law | Scope of application | Key legal instruments | Main protection provisions/ principles |
|---|---|---|---|
| Refugee law | Applies to persons who, owing to well-founded fear of being persecuted, are outside their country of nationality or habitual residence | 1951 Refugee Convention<br>1967 Protocol | Principle of non-refoulement |
| International humanitarian law (IHL) | Applies to armed conflict settings. IHL protects civilians and combatants/fighters no longer taking part in hostilities | 1949 Geneva Conventions<br>1977 Additional Protocols | Principle of distinction between civilian and military targets |
| International human rights law (IHRL) | Applies in all circumstances but can be curtailed during public emergencies, including during armed conflicts<br>Applies to all persons within jurisdiction of a state party to international human rights law | 1966 International Covenant on Civil and Political Rights<br>1966 International Covenant on Economic, Social and Cultural Rights | Right to life<br>Right to liberty of movement<br>Right to adequate housing |
| Legal mechanisms relating to internal displacement | Persons or groups of persons who have been forced or obliged to flee or to leave their homes or places of habitual residence, in particular as a result of or in order to avoid the effects of armed conflict, situations of generalised violence, violations of human rights, or natural or human-made disasters, and who have not crossed an internationally recognised state border (Guiding Principle 2) | Guiding Principles on Internal Displacement<br>Kampala Convention on internally displaced persons<br>A range of domestic bodies of law mitigate displacement and protect internally displaced persons, including criminal law | Guiding Principles incorporate elements of IHL and IHRL |

justiciable rights within the relevant legal systems, the capacity to hold duty bearers accountable for protecting these rights, as well as the freedom for civil society to advocate for political and legal change (Sen 2004; Hennink et al. 2012, p. 210).

Against this backdrop, there has been an increasing emphasis on the promotion of a rights-based approach to deal with displacement. The rights-based approach has been criticised, within humanitarian settings, as being superficial (Uvin 2002). However, when treated seriously, the approach represents a strategy for translating the various protection norms into practical reality. It potentially shifts the portrayal of displaced persons from helpless victims to actors in their own right who possess an agency that may well be limited by circumstances, but which is often also asserted in displacement contexts (Gibbons et al. 2017; Isayev 2017; London 2008, p. 68; Slim 2002, p. 3).

The rights-based approach can be distinguished from other approaches to progressive change on the basis that it involves the explicit identification of rights holders and duty bearers. By recognising displaced persons as rights holders, their potential agency is acknowledged, and their empowerment is promoted. Concomitantly, duty bearers such as national governments and non-state armed actors are appraised of their duties and their activities in light of their duties are monitored. An avenue is also provided for the remedy of violations. Claim–duty relationships are affixed to all relevant subjects and objects at subnational, (host) community, and household level. In so doing, the balance between rights and duties at each societal level is emphasised. This recognises that the links to social institutions such as the (extended) family, community, civil society, and state that displaced persons may have previously relied upon for protection and assistance are often highly disrupted or severed (Mooney 2016). Figure 8.1 shows how actors at different levels have a range of complementary roles in creating an enabling environment for the protection of the rights of individuals.

## Conclusion

Displaced persons are protected by a range of international, regional, and national legal frameworks that are rooted in a concern with human dignity. Depending on their respective scope of application, these bodies of law provide protection to those at risk of displacement, protect those who have been displaced, and provide durable solutions. Through their legal effect a powerful tool is provided to displaced persons and those who advocate on their behalf.

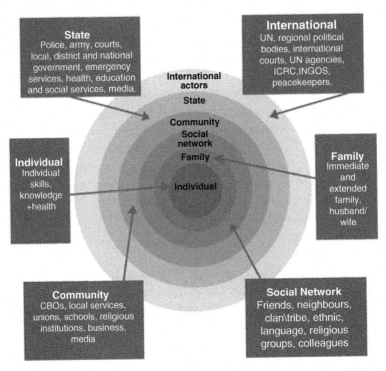

**Fig. 8.1** The different societal levels that influence protection (Action Aid 2010)

Rights-based approaches to the protection of displaced persons recognise the need for the creation of an enabling environment that affixes rights and duties to various actors at various societal levels. Nonetheless, such rights-based approaches should not be viewed as a means of de-responsibilising actors at higher societal levels, including states and international organisations, who carry clear, legally binding protection obligations. Despite scepticism from some quarters as to how the rhetoric of the rights-based approach is implemented in practice, in its ideal form it facilitates the agency of displaced persons and draws attention to the obligations of duty bearers to promote their protection.

**Acknowledgements** This chapter has been written as part of the Preparedness and Resilience to address Urban Vulnerabilities (PRUV) project. The project has received funding from the European Union's Horizon 2020 research and innovation programme under the Marie Skłodowska-Curie grant agreement no. 691060.

## Notes

1. Inter-Agency Standing Committee definition, Protection of Internally Displaced Persons (IASC 1999, p. 4).
2. Geneva Convention IV, Arts 49 and 147; AP I Art. 85(4)(a); AP II Art. 17; Customary IHL Rule 129. See also Additional Protocol I Arts 51(7) and 78(1) and Additional Protocol II Art. 4(3)(e).
3. Geneva Convention IV Art. 49 and Customary IHL Rule 132.
4. For example, see Paragraph 11, UN Committee on Economic, Social and Cultural Rights, *Concluding observations of the Committee on Economic, Social and Cultural Rights: Honduras*, 21 May 2001, E/C.12/1/Add.57.
5. UN Commission on Human Rights, *Resolution 1997/39 on Internally Displaced Persons*, 11 April 1997, E/CN.4/RES/1997/39.
6. UN Commission on Human Rights, *Report of the Representative of the Secretary-General, Mr. Francis M. Deng, submitted pursuant to Commission resolution 1997/39. Addendum: Guiding Principles on Internal Displacement*, 11 February 1998, E/CN.4/1998/53/Add.2.
7. See UNHCR's *Global Report for 2018*: http://reporting.unhcr.org/sites/default/files/gr2018/pdf/GR2018_English_Full_lowres.pdf Accessed on 25 February 2020.

## References

Abebe, A. (2011). Special report—Human rights in the context of disasters: The special session of the UN Human Rights Council on Haiti. *Journal of Human Rights, 10*(1), 99–111.

Action Aid. (2010). *Safety with dignity—A field-based manual for integrating community-based protection across humanitarian programs*. Retrieved October 23, 2018, from https://actionaid.org/publications/2010/safety-dignity

Akande, D. (2012). Classification of armed conflicts: Relevant legal concepts. In E. Wilmshurst (Ed.), *International law and the classification of conflicts* (pp. 32–49). Oxford, UK: OUP.

Barber, R. (2008). Protecting the right to housing in the aftermath of natural disaster: Standards in international human rights law. *International Journal of Refugee Law, 20*(3), 432–468.

Finnemore, M., & Sikkink, K. (1998). International norm dynamics and political change. *International Organisation, 2*(4), 887–917.

Gibbons, P., McDermott, R., Maitra, S., & Herman, J. (2017). Building on the capacities of crisis-affected populations: From victims to actors. *Development Policy Review, 36*(5), 547–560.

Hennink, M., Ndunge, K., Pillinger, M., & Jayakaran, R. (2012). Defining empowerment: Perspectives from international development organisations. *Development in Practice, 22*(2), 202–215.

IASC (Inter-Agency Standing Committee). (1999, December). *Protection of internally displaced persons* (Inter-Agency Standing Committee Policy Paper). Retrieved from https://interagencystandingcommittee.org/focal-points/documents-public/iasc-policy-paper-protection-internally-displaced-persons-1999

ICRC. (2015). *How does IHL protect refugees and internally displaced persons?* Retrieved October 23, 2018, from https://www.icrc.org/en/document/how-does-humanitarian-law-protect-refugees-and-internally-displaced-persons-0

Isayev, E. (2017). Between hospitality and asylum: A historical perspective on displaced agency. *International Review of the Red Cross, 99*(1), 75–98.

Koser, K. (2007). *International migration: A very short introduction.* Oxford, UK: OUP.

London, L. (2008). What is a human-rights based approach to health and does it matter? *Health and Human Rights, 10*(1), 65–80.

McDermott, R., & Gibbons, P. (2017). Human rights and proactive displacement: Determining the appropriate balance between the duty to protect and the right to remain. *Disasters, 41*(3), 587–605.

Mooney, E. (2016). Displacement and the protection of civilians under international law. In H. Willmot, R. Mamiya, S. Sheeran, & M. Weller (Eds.), *Protection of civilians* (pp. 177–204). Oxford, UK: OUP.

Sen, A. (2004). Elements of a theory of human rights. *Philosophy and Public Affairs, 32*(4), 315–356.

Slim, H. (2002). A response to Peter Uvin: Making moral low ground: Rights as the struggle for justice and the abolition of development. *PRAXIS: The Fletcher Journal of Development Studies, 17*, 1–5.

Sommario, E. (2018). Limitation and derogation provisions in international human rights law treaties and their use in disaster settings. In F. Giustiniani, E. Sommario, F. Casolari, & G. Bartolini (Eds.), *International disaster response law.* The Hague, Netherlands: Springer.

UNHCR. (2018). *Global Report 2018.* Retrieved February 25, 2020, from http://reporting.unhcr.org/sites/default/files/gr2018/pdf/GR2018_English_Full_lowres.pdf

Uvin, P. (2002). On high moral ground: The incorporation of human rights by the development enterprise. *PRAXIS: The Fletcher Journal of Development Studies, 17*, 1–11.

Worster, W. (2012). The evolving definition of the refugee in contemporary international law. *Berkeley Journal of International Law, 30*, 94–160.

# 9

# Queering Displacement/The Displacement of Queers

Scott McKinnon

In Australia, as I write this chapter, refugees are again front-page news. The nation appears poised for another election fought over which of the two major parties can best be trusted to 'protect our borders.' The current government trumpets its 'success' in 'stopping the boats,' that is, halting the movement by sea of displaced people who have managed to reach our northern neighbours and now hope to make it to Australia in search of safety (Davidson 2018). A key strategy behind the government's purported policy triumph is the indefinite incarceration of asylum seekers in 'offshore' detention camps located on Manus Island (part of Manus Province in Papua New Guinea) and Nauru (Seuffert 2013). The conditions in these camps have been heavily criticised by the United Nations and a range of human rights and medical organisations (Tingle 2018).

This frequently recurring political moment has deeply informed my writing of this chapter on queer displacement, in which I deploy 'queer' as both an umbrella term for a multiplicity of non-heterosexual and non-cisgender identities and as a process which resists normativity and the political, cultural, and social centring of heterosexual and cisgender lives (Gorman-Murray 2007; Patrick 2014). I attend first, to the ways in which subjects with non-normative sexual and gender identities are displaced by discriminatory policies and practices and, second, to the uneven impacts of crisis on displaced sexual and gender minority populations. Drawing on definitions of

S. McKinnon (✉)
University of Wollongong, Wollongong, NSW, Australia
e-mail: scottmck@uow.edu.au

© The Author(s) 2020
P. Adey et al. (eds.), *The Handbook of Displacement*,
https://doi.org/10.1007/978-3-030-47178-1_9

displacement as both process and condition (Bakewell 2011; Winton 2019), I argue for the importance of revealing the heteronormative and homonormative structures that form queer experiences of migration and mobility. Ultimately, I am interested in how people who might be understood as 'queer' experience displacement, as well as the value of 'queering' displacement in order to reveal normative understandings that silence or erase difference.

The critical importance of making queerness visible in our understandings of displacement has been brought into stark focus by Australian refugee policy. Globally, around seventy-five countries continue to criminalise same-sex acts. Among the asylum seekers fleeing to Australia are individuals displaced from their country of origin by the criminalisation of their sexual and romantic lives (Laughland 2014). Although Australia offers substantial legal protections to lesbian, gay, bisexual, transgender, and queer (LGBTQ) citizens and residents, queer refugees who have sought asylum in Australia have nonetheless been sent to detention camps in Manus, where same-sex acts are illegal and punishable by up to fourteen years of imprisonment. In other words, displaced individuals who have sought safety in Australia have been detained in circumstances that replicate the legal structures through which they were initially displaced. The Australian government's advice to any detainees concerned about arrest on Manus has been to refrain from same-sex acts. Openly gay asylum seekers detained on Manus have reported harassment, abuse, and violence from other detainees and members of the local population (Seuffert 2013).

The already traumatising circumstances of offshore detention are therefore exacerbated by heteronormative systems that place queer people in danger. The consequences on individual lives are devastating. In 2014, *The Guardian* newspaper published letters written anonymously by asylum seekers. One stated, 'I am so sorry that I was born gay. … I wish our boat had sunk in the ocean and stopped me living the most painful year in my life. I thought Australia and its people would be my protector, but they taught me otherwise' (quoted in Laughland 2014). For displaced individuals, displacement is not an easily defined movement from danger to safety but instead a process of constantly re-making place in contexts which define non-heterosexual and non-cisgender identities as either invisible or as a target for punishment. The displacement of sexual dissidents operates, in the words of Winton, as 'temporally and spatially contingent not on movement per se, but rather on being in or out of place' (2019, p. 97).

Below, I draw on the work of a range of scholars to understand queer experiences of displacement at multiple scales, ranging from homeless LGBTQ youth in large cities, to LGBTQ families displaced by disaster, to queer asylum seekers searching across and within national borders for places of safety. These diverse examples allow us to understand the complex and embodied experiences of individuals forced into movement by discriminatory structures, policies, and practices.

## The Displacement of Queers

Since the 1980s, scholars in a range of disciplines have explored processes of place-making by sexual and gender minority populations, initially focussing on gay men in urban spaces. In much of this research, the idea of displacement was implicit. Inner-city neighbourhoods in large cities of the Global North were identified as the home of gay men and other queer people who had gathered in these spaces in order to find community and participate in the co-creation of spaces of acceptance and inclusivity. These spaces were often seen as the end of a migratory journey—what Knopp has called 'queer quests for identity' (2004, p. 121)—that began in unwelcoming homes and towns. Journeys in search of safety imply the need to leave spaces that are dangerous to physical or mental health, emotional well-being, and ontological security.

Into the 1990s, and under the influence of queer theory, increasing complexity has been acknowledged in the fluidity of identities, the spatial and temporal aspects of journeys undertaken, and the intersectional factors that lead to and alter the experience of displacement across multiple scales. Broadly acknowledged is the fact that discrimination against sexual and gender minority populations often forces queer people to leave their home, town, or nation. As stated by Gorman-Murray, 'queer sexualities often appear "out of place" in communities of origin, and are frequently only enabled by relocation elsewhere' (2007, p. 105). Gorman-Murray argues for the need to understand queer migrations at the scale of the body, stating, 'a focus on embodied displacement … enables us to contemplate *peripatetic* migrations by queers as progressive quests for sexual identity' (2007, p. 105—emphasis in original).

The experiences of many LGBTQ youth reveal the ways in which queer people are displaced by discriminatory forces. Spaces including homes and schools are commonly represented as 'safe' for children and young people. Although many queer youth do find support within the home (Gorman-Murray 2008), geographers and other scholars have shown how these spaces are also often defined as heterosexual and cisgender (Rodó-de-Zárate 2015; McKinnon et al. 2017). Unlike other minority groups, queer youth commonly do not share the identities of their parents and other family members and thus are 'coming out' into atmospheres that are potentially hostile to queerness.

The consequences for LGBTQ youth displaced from purported spaces of care can include homelessness and exposure to violence (Rosenberg 2017; Tunåker 2015; Valentine et al. 2003). Displaced queer youth are forced into processes of locating and re-making 'home' in spaces including the homes of friends, youth shelters, and, potentially, the street. The result is an ongoing sense of spatial and ontological precarity. As argued by Tunåker, 'Being "homeless," to LGBT youth, implies a whole set of complex ideas of living outside the boundaries of what they perceive as the norm in the society that surrounds them' (2015, p. 254).

Intersecting factors, including race, mental health, and bodily ability, define the process and condition of displacement for queer youth. Rae Rosenberg, exploring the experiences of LBGTQ youth of colour in the United States city of Chicago, describes systems of community policing which foster 'a White queer politics of belonging that encourages the mistrust and consequent criminalizing of people of color' (2017, p. 146). Rosenberg's research indicates the role of heteronormative and homonormative (Duggan 2002) practices which define LGBTQ youth of colour as 'out of place' even within spaces otherwise defined as 'queer.'

At broader scales, and as discussed above, the displacement of queer people can also force inter- or intra-country migration. The field of migration studies initially emerged with an implicit assumption of uniform heterosexuality among migrants (Lewis and Naples 2014). More recently, a growing body of literature has explored the pivotal role of minority sexual and gender identity in forcing displacement and in shaping (or limiting) the possibilities for mobility across and within borders (Güler et al. 2019; Lewis and Naples 2014; Spijkerboer 2013). According to Lewis and Naples, acknowledging 'the experiences of LGBTQI immigrants, refugees and asylum seekers brings into view the way in which movement across borders reinscribes heterosexuality, regulates homosexual expression, and renders invisible the bodies and self-identities of those who dare to cross' (2014, p. 912).

In her investigation of queer displacement in northern Central America and Central Mexico, Ailsa Winton (2019) makes clear how multiple scales of displacement become entwined in individual lives. Winton gives the example of George, a gay 25-year-old El Salvadorian man who faced harassment and abuse from his family at home, and who was frequently stopped, harassed, and beaten by police as he walked to work. Persecution by the state and rejection by family, occurring both in public and private spaces, meant that George 'was unable to make any space his … and so he had to "go somewhere"'

(Winton 2019, p. 109). With only fifty dollars and one change of clothes, he fled to Mexico and began the asylum application process. According to Winton, 'The intersection particularly between gender and sexual transgression and economic and social marginalization translates into pervasive precarity' (2019, p. 111).

As the examples from Australia with which I began the chapter make clear, processes of applying for asylum or navigating complex migration systems are further complicated by normative structures that erase sexual and gender difference. Individuals seeking asylum based on persecution related to their sexuality, for example, may be required to prove that they are homosexual (Brazil and Arnold 2019; Jansen and Spijkerboer 2011). This raises a series of complex questions over how sexual identity can be proved, particularly when the applicant may have a different understanding of gay or queer identity—based on cultural differences—than the definitions used by the host country. A Nigerian applicant for asylum in Ireland, for example, was initially rejected as uncredible based, in part, on his inability to name a gay bar he had attended or to recall the address of Dublin's most popular gay venue (Brazil and Arnold 2019). The assumption on the part of investigators that a man who has not visited a gay bar is not, therefore, likely to be gay is based more on a series of stereotypes than the lived experience of each queer person seeking asylum.

## Displaced Queers

As the above examples attest, queer readings of displacement have revealed how sexual and gender minority populations experience forced mobility as a result of anti-queer discrimination. Equally, queering displacement has offered insights into how similar forces shape the experiences of queer minorities within broader populations displaced by various forms of crisis, including war and disaster. As with migration studies, the field of disaster research has, until recently, operated with an implicit assumption that all individuals struck by disaster are heterosexual and cisgender (Dominey-Howes et al. 2014). Equally, heteronormative and cisnormative emergency management policy and practice has left queer people invisible, resulting in complex challenges within disaster-impacted communities. Recent research across a number of global contexts has revealed the need to develop policies that better support queer individuals and families displaced by catastrophe.

Research by Balgos et al. (2012), for example, has examined the impacts of disaster on *waria* communities following the 2010 volcanic eruption of Mt Merapi, in Central Java. The word *waria* is derived from the Indonesian words *wanita* (woman) and *pria* (man). It describes a subculture of people identified as male at birth, who dress in female attire and do not identify as either male or female. In predominantly Islamic Indonesian society, *waria* are subjected to discrimination, violence, and marginalisation. The eruption forced more than 400,000 people to evacuate their homes and resulted in the deaths of an estimated 300 people. In official examinations of the disaster's impacts, however, *waria* were invisible. Details of evacuees in temporary shelters were collected on a binary gender (male/female) model, meaning that *waria* were not included. According to Balgos et al. (2012), some *waria* displaced by the disaster chose not to go to shelters for fear of discrimination, instead attempting to find shelter in the homes of friends.

Similarly, research by Dominey-Howes et al. (2016) has revealed the role of faith-based organisations in Australian emergency management and the consequent impacts on LGBTQ disaster survivors. Australian state governments outsource a number of post-disaster support services (including the provision of food and clothing to survivors) to faith-based groups. Many of the churches that run these groups teach that LGBTQ behaviours and identities are sinful and perverse. In states including New South Wales, faith-based organisations also enjoy exemptions under anti-discrimination legislation meaning that they are legally permitted to discriminate against LGBTQ people in employment and provision of services. LGBTQ individuals, couples, and families displaced from their homes by disaster are thus required to seek support from organisations who have the right to deny service and who actively lobby against LGBTQ equality.

## Conclusion

While drawing this chapter to a close, a new report appeared in the local media. The government is now arguing against legislation that would make it easier for refugees detained on Manus and Nauru to be brought to Australia for medical treatment. Providing care for sick people, it would seem, would 'weaken our borders.' What's more, argues Prime Minister Scott Morrison, Australians 'who need medical services are going to be displaced from those services, because if you bring hundreds and hundreds of people from Nauru and Manus down to our country, they are going to go into the health network' (quoted in Davidson 2019). Refugees are positioned here as a threat to the people of 'our country.' They are not the displaced but instead threaten to displace 'us.'

Our government, it seems, works in binaries—us/them, citizen/refugee, male/female, and straight/gay. What queer readings of displacement offer is an opportunity to break down these binaries. In doing so, what comes into focus are, first, the ways in which normative systems operate to displace queer people; second, the ways in which displaced queer people are further marginalised by those systems; and, third, the ways in which intersectional factors including race, gender, and socioeconomic status, further shape the (im) mobility of queer people. By applying a queer lens to displacement, ultimately what was revealed is the need to build more inclusive systems that allow queer people to be 'in place' and which provide greater support for those forced 'out of place' by crisis.

## References

Bakewell, O. (2011). Conceptualising displacement and migration: Processes, conditions, and categories. In K. Koser & S. Martin (Eds.), *The migration-displacement nexus: Patterns, processes, and policies* (pp. 14–28). Oxford: Berghahn Books.

Balgos, B., Gaillard, J. C., & Sanz, K. (2012). The *warias* of Indonesia in disaster risk reduction: The case of the 2010 Mt Merapi eruption in Indonesia. *Gender & Development, 20*(2), 337–348.

Brazil, P., & Arnold, S. (2019). LGBTI asylum applications in Ireland: Status determination and barriers to protection. In A. Güler, M. Shevtsova, & D. Venturi (Eds.), *LGBTI asylum seekers and refugees from a legal and political perspective* (pp. 95–113). Cham, Switzerland: Springer.

Davidson, H. (2018, September 19). 'I stopped these': Scott Morrison keeps migrant boat trophy in office. *The Guardian.* Retrieved from https://www.theguardian.com/australia-news/2018/sep/19/i-stopped-these-scott-morrison-keeps-migrant-boat-trophy-in-office

Davidson, H. (2019, March 1). Morrison backs Dutton claim refugees' medical care will 'displace' Australians. *The Guardian.* Retrieved from https://www.theguardian.com/australia-news/2019/mar/01/morrison-backs-dutton-claim-refugees-medical-care-will-displace-australians

Dominey-Howes, D., Gorman-Murray, A., & McKinnon, S. (2014). Queering disasters: On the need to account for LGBTI experiences in natural disaster contexts. *Gender, Place & Culture, 21*(7), 905–918.

Dominey-Howes, D., Gorman-Murray, A., & McKinnon, S. (2016). Emergency management response and recovery plans in relation to sexual and gender minorities in New South Wales, Australia. *International Journal of Disaster Risk Reduction, 16,* 1–11.

Duggan, L. (2002). The new homonormativity: The sexual politics of neoliberalism. In R. Castronovo & D. D. Nelson (Eds.), *Materializing democracy: Toward a revitalized cultural politics* (pp. 175–194). Durham, NC: Duke University Press.

Gorman-Murray, A. (2007). Rethinking queer migration through the body. *Social and Cultural Geography, 8*(1), 105–121.

Gorman-Murray, A. (2008). Queering the family home: Narratives from gay, lesbian and bisexual youth coming out in supportive family homes in Australia. *Gender, Place and Culture, 15*(1), 31–44.

Güler, A., Shevstova, M., & Venturi, D. (2019). *LGBTI asylum seekers and refugees from a legal and political perspective: Persecution, asylum and integration.* Cham, Switzerland: Springer.

Jansen, S., & Spijkerboer, T. (2011). *Fleeing homophobia: Seeking safety in Europe.* Amsterdam: COC Nederland & VU University Amsterdam.

Knopp, L. (2004). Ontologies of place, placelessness, and movement: Queer quests for identity and their impacts on contemporary geographic thought. *Gender, Place & Culture, 11*(1), 121–134.

Laughland, O. (2014, September 24). Gay asylum seekers on Manus Island write of fear of persecution in PNG. *The Guardian.* Retrieved from https://www.theguardian.com/world/2014/sep/24/gay-asylum-seekers-manus-island-fear-persecution-png

Lewis, R. A., & Naples, N. A. (2014). Introduction: Queer migration, asylum, and displacement. *Sexualities, 17*(8), 911–918.

McKinnon, S., Waitt, G., & Gorman-Murray, A. (2017). The Safe Schools Program and young people's sexed and gendered geographies. *Australian Geographer, 48*(2), 145–152.

Patrick, D. J. (2014). The matter of displacement: A queer urban ecology of New York City's High Line. *Social & Cultural Geography, 15*(8), 920–941.

Rodó-de-Zárate, M. (2015). Young lesbians negotiating public space: An intersectional approach through places. *Children's Geographies, 13*(4), 413–434.

Rosenberg, R. (2017). The whiteness of gay urban belonging: Criminalizing LGBTQ youth of color in queer spaces of care. *Urban Geography, 38*(1), 137–148.

Seuffert, N. (2013). Haunting national boundaries: LBGTI asylum seekers. *Griffith Law Review, 22*(3), 752–784.

Spijkerboer, T. (Ed.). (2013). *Fleeing homophobia: Sexual orientation, gender identity and asylum.* Abingdon: Routledge.

Tingle, L. (2018, December 6). Offshore detention case to be brought to UN against Australian Government as families hope for reunion. *ABC News.* Retrieved from https://www.abc.net.au/news/2018-10-15/offshore-detention-case-against-government-in-un/10379218

Tunåker, C. (2015). 'No place like home?' Locating homeless LGBT youth. *Home Cultures, 12*(2), 241–259.

Valentine, G., Skelton, T., & Butler, R. (2003). Coming out and outcomes: Negotiating lesbian and gay identities with, and in, the family. *Environment and Planning D: Society and Space, 21*(4), 479–499.

Winton, A. (2019). 'I've got to go somewhere': Queer displacement in northern Central America and southern Mexico. In A. Güler, M. Shevtsova, & D. Venturi (Eds.), *LGBTI asylum seekers and refugees from a legal and political perspective* (pp. 95–113). Cham, Switzerland: Springer.

# 10

# Gendered and Feminist Approaches to Displacement

Katherine Brickell and Jessie Speer

## Introduction

In this chapter, we explore what gendered and feminist approaches bring to research on displacement. Current scholarship on displacement substantively deals with, but also goes beyond, the study of the refugee crisis. As Horn and Parekh (2018, p. 503) identify in their introductory editorial to a special issue of *Signs: Journal of Women in Culture and Society* on the topic, 'Displacement is not the same as the global refugee crisis, though that is likely to come to mind for many readers.' Indeed, gender and feminist-oriented research is varied yet is typically grouped into work on displacement and resettlement which is induced by conflict (Brun 2000; Ensor 2017; Manchanda 2004; Meertens and Segura-Escobar 1996; Suerbaum 2018), development (Brickell 2014; Fernández Arrigoitia 2017; Mehta 2009; Perry 2013), gentrification (Lyons et al. 2017; Mirabal 2009; Sakizhoglu 2018), the extractive industries (Ahmad and Lahiri-Dutt 2006), and natural disasters (de Mel 2017; Gorman-Murray et al. 2014; Juran 2012; Tanyag 2018).

A gendered lens tends to draw on ideas of social construction to understand how culturally shaped ideals and norms associated with masculinity and

---

K. Brickell (✉)
Department of Geography, Royal Holloway, University of London, Egham, UK
e-mail: katherine.brickell@rhul.ac.uk

J. Speer
School of Geography, Queen Mary University of London, London, UK
e-mail: j.speer@qmul.ac.uk

© The Author(s) 2020
P. Adey et al. (eds.), *The Handbook of Displacement*,
https://doi.org/10.1007/978-3-030-47178-1_10

femininity are relational, are performed (Butler 1990), and contextually vary in private and public life. There is now a consolidated body of literature which evidences how the effects of displacement are not gender-neutral but rather impact men and women differently (Indra 2001; Nolin 2006). Intersectional politics are also significant to such gendered displacements. As Roy (2017, p. A3) asserts in the US context, evictions are 'part of broader processes of racial banishment' that negate any claim to displacement as a sanitised process. Gendered and feminist approaches to the study of displacement attend to such intersecting contours of gender, sexuality, class, race, and ethnicity that produce and mediate these banishments (Schrijvers 1999).

In academic and policy literature, in the past, displacement was framed as a gender-neutral phenomenon. There are continuing issues, however, with a lack of gender-disaggregated data collected on displaced populations (e.g. by the International Organization for Migration regarding the tracking of internal displacement) which limits the ability to undertake gender-sensitive analyses.[1] In their absence, policies targeted towards displaced groups cannot be effectively designed to take into account deeply gendered processes and consequences of displacement (O'Neil et al. 2016). This matters, given that women and girls make up nearly 50 per cent of any refugee, internally displaced, or stateless population (UNHCR 2016). Those who are unaccompanied, pregnant, heads of households, disabled, or elderly are especially vulnerable (United Nations General Assembly 2016). For example, for women and girls in transit, challenges include family separation, psychosocial stress and trauma, health complications, physical harm and injury, and risks of exploitation (United Nations General Assembly 2016). Furthermore, national governments across the globe have been shown to criminalise and brutally punish gender-variant behaviour and same-sex sexualities, leading to widespread survival migration. Such migrants, in turn, are doubly impacted by heteronormative state approaches to refuge and asylum in receiving countries (Lewis and Naples 2014). Indeed, writing an editorial to a special issue on gender and displacement in the early 2000s, El-Bushra (2000, p. 4) notes:

> Assumptions about gender differences operate not only between individuals but also within institutions, including the household, the community, the state, schools and places of employment, including development agencies themselves. It is these institutions which reinforce and perpetuate gender discrimination, and it is these which must be challenged if gender injustice is to be transformed.

A feminist approach is underpinned by the commitment to theorising and addressing masculinist practices and connected drivers and social relations of oppression (e.g. patriarchy and capitalism) which structure gender inequalities and differences of power and status. 'In committing itself to gender as a category of analysis, contemporary feminism also commits itself to gender equality as a social goal,' Tickner (1992, p. 8) explains. This sentiment highlights a key point; gendered analysis is necessary but not sufficient for feminist analyses which seek not only to describe gendered identities, roles, and practices but also to intervene to change systems and patterns of discrimination and inequality (Sharp 2009).

Perhaps two of the most consistently used conceptual frames in the feminist study of displacement currently are feminist political ecology (Doshi 2017; Vaz-Jones 2018) and feminist geopolitics (Casolo and Doshi 2013; Culcasi 2019; Hyndman 2019). Both are driven by the production of situated, grounded, and embodied knowledges of displacement which take into consideration the multi-scalar politics of (im)mobility and its violences on differently positioned gendered bodies. With this in mind, the chapter is divided into two parts, the first on the physical and emotional labour involved in responding to and living with displacement and the second on neglected temporalities and hidden geographies of displacement which are commonly missing in mainstream writing. Together they show the value of gendered and feminist approaches to the study of displacement.

## Labours of Displacement

An emerging focus in writing on displacement is labour, an issue that has been skirted around rather than having received concerted engagement. Feminist writing has long established the significance of women's labour, both paid and unpaid, and there is, in relation to gender and migration literature specifically, a particular onus on its emotional as well as physical nature. Emotional or caring labour is typically involved in the care and nurturance of children and other family members, be this a woman's own or those of her employer (Hochschild 1983). Given the multiple financial and corporeal traumas of displacement, the emotional labour required to support family members through, and on from, it can be considerable. Displacement, after all, 'can be seen as a traumatic rupture in time and space of domestic and social reproduction in all their dimensions, including the symbolic dimensions of identity and representation' (Meertens and Segura-Escobar 1996, p. 171). As Brickell (2020, np) argues, there is an acute need then to focus on 'survival-*work*' to not lose sight of the range of labour that women undertake in the context of displacement (emphasis in original). There are major 'fallouts' of displacement, tolls on a person's spirit, which can disproportionately impact on women (Desmond 2016).

Women's management of this displacement ordinariness is an important concern in academic work and is diverse in its geographical origins of study. The ordinariness of displacement, Baviskar (2009, p. 72) comments, 'is an inescapable part of being female,' for example, in North India where marriage is understood as a displacement from a woman's natal home to that of her husband's. Here 'Women's subjective experience of marriage as a form of displacement, and their resigned acceptance of the norms of village exogamy and virilocality, is recognised all over north India through numerous songs that voice sorrow, yearning and reproach' (Baviskar 2009, p. 72). Other research questions the norms of female altruism which are evoked in response to displacement. In the context of Typhoon Haiyan, which left its devastating wake in 2013, Tanyag (2018) looks to the Philippine case to argue that disaster resilience has further divested responsibility for disaster response from the state to the household and community levels. In so doing, she argues that the assumed elasticity of women's unpaid labour, and propensity for self-sacrificing behaviour, 'has served to reinforce the structural roots of gendered vulnerability' (Tanyag 2018, p. 566). The onus on women's 'coping labour' and its potentially harmful implications is something increasingly spoken of in this recent literature. Culcasi's (2019) research, for example, focuses on the experience of Syrian women refugees in Jordan using ideas of coping and coping labour as a framework to explore intimate spaces of displacement. She describes how 'they generally take on new performances within the household, including becoming a provider, head of household and decision maker, all while maintaining their existing roles as caretakers' (Culcasi 2019, p. 5). As a result, there are immediate and longer-term changes to gender relations and women's bodily sustainability which require attention in the study of displacement.

In combination with this coping labour is the labour involved in resisting displacement. Prevailing gender norms are commonly challenged because it is women who have largely been at the forefront of contestations over their displacement; anti-displacement tactics, protects, and measures are thus thoroughly gendered (Casolo and Doshi 2013). The lost, threatened, or home made in the 'meanwhile' is a key site in which this happens. Culcasi (2019) critiques, for example, how 'mainstream migration and refugee discourses often frame refugees as living in "limbo" and merely waiting to return home.' Yet homemaking practices continue to be practised in these spaces (Brun and Fábos 2015). Therefore examining 'displaced people's intimate spaces of home and household can help reveal the rather invisible yet complex ways that macro-scaled geopolitics of forced displacement is entangled with daily life, coping practices and gender relations' (Culcasi 2019, p. 4). Women's labour in the domestic sphere works to establish the right to inhabit the abstract space of 'home.' Therefore, the assertion of the right to dwell echoes the wider view taken by Davidson (2009, p. 226) that there needs to be a move to 'reassert the place in dis*place*ment' (emphasis in original).

## Neglected Temporalities and Hidden Geographies of Displacement

In this second section of the chapter, we turn to neglected temporalities and hidden geographies of displacement which feminist research reveals. In terms of temporality, a keen focus of such work is on 'slow violence' (Hyndman 2019, p. 11), and the 'banal and ubiquitous practice' of waiting which is linked 'in myriad ways to mobility and (im)mobility in the contemporary era' (Conlon 2011, p. 353). This focus on the gradual and prolonged sidesteps many of the criticisms made on the 'eventfulness' of current literatures on displacement. Legitimate concerns speak to wider blind spots in the social sciences on suffering, with the vital importance of 'non-eventful geographies' (Wilkinson and Ortega-Alcázar 2018, p. 8) of displacement too often neglected. Such 'quasi-events' run the risk of never quite achieving 'the status of having occurred or taken place' (Povinelli 2011, p. 13) as a result. An apposite example in this regard is women's vulnerability to familial forms of displacement, including forced marriage, sex trafficking, and domestic violence; all intricately linked to the politics of sexuality, reproduction, and gender. Challenging neglected temporalities, therefore, opens up the possibility of exploring more hidden geographies of displacement which do not receive the same level of attention as other more visible and focused-upon forms of displacement.

Gender-based violence, in particular, is dramatically under-acknowledged in studies of conflict and displacement, in contrast to a robust literature on questions of global security and macro-level violence (Brickell 2015, 2020; Pain 2015). Indeed, the ways in which gender-based violence causes displacement at the scale of the home, city, nation, and globe is an emerging area of focus that requires greater attention. Domestic violence, in particular, is linked to one of the most overlooked forms of displacement, as it takes place during everyday life in ostensible times of peace. In the UK—a country which has reported to the United Nations that it only has people displaced by 'disaster,' that is, natural disasters (see IDMC 2018)—tens of thousands of forced migrations occur each year as a result of domestic violence (Bowstead 2015, 2017). Women are often forced to leave their homes suddenly, without their possessions, to an unknown and unfamiliar place (Graham and Brickell 2019; Warrington 2002). Those seeking shelter tend to undertake individual, isolated journeys, and move multiple times before they are able to find a safe, settled home (Bowstead 2015). For many women, the public sphere—typically framed as masculine and dangerous—becomes a refuge from the kinds of intimate violence that take place in private (Meth 2003). A recent study in the UK analysed casework data related to 264 women and found that 12 per cent of domestic violence survivors ended up sleeping rough to escape abuse, while 46 per cent 'sofa surfed' to survive while waiting for a refuge space (Women's Aid 2018). Such displacement also profoundly impacts LGBT youth, as hetero patriarchal violence in the home has resulted in what Ray (2007) calls 'an epidemic of youth homelessness' in the United States. Many women and queer youth displaced by domestic violence are at risk of being abused again, such that they experience a kind of invisible homelessness that involves cycling between unsafe domestic spaces (Speer 2018).

While displacement is fundamental to the experience of domestic violence, such patterns are also connected to geographies of containment and isolation (Bowstead 2017; Murray et al. 2019; Warrington 2002). Home can become a prison-like site of confinement, as abuse is often accompanied with enforced social isolation and regulation of movement. Upon escape, women often seek to move as far away from an abuser as possible, which further reinforces the experience of isolation and social disruption. In this way, many women are not only displaced from housing and from the space of the home but from their community at large. Such spatial isolation impacts immigrant women in particular, whose status may be dependent upon their husbands, or who may be denied access to their immigration paperwork. In some cases, domestic violence not only results in domestic displacement but in immigration detention or deportation (Cassidy 2018). For women living in rural communities—and Indigenous women in particular—social isolation and a lack of services have contributed to much higher rates of severe injury and death in cases of domestic of domestic violence in the United States (Pruitt 2008). This dynamic sheds light on the ways in which gender-based violence can enforce both mobility and fixity simultaneously, and how these paradoxical geographies are intricately tied to class, gender, ethnicity, and citizenship.

## Concluding Thoughts

Gendered and feminist approaches to the problem of displacement illuminate a range of concerns. Crucially, such research reveals that managing displacement is itself a unique kind of labour, one which is unpaid and profoundly gendered. Women bear the brunt of efforts to make do with precarious domesticities, prepare for future displacement, remake the space of the home, and cope with its loss. The work of managing displacement is not only a necessary part of social reproduction but also a mode of resistance that challenges the boundaries between spectacular and everyday politics. This, in turn, calls attention to struggles for emplacement, and the creation of a complex range of interrupted, precarious, and adaptive domesticities. Such insights highlight the importance of understanding how place is experienced and remade in the face of ongoing dislocation.

Another strand of feminist research reveals the slow, banal nature of women's familial and sexual displacement, integrated as it is into the fabric of everyday gender relations. Forced marriage, trafficking, domestic violence, and reproductive restrictions have all resulted in private and intimate kinds of displacement that are often overlooked in literature on the subject. Bringing such dynamics to light reveals a hidden temporality of displacement—one that is cyclical, ongoing, and linked to a paradoxical geography of containment and isolation. Overall, the approaches discussed in this chapter should not be framed as peripheral to the problem of displacement, or relevant only to questions of gender. Rather, feminist approaches speak to the central problem of how embodied struggles against displacement play out in everyday life.

## Note

1. https://displacement.iom.int.

## References

Ahmad, N., & Lahiri-Dutt, K. (2006). Engendering mining communities: Examining the missing gender concerns in coal mining displacement and rehabilitation in India. *Gender, Technology and Development, 10*(3), 313–339.

Baviskar, A. (2009). Breaking homes, making cities: Class and gender in the politics of urban displacement. In L. Mehta (Ed.), *Displaced by development: Confronting marginalization and gender injustice* (pp. 59–81). New Delhi, India: Sage.

Bowstead, J. C. (2015). Forced migration in the United Kingdom: Women's journeys to escape domestic violence. *Transactions of the Institute of British Geographers, 40*(3), 307–320.

Bowstead, J. C. (2017). Segmented journeys, fragmented lives: Women's forced migration to escape domestic violence. *Journal of Gender-Based Violence, 1*(1), 43–58.

Brickell, K. (2014). 'The whole world is watching': Intimate geopolitics of forced eviction and women's activism in Cambodia. *Annals of the Association of American Geographers, 104*(6), 1256–1272.

Brickell, K. (2015). Towards intimate geographies of peace? Local reconciliation of domestic violence in Cambodia. *Transactions of the Institute of British Geographers, 40*(3), 321–333.

Brickell, K. (2020). *Home SOS: Gender, violence and survival in crisis ordinary Cambodia*. Oxford, England: Wiley.

Brun, C. (2000). Making young displaced men visible. *Forced Migration Review, 10*, 8–12.

Brun, C., & Fábos, A. (2015). Making homes in limbo? A conceptual framework. *Refuge, 31*(1), 5–17.

Butler, J. (1990). *Gender trouble: Feminism and the subversion of identity*. London, England: Routledge.

Casolo, J., & Doshi, S. (2013). Domesticated dispossessions? Towards a transnational feminist geopolitics of development. *Geopolitics, 18*(4), 800–834.

Cassidy, K. (2018). Where can I get free? Everyday bordering, everyday incarceration. *Transactions of the Institute of British Geographers, 44*(1), 48–62.

Conlon, D. (2011). Waiting: Feminist perspectives on the spacings/timings of migrant (im)mobility. *Gender, Place & Culture, 18*(3), 353–360.

Culcasi, K. (2019). 'We are women and men now': Intimate spaces and coping labour for Syrian women refugees in Jordan. *Transactions of the Institute of British Geographers*. Advance online publication. https://doi.org/10.1111/tran.12292.

Davidson, M. (2009). Displacement, space and dwelling: Placing gentrification debate. *Ethics, Place and Environment, 12*(2), 219–234.

Desmond, M. (2016). *Eviction: Poverty and profit in the American city*. London, England: Allen Lane.

Doshi, S. (2017). Embodied urban political ecology: Five propositions. *Area, 49*(1), 125–128.

El-Bushra, J. (2000). Gender and forced migration: Editorial. *Forced Migration Review, 9*, 4–7.

Ensor, M. (2017). Lost boys, invisible girls: Children, gendered violence and wartime displacement in South Sudan. In S. Buckley-Zistel & U. Krause (Eds.), *Gender, violence, refugees* (pp. 197–218). New York, NY: Berghahn Books.

Fernández Arrigoitia, M. (2017). Unsettling resettlements: Community, belonging and livelihood in Rio de Janeiro's Minha Casa Minha Vida. In K. Brickell,

M. Fernández Arrigoitia, & A. Vasudevan (Eds.), *Geographies of forced eviction: Dispossession, violence, resistance* (pp. 71–96). Basingstoke, England: Palgrave.

Gorman-Murray, A., McKinnon, S., & Dominey-Howes, D. (2014). Queer domicide? LGBT displacement and home loss in natural disaster impact, response and recovery. *Home Cultures, 11*(2), 237–262.

Graham, N., & Brickell, K. (2019). Sheltering from domestic violence: Women's experiences of punitive safety and unfreedom in Cambodian safe shelters. *Gender, Place and Culture*. Advance online publication. https://doi.org/10.1080/0966369X.2018.1557603.

Hochschild, A. R. (1983). *The managed heart: Commercialization of human feeling*. Berkeley, CA: University of California Press.

Horn, D. M., & Parekh, S. (2018). Introduction to 'displacement'. *Signs: Journal of Women in Culture and Society, 43*(3), 503–514.

Hyndman, J. (2019). Unsettling feminist geopolitics: Forging feminist political geographies of violence and displacement. *Gender, Place and Culture, 26*(1), 3–29.

IDMC. (2018). 2018 internal displacement figures by country. Internal Displacement Monitoring Centre. Retrieved from http://www.internal-displacement.org/database/displacement-data.

Indra, D. (Ed.). (2001). *Engendering forced migration: Theory and practice*. New York, NY: Berghahn Books.

Juran, L. (2012). The gendered nature of disasters: Women survivors in post-tsunami Tamil Nadu. *Indian Journal of Gender Studies, 19*(1), 1–29.

Lewis, R. A., & Naples, N. A. (2014). Introduction: Queer migration, asylum, and displacement. *Sexualities, 17*(8), 911–918.

Lyons, T., Krusi, A., Pierre, L., Small, W., & Shannon, K. (2017). The impact of construction and gentrification on an outdoor trans sex work environment: Violence, displacement and policing. *Sexualities, 20*(8), 881–903.

Manchanda, R. (2004). Gender conflict and displacement: Contesting 'infantilisation' of forced migrant women. *Economic and Political Weekly, 39*(37), 4179–4186.

Meertens, D., & Segura-Escobar, N. (1996). Uprooted lives: Gender, violence and displacement in Columbia. *Singapore Journal of Tropical Geography, 17*(2), 165–178.

Mehta, L. (Ed.). (2009). *Displaced by development: Confronting marginalization and gender injustice*. Los Angeles, CA: Sage.

de Mel, N. (2017). A grammar of emergence: Culture and the state in the post-tsunami resettlement of Burgher women of Batticaloa, Sri Lanka. *Critical Asian Studies, 49*(1), 73–91.

Meth, P. (2003). Rethinking the 'domus' in domestic violence: Homelessness, space and domestic violence in South Africa. *Geoforum, 34*(3), 317–327.

Mirabal, N. R. (2009). Geographies of displacement: Latina/os, oral history, and the politics of gentrification in San Francisco's Mission District. *The Public Historian, 31*(2), 7–31.

Murray, L., Warr, D., Chen, J., Block, K., Murdolo, A., Quiazon, R., et al. (2019). Between 'here' and 'there': Family violence against immigrant and refugee women in urban and rural Southern Australia. *Gender, Place & Culture, 26*(1), 91–110.

Nolin, C. (2006). *Transnational ruptures: Gender and forced migration (gender in a global/local world)*. London, England: Routledge.

O'Neil, T., Fleury, A., & Foresti, M. (2016). Women on the move: Migration, gender equality and the 2030 Agenda for Sustainable Development. London, England: Overseas Development Institute. Retrieved from https://www.odi.org/sites/odi.org.uk/files/resource-documents/10731.pdf.

Pain, R. (2015). Intimate war. *Political Geography, 44*, 64–73.

Perry, K.-K. Y. (2013). *Black women against the land grab: The fight for racial justice in Brazil*. Minneapolis, MN: University of Minnesota Press.

Povinelli, E. A. (2011). *Economies of abandonment: Social belonging and endurance in late liberalism*. Durham, NC: Duke University Press.

Pruitt, L. R. (2008). Place matters: Domestic violence and rural difference. *Wisconsin Journal of Law, Gender, and Society, 23*(2), 347–416.

Ray, N. (2007). *Lesbian, gay, bisexual, and transgender youth: An epidemic of homelessness*. Washington, DC: National Gay and Lesbian Task Force Policy Institute and National Coalition for the Homeless.

Roy, A. (2017). Dis/possessive collectivism: Property and personhood at city's end. *Geoforum, 80*, A1–A11.

Sakizhoglu, B. (2018). Rethinking the gender-gentrification nexus. In L. Lees & M. Philips (Eds.), *Handbook of gentrification studies* (pp. 205–224). London, England: Routledge.

Schrijvers, J. (1999). Fighters, victims, and survivors: Constructions of ethnicity, gender, and refugeeness among Tamils in Sri Lanka. *Journal of Refugee Studies, 12*(3), 307–333.

Sharp, J. (2009). Geography and gender: What belongs to feminist geography? Emotion, power and change. *Progress in Human Geography, 33*(1), 74–80.

Speer, J. (2018). *Losing home: Housing, displacement, and the American Dream* (Doctoral dissertation). Retrieved from ProQuest Dissertations Publishing (10838329).

Suerbaum, M. (2018). Defining the other to masculinize oneself: Syrian men's negotiations of masculinity during displacement in Egypt. *Signs: Journal of Women in Culture and Society, 43*(3), 665–686.

Tanyag, M. (2018). Resilience, female altruism, and bodily autonomy: Disaster-induced displacement in post-Haiyan Philippines. *Signs: Journal of Women in Culture and Society, 43*(3), 563–585.

Tickner, J. A. (1992). *Gender in international relations: Feminist perspectives on achieving global security*. New York, NY: Columbia University Press.

UNHCR (United Nations High Commissioner for Refugees). (2016). *Global trends: Forced displacement in 2015*. Geneva: United Nations High Commissioner for Refugees.

United Nations General Assembly. (2016). *In safety and dignity: Addressing large movements of refugees and migrants*. Retrieved from https://refugeesmigrants.un.org/sites/default/files/in_safety_and_dignity_-_addressing_large_movements_of_refugees_and_migrants.pdf.

Vaz-Jones, L. (2018). Struggles over land, livelihood, and future possibilities: Reframing displacement through feminist political ecology. *Signs: Journal of Women in Culture and Society, 43*(3), 711–735.

Warrington, M. (2002). 'I must get out': The geographies of domestic violence. *Transactions of the Institute of British Geographers, 26*(3), 365–382.

Wilkinson, E., & Ortega-Alcázar, I. (2018). The right to be weary? Endurance and exhaustion in austere times. *Transactions of the Institute of British Geographers, 44*(1), 155–167.

Women's Aid. (2018) *Nowhere to Turn: Findings from the second year of the No Woman Turned Away Project*. Retrieved from https://1q7dqy2unor827bqjls0c4rn-wpengine.netdna-ssl.com/wp-content/uploads/2018/06/NWTA-2018-FINAL.pdf.

# 11

# 'Race,' Ethnicity, and Forced Displacement

Luisa F. Freier, Matthew D. Bird, and Soledad Castillo Jara

## Introduction

Biologically, 'race' does not exist, in the sense that it is not a scientifically reliable measure of human genetic variation (Hunley et al. 2009; Templeton 2013). Rather, it is an essentialising sociocultural construct, which social scientists interpret in different ways (Ifekwunigwe et al. 2017). Similarly, ethnicity is not a series of objective physical and cultural characteristics, but the result of complex processes of social classification (Barth 1970; Jenkins 1997; Brubaker et al. 2004). As such, some authors have grown critical of the usefulness of both concepts in understanding human displacement (Goldberg 2015), arguing that linguistic skills, residence status, and economic and geopolitical issues matter more than racism when explaining migratory experiences (Ahmed 2004; Amisi et al. 2011; Landau 2012; Cortina 2017).

Although most societies no longer defend systematic theories based on an allegedly objective hierarchy of 'races,' a sense of division remains between many ethnic groups, especially towards foreigners of different phenotypes. In other words, race and ethnicity as social categories are real (Smedley and Smedley 2005). Since societies think with and through them, such categories systematically structure practices that shape access to resources and opportunities (Kibria et al. 2013). Although the two constructs correlate with other characteristics shaping the reception of displaced people, they serve as markers for discrimination, which in turn objectively and subjectively affect

---

L. F. Freier (✉) • M. D. Bird • S. Castillo Jara
Universidad del Pacífico, Lima, Peru
e-mail: lf.freierd@up.edu.pe; bird_md@up.edu.pe; sr.castilloj@up.edu.pe

displaced people worldwide. The study of forced displacement thus remains incomplete without including these categories. Yet as sociocultural constructs with particular histories, the processes through which race and ethnicity determine and condition the reception of displaced people vary.

This chapter takes a critical look at the literature on the role of race and ethnicity as a conditioning factor of forced displacement, including in contemporary immigration and refugee policies, and argues that it suffers from a geographical bias, which impedes a broader historic, global, and thus theoretical understanding of the triangular relationship between racism, citizenship, and displacement. While the literature on the role of racism in the reception of displaced people focuses on receiving countries in North America, Europe, and Australia, there is less evidence on reception in the Global South. This bias is problematic since processes of racial discrimination and ethnic violence play out differently in postcolonial societies. We demonstrate this by sharing concrete examples from Latin America which highlight existing theories' limitations in handling empirical cases outside the Global North. Our goal in this chapter is to highlight avenues for further research, which could lead to a more integrative framework.

Throughout the chapter, when we refer to racism and xenophobia, we are concerned with discrimination based on race *and* ethnicity—which we understand as (self) identification based on shared, albeit constructed, physical traits, culture, religion, language, or alike that are perceived as distinctive from an out-group or the 'other.' We thus understand discrimination as conscious or unconscious, agentic or institutionalised acts taken to exclude displaced people who are defined by incompatible physical characteristics or inferior cultural elements, which threaten the host society or dominant social group's makeup or cohesion.

## Racism and the Integration of Displaced People

Racial discrimination significantly impacts the treatment of voluntary migrants and displaced people—as well as the opportunities that they have once arriving in destination countries. Scholarship on integration processes focuses mostly on receiving countries in the Global North, especially the United States (Ong et al. 1996; De Greiff and Gracia 2000; Quesada et al. 2011; Oh et al. 2019), Europe (Wrench et al. 1999; Silverstein 2005; Virdee and McGeever 2018), and Australia (Jakubowicz 1989; Jones 1998; Onsando and Billett 2017).

The unevenness of this evidence base impedes further theoretical development, especially with respect to the operations of racial discrimination. The literature on racism in the context of the reception of displaced people mostly deals with the issue of negative social, political, and legal discrimination. Yet not all discrimination is negative. Some groups are discriminated against, while others enjoy the privilege of positive racial prejudice.

Regarding the reasons for ethnic discrimination against forced migrants in the context of Western receiving countries, some authors conceptualise the negative discrimination of newly arrived migrants as a continuity of the racialisation of citizens from former colonial territories. McGhee (2009), for example, draws a parallel between the preference for EU citizens in the United Kingdom and the legal restrictions that were set against non-white citizens of Commonwealth countries in the postwar period. In the case of Germany, Schönwälder (2004) shows that foreign workers from Europe are also preferred to those from African or Asian countries. The same is seen in Italy, where Kovačič and Erjavec (2010) study the racialisation and exclusion of immigrants from African countries such as Nigeria, Somalia, and Eritrea.

Rather than seeing a link between present-day racism and the racialisation of groups that were subjugated in colonial times, other authors draw a more complex connection between racism and culture, religion, security, and national sovereignty. In the context of the 'war on terror,' some argue that xenophobia and racism are not the result of any 'culture clash' but are reflective of an appeal to the pact of solidarity and the political struggle about whom the state must care for (Wimmer 1997). Marz (2017, 54) argues that political actors increasingly justify divisions based on cultural or civilisational grounds, and that in Europe and the US, the concept of 'white race' has been replaced by 'Western culture.' In a context of the contemporary crises of the nation state (Roshwald 2015), xenophobic discourses are politically useful to reassure people of their imagined shared identity in cultural terms.

Yet there are fewer studies on how racism and xenophobia condition the integration of forced migrants in the Global South, with exceptions including South Africa (Beyers 2008; Vromans et al. 2011; Chigeza et al. 2013; Klotz 2016; Dube 2017) and East Asia (Lok-Sun Ngan and Chan 2013; Hoffstaedter 2014; Lee et al. 2017). The literature on Latin America has mostly focused on how displacement in the context of internal armed conflicts both resulted from and perpetuated racism and social exclusion, for example, in Colombia (Escobar 2017; Meertens 2008) and Peru (Barrantes 2012; Diez 2003). However, despite increasing extra-continental immigration from Asia, Africa, and the Caribbean (IOM 2013) and the displacement crises of Colombian (Carreño 2012) and Venezuelan citizens (Freier and Parent 2019), the literature on xenophobia in the region remains underdeveloped (e.g. Dutra 2018; IOP and IDEHPUCP 2019).

To exemplify what is at stake in relying on the evidence base from the Global North and related theory, we will deepen the analysis of Latin America where, in our experience, the existing literature is ill-equipped to understand how issues of race and ethnicity shape the integration of displaced people, such as Venezuelan forced migrants in Peru. In contrast to categorical concepts of race (Elias and Feagin 2016), race and ethnicity in much of Latin America are *relative* and interlinked with gender, education, and class, assuming the form of a mosaic. When examined from afar, the socio-racial differences operate more as a relational continuum and blend into one another rather than presenting sharp categories, a point also made by Bourdieu and Wacquant (1999) in their critique of the implicit importation of 'imperial' theories of bimodal race to the Brazilian case.

The relational operation of race has roots in the colonial era and is what prompted attempts to legislate caste differences (Mörner 1967). In the Republican era, the concept of race became entangled with nation-building projects (Appelbaum et al. 2003). While some countries explicitly sought to encourage international migration in the late nineteenth and early twentieth centuries to 'whiten' the population, a counter discourse of *mestizaje* emerged and spread through most Latin American countries in the early twentieth century, seeking to generate a unitary sense of nationality and belonging (de la Cadena 2000).

*Mestizaje* refused to embrace the Indigenous or other marginalised groups and ignored the social disparities rooted in racism. Although the discourse and construction of *el mestizo* sought the negation of race and racism in many Latin American societies, discrimination continues in different forms across the region (Wade 1997; Beck et al. 2011). In some countries, such as Ecuador and Brazil, racism against people of African descent is prominent and considered a contentious social and policy issue (Beck et al. 2011; Telles 2014), whereas Afro populations in other countries, such as Argentina, have historically been treated as invisible (Cottrol 2007; Frigerio 2008). These differences impact the reception and socioeconomic integration of the growing numbers of African immigrants in different ways, with more violent racially motivated xenophobia in countries where racism against people of African descent is prominent (Freier and Zubrzycki 2019).

The structuring role of race and ethnicity in many Latin American countries is extremely complex. To give one detailed example, in Peru racial concepts were shaped by the relationships between Indigenous Andean and Amazonian populations, the Spanish conquest, the forced arrival of African slaves, and European and Asian immigration (e.g. Poole 1997; de la Cadena 2000; Gotkowitz 2007). These historic 'ingredients' make it difficult to define race categorically, except at the phenotypic Indigenous, African, and white extremes. Instead of a colour wheel, one can imagine a colour triangle, with different combinations as one moves towards one pole or the other. These combinations represent variations of class and *mestizaje*.

The massive rural-to-urban migration in Peru in the second half of the twentieth century gave way to ever more complex forms of discrimination (Bird 2010; Santos 2014). In practice, one is not white, black, Indigenous, or even mestizo—and money is not the main marker of class; it is contextual (Nugent 1992; Sulmont and Callirgos 2014). Here, physical, cultural, linguistic, occupational, and geographic characteristics linked to being 'Indigenous' remain conceptually linked to social stagnation and underdevelopment, whereas being 'white' pertains to the country's 'upper class' (Telles 2014). However, one can be more or less white, black, Indigenous, and even Asian depending on whether you have more or less formal education, a lower or higher status occupation, the relative status of the region or district of residence, your way of dressing, and certain manners.

Since the majority of the population does not occupy one of the three extreme poles, citizens learn to play games of social distinction that are socio-racially marked. For example, someone who is socio-racially higher in the hierarchy may have less income than the person below them, but can still discriminate, not allowing them to rent homes or access clubs in certain districts. You can be more or less white or more or less Indigenous depending upon the context, and with whom you are compared. In the process of socialisation, Peruvians are thus trained to recognise and simultaneously hide racial differences (Portocarrero 1993, p. 31).

Venezuela, to give a different example, is one of Latin America's allegedly 'colourblind' societies, which has led to a national belief that neither racism nor 'race' exist in the country (Herrera Salas 2005; Ishibashi 2003; Nichols 2013). At the same time, Venezuela's racial heritage has fostered cultural concern with physical beauty that privileges 'white' or 'European' physical characteristics, while the perceived problematic characteristics are linked to marginalised groups (Nichols 2013). In Venezuela, the conversation about race and discrimination was thus replaced by a discussion about beauty, physical attributes (hair, lips, the shape of the nose), and 'decency' (Ishibashi 2003).

These two contexts are critical for understanding how Venezuelan forced migrants position themselves racially in Latin American host societies, such as Peru, and how race, ethnicity, and both positive and negative discrimination affect their integration experience. Based on their physical and in relative terms more European traits, Venezuelans were initially positively received in Peruvian society, before these same phenotypes became linked to criminalisation and hypersexualisation processes (Freier and Pérez 2020). At the same time, many Venezuelans see themselves as racially superior to Peruvians, an understanding grounded not only on physical traits but also on relative educational levels and former class position in Venezuela (Fernández 2019).

When studying the relationship of race and displacement in postcolonial contexts, some groups of migrants are victims of negative racial prejudice, while others enjoy privilege based on their physical traits. Historically, in southern world regions, such privilege has been linked to being 'white.' Studies on this issue in Latin America focus on the historical analysis of how political elites decided not only to allow, but to actively encourage, European immigration in order to populate rural areas with people of an allegedly superior race and civilisation towards the end of the nineteenth century (Bastia and Vom Hau 2014; FitzGerald 2013; Gomes 2007; Gregory 2007; Coletta 2011). However, we are not aware of any studies of the privilege that comes with being perceived as white migrants in present-day Latin America.

## Racism in Contemporary Migration Laws and Policies

Although historically many migration laws and policies contained explicit discrimination or quotas limiting the entry of groups of different nationalities, race, and ethnicity, contemporary refugee policies and laws usually postulate non-discrimination, including on grounds of 'race and colour.'[1] In this context, especially noteworthy is the focus on non-discrimination—on all grounds—of a new generation of refugee laws in Latin America, which in some cases are more progressive than European protection standards regarding explicit non-discrimination on the bases of gender and sexual orientation and the extension of social and economic rights to refugees (Freier and Gauci 2020).

Critically, forcibly displaced people do not always neatly fit the category of refugee of the 1951 Convention, or the expanded regional refugee definitions of Latin America and Africa, and often move in mixed migration flows. Mixed migration refers to the cross-border movements of both forcibly displaced and economic migrants who are motivated by a multiplicity of factors and have different legal statuses and vulnerabilities but travel along similar routes, often assisted by migrant smugglers. Immigration laws and policies, for and foremost those regulating access to receiving countries, thus affect victims of forced displacement and are used to limit the entry of asylum seekers (FitzGerald 2019; Freier and Luzes 2020).

Similar to the literature on racism as a determinant of the migratory experience and integration of displaced people, the scholarship on racism in contemporary migration laws and policies focuses on Western receiving countries. The racialisation of citizenship, which today finds its expression in admission and naturalisation laws, has deep roots in European history. Examples include the proto-racial notion of citizenship in ancient Greece and the presence of anti-Jewish and anti-Moorish norms in the Iberian territories in the sixteenth century (FitzGerald 2017).

Scholars have also studied how discrimination and racist immigration policies in the Americas and Europe in more recent history targeted people of African and Asian descent (Geddes 1995; Schönwälder 2004; Gilbert 2008; Johnson 2009; FitzGerald and Cook-Martín 2014) and found that Western immigration policies today especially discriminate against Arabs and Muslims (Sivanandan 2007; Lentin 2016; Hammerstad 2014). Contrary to a broad consensus that describes a sustainable shift from prevalent negative ethnic selectivity towards widespread ethnic and racial neutrality since 1945 in *de jure* immigration policies (Freeman 1994; Hansen 2002; Joppke 2005; Cook-Martín and FitzGerald 2010; FitzGerald and Cook-Martín 2014), some authors have recently suggested that the ongoing racialised securitisation of migration finds expression in the tourist visa regimes of Europe.

Ethnic selectivity in this context is expressed through discrimination based on nationality or country of birth. For example, Hobolth (2012) shows that travellers from poor, Muslim, and asylum-producing countries tend to be refused tourist visas to European territory. FitzGerald et al. (2018) compare immigrant admission requirements across the Americas and Europe to identify and explain the historical evolution of assimilability requirements. They find that, even though explicit ethnic restrictions are now delegitimised, admission tests applied to immigrants still exhibit preference for some ethnic groups and the exclusion of others based on the argument that some are more likely to integrate to the majoritarian culture. FitzGerald (2019) further points out cases where preference has been given to white asylum seekers, such as Eastern Europeans during the Cold War.

Many countries in Latin America mirror American and European policies in terms of the ethnic groups negatively targeted (FitzGerald and Cook-Martín 2014; Freier and Holloway 2018). At the same time, a growing literature highlights the unique characteristics of Latin American immigration and refugee policies, where there is a reverse gap between liberal political discourses that speak of non-discrimination and even the 'right to migrate', but de facto discrimination against African, Asian, and Caribbean migrants of colour (Acosta and Freier 2015; Freier 2016; Cantor et al. 2015; Freier and Holloway 2018). As alluded to above, racial selectivity in Latin America is based on criteria that have its roots in the colonial history of the region, but also goes back to policy diffusion, especially from the United States.

FitzGerald and Cook-Martín (2014) show how Latin American countries historically mimicked the immigration policies of the United States, which led the way in using legal means to exclude 'inferior' ethnic groups in the Americas. Freier (2016) argues that racism and perceived security threats of domestic and international political actors, including first and foremost the United States, constrained immigration policy liberalisation in the case of Ecuador's 2008 policy of open borders. Further studies should examine how the circumstances and histories in other regions, including Africa, Asia, and the Middle East, impact the forms migration laws and policies take *de jure* and de facto.

## Conclusion

In this chapter we critically reviewed the literature on racism—broadly defined—as conditions for the integration of displaced people, and factors in contemporary refugee laws and policies. This literature shows a clear geographical bias towards Western receiving states. Scholars working in these fields should broaden their geographical scope to include case studies in the Global South. We introduced several concrete examples from Latin America to demonstrate the promise a new evidence base could bring to our understanding. In general, the literature on race, ethnicity, and displacement would significantly benefit from cross-regional and cross-temporal studies for the purpose of theory building. At the same time, there is much room for further research on the process of international diffusion of public attitudes and policies as they concern issues of race and forced displacement.

## Note

1. Art. 3 Geneva Refugee Convention.

# References

Acosta, D., & Freier, L. F. (2015). Turning the immigration policy paradox upside down? Populist liberalism and discursive gaps in South America. *International Migration Review, 49*, 659–696.

Ahmed, S. (2004). *The cultural politics of emotion*. Edinburgh: Edinburgh University Press.

Amisi, B., Bond, P., Cele, N., & Ngwane, T. (2011). Xenophobia and civil society: Durban's structured social divisions. *Politikon: South African Journal of Political Studies, 38*(1), 59–83.

Appelbaum, N., Macpherson, A., & Rosemblatt, K. A. (Eds.). (2003). *Race and nation in modern Latin America*. Chapel Hill, NC: University of North Carolina Press.

Barrantes, R. (2012). *Reparations and displacement in Peru*. New York: ICTJ/Brookings.

Barth, F. (1970). *Ethnic groups and boundaries: The social organization of culture difference*. Oslo: Universitats Forlaset.

Bastia, T., & Vom Hau, M. (2014). Migration, race and nationhood in Argentina. *Journal of Ethnic & Migration Studies, 40*(3), 475–492.

Beck, S., Mijeski, K. J., & Stark, M. M. (2011). Qué es Racismo? Awareness of racism and discrimination in Ecuador. *Latin American Research Review, 46*(1), 102–125.

Beyers, C. (2008). Memory, displacement and post-Apartheid restitution. *Journal of Southern African Studies, 34*(3), 731.

Bird, M. (2010). *The other paths: Cultural economics, comparative economic performance, and the formation of capitalist mentalities in Lima, Peru* (Doctoral dissertation). University of Chicago, Chicago, IL.

Bourdieu, P., & Wacquant, L. (1999). On the cunning of imperialist reason. *Theory, Culture & Society, 16*(1), 41–58.

Brubaker, R., Loveman, M., & Stamatov, P. (2004). Ethnicity as cognition. *Theory and Society, 33*, 31–64.

Cantor, D. J., Freier, L. F., & Gauci, J. P. (Eds.). (2015). *A liberal tide? Immigration and asylum law and policy in Latin America*. London: University of London Institute of Latin American Studies.

Carreño, A. (2012). Situación de refugiados colombianos en Ecuador. *Justicia, 17*(22), 20–46.

Chigeza, S., De Wet, A., Roos, V., & Vorster, C. (2013). African migrants' experiences of xenophobic violence in South Africa: A relational approach. *Journal of Psychology in Africa, 23*(3), 501–505.

Coletta, M. (2011). The role of degeneration theory in Spanish American public discourse at the fin de siècle: Raza latina and immigration in Chile and Argentina. *Bulletin of Latin American Research, 30*, 87–103.

Cook-Martín, D., & FitzGerald, D. (2010). Liberalism and the limits of inclusion: Racialized preferences in immigration laws of the Americas, 1850–2000. *Journal of Interdisciplinary History, 16*, 7–25.

Cortina, A. (2017). *Aporofobia, el rechazo al pobre: Un desafío para la sociedad democrática*. Barcelona: Paidós.

Cottrol, R. J. (2007). Beyond invisibility: Afro-Argentines in their nation's culture and memory. *Latin American Research Review, 42*(1), 139–156.

De Greiff, P., & Gracia, J. J. E. (2000). *Hispanics/Latinos in the United States: Ethnicity, race, and rights*. New York and London: Routledge.

de la Cadena, M. (2000). *Indigenous mestizos: The politics of race and culture in Cuzco, Peru (1919–1991)*. Durham, NC: Duke University Press.

Diez, A. (2003). *Los desplazados en el Perú*. Lima: Comité Internacional de la Cruz Roja.

Dube, G. (2017). Levels of othering—The case of Zimbabwean migrants in South Africa. *Nationalism and Ethnic Politics, 23*(4), 391–412.

Dutra, D. (2018). Experiências de racismo desde a imigração haitiana e africana no Brasil. *Revista Interdisciplinar da Mobilidade Human, 26*(53), 99–113.

Elias, S., & Feagin, J. (2016). *Racial theories in social science: A systemic racism critique*. New York: Routledge.

Escobar, G. (2017). Colombia's displaced indigenous women. *Forced Migration Review, 56*, 37–38.

Fernández Rodriguez, N. (2019). La migracion venezolana en Perú (2018–2019): Derechos Humanos, percepciones mutuas y xenofobia. *Master's Thesis*. Estudios Avanzados de Derechos Humanos, Universidad Carlos III de Madrid.

FitzGerald, D. (2013). Ethnic selection in immigration to Latin America. In I. Ness (Ed.), *The Encyclopedia of Global Human Migration*. https://doi.org/10.1002/9781444351071.

FitzGerald, D. (2017). The history of racialized citizenship. In A. Shachar, R. Bauböck, I. Bloemraad, & M. Vink (Eds.), *The Oxford handbook of citizenship* (pp. 129–148). Oxford, England: Oxford University Press.

FitzGerald, D. (2019). *Refuge beyond Reach: How Rich Democracies Repel Asylum Seekers*. Oxford University Press.

FitzGerald, D., & Cook-Martín, D. (2014). *Culling the masses: The democratic origins of racist immigration policy in the Americas*. Cambridge, MA: Harvard University Press.

FitzGerald, D. S., Cook-Martín, D., García, A. S., & Arar, R. (2018). Can you become one of us? A historical comparison of legal selection of 'assimilable' immigrants in Europe and the Americas. *Journal of Ethnic & Migration Studies, 44*(1), 27–47.

Freeman, G. P. (1994). Can liberal states control unwanted immigration? *Annals of the American Academy of Political and Social Science, 534*, 17–30.

Freier, L. F. (2016). A reverse migration paradox? Policy liberalisation and new South-South migration to Latin America (Doctoral dissertation, LSE). Retrieved from http://etheses.lse.ac.uk/3455/.

Freier, L. F., & Gauci, J. P. (2020). Latin America: The new avant-garde of international refugee protection? *Refugee Studies Quarterly*. Rights across regions: a comparative overview of legislative good practices in latin america and the eu.

Freier, L. F., & Luzes, M. (2020). How humanitarian are humanitarian visas? An analysis of theory and practice in Latin America. In L. Jubilut, G. Mezzanotti, and M. Vera Espinoza (Eds.), *Latin America and Refugee Protection: regimes, logics and challenges*. Oxford, New York: Berghahn.

Freier, L. F., & Pérez, L. (2020). "Watch out, they'll quarter you!": Experiences of Nationality-based Criminalisation of Venezuelan Immigrants in Peru. Working Paper.

Freier, L. F., & Holloway, K. (2018). The impact of tourist visas on intercontinental South-South migration: Ecuador's policy of 'open doors' as a quasi-experiment. *International Migration Review*, 53(4), 1171–1208. https://doi.org/10.1177/0197918318801068.

Freier, L. F., & Parent, N. (2019). The regional response to the Venezuelan exodus. *Current History*, 118(805), 56–61.

Freier, L. F., & Zubrzycki, B. (2019). How do immigrant legalization programs play out in informal labor markets? The case of Senegalese street hawkers in Argentina. *Migration Studies*. https://doi.org/10.1093/migration/mnz044.

Frigerio, A. (2008). *De la desaparición de los negros a la reaparición de los afrodescendientes: comprendiendo las políticas de las identidades negras, las clasificaciones raciales y de su estudio en Argentina*. Cordoba and Buenos Aires: CLACSO.

Geddes, A. (1995). *The politics of immigration and race*. Manchester: Basiline Books.

Gilbert, L. (2008). National identity and immigration policy in the U.S. and the European Union. *Columbia Journal of European Law*, 14, 99–145.

Goldberg, D. T. (2015). *Are we all postracial yet?* Cambridge: Polity.

Gomes, A. (2007). Imigrantes italianos: Entre a italianitá e a brasilidade. In Instituto Brasileiro de Geografia e Estatística, *Brasil: 500 anos de povoamento*. Rio de Janeiro: IBGE, Centro de Documentação e Disseminação de Informações.

Gotkowitz, L. (2007). A revolution for our rights. In *Indigenous struggles for land and justice in Bolivia, 1880–1952*. Durham and London: Duke University Press.

Gregory, V. (2007). Imigração alemã: formação de uma comunidade teuto-brasileira. In Instituto Brasileiro de Geografia e Estatística, *Brasil: 500 anos de povoamento*. Rio de Janeiro: IBGE, Centro de Documentação e Disseminação de Informações.

Hammerstad, A. (2014). *The rise and decline of a global security actor: UNHCR, refugee protection and security*. Oxford, England: Oxford University Press.

Hansen, R. (2002). Globalization, embedded realism, and path dependence: The other immigrants to Europe. *Comparative Political Studies*, 35, 259–283.

Herrera Salas, J. M. (2005). Ethnicity and revolution: The political economy of racism in Venezuela. *Latin American Perspectives, 32*, (2), Venezuelan exceptionalism revisited: The unraveling of Venezuela's model, 72–91.

Hunley, K., Healy, M., & Long, J. (2009). The global pattern of gene identity variation reveals a history of long-range migrations, bottlenecks, and local mate exchange: Implications for biological race. *American Journal of Physical Anthropology, 139*(1), 35–46.

Hobolth, M. (2012). *Wanted and unwanted travellers: Explaining similarities and differences in European tourist visa practice* (London Migration Research Group (LMRG) Seminar Series). London.

Hoffstaedter, G. (2014). Place-making: Chin refugees, citizenship and the state in Malaysia. *Citizenship Studies, 18*(8), 871–884.

IOM. (2013). *Migrantes extra-continentales en América del Sur: Estudio de caso* (OIM Cuadernos Migratorios N°5). Retrieved from https://publications.iom.int/books/cuadernos-migratorios-ndeg5-migrantes-extracontinentales-en-america-del-sur-estudio-de-casos.

IOP & IDEHPUCP. (2019). *Creencias y actitudes hacia los inmigrantes venezolanos en el Perú* (Boletín N° 157). Estado de la Opinión Pública. Retrieved from http://repositorio.pucp.edu.pe/index/handle/123456789/134548.

Ifekwunigwe, J. O., Wagner, J. K., Yu, J.-H., Harrell, T. M., Bamshad, M. J., & Royal, C. D. (2017). A qualitative analysis of how anthropologists interpret the race construct. *American Anthropolist, 119*(3), 422–434.

Ishibashi, J. (2003). Hacia una apertura del debate sobre el racismo en Venezuela: Exclusión einclusión estereotipada de personas 'negras' en los medios de comunicación. In D. Mato (Ed.), *Políticas de identidades y diferencias sociales en tiempos de globalización* (pp. 33–61). Caracas: FACES-UCV.

Jakubowicz, A. (1989). The state and the welfare of immigrants in Australia. *Ethnic and Racial Studies, 12*(1), 1–35.

Jenkins, R. (1997). *Rethinking ethnicity: Arguments and explorations*. Thousand Oaks, CA: Sage Publications.

Johnson, K. (2009). The intersection of race and class in U.S. immigration law and enforcement. *Law and Contemporary Problems, 72*(4), 1–35.

Jones, F. L. (1998). Recent trends in labour market disadvantage among immigrants in Australia. *Journal of Ethnic and Migration Studies, 24*(1), 73–95.

Joppke, C. (2005). *Selecting by origin: Ethnic migration in the liberal state*. Cambridge and London: Harvard University Press.

Kibria, N., Bowman, C., & O'Leary, M. (2013). *Race and immigration*. Cambridge: Polity.

Klotz, A. (2016). Borders and the roots of xenophobia in South Africa. *South African Historical Journal, 68*(2), 180–194.

Kovačič, M., & Erjavec, K. (2010). The European Union—A new homeland for illegal immigrants? A study of imaginaries of the European Union. *Dve Domovini, 31*, 169–183.

Landau, L. (Ed.). (2012). *Exorcising the demons within: Xenophobia, violence and statecraft in contemporary South Africa*. Tokyo, New York and Paris: United Nations University Press.

Lee, J., Jon, J.-E., & Byun, K. (2017). Neo-racism and neo-nationalism within East Asia. *Journal of Studies in International Education, 21*(2), 136–155.

Lentin, A. (2016). Racism in public or public racism: Doing anti-racism in 'post racial' times. *Ethnic and Racial Studies, 39*(1), 33–48.

Lok-Sun Ngan, L., & Chan, K. W. (2013). An outsider is always an outsider: Migration, social policy and social exclusion in East Asia. *Journal of Comparative Asian Development, 12*(2), 316–350.

Marz, U. (2017). Annäherungen an eine Kritische Theorie des Rassismus. *Peripherie, 37*(146/147), 250–270.

McGhee, D. (2009). The paths to citizenship: A critical examination of immigration policy in Britain since 2001. *Patterns of Prejudice, 43*(1), 41–64.

Meertens, D. (2008). Discriminación racial, desplazamiento y género en las sentencias de la Corte Constitucional. El racismo cotidiano en el banquillo. *Universitas Humanística, 66*(66), 83–106.

Mörner, M. (1967). *Race mixture in the history of Latin America*. Boston: Little, Brown & Company.

Nichols, E. (2013). 'Decent girls with good hair': Beauty, morality and race in Venezuela. *Feminist Theory, 14*(2), 171–185.

Nugent, G. (1992). *El laberinto de la choledad*. Lima: Universidad Peruana de Ciencias Aplicadas.

Oh, H., Stickley, A., Koyanagi, A., Yau, R., & DeVylder, J. E. (2019). Discrimination and suicidality among racial and ethnic minorities in the United States. *Journal of Affective Disorders, 245*, 517–523.

Ong, A., Dominguez, V. R., Friedman, J., Schiller, N. G., Stolcke, V., & Ying, H. (1996). Cultural citizenship as subject-making: Immigrants negotiate racial and cultural boundaries in the United States. *Current Anthropology, 37*(5), 737–762.

Onsando, G., & Billett, S. (2017). Refugee immigrants' experiences of racism and racial discrimination at Australian TAFE institutes: A transformative psychosocial approach. *Journal of Vocational Education & Training, 69*(3), 333–350.

Poole, D. (1997). *Vision, race, and modernity: A visual economy of the Andean image world*. Princeton: Princeton University Press.

Portocarrero, G. (1993). La cuestión racial: Espejismo y realidad. In G. Portocarrero (Ed.), *Racismo y mestizaje* (pp. 181–223). Lima: Sur Casa de Estudios del Socialismo.

Quesada, J., Hart, L. K., & Bourgois, P. (2011). Structural vulnerability and health: Latino migrant laborers in the United States. *Medical Anthropology, 30*(4), 339–362.

Roshwald, A. (2015). The global crisis of the nation state. *Current History, 114*(768), 3–8.

Santos, M. (2014). La discriminación racial, étnica y social en el Perú: Balance crítico de la evidencia empírica reciente. *Debates En Sociología, 39*, 5–37.

Schönwälder, K. (2004). Why Germany's guestworkers were largely Europeans: The selective principles of post-war labourrecruitment policy. *Ethnic & Racial Studies, 27*(2), 248–265.

Silverstein, P. (2005). Immigrant racialization and the new savage slot: Race, migration, and immigration in the new Europe. *Annual Review of Anthropology, 34*(1), 363–384.

Sivanandan, A. (2007). Racism, liberty and the War on Terror. *Race & Class, 48*, 45–96.

Smedley, A., & Smedley, B. D. (2005). Race as biology is fiction, racism as a social problem is real: Anthropological and historical perspectives on the social construction of race. *American Psychologist, 60*(1), 16–26.

Sulmont, D., & Callirgos, J. (2014). ¿El país de todas las sangres? Race and ethnicity in contemporary Peru. In E. Telles (Ed.), *Pigmentocracies. Ethnicity, race and color in Latin America* (pp. 126–171). Chapel Hill, NC: University of North Carolina Press.

Templeton, A. R. (2013). Biological races in humans. *Studies in History and Philosophy of Biological and Biomedical Sciences, 44*(3), 262–271.

Telles, E. (2014). *Pigmentocracies: Ethnicity, race and color in Latin America*. Chapel Hill, NC: University of North Carolina Press.

Virdee, S., & McGeever, B. (2018). Racism, crisis, Brexit. *Ethnic and Racial Studies, 41*(10), 1802–1819.

Vromans, L., Schweitzer, R. D., Knoetze, K., & Kagee, A. (2011). The experience of xenophobia in South Africa. *American Journal of Orthopsychiatry, 81*(1), 90–93.

Wade, P. (1997). *Race and ethnicity in Latin America*. London: Pluto Press.

Wimmer, A. (1997). Explaining xenophobia and racism: A critical review of current research approaches. *Ethnic and Racial Studies, 20*(1), 17–41.

Wrench, J., Rea, A., & Ouali, N. (Eds.). (1999). *Migrants, ethnic minorities and the labour market. Integration and exclusion in Europe*. London: Macmillan Press.

# 12

## Conceptualising Postcolonial Displacement Beyond Aid and Protection

Jose Jowel Canuday

In laying down the concept of forced migration nearly two decades ago, David Turton underscored that the 'the metaphorical language we use to talk about migration carries with it certain implications for the way we think about, and therefore act towards, migrant' (Turton 2003a, p. 2). Forced migration, as articulated by Turton in reference to the displacement experience, emanates from a point of view and geopolitical position of the non-migrant towards a migrant. This position indicates that thinking about and taking action over human suffering behind the displacement experience are not only value-free exercises of knowing. Conceptualising displacement is a project of prescribing an understanding from a relative position of power and in geopolitical location.

The number of the world's displaced peoples stands at sixty-five million, situated mostly in the strife-torn and disaster-prone regions covering the Horn of Africa, the Middle East, South Asia, and South East Asia, according to recent estimates. These regions cover social worlds that were viewed as objects of North American and European civilising missions during the colonial centuries, and subsequently as recipients of military, development, and civilian aid in the bifurcated politics of the Cold War years (Weiner 1995; Cohen and Layton-Henry 1997; Taithe and Borton 2016).

---

J. J. Canuday (✉)
Ateneo de Manila University, Quezon City, Philippines
e-mail: jcanuday@ateneo.edu

In the past few decades, several governments across the region have reeled or caved under pressure from geopolitical realignments, rebel uprisings, civil wars, the unravelling of colonially suppressed and precipitated ethnic divides, and poorer state responses to devastating hazard-induced disaster events (Nair 1997; Jacobsen 2010; Redclift 2016; Siddiqi 2017). This explosive mix of events in turn induced massive episodes of population displacements that are partly but also unavoidably entangled with the politics of postcolonial transitions. Such entanglements suggest that colonisation and decolonisation experiences are equally crucial in conceptualising the displacement experience rather than merely being set as backdrop to it.

Contemporary institutional understanding, however, regarded the displaced primarily as beneficiaries of international humanitarian and human rights intervention from inter-governmental and non-governmental organisations (NGOs) headquartered mainly in North America and Europe. These assessments in turn catalysed global actions and thinking that relegated the displaced as suffering populations needing immediate protection and aid. Consequently, the displaced are referred in legal, analytical, and prescriptive categories as 'persons of concerns' (Kendle 1998; Scheel and Ratfisch 2014; Stevens 2016; Prosperi 2017). The United Nations High Commissioner on Refugees breaks down this category rather sharply into internally displaced persons (IDPs), refugees and resettled refugees, asylum seekers, and stateless persons (Turton 2003b; Zetter 2007; Cameron 2014). These categories relegate refugees and other displaced peoples as subjects of international humanitarian and human rights assistance, inadvertently subverting the colonial links and the geopolitically imbued play for power that spur displacement events in the decolonisation process.

The displacement experience constitutes a discursive field encompassing legal instrumentalities, articulations of history, ideologies, and cultural constructs that are simultaneously incoherent and distinguishable. Tracking these constructs demands an ontological view, one that pays close attention to the reconfigurations of the past, continuities, and trajectories of the colonial and postcolonial encounters (cf. Country et al. 2018). The aim is not necessarily to strike coherence in articulating a notion of displacement but, rather, understand the outlines of the complex relations underpinning the displacement experience.

## Discursive Fields of Displacement

The English term refugee, a term etymologically traceable to the seventeenth-century French word *réfugié*, serves as descriptor of persons displaced by disruptive circumstances to seek refuge. Varying social worlds beyond Europe, however, articulate the displacement experience in differential terms. The word's rough translation in Afghani is *mujahin*, which means 'those who leave their homes in the cause of Allah, after suffering oppression' (Daniel 2002, p. 274). Among the Tamil, the notion of refugee is encapsulated in two terms—the *saranadaivor*, which comes from the word *saranam* that means refuge, shelter, and a place for 'one's true or ultimate home.' The second term is encapsulated in the word *ahati*, connoting 'victimhood' (Daniel 2002). The notion of 'refugee' and non-European lexicons of displacements are embedded in an encompassing discourse of attendant terms anchored in specific cultural and geopolitical relations; however, this has been substantially recast into an indexical category of people encumbered by the postcolonial geopolitics of nation-state borders.

From its old everyday meaning of persons fleeing to seek refuge, the 1951 Convention Relating to the Status of Refugees transformed the word refugee into a value-laden term that exclusively applies to 'any person [that] owing to well-founded fear of being persecuted for reasons of race, religion, nationality membership of a particular social group or political opinion, is outside the country of his nationality.' In this articulation, the convention invoked a 'strategic vision which dictated that refugees possessed ideological and political value' (Chimni 1998, p. 355) linked to the politics of the Cold War divide, a 'pivotal issue' in the arena of national and international politics (Castles 2003). The 1951 Convention and subsequent amendments framed the notion of the 'refugee' in response to the political conditions arising from the Cold War in the 1950s, highlighting the ideological nature of the instrument (Weiner 1995; Cohen and Layton-Henry 1997). Some governments moved to control the international flow of populations by restricting emigration (Weiner 1995, p. 34). Consequently, international humanitarian norms of intervention materialised and were shaped 'in the vacuum led by the Cold War' (Taithe and Borton 2016). Following the collapse of the European communist regimes, the strategic political value of the term refugee shifted and waned. In other words, several states across the Global North mobilised the protocols set under the Refugee Conventions as an instrument of managing population mobilities coming into and moving out of their territories.

The contested frames of displacement also play out in the conflicted zone of the Global South, including my home region in the Southern Philippines, Mindanao, where I also do fieldwork (Canuday 2009). In this area, the idea of the person displaced by repeated violence is generally encapsulated by the term *bakwit*. Etymologically, *bakwit* was a contraction of the English term *evacuate*, which gained currency at the time of the American colonial military occupation of the Philippines from the turn of the nineteenth century through the end of the Second World War in 1946. Nonetheless, the term *bakwit* was further used in referring to people displaced by disasters induced by hazards. *Bakwit* in a sense was a legacy arising from the horrific experiences of colonial wars but later deployed to conjure both the people and the experience of being displaced under circumstances arising from the Philippine decolonisation experience. Whenever I interchange the notion of the *bakwit* with refugees, friends in the local humanitarian community argue that the locally displaced did not move national borders and therefore do not qualify as refugees. Local aid workers align the local displacement experience to normative differentiation of the world's displaced peoples, reappropriating a colonial discourse into a contemporaneous postcolonial experience with global humanitarian and human rights implications.

The 1997 United Nations Guiding Principles on Internal Displacement refers to persons displaced from home communities by conflict and hazard but unable to cross international borders. Following the tenets of the guiding principles, humanitarian movements working in the Philippines dissociate the *bakwit* experience from an Anglo–American colonial lexicon of violence. In the contemporary humanitarian discursive field, the *bakwit* are IDPs covered by the mandate of the UN guiding principle on internal displacement, consequently extending the historical but localised idea of displacement beyond the local sociocultural, political, and historical contexts of conflicts, tensions, and displacement. The guiding principle sums up an idea of the displaced within a universalising worldview of protection framed by governments, international NGOs, and scholars. Refugees, IDPs, Mindanao's *bakwit*, Afghani's *mujahin*, or the Tamil's *saranam*, noted earlier, constitute a sprawling field of articulations of displacement that are of strategic value to legal-institutional and political regimes, societies, and communities. Humanitarian and human rights principles primarily framed the responses to the horrific conditions unravelling behind these displacement events. These frames paradoxically circumscribe the experiences of suffering and brutality into identifiable markers over which the donor community decides and draws the terms of their intervention as they continue to dominate the global humanitarian frontline (De Cordier 2009). These frames resonate with, overlap, and extend the Western decolonising mission of raising the 'capacity' of what are essentially former colonies of old colonial empires in navigating the geopolitics of post-Second World War postcolonial order (cf. Kingsbury 2011). Under such circumstances, 'new arrangements of knowledge and power, new practices, theories, and strategies' are constructed in ways that inadvertently render struggling communities, such as the displaced, as a 'space for subject peoples' of aid (cf. Escobar 1995, p. 9).

## Humanitarian Standards of Practice, Technology, and Surveillance

The 1990s marked the establishment of 'standards of practice' in humanitarian action, which in turn led to the reconstruction of the idea of aid as a normative engagement marked by qualitative and quantitative measures over which the adequacy and politics of interventions are appraised (Apthorpe 2012; Sphere Association 2018; Taithe and Borton 2016, p. 213; Bridges 2010). In recent years, aid groups have upgraded the instruments of intervention through the advancement of information and communication technology with crisis mapping tools, data informatics, crowd sourcing, bank transfers, and other forms of digital infrastructure to inform the management of humanitarian intervention (Franke 2016). These initiatives raised hopes of tapping the democratising potential of technology in breaking down the top-down approach to humanitarian intervention and empower beneficiaries (Flint 2014). Such an approach, however, is not without complications with some scholars calling for a critical attention to the neoliberal financial economy shaping the politics behind the deployment of technology (Jacobsen and Sandvik 2018).

In Kenya, digital technology precipitated the integration of the displaced into the global financial system through mobile phone and online banking. Electronic methods of cash transfer and biometric scans to verify the authenticity of beneficiaries to access aid intervention inadvertently turn human bodies into loci of legitimation and set the era of biopolitics in this field of human relations (McKee 2011; Human and Robins 2014; Donovan 2015, p. 744). Rather than empowerment, certain technological intervention produces unequal 'networks of power' (Hughes 1993) and recalibrates humanitarian action into an instrument of market integration, individual regulation, and population surveillance (Duffield 2016). The shift towards electronic and biometric scans in Kenyan aid delivery illustrates that the displacement experience is not exclusively associated to the humanitarian and human rights mission but integrated into the postcolonial global financial order, markets, and connectivities.

Other forms of humanitarian responses, however, continue to run along paternalistic and hegemonic forms of relations (Dale and Kyle 2016) with the donors and aid agents referencing the displaced as 'child-like victims' of armed conflicts, desperately needing immediate assistance from the international community in ways that undermine their capacity to take action (I'Anson and Pfeifer 2013, p. 59). Cunningly, such stereotypical portrayal of the displaced privileges donors to imagine their beneficiaries as powerless victims of disasters to attract aid support, exposing how relations of power mediate both ends of the aid continuum (Vestergaard 2013; Smirl 2015).

To some degree, humanitarian issues are also implicated by the Christian missionary value of compassion and the ethical claim of bringing salvation to blighted communities of the world in accordance to the benevolent role that they think they ought to play to lift up the developing world (O'Sullivan 2014; De Cordier 2009). Humanitarian response is primarily viewed as a moral crusade locating the givers on the good side of humanity and the beneficiaries on the receiving end of that 'kindness' (Apthorpe 2012, pp. 1554, 1557; Faist 2018). These categorisations prescribe a normative approach of understanding displacement with little consideration of the geopolitical and social contexts shaping the wider postcolonial reconfigurations of mobility, borders, ethics, markets, history, and surveillance.

## Displacement and State Orders

States refer to the displaced not necessarily as people needing protection but a constituency with specific needs that governments are mandated to address. To this end, some states sought to narrowly define who among the displaced deserve state assistance by setting criteria for qualification and disqualification of IDP status (Shutzer 2012). In some of the violent events in Mindanao, the government refers to the displaced as fire victims rather than IDPs in a manner that tacitly subverts the characterisation of the displacement experience as an international concern to, rather, primarily a domestic affair covered by the terms and conditions set by a sovereign Philippine state and that global humanitarian action must contend with. In Kenya, the return of IDPs to their homes was entangled in the domestic politics aid rooted in the country's 'troubled land history' (Shutzer 2012). Housing and resettlement aid provided for by UN aid programmes and NGOs was mainly given to IDPs registered as landowners under colonial and postcolonial land laws, while forcing those without land registration to return to unspecified 'homes.' The scheme was primarily aimed at pre-empting efforts of mobilising the displaced to gain property rights and alter colonial land relations, while underscoring the intricate link between contemporary displacement experience and the continuity of colonial and postcolonial issues.

Moreover, the displacement experience creates conditions that breed criminal gangs, perpetration of corruption, and an increased number of rebels, especially in highly contested relations between communities and state authorities (Gill et al. 2011; Perera-Mubarak 2012). Tensions, political uncertainties, and vacuums of power in areas of displacement, such as camps, host cities, and regions, enabled criminal networks, gangs, thugs, and newly recruited rebels—associated or not to the state military—to emerge (Loescher and Milner 2005; Sindre 2014; Glassman 2006, p. 9). Crimes alongside border issues, sovereignty concerns, and territorial questions indicate that displacement is in many ways part of the underlying problems affecting the 'national order of things' (Malkki 1995, p. 496). States, inevitably, were often viewed as the entity most accountable for the displacement experience (Gill 2010).

## Displacement and Humanitarian Action

The stereotypical representations of the displaced, however, are not shaped solely by the dominant agents of the humanitarian and human rights divide. Sometime the displaced actively participated by playing the roles expected of them by aid providers and givers as a tactical way of dealing with the horrific experiences of displacement (Harrell-Bond 2002). Moreover, the economic ascendancy of former colonies such as India, Brazil, and the emirates around the Gulf States facilitated their rise as aid donors (Hammerstad 2015). This development underscores the problem with conceiving displacement along human rights and humanitarian frames of thought. Julieta Lemaitre points that part of the problem of human rights and humanitarian conception of displacement is the broader tendency of their advocates to generalise all conflict-ridden states as inadequate in addressing and advancing both issues (Lemaitre 2018).

States such as Colombia, Lebanon, Jordan, Honduras, Nicaragua, and Mexico have strong centres even though the capacity of state structures to implement their humanitarian and human rights commitments at the outer regions of these countries is weak (Lemaitre 2018, p. 545). Questions over the costs, politicisation, marketing audience, and fundraising practices of humanitarian media campaigns have been raised (Vestergaard 2013), in as much as humanitarian action has been blamed for prolonging, exacerbating, and even harming aid recipients (Efuk 2000; Goodhand 2002; Džuverović and Vidojević 2018) in hazard and conflict-prone countries such as Sudan, Ethiopia, Angola, and Somalia (Diprizio 1999, pp. 97, 99).

In certain situation, rebels manipulate and take control of humanitarian aid, for example, the 'predatory behaviour' of the Free Aceh Movement (GAM) in Indonesia and the Tamil Tigers (LTTE) in Sri Lanka in the wake of the outpouring of assistance supporting the peace processes and deals struck with the governments in these respective countries following the 2004 Indian Ocean Tsunami (Sindre 2014). Such rebel predatory acts and aid manipulation, however, were not necessarily disastrous but, rather, instrumental in bringing about political stability, transforming the militarist structure of rebel groups into productive public movements, and setting peacebuilding on much more solid ground (Sindre 2014). The complexity of humanitarian action and its outcomes highlights the difficulty in establishing a normative approach in conceptualising displacement. A displacement event is not a singular experience definable by a universal epistemological frame. These variant understandings of the displaced indicate that the categories of displacement do not simply reflect the human condition they purportedly represent but are themselves outcomes of the complicated consequences of violence and hazard-induced disasters in a postcolonial setting.

## The Trouble with Conceptualising Displacement

Indeed, a few observers remarked that categories are less useful in understanding the intricacies of displacement and the relevance of such phenomenon in knowing the human condition and even 'downright unhelpful' in building epistemological understanding of displacement in a globalising world (Turton 2003a; Zetter 2015). These categories emerged as a product of the broader asymmetry of power relations enmeshed with the tangled politics of nation states, communities, and social forces that include humanitarian and human rights advocacy groups responding to these conditions (cf. Agier 2002).

David Turton has critiqued the term 'forced migration' as conceptually paradoxical even as he cautions not to discard it as yet, noting that it is 'probably the best term available' (Turton 2003a; p. 11) in understanding displacement. He outlines and concludes that the discursive formation of the language of forced migration has a 'dehumanizing effect' as the term actually implies that refugees 'have little or no scope for independent rational decision making; that they are simply passive victims of circumstances' (Turton 2003a, p. 10). Turton argued that as a consequence the 'dehumanizing effect' of the forced migration discourse is 'carried through the practice, intervention, and responses of states and international organizations as they try to control and manage forced migration' and responds 'to forced migrants not as individual human beings, people like us, embedded in contingent social and historical circumstances, but as anonymous and dehumanised masses' (Turton 2003a, p. 10).

Turton eloquently argued for a concept of displacement that does not reduce people at the receiving end of intervention into helpless and powerless victims of forced migration but also yearned to establish some form of coherence in understanding such a horrific experience. By aiming for coherence, however, other enduring and mercurial forces at play in colonial and postcolonial transitions tend to be left out in the analysis. This underscores the trouble with attempting to conceptualise displacement as a coherent category of people and offering less attention to the underlying political, historical, and cultural experiences that implicate not only the geopolitical contingencies of the contemporary world but also the lasting impact of the colonial experience.

Constructing the categories of displacement forms part of the outcome of the postcolonial reordering. Such an endeavour calls not only for pluralist analysis and a more sensitive consideration of the agency of the displaced but, rather, the decentring of the project of knowledge production of the displacement experience. These events open up various worlds of simultaneously differentiated and shared experiences that demand ontological examination and take into consideration their historical continuities to know how these categories were constructed, established, and mobilised as social facts. The complexity of the displacement discourse demands a decentred locus of analysis by factoring the politics of knowledge and its relation to the politics of reality (Savransky 2017, p. 13). By decentring the analysis of displacement, the project ceases to be an enterprise of finding coherence in knowing about the human condition but rather an engagement that unravels the contours of continuing inequality at play in the discursive formation of the displacement experience.

# References

Agier, M. (2002). Between war and city: Towards an urban anthropology of refugee camps (R. Nice & L. Wacquant, Trans.). *Ethnography, 3*(3), 317–341.

Apthorpe, R. (2012). Effective aid: The poetics of some aid workers' angles on how humanitarian aid 'works'. *Third World Quarterly, 33*(8), 1545–1559.

Bridges, K. M. (2010). Between aid and politics: Diagnosing the challenge of humanitarian advocacy in politically complex environments—The case of Darfur, Sudan. *Third World Quarterly, 31*(8), 1251–1269.

Cameron, B. T. (2014). Reflections on refugee studies and the study of refugees: Implications for policy analysts. *Journal of Management & Public Policy, 6*, 1–13.

Canuday, J. J. (2009). *Bakwit: The power of the displaced*. Quezon City, Philippines: Ateneo de Manila University Press.

Castles, S. (2003). The international politics of forced migration. *Development, 46*(3), 11–20.

Chimni, B. S. (1998). The geopolitics of refugee studies: A view from the south. *Journal of Refugee Studies, 11*(4), 50373.

Cohen, R., & Layton-Henry, Z. (1997). Introduction to the politics of migration. In R. Cohen & Z. Layton-Henry (Eds.), *The politics of migration* (pp. ix–xvi). Cheltenham, England: Edward Elgar.

Country, B., Wright, S., Suchet-Pearson, S., Lloyd, K., Burarrwanga, L., Ganambarr, R., et al. (2018). The politics of ontology and ontological politics. *Dialogues in Human Geography, 6*(1), 23–27.

Dale, J., & Kyle, D. (2016). Smart humanitarianism: Re-imagining human rights in the age of enterprise. *Critical Sociology, 42*(6), 783–779.

Daniel, E. V. (2002). The refugee: A discourse on displacement. In J. MacClancy (Ed.), *Exotic no more: Anthropology on the front lines* (pp. 270–286). Chicago and London: The University of Chicago Press.

De Cordier, B. (2009). The 'humanitarian frontline,' development and relief, and religion: What context, which threats and which opportunities? *Third World Quarterly, 30*(4), 663–684.

Diprizio, R. (1999). Adverse effects of humanitarian aid in complex emergencies. *Small Wars & Insurgencies, 10*(1), 97–106.

Donovan, K. K. (2015). Infrastructuring aid: Materialising humanitarianism in Northern Kenya. *Environment and Planning D: Society and Space, 33*, 732–748.

Duffield, M. (2016). The resilience of the ruins: Towards a critique of digital humanitarianism. *Resilience, 4*(3), 147–165.

Džuverović, N., & Vidojević, J. (2018). Peacebuilding or 'peacedelaying': Social exclusion of refugees and internally displaced persons in post-war Serbia. *Ethnopolitics, 17*(1), 55–70.

Efuk, S. (2000). 'Humanitarianism that harms': A critique of NGO charity in Southern Sudan. *Civil Wars, 3*(3), 45–73.

Escobar, A. (1995). *Encountering development: The making and unmaking of the Third World*. Princeton, NJ & Oxford, England: Princeton University Press.

Faist, T. (2018). The moral polity of forced migration. *Ethnic and Racial Studies, 41*(3), 412–423.

Flint, A. (2014). Ownership and participation: Toward a development paradigm based on beneficiary-led aid. *Journal of Developing Societies, 30*(3), 273–295.

Franke, M. F. N. (2016). UNHCR's territorial depoliticization of forced displacement through the governance mechanisms of participatory geographical information systems. *Territory, Politics, Governance, 4*(4), 421–437.

Gill, N. (2010). New state-theoretic approaches to asylum and refugee geographies. *Progress in Human Geography, 34*, 626–645.

Gill, N., Caletrío, J., & Mason, V. (2011). Introduction: Mobilities and forced migration. *Mobilities, 6*(3), 301–316.

Glassman, J. (2006). Imperialism imposed and invited: The 'War on Terror' comes to Southeast Asia. In D. Gregory & A. Pred (Eds.), *Violent geographies: Fear, terror, and political violence* (pp. 93–110). New York, London: Routledge.

Goodhand, J. (2002). Aiding violence or building peace? The role of international aid in Afghanistan. *Third World Quarterly, 23*(5), 837–859.

Hammerstad, A. (2015). The international humanitarian regime and its discontents: India's challenge. *The Round Table, 104*(4), 457–471.

Harrell-Bond, B. (2002). Can humanitarian work with refugees be humane? *Human Rights Quarterly, 24*(1), 51–85.

Hughes, T. (1993). *Networks of power: Electrification in Western society, 1880–1930*. Baltimore, MD: JHU Press.

Human, O., & Robins, S. (2014). Social movements and biopolitical states: A study of humanitarian aid in Cape Town 2008. *Culture and Organization, 20*(2), 121–134.

I'Anson, C., & Pfeifer, C. (2013). A critique of humanitarian reason: Agency, power, and privilege. *Journal of Global Ethics, 9*(1), 49–63.

Jacobsen, K. L. (2010). Making design safe for citizens: A hidden history of humanitarian experimentation. *Citizenship Studies, 14*(1), 89–103.

Jacobsen, K. L., & Sandvik, K. B. (2018). UNHCR and the pursuit of international protection: Accountability through technology? *Third World Quarterly, 39*(8), 1508–1524.

Kendle, A. B. (1998). Protecting whom? The UNHCR in Sri Lanka, 1987–1997. *The Round Table, 87*(348), 521–541. https://doi.org/10.1080/0035853980845444.

Kingsbury, D. (2011). Post-colonial states, ethnic minorities and separatist conflicts: Case studies from Southeast and South Asia. *Ethnic and Racial Studies, 34*(5), 762–778. https://doi.org/10.1080/01419870.2010.537357.

Lemaitre, J. (2018). Humanitarian aid and host state capacity: The challenges of the Norwegian Refugee Council in Colombia. *Third World Quarterly, 39*(3), 544–559. https://doi.org/10.1080/01436597.2017.1368381.

Loescher, G., & Milner, J. (2005). Security implications of protracted refugee situations. *The Adelphi Papers, 45*(375), 23–34. https://doi.org/10.1080/05679320500212130.

Malkki, L. (1995). Refugees and Exile: From "Refugee Studies" to the National Order of Things. *Annual Review of Anthropology, 24,* 495–523.

McKee, Y. (2011). On climate refugees: Biopolitics, aesthetics, and critical climate change. *Qui Parle, 19*(2), 309–325.

Nair, R. B. (1997). Acts of agency and acts of God: Discourse of disaster in a post-colonial society. *Economic and Political Weekly, 32*(11), 535–542.

O'Sullivan, K. (2014). Humanitarian encounters: Biafra, NGOs and imaginings of the Third World in Britain and Ireland, 1967–70. *Journal of Genocide Research, 16*(2–3), 299–315.

Perera-Mubarak, K. N. (2012). Reading 'stories' of corruption: Practices and perceptions of everyday corruption in post-tsunami Sri Lanka. *Political Geography, 31*(6), 368–378.

Prosperi, L. (2017). U.N.H.C.R. as a bulwark against violations of migrants' rights. Dialogue with Stephane Jacquemet. *International Review of Sociology, 27*(2), 287–290.

Redclift, V. (2016). Displacement, integration and identity in the postcolonial world. *Identities, 23*(2), 117–135.

Savransky, M. (2017). A decolonial imagination: Sociology, anthropology and the politics of reality. *Sociology, 51*(1), 11–26.

Scheel, S., & Ratfisch, P. (2014). Refugee protection meets migration management: UNHCR as a global police of populations. *Journal of Ethnic and Migration Studies, 40*(6), 924–941.

Shutzer, M. A. (2012). The politics of home: Displacement and resettlement in postcolonial Kenya. *African Studies, 71*(3), 346–360.

Siddiqi, A. (2017). 'Disaster citizenship': An emerging framework for understanding the depth of digital citizenship in Pakistan. *Contemporary South Asia, 26*(2), 157–174.

Sindre, G. M. (2014). Rebels and aid in the context of peacebuilding and humanitarian disaster: A comparison of the Free Aceh Movement (GAM) and the Tamil Tigers (LTTE). *Forum for Development Studies, 41*(1), 1–21.

Smirl, L. (2015). *Spaces of aid: How cars, compounds and hotels shape humanitarianism.* London: Zed Books.

Stevens, D. (2016). Rights, needs or assistance? The role of the UNHCR in refugee protection in the Middle East. *The International Journal of Human Rights, 20*(2), 264–283.

Taithe, B., & Borton, J. (2016). History, memory and 'lessons learnt' for humanitarian practitioners. *European Review of History: Revue européenne d'histoire, 23*(1–2), 210–224.

Sphere Association. (2018). *The Sphere handbook: Humanitarian charter and minimum standards in humanitarian response, Fourth edition.* Geneva, Switzerland; Rugby, UK: Sphere Association and Practical Action Publishing. www.spherestandards.org/handbook.

Turton, D. (2003a). *Conceptualising forced migration* (RSC Working Paper Series, No. 12). Oxford: Refugee Studies Centre, University of Oxford.

Turton, D. (2003b). *Refugees and 'other forced migrants'* (RSC Working Paper Series. No. 13). Oxford: Refugee Studies Centre, University of Oxford.

Vestergaard, A. (2013). Humanitarian appeal and the paradox of power. *Critical Discourse Studies, 10*(4), 444–467

Weiner, W. (1995). *The global migration crisis: Challenge to states and to human rights.* Longman, GL: Addison-Wesley Education Publishers.

Zetter, R. (2007). More labels, fewer refugees: Remaking the refugee label in an era of globalization. *Journal of Refugee Studies, 20*(2), 172–192.

Zetter, R. (2015). *Protection in crisis: Forced migration and protection in a global era.* Washington, DC.: Translantic Council on Migration, Migration Policy Institute.

# Part II

## Section Two: Technologies of Displacement

# 13

## Intervention: Displacement Aesthetics

### Kaya Barry and Peter Adey

Emergency diagrams, instructions for evacuation, safety cards, and pictographic displays inform how and where mobile bodies should move in exceptional circumstances. During times of danger, insecurity, panic, and disruption, these informational diagrams guide the movements and behaviours of the general public. Yet in daily life, these diagrams go largely unnoticed, sitting in the background of transit spaces, vehicles, vessels, or aircrafts. We might encounter them through the swift gesture during a safety induction, in a presentation from a flight attendant, or an automated security announcement. Bright colours, large arrows, and pictographics attempt to stimulate our attention, alongside generic shapes of human bodies that enacting complicated evacuation and safety procedures. The emergency diagram, airline safety card, and the ridiculous safety video attempt to stimulate passenger/audiences to do something in order to care for their safety in emergency. They have increasingly been drawn to humour, with strange juxtapositions and metaphors to encourage the public's response (Bissell et al. 2012). In this intervention we reflect on a series of artworks produced by Kaya Barry in response to the

---

K. Barry (✉)
Griffith University, Nathan, QLD, Australia
e-mail: k.barry@griffith.edu.au

P. Adey
Department of Geography, Royal Holloway, University of London, Egham, UK
e-mail: Peter.Adey@rhul.ac.uk

abstracted designs and obscure spatial depictions of emergency instructions. The artworks are adapted from examples and illustrations found in aircraft 'safety cards'—the laminated cards that are found in the pocket of the seat in front during air travel. These creative responses are intended to provide what Barry has elsewhere called 'alternative ways of knowing and accessing affective experiences [and] techniques to engage and complement the wide range of mobilities encountered in air travel, as well as everyday experiences of movement' (Barry 2017, p. 367).

The images are material-representational discursive compositions of bodies, movements, anticipatory affects, and expressivities, which are diagrammed into prescriptive sequences of action and comportment. They conjure up emergency situations that are made unimaginable for most, but in the most subduing, glossy kinds of ways. For the majority of the travelling public, these 'what if' speculative scenarios of evacuation, forceful eviction, or disaster response are an 'aesthetics of transit' (Barry 2017) that lie in the peripheries of our attention. They continue, to some extent, a classed and romantic aesthetic of travel and mobility, reinforcing the ideals of the 'right' and 'wrong' types of persons undertaking a journey. Bodies take the form of either generic figures of passengers or the glamorous attendants who effortlessly demonstrate how to deploy lifeboats, inflate jackets, and open emergency doors (Fig. 13.1).

However, these emergency diagrams have gained a new meaning through contemporary representations of forced migration and displacement. No longer are these passive objects and images for possible future 'worst case' scenarios of evacuation and emergencies. The proliferation of materials from the 2015–2016 'global refugee crisis' has been collected, documented, and exhibited by artists around the world. Artworks that take form through arrangements of life jackets, rafts, foil blankets, and washed-up passports and belongings appear alongside photographs and documentation of these perilous journeys in galleries, exhibitions, and cultural institutions, as well as extensive media coverage of these visuals. The result is a charged aesthetics of mobilities—where the textures, colours, consistencies, and surfaces of such materials are amplified (Fig. 13.2). These materials 'destabilise and unsettle the unusual public passivity to seeing these material traces of migration. … They are right here-and-now, tangible and forcibly present' (Barry 2019, p. 209).

The oversaturation of these objects, and the hypervisibility of such emergency diagrams, now pervades public attention. They signify far more precarious situations and journeys. These are new contexts where the experiences hinted at in the emergency diagrams are now commonplace. The stylisation

**Fig. 13.1** *Inflation (night)* (Kaya Barry 2019)

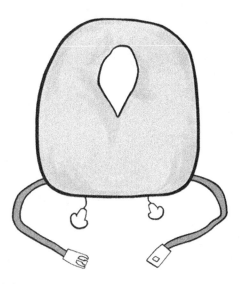

**Fig. 13.2** *Inflation (life vest)* (Kaya Barry 2016)

of emergency diagrams becomes a kind of bridge between contexts of luxury and safety in travel to those experiences of precarity and displacement.

The mutation from preemptive and passive objects of evacuation and disaster mitigation to the now highly charged association with perilous refugee journeys recasts emergency diagrams into a new aesthetics of displacement mobilities. These representations of mobility, secured through the depiction of materials of transit and emergency, are now 'enfolded with the threat' of the unknown traveller (Hall 2015, p. 19). Here, im/mobilities are recast through a pervasive aesthetic of the desirable, permitted, and 'good' traveller that lies in opposition to the illegal, suspicious, and undesirably anonymous body of the migrant.

At the same time, there may be some hope. A different kind of displacement and mobility politics is possible within this kind of aesthetic. On the one hand the diagrams bifurcate emergency displacement mobility into oppositions relying upon expected and structural orientations by way of wealth and class, race, (non)citizenship, gender, ableism, and so on. On the other, at least imaginatively, and even if they are relatively abstract—the bodies occupying the rubber inflatable rafts (Fig. 13.3) are silhouetted figures—they

**Fig. 13.3** *Emergency exit (water)* (Kaya Barry 2019)

may erode those distinctions by placing travellers into intimate proximity with the experiences of others.

Creative approaches that appropriate transit aesthetics in this manner may also, paradoxically, prove more accessible than the official airline safety card or emergency instruction diagrams. Although the artworks depict people appearing calm, complacent, and willing to obey in the most extreme circumstances, the spatialisation of bodies and materials in such abstracted environments amplifies the universal nature of these possible scenarios. We wonder whether, just in the way emergency politics may be harnessed more democratically (Honig 2009), that instead of the articulation of migrant displacement as an emergency for the sovereignty of bordered states, the diagrams present those travels as in common and as possible. They are not only an emergency situation but a durable and normal condition for others. The emergency *is* that this is happening to anyone at all and that urgent action is necessary to protect and care for those precarious journeys at all costs.

## References

Barry, K. (2017). The aesthetics of aircraft safety cards: Spatial negotiations and affective mobilities in diagrammatic instructions. *Mobilities, 12*(3), 365–385.

Barry, K. (2019). Art and materiality of the global refugee crisis: Ai Weiwei's artworks and the emerging aesthetics of mobilities. *Mobilities, 14*(2), 204–217.

Bissell, D., Hynes, M., & Sharpe, S. (2012). Unveiling seductions beyond societies of control: Affect, security, and humour in spaces of aeromobility. *Environment and Planning D: Society and Space, 30*(4), 694–710.

Hall, R. (2015). *The transparent traveller*. Durham, NC: Duke University Press.

Honig, B. (2009). *Emergency politics: Paradox, law, democracy*. Princeton, NJ: Princeton University Press.

# 14

# The Artwashing of Gentrification and Social Cleansing

Stephen Pritchard

## Introduction

Art is increasingly used to smooth and gloss over social cleansing and gentrification, functioning as a 'social licence,' a public relations tool, and a means of pacifying local communities. This practice is known by activists as 'artwashing.' Artwashing lacks an academic definition but is perhaps best defined as 'a process that uses artistic practices unwittingly (or not) in the service of private capital' in which art is intentionally employed as a tool designed 'to make a place more "amenable" for private capital and the aesthetics that it currently desires' (Mould 2017).

It is commonly accepted that regeneration and gentrification are driven by a process known as 'accumulation by dispossession' (Harvey 2008). Artists and arts institutions are instrumentalised by state and corporate interests alike to become pawns in the social cleansing process—the 'gentrifying foot soldiers of capitalism' (Pritchard 2016). From public art to artists' studios, carefully sited new galleries and museums to carefully depoliticised spectacles, art has proved the perfect foil for the vengeful ideology of neoliberalism—a smokescreen for dispossession and displacement. Yet artists are also employed to perform a different role, that of engendering trust and 'building' or 'growing' social capital. Artists engaged in community contexts are perfectly suited

S. Pritchard (✉)
Independent Academic, Newcastle-upon-Tyne, UK

to harnessing social capital because, unlike corporate consultants, they are frequently able to earn the trust of local people and community groups, and trust is perhaps the single most important element of social capital. This 'community artwashing' is particularly exploitative and deceitful, leading artists and creative placemakers to become what is termed 'social capital artists' (Pritchard 2017a). However, other artists stand in opposition to the dispossession of homes and displacement of people and to the use of art as a veneer for social and ethnic cleansing (Fig. 14.1).

This chapter examines the histories and narratives that have developed alongside our understanding of the term 'artwashing' and how this knowledge can provide new ways of thinking about art's role in assisting, resisting, and opposing gentrification and social cleansing. Focusing on state-led artwashing, the chapter considers how the V&A—a UK state arts institution—artwashed the displacement of working-class families from the Robin Hood Gardens estate in East London by excavating a 'fragment' of the iconic ex-council housing estate as design 'salvage' for display in Venice and its new museum in neighbouring Stratford. The chapter begins by exploring the ways in which artwashing functions as a tool for dispossession and displacement, before moving on to investigate how artists and arts organisations have

**Fig. 14.1** Montage of demolition of Robin Hood Gardens estate and Blackwall Reach promotional video. (Film still from *A Cacophony of Crows* by S. Pritchard 2018d)

interacted with and depicted Robin Hood Gardens. It then looks at the V&A Museum's interventions on the estate as a case study before concluding.

## Artwashing, Dispossession, and Displacement

The art world has become increasingly 'entrenched within cycles of urban change' (Mathews 2010, p. 660). Urban change is commonly dressed-up with terms such as 'renaissance,' 'regeneration,' 'revitalisation,' and 'renewal' which, in turn, serve to disguise gentrification and the processes of dispossession and displacement that accompany it. In the UK, council estates and social housing are, particularly in the post-2008 crash era of austerity, soft targets for dispossession and displacement, with many being demolished or earmarked for demolition; a select few (e.g., Balfron Tower in Poplar, London, and the Park Hill estate in Sheffield) are 'preserved' by displacing social housing tenants then 'refurbishing' the properties for middle-class buyers. This process of demolishing or refurbishing council homes so that new private accommodation can be built for new, wealthy incomers has been labelled 'state-led gentrification' (Lees et al. 2014, p. 5). Sociologist David Madden sums up this process succinctly:

> Here's how gentrification talk typically goes: poor neighborhoods are said to need 'regeneration' or 'revitalization,' as if lifelessness and torpor—as opposed to impoverishment and disempowerment—were the problem. Exclusion is rebranded as creative 'renewal.' The liberal mission to 'increase diversity' is perversely used as an excuse to turn residents out of their homes in places like Harlem or Brixton—areas famous for their long histories of independent political and cultural scenes. (2013)

Once gentrified, neighbourhoods are celebrated for having 'bounced back,' whilst the fact that poverty has simply been dispersed or relocated is ignored and, sometimes, celebrated. For Madden, this 'urban renaissance' narrative, based as it is on 'heroic elites' saving inner cities from the 'dangerous classes,' is both 'condescending and often racist' (Madden 2013) (Fig. 14.2).

Artists are drawn to the constantly shifting urban frontiers which immediately prefigure gentrification. And they are increasingly vilified in some contexts for their role in the displacement of lower-class people and ethnic minorities—for example, in Boyle Heights, Los Angeles (Boyle Heights Alliance Against Artwashing and Displacement 2019)—and, simultaneously, celebrated for their ability to (temporarily) inhabit run down urban

**Fig. 14.2** Montage of demolition of Robin Hood Gardens estate and Blackwall Reach promotional video. (Film still from *A Cacophony of Crows* by S. Pritchard, 2018d)

areas—for example, Hotel Elephant in Southwark which once occupied a space on the condemned Heygate Estate (Southwark Notes 2016). In this sense, artists become both 'victims and aggressors' (Mathews 2010, p. 672). For Rebecca Solnit, gentrification is 'the fin above water,' betraying 'the rest of the shark' below it that devours 'cultural diversity' (2000, pp. 13–18). Certain forms of art—for example public art and socially engaged art—have secured a place within the global urban economy as powerful placemaking tools, strategic policy devices capable of 'improving' places, people, and, ultimately, economies. I argue that these artistic practices, with their claims of community empowerment and social engagement, when deployed in areas undergoing or under threat of displacement of working-class and ethnic minority residents, become the artistic fin above the water, while the creative industries, the state, and the corporate investors become the rest of the shark lurking below. In such circumstances, art becomes artwashing.

The artwashing of dispossession and displacement is a global practice, increasingly undertaken by property developers, local and national authorities, and other corporate and financial institutions. It is often supported and (at least partly) financed by state agencies responsible for arts and culture. In Vancouver, property developer Westbank has used art to sell luxury

## 14 The Artwashing of Gentrification and Social Cleansing

apartments. It uses public art installations and exhibitions such as *Fight for Beauty* (2018). Interestingly, the developer labels itself as a 'cultural practice' that uses 'beauty' to 'build cities and culture' (Westbank 2018). Westbank also installed the 2016 Serpentine Pavilion on the site of a new development. *Fight for Beauty* has been widely criticised with ex-Vancouver arts and cultural policy panel member Melody Ma accusing Westbank of artwashing:

> What we're seeing here is a real estate company ... co-opting the arts and culture to market luxury condos in neighbourhoods like Chinatown, which in fact at the end of the day economically and physically displace people and culture that's already there. (Zeidler 2017)

Similarly, developers have used artwashing in Hamburg (Pritchard 2018c), Melbourne (Davey 2017), Paris (Blanchard 2017), Mumbai (Bas 2015), Beijing (McCarthey 2018), Bilbao (Vicario and Martinez Monje 2003), and in many other cities around the world. Artwashing serves to mask the exploitation of people and property, underwriting the accumulation of capital by dispossession (Fig. 14.3).

Sociologists Sharon Zukin and Laura Braslow argue that property developers and local authorities often utilise the 'artistic mode of production' as

**Fig. 14.3** People of Southwark mural at The Artworks Elephant box park, Southwark, London. (Photograph by S. Pritchard 2016)

symbolic capital with which to construct 'new place-identities' that increase their economic value by rebranding them as 'creative' (2011, p. 131). Typically, gentrification and displacement follow. It can be argued that Zukin and Braslow were here describing what later became known as 'artwashing' and its urbanist alter ego, 'creative placemaking.' Artists have increasingly been tasked with ameliorating lower-class '"poverty of aspiration" and "low expectations"' (Seymour 2009, p. 34) whilst simultaneously providing, what Benedict Seymour describes as, 'stimulus to and communitarian credibility for the process of privatisation and gentrification' (2009, p. 34). Socially engaged art and creative placemaking have become commonly used means of collecting 'memories of displacement' and mapping urban decline. In these circumstances, artists can be considered as 'surrogate and simulacral service providers' delivering 'cheap but cosmetic substitutes for welfare provision'—models of 'regenerate citizenship' who often work for free (or comparatively cheaply) and create a 'marketable "buzz"' (Seymour 2009, p. 34).[1]

Art and other creative practices serve as the perfect foils for what has become known as 'austerity urbanism' (Peck 2012). Artists involved in artwashing do not only collect, or 'harvest,' people's stories, they also collect objects from communities awaiting or undergoing displacement (Pritchard 2017b). Artwashing gentrification occurs when developers (whether corporate, or local authority, or state) use art (such as commissioned street art, artists' studios in meanwhile spaces, pop-up art galleries, and socially engaged art projects) as a means of attracting a creative class as a magnet for the middle class. Artwashing epitomises Schumpeterian neoliberalism of 'creative destruction' with artists cast as the future bringers of prosperity for some and the harbingers of poverty and displacement for many others (Pritchard 2018a).

## Artists and Robin Hood Gardens

Artists have been involved in the planned redevelopment of Robin Hood Gardens (RHG) since 2006, when Erect Architecture delivered a community engagement project on the estate that culminated in an event held as part of that year's London Open House programme. *Streetlife* claimed to 'aid the regeneration of the estate from within' by inviting 'residents to reconsider their environment in order to imagine different sorts of living' in a series of workshops that used art to explore 'built aspects of the estate as well as multicultural issues' (Erect Architecture 2006). The project utilised techniques that today would be considered socially engaged art and creative placemaking. It culminated in a fair on one of RHG's 'streets in the sky' which included an exhibition of art produced with some of the residents and the opportunity for visitors to look inside one of the flats (Beech 2006).

Transitions Project—a collective of seven photographers—produced the exhibition *Robin Hood Gardens: Out of Time* in 2016 which was, like *Streetlife*, part of London Open House. Similarly, the event included guided tours of the building (Transitions Project 2016a). The collective's previous exhibitions had focused on other contested spaces in Poplar, such as Balfron Tower in 2014 and Chrisp Street Market in 2015. It is important to note that the Open House event was facilitated by Tower Hamlets Council, Tower Hamlets Homes, and Swan Housing Association (Transitions Project 2016b).

During the course of 2016 and 2017, painter Charlie Warde produced a series of works including *Deposition (Robin Hood Gardens)*—a '3D "Plastic Painting"' that recreated elements of the estate's East Block (Axisweb 2017)—and *Memorandum (Robin Hood Gardens)*, a limited edition set of gold-plated etched copper plates featuring the estate, the first of which was purchased by the V&A (V&A 2017). Socially engaged artist Jessie Brennan also worked on the estate from 2014 to 2016. Brennan was commissioned by the Foundling Museum to produce *A Fall of Ordinariness and Light* in 2014. The series of drawings depicted RHG as a squashed, flattened, and crumpled ruin. For Brennan, the drawings visualised 'an apparently failed utopian ideal' ahead of the estate's 'imminent demise' (Foundling Museum 2014). A review of the exhibition described how the artist's practice 'confronts the politics of contemporary urban living, the impact of regeneration on people and places' (Nettell 2014). As with Warde, the V&A now own one of Brennan's works (V&A 2018).

In 2015, HS Projects commissioned Brennan to produce another exhibition and book about RHG entitled *Regeneration!* HS Projects is an arts consultancy that produces 'public art commissions for institutions, developers and local authorities,' develops 'exhibition programmes for private and corporate collections,' and brokers 'mutually beneficial partnerships working collaboratively with arts organisations, charities, government and the private sector' (Sotiriadi 2018). For Brennan, *Regeneration!* was an exploration of what it was like for residents of RHG to live with impending displacement, with doormat rubbings forming the basis for conversations with tenants. Brennan seemed to oppose the impending demolition, writing that it 'doesn't merely sweep away an entire social housing estate, it threatens to crush those collective dreams' (2015a). For her, RHG symbolised 'political struggle' (Brennan, 2015a). Nevertheless, it is interesting to note that HS Projects brokered support for *Regeneration!* from the Insight Community Arts Programme (Brennan 2015b). The programme was part of Insight Investment

Management's corporate social responsibility commitment. Its purpose was described as 'supplying regular photographic artwork for the Insight offices via disadvantaged community groups' (Nutter 2018). Part of the Bank of New York Mellon corporation—one of the world's largest financial institutions—the company is also a pension provider for Tower Hamlets Borough Council (Insight Investment 2018).

With the demolition of RHG looming, artists were attracted to the estate to undertake various attempts to document its last days. Yet, whilst the intentions of Brennan and Warde appear to have been benign, their work was collected by the V&A in advance of its intervention at RHG. However, it can be argued that the work of Erect Architecture was a precursor to the planned demolition of the estate, utilising community engagement as a form of community consultation and London Open House tours as a means of attracting middle-class interest in the buildings. Similarly, the Transition Project utilised London Open House tours—this time backed by the partners involved in the redevelopment of RHG. This project appears to have been a form of state- and local authority-led artwashing, although it is worth noting that its impact was relatively low-key (Fig. 14.4).

## The V&A Museum and State-Led Artwashing

Much has been written about the V&A's decision to purchase a part of Robin Hood Gardens (RHG): an ex-council estate, more recently social housing. The estate is currently being demolished to make way for Blackwall Reach—a luxury property development. Campaigners fought to save Robin Hood Gardens: some because of its architectural significance, others because they believed in maintaining social housing. The estate was not saved. Its tenants were and will be displaced. As such, the V&A's acquisition can be considered 'artwashing.' A large proportion of the press coverage and broader debate about the V&A's acquisition of a slice of RHG was positive: the museum portrayed as 'saving,' 'salvaging,' and 'preserving' an important part of the UK's Brutalist architectural heritage. There is little mention of the dispossession and displacement of hundreds of people, families, and children.

RHG was an experiment—part of the utopian dream of providing those with the least in our society with new, spacious, comfortable, communal homes; a dream for better lives for everyone. However, council estates were quickly run down and demonised as 'sink estates,' 'slums,' 'ghettos,' 'no-go-zones,' and so on. This was part of the 'regeneration game,' part of the state-led cycle of gentrification—of planned disinvestment, planned dispersal of

# 14 The Artwashing of Gentrification and Social Cleansing

**Fig. 14.4** One of the walkways at Robin Hood Gardens before demolition. (Photograph by S. Pritchard 2016)

working-class people, planned demolition, and planned reinvestment. RHG is part of this cycle. Sociologist Nicholas Thoburn described the estate as lying 'on the fault-line of class and inequality, as London's Docklands meets and ejects one of the city's most economically deprived communities. More than a cleavage between rich and poor, this fault-line is an experience of destruction and restructuring intrinsic to the life of the estate' (Thoburn 2018, p. 612). Indeed, RHG has often featured as an example of the decline of the modernist vision of high-rise council housing as part of government masking and repackaging of social cleansing (Thoburn 2018, p. 613).

The V&A inserted itself into this class fault line via an Andrew W. Mellon Foundation residency during 2016–2017 in which muf architecture/art were commissioned to explore the museum's relationship with East London and its forthcoming V&A East museum on Stratford Waterfront in the Queen Elizabeth Olympic Park (Heathcote 2017). The result of the residency was that muf suggested that the V&A should acquire some sections of RHG ahead of its demolition (Pritchard 2018b). muf's 'provocation' to the museum was: 'Given the museum engages with the world through objects, can RHG prompt engagement with east London?' (Wainwright 2018). The proposal was unanimously supported by the V&A (Sayer 2017). The idea of taking sections of the condemned ex-council housing estate and turning them into objects—museum objects and art objects—generated a great deal of media attention and debate (Braidwood and Hurst 2017; Brown 2017; O'Sullivan 2017). muf had argued that it was the function of the V&A 'to provide a platform for … difficult conversations,' pointing out that the museum was 'full of bits of buildings that were victims of regeneration, or changes in liturgical fashion and administrative power over the centuries' (Wainwright 2018).

Unsurprisingly, the acquisition of sections of the estate was enthusiastically supported by Blackwall Reach development partners Swan Housing Association, Tower Hamlets Borough Council, and the Mayor of London (Swan Housing Association 2017; HaworthTompkins 2018). The V&A residency became a means of advertising two major investments for the Mayor of London: the £1.3 billion investment in what has now been rebranded as East Bank, in the city's rapidly gentrifying borough of Newham, and the £300 million investment in Blackwall Reach. The 'salvaged' chunks of what were once people's homes, once council homes for those most in need of housing, became objects for collection and preservation: 'objects' designed to 'prompt engagement with east London' (Wainwright 2018). In this sense, the sections of RHG could be considered as anchors for the gentrification of East London and palimpsests of erased working-class homes and of dispossession and displacement. When fragments of council housing estates are reduced to objects, they become 'museumified.' Museumification happens when 'a living city' becomes 'an idealized re-presentation of itself': usefulness is displaced by the possibility of everything (tangible and intangible) becoming a museum artefact (Di Giovine 2009, p. 261).

The sections of RHG are significant objects for the V&A and particularly for its forthcoming museum on the East Bank. The segment destined for the new museum is the biggest object in the V&A's collection. It is no coincidence that V&A East is less than one mile away from RHG. According to the V&A, the three-storey section taken from the estate will become an immersive exhibition in the new museum, complete with 'cherry-picked' items from other flats (Sayer 2017). One of its curators described the acquisition as a 'representative portion' of the ex-council housing estate (Wainwright 2018). It was to become an 'anchor' for the new museum (Heathcote 2017).

RHG has been a contested space since at least 2006 when initial forays into 'community engagement' (as per the example of Erect Architecture above) began on the estate. Whilst many architects and residents fought against planned demolition and regeneration, many other experts and those with vested interests in regeneration sought to portray RHG as a sink estate. The consultancy commissioned by Tower Hamlets Borough Council to advise on whether RHG should be listed described the estate as 'a heroic failure—monumental but inhumane,' adding that not only was it inadequate as a dwelling place but that it was architecturally below par. The report concluded that it stood 'in the way of providing a better, more humane environment in this problematic part of East London,' but recommended that the estate should be 'carefully' recorded and documented so 'lessons could be learnt' (Peter Stewart Consultancy 2007, pp. 19–20) (Fig. 14.5).

The language in the report was widely replicated by those who supported the estate's regeneration. Laden with codified meanings, the message is that the area should be redeveloped to replace the grimly inhumane slum of the failed RHG estate and its problematic inhabitants with something better. The message signals the displacement of working-class social housing tenants with middle-class tenants and homeowners. Yet the idea of documenting the estate and its inhabitants before demolition and displacement is particularly telling—an open invitation to cultural and heritage institutions like the V&A and to artists like those mentioned above. Terms such as documenting, recording, collecting, and preserving are all too frequently deployed as code for artwashing. For Thoburn, the Brutalism of working-class social crisis—'a condition of exploitation, insecurity, displacement, and dispossession'—is

**Fig. 14.5** View of Robin Hood Gardens with some units boarded up. (Photograph by S. Pritchard 2016)

transformed into one of two middle-class, clearly related, aesthetic forms: 'concrete monstrosity' or the 'middle-class Brutalism' of a 'class-cleansed "modernist masterpiece"' (2018, p. 613).

It is this middle-class aestheticisation and fetishisation of Brutalism that gives the codified language of artwashing its visual form. The developers of RHG employed the first of these middle-class aesthetic tropes—the image of a vandalised, decrepit working-class ghetto; the V&A employed the second form—portraying the estate as an example of Brutalist architecture that needed rescuing and preserving from its working-class tenants (Thoburn 2018, pp. 615–20). The museum is reproducing its own history of class privilege and colonialism: collecting trophies of unexpected architectural 'beauty' from 'uncivilised' urban 'savages.' Thoburn's description of the V&A's acquisition of a slice of RHG comes very close to artwashing:

> Here is a symbolic distillation of the truth of middle-class Brutalism, the usurping of working-class housing for middle-class pleasure. But the acquisition serves also to symbolically cleanse the social violence of demolition, by lifting the estate—and by association the social and aesthetic form of modernist council housing—out of its crisis in the present and into the sealed and sanitized past of a museum artifact. (2018, p. 620)

My own research on the estate and discussions with residents and activist groups campaigning to save RHG found that people seemed to like living on the estate, although they were concerned about the poor maintenance and lack of investment by the council. The estate was certainly not a 'slum' and it did not feel unsafe. The playground on the estate had been decorated with colourful children's murals celebrating the 2012 London Olympics (Fig. 14.6). Looming over the murals: Blackwall Reach hoardings with smiling middle-class people peering into the condemned estate. Behind them, glass towers and the hum of construction. This was 2016. When I returned in 2017, part of one of the blocks had already been demolished, the towers of HSBC and other global financial corporations now entirely visible. Residents were still living in the other block. Their stay of execution prolonged, they became spectators watching the wrecking ball of dispossession and displacement from their front-row seats (Pritchard 2014–2018).

My research is echoed by architecture and design critic Edwin Heathcote:

> Walking around the still-inhabited half of the estate, you do not get the sense of a failed project. The elevated walkways are not, perhaps, the lively community spaces envisaged by their architects. Yet neither are they the crime-ravaged,

# 14 The Artwashing of Gentrification and Social Cleansing

**Fig. 14.6** London Olympics mural in playground at Robin Hood Gardens with the Blackwall Reach development on other side of road. (Photograph by S. Pritchard 2016)

graffiti-spattered labyrinths of urban myth. Front doors have been customised, frilly net curtains hung at the windows. It is a part of the city, albeit a part that has been radically transformed. Heavy industry has given way to the property industry as this area is repackaged, rebuilt, rebranded and re-sold as 'East London.' Hotels in Hong Kong and Shanghai are hosting events and marketing the new development as 'nothing but excellence.' (2017)

Nonetheless, for Heathcote, the V&A's acquisition of a section of RHG—'a pivotal architectural landmark'—should be welcomed as a 'simultaneously a brilliant and a problematic and political acquisition' (2017). This echoes the middle-class fetishisation of Brutalism. A chunk of ex-council housing becomes a 'conversation piece' that can help 'anchor' the V&A to its new site in a part of London—Newham—that is undergoing extreme culture-led, state gentrification. Anna Minton argues that the V&A is seeking to generate cultural capital from 'an iconic building which was not deemed worth saving as housing for people to live in' (2018). The demolition of RHG and its replacement with the Blackwall Reach development exemplifies what is commonly now referred to as 'placemaking'—a process repeated across London and other UK and US cities, including in the Newham 'opportunity' area—the site of the forthcoming V&A East development (Minton 2018). The use of art as a means of placemaking is known in the UK and US as 'creative placemaking' (Fig. 14.7).

The artwashing of the displacement of residents of RHG was reinforced by V&A's decision to float one of the pieces of the estate for the 2018 Venice Architecture Biennale. It became clear that the museum had purchased more than one piece of the estate. One V&A curator described the piece sent to Venice as 'a small section of the ruin that RHG has now become,' stating that the museum was bringing the fragment 'back to Venice to look at the original vision and what we might learn for the future of social housing from the Smithsons' long engagement with the issue' (Brown 2018). This was a reference to the Smithsons' exhibition of a photograph of the estate and a bench at the biennale in 1976 in which they called the estate a 'ruin in reverse' (Brown 2018). Exhibiting the segment at the 2018 Venice Architecture Biennale led to further media attention. The concrete section was shown alongside an exhibition by artist Do Ho Suh which included film from inside flats that, at the time of filming, were mostly still inhabited.

**Fig. 14.7** Montage of section retrieved from Robin Hood Gardens estate installed at V&A's *Robin Hood Gardens: A Ruin in Reverse* exhibition at the 2018 Venice Architecture Biennale and footage from V&A's Biennale promotional video. (Film still from *A Cacophony of Crows* by S. Pritchard 2018d)

## Conclusion

Whilst the V&A's RHG 'exhibit' may seem like a well-intentioned act of historic preservation, it can also considered as a headstone for the displacement of Poplar's working-class people; the estate's smashed council homes 'recycled' to become 'London rocks.' The Blackwall Reach development will sit atop of another of the shallow graves of council housing, with rumours that other crushed and 'preserved' pieces from RHG are being used for 'memorial' paths and public art. You can imagine the exclusive exhibition preview evening when V&A East opens: just like the Venice Biennale opening, London's middle class will celebrate this council-home-turned-art-object with chinking champagne flutes. The V&A paid a contribution towards the cost of extraction to the developers of Blackwall Reach (Mairs 2017). As the development partnership involves the local council, the Mayor of London, and a housing association, and the V&A is a state-funded cultural institution, this is a strange transaction involving public money. One in which the museum benefits from an acquisition that will generate significant interest in its new site in rapidly gentrifying East London; the local council and housing association benefit from brand association and free advertising; and the Mayor of London and the UK government benefit from the culture-led gentrification of the boroughs of Tower Hamlets and Newham.

Whilst all these 'benefits' have been achieved using the cultural capital of the V&A Museum, the tenants of RHG have not been so fortunate. They have been decanted and displaced. This is how state-led artwashing works: the middle class and wealthy benefit at the expense of the working class and poor (Fig. 14.8).

## Note

1. Benedict Seymour described the function of participating in socially engaged art under New Labour as follows: 'Participation in the valorisation of life/labour—whether helping run your block of flats or talking to a concerned artist about your memories of displacement—is not so much solicited as compulsory. Consequently, in a regeneration regime it becomes easier to get your experience of urban blight plotted on a psychogeographic map of your area than to obtain hospital treatment, housing or a day off work' (2009, p. 34).

**Fig. 14.8** Robin Hood Gardens awaiting demolition. (Photograph by S. Pritchard 2017b)

## References

Erect Architecture. (2006). Streetlife. *Erect Architecture*. Retrieved December 1, 2018, from https://www.erectarchitecture.co.uk/projects/street-life-robin-hood-gardens/.

Axisweb. (2017). Deposition (Robin Hood Gardens). Retrieved December 2, 2018, from https://www.axisweb.org/p/charliewarde/workset/227847-deposition-robin-hood-gardens.

Bas, S. (2015, July 10). Kala Ghoda 2.0: Mumbai's art diaspora shifts away from SoBo. *Hindustan Times*. Retrieved February 14, 2019, from https://www.hindustantimes.com/art-and-culture/kala-ghoda-2-0-mumbai-s-art-diaspora-shifts-away-from-sobo/story-66uWalhQLJmMrBBDHPzYEN.html.

Beech, N. (2006). Problems with easy street: At the Smithsons' 'Robin Hood Gardens Estate' for open house weekend, Sunday 17 September 2006. Opticon, *1826*(1).

Blanchard, S. (2017, September 29). Street art, rénovation urbaine et gentrification dans le Nord-Est parisien: Entre marketing urbain et gender mainstreaming. *Urbanités*. Retrieved February 14, 2019, from http://www.revue-urbanites.fr/9-street-art-renovation-urbaine-et-gentrification-dans-le-nord-est-parisien-entre-marketing-urbain-et-gender-mainstreaming.

Boyle Heights Alliance Against Artwashing and Displacement. (2019). Home. Retrieved February 14, 2019, from http://alianzacontraartwashing.org/en/bhaaad.

Braidwood, E., & Hurst, W. (2017, November 13). Critics round on V&A's acquisition of Robin Hood gardens section. *Architects' Journal*. Retrieved December 20, 2018, from https://www.architectsjournal.co.uk/news/critics-round-on-vas-acquisition-of-robin-hood-gardens-section/10025178.article.

Brennan, J. (2015a, August 11). Robin Hood Gardens and the politics of regeneration. *Apollo*. Retrieved December 2, 2018, from https://www.apollo-magazine.com/robin-hood-gardens-and-the-politics-of-regeneration.

Brennan, J. (2015b). *Regeneration!* London: Silent Grid.

Brown, M. (2017, November 9). V&A acquires segment of Robin Hood Gardens council estate. *The Guardian*. Retrieved December 20, 2018, from https://www.theguardian.com/artanddesign/2017/nov/09/va-buys-three-storey-chunk-robin-hood-gardens-council-estate-alison-peter-smithson-brutalist-architecture.

Brown, M. (2018, March 6). Chunk of London housing estate to star at architecture festival. *The Guardian*. Retrieved March 7, 2018, from https://www.theguardian.com/artanddesign/2018/mar/06/chunk-of-london-housing-estate-to-star-at-architecture-festival.

Davey, M. (2017, October 5). Gentrification, street art and the rise of the developer-sponsored block party. *The Guardian*. Retrieved February 14, 2019, from https://www.theguardian.com/australia-news/2017/oct/05/gentrification-street-art-and-the-rise-of-the-bogus-block-party.

Di Giovine, M. (2009). *The heritage-scape*. Lanham, MD: Lexington Books.

Foundling Museum. (2014). Progress. *Foundling Museum*. Retrieved December 2, 2018, from https://foundlingmuseum.org.uk/events/progress-2014.

Harvey, D. (2008, September/October). The right to the city. *New Left Review, 53*. Retrieved from http://newleftreview.org/II/53/david-harvey-the-right-to-the-city.

HaworthTompkins. (2018). *Blackwall Reach, 2020*. Retrieved December 1, 2018, from http://www.haworthtompkins.com/work/blackwall-reach.

Heathcote, E. (2017, November 24). Robin Hood Gardens travels to Venice. *Financial Times*. Retrieved September 21, 2018, from https://www.ft.com/content/b9d4bf88-ca2a-11e7-8536-d321d0d897a3.

Insight Investment. (2018). *London Borough of Tower Hamlets pension fund: BNY Mellon absolute return bond fund*. London: Insight Investment.

Lees, L., & Just Space, & SNAG. (2014). *Staying put: An anti-gentrification handbook for council estates in London*. London: Antipode Foundation.

Madden, D. (2013, October 10). Gentrification doesn't trickle down to help everyone. *The Guardian*. Retrieved December 20, 2015, from http://www.theguardian.com/commentisfree/2013/oct/10/gentrification-not-urban-renaissance.

Mairs, J. (2017, November 9). V&A acquires entire one-bedroom flat from Robin Hood Gardens. *Dezeen*. Retrieved December 7, 2018, from https://www.dezeen.com/2017/11/09/va-museum-acquires-robin-hood-gardens-flat-alison-peter-smithson-news-london-uk-architecture.

Mathews, V. (2010). Aestheticizing space: Art, gentrification and the city. *Geography Compass, 6*(4), 660–675.

McCarthey, J. (2018, July 6). Beijing's 798 Art Zone: Art, regeneration and gentrification. *Asia Dialogue*. Retrieved February 14, 2019, from http://theasiadialogue.com/2018/07/06/beijings-798-art-zone-art-regeneration-and-gentrification/.

Minton, A. (2018, September). The price of regeneration. *Places Journal*. Retrieved September 30, 2018, from https://placesjournal.org/article/the-price-of-regeneration-in-london.

Mould, O. (2017, September 14). Why culture competitions and 'artwashing' drive urban inequality. *OpenDemocracy*. Retrieved January 26, 2018, from https://www.opendemocracy.net/uk/oli-mould/why-culture-competitions-and-artwashing-drive-urban-inequality.

Nettell, M. (2014, June 3). A response to Hogarth: 'I'm thinking about progress as a concept.' *A-N*. Retrieved December 1, 2018, from https://www.a-n.co.uk/news/a-response-to-hogarth-thinking-about-progress-as-a-concept.

Nutter, G. (2018). Geoff Nutter [Linked in profile]. Retrieved December 2, 2018, from https://www.linkedin.com/in/geoffnutter.

O'Sullivan, F. (2017, November 10). A duplex of London's public housing will become a museum exhibit. *Citylab*. Retrieved December 20, 2018, from https://www.citylab.com/design/2017/11/a-duplex-of-londons-public-housing-will-become-a-museum-exhibit/545528.

Peck, J. (2012). Austerity urbanism: American cities under extreme economy. *City, 16*(6), 626–655.

Peter Stewart Consultancy. (2007). *Robin Hood Gardens: Report on potential listing*. London: London Borough of Tower Hamlets.

Pritchard, S. (2014–2018). *A critical ethnography of the social cleansing of Robin Hood Gardens* (Unpublished case study). Newcastle, England.

Pritchard, S. (2016, September 13). Hipsters and artists are the gentrifying foot soldiers of capitalism. *The Guardian*. Retrieved December 20, 2018, from https://www.theguardian.com/commentisfree/2016/sep/13/hipsters-artists-gentrifying-capitalism.

Pritchard, S. (2017a, September). Extracting new cultural value from urban regeneration: The intangible rise of the social capital artist. Sluice [Online].

Pritchard, S. (2017b, November 13). Rethinking the role of artists in urban regeneration contexts [Web log post]. Retrieved December 20, 2018, from http://colouringinculture.org/blog/rethinkingartistsinurbanregen.

Pritchard, S. (2018a, July 16). New Bohemias: Artists, hipsters and gentrification [Web log post]. Retrieved December 20, 2018, from http://colouringinculture.org/blog/oldtonewbohemias-gentrification.

Pritchard, S. (2018b, June 1). No breathing space—V&A, artwashing and the theft of Robin Hood Gardens. *Bella Caledonia*. Retrieved December 1, 2018, from https://bellacaledonia.org.uk/2018/06/01/no-breathing-space-va-artwashing-and-the-theft-of-robin-hood-gardens.

Pritchard, S. (2018c, April 27). Artwashing and gentrification. *Bella Caledonia*. Retrieved February 14, 2019, from https://bellacaledonia.org.uk/2018/04/27/artwashing-and-gentrification.

Pritchard, S. (2018d). A Cacophony of Crows [Film]. Retrieved December 20, 2018, from https://vimeo.com/300303157.

Sayer, J. (2017, November 20). Publicity stunt? 'Working-class theme park?' Questions loom for the V&A's three-story Robin Hood Gardens fragment. *Metropolis*. Retrieved December 1, 2018, from https://www.metropolismag.com/architecture/victoria-albert-museum-robin-hood-gardens.

Seymour, B. (2009). Shoreditch and the creative destruction of the inner city. *Variant, 34*, 32–34.

Solnit, R. (2000). *Hollow city: The siege of San Francisco and the crisis of American urbanism*. New York, NY: Verso.

Sotiriadi, T. (2018). Tina Sotiriadi [Linked in profile]. Retrieved December 1, 2018, from https://www.linkedin.com/in/tina-sotiriadi-a4432b31.

Southwark Notes. (2016, September). The fine art of regeneration in Southwark. *Southwark Notes*. Retrieved February 14, 2019, from https://southwarknotes.wordpress.com/art-and-regeneration/.

Swan Housing Association. (2017). *Swan is delighted to support V&A to acquire a fragment of Robin Hood Gardens*. Retrieved December 2, 2018, from https://www.swan.org.uk/home/news/swan-is-delighted-to-support-va-to-acquire-a-fragment-of-robin-hood-gardens.aspx.

Thoburn, N. (2018). Concrete and council housing: The class architecture of brutalism 'as found' at Robin Hood Gardens. *City, 22*(5–6), 612–632.

Transitions Project. (2016a). Robin Hood Gardens: Out of time. Retrieved June 13, 2018, from http://www.robinhoodgardens.london.

Transitions Project. (2016b). Open House London 2016. Retrieved December 2, 2018, from https://www.transition.photography/robin-hood-gardens.

V&A. (2017). *Memorandum (Robin Hood Gardens)*. Retrieved December 2, 2018, from http://collections.vam.ac.uk/item/O1419998/memorandum-robin-hood-gardens-artists-multiple-warde-charlie.

V&A. (2018). *A fall of ordinariness and light: The justification*. Retrieved December 3, 2018, from http://collections.vam.ac.uk/item/O1329223/a-fall-of-ordinariness-and-drawing-brennan-jessie.

Vicario, L., & Martınez Monje, P. (2003). Another 'Guggenheim Effect'? The generation of a potentially gentrifiable neighbourhood in Bilbao. *Urban Studies, 40*(12), 2383–2400.

Wainwright, O. (2018, May 15). Will this three-storey slice of British brutalism be the toast of Venice? *The Guardian*. Retrieved May 16, 2018, from https://www.theguardian.com/artanddesign/2018/may/15/robin-hood-gardens-three-storeys-british-brutalism-venice-biennale.

Westbank. (2018). Fight for beauty. *Westbank*. Retrieved February 14, 2019, from http://fightforbeauty.westbankcorp.com/overview.

Zeidler, M. (2017, October 22). 'Displacement is not beautiful': Critics slam Westbank's Fight for Beauty exhibition. *CBC News*. Retrieved from https://www.cbc.ca/news/canada/british-columbia/displacement-is-not-beautiful-critics-slam-real-estate-developer-s-art-exhibition-1.4366125.

Zukin, S., & Braslow, L. (2011). The lifecycle of New York's creative districts. *City, Culture and Society, 2*(3), 131–140.

# 15

# Taking the Weather with You: Remittances, Translocality, and the Climate Migrant Within

Laurie Parsons

As climate change becomes more prominent in political discourse, the nature and extent of human mobility in response to it is a question of growing relevance. In particular, the issue of displacement—migration in response to vulnerability—has become increasingly prominent, as governments and international organisations attempt to plan for the growing mobility of global southern populations. Yet whilst academics and policymakers alike have devoted much ink to the question, it is one that poses significant challenges both methodologically and conceptually. Policy and media discourse both reflect a persistent othering of the climate migrant, engendering an implicit but almost ubiquitous exceptionalism that greatly impedes analysis.

Yet, in reality, the displaced are not a cleanly divisible category. The impacts of the climate are both physical and portable, carried with or thrust upon migrants in the form of commitments, connections, and duties. Bound by these obligations, migrants take the weather with them when they travel. Whether internally or internationally migrating, social and financial flows in the form of remittances engender close-knit interlinkages between migrants and their sender communities, structuring behaviour and engendering a durable connection between migrants of all descriptions and spatially distant ecologies.

---

L. Parsons (✉)
Royal Holloway, University of London, Egham, UK
e-mail: Laurie.Parsons@rhul.ac.uk

© The Author(s) 2020
P. Adey et al. (eds.), *The Handbook of Displacement*,
https://doi.org/10.1007/978-3-030-47178-1_15

Taking this perspective as its starting point, this chapter explores the dynamic interrelationship of mobility and climate, outlining what a mobile lens on climate change can help to explain about human response to the environment. After highlighting the role of exceptionalist assumptions in shaping analysis around displacement and migration, the chapter will then proceed to highlight how translocal analysis may be used to reshape the analysis of climate migration. In this way, it will make a case for the need to consider the climate as an integrated component within wider systems of migration, linked into and co-constitutive of socioeconomic structures at multiple scales.

## Climate Change and Migration in Public Discourse

Herds, hoards, and horror stories; catastrophe and cataclysm. Ever a concept to capture the imagination, the topic of climate migration is inevitably colourful in its discussion. Newspapers proclaim a coming 'wave of climate migration' (Thompson 2018, p. 1) and that 'climate change will create the world's biggest refugee crisis' (Taylor 2017), whilst advocacy groups proclaim 'a growing global crisis' (Arcanjol 2018, p. 1) in which 'every second, one person is displaced by disaster' (IDMC 2017). Drawn from across the political spectrum, these views infuse the rhetoric of climate change communication with an apocalyptic character drawn from the 'weaponising' (Misra 2018, p. 1) of migration more broadly, defining the boundaries of a field by its 'quest for ever larger numbers' (Jakobeit and Methmann 2012, p. 301).

Alarmist perspectives 'seem to dominate the debates at the beginning of the 21st century' (Klepp 2017, p. 3). Yet in both their growth and range, the numbers belie the uncertainty underpinning the blurred boundary between migration, viewed generally as a form of adaptation, and displacement, which is viewed conversely as a manifestation of vulnerability (Adger et al. 2018). Is every migrant in a drought prone area a climate migrant, for example? Is somebody who migrates annually to avoid floods, but returns each year? Do existing migrants whose behaviour is changed by conditions on their family farm become climate migrants and if so, when? The permutations are endless, yet the importance of the issue keeps it firmly in public and academic consciousness, the quest for clarity subsumed beneath the fixity of the displacement category.

Perhaps as a result of this uncertain border between migration and displacement, calls are growing globally for greater recognition of climate refugees, as a group, a reality, and a political agenda. It has been a rocket of a concept. Since its coinage in the course of early re-examinations of the climate–migration nexus (see El-Hinnawi 1985; Brown 1976), it has featured in news outlets from *The New York Times* to the *Daily Mail*, from Fox News to the BBC, as a means to represent the human dimension of climate change. Moreover, it has in many ways been effective, bringing the topic to bear across a range of influence forums. The United Nations (UN 2018) and European Parliament (European Parliamentary Research Service 2018) have both issued statements on climate refugees in recent years, placing the issue at the forefront of global discourse.

Though such growing awareness is undeniably positive, there is a sententious underlying character to the issue's presentation. Climate refugees—like refugees more generally (Bleiker et al. 2013)—are presented as invaders, labelled by politicians and policymakers as 'swarms, floods and marauders' (Shariatmadari 2015, p. 1) to be managed and controlled, rather than people or citizens to be accommodated. As outlined by Baldwin (2013, p. 1474), for example, 'the figure of the climate-change migrant designates a form of racial Other prevalent throughout climate change discourse.' Whether characterised as threat, or as victim, the narratives that frame this discourse present the object as 'different from some purportedly normal, unmarked body' (Baldwin 2013, p. 1475).

This implicit distinction underpins analysis, as well as policy. Despite growing awareness of slow onset climate impacts (World Bank 2018; Stojanov et al. 2017), this typology reflects the implicit influence of classic economic models (e.g. Todaro 1969; Rostow 1960; Lewis 1954) on migration scholarship, segregating as voluntary that dimension of migration that is 'pulled' by wage differentials rather than 'pushed' by broader livelihood factors at origin. Consequently, whilst migration and displacement are recognised as existing along a continuum (Hugo 1996), there remains an implicit distinction between migration that is enforced by the climate on the one hand and 'the voluntary nature of movement and the ability to exercise choice,' which in practice is attributed to economic rationale (Adger et al. 2018).

Nevertheless, whilst standing as the basis for a number of influential studies on the human impact of the environment (e.g. Biermann and Boas 2008; Myers 2002; Jacobson 1988), this distinction is a controversial one. Critical scholars of the environment (e.g. Hulme 2008; Black 2001) have questioned the very concept of the 'environmental refugee' or 'climate refugee,' pointing instead to the inherent interrelationship between environmental and economic factors. Yet its persistence and growth indicates a tendentious exceptionalist trend in policy discourse on climate impacts. Whether labelled as refugees, or simply migrants, those responding to changes in their environment are firmly excluded from the business as usual of labour migration.

From this standpoint, therefore, this chapter will proceed in two parts. In the next section, it will highlight how the discursive othering of the climate migrant in public discourse feeds in both to the persistently dichotomous policy framings and unilinear analysis of environmental migration and displacement. The following section will thereafter demonstrate the potential of a translocal perspective on climate migration to provide insight into the complex interlinkages of climatic and economic factors in promulgating migration decisions.

## The Climate and Policy: Implicit Exceptionalism and the Other

Climate migrants have an image problem. Pictures of 'war, flight and human ruin' have become so ubiquitous as to challenge the iconic 'stranded polar bear' as the face of climate change (Whibey 2015, p. 1). This is in the first instance a dehumanising narrative, but more broadly it is an isolating one, contributing to the conceptual dislocation of climate migration from other forms of human mobility. Like the arctic megafauna with whom they vie for visual supremacy, climate migrants are left stranded in dismal isolation, a position as deleterious to the formulation of policy as to the formation of public opinion. Thus, regardless of the particular typology of climate and migration employed, climate migrants are invariably portrayed as 'climate barbarians at the gate' (Bettini 2013, p. 63). Rather than being the bearers of tangible contemporary misfortune, they become portents of a shadowy and uncertain future doom; their statistical aggregation compounded and reinforced by media discourse.

Yet beyond her deeply ingrained public image problem, the climate migrant's problems run deeper still. Media representations are not direct influencers of policy, but in reflecting and shaping how mobility in response to the climate is conceived, they have a wider resonance. In particular, the international framing of the issue—combined with its inherent problematisation—ensures that 'methodological nationalism' (Kalir 2013) pervades much migration policy. Consequently, where the impact of the climate on human mobility is concerned, the debate is dominated by a securitisation narrative, which views climate migrants either as requiring assistance themselves or as engendering a need for protection by others (White 2012).

Policy discourse surrounding the development of protective frameworks for climate refugees is a key example of this, with international organisations such as the UN and the EU concerned increasingly with the security of climate migrants' international mobility. The European Parliamentary Research Service (2018, p. 2), for example, has adopted the refugee categorisation for this explicit purpose, stating that 'there is a protection gap with regard to "climate refugees," who are neither clearly defined as a category nor covered by the 1951 Convention relating to the Status of Refugees (the 1951 Refugee Convention).' As they outline, the lack of political persecution hinders the attainment of refugee status for those displaced by the climate, resulting in a reliance on human rights frameworks, which are non-binding in many cases, or include little direct provision for environmental rights.

The implication, in other words, is that climate migrants are a special case regardless of their treatment by their national government, an extraordinary outlier for whom the usual protections have failed. Consequently, the noting of environmental issues in the recently agreed UN Global Compact for Migration has been acclaimed as 'a real breakthrough' in its recognition of the role of environmental pressures in shaping migration patterns (Kaelin, cited in Taylor 2019). The compact designates that signatories will:

> Cooperate to identify, develop and strengthen solutions for migrants compelled to leave their countries of origin due to slow-onset natural disasters, the adverse effects of climate change, and environmental degradation, such as desertification, land degradation, drought and sea level rise, including by devising planned relocation and visa options, in cases where adaptation in or return to their country of origin is not possible. (UN 2018, p. 9)

Nevertheless, although it provides the basis of a policy platform, the compact is in reality little more attuned to the nexus of climate and migration than its agglomeration of predecessors. In particular, the fundamental line of analytical distinction here is a national one, drawn between migrants who are able to adapt to their changing environment—that is, those who do not leave the country—and 'those who are compelled to leave their countries of origin … in cases where adaptation in or return to their country of origin is not possible' (UN 2018, p. 9). In other words, adaptation gives way to problematic migration the moment a border is crossed, a conceptualisation that fits poorly both with the international nature of many adaptation strategies and the domestically contained nature of the majority of struggles with the environment.

Indeed, this is something which even the World Bank—a noted purveyor of securitisation discourse (Taylor 2009)—has come recently to recognise, dedicating a high-profile report, *Groundswell*, to 'preparing for internal climate migration' (World Bank 2018, p. i). Setting aside the implicit futurism of the report's mandate, this is a development that begins to draw climate migration back into the fold of ordinary human behaviour by decoupling it from illegality. However, though such developments are welcome, analytical concerns persist. In particular, despite confounding the rather simplistic narrative of the climate refugee, much of the implicit exceptionalism persists through the report's focus on the marginal impact of climate change on human mobility.

This is in many ways an entirely valid concern. By focussing on the margins of the climate's impact, climate migration scholarship is aiming to establish a body of irrefutable evidence, or the sort that since the 1990s has succeeded in bringing 97 per cent of the scientific establishment around to the extraordinary claim of climate change (Cook et al. 2013). Consequently, though possible only in the rarefied realm of aggregate statistics, isolating the climate change margin of migration remains at the forefront of the agenda. A suite of high-profile studies have emerged in recent years (McGranahan et al. 2007; IPCC 2007; Anthoff et al. 2006) seeking to link sea level rises to migration at a national scale or greater.

Yet the cost of chasing clarity is detail. Confidence in the data on sea level rises may be relatively high—after all, they estimate displacement on the basis of the actual disappearance of inhabited land—but in reality, 'these territories represent a comparatively small fraction of a much greater problem' (Arcanjol 2018, p. 1). The majority of people whose mobility is influenced by climate change will not be fleeing rising seas, but moving into or out of rainfed croplands (World Bank 2018), patterns of behaviour that are more complex as well as more extensive.

Consequently, where attempts are made to explore climate migration in a broader sense, confidence in the projections drops significantly (IPCC 2014), undermined by significant data challenges (ODI 2017). The wider reality of climate migration is messy and complex, bound up so completely in the wider milieu of socioeconomic structures and forces as to be, in analytical terms, almost inseparable. Indeed, despite the temptation to affect a 'simplistic linkage of climate change and mass displacement and migration,' the climate does not influence behaviour as an independent variable, but as a 'complex interplay of factors beyond climate-related hazards that contribute to people's decision to move or stay in place' (Stapleton et al. 2017, p. 10).

Studies recognising such bottom-up complexity constitute what has been described as a 'third wave' (Arnall and Kothari 2015, p. 1) of climate migration research—following an initial focus on climate refugees and subsequent attention to 'drivers of climate migration'—which views climate migration as part of the socioeconomic milieu in which it is embedded. The approach has proved successful. Not only have such studies succeeded in highlighting the embeddedness of climate migrants in their environment, but, at long last, policymakers are beginning to come round to this idea. The most recent policy reports (e.g. World Bank 2018; Stapleton et al. 2017) have gone some way to recognising this complexity, acknowledging, in particular, that 'there is a lack of clarity as to the direct influence of climate change on human mobility' (Stapleton et al. 2017, p. 7).

Yet key problems endure. Though noting the absence of a 'straight line of causation from environmental stress to the movement of people' (World Bank 2018, p. 1), this latest World Bank report retains the explicit aim to 'isolate the portion of future changes in population distribution that can be attributed to internal climate migration' (World Bank 2018, p. 1), thereby erecting an analytical target that runs counter to its purported methodological nuance. The key questions are, as before: Does somebody move or not? and How many others also move?

Alternative questions, concerning the direction and—of perhaps even greater importance—the interrelatedness of movement, have been left relatively little attended, lost in the implicit binaries of adaptation and (im)mobility. Yet, in reality, the climate need not be the preeminent factor in mobility. A climate migrant is an above all a migrant, with the same suite of social, familial, and economic concerns as any other. Nevertheless, the deeply ingrained exceptionalist discourse attributed to climate migrants often obscures this recognition, resulting in a tautologous definitional logic wherein a climate migrant is somebody who is moved by the climate.

This is an anachronistic shortcoming. Indeed, as Potts (2010, p. 1) argues, 'the experiences of the last three decades have entirely undermined the utility of conceptualizing [migratory] processes in relation to unilinear models.' Yet, in the absence of a coherent alternative, the sense nevertheless persists that the only genuine impact of the climate on migration must, almost by definition, be a one-way journey. Clearly, this is an impediment to good research in the area. What is required is a methodological and analytical frame capable of determining how structural factors shape behaviour in response to the climate and how they in turn are shaped by climate response. Focusing on the translocal impacts of remittances, what follows will outline such a response.

## Reintegrating the Climate Migrant: Towards a Translocal Perspective

Just as the human experience of the climate does not begin with migration, neither does it end there. Migrants are never independent entities, but bring with them cultural (Erel 2010; Cederberg 2015) and socioeconomic (Parsons 2017) baggage that structures their behaviour long after their movement has ceased. Consequently, just as socioeconomic structures shape mobility, so too does mobility itself feed back into those same structures. As has long been clear in the literature both on seasonal labour migration (e.g. Shah 2006; Gardner and Osella 2003; de Haan and Rogaly 2002; Rogaly et al. 2001) and gender and migration (e.g. Jacka 2006; Curran and Saguy 2001; Chant and Radcliffe 1992; Tienda and Booth 1991), they are moulders and reshapers of their environment in both origin and destination.

In addition, this is a two-way process. Migrants' own experiences are continually constituted by conditions at origin, as well as destination, via two-way flows of remitted resources (e.g. Lindley 2013; Anh et al. 2012; Rigg et al. 2014) which reshape 'the socio-spatial context of migration' (Silvey 2001, p. 43). Migrants from the poorest households make burdensome 'sacrifices' in the urban space (Lindley 2009, p. 1326), whilst the linkages produced by social behaviour such as this appear to be one of the means by which wealthier migrants gain access to urban opportunity (Yu and Berryman 1996).

Nevertheless, analysis of migratory networks is one-sided in relation to the climate. Whilst it is generally accepted, for example, that social networks are a crucial dimension of climate migrants' adaptive strategies (Scheffran et al. 2012), analyses of such social resources tend to focus either on the negative dimensions of their breakdown (Torres and Casey 2017) or on the positive impacts of their persistence (Scheffran et al. 2012). Thus, evidence that 'the process of fleeing acute disasters and conflict may disperse family and community members' (Torres and Casey 2017, p. 4) is linked to outcomes wherein 'migrants and displaced persons often face societal stigma and marginalization as well as social isolation' (Torres and Casey 2017, p. 45).

Yet, rarely or never is the obverse considered: how might the maintenance of climate migrants' social ties result in isolation in stigma in their destination? After all, the scale and importance of remittances—a key indicator of these ongoing linkages—has long been recognised. The World Bank estimates that $601 billion was remitted in 2016 (World Bank 2016) and in certain countries, such as Cambodia, remittances sent by internal migrants may consistently reach almost 50 per cent of income (CARE International 2017).

There, as in many countries, modern sector income is the lifeline keeping rural people afloat, facilitating adaptations and transformations in responses to the changing climate.

Moreover, this is not solely a domestic issue. In an age where foreign remittances account for almost 6 per cent of GDP in low-income countries (World Bank 2017), information and capital flows have rendered major urban economies as central a site of climate change as low-lying atolls. Nevertheless, despite mounting evidence of the role of remittances in shaping responses to climate change at the local level (Musah-Surugu et al. 2018; Banerjee et al. 2017), studies in the academic literature have addressed the topic only in passing (Bendandi and Pauw 2016), leaving adaptation frameworks rooted in assumptions of unilinearity and discrete sender and destination environments.

Otherwise put, migration has long been a subject addressed with strong preconceptions and weak tools by climate scholars, whilst the changing climate has until recently figured only peripherally in migration analysis (Black et al. 2011). Nevertheless, the tools are available for a fruitful reconciliation. From the influence of rural environmental change on urban migrant behaviour to its intersection with questions of inequality and economic development, there remains great untapped value in the application of such frameworks to a more specific focus on environmental and climactic factors.

In particular, frameworks emphasising the environmental dimensions of, first, circularity (Castles and Ozkul 2014; Vertovec 2007) and, latterly, translocality (Greiner and Sakdapolrak 2013; Brickell and Datta 2011) offer a useful means to integrate a concern for environmental themes into the study of mobility in all its wide-ranging and nuanced forms. Not only do these frameworks offer a route into the cutting edge of migration scholarship more broadly, but a concrete and proven means by which to move away from persistently unilinear conceptions of climate migration.

In simple terms, translocal approaches crosscut the traditional binaries of rural/urban, migrant/non-migrant, and human/non-human, viewing origin and destination as linked in so fundamental a manner as to be mutually co-constitutive. Their analysis is therefore predicated on two recognitions: first, that environmental changes may motivate forms of mobility to operate translocally—that is, across multiple linked places simultaneously—and, second, that the climate's impact on mobility is both structured by and feeds back to the unequal distribution of resources.

Certainly, this is a concept which complicates, rather than simplifies, the analysis of climate migration. Nevertheless, it is one which is increasingly necessary to capture the complexities of mobility in the age of widespread mobile telephony and Internet access (Dekker and Engbersen 2014). Whilst migrant behaviour has always been structured by sender-side conditions, the current era offers something new and unavoidable in the ease, speed, and availability of long-distance communication, facilitating repetition of, reaction to, and anticipation of communication in a manner previously unseen (Dekker and Engbersen 2014).

Overarchingly, the result of such developments is a greatly increased mutual sensitivity to conditions in origins and destination; yet remittances, in particular, play a far more finely attuned role than in the past. A medical emergency, pressing debt, or failed crop is not something to be discovered upon a migrant's return, but a phenomenon experienced in real time. Via the catalyst of technological development, social and familial ties communicate acute and chronic stresses over hundreds or even thousands of miles, ensuring their effects are felt in origin and destination alike. Migrant factory workers experience the greatest economic stress during their rural households' planting season (Lawreniuk and Parsons 2018), whilst economic migrants settled in far-off Western cities bear the burden of disaster relief in the Philippines (Dalgas 2018).

Clearly then, these relationships have implications for how climate change and migration are linked. Rather than being merely a case of those displaced by the immediate impact of disasters, the durable linkages than connect people and places post-migration mean that even existing migrants may become climate migrants over time, as they begin to adopt responsibility for environmental impacts on their sender household. In a related manner, moreover, those same migrations may structure how the climate is experienced on the sender side, potentially providing the resources to supersede, or even benefit from, the climactic changes underway (Sakdapolrak et al. 2016).

Indeed, this is the key point at hand. The climate's impact on human behaviour is not a direct one, but is articulated through the lens of positionality (Kaijser and Kronsell 2014) and resources (Raleigh et al. 2008). Depending on one's livelihood, an extended period of rain or dry weather may be a great boon, an irrelevance, or the harbinger of ruin. A storm, similarly, may destroy a palm leaf house, but leave a brick one entirely unaffected. Equally, the ability to borrow money from a migrant relative may mean the difference between abandoning and rebuilding the family farm in the wake of a disaster. Migration, in other words, structures both disaster impacts and the nature of future migrations.

Yet crucially, the reverse is also true: disasters have shock waves far beyond their immediate impact, leaving a longstanding mark on livelihoods as they manifest across both time and space. The pressures borne by current and prospective migrants to compensate for the impacts of the climate on their translocal households are not costless, but structure decision-making in destination areas, through the medium of remittances. In the short term, this places considerable strain on the migrant livelihoods of those whose households face the greatest demands (Lindley 2013), but it is over the longer run that the impacts of environmental precarity are most strongly felt. Those whose sender households demand the highest levels of remittance—a function often of their endurance of the greatest struggles with the climate—have less available to invest in integrating into destination areas both economically and socially (Torres and Casey 2017) and thus less ability to advance their livelihoods over time (Parsons 2016).

Left to subsist on what their sender households can spare, migrant workers like these are therefore a key site of environmental inequality. They have never fled disaster and may well have had economic sums in mind when they set off on their journeys initially; indeed, the environmental dimension of their behaviour and choices is in many ways subsumed within a wider set of constraints. There is no clearly definable group, behaviour, or margin of activity to attribute to the changing climate. Nevertheless, their ongoing connection to areas vulnerable to climactic instability ensures that they remain enmeshed within processes of environmental change.

This is in some ways a daunting task, yet the analytical tools are already in place to bring far greater nuance to our understanding of mobility in response to the climate. What needs to change is the exceptionalism attributed to this form of movement. Climate migration is not only that which follows disasters, but that which emerges from slower and far more complex pressures. Crucially, moreover, it is also that which begins to exert an influence on existing migrations over time, structuring behaviour and shaping outcomes for migrants of all descriptions. Those affected in this way are not hoards, swarms, nor even groups; they are not coming, but in many cases have already arrived. Thus, what is necessary to understand their behaviour is not a discrete field of inquiry, but the integration of a climate dimension into the study of climate migration as a whole.

In simple terms, this means that the climate migrant as an analytical object must be dispensed with, in favour of the climate as a mobile analytical theme. The aim must be not to ask 'Who is a climate migrant?,' but rather to ask of any subject: How has the climate affected this person's mobility? Moreover, this is a question to be asked without the constraint of geographical boundaries. As climate change and migration—both domestic and international—continue to proceed in parallel, greater attention must be paid to the linkages and overlaps that conjoin these two processes. In place of a deeply rooted exceptionalism in a media discourse of 'awe and dread' (O'Neill and Smith 2014), what must be recognised is 'the normality of translocal lives' (Etzold and Mallick 2016, p. 105) and thus the everyday nature of the climate migrant within.

## Conclusion

Combining doom-laden fears of a dying planet with the intensifying politicisation of human mobility, climate migration was always set to be a hotly contested topic. Increasingly labelled refugees to denote both placelessness and political abandonment, climate migrants are persistently othered by the language of displacement, segregated from the everyday business of economic migration by media narratives and the microcosmic formulation of policy around those cases most analytically distinct. Yet whilst this approach has long helped to draw attention to an issue of growing importance, it is an impediment to effective analysis. The reality of climate migration is messy and confusing. Stripped of her refugee label, a climate migrant looks like any other mobile person: a product of economic and social, as well as climactic, conditions.

To exemplify this, it is useful to abandon, for a moment, the concept of climate change migration entirely. Rising, carbon induced, tides over low-lying island atolls may capture the imagination, but in purely numerical terms such scenarios are outliers. In most cases, there is nothing inherently different about the mobility that occurs in response to climate change—defined as the anthropocentric patterns of change currently underway as a result of human activity—and that driven merely by the climate in general.

Indeed, although climate change will engender some environmental changes that have not been seen before, it is in general a process of change by degree, not kind. Changes in the climate are shifting the probability of certain events, their geography, location, and scale. Yet, they are not reinventing the weather. As a result, much of what we need to know about how people respond to changes in the climate is already exemplified, recorded, and available. Case studies in Africa (Musah-Surugu et al. 2018) and Asia (Etzold and Mallick 2016) demonstrate clearly the interaction of ecological with economic and social factors, engendering a complex milieu whose analysis necessitates utilising the varied tools of social scientific analysis, rather than any attempt to bring the influence of the climate down to a single discrete and non-fungible variable.

Yet, the implicitly discrete analytical character attributed to displacement continues to pervade analysis, enforcing a unilinear perspective on a highly complex and multifaceted issue. Perhaps most damaging of all, it introduces the persistent fallacy that the climate acts on a person once and once only, a perspective deeply intertwined with historical assumptions in migration literature, but increasingly recognised as flawed. With respect to climate migration, it is especially so. Far from being dim and distant individuals, the climate migrant of greatest relevance is the climate migrant within: ordinary, integrated, and above all local—the mobile bearers of ecologies far from their physical selves.

# References

Adger, W. N., de Campos, R. S., & Mortreux, C. (2018). Mobility, displacement and migration, and their interactions with vulnerability and adaptation to environmental risks. In R. McLeman & F. Gemenne (Eds.), *Routledge handbook of environmental displacement and migration* (pp. 29–41). London: Routledge.

Anh, N. T., Rigg, J., Huong, L. T. T., & Dieu, D. T. (2012). Becoming and being urban in Hanoi: Rural–urban migration and relations in Viet Nam. *The Journal of Peasant Studies, 39*(5), 1103–1131.

Anthoff, D., Nicholls, R. J., Tol, R. S., & Vafeidis, A. T. (2006). *Global and regional exposure to large rises in sea-level: A sensitivity analysis*. Tyndall Centre for Climate Change Research. Working Paper 96. Norwich, UK: Tyndall Centre for Climate Change Research.

Arcanjol, M. (2018). *Climate migration: A growing global crisis*. Climate Institute. Retrieved February 14, 2019, from http://climate.org.

Arnall, A., & Kothari, U. (2015). Challenging climate change and migration discourse: Different understandings of timescale and temporality in the Maldives. *Global Environmental Change, 31*, 199–206.

Baldwin, A. (2013). Racialisation and the figure of the climate-change migrant. *Environment and Planning A, 45*(6), 1474–1490.

Banerjee, S., Kniveton, D., Black, R., & Bisht, S. (2017). *Do financial remittances build household-level adaptive capacity? A case study of flood-affected households in India*. KNOMAD Working Paper 18. Retrieved April 29, 2019, from https://www.knomad.org.

Bendandi, B., & Pauw, P. (2016). Remittances for adaptation: An 'alternative source' of international climate finance? In A. Milan, B. Schraven, K. Warner, & N. Cascone (Eds.), *Migration, risk management and climate change: Evidence and policy responses* (pp. 195–211). Cham, Switzerland: Springer.

Bettini, G. (2013). Climate barbarians at the gate? A critique of apocalyptic narratives on 'climate refugees'. *Geoforum, 45*, 63–72.

Biermann, F., & Boas, I. (2008). Protecting climate refugees: The case for a global protocol. *Environment: Science and Policy for Sustainable Development, 50*(6), 8–17.

Black, R. (2001). *Environmental refugees: Myth or reality?* United Nations High Commissioner for Refugees (UNHCR) Working Paper Number 4. Retrieved April 29, 2019, from https://www.unhcr.org.

Black, R., Adger, W. N., Arnell, N. W., Dercon, S., Geddes, A., & Thomas, D. (2011). The effect of environmental change on human migration. *Global Environmental Change, 21*, S3–S11.

Bleiker, R., Campbell, D., Hutchison, E., & Nicholson, X. (2013). The visual dehumanisation of refugees. *Australian Journal of Political Science, 48*(4), 398–416.

Brickell, K., & Datta, A. (Eds.). (2011). *Translocal geographies*. London: Ashgate Publishing, Ltd.

Brown, L. R. (1976). *World population trends: Signs of hope, signs of stress*. Worldwatch Paper 8. Washington, DC: Worldwatch Institute.

CARE International. (2017). *'I know I cannot quit': The Prevalence and Productivity Cost of Sexual Harassment to the Cambodian Garment Industry*. Canberra: CARE Australia.

Castles, S., & Ozkul, D. (2014). Circular migration: Triple win, or a new label for temporary migration? In G. Batistella (Ed.), *Global and Asian perspectives on international migration* (pp. 27–49). Cham, Switzerland: Springer.

Cederberg, M. (2015). Embodied cultural capital and the study of ethnic inequalities. In L. Ryan, U. Erel, & A. D'Angelo (Eds.), *Migrant capital* (pp. 33–47). London: Palgrave Macmillan.

Chant, S., & Radcliffe, S. A. (1992). Migration and development: The importance of gender. In S. Chant (Ed.), *Gender and migration in developing countries* (pp. 1–29). London: Belhaven Press.

Cook, J., Nuccitelli, D., Green, S. A., Richardson, M., Winkler, B., Painting, R., et al. (2013). Quantifying the consensus on anthropogenic global warming in the scientific literature. *Environmental Research Letters, 8*(2), 1–7.

Curran, S. R., & Saguy, A. C. (2001). Migration and cultural change: A role for gender and social networks? *Journal of International Women's Studies, 2*(3), 54–77.

Dalgas, K. (2018). Translocal disaster interventions: The role of individual relief channels in Philippine disasters. *Journal of Contingencies and Crisis Management, 26*(3), 377–384.

Dekker, R., & Engbersen, G. (2014). How social media transform migrant networks and facilitate migration. *Global Networks, 14*(4), 401–418.

El-Hinnawi, E. (1985). *Environmental refugees*. Nairobi, Kenya: United Nations Environmental Programme.

Erel, U. (2010). Migrating cultural capital: Bourdieu in migration studies. *Sociology, 44*(4), 642–660.

Etzold, B., & Mallick, B. (2016). Moving beyond the focus on environmental migration towards recognizing the normality of translocal lives: Insights from Bangladesh. In A. Milan, B. Schraven, K. Warner, & N. Cascone (Eds.), *Migration, risk management and climate change: Evidence and policy responses* (pp. 105–128). Cham, Switzerland: Springer.

European Parliamentary Research Service. (2018). *The concept of 'climate refugees': Towards a possible definition*. Retrieved April 29, 2019, from http://www.eprs.ep.parl.union.eu.

Gardner, K., & Osella, F. (2003). Migration, modernity and social transformation in South Asia: An overview. *Contributions to Indian Sociology, 37*(1–2), v–xxviii.

Greiner, C., & Sakdapolrak, P. (2013). Translocality: Concepts, applications and emerging research perspectives. *Geography Compass, 7*(5), 373–384.

de Haan, A., & Rogaly, B. (2002). Introduction: Migrant workers and their role in rural change. *Journal of Development Studies, 38*(5), 1–14.

Hugo, G. (1996). Environmental concerns and international migration. *International Migration Review, 30*, 105–131.

Hulme, M. (2008). The conquering of climate: Discourses of fear and their dissolution. *Geographical Journal, 174*(1), 5–16.

IDMC (International Displacement Monitoring Centre). (2017). *Global report on internal displacement*. Geneva, Switzerland: IDMC.

IPCC (Intergovernmental Panel on Climate Change). (2007). *Climate change 2007: Impacts, adaptation, and vulnerability*. Geneva, Switzerland: IPCC.

IPCC (Intergovernmental Panel on Climate Change). (2014). *Climate change 2014: Synthesis report. Contribution of working groups I, II and III to the fifth assessment report of the intergovernmental panel on climate change*. Geneva, Switzerland: IPCC.

Jacka, T. (2006). *Rural women in urban China: Gender, migration and social change*. Armonk, NY: ME Sharpe.

Jacobson, J. L. (1988). *Environmental refugees: A yardstick of habitability*. Worldwatch Paper 86. Washington, DC: Worldwatch Institute.

Jakobeit, C., & Methmann, C. (2012). 'Climate refugees' as dawning catastrophe? A critique of the dominant quest for numbers. In H. Brach, J. Scheffran, & M. Brzoska (Eds.), *Climate change, human security and violent conflict* (pp. 301–314). Berlin: Springer.

Kaijser, A., & Kronsell, A. (2014). Climate change through the lens of intersectionality. *Environmental Politics, 23*(3), 417–433.

Kalir, B. (2013). Moving subjects, stagnant paradigms: Can the 'mobilities paradigm' transcend methodological nationalism? *Journal of Ethnic and Migration Studies, 39*(2), 311–327.

Klepp, S. (2017). Climate change and migration. In *Oxford research encyclopaedia, climate science* (pp. 1–35). Oxford, UK: Oxford University Press.

Lawreniuk, S., & Parsons, L. (2018). For a few dollars more: Towards a translocal mobilities of labour activism in Cambodia. *Geoforum, 92*, 26–35.

Lewis, W. A. (1954). Economic development with unlimited supplies of labour. *The Manchester School, 22*(2), 139–191.

Lindley, A. (2009). The early-morning phonecall: Remittances from a refugee diaspora perspective. *Journal of Ethnic and Migration Studies, 35*(8), 1315–1334.

Lindley, A. (2013). *The early morning phone call: Somali refugees' remittances.* Oxford: Berghahn Books.

McGranahan, G., Balk, D., & Anderson, B. (2007). The rising tide: Assessing the risks of climate change and human settlements in low elevation coastal zones. *Environment and Urbanization, 19*(1), 17–37.

Misra, T. (2018, 2 November). *On weaponising migration.* City Lab. Retrieved February 14, 2019, from http://www.citylab.com.

Musah-Surugu, J. I., Owusu, K., Yankson, P. W. K., & Ayisi, E. K. (2018). Mainstreaming climate change into local governance: Financing and budgetary compliance in selected local governments in Ghana. *Development in Practice, 28*(1), 65–80.

Myers, N. (2002). Environmental refugees: A growing phenomenon of the 21st century. *Philosophical Transactions of the Royal Society of London. Series B: Biological Sciences, 357*(1420), 609–613.

O'Neill, S. J., & Smith, N. (2014). Climate change and visual imagery. *Wiley Interdisciplinary Reviews: Climate Change, 5*(1), 73–87.

ODI (Overseas Development Institute). (2017). *Climate change, migration and displacement: The need for a risk-informed and coherent approach.* London: ODI.

Parsons, L. (2016). Mobile inequality: Remittances and social network centrality in Cambodian migrant livelihoods. *Migration Studies, 4*(2), 154–181.

Parsons, L. (2017). Multi-scalar inequality: Structured mobility and the narrative construction of scale in translocal Cambodia. *Geoforum, 85*, 187–196.

Potts, D. (2010). *Circular migration in Zimbabwe and contemporary sub-Saharan Africa.* Rochester, NY: Boydell & Brewer.

Raleigh, C., Jordan, L., & Salehyan, I. (2008). *Assessing the impact of climate change on migration and conflict.* Paper commissioned by the World Bank Group for the Social Dimensions of Climate Change Workshop, Washington, DC.

Rigg, J., Nguyen, T. A., & Luong, T. T. H. (2014). The texture of livelihoods: Migration and making a living in Hanoi. *Journal of Development Studies, 50*(3), 368–382.

Rogaly, B., Biswas, J., Coppard, D., Rafique, A., Rana, K., & Sengupta, A. (2001). Seasonal migration, social change and migrants' rights: Lessons from West Bengal. *Economic and Political Weekly, 36*(49), 4547–4559.

Rostow, W. W. (1960). *The stages of growth: A non-communist manifesto.* Cambridge: Cambridge University Press.

Sakdapolrak, P., Naruchaikusol, S., Ober, K., Peth, S., Porst, L., Rockenbauch, T., & Tolo, V. (2016). Migration in a changing climate. Towards a translocal social resilience approach. *DIE ERDE – Journal of the Geographical Society of Berlin, 147*(2), 81–94.

Scheffran, J., Marmer, E., & Sow, P. (2012). Migration as a contribution to resilience and innovation in climate adaptation: Social networks and co-development in Northwest Africa. *Applied Geography, 33*, 119–127.

Shah, A. (2006). The labour of love: Seasonal migration from Jharkhand to the brick kilns of other states in India. *Contributions to Indian Sociology, 40*(1), 91–118.

Shariatmadari, D. (2015, 10 August). Swarms, floods and marauders: The toxic metaphors of the migration debate. *The Guardian*. November 15, 2018, from http://www.theguardian.com.

Silvey, R. M. (2001). Migration under crisis; Household safety nets in Indonesia's economic collapse. *Geoforum, 32*(1), 33–45.

Stapleton, S. O., Nadin, R., Watson, C., & Kellett, J. (2017). *Climate change, migration and displacement: The need for a risk-informed and coherent approach*. London: Overseas Development Institute.

Stojanov, R., Duží, B., Němec, D., & Procházka, D. (2017). *Slow onset climate change impacts in Maldives and population movement from islanders' perspective*. KNOMAD Working Paper 20. Retrieved April 29, 2019, from https://www.knomad.org.

Taylor, M. (2009). Displacing insecurity in a divided world: Global security, international development and the endless accumulation of capital. *Third World Quarterly, 30*(1), 147–162.

Taylor, M. (2017, 2 November). Climate change 'will create world's biggest refugee crisis'. *The Guardian*. Retrieved February 14, 2019, from http://www.theguardian.com.

Taylor, L. (2019, 19 July). UN's New Migration Pact Offers Hope for Climate Refugees. *Global Citizen*. Retrieved April 29, 2019, from https://www.globalcitizen.org.

Thompson, A. (2018, 23 March). Wave of climate migration looms, but it 'doesn't have to be a crisis'. *Scientific American*. Retrieved April 26, 2019, from https://www.scientificamerican.com/.

Tienda, M., & Booth, K. (1991). Gender, migration and social change. *International Sociology, 6*(1), 51–72.

Todaro, M. P. (1969). A model of labor migration and urban unemployment in less developed countries. *The American Economic Review, 59*(1), 138–148.

Torres, J. M., & Casey, J. A. (2017). The centrality of social ties to climate migration and mental health. *BMC Public Health, 17*(1), 600.

UN (United Nations). (2018). *Report of the United Nations high commissioner for refugees*. New York: United Nations.

Vertovec, S. (2007). *Circular migration: The way forward in global policy?* Working Paper Number 4. International Migration Institute, University of Oxford. Retrieved April 29, 2019, from https://ora.ox.ac.uk.

Whibey, J. (2015, 14 December). Nuancing 'climate refugee language and images'. *Yale Climate Connections*. Retrieved January 31, 2020, from https://www.yaleclimateconnections.org.

White, G. (2012). The 'securitization' of climate-induced migration. In J. H. Khory (Ed.), *Global migration: Challenges in the 21st century* (pp. 17–44). New York: Palgrave Macmillan.

World Bank. (2016). *Migration and remittances factbook 2016*. Washington, DC: World Bank.

World Bank. (2017). *Migration and remittances: Recent developments and outlook; Special topic: Global compact on migration*. Migration and Development Brief 27. Washington, DC: World Bank.

World Bank. (2018). *Groundswell: Preparing for internal climate migration*. Washington, DC: World Bank Press.

Yu, P., & Berryman, D. L. (1996). The relationship among self-esteem, acculturation, and recreation participation of recently arrived Chinese immigrant adolescents. *Journal of Leisure Research, 28*, 251–273.

# 16

## Barbed Displacement: Walls to the Disciplined Migrant

### Bénédicte Michalon

In 2016, the French state hired a private jet to transport migrants arrested at the Calais Jungle to several detention centres and dozens of buses to convey those who had agreed to stay in the various shelters, hastily opened across the country, offered by the state. In August 2018, people rescued at sea and transferred by the naval vessel *Di Ciotti* were refused entry on to Italian soil. The vessel, which docked at the Port of Catania, became a de facto place of detention for ten days. The Italian Ministry of the Interior tried to pressure other countries within the European Union (EU) to take in some of the retained migrants. In October 2018, the shareholders of the Australian airline company Qantas voted on a motion requiring the company to cease transporting people deported by the Australian government, for fear of being accused of not respecting the rights of asylum seekers and refugees. These three episodes illustrate the close relationship between migration policy and the more or less coerced displacement of migrants.

The link between migration policy and forced displacement may surprise. To date, research in the social sciences has more commonly overlooked procedures for immobilising migrants, particularly in detention centres. Indeed, the 'ostensible rationale' of places of confinement may be considered to be to halt mobility (Philo 2014, p. 495). Thus designed, confinement constitutes the exact opposite of mobility understood as the manifestation of a subject's autonomy and freedom (Bauman 2000). Yet the particular form of biopower

---

B. Michalon (✉)
French CNRS, Bordeaux, France
e-mail: benedicte.michalon@cnrs.fr

established by confinement cannot exist without multiple kinds of mobility—of people, ideas, knowledge, and technology.

Displacements triggered by migration policy give way to further questioning of the relationship between mobility and power, examining the way mobility is used to better control society. The restrictive nature of these displacements is, nevertheless, anything but obvious to deal with. Just like expulsions, they can be perceived as 'a liberal-statist world order' (Gibney 2013, p. 116) from Western states. They can be seen as part of the state's legitimate and rightful power. The existing works on this issue nonetheless coincide with the idea that these mobilities lead to limiting an individual's freedom. They are considered a political technology that helps govern populations; the very efficient term 'governmental mobility' (Gill 2009) expresses this assumption, using a Foucauldian approach to power, government, and governmentality.

This chapter aims to synthesise the present state of knowledge on the governmental power of displacements carried out within the framework of migration control. Section "Dialogues: The Carceral, Mobility, and Borders" places this theme at the intersection of studies on the carceral, mobility, and borders. Section "Dimensions of a Technology of Power" outlines the main themes of study about the power and the restriction of individual freedom through governmental mobility. Section "Variations of Power" is dedicated to a typology of governmental mobility, showing how the barbed wire surrounding detention places also characterises, materially or metaphorically, the displacements of foreign nationals.

## Dialogues: The Carceral, Mobility, and Borders

Governmental mobility of foreign nationals today ranks highly among the tools used by states to control international migration. Their assessment by social scientists results from a dialogue between debates about the 'de-carceralisation' of the readings of the carceral, the political power of mobility, and the recomposition of borders.

### Mobility at the Heart of the Carceral

The modern era is said to be characterised both by an increase in the use of confinement (Simon 1998) and by reinforcement of constraints imposed on detainees, symptomatic of a 'punitive turn' (Garland 2001). It is also said to be characterised by a demand for 'mitigation' and transparency (Darley et al.

2013), which is supposed to lead to the opening of carceral settings to society and to 'normalisation' of life inside such places. This tension has given rise to heightened interest among researchers since the 2000s, translated, in particular, by the 'de-carceralisation' of the sociological approach of the carceral and a growing interest in mobility induced by carceral policies.

This opening of thinking on confinement involves the heightened interest in the links between carceral institutions and the society to which they belong, as in the centrality of the dialectics of opening/closure in the conception, justification, and implementation of the carceral (Moran 2015; Turner 2016). Networks of family, friends, and neighbours or even gangs of detainees and the flows they generate between the inside and the outside have been analysed (Comfort 2008; da Cunha 2008). The broadening of the scientific gaze has also made it possible to question the importance of confinement in biography and to bring to light 'institutional journeys' (Schliehe 2016). In addition, 'community control' restricts but doesn't fully prevent mobility; it suggests the extension of the 'carceral net' (de Larminat 2014).

Carceral policy increasingly relies on mobility, which tends to emerge as one of the conditions for maintaining carceral institutions (Mincke 2017). Indeed, certain measures aim to establish mobility as a mean to improve detention conditions. It involves giving greater autonomy to the detainees. However, this comes at a cost since in many cases it is accompanied by technological tools to strengthen control and to trace movements (Akrich and Callon 2004; Scheer 2016). The process seems to intensify in 'the ideology of mobility' (Mincke 2013). In carceral settings, as in many parts of social life, mobility appears as an imperative, an objective in itself: the 'good' detainee should not only be able to deal with deprivation of freedom but also to meet the requirement of mobility and to display their will and ability to set themselves in motion.

## Mobility as an Instrument of Power

The interest in governmental displacement of foreign nationals stems from another body of work devoted to mobility. The current era is divided between the increase of mobility through social life as a whole within the context of globalisation and the reinforcement of immobilisation processes (Hannam et al. 2006). This invites us to study relations between mobility and power (Cresswell 2006) and to consider the resulting changes in rationality, targets, and methods of power.

In a Foucauldian theoretical frame, mobility can be understood as the embodiment of individual and collective disciplinary norms, a notion conveyed by the expression 'disciplined mobility' (Packer 2003). The key issue here is understanding how disciplinary mechanisms transform the 'docile bodies,' dear to the French philosopher, into 'docile mobile bodies' (Packer 2003, p. 141). Other authors have approached mobility as a tool for authorities (Bærenholdt 2013). For instance, the colonial state was only made possible by means of transporting convicted populations or those reduced to slavery, as well as of the displacement of the native populations (Nicholson and Sheller 2016). The mobility imagined and organised by the state is one variant of forced mobility which can also be organised by local authorities, for example when bringing occupation of land to an end or when eliminating shanty towns or camps by dispersing their occupants (Blot and Spire 2014; Cousin and Legros 2014).

These kinds of setting in motion can be justified by protective aims. As such, disabled children placed in orphanages in Russia are moved by the institution. Their forced displacements, presented as integral to their care, are in fact disciplinary or even punitive (Disney 2017). Modern transformations of the state play a role in such uses of mobility. *New poverty management* has, since the 1970s, pushed both public and charity institutions and services providing aid to the homeless in the US to reduce their operating costs via quicker turnover of their beneficiaries, leading to an increase in unwanted mobility between such schemes' various facilities (DeVerteuil 2003).

Research on governmental mobility is partly grounded on understanding mobility as an instrument of power, especially in a preeminent domain of sovereign power: border control.

## Mobile Borders

A third area of debate focuses attention on governmental mobility. It is dedicated to the borders of states and their control and more specifically to attempts to resolve the tension between erasing borders and strengthening them. Many critical border studies argue that, far from disappearing, borders are recomposed and reappear in multiple forms and unexpected spaces depending on the evolution of flows, policies, and control techniques.

Borders take the form of places, of points in space (e.g., places of confinement of strangers, ports of entry, Mountz 2011a). They are also shaped in networks of border control professionals, but also and more and more people without a direct connection to this particular activity (Burridge et al. 2017). Feminist geographers, for their part, have highlighted the multi-scalar nature of borders and have examined the body of people subjected to the bordering process (Schmoll 2014). The use of biometrics and of 'smart borders' makes it possible to trace people through information about their body and personal life (Jeandesboz 2016).

Other studies argue that borders must be understood via their own mobility—mobility partly generated to respond to the mobility of those whom the state intends to control. The control of migration paths is translated into the relocation of the border on migration roads (Hess 2012). The construction of border-crossing points far from the national territory, such as islands, or the outsourcing of migration policies, testifies to the mobility of the state (Mountz 2011b). They participate in the transnationalisation of the borderwork of the state—'border enforcement never begins or ends at national boundaries' (Loyd and Mountz 2018, p. 17). The functions and the spaces of borders are more and more dissociated. Considering the border as mobile would allow one to break with 'its traditional fixity in time and space' (Amilhat-Szary and Giraut 2015, p. 6) and take into account its spatial instability.

Debates on the carceral, mobility, and borders intersect to define a framework of analysis. The governmental mobility of foreign nationals results from the emergence of mobility as a political and behavioural imperative in carceral institutions. It is a powerful instrument for the authorities as part of population control policies. Finally, governmental mobility contributes to transforming the forms of the border and its governance.

## Dimensions of a Technology of Power

The studies related to governmental mobility coalesce around how the displacements implement the state's constraint over foreign nationals. This main concern breaks down into four dimensions of power, related to theoretical considerations, means, and actors of mobility, as well as their effects on the displaced persons.

## Theoretical Readings of the Power of Governmental Mobility

Forced displacement is often conceptualised as a part of the carceral. Detention of foreign nationals has been theorised as a process stemming from the interdependence between confinement and mobility (Mountz et al. 2012). Other works have reinforced the reading of detention settings as places of flow production and management (Moran et al. 2012; Turner and Peters 2016), a point made in 1975 by Michel Foucault in qualifying prison as an 'inevitable motel' (Foucault 2001). Some authors view carceral spaces as circuits, to 'give priority to the connections between, around, within, and beyond carceral institutions' (Gill et al. 2016, p. 2).

Other research emphasises the continuous movement caused by governmental mobility. It supplements older analysis which had already underlined the capacity of mobility to generate mobility. Michel Foucault, once again, said in 1979 that those who were subjected to the vagrancy of repeated forced displacements were 'detained outside.' His formula has been widely disseminated in French-language research (Agier 2008), particularly in theorising links between alienation, confinement, and circulation (Makaremi and Kobelinsky 2009). The expression 'containment through mobility' highlights the role of forced mobility in the control over migrants and indicates that 'migration movements are obstructed in their autonomy not only by generating immobility and conditions of strandedness, nor through constant surveillance, but through administrative, political and legal measures that use (forced) mobility as a technique of government' (Tazzioli 2017, p. 2).

Finally, other authors underline the spatial, temporal, social, and political complexity of governmental mobility. This is what 'deportation studies' illustrate, devoted to the most well-known form of forced mobility. Their complexity has been highlighted by the notion of 'deportation corridors' (Drotbohm and Hasselberg 2015). It aims to broaden understanding of the phenomenon to avoid reducing it to the sole action of expulsion. It incorporates its historicity, the diversity of places and multi-directionality of the displacements it implies, the large array of actors it involves, as well as takes account of its normative and emotional power (Peutz 2006).

## Means of Transport and the Space-Time of Power

The space-time of displacement is characterised by the issues of power that are at stake (Blue 2015; Moran et al. 2012; Walters 2015). All means of transport are used in governmental mobility. They are generally originally designed for unconsented mobility. Their use in a restrictive framework raises the issue of using 'civil' means to impose power. They then become space-times for the surveillance of displaced persons. They materialise a 'vernacular violence,' 'a violence that is masked by its very ordinariness; [...] a violence made invisible by its very vernacularity' (Pugliese 2009, p. 153). In this respect, the airplane and the boat pose different technical challenges. In response to the difficulties in deporting people on commercial flights, states have created specific charter flights. This has made it possible to remove deportation operations from the public eye and confine them within an opaqueness (Walters 2017). With regard to boats, their materiality is transformed to fulfil new purposes of confinement and displacement. For instance, the boats of the Customs Maritime Enforcement Officers, requisitioned to intercept asylum seekers before their arrival on Australian shores and to take them to 'offshore processing centres,' were refitted to accommodate people rescued at sea for several days (Pickering 2014). Although initially foreign to any customs or border control purposes, merchant navy boats are witnessing a similar development in response to the construction of maritime stowaways as a 'risk' and the policy of confining this 'risk' (Maquet and Burtin 2012). Not only are ports secured to prevent boarding by stowaways, but the boats themselves are transformed into places of confinement—by the construction of cells, the dependence on personnel for primary needs, and surveillance by sailors (Clochard 2015).

Research has therefore demonstrated how the transport space-time materialises the power relations inherent to governmental mobility. It has also examined how transports materialise the delegation of migration control.

## Delegation of Power and Diversification of Rationalities

The delegation of migration control is especially marked in governmental mobility procedures and leads to in-depth changes in its methods and rationalities. Delegation of control firstly occurs between various levels of power. Governmental mobility is frequently decided upon and implemented outside the realms of national policy, for instance by local actors of migration policy, such as directors of asylum seeker centres or detention centres (Michalon 2013). In the UK, the Detention Escorting and Population Management Unit is responsible for organising transfers to and from detention centres, but performance of such transfers is entrusted to private sector companies who consequently negotiate with the state (Gill 2009). Subcontracting to private sector actors modifies the methods of displacements in as much as the priorities of state service providers may override concerns of the state. Delegation to economic players takes on an extreme form in the case of stowaways on board merchant navy ships. The ship's captain is responsible for finding a port of disembarkation, which is chosen according to the commercial route taken. The stowaways can be transferred to another boat belonging to the same company or sent back by plane in order to avoid slowing down maritime activity (Maquet 2014).

Displacement decisions may be taken to respond to the economic concerns of private partners rather than those of public authorities, especially in countries that massively subcontract control of foreign nationals. In the US, many transfers between detention centres are justified by the need to fill the establishments (Hiemstra 2013). Such displacements may also be grounded on humanitarian rhetoric: the dispersion of asylum seekers in the UK is justified by the necessity to provide them with housing and to process asylum more quickly (Darling 2011). The intertwining of control and humanitarian concerns also relies on the involvement of non-governmental organisations. Displacement of migrants from the Calais Jungle in 2015–2016 was only possible due to certain associations agreeing to take part in the operations decided upon by the French authorities (Agier et al. 2018).

Research into governmental mobility confirms that the diversification of migration policies' actors increases the confrontation or hybridisation of rationalities (Pickering 2014). Security-based, economic, but also humanitarian logic all contribute to the forced mobility of foreign nationals.

## Displacing to Render 'Deportable'

The geographical and social effects of governmental mobility have been addressed in the literature. They produce their specific channels, networks, and hubs (Walters 2017) and structure the physical and social space. The confinement of foreign nationals is often justified by the preparation and/or implementation of such forced mobility. In turn, forced mobility is a condition for confinement. The resulting complex journeys are made up of phases of confinement and phases of mobility, a sort of 'confinement in movement' (Makaremi and Kobelinsky 2008, p. 3). They may adopt a circular form (Michalon 2013) or be chaotic in their geography (Hiemstra 2013). Such geography is also hidden: the forced displacements of foreign nationals are generally not recorded or made public (Conlon and Hiemstra 2014). Their invisibility comes in addition to that of the detention centres. Lastly, such mobility produces a specific experience of dispossession of the migration project, which is not any longer decided by the migrant himself/herself but by other stakeholders. It contributes to a broader feeling of self-dispossession (Michalon 2013).

Governmental mobility deeply affects the social relations of the people displaced, as they sometimes aim at normalising behaviours. The fragile social balances that prevail in carceral places are modified by the departure and/or arrival of new detainees: as places of social hierarchies, the relationships between detainees on the one hand, and between detainees and surveillance agents on the other, are disrupted. The transferred person must discover and adapt to a new social environment; for some time, it isolates and weakens him/her. Moreover, governmental displacement affects more than the foreign nationals; it impacts the various professional bodies that interact with them, such as surveillance officers, lawyers, and associations (Gill 2009, 2016).

Governmental mobility has an isolating effect. It separates foreign nationals from their entourage and from any social support. It may marginalise them, be it in camps of all types as in depressed urban zones (Darling 2016). It accentuates their *deportability* (De Genova 2002). Analysis of the vectors, actors, and effects of governmental mobility draw the contours of this technology of power. They also bring into the fore its common features with other tools of migration policy: integration of 'civilian' techniques into police objectives, transformation of the state, and exclusion of foreign nationals. In addition, they illuminate another change of migration policy: their increasingly strong hold on the subject.

## Variations of Power

Governmental mobility of foreign nationals can be split into several categories according to the intensity and methods of the constraint. The subject is deprived of the possibility to decide whether to move or not, or, inversely, is forced to make this decision. Between the two, he/she is (forcefully) guided with regard to the choices to be made. The less the constraint to move is materially driven, the more the subject is transformed into the technology of his/her constraint, in a disciplinary logic. These various procedures correspond to different objectives: to displace, channel, or discipline or even to punish.

### Displacing: Imposed Mobility

Governmental mobility policies have initially been informed by scientists in their most restrictive versions, based on deprivation of freedom. They are decided upon and implemented by the authorities in charge of migration control, but also by directors of detention centres. The detainees only have a very limited power over the choice and methods of their displacement.

Deportations are the most radical type of governmental mobility. Their goal is to return the migrant to his/her country of origin. They have sharply increased since the 'deportation turn' in the 2000s, made possible by the growing human and material resources and legal provisions allocated thereto (Gibney 2008). They most often involve a period of detention, which is justified by the authorities by the need to prevent the person to be deported from fleeing.

Expulsion is often only a step in a path of constrained mobility. It is articulated with other forced displacements, towards detention places after arrest, or during transfers between facilities, whether these relate specifically to foreign nationals (Michalon 2013), are penal (Peutz 2006), or even medical. Similarly, such transfers can be organised towards the diplomatic missions of the supposed country of origin, as well as to or from 'ports of entry.'

Bilateral agreements lead to forced mobility through readmissions between states, such as the readmission agreement or Dublin Regulation in the EU, or the Safe Third Country Agreement between the US and Canada. Australia also resorts to such displacements within the framework of its Regional Cooperation Model with Indonesia, under which it can send migrants intercepted at sea to Indonesia and deliver them to the Indonesian authorities. The controversies surrounding such forced mobility have led states, the United Nations High Commissioner for Refugees (UNHCR), and the International Organization for Migration to promote returns. These improperly named 'voluntary returns' offer no alternative but return to the country of origin; their only 'advantage' for the migrant is the less restrictive conditions of return (Webber 2011). They attempt to make the subject responsible for the modalities of his/her displacement, a tendency that is found in other forms of governmental mobility.

## Channelling: Monitored Mobility

Other schemes allow the subject slightly greater room for manoeuvre regarding his/her mobility, which however remains significantly controlled. His/her displacement is organised and implemented by other actors. Such mobility differs from the previous ones through greater geographical diversity, because it generally takes place between non-carceral spaces.

Monitored mobility occurs through programmes involving the dispersion of migrants on an international scale. The UNHCR's *refugee resettlement programme* and the EU's *relocation scheme* are based on selection conducted in the first country of reception. They are grounded in the distinction between 'refugees' and 'economic migrants,' with only the first category being eligible. The UNHCR's programme concerns people deemed to be 'vulnerable,' selected jointly by the UNHCR and a receiving country. The selected persons are then displaced to that country, which will provide them with lasting legal protection (Tissier Raffin 2018). The EU's relocation scheme selects according to nationality. Hotspots were opened in summer 2015 in Italy and Greece to identify migrants in order to facilitate the resettlement of refugees in other Member States and to coordinate the expulsion of other migrants. The selection is first made by the Greek and Italian authorities, after which potential host states proceed to the selection. Asylum seekers are finally offered a destination. If they accept it, they are transferred to the mainland by UNHCR and then to their place of installation by the authorities of the host country.[1]

The European relocation scheme is worth attention given the many movements to which it gives rise. Greece and Italy indeed use hotspots to control mobility of foreign nationals on their respective territories. In Greece, stays in hotspots were considerably lengthened in 2016 in order to slow the migrant traffic through the country and to limit the arrival of foreign nationals at the border with Macedonia, where thousands of people were blocked on the Greek side. In Italy, the hotspot of Taranto has served to forcibly displace migrants blocked in Ventimiglia and Como to the south; the transfers, repeated each week, were carried out by bus. Whilst some of these displaced migrants were deported, most of them were released and returned to the border by train a few days later (Tazzioli 2017).

These monitored mobility policies 'force them [migrants] to manifold rerouting and to a condition of permanent mobility' (Tazzioli 2017, p. 10). They are more about managing migration than preventing it completely. This idea about border sorting had already been advanced between the US and Mexico in the early 1990s (Kearney 1991) and reemerged in the control of so-called transit migration in Europe (Hess 2012). Governmental mobility reactivates this work of channelling, slowing down, or accelerating migration paths.

## Disciplining: The Subject and the Mobility

Governmental mobility includes a third profile. Here, the foreign national is authorised to move freely but within a frame delimited by the state.

Asylum seekers' processing and accommodation schemes have for a long time made asylum seekers responsible to the state for their mobility and immobility. Many countries use a policy of dispersing exiles throughout their territory (Robinson et al. 2003; Darling 2011). Asylum seekers generally cannot choose the place where they are assigned housing—if housing can be provided. A refusal implies loss of social rights. Furthermore, their right to mobility is limited by the state. For instance, in Romania, asylum seekers are not authorised to leave the municipality in which they are domiciled. Any movement beyond its confines requires permission from the authorities. This country is still perceived as a space of transit and the mobility of exiles tend to be interpreted as a proof that the request for asylum is unjustified (Michalon 2017). As such, whilst asylum seekers are not forbidden from travelling, it falls to them to restrict their movement in order to be perceived as credible in their asylum claim.

So-called alternative schemes to detention also operate through the discipline of foreign nationals. They are applied in environments referred to as 'open' and authorise mobility within certain limits. In France, house arrest goes hand in hand with restriction of movement within a given zone, which can only be left with authorisation from the local public authority or police headquarters. They are obliged to report to law enforcement agencies or police headquarters at regular intervals. These constraints give rise to many obstacles. Restriction of the right to mobility makes access to lawyers and associations complicated; the obligation of travelling to report to the authorities gives rise to difficulties in terms of transport but also with regard to carrying out daily activities. Any failure to observe these obligations can be punished by detention or even imprisonment for up to three years. In the UK, foreigners released from detention and placed under 'community control' are subject to similar constraints. They may also have to wear an electronic tag and to obey a curfew, so as to limit their movements (Klein and Williams 2012).

Through an obligation to restricted movement within the authorised zones and through movements resulting from the obligation to report to the authorities, the subject is expected to discipline his/her mobility. The state's control extends over the private, daily, and even intimate sphere.

## Punishing Through Mobility

Unlike penal instruments, foreign nationals' control schemes do not officially have a punitive purpose. However, many authors claim that their action depends on punitive mechanisms, that the persons under control experience them as punishment, and that this alters the notion of justice (Bosworth 2012, 2017). The increase in governmental mobility supports this argument. Indeed, they are frequently implemented for sanctioning acts considered to violate migration policies.

The failure to respect the legal requirements to which foreign nationals are subjected can lead to a variety of sanctions: loss of related rights, payment of fines, or, in detention centres, solitary confinement. Forced mobility is frequent, either via expulsion from the accommodation (e.g., in centres for asylum seekers) or by detention or even imprisonment for people who have not respected the conditions of their 'community control' (Michalon 2017).

Displacements may also be implemented due to behaviour deemed to be problematic from the viewpoint of the control officers. Those who disobey the written or tacit rules governing community life in a centre for asylum seekers or a detention centre may be transferred to another establishment. Failure to follow rotas detailing cleaning of communal spaces, contravening times and durations of authorised absences, and creating conflicts with cohabitants are all reasons for transfers invoked by the directors and personnel of these establishments.

Finally, the behaviours that are most likely to be punished by forced displacements are collective uprisings. Foreign nationals are not passive in the face of their conditions and the different forms of control to which they are subjected can give rise to protests or collective revolts (Clochard 2016). The personnel of detention centres are especially mindful of such events (Gill 2016; Hiemstra 2013). In such a case, the aim of displacement is not only to bring a halt to a movement of contestation but also to split up groups of protesters: punitive mobility is also a means of managing 'migrant mobs' (Tazzioli 2016).

The punitive dimension of migration control schemes is always lurking beneath the surface. Even if it is not their main purpose, it acts as a guarantee of their durability by imposing foreign nationals to adopt the 'right' behaviour expected of them.

## Conclusion

Mobility has long benefited from a positive view of researchers, who, as Peter Adey recalls, have made it 'a desperate passageway full of hope' or something 'banal and forgettable' (Adey 2017, p. XV). Work on the forced displacement of foreigners for migration policy brings a different light to this major social and political issue. It is located at the intersection of carceral policy, border control, and the use of mobility in public policies. Not only does it increase mobility but it also diversifies its causes, as well as its spatial, temporal, social, and political scales and modalities. Finally, governmental mobility responds to multiple and sometimes contradictory rationalities that all challenge the subject's freedom.

While the existing body of work has laid solid foundations, much still remains to be researched to fully understand how governmental mobility implements power on foreign nationals. Its most restrictive forms have been the most studied. The use of mobility as an instrument of discipline remains to be further documented. For example, the actual organisation of travel by asylum seekers or by foreigners under 'community control' could shed light on the relations between autonomy and constraint.

Governmental mobility is most often apprehended as an instrument of the reinforced control of foreign nationals. The increasing use of confinement goes hand in hand with the growing use of that special type of mobility. But in turn it generates surveillance practices of the foreign nationals to be moved. Some evolutions in Europe since 2015 bear witness to this. The *Ankerzentren*, created in 2018 in Germany, are intended for the accommodation of asylum seekers waiting to be moved either to be relocated in Germany or to be expelled. In France, the *Prahda* (one of the many Reception and Accommodation Programs for Asylum Seekers) was created in 2016 to accommodate asylum seekers to be transferred to another Member State under the Dublin agreement. Understanding these changes engages researchers to question how governmental mobility produces policies and practices of control of foreign nationals.

## Note

1. The information concerning the European procedure was collected by Emilie Hétreau within the ARRECO project on relocation of refugees in Europe (https://arreco.hypotheses.org/). I thank her and Estelle d'Halluin for kindly sharing it with me.

## References

Adey, P. (2017). *Mobility*. Abingdon, UK: Routledge.
Agier, M. (2008). *Gérer les indésirables. Des camps de réfugiés au gouvernement humanitaire*. Paris: Flammarion.

Agier, M., Bouagga, Y., Galisson, M., Hanappe, C., Pette, M., & Wannesson, P. (2018). *La jungle de Calais*. Paris: Presses Universitaires de France.

Akrich, M., & Callon, M. (2004). L'intrusion des entreprises privées ans le monde carcéral français : le Programme 13 000. In P. Artières & P. Lascoumes (Eds.), *Gouverner, enfermer. La prison, un modèle indépassable ?* (pp. 295–317). Paris: Press de Sciences Po.

Amilhat-Szary, A.-L., & Giraut, F. (2015). *Borderities and the politics of contemporary mobile borders*. Basingstoke, UK: Palgrave Macmillan.

Bærenholdt, J. O. (2013). Governmobility: The powers of mobility. *Mobilities, 8*(1), 20–34.

Bauman, Z. (2000). *Liquid modernity*. Cambridge, UK: Polity Press.

Blot, J., & Spire, A. (2014). Déguerpissements et conflits autour des légitimités citadines dans les villes du Sud. *L'Espace Politique, 22*(1). https://doi.org/10.4000/espacepolitique.2893.

Blue, E. (2015). Strange passages: Carceral mobility and the liminal in the catastrophic history of American deportation. *National Identities, 17*(2), 175–194.

Bosworth, M. (2012). Subjectivity and identity in detention: Punishment and society in a global age. *Theoretical Criminology, 16*(2), 123–140. https://doi.org/10.1177/1362480612441116.

Bosworth, M. (2017). Immigration detention, punishment and the transformation of justice. *Social & Legal Studies, 28*(1), 81–99. https://doi.org/10.1177/0964663917747341.

Burridge, A., Gill, N., Kocher, A., & Martin, L. (2017). Polymorphic borders. *Territory, Politics, Governance, 5*(3), 239–251. https://doi.org/10.1080/21622671.2017.1297253.

Clochard, O. (2015). Enfermés à bord des navires de la marine marchande. *Annales de géographie, 702–703* (2–3), 185–207. https://doi.org/10.3917/ag.702.0185.

Clochard, O. (2016). Révoltes, protestations et 'résistances du quotidien': Des étrangers à l'épreuve de la détention. *Migrations Société, 164*(2), 57–72.

Comfort, M. (2008). *Doing time together. Love and family in the shadow of the prison*. Chicago, IL: University of Chicago Press.

Conlon, D., & Hiemstra, N. (2014). Examining the everyday micro-economies of migrant detention in the United States. *Geographica Helvetica, 69*(5), 335–344. https://doi.org/10.5194/gh-69-335-2014.

Cousin, G., & Legros, O. (2014). Gouverner par l'évacuation ? L'exemple des 'campements illicites' en Seine-Saint-Denis. *Annales de géographie, 700*(6), 1262–1284. https://doi.org/10.3917/ag.700.1262.

Cresswell, T. (2006). *On the move. Mobility in the modern Western world*. New York and London: Routledge.

da Cunha, M. I. P. (2008). Closed circuits: Kinship, neighborhood and incarceration in urban Portugal. *Ethnography, 9*(3), 325–350. https://doi.org/10.1177/1466138108094974.

Darley, M., Lancelevée, C., & Michalon, B. (2013). Où sont les murs? Penser l'enfermement en sciences sociales. *Cultures & Conflits, 90*(2), 7–20.

Darling, J. (2011). Domopolitics, governmentality and the regulation of asylum accommodation. *Political Geography, 30*(5), 263–271. https://doi.org/10.1016/j.polgeo.2011.04.011.

Darling, J. (2016). Forced migration and the city: Irregularity, informality, and the politics of presence. *Progress in Human Geography, 41*(2), 178–198. https://doi.org/10.1177/0309132516629004.

De Genova, N. (2002). Migrant 'illegality' and deportability in everyday life. *Annual Review of Anthropology, 31*, 419–447.

DeVerteuil, G. (2003). Homeless mobility, institutional settings, and the new poverty management. *Environment and Planning A: Economy and Space, 35*(2), 361–379. https://doi.org/10.1068/a35205.

Disney, T. (2017). The orphanage as an institution of coercive mobility. *Environment and Planning A: Economy and Space, 49*(8), 1905–1921. https://doi.org/10.1177/0308518X17711181.

Drotbohm, H., & Hasselberg, I. (2015). Introduction. Deportation, anxiety, justice: New ethnographic perspectives. *Journal of Ethnic and Migration Studies, 41*(4), 551–562.

Foucault, M. (2001). 'Préface' in Jackson (B.), Leurs prisons. Autobiographies de prisonniers américains, Paris, Plon, 1975 (texte n°144). In D. Defert, F. Ewald, & J. Lgrange (Eds.), *Dits et écrits I. 1954–1975* (pp. 1556–1559). Paris: Gallimard.

Garland, D. (2001). *The culture of control: Crime and social order in contemporary society*. Chicago, IL: University of Chicago Press.

Gibney, M. J. (2008). Asylum and the expansion of deportation in the United Kingdom. *Government and Opposition, 43*(2), 146–167. https://doi.org/10.1111/j.1477-7053.2007.00249.x.

Gibney, M. J. (2013). Is deportation a form of forced migration? *Refugee Survey Quarterly, 32*(2), 116–129.

Gill, N. (2009). Governmental mobility: The power effects of the movement of detained asylum seekers around Britain's detention estate. *Political Geography, 28*(3), 186–196. https://doi.org/10.1016/j.polgeo.2009.05.003.

Gill, N. (2016). *Nothing personal? Geographies of governing and activism in the British asylum system*. Oxford, UK: Wiley Blackwell.

Gill, N., Conlon, D., Moran, D., & Burridge, A. (2016). Carceral circuitry: New directions in carceral geography. *Progress in Human Geography, 42*(2), 183–204. https://doi.org/10.1177/0309132516671823.

Hannam, K., Sheller, M., & Urry, J. (2006). Editorial: Mobilities, immobilities and moorings. *Mobilities, 1*(1), 1–22. https://doi.org/10.1080/17450100500489189.

Hess, S. (2012). Gefangen in der Mobilität. Prekäre Zonierungsprozesse an den Rändern Europas. *Behemoth. A Journal on Civilisation, 5*(1), 8–29.

Hiemstra, N. (2013). 'You don't even know where you are': Chaotic geographies of US migrant detention and deportation. In D. Moran, N. Gill, & D. Colon (Eds.),

*Carceral geographies: Mobility and agency in imprisonment and migrant detention* (pp. 57–75). Farnham, UK: Ashgate.

Jeandesboz, J. (2016). Smartening border security in the European Union: An associational inquiry. *Security Dialogue, 47*(4), 292–309. https://doi.org/10.1177/0967010616650226.

Kearney, M. (1991). Borders and boundaries of state and self at the end of empire. *Journal of Historical Sociology, 4*(1), 52–74. https://doi.org/10.1111/j.1467-6443.1991.tb00116.x.

Klein, A., & Williams, L. (2012). Immigration detention in the community: Research on the experiences of migrants released from detention centres in the UK. *Population, Space and Place, 18*(6), 741–753. https://doi.org/10.1002/psp.1725.

de Larminat, X. (2014). *Hors des murs. L'exécution des peines en milieu ouvert*. Paris: Presses Universitaires de France.

Loyd, J. M., & Mountz, A. (2018). *Boats, borders, and bases*. Oakland: University of California Press.

Makaremi, C., & Kobelinsky, C. (2008). Editorial. Confinement des étrangers: Entre circulation et enfermement. *Cultures & Conflits, 71*(3), 7–11.

Makaremi, C., & Kobelinsky, C. (2009). *Enfermés dehors. Enquêtes sur le confinement des étrangers*. Bellecombe-en-Bauge, France: Editions du Croquant.

Maquet, P. (2014). Passages clandestins, les assureurs s'en mêlent. *Plein Droit, 2*(101), 19–22.

Maquet, P., & Burtin, J. (2012). Les passagers clandestins sur les navires de marine marchande : de la gestion d'un événement à la production d'un risque. *L'Espace Politique, 16*(1). Retrieved from http://journals.openedition.org/espacepolitique/2279.

Michalon, B. (2013). Mobility and power in detention: The management of internal movement and governmental mobility in Romania. In D. Moran, N. Gill, & D. Colon (Eds.), *Carceral geographies: Mobility and agency in imprisonment and migrant detention* (pp. 37–55). Farnham, UK: Ashgate.

Michalon, B. (2017). Accommodation for asylum seekers and 'tolerance' in Romania. Governing foreigners by mobility? In J. Turner & K. Peters (Eds.), *Carceral mobilities: Interrogating movement in incarceration* (pp. 57–70). Abingdon, UK: Routledge.

Mincke, C. (2013). Mobilité et justice pénale. L'idéologie mobilitaire comme soubassements du managérialisme. *Droit et société, 2*(84), 359–389.

Mincke, C. (2017). Prison. Legitimacy through mobility? In J. Turner & K. Peters (Eds.), *Carceral mobilities: Interrogating movement in incarceration* (pp. 236–249). Abingdon, UK: Routledge.

Moran, D. (2015). *Carceral geography: Spaces and practices of incarceration*. Farnham, UK: Ashgate.

Moran, D., Piacentini, L., & Pallot, J. (2012). Disciplined mobility and carceral geography: Prisoner transport in Russia. *Transactions of the Institute of British Geographers, 37*(3), 446–460.

Mountz, A. (2011a). Specters at the port of entry: Understanding state mobilities through an ontology of exclusion. *Mobilities, 6*(3), 317–334.

Mountz, A. (2011b). The enforcement archipelago: Detention, haunting, and asylum on islands. *Political Geography, 30*(3), 118–128. https://doi.org/10.1016/j.polgeo.2011.01.005.

Mountz, A., Coddington, K., Catania, R. T., & Loyd, J. M. (2012). Conceptualizing detention: Mobility, containment, bordering, and exclusion. *Progress in Human Geography, 37*(4), 522–541. https://doi.org/10.1177/0309132512460903.

Nicholson, J. A., & Sheller, M. (2016). Race and the politics of mobility. Introduction. *Transfers, 6*(1). Retrieved from https://www.berghahnjournals.com/view/journals/transfers/6/1/trans060102.xml.

Packer, J. (2003). Disciplining mobility. Governing and Safety. In J. Bratich, J. Packer, & C. McCarthy (Eds.), *Foucault, cultural studies, and governmentality* (pp. 135–161). Albany, NY: State University of New York Press.

Peutz, N. (2006). Embarking on an anthropology of removal. *Current Anthropology, 47*(2), 217–241.

Philo, C. (2014). 'One must eliminate the effects of … diffuse circulation [and] their unstable and dangerous coagulation': Foucault and beyond the stopping of mobilities. *Mobilities, 9*(4), 493–511.

Pickering, S. (2014). Floating carceral spaces: Border enforcement and gender on the high seas. *Punishment & Society, 16*(2), 187–205. https://doi.org/10.1177/1462474513517018.

Pugliese, J. (2009). Civil modalities of refugee trauma, death and necrological transport. *Social Identities, 15*(1), 149–165. https://doi.org/10.1080/13504630802693687.

Robinson, V., Andersson, R., & Musterd, S. (2003). *Spreading the 'burden'?* (1st ed.). Bristol, UK: Bristol University Press. https://doi.org/10.2307/j.ctt1t894cb.

Scheer, D. (2016). Condamnés à l'immobilité. La prison contemporaine en quête de discipline. *SociologieS*. Retrieved from https://journals.openedition.org/sociologics/5176.

Schliehe, A. K. (2016). *Tracing outsideness: Young women's institutional journeys and the geographies of closed space*. Doctoral dissertation, University of Glasgow, Glasgow.

Schmoll, C. (2014). Gendered spatialities of power in 'borderland' Europe: An approach through mobile and immobilised bodies. *International Journal of Migration and Border Studies, 1*(2), 173–189.

Simon, J. (1998). Refugees in a carceral age: The rebirth of immigration prisons in the United States. *Public Culture, 10*(3), 577–607.

Tazzioli, M. (2016). The government of migrant mobs: Temporary divisible multiplicities in border zones. *European Journal of Social Theory, 20*(4), 473–490. https://doi.org/10.1177/1368431016658894.

Tazzioli, M. (2017). Containment through mobility: Migrants' spatial disobediences and the reshaping of control through the hotspot system. *Journal of Ethnic and Migration Studies, 44*(16), 2764–2779. https://doi.org/10.1080/1369183X.2017.1401514.

Tissier Raffin, M. (2018). Entretien avec Pascal Brice, Directeur général de l'OFPRA : 'Entre continuité et modernisation · la diversification des missions de l'OFPRA'. *La Revue des droits de l'homme, 13*. Retrieved from http://journals.openedition.org/revdh/3681.

Turner, J. (2016). *The prison boundary. Between society and carceral space*. London: Palgrave Macmillan.

Turner, J., & Peters, K. (Eds.). (2016). *Carceral mobilities: Interrogating movement in incarceration*. Abingdon, UK: Routledge.

Walters, W. (2015). Migration, vehicles, and politics: Three theses on viapolitics. *European Journal of Social Theory, 18*(4), 469–488. https://doi.org/10.1177/1368431014554859.

Walters, W. (2017). Aviation as deportation infrastructure: Airports, planes, and expulsion. *Journal of Ethnic and Migration Studies, 44*(16), 2796–2817. https://doi.org/10.1080/1369183X.2017.1401517.

Webber, F. (2011). How voluntary are voluntary returns? *Race & Class, 52*(4), 98–107. https://doi.org/10.1177/0306396810396606.

# 17

# Technologies of Deportation

## William Walters

Today the controversial practice of deportation casts a long shadow across the turbulent landscape of migration policies, border regimes, refugee protection, and population flows. For many decades deportation operated as merely one amongst a range of tools states had at their disposal for the purposes of immigration enforcement (Collyer 2012). Since the early 1990s it has acquired much greater prominence. In many countries the deportation of illegalised migrants, rejected asylum seekers, non-citizen offenders, and other categories of unwanted foreigners (whose number includes many long-settled residents) has become a large-scale operation, accounting for significant levels of state funding, and employing specialised agencies and authorities. It is also a mode of human displacement that provokes widespread concern on the part of human rights groups and activists who see in deportation a violent activity that offers further evidence of the erosion of refugee and migrant protection regimes (Kanstroom 2007; Bloch and Schuster 2005; Khosravi 2009; Weber 2015; De Genova and Peutz 2010; Walters 2002). Scholarly interest in this deportation 'turn' (Gibney 2008) has grown accordingly to the point where deportation studies has emerged as a subfield of migration and security

---

I thank Peter Adey for helpful comments on an earlier draft of this chapter, as well as Nicholas De Genova, Huub Dijstelbloem, Clara Lecadet, and Martina Tazzioli for our ongoing conversation about borders, migration, and politics.

---

W. Walters (✉)
Carleton University, Ottawa, ON, Canada
e-mail: william.walters@carleton.ca

© The Author(s) 2020
P. Adey et al. (eds.), *The Handbook of Displacement*,
https://doi.org/10.1007/978-3-030-47178-1_17

research (Coutin 2015). During the 1990s migrations and mobilities were often debated in terms of globalisation theories that emphasised liquid flows and transnational citizenships. By comparison, contemporary interest in deportation is contributing to a picture, in some ways much more sobering, in which global processes are studied in conjunction with new forms of bordering, stratifying, and policing; a picture in which the very notion of the global is no longer synonymous with forces of transgression and unbounding but includes the multiplication of boundaries and controls (Golash-Boza 2015; Mezzadra and Neilson 2013).

This chapter offers a partial survey of recent interdisciplinary work on deportation. It is partial because it reads this literature in terms of a particular thematic focus on technologies of deportation. While the reference to technologies may suggest a Foucauldian interpretation, I understand the term somewhat ecumenically. For the present purposes it entails an emphasis less on the causes of this turn towards deportation, as important as they may be, and more a critical engagement with the practices, authorities, spaces, laws, procedures, devices that have been fashioned to give effect to deportation, as well as the political, economic, cultural effects generated by the prominence that deportation now has. My core argument is that the compulsory removal of people from one territory to another is legally, ethically, politically, and logistically an inherently difficult business. It is made much more tricky when states claim, as they generally do, to be respecting the fundamental human rights of those being removed, as well as the sovereignty of the states where deportees are being transported. Squaring this circle makes contemporary deportation a highly dynamic, volatile, and, in some ways, experimental domain. A focus on technologies offers a promising way to open the black box of deportation and capture this dynamism. As such, it reveals that deportation is not something fixed but a shifting and contradictory field of practices, programmes, strategies, and struggles.

A focus on technologies also brings an element of symmetry to the study of deportation. A growing body of scholarship is exploring 'technological borders' (Dijstelbloem and Meijer 2011), advancing a 'more-than-human' account of borders which highlights the various objects, knowledges, databases, and technical systems that mediate the control of mobility. In this chapter I suggest aspects of this technological turn can be fruitfully extended to our accounts of deportation.

The chapter is organised into four sections. First, I make the case for an expanded understanding of deportation looking to the concept of machine to move beyond a legal or state-centric view. Thinking in terms of machines builds on scholarship on border regimes, infrastructures, and systems. It is consistent with a focus on technologies, since it concerns questions about how deportation is effected, and captures the heterogeneous actors, elements, and processes which subtend the involuntary movement of illegalised people. Second, I move from considering the technologies which make up the machine to the question of deportation *as* a technology. What does deportation do? What social, economic, symbolic, and political functions has it come to be associated with? Third, I examine the rise of an ethnographic perspective on deportation, showing that this turn promotes understanding of the multiple spatialities and temporalities, and the various scales and dispersed sites of expulsion. Finally, I highlight some possible new directions for critical research on technologies of deportation.

A caveat is in order before proceeding. Today deportation is largely understood as a legalised and routinised form of expulsion from a state territory and an operation of immigration enforcement. I shall confine myself mostly to literature that largely shares this view. Nevertheless, it is important to register a point about expulsion in history. A wider view reveals the existence of multiple forms and logics of expulsion (Walters 2002; Kedar 1996). Historically expulsion has been used by states and other polities on racial and religious grounds as a weapon against political enemies, as an instrument of ethnic or social cleansing, as a practice of labour regulation, and as a mode of punishment and support for colonisation. We should be wary of a form of presentism that loses sight of these other expulsions and displacements. A historical sensibility is valuable both to attune us to the ways in which logics of race, religion, colonialism, penality, political partisanship, and economy continue to structure deportation today, but equally as a way to understand what is distinctive and novel about the deportations of our time.

## Mapping the Deportation Machine

To move beyond the view that deportation is merely one immigration enforcement policy amongst many, and to challenge the assumption that deportation is simply an instrument wielded by the state, scholars have looked to a range of terms, describing it as a system (Kanstroom 2017), regime (De Genova and Peutz 2010), continuum (Kalir and Wissink 2015), and machine (Fekete 2005). If I opt for the term machine here, I give it a Deleuzian inflection, using this metaphor to emphasise a phenomenon that is assembled, heterogeneous, dynamic, and generative. But irrespective of the particular term they privilege, what all these arguments share is this: we are dealing not just with a policy but an entire infrastructure (Xiang and Lindquist 2014) with irreducible features, elements, and dynamics that call for scholarly research in their own right. In an age when politicians promise accelerated removals to their electorates and set deportation targets for their bureaucracies (Fekete 2011), when a detention system is growing (Bosworth 2014; Weber 2015), and when a sizeable market of providers has grown up to service this infrastructure (Martin 2017), it is not hard to grasp that a deportation machine now exists. It is a machine that extends its reach far beyond the deportee or the immigration and interior ministries: development aid, social welfare services, local police agencies, transportation markets, medical and psychological authorities, a whole sector of humanitarian charities and NGOs, and various specialised international organisations (Geiger and Pécoud 2014) all find themselves imbricated in a form of forced population movement that is both routinised and contested.

Deportation cannot be grasped solely from the angle of the state (Peutz 2006). To do so would downplay the significance of the kinds of non-state actors and agencies I just mentioned. It would also overlook situations and practices where the forcible and legalised movement of people is carried out not under the formal auspices of state agents but by private actors. For example, the repatriation of stowaways discovered on ocean-going ships is largely conducted by a network of maritime insurance and security agencies (Walters 2008; Senu 2018). Nevertheless, individual states as well as the interstate system profoundly contour the deportation machine. For some observers deportation illustrates the limits of the *liberal* state within the realm of immigration enforcement (Gibney and Hansen 2003). The starting point for such analyses is the notion of an enforcement or implementation gap, namely the disjuncture between the numbers of deportees removed in a given year and the far larger number of illegalised people who, based on estimates, *could* be subjected to removal procedures (Paoletti 2010). The US is often cited as a classic example of the enforcement gap. Despite serving as a kind of vanguard to the deportation turn, millions of non-status people reside within its borders. If deportation has become so central to politics, why aren't more people being deported?

To answer this question, scholars have explored a range of factors. These include the power of rights, the mobilisation of public opinion in support of particular cases and communities, the struggles of deportees and their supporters to resist their removal by employing an ever-shifting array of tactics (Nyers 2003; Gill et al. 2014), the considerable cost of detaining and removing people, and the difficulty of ascertaining the nationality of people who lack identity papers (Ellermann 2008; Gibney and Hansen 2003). The latter point touches on one of the most formidable challenges states face in pursuing their deportation agendas: in a world divided into sovereign nation-states, the act of deportation usually requires some degree of compliance on the part of receiving states (Walters 2002; Collyer 2012). This means we have to embed our discussion of deportation in a global account of power relations. While formally deportation can operate between any two states, as a social phenomenon it is inscribed within a global hierarchy of states and regions, a power structure which it both reflects and enacts. Major flows of expulsion move along a gradient from Global North to Global South (Bialasiewicz 2012) and often reinscribe colonial and postcolonial geographies (El Qadim 2014; Bosworth 2017). Yet there are significant issues that governments in the Global South face when pressured to accept deportees. For example, many of them rely heavily on the remittances of their emigrants abroad. They are also wary of the negative public image of large numbers of their nationals being returned. These and other factors mean many governments will play the role of countries of transit and readmission only with great reluctance (El Qadim 2014; Ellermann 2008).

A focus on these constraints is important not least to avoid the image of an overly potent state that moves people at will. But whatever the nature of these constraints, they should not be viewed as fixed or absolute. On the contrary, each constraint might be regarded as a limit, a frontier zone where state actors and international agencies engage in ongoing projects to work around or overcome these obstacles. Laws and regulations certainly constrain a government's capacity to quickly remove people from its territory, yet such laws and regulations are subject to constant revision and restructuring to extend the grasp of and accelerate the deportation machine (Fekete 2011). Faced with people of unknown nationality, states have harnessed a range of practices including language analysis, DNA testing, legal sanctions, and consultations with foreign embassies and ministries in order to fix an identity and render the subject deportable (Griffiths 2012; Ellermann 2008). Finally, states have looked to readmission agreements, development aid, bilateral negotiations, a network of immigration liaison officers, and other threats and inducements to encourage specific countries to open their borders to their repatriated citizens, and sometimes deportees of other nationalities as well (El Qadim 2014; Bialasiewicz 2012; Collyer 2012; Bosworth 2017). And on this point, we should stress that targeting is a technology of deportation as well (Fekete 2011). Deportation does not proceed uniformly and evenly. Instead, particular groups, demographics, and statuses get prioritised for removal. For example, very recently Germany has made concerted efforts to deport rejected asylum seekers who are identified with Afghanistan, it being speculated that this is a 'test run' for a wider programme of deportations that will form the counterpart to the opening of German borders to refugees that happened in 2015 ('Germany Ramps Up deportations,' 2018). So, all these struggles and experiments make the deportation machine a dynamic ensemble. It unfolds not according to a linear logic but in a way that is powerfully shaped by struggles.

There are other mechanisms which have become prominent features of the deportation machine. Especially important are the tools which states and other agencies have fashioned to elicit compliance from people targeted with deportation. These include the mechanisms scholars have called 'soft deportation' (Leerkes et al. 2017; Kalir 2017). Operating under such administrative labels as 'assisted return' or 'voluntary return,' we see the use of cash incentives, help with travel documents and plane tickets, resettlement allowances, as well as the involvement of counsellors, social workers, doctors, and other care workers. In such ways a host of technologies of care, along with many NGOs and charities who specialise in these services, are becoming prominent elements within the deportation scene. But there is nothing smooth about this game of co-option: the decision to cooperate with the state induces friction and division within the NGO sector.

If deportation practices exist on a spectrum, then at the opposite end of the spectrum from soft deportation we find forced removals. People who refuse to leave under assisted return schemes may well find themselves subjected to forced removals, leading scholars to question the voluntariness implied by 'voluntary return' (Webber 2011). Yet in turning our attention to the most coercive forms and practices of expulsion we should not regard the exercise of force as somehow self-evident. Nor should we overlook the way force takes an embodied, corporeal subject as its target (Khosravi 2009). If we can identify various technologies of soft deportation, we should also speak of the technologies of force which deportation has utilised in different situations and combinations (Makaremi 2018). For force is not a generic capacity. Instead, authorities have asked themselves a whole series of questions about how to exercise force over the body of the deportee (Fekete 2011). What kinds of holds and manoeuvres are legally and medically legitimate in the policing of deportees—a question that has become all the more urgent following incidents in which migrants have died from 'positional asphyxia' (a form of suffocation) occurring while being restrained on planes and in airports and detention centres (Council of Europe 2013, p. 9). When, how, and by whom can pain be inflicted to elicit compliance from people resisting their own removal? What kinds of handcuffs, belts, and manacles are acceptable? Can we use particular drugs to sedate people in transit (IATA 2010, § 6.11–6.13)? Is it ethical to use deception and surprise, to wake people in the middle of the night and bundle them towards a waiting plane (HM Chief Inspector of Prisons 2018, § 3.3)? These questions of force and confinement highlight the role which the expertise of doctors, security experts, aviation specialists, and technicians of violence plays within the deportation machine. At stake there is a whole political anatomy of the body of the deportee oriented by the problem of how to exercise force while minimising risk of serious harm (Walters 2017). Technologies of force are eminently microphysical in the way they circulate and engage the body. But they are hardly marginal: without this microphysics of deportation it is hard to imagine how states and international agencies could possibly move deportees on the macro-scale of the interstate system.

## Deportation as a Technology

I shall return to the theme of deportation and scale in a moment. However, to nuance this brief mapping of the deportation machine I want to make a point about how we might regard deportation not just in terms of its constitutive technologies but as a technology in its own right. To do so is to pose questions about its functions, its effects, its strategic relationships with political economy, population, and sovereignty. Returning to the theme of the enforcement gap, scholars have asked: if deportation policy generally fails to meet its targets, if the potential scope of deportation typically seems to exceed its actual employment, then what other factors might explain its persistence? There are at least three kinds of answer here. In practice, they often overlap but for analytical purposes I differentiate them here. First, there is the argument that the current deportation turn can be understood as part of a wider crisis and mutation in the exercise of state sovereignty (Stumpf 2006; Bosworth 2017). When politicians promise to round up and deport they are projecting and enacting sovereign power (Tyler 2018). They are, in other words, performing the fantasy of a state which, faced with the reality of permeable boundaries, actually can control who resides on its territory.

The second argument about deportation as a technology centres on the idea of deportability (De Genova 2002). Here it is the theme of neoliberal capitalism and flexible accumulation strategies that provides the backdrop (Golash-Boza 2015). Deportation policy, like border control, might be considered to 'fail' when judged on the targets it falls short of. But it succeeds in other ways, the most significant being the degree of fear, insecurity, and anxiety it spreads not just within the lives of undocumented people, but the much wider family and community networks they inhabit. In this iteration the threat of deportation, alongside other forms of conditional residence status which Moffette (2014) calls 'probation,' casts a long shadow and fosters a kind of vulnerable subject which businesses harness and exploit. Moulier Boutang reminds us that unfree labour is not something that belongs only to capitalism's 'primitive' origins. On the contrary capitalism has *always* required unfree labour which has taken a variety of forms including enslavement, indenture, and guest worker schemes (Moulier Boutang 1998; see also McKeown 2012). Deportability is merely one of the most recent of these.

A third view builds on Foucault's concept of police in order to understand deportation as a form of international population management (Walters 2002). To speak of an international police of population is not, of course, to suggest the presence or tendency of a world state, but rather to identify a dispersed network of state organisations and international agencies (Geiger and Pécoud 2014). As noted earlier, modern deportation presumes a world divided into discrete nation-states, each of which is the proper 'home' to its citizens. Seen at the level of a given state, deportation represents expulsion and exclusion. But seen at the level of the world's population, it represents a form of division and involuntary reallocation of population with citizenship operating as a kind of location marker (Hindess 2000). This international police takes other forms as well, including the resettlement of refugees and displaced persons, the extradition of criminal or terror suspects, and the partitioning of whole countries. What these practices have in common is the authoritative allocation of individuals and populations to sovereigns and territories. Unfreedom was an extremely common condition of mobility in previous centuries; only in the twentieth century did society come to equate immigration so thoroughly with 'free passengers' (Fogleman 1998) and with a 'free,' 'atomized, self-motivated individual' (McKeown 2012, p. 21). An analytic of international police offers a lens on aspects of the governance of international mobility, including deportation, that does not naturalise this free, liberal subject of movement. As a theme, international police cautions us against treating forceful or collective movements of population as exceptional since they are in many ways woven into the constitution of social order, and the way population has always been distributed.

Having addressed some of the 'big picture' ways in which the technology of deportation can be understood, let us now shift focus and consider how it appears in terms of other scales and sites.

## Scales, Sites, Spaces

In media representations and political talk, deportation often appears as a legitimate act of resettlement. It is the image of non-citizens who are not rightfully 'here' and the muscular act of sending them 'home.' A particular imagined temporality features in this deportation imaginary as much as space. Deportation is typically depicted as a single, decisive event.

Recent scholarship complicates this image of a singular, decisive relocation, a 'repatriation' from 'here' to 'there' (Coutin 2015, p. 4). A multi-sited and multi-temporal deportation topography is emerging through careful accounts of social life within and across many sites and spaces connected to the deportation machine (Peutz 2006). These sites include detention centres (Bosworth 2014; Hall 2012; Mountz 2011a), airports (Alpes 2015), places of sanctuary (Squire and Bagelman 2012), immigration offices (Kalir and Wissink 2015; Nyers 2003), courts and hearings (Burridge and Gill 2017), and drop-in centres (Darling 2014). As well as challenging the compressed, evental image of deportation, this broadening of focus brings many actors besides state officials and deportees into the picture, reflecting calls to move migration studies beyond a singular and sometimes fetishistic focus on the lives of 'migrants' (Garelli and Tazzioli 2013; Xiang and Lindquist 2014). Ethnographic research in many of these sites is often difficult since many are restricted settings, and the subjects are frequently vulnerable people. Nevertheless, research offers many insights. For example, it stresses that as much as deportation might appear as a spectacle or an event at the level of popular media, and as much as it might be experienced as a most dreadful trauma, there is also a banality of deportation, a world of routines and practices (Kalir and Wissink 2015). To echo my earlier point: forcibly removing people is a nasty business. As Kalir and Wissink (2015) demonstrate, understanding the way in which deportees are translated into files, cases, seats on a plane, or 'clients' to be 'helped' by bureaucrats goes some way to explaining how these bureaucrats justify their work to themselves and others, a form of normalisation that underwrites the viability of the deportation process.

Second, these ethnographies of specific sites and experiences are important because they register a whole affective dimension of deportation (Drotbohm and Hasselberg 2014). They move us beyond a narrowly legal or structural understanding of the activity of forced relocation, as well as a state-centric view (Peutz 2006). In addition, they capture the ways in which feelings, emotions, hopes and fears, defiance and resignation suffuse the deportation scene. For example, Fischer (2014) examines the acute anxiety which comes to afflict persons held in the French detention system where expulsion looms over them. He shows how the possibility of self-harm becomes a risk which authorities seek to manage and thus how mental health and emotional life become territories within the deportation machine. On the other hand, Drotbohm (2012) highlights the work of surviving. She examines everyday ways in which people threatened with deportation develop coping strategies: the methods they use to negotiate a condition of precarity, anxiety, and uncertainty. For example, she finds that compliance with regulations itself can become a form of resistance under restricted circumstances. As one informant, Tony, puts it to her, it's as though the authorities want him to resort to crime to survive, thus making him all the more deportable: 'They just want to squeeze me into the criminal side. … But they are not going to get me. I stay clean' (Drotbohm 2012, p. 149).

If recent research has multiplied the sites, scales, and identities pertaining to deportation, a particularly important development in this vein is the study of post deportation (Lecadet 2013; Alpes 2015; Khosravi 2018; Peutz 2006; Schuster and Majidi 2013). Until quite recently most scholarly attention focused on the countries of the Global North from which expulsions were emanating. Little was known about the social impact deportation was having on the places to which people were being expelled, nor the entanglement of deportation issues within the politics of these states (El Qadim 2014). Growing scholarly interest in post deportation examines the challenges people face after deportation where relationships of gender, ethnicity, region, age, and experiences of dislocation and estrangement all shape their social worlds (Khosravi 2018). It reveals the efforts many deportees make to sustain networks of connection with the social worlds they have been separated from, as well as the fact many remigrate, leading some to speak of an emerging deportspora (Nyers 2003). It also reveals how former deportees can face stigma, shame, and persecution upon being returned. People often manage the risks and stigma attached to deportation by making themselves less visible socially. However, in some countries deportees have managed to self-organise, forging political and civic space. Can the deportee speak? Can they add their own voice to the debate about the borders and migration policies that define, say, Europe's relationship with its neighbours? Lecadet documents how in several West African and Central African countries this is precisely what has happened (Lecadet 2013; Lecadet 2018).

This complication of the deportation landscape which recent research is providing is not limited to multiplying accounts of the sites or harnessing the power of ethnography to zoom in on telling little details. It also involves explorations of how deportation and its contestation remake scales, spaces, and temporalities of governance. One account of this generative power comes from geographers who explore the changing relationships between states and commercial actors like security companies and prison providers, as well as the particular ways in which national, local and regional, and international governments take on new relationships around deportation (Martin 2017; Coleman 2012; Bialasiewicz 2012; Geiger and Pécoud 2014). Another version comes from scholars who have fashioned concepts like 'deportation corridor' (Drotbohm and Hasselberg 2014) to make sense of the ways in which practices of detention and transportation not only confine people in a physical sense but foster feelings and experiences of disconnection, disorientation, and despair. These practices engineer deportability through the manipulation of time-space and affect (Gill 2009; Hiemstra 2013).

Finally, a third kind of emergent space should be mentioned: a space of communication and publicity. Tyler speaks of a 'deportation propaganda machine,' referring to the various ways in which governments seek to orchestrate media coverage of police raids and deportation operations or use public information campaigns that warn illegalised people it is time to 'go home' (Tyler 2018; see also Jones et al. 2017). At the same time, refugees and migrants have long understood that sometimes publicity and, at other times, invisibility can be weapons in frustrating their own deportations (Nyers 2003). The point here is perhaps that we should not see media visibility as a second-order phenomenon that results in struggles over deportation after the fact. Instead, we need to recognise that technologies of visibility and invisibility and dialectics of publicity and secrecy are woven into the deportation field (Fischer 2005).

## New Directions

While scholarship examining technologies of deportation has grown remarkably, there are some areas that remain undeveloped. In this final section I highlight three where useful advances could be made. First, a point about mobilities. There has been rather little attention paid to the ways in which deportation actually moves people. This is paradoxical given that there can be no deportation without transportation. Scholars are only now beginning to examine technologies of mobility in the context of detention (Gill 2009), how a system of highways and a culture of automobility shapes immigration policing as well as its resistance (Stuesse and Coleman 2014), or the difference that railways (Blue 2015) or aircraft (Walters 2016; Hiemstra 2013) have made to the technology of deportation. There would seem to be ample scope for forging closer interaction between research on carceral geographies, mobility infrastructures, and deportations. One benefit of such a move would be a more mobile account of state power itself (Mountz 2011b). Why is it we speak of sea power and air power only in the context of military affairs; why not extend these notions to the international police of population?

Second, a point about the geography of deportation scholarship. The English-speaking literature on deportation focuses overwhelmingly on expulsions from Europe, North America, and Australia to the Global South (Drotbohm and Hasselberg 2014). Important work is rectifying this situation (Lecadet 2013; Galvin 2015; Lindquist 2018), but still we know relatively less about deportation between the countries of the Global South. In some situations, these South–South deportations appear as a ripple effect of forced movements that originate in the wealthier regions. This was the case with Italy's infamous 'push-backs' that in 2009 returned refugees to Libya, a state that was in turn conducting mass expulsions towards its African neighbours (Bialasiewicz 2012). Yet a properly global perspective on the deportation machine requires that we do not see deportation as though its geography was fully contained and explained by a centre/periphery hierarchy (Garelli and Tazzioli 2013). Paradoxically, a global understanding requires appreciation for the local contexts and dynamics which shape any given process of expulsion and give it a specific character. For example, Galvin looks at the ways in which undocumented Zimbabwean workers experience work and deportation from Botswana (Galvin 2015). She shows that while disruptive and stressful, these deportations are experienced not as absolute or final but as a relatively normal feature of cross-border mobility requiring negotiation and management. In this way she deepens understanding of the different temporalities and moods that mediate the experience of deportation.

Third, and finally, a point about the theoretical tools on which scholarship on technologies of deportation draws. While thinkers like Foucault, Arendt, Rancière, and Agamben have been used productively to think beyond the limits of liberalism's juridical and institutional view of power, there are still dimensions and registers of power that are currently off our maps. There is a case for widening the repertoire of critical thinkers when it comes to studying technologies of deportation. For example, to forcibly remove people, states sometimes employ techniques of secrecy and surprise. Here the work of Canetti on secrecy and speed as tools of predation could be useful in capturing some of the more sinister modalities while bringing greater corporeality into the picture (Canetti 1973; see also Chamayou 2012). Deportation research might also profit from Sloterdijk's philosophical work on atmosphere. Sloterdijk thinks about security in 'morphological' and 'spherological' terms, highlighting the ways in which architectures, rooms, enclosures, and other 'conditioning systems' are a fundamental condition of our human being (Klauser 2010). When governments seek to restrict housing, healthcare, schooling, and other social provisions in a bid to 'deter' migration; when they threaten to criminalise civil society actors for offering the most elementary forms of solidarity and support, like a bed for the night or a bowl of soup; when they ally with media organisations to foster a public climate of hatred and rejection, then it seems that all the mechanisms that might normally constitute the life-support systems of society are being weaponised and turned into instruments of hostility (Dijstelbloem and Walters 2019). Might Sloterdijk's spherology be useful for thinking about this new scenario in which states seek to effect expulsion not only through the formal mechanisms of deportation but by destroying migrants' social worlds as a way to effect 'voluntary departure' (Tyler 2018)?

## Conclusion

In very broad brush strokes I have sketched some lines of a deportation machine paying special attention to the level of its technologies. The risk associated with the machine metaphor is to accord too much consistency and orderliness to a field of activity that exhibits a great deal of ad hocery, dissensus, and incoherence (Kalir and Wissink 2015). To ward off this impression it should be stressed that this machine is shot through with struggles. There would be no multiplication and experimentation in technologies of deportation were it the case that illegalised people meekly departed on their own accord or had not themselves learnt ways to render themselves undeportable. As much as deportation might be designed to police boundaries of citizenship and belonging, let us not forget that everyday resistance, campaigns for rights and recognition, and anti-deportation movements mean that deportation is also sometimes a site where new forms of solidarity, affiliation, and even citizenship are emergent (Anderson et al. 2011; Nyers 2003; Lecadet 2013; Garelli and Tazzioli 2013).

## References

Alpes, J. (2015). Airport casualties: Non-admission and return risks at times of internalized/externalized border controls. *Social Sciences, 4*, 742–757.

Anderson, B., Gibney, M., & Paoletti, E. (2011). Citizenship, deportation and the boundaries of belonging. *Citizenship Studies, 15*(5), 547–563.

Bialasiewicz, L. (2012). Off-shoring and out-sourcing the borders of EUrope: Libya and EU border work in the Mediterranean. *Geopolitics, 17*(4), 843–866.

Bloch, A., & Schuster, L. (2005). At the extremes of exclusion: Deportation, detention and dispersal. *Ethnic and Racial Studies, 28*(3), 491–512.

Blue, E. (2015). Strange passages: Carceral mobility and the liminal in the catastrophic history of American deportation. *National Identities, 17*(2), 175–194.

Bosworth, M. (2014). *Inside immigration detention.* Oxford, England: Oxford University Press.

Bosworth, M. (2017). Penal humanitarianism? Sovereign power in an era of mass migration. *New Criminal Law Review, 20*(1), 39–65.

Burridge, A., & Gill, N. (2017). Conveyor-belt justice: Precarity, access to justice, and uneven geographies of legal aid in UK asylum appeals. *Antipode, 49*(1), 23–42.

Canetti, E. (1973). *Crowds and power.* Harmondsworth, England: Penguin Books.

Chamayou, G. (2012). *Manhunts: A philosophical history* (S. Rendall, Trans.). Princeton, NJ: Princeton University Press.

Coleman, M. (2012). The 'local' migration state: The site-specific devolution of immigration enforcement in the US South. *Law and Policy, 34*(2), 159–190.

Collyer, M. (2012). Deportation and the micropolitics of exclusion: The rise of removals from the UK to Sri Lanka. *Geopolitics, 17*(2), 276–292.

Council of Europe. Parliamentary Assembly. (2013, 7 November). *Monitoring the return of irregular migrants and failed asylum seekers by land, sea and air* (Doc. 13351). Committee of Migration, Refugees and Displaced Persons.

Coutin, S. (2015). Deportation studies: Origins, themes and directions. *Citizenship Studies, 41*(4), 671–681.

Darling, J. (2014). Another letter from the Home Office: Reading the material politics of asylum. *Environment and Planning D: Society and Space, 32*, 484–500.

De Genova, N. (2002). Migrant 'illegality' and deportability in everyday life. *Annual Review of Anthropology, 31*, 419–447.

De Genova, N., & Peutz, N. (Eds.). (2010). *The deportation regime: Sovereignty, space and the freedom of movement.* Durham, NC: Duke University Press.

Dijstelbloem, H., & Meijer, A. (Eds.). (2011). *Migration and the new technological borders of Europe.* New York, NY: Palgrave Macmillan.

Dijstelbloem, H., & Walters, W. (2019). Atmospheric border politics: The morphology of migration and solidarity practices in Europe. *Geopolitics.* https://doi.org/10.1080/14650045.2019.1577826.

Drotbohm, H. (2012). *An ethnography of deportation from Britain* (Doctoral dissertation). University of Sussex, Falmer, England.

Drotbohm, H., & Hasselberg, I. (2014). Deportation, anxiety, justice: New ethnographic perspectives. *Journal of Ethnic and Migration Studies, 41*(4), 551–562.

El Qadim, N. (2014). Postcolonial challenges to migration control: French–Moroccan cooperation practices on forced return. *Security Dialogue, 45*(3), 242–261.

Ellermann, A. (2008). The limits of unilateral migration control: Deportation and inter-state cooperation. *Government and Opposition, 43*(2), 168–189.

Fekete, L. (2005). The deportation machine: Europe, asylum and human rights. *Race and Class, 47*(1), 64–91.

Fekete, L. (2011). Accelerated removals: The human cost of EU deportation policies. *Race & Class, 52*(4), 89–97.

Fischer, N. (2005). Clandestins au secret: Contrôle et circulation de l'Information dans les Centres de Rétention Administrative Français. *Cultures & Conflits, 57*, 91–118.

Fischer, N. (2014). The management of anxiety: An ethnographical look at self-mutilations in a French immigration detention centre. *Journal of Ethnic and Migration Studies, 41*(4), 599–614.

Fogleman, A. (1998). From slaves, convicts, and servants to free passengers: The transformation of immigration in the era of the American Revolution. *Journal of American History, 85*(1), 43–76.

Galvin, T. (2015). 'We deport them but they keep coming back': The normalcy of deportation in the daily life of 'undocumented' Zimbabwean migrant workers in Botswana. *Journal of Ethnic and Migration Studies, 41*(4), 617–634.

Garelli, G., & Tazzioli, M. (2013). Challenging the discipline of migration: Militant research in migration studies, an introduction. *Postcolonial Studies, 16*(3), 245–249.

Geiger, M., & Pécoud, A. (2014). International organisations and the politics of migration. *Journal of Ethnic and Migration Studies, 40*(6), 865–887.

Germany ramps up deportations of failed Afghan asylum seekers. (2018, 12 July). *CNN.com.* https://www.cnn.com/2018/07/12/europe/germany-deportations-afghanistan-intl/index.html.

Gibney, M. (2008). Asylum and the expansion of deportation in the United Kingdom. *Government and Opposition, 43*(2), 146–167.

Gibney, M., & Hansen, R. (2003). *Deportation and the liberal state: The forcible return of asylum seekers and unlawful migrants in Canada, Germany and the United Kingdom* (New Issues in Refugee Research Working Paper 77). Geneva, Switzerland: UNHCR.

Gill, N. (2009). Governmental mobility: The power effects of the movement of detained asylum seekers around Britain's detention estate. *Political Geography, 28*(3), 186–196.

Gill, N., Conlon, D., Tyler, I., & Oeppen, C. (2014). The tactics of asylum and irregular migrant support groups: Disrupting bodily, technological, and neoliberal strategies of control. *Annals of the Association of American Geographers, 104*(2), 373–381.

Golash-Boza, T. M. (2015). *Deported: Immigrant policing, disposable labor, and global capitalism.* New York, NY: New York University Press.

Griffiths, M. (2012). Anonymous aliens? Questions of identification in the detention and deportation of failed asylum seekers. *Population, Space & Place, 18*(6), 715–727.

Hall, A. (2012). *Border watch: Cultures of immigration, detention and control.* London: Pluto Press.

Hiemstra, N. (2013). 'You don't even know where you are': Chaotic geographies of US migrant detention and deportation. In D. Moran, N. Gill, & D. Conlon

(Eds.), *Carceral spaces: Mobility and agency in imprisonment and migrant detention*. Bristol, England: Ashgate.

Hindess, B. (2000). Citizenship in the international management of population. *American Behavioral Scientist, 43*(9), 1486–1497.

HM Chief Inspector of Prisons. (2018). *Detainees under escort: Inspection of escort and removals to Pakistan, February 14–15, 2018*. London: HMIP.

IATA (International Air Transport Association). (2010). *Guidelines for the removal of deportees* (Version 3.0).

Jones, H., Gunaratnam, Y., Bhattacharyya, G., Davies, W., Dhaliwal, S., Forkert, K., et al. (2017). *Go home? The politics of immigration controversies*. Manchester, England: Manchester University Press.

Kalir, B. (2017). Between 'voluntary' return programs and soft deportation: Sending vulnerable migrants in Spain back 'home.'. In Z. Vathi & R. King (Eds.), *Return migration and psychosocial wellbeing* (pp. 56–71). New York, NY: Routledge.

Kalir, B., & Wissink, L. (2015). The deportation continuum: Convergences between state agents and NGO workers in the Dutch deportation field. *Citizenship Studies, 20*(1), 1–16.

Kanstroom, D. (2007). *Deportation nation: Outsiders in American history*. London: Harvard University Press.

Kanstroom, D. (2017). The 'right to remain here' as an evolving component of global refugee protection: Current initiatives and critical questions. *Journal on Migration and Human Security, 5*(3), 614–644.

Kedar, B. (1996). Expulsion as an issue of world history. *Journal of World History, 7*(2), 165–180.

Khosravi, S. (2009). Sweden: Detention and deportation of asylum seekers. *Race & Class, 50*(4), 38–56.

Khosravi, S. (Ed.). (2018). *After deportation: Ethnographic perspectives*. New York, NY: Palgrave.

Klauser, F. (2010). Splintering spheres of security: Peter Sloterdijk and the contemporary fortress city. *Environment and Planning D: Society and Space, 28*(2), 326–340.

Lecadet, C. (2013). From migrant destitution to self-organization into transitory national communities: The revival of citizenship in post-deportation experience in Mali. In B. Anderson, M. Gibney, & E. Paoletti (Eds.), *The social, political and historical contours of deportation* (pp. 143–158). New York, NY: Springer.

Lecadet, C. (2018). Post-deportation movements: Forms and conditions of the struggle amongst self-organising expelled migrants in Mali and Togo. In S. Khosravi (Ed.), *After deportation: Ethnographic perspectives* (pp. 187–204). New York, NY: Palgrave.

Leerkes, A., van Os, R., & Boersma, E. (2017). What drives soft deportation? Understanding the rise in assisted voluntary return among rejected asylum seekers in the Netherlands. *Population, Space & Place, 23*(8), 1–11.

Lindquist, J. (2018). Infrastructures of escort: Transnational migration and economies of connection in Indonesia. *Indonesia, 105*, 1–19.

Makaremi, C. (2018, 15 January). Deportation and the technification of force: Violence in democracy. *Technosphere Magazine.* https://technosphere-magazine.hkw.de/p/Deportation-and-the-Technification-of-Force-Violence-in-Democracy-ngjuok8Rb5q1U7ae54VHwp.

Martin, L. (2017). Discretion, contracting, and commodification: Privatisation of US immigration detention as a technology of government. In D. Conlon & N. Hiemstra (Eds.), *Intimate economies of immigration detention: Critical perspectives* (pp. 32–50). Abingdon, England: Routledge.

McKeown, A. (2012). How the box became black: Brokers and the creation of the free migrant. *Pacific Affairs, 85*(1), 21–45.

Mezzadra, S., & Neilson, B. (2013). *Border as method, or, the multiplication of Labor.* Durham, NC: Duke University Press.

Moffette, D. (2014). Governing immigration through probation: The displacement of borderwork and the assessment of desirability in Spain. *Security Dialogue, 45*(3), 262–278.

Moulier Boutang, Y. (1998). *De l'esclavage au salariat: Économie historique du salariat bride.* Paris: Presses universitaires de France.

Mountz, A. (2011a). The enforcement archipelago: Detention, haunting, and asylum on islands. *Political Geography, 30*(3), 118–128.

Mountz, A. (2011b). Understanding state mobilities through an ontology of exclusion. *Mobilities, 6*(3), 317–334.

Nyers, P. (2003). Abject cosmopolitanism: The politics of protection in the anti-deportation movement. *Third World Quarterly, 24*(6), 1069–1093.

Paoletti, E. (2010). *Deportation, non-deportability and ideas of membership* (Refugee Studies Centre Working Paper Series No. 65). Oxford, England: Refugee Studies Centre, University of Oxford.

Peutz, N. (2006). Embarking on an anthropology of removal. *Current Anthropology, 47*(2), 217–231.

Schuster, L., & Majidi, N. (2013). What happens post-deportation? The experience of deported Afghans. *Migration Studies, 1*(2), 221–240.

Senu, A. (2018). *The global assemblage of multi-centred stowaway governance* (Doctoral dissertation). Cardiff University, Cardiff, Wales.

Squire, V., & Bagelman, J. (2012). Taking not waiting: Space, temporality and politics in the city of sanctuary movement. In P. Nyers & K. Rygiel (Eds.), *Citizenship, migrant activism and the politics of movement* (pp. 146–164). New York, NY: Routledge.

Stuesse, A., & Coleman, M. (2014). Automobility, immobility, altermobility: Surviving and resisting the intensification of immigrant policing. *City & Society, 26*(1), 51–72.

Stumpf, J. (2006). The crimmigration crisis: Immigrants, crime, and sovereign power. *American University Law Review, 56*(2), 367–419.

Tyler, I. (2018). Deportation nation. *Journal for the Study of British Cultures, 25*(1), 1.

Walters, W. (2002). Deportation, expulsion, and the international police of aliens. *Citizenship Studies, 6*(3), 265–292.

Walters, W. (2008). Bordering the sea; shipping industries and the policing of stowaways. *Borderlands e-journal, 7*(3), 1–25.

Walters, W. (2016). Flight of the deported: Aircraft, deportation, and politics. *Geopolitics, 21*(2), 435–458.

Walters, W. (2017). The microphysics of power redux. In P. Bonditti, D. Bigo, & F. Gros (Eds.), *Foucault and the modern international* (pp. 57–76). New York, NY: Palgrave.

Webber, F. (2011). How voluntary are voluntary returns? *Race & Class, 52*(4), 98–107.

Weber, L. (2015). Deciphering deportation practices across the Global North. In S. Pickering & J. Ham (Eds.), *The Routledge handbook on crime and international migration* (pp. 155–178). New York, NY: Routledge.

Xiang, B., & Lindquist, J. (2014). Migration infrastructure. *International Migration Review, 48*(s1), S122–S148.

# 18

## Street Technologies of Displacement: Disposable Bodies, Dispossessed Space

Elijah Adiv Edelman

In this chapter I explore exclusionary space and practices of embodied, social, and political erasure within cities across geopolitical landscapes. I approach displacement in this chapter as dynamic and existing across 'diverse scales and temporalities … so that their negative emotional, psychosocial and material impacts can be more fully documented, and resisted' (Elliott-Cooper et al. 2019, p. 1). Historically, practices of urban displacement have served a larger structural role to delineate and segregate those bodies deemed pathological, undesirable, and, in some cases, disposable. That is, these bodies come to serve as necropolitical anchoring points, indexing that which is morally suspect and intrinsically disposable, representing 'the condition for the acceptability of putting to death' (Mbembe 2003, p. 17). Zones of exclusion, zero tolerance zones, and other similar kinds of 'exclusionary regimes' emphasise 'the undeserving and the unreformable nature of deviants' (Beckett and Weston 2001, p. 44). I focus here on how values associated with devalued subjectivities and bodies—sex workers and trans communities of practice—become codified features of urban displacement.

Cityscapes and urban spheres provide significant and productive grounds in which to visualise and track processes of neoliberalism, nationalisms, and bio/necropolitics as technologies of displacement. Zones of exclusion and

E. A. Edelman (✉)
Rhode Island College, Providence, RI, USA
e-mail: eedelman@ric.edu

concomitant projects of 'urban renewal' and 'development' work symbiotically to physically and sociopolitically carve out bodies spatially from space. The exclusion of 'undesirables' from the urban terrain 'must be seen as part of a broader process by which the law includes, weighs and assesses all urban denizens' (Carr et al. 2009, p. 1962). Thus, if we orient 'citizenship' as the 'right to access and use specific kinds of space' (Hubbard 2001, p. 54), zones of exclusion and practices of urban displacement operate to delineate between those that qualify as potential citizens—those that can maintain a claim of belonging to space or place—and those that do not. In other words, the value of a body and the 'capacity and potential of individuals and the population as living resources' (Ong 2006, p. 6) become collapsible features in the possession or dispossession of space and place. In this chapter I explore the material implications of displacement and spatial exclusion for those that are viewed as 'undesirable' through my own work on 'Prostitution Free Zones' and in trans community map-making of Washington, D.C.

## Displacement and States of Exception

While there are various paradigms in which to theorise how space and place function, space is often situated as 'a natural fact—a collection of properties that define essential reality of settings of action' whereas place refers to 'a social product, a set of understandings that come about only after spaces have been encountered by individuals and groups' (Dourish 2006, p. 2). However, this 'myth of spatial immanence and a fallacy of spatial relativism' (Keith and Pile 1993, p. 6) operates to obscure how 'pace and place are heavily regulated by and through power. However, because space is "fragmented, multidimensional, contradictory and provisional" (Blunt and Rose 1994, p. 7) we can understand any totalising claims or management of space to be inherently political. That is, space and place can be articulated in ways that are categorically different. Spaces are constantly in flux and their meanings, however hegemonic or arbitrary, are unstable in meaning.

Moreover, who ultimately belongs or can claim 'belonging' to a space or place, and thus become displaceable or 'un-homed,' is 'inevitably based around the anthropophagic erasure or anthropoemic exclusion of difference, rather than its celebration' (Hubbard 2001, p. 60). In short, the codification of difference preconfigures space to function as either inclusive or exclusive.

The production of exclusionary space serves a broader societal role—to delineate and segregate those bodies deemed sick, pathological, undesirable, and, in some cases, disposable. Globally, zones of exclusion, zero tolerance zones, and other similar geospatial policies are commonly understood to identify practices that are understood as 'not belonging' in these spaces. Class, race, ethnic, gendered, sexualised, and other 'marked' forms of difference are all implicated in the ways in which bodies are demarcated as 'valuable' from those which are not. States of exception, wherein what may be legal, or even illegal, are subsumed by the desires of those in power. Agamben's 'homo sacer,' a formulation of the body that may be killed with impunity but not in sacrifice (Agamben 1998), is designed. Those bodies marked as ideologically suspect through biopolitical evaluation occupy a state wherein value can only be found within death—occupying a subjectivity that promises neither death nor life. Importantly, states of exception are 'neither external nor internal to the juridical order' (Agamben 2005, p. 23) further complicating the ways in which the violence of displacement is either erased or rendered visible. In other words, the ways in which our lives do, or do not, matter are directly linked to who we are understood to be.

For example, laws regarding urban-based street art 'graffiti' may function less to manage the presentation of 'street art' and more to do with establishing moral economies based in anti-black racism and class aesthetics in value production (Young 2010). Indeed, 'the sometimes arbitrary separation of graffiti from street art by metropolitan agencies has allowed an embrace and even valorisation of the power of "street art" to activate space, at a time of increasing criminalisation of "graffiti"' (McAuliffe 2012, p. 190). This seemingly innocuous example of the public management of art underlies how bodies understood to violating a space of exclusion are then managed by the juridico-legal state.

The urban context is a particularly salient site in which to view how the combination of limited space, fluctuating economies, and shifting cash flows transforms physical landscapes into nearly unrecognisable forms of redevelopment. While the 'revitalisation' of an abandoned building with multi-million dollar condos may serve to make some persons feel comfortable and welcome in that geographic locality, others, many of whom cannot afford such accommodations, may suffer displacement and henceforth regard this development as alienation. In this example, the role of gentrification, as a tool of both displacement and enrichment, cannot be regarded as merely another element of multiple intersecting bodies. Rather, it should be situated in the sociopolitical environment in which it comes to action, wherein particular kinds of bodies or practices are negatively evaluated and displaced by those practices and persons of value to city planners and developers. In the same manner that who we are impacts how much value our life or death has, who we are is also viewed as facilitating—or standing in the way of—profit.

## Urban Space, Cityscapes, and 'Public' Space

Spatially and geographically defined, the 'city,' and how bodies come to be regulated by its terrain, is a powerful site of ideological work. In thinking about the particular spaces in which the regulation of bodies at work and place can be visualised, the 'inner city' becomes a 'soft spot for the implementation of neoliberal ideals' (Hackworth 2007, p. 13). Most broadly, the 'city' should be regarded as dynamic, as both produced and consumed by its inhabitants and visitors. Building on the concept that the city is dialectically linked to 'very physical expressions of social relations, movements and ideologies' (Hackworth 2007, p. 79). One's experience of the city 'is the product both of immediate sensation and of the memory of past experience, and it is used to interpret information and to guide action' (Lynch 1960, p. 4).

When framing the 'public sphere' as being constituted by the freedom of movement and habitation, the treatment and 'involuntary publicity of the homeless' (Mitchell 1995) calls into question the scale and scope of that freedom. Additionally, processes such as gentrification carve out literal geographic spaces of exceptionality, wherein the management of sovereignty and sovereign bodies does not sit within the nation-state but rather is co-managed by the nation-state and capital investors. It is this relationship between the nation-state and the land developers that creates zones of exceptionality wherein the loss of human life is acceptable.

Gentrification, in addition to the destruction of public services, including affordable housing, clinics, and community meeting spaces in deference to corporate development, 'can be seen as the material and symbolic knife-edge of neoliberal urbanism representing the erosion of the physical and symbolic embodiment of neoliberal urbanism's putative other—the Keynesian activist state' (Hackworth 2007, p. 98). This is particularly true within cityscapes, wherein the combination of limited space, fluctuating economies, and shifting cash flows literally transforms the physical landscapes into nearly unrecognisable forms of redevelopment. The way necropolitics articulates with bodies in space in gentrifying spaces represents the expression of 'necrocapitalism' (Banerjee 2008). Gentrification, as a kind of necrocapitalistic reformation of space, renders bodies that stand in the way of capital productivity as pathological and malignant tumours in an otherwise healthy expansion of capitally productive landscapes. Specifically, the necrocapitalist 'practices of organizational accumulation that involve violence, dispossession, and death' provide the logic that buttresses the destruction of public housing and low-income neighbourhoods, as well as the bodies that once occupied that space (Banerjee 2008, p. 1543).

We can conceptualise the urban policies of exclusion and displacement, such as gentrification, as a way in which necropolitics articulates with space. Perhaps not surprisingly, the ethos of 'zero tolerance' does not function to remove but rather displace. The social practice of 'zero tolerance' within a limited space, such as public schools in the United States, has reflected that not only do these policies not lower rates of suspension or other negative outcomes for students but rather exacerbate already existing ones. Policies of exclusion, spatial or otherwise, do not reflect an overly criminalising state but rather is inherent to how 'inclusion' is understood. Exclusion is 'necessarily constitutive of politics itself. In other words, the policing of what must remain "outside" the state is very much an "inside" activity' (Stevenson 2007, p. 141). Biopower as 'a constitutive form of power that takes as its object human life' (Foucault 1977, p. 212) underlies the 'capacity and potential of individuals and the population as living resources' (Ong 2006, p. 6). In other words, the potential for the productivity of the body hinges on the cooperation and investment on the part of the subject insofar as it is *permitted* to engage in projects of productivity.

## Mapping Spaces of Exception

Bodies do not move through vacuums of space but rather are always already understood through discourses of power. These discourses of power are visible in the depictions of space we use in order to understand and navigate space. Maps serve an additional means towards understanding how the regulation of space serves as a way to make visible the felt experiences of negotiating the world as an embodied subject. Lefebvre discusses this dialectic between space and the body, noting that 'the capacity of bodies that defy visual and behavioral expectation to disrupt the shared meaning of public space' reflects the multi-directionality of meaning-making (cited in Brown and Knopp 2003, p. 315). Moreover, I would agree with De Certeau that 'every story is a travel story—a spatial practice' (1984, p. 115) and within those space and place-based practices (however they may be defined) are yet more 'stories' about power, practice, and somatic experience. Space is as an active constitutive component of hegemonic power: 'it tells you where you are and it puts you there' (Keith and Pile 1993, p. 37). Elizabeth Grosz stresses the importance of this body/city dynamic as 'complex feedback relation' wherein '[t]he body and its environment, rather, produce each other as forms of the hyperreal, as modes of simulation which have overtaken and transformed. … the city is made and made over into the simulacrum of the body, and the body, in its turn, is transformed, "citified," embraced as a distinctly metropolitan body' (Grosz 1998, p. 43). This matters in the sense that in any context in which we turn to a map to tell us where we are—or where we might go—we reproduce the discursive power of maps.

## Washington, D.C.: Case Study in Spaces of Exclusion

Washington, D.C., as the capital city of the United States, metonymically indexes the highest power of the nation-state and is, as such, a particularly poignant example of how policies of exclusion and displacement impact marginalised communities of practice. In this section I focus on both laws that regulate space formally, such as through the Prostitution Free Zones, as well as informal forms of regulation, as expressed in trans community-produced maps of DC.

Following Hubbard, and noting the utility of community map-making, I argue here for the need to approach spaces as 'becoming' (2001). While some features of a space or place can be more static in experience, such as a building or other 'permanent' structures, the ways in which these spaces are regulated, imagined, or experienced are contingent on inherently dynamic features: time, bodies present in the space, climate, and so on. As such, we must also regard the laws and policies that regulate public space as emerging out of a dynamics of cultural practice that is also 'becoming.' In many ways, pedestrian and other exclusionary policies regulating public space do so in the interest of the discourses about space, rather than the space itself. Moreover, as with the production of laws, the application of laws is also an inherently dynamic practice; laws can only be applied when an authority is present, aware of the legal infraction, and choosing to apply the law. Similarly, and as discussed in this chapter, the regulation of bodies in space (where one is expected to go or not go) does not rely on law. Rather, laws provide a legal (and therefore 'valid') basis for what has otherwise been referred to as legalised racism, anti-sex work, and a form of criminalisation of poverty, and other cultural practices that seek to organise and situate specific bodies in space.

As I discuss elsewhere (Edelman 2014; Edelman et al. 2015) Prostitution Free Zones (PFZs) in Washington, D.C., exemplify the role of exclusionary space in managing codified difference. Prostitution Free Zones sit at the intersection of the 'juridico-political and the biopolitical' (Mitchell 2006, p. 102). PFZs operate in line with what is expected of a 'post justice city' in which urban policies are emerging 'based on social and racial containment, the purification of public spaces, the subsidization of elite consumption, the privatization of social reproduction, the normalization of economic insecurity and preemptive crime control' (Peck 2004, p. 225). This kind of spatial governmentality is where the nation-state's policies work to 'manipulate the spatial order of a regions or community' to materialise this neoliberal ethos (Sanchez 2004, p. 262). Thus, PFZs do not actually attend to the crux of the 'crime' or the 'criminal' but rather merely shift the practices to a different space not deemed as valuable as that within a PFZ and, in this case, this implicitly refers to racial and gendered practices.

Within Washington, D.C., the kinds of class and race stratification marking the broader landscapes of inequality in the city are also clear in the navigations and understandings of space within different trans experiences. Marginalised groups such as transgender communities of practice—the groups that I work with—are found across these sub-cities and are members of various communities of practice. Trans communities of practice in Washington, D.C., are brought together through both proximity within the city and healthcare and social and political needs as they relate to trans lived experience. Particularly within the context of limited healthcare resources in DC, trans persons living in the district who seek any kind of trans-sensitive or aware medical care, whether transition related or otherwise, often must utilise the same resources, bridging together a radically diverse, and otherwise entirely unrelated community, in unexpected ways.

## Methodology: Trans Community Map-Making

Gentrification-induced displacement is not simply a matter of 'out-migration' and 'in-migration' but rather involves 'multiple processes of un-homing' (Elliott-Cooper et al. 2019, p. 6). Specifically, gentrified space, as a racial and class-based form of displacement, is expressed as places where trans communities are not expected to belong. The data I discuss in this section emerges from research I conducted between 2006 and 2016 on transgender experiences in Washington, D.C. As an anthropologist I employed both 'traditional' modes of anthropological research—participant observation, semi-structured interviews, and survey analysis—along with 'non-traditional' techniques of community map-making. The maps discussed here were produced through both one-on-one interviews and group discussions. As an element of the one-on-one semi-structured interviews, I asked interviewees to draw and describe a map of Washington, D.C., according to what they describe as a 'trans DC.'

These community-produced maps reflect how spatial and ideological displacement are also lived experiences of gentrification and 'un-homing.'

Mainstream maps of Washington, D.C., generally frame the city as a space for national interest and consumption. These maps are often limited in range spatially and depict only the 'downtown' portions of the city, typically including national monuments, government offices, and other historical points of interest. Community map-making, such as that I discuss here, transcends normative cartographic methods of aligning space according to objective scientific means and, instead, utilises community conceptualisations of space and place in which to visualise the city (Geltmaker 1997, p. 234; Bhagat and Mogel 2007, p. 6). Jan's map (Fig. 18.1) notably leaves blank the areas of the city which house national monuments and instead depicts the city as defined by spaces of wealth, trans exclusion, and trans inclusion.

In this context, dollar signs in the north-western region of the city reflect the relative wealth found in this area as a contrast to the poverty across the south-eastern areas of the city. Jan identifies trans exclusion as in Virginia, which sits across the Potomac River. Finally, trans inclusion is focused on where they live (a group house known as 'Fireswap') and within a sex-worker support agency (HIPS, Helping Individual Prostitutes Survive). While these spaces are depicted as in close proximity to each other, this spatiality is not on a normative geographic scale but rather an affective one.

Brett's map (Fig. 18.2) reflects the city not through geographic form but rather experiences and places.

Brett includes trans inclusionary areas of the city where they find friends, their own home and friend's homes, along with exclusionary areas as populated by briefcase-holding 'assholes' or the mixed experiences of the area of the city known as Anacostia. As with Jan's map, these elements are not depicted through geographic spatiality, but rather as linked both through affect and temporality alongside the un-homing threats of gentrification and class inequality. Cameron's map (Fig. 18.3) reflects approaches both Jan and Brett used in their maps by their depiction of the city in geographic form as well as experience. Cameron utilises the well-known landmarks of the White House, the Capitol Building, and Whitman Walker (a low-income and LGBT health clinic), as well as waterways to establish the geographic specificity of DC. The reminder of the map reflects concentrations of where trans and gender non-conforming people live alongside 'all others.' Importantly, Cameron identifies the greatest concentration of trans and gender non-conforming people as existing along 'K St.' This road does not house a large number of trans persons but rather is a space where Cameron knows they can find other trans people. This is significant both in understanding trans experiences of DC and in how

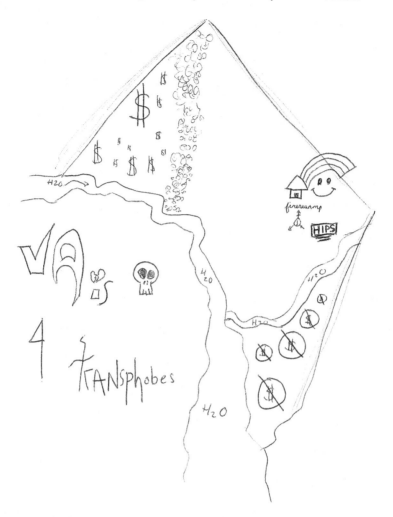

**Fig. 18.1** *Jan's map* (Example 1): Participant in 2010 Trans DC Panel, consent for reproduction provided by producer

exclusionary policies and laws such as the Prostitution Free Zone come to disproportionately impact trans women of colour. This road sits at the confluence of gentrification, sex work(er) spaces, and the criminalisation of trans bodies of colour. Empowered by the discriminatory practices of the PFZ, of which are enacted along spaces of gentrification, police officers often assume trans women of colour along this road are engaging in sex work. However, as evidenced both in Cameron's map and across the research on this area, trans persons in this area congregate to be alongside other trans persons which is not, and cannot be used as, evidence of sexual solicitation.

**Fig. 18.2** *Brett's map* (Example 2): Participant in 2010 Trans DC Panel, consent for reproduction provided by producer

Importantly, Washington, D.C., like many major cities in the United States, is home to a number of organisations, support groups, and other 'activist'-oriented projects that claim to be salient and accessible resources for trans persons living in DC. But, focusing only on the organisations and groups included in these maps, those who participated identified resources as

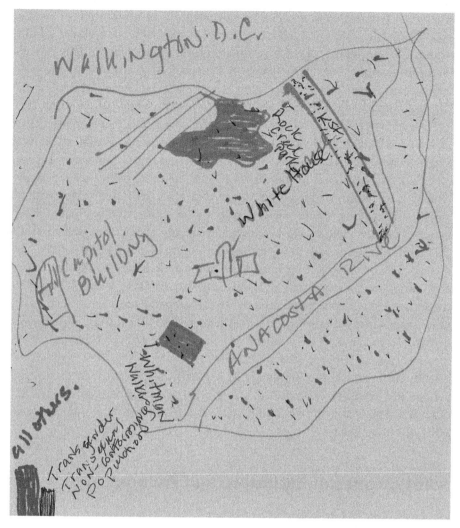

**Fig. 18.3** *Cameron's map* (Example 3): Participant in 2010 Trans DC Panel, consent for reproduction provided by producer

emerging through spaces with similar social justice, political, or religious practices (such as a food cooperative or church group). In other words, these maps reflect that the spaces that 'matter' the most as where to access trans-specific support are also spaces that are heavily regulated by policies of exclusion and displacement, like PFZs, and other expressions of gentrification.

'Safety' links the major themes in these community-produced maps, as well as raises a question of how we conceptualise what 'exclusionary space' might actually feel like. Safety was described in the community-produced maps as where one feels safe, where one does not feel safe, and how, even in areas of potential criminalisation, such as the sex work(er) stroll, one seeks, and finds, support. What requires closer investigation is the ways 'safety' is differentially understood among the participants of this project in contrast to the ways mainstream efforts define 'safe.' Safety is a phenomenological experience (i.e. felt and embodied). The relationships between ourselves—the 'feeling subject'—and what is being felt work in tandem to produce one's experience of materiality. This dialectic also emerges as a core organising subtext binding together the features included in the maps collected in this project. The spaces included, and excluded, in the maps of those collected in this project represent where project participants *experienced* safety, even in objectively 'unsafe areas,' such as along the intense liminality of the strolls. In many ways, spaces that participants identify as where they have or can access care stand in direct contrast to the kinds of 'safety' promised by organisations that ostensibly exist to serve gender marginal persons. However, safety, in the context of spaces and places that are specifically demarcated as 'gay-friendly,' is thus best understood as catering to predominantly white and non-trans consuming classes. In other words, the sexual or gender marginal person who can or is willing to perform a specific kind of racialised and classed personhood is the one who is the least impacted by policies of exclusion. That is to say, these spaces only provide support, whether implicitly or explicitly, to particular formations of gender and sexuality. In contrast, 'safety' for the participants in this project often refers instead to areas wherein one's trans history or present is *not* necessarily of public issue.

## Safety, Zones of Exclusion, and Friction

Contrasted to mainstream maps, the community-produced maps discussed here focus primarily on spaces that are commonly organised around broader concerns of safety, risk, and where to access trans-specific care. These maps, as situated in the embodied knowledge of the subject, can, in turn, then be used to produce a far richer understanding of those who participated here and depicted a city heavily demarcated by spatial policies. This becomes of particular importance to gender and sexuality rights organisations and other advocates when critiques or notable exclusions of particular groups over others emerge. Notions of 'safe space' that circulate in the mainstream imaginary community (such as the gay bar or the 'LGBT' community centre) are often only referenced in the maps collected in this project for their lack of actual safety. Indeed, the processes of deeming safe versus unsafe space are not purely individualistic as much as they are linked to broader discourses circulating within the trans community and larger organisational efforts with regard to where 'safety' exists. The inflexibility of the margins of text in this case keeps trans community members going to the spaces offering the greatest degrees of safety, or comfort, even in moments wherein that safety is fleeting. As highlighted by the maps collected in this project, space is a contextualised experience that is reliant upon broader social and political interrogations of power rather than simplistic categories of gay space as 'safe space.'

Importantly, while this chapter considers laws and practices that may emerge within or across nation-states and jurisdictions in which there are different cultural practices, historical practices, and legal frameworks, the focus of this chapter is on the regulations of bodies in urban spaces as enacted through a legal framework of 'exclusion.' Importantly, zones of exclusion are not all equal in their potential violence. When placed in broader social contexts, we can then understand exclusionary laws and policies as focused more on identifying bodies that are seen as out of space than as enactments of laws that are focused on space itself.

Finally, as a means of thinking about the 'becoming' offered by 'zones of exclusion,' I end here calling for a consideration of how displacement can provide the 'friction' necessary for effecting change. The ways in which various social actors enforce, or resist, exclusionary policies or practices illuminate the productive potential of what is generally understood as a form of social death or erasure. Anthropologist Anna Tsing considers 'friction' a productive force (Tsing 2011) wherein spaces of conflict or contrasting needs produce the very spaces in which mobilising for change is possible. Tsing refers to this relationship as symbiotic, that the exclusions of a social, economic, or political practice are the inclusions of another. The process weaves together key stakeholders. In thinking about the increased political polarisation of the global north, with hyper-nationalistic rhetoric offering up new opportunities for committing violence, the productive moment is important to consider: how will spaces that once performed inclusion, that now perform hyper-exclusion, attend to this shift?

# References

Agamben, G. (1998). *Homo sacer: Sovereign power and bare life*. Stanford, CA: Stanford University Press.

Agamben, G. (2005). *State of exception* (K. Attell, Trans.). Chicago, IL: The University of Chicago Press.

Banerjee, S. B. (2008). Necrocapitalism. *Organization Studies, 12*, 1541–1563.

Beckett, K., & Western, B. (2001). Governing social marginality: Welfare, incarceration, and the transformation of state policy. *Punishment & Society, 3*(1), 43–59.

Bhagat, A., & Mogel, L. (2007). *An atlas of radical cartography*. Los Angeles, CA: Journal of Aesthetics and Protest Press.

Blunt, A., & Rose, G. (1994). *Writing women and space: Colonial and postcolonial geographies*. New York, NY: Guilford Press.

Brown, M., & Knopp, L. (2003). Queer cultural geographies – We're here! We're queer! We're over there, too. In K. Anderson, M. Domosh, S. Pile, & N. Thrift (Eds.), *Handbook of cultural geography* (pp. 313–324). London: Sage Publications.

Carr, J., Brown, E., & Herbert, S. (2009). Inclusion under the law as exclusion from the city: Negotiating the spatial limitation of citizenship in Seattle. *Environment and Planning A, 41*, 1962–1978.

De Certeau, M. (1984). *The practice of everyday life*. Berkeley, CA: University of California Press.

Dourish, P. (2006). Re-space-ing place: Place and space ten years on. In *Proceedings of the 2006 20th anniversary conference on computer supported cooperative work* (pp. 299–308). ACM.

Edelman, E. A. (2014). Walking while transgender: Necropolitical regulations of trans feminine bodies of colour in the nation's capital. In J. Haritaworn, A. Kuntsman, & S. Posocco (Eds.), *Queer necropolitics* (pp. 172–190). London: Routledge.

Edelman, E. A., Corado, R., Lumby, E. C., Gills, R. H., Elwell, J., Terry, J. A., & Emperador Dyer, J. (2015). *Access denied: Washington DC trans needs assessment report*. Washington, DC: DC Trans Coalition.

Elliott-Cooper, A., Hubbard, P., & Lees, L. (2019). Moving beyond Marcuse: Gentrification, displacement and the violence of un-homing. *Progress in Human Geography, 30*, 1–18. https://doi.org/10.1177/0309132519830511.

Foucault, M. (1977). *Discipline and punish: The birth of the prison* (A. Sheridan, Trans.). New York, NY: Vintage.

Geltmaker, T. (1997). The queer nation acts up: Health care, politics, and sexual diversity in the county of angels, 1990–92. In G. Ingram, A. M. Bouthillette, & Y. Retter (Eds.), *Queers in space: Communities, public places, sites of resistance* (pp. 233–274). Seattle, WA: Bay Press.

Grosz, E. (1998). Bodies–cities. In H. Nast & S. Pile (Eds.), *Places through the body* (pp. 42–51). London: Routledge.

Hackworth, J. (2007). *The neoliberal city: Governance, ideology, and development in American urbanism*. Ithaca, NY: Cornell University Press.

Hubbard, P. (2001). Sex zones: Intimacy, citizenship and public space. *Sexualities, 4*(1), 51–71.

Keith, M., & Pile, S. (1993). *Place and the politics of identity*. London: Routledge.

Lynch, K. (1960). *The image of the city*. Cambridge, MA: MIT press.

Mbembe, A. (2003). Necropolitics. *Public Culture, 15*(1), 11–40.

McAuliffe, C. (2012). Graffiti or street art? Negotiating the moral geographies of the creative city. *Journal of urban affairs, 34*(2), 189–206.

Mitchell, D. (1995). The end of public space? People's Park, definitions of the public, and democracy. *Annals of the association of American geographers, 85*(1), 108–133.

Mitchell, K. (2006). Geographies of identity: The new exceptionalism. *Progress in Human Geography, 30*(1), 95–106.

Ong, A. (2006). *Neoliberalism as exception: Mutations in citizenship and sovereignty.* Durham, NC: Duke University Press.

Peck, J. (2004). Geography and public policy: Constructions of neoliberalism. *Progress in Human Geography, 28*(3), 392–405.

Sanchez, L. E. (2004). The global e-rotic subject, the ban, and the prostitute-free zone: Sex work and the theory of differential exclusion. *Environment and Planning D: Society and Space, 22*(6), 861–883.

Stevenson, R. B. (2007). Schooling and environmental education: Contradictions in purpose and practice. *Environmental Education Research, 13*(2), 139–153.

Tsing, A. L. (2011). *Friction: An ethnography of global connection.* Princeton, NJ: Princeton University Press.

Young, A. (2010). Negotiated consent or zero tolerance? Responding to graffiti and street art in Melbourne. *City, 14*(1–2), 99–114.

# 19

# Olympic Favela Evictions in Rio de Janeiro: The Consolidation of a Neoliberal Displacement Regime

James Freeman

## Introduction

'We are afraid of Morar Carioca,' Cristiane explained. 'You never know when you are going to come home and find your house has been marked' (Author interview, 21 July 2011). Cristiane was born and raised in the favela of Chapéu Mangueira, a dense jumble of modest cinderblock houses covering a hillside behind the middle-class apartment buildings of Copacabana. Rio's favela residents, like Brazil's poor in general, are the inheritors of a history of multiple displacements, from the displacements of Brazil's Indigenous populations, to the displacements of the Atlantic slave trade, to the displacements of rural–urban migration that created the first favelas, to repeated state-mandated displacements within and between Rio's tenements and favelas from late nineteenth century onwards.

Cristiane was worried about Morar Carioca (Carioca Living), an ambitious slum upgrading programme, with the goal of urbanising all of Rio's more than 1000 favelas by 2020 (Steiker-Ginzberg 2014). According to residents and activists, Morar Carioca intentionally displaced people in the process of building roads and other infrastructure through dense favela neighbourhoods. Morar Carioca was part of a set of favela policies that were developed and

J. Freeman (✉)
Department of Geography, Planning and Environment, Concordia University, Montreal, QC, Canada
e-mail: james.freeman@concordia.ca

© The Author(s) 2020
P. Adey et al. (eds.), *The Handbook of Displacement*,
https://doi.org/10.1007/978-3-030-47178-1_19

became dominant between 2007 and 2016 as Rio prepared for a series of mega-events culminating in the 2016 Olympic Games. This was a period when Rio's favelas got a lot of attention from the authorities, much of it seemingly positive. Morar Carioca was a municipal programme with matching funds from the Inter-American Development Bank. There was also PAC (Program for Accelerated Growth), a federal infrastructure programme, that included urbanisation and transportation projects in Rio's favelas (Dias Simpson 2013). There was a major push to build housing for the poor and lower middle class through the federal Minha Casa Minha Vida (My House My Life) programme (Rolnik 2015). There was the UPP (Police Pacification Unit) programme that sought to assert state control over favelas, which had long been dominated by drug gangs, through 'proximity policing' (Cunha 2011; Freeman 2012; Müller 2018). Rio's favelas also got increased attention from GEO-Rio, a municipal agency that seeks to protect favela residents from mudslides, following the rains of April 2010 (Freeman and Burgos 2016).

In many ways, these programmes were the culmination of the democratisation movement at the end of the 1964–1985 military dictatorship, where authoritarian policies towards favelas that resulted in mass evictions were replaced by a legal regime that protected the rights of favela residents to stay in their communities, and promoted state investment in favelas coupled with community participation (Caldeira and Holston 2014; Gonçalves 2013; McCann 2014). But they were also created in a period when cities had become increasingly dependent on attracting mobile capital, when 'business climate' and city image had become key projects for city officials (Harvey 1989, 2005). While many of these programmes had idealistic roots, they ended up lending their services to a mega-event development strategy that hinged on subduing Rio's favelas.

So, while mass displacements by an authoritarian state were no longer politically possible, these programmes collaborated to pacify Rio's famously unruly favelas, in part by thinning them out and selectively displacing a portion of their residents to the urban periphery, thus creating new opportunities for capital accumulation (Faulhaber and Azevedo 2015; Freeman 2012; Freeman and Burgos 2016). In this chapter I place Rio's mega-event regime of displacement in the context of the history of favela evictions. I argue that the world has gone through a series of displacement regimes with the evolution of global capitalism, each accompanied by a new set of internationally circulating technologies and ideas.

## Primitive Accumulation and Regimes of Displacement

Capitalism is a system that regularly creates displacements of human populations. While capital accumulation through commodity production is in principle a non-violent process, an initial violence was necessary to get this system off the ground, what Marx calls primitive accumulation (Marx 1976, p. 915). This initial violence involved a series of major displacements that were fundamental for the establishment of modern capitalism. The enclosures movement in England leading up to the industrial revolution displaced large numbers of people, as did the commodification of land everywhere as capitalism spread. Africans were enslaved and displaced to the Americas to produce the raw materials that fuelled commodity production in Europe: cotton, tobacco, and sugar (Hobsbawm 1996; Williams 1994). At the same time, Indigenous populations of the Americas experienced repeated displacements as Europeans expanded into the continent (Mann 2012). Capital must continue to expand, but non-violent expanded reproduction through commodity production has limits, and further primitive accumulation—or what Harvey calls accumulation by dispossession—can allow capital to temporarily overcome those limits (Harvey 2003; Jeffrey et al. 2012).

## Haussmannisation

As capitalist production concentrated in cities, urban displacement became a regular feature of the system (Harvey 2008). The industrial revolution created massive urban slum populations of industrial workers that became a problem for the dominant classes throughout the nineteenth-century industrialised world (Engels 1968; Hall 1996). Emperor Louis Napoleon and his prefect Baron Haussmann came up with a solution in Paris that would become a model for the rest of the world. In the 1850s and 1860s, broad avenues were carved through the old central city. These avenues served the needs of capital accumulation: absorbing surplus capital, providing for more efficient movement of goods and people through the city, and facilitating social control of the working classes (Berman 1982; Hall 1996; Harvey 1985, 2006; Kirkland 2013; Scott 1998). But once again the needs of capital meant the displacement of vulnerable populations. Haussmann himself estimated that his boulevards displaced 350,000 people (Clark 1985, p.37).

Haussmann's model was copied extensively in Europe, and by the late nineteenth century cities in the developing world such as Buenos Aires (Needell 1995), Cairo (AlSayyad 2011), Mexico City (Tenorio 1996), and Rio de Janeiro were displacing poor populations to make way for Parisian-style boulevards fronted by Beaux Arts-style façades. Rio's Haussmann-style reforms were part of the first large-scale displacement of the urban poor in that city, marking the beginning of a long history of removing the unsightly, dangerous poor from valuable real estate in central areas of the city. The displacements carried out at the end of the nineteenth century and beginning of the twentieth established a pattern of socio-spatial segregation that persists to this day, dividing the wealthy seaside enclave of the South Zone from the poor North Zone of the city and creating the first significant favela population (Chalhoub 1996; Vaz 1994). Urban reforms during the period targeted tenements, known as *cortiços* or bee hives, which housed Rio's poor, notably a large population of former African slaves, leading up to and following the abolition of slavery in 1888 (Macedo 1955). The anti-*cortiço* campaign led by Mayor Barata Ribeiro and the Parisian-style boulevards created by Mayor Pereira Passos together may have displaced 100,000 people or 25 per cent of Rio's population at the time (Abreu 1997, p. 67; Ribeiro 1997, p. 175; Teixeira 1994, p. 574).

Rio's turn of the century reforms served the interests of capital during a period of economic transition. Rio was the main port for a booming coffee export economy and was experiencing the beginnings of industrialisation (Abreu 1997). But the abolition of slavery led to economic crisis (Speck 1982). The massive intervention in the built environment absorbed a large amount of capital and led to a speculative real estate boom in areas adjacent to the new avenues (Ribeiro 1997). At the same time the reforms represented a transition in the regime of social control, imposing a new kind of power on a potentially threatening urban working class, swelled by the influx of former slaves from the countryside.

## Urban Renewal

The reforms carried out by Haussmann and his imitators represented a first wave of urban displacements in the process of restructuring cities for the next phase of capital accumulation. But cities have required regular creative destruction, as capital reaches the limits of a particular built environment with its associated technologies, products, institutions, and social relations (Walker 1981). The United States provided a model for the next phase of urban restructuring. The Great Depression was the result of capital reaching growth limits within the structures it had created, and in the United States, car-oriented suburbanisation became an important outlet for further growth. The remaking of US cities for an economy centred around the automobile in the post-World War II period entailed another major wave of displacement. Following Robert Moses's pioneering highway building and slum clearance programme in New York (Caro 1974; Chronopoulos 2013), the US Federal Government mandated an analogous nationwide policy (Collins and Shester 2013; Hall 1996; Platt 2015). Under urban renewal, 'Approximately one million people were displaced in 2500 projects carried out in 993 American cities; 75% of those displaced were people of color' (Fullilove and Wallace 2011, p. 382).

The model of slum clearance and replacement government-built housing spread around the world (Hall 1996). In Asia, Africa, and Latin America, import substitution industrialisation policies and the modernisation of agriculture in the post-World War II period led to a massive rural–urban migration, causing the mushrooming of slum populations (Roberts 1978). Third World governments responded with a new wave of slum clearance, inspired by the US example and making use of internationally circulating ideas about the relationship between disorderly housing on the one hand and dysfunctional people on the other (Lewis 1959; Perlman 1976; Platt 2015). Gilbert cites public housing projects built to receive displaced slum populations in San Salvador, Bogotá, Rio de Janeiro, São Paulo, Mexico City, Lagos, Lusaka, Nairobi, Kuala Lumpur, Manila, Tunis, and Rabat during the period (Gilbert and Gugler 1992, p. 137).

Rio de Janeiro experienced a particularly brutal period of forced slum removal along urban renewal lines in the 1960s and 1970s. The political project of favela removal was consolidated with the election of Carlos Lacerda, a centre-right technocratic moderniser, as governor of the State of Guanabara in 1960.[1] Gonçalves writes, 'The final tally of the Lacerda government in this account was the removal of approximately 42,000 people, the demolition of 8078 shacks and the total or partial eradication of 27 favelas between 1962 and 1965' (Gonçalves 2013, p. 218). After the military coup of April 1964, Lacerda's pilot project was scaled up to a national policy of systematic slum removal. Within a few months of taking power, the military government created the National Housing Bank (BNH) and a new national pension fund, the FGTS, so that workers' forced retirement savings would fund a massive slum removal and replacement public housing programme (Rolnik 2015). CHISAM was the agency tasked with carrying out removals in Rio de Janeiro. Perlman writes, 'From its inception in 1968 until its demise in 1975

[CHISAM] removed over 100 favelas, destroying more than 100,000 dwellings and leaving at least half a million poor people without their homes. The eradications were systematic and relentless, targeting the South Zone favelas first, where land values were the highest, and then continuing into the North Zone' (Perlman 2010, p. 271).

Favela leaders who sought to organise resistance were imprisoned and tortured (Gonçalves 2013). One favela was infamously burned to the ground after residents refused to move (Perlman 1976, pp. 206–209). Favela residents were kept in the dark about plans for removal, often receiving notice of the final move only the night before. They were piled in the back of dump trucks with their belongings and deposited in remote BNH *conjuntos*. Residents had no say about where they would end up, and there was little attempt to keep families and communities together. Separated from jobs, schools, social networks, commerce, and transportation connections, residents suffered depression and several older people died (Perlman 1976). Burgos writes, 'The history of these removals, carried out mostly between 1968 and 1975, represents one of the most violent chapters in a long history of Brazilian state repression and exclusion' (1998, p. 36).

There was a strong economic motive behind these policies. Capital accumulation within a certain set of institutional structures and a certain built environment associated with the early phase of import substitution industrialisation had reached its limits in Brazil by 1964 (O'Donnell 1973). Further capital accumulation required the seizure of assets and the colonisation of territory by force. Workers were dispossessed of 8 per cent of their paycheques, so that the urban periphery could be colonised for low-income housing production. That housing boosted the construction industry, encouraged real estate speculation, and created demand for national production of construction materials (Dias Simpson 2013, p. 7). At the same time favela residents were dispossessed of the use value of their centrally located land so that capital could be accumulated through the construction of high-rise apartment buildings for the middle classes. The period of most intense favela removal coincided with the Brazilian economic miracle in which GDP grew at a rate of 10 per cent per year, and the construction and real estate industries boosted by the favela removal programme were a significant part of the boom (Gonçalves 2013, p. 223).

## Neoliberal Displacements

Urban renewal and the carving of elevated highways through central cities in North America were strongly contested and began losing hegemony in the late 1960s. Jane Jacobs famously argued that the poor were better off staying in their communities rather than being transferred to soulless vertical slums (Jacobs 1961). At the same time the World Bank led a shift in Third World slum policy from removal to various forms of *in situ* upgrading (Dias Simpson 2013; Gilbert and Gugler 1992; Goldman 2005; Pugh 2001; UN-Habitat 2003; Wakely 1988). In Rio the tide turned away from slum removal. Favela residents were important protagonists in the democratisation process, and their rights were enshrined in the post-dictatorship legal regime, particularly the 1988 Constitution, the 2001 Statute of the City, and a series of state and municipal laws (Caldeira and Holston 2014; Gonçalves 2013; McCann 2014; Rolnik 2011).

Ironically, the ideas Jacobs articulated helped pave the ideological way for the next phase of urban capital accumulation and its associated displacements. Again, capital needed to remake landscapes and displace people for further growth. As the Fordist economy—based on suburban housing, highways, automobiles, and First World urban manufacturing—reached saturation, the economic role of cities changed. Manufacturing shifted to ex-urban export processing zones in Asia, so that cities lost the manufacturing role that supported large working-class populations. A handful of global cities were able to specialise in finance, corporate headquarters, and producer services (Sassen 1991). Other cities were able to thrive as agglomerations of technological and creative innovation (Saxenian 1994; Storper and Christopherson 1987). But most cities have spent the period since the 1980s trying to replace manufacturing jobs with consumption and real estate development—promoting tourism, building shopping malls and sport stadiums, and facilitating real estate speculation (Harvey 1989).

This restructuring of cities with the restructuring of the global economy has led to a new wave of displacements. As jobs disappeared, urban working-class neighbourhoods deteriorated, and capital eventually saw an opportunity to return (Smith 1987, 1996). The arguments of ideologues like Richard Florida, repurposing the ideas of Jane Jacobs, became a pretext for speculative real estate development, once again displacing the poor from central neighbourhoods to the edge of the city (Peck 2005; Tochterman 2012).

Third World cities have also experienced displacements due to something called 'gentrification' during the neoliberal period. There is a growing literature on gentrification in Asia (Ley and Teo 2014; Shin 2009), Africa (Lees et al. 2015; Winkler 2009), and Latin America (Inzulza-Contardo 2016; Janoschka et al. 2014; López-Morales et al. 2016). But while gentrification in the First World is often discussed in terms of market-led displacement, Third World gentrification and displacement is usually the result of direct state action to upgrade historic districts or promote real estate development projects (Betancur 2014). Mega-events and their associated mega-projects are often part of the process, as with the massive clearing of hutong housing in Beijing leading up to the 2008 Olympics. Cities throughout Asia, Africa, and Latin America are clearing street vendors from central areas, investing in redesigning and rebranding of urban spaces, and promoting real estate development that displaces the poor from jobs and from homes.

## Rio's Mega-event Development and Displacement Regime

The 1980s in Brazil was a period of hyperinflation, structural adjustment, austerity, and the growth of favelas. By the 1990s inflation was tamed, Brazil was no longer beholden to the IMF, and there was a new wave of urban investment. Like in North American cities, capital had retreated, leaving Rio devalued to the point that it now offered investment opportunities.

Mayor Cesar Maia (1993–1996; 2001–2008) led Rio on an urban entrepreneurial development programme that focused on urban design and the hosting of mega-events, following the example of the 1992 Barcelona Olympics (Vainer 2011). Maia took office in the wake of Rio's hosting of the 1992 UN Conference on Environment and Development. He led unsuccessful bids for the 2004 and 2012 Summer Olympics, hosted the 2007 Pan-American Games, redesigned favelas through the Favela-Bairro programme, redesigned formal neighbourhoods through the Rio Cidade programme, and unsuccessfully attempted to bring a Guggenheim museum to Rio. This mega-event city project would be more fully realised, during the next political regime, in the form of an alliance between Mayor Eduardo Paes (2009–2016), Governor Sergio Cabral (2007–2014), and President Luiz Inácio Lula da Silva (2003–2010), with support from FIRJAN (Rio's Chamber of Commerce) and NGOs such as Viva Rio and the UN Development Programme (Freeman 2018; Moraes et al. 2015).

Between the time Sergio Cabral took office in January 2007 and the August 2016 Summer Olympics, Rio went through a major transformation that threatened to displace hundreds of thousands of people. Evictions in Rio resulted from transportation and stadium projects, beautification near Olympic sites, favela upgrading under PAC and Morar Carioca, and particularly GEO-Rio's designation of 'areas of risk' in numerous favelas following a series of landslides in April 2010. Morar Carioca and PAC, unlike the earlier slum upgrading programme Favela-Bairro (Brakarz and Aduan, 2004; Dias Simpson 2013, p. 13; Perlman 2010, p. 276), had a significant 'mobility' component, meaning carving roads through dense neighbourhoods and building cable car systems with large footprints. These programmes also had a mandate to 'oxygenate' and 'de-densify' (Freeman and Burgos 2016). Thus, Morar Carioca and PAC tended to displace people in a way Favela-Bairro did not (Dias Simpson 2013, pp. 17–18; Gomes and Motta 2013, p. 10). If these programmes had continued as planned, 'urbanising' all of Rio's more than one thousand favelas, massive evictions would have been the result.

Likewise, GEO-Rio arbitrarily designated large sections of over one hundred favelas as 'areas of risk' and thus subject to removal (Freeman and Burgos 2016). Many more houses were marked with eviction serial numbers than were actually removed. Community activists leveraged post-dictatorship legislation to block some of the evictions and mobilised independent technical experts to counter claims of landslide risk in several cases. Between 2009 and 2015, according to the municipal government, 22,059 families or 77,206 individuals had been evicted (Comité Popular da Copa e Olimpíadas do Rio de Janeiro 2015; Faulhaber and Azevedo 2015; Freeman and Burgos 2016). By comparison, 1.5 million people were displaced leading up to the 2008 Beijing Olympics, and 200,000 people were evicted in the context of the 2010 Commonwealth Games in New Delhi (Centre on Housing Rights and Evictions 2008; Housing and Land Rights Network 2011). Likely the collapse of the mega-event development coalition alongside the political and economic crisis after 2014 was an important factor in preventing more evictions. Evictions were mostly not wholesale, unlike during the military dictatorship, when entire communities were razed. Rather the objective seemed to be to thin out centrally located communities through public works, as part of a project of pacification and rebranding. While residents theoretically benefited from legal protections not available to their counterparts in the 1960s and 1970s, they suffered much of the same lack of information and disrespect (Freeman and Burgos 2016).

The dispossession of the poor leading up to the 2016 Olympics served the needs of capital accumulation (Freeman 2012). That dispossession took place in the context of an entrepreneurial city marketing programme that saw favelas as an image problem. The pacification of strategically located favelas—through UPP, PAC, Morar Carioca, GEO-Rio, and MCMV—was an attempt to change the image of the city by changing the image of key favelas from dangerous outlaw territories to the model projects of an enlightened government. This transformation allowed FIFA and the International Olympic Committee to associate their brands with the exotic but safe brand of Rio de Janeiro, selling billions of dollars in broadcasting rights and sponsorship agreements, so that a series of global brands could in turn be associated with those of Rio de Janeiro, FIFA, and the Olympic movement. The pacification of key favelas and the displacement of populations allowed Brazilian multinational engineering and construction companies, such as Odebrecht and OAS, to profit greatly by building the mega-event city.[2] These companies built the stadiums, the bus rapid transit lines, and the subway extension. They carried out the transportation and architectural interventions in favelas, built the MCMV housing that received displaced favela residents, and managed the highly speculative Porto Maravilha port revitalisation project, that depended directly on the pacification of the Morro da Providência favela (Oliveira 2015). In a general climate of real estate speculation, property values in areas near pacified favelas increased dramatically. Pacified favelas themselves became new markets for commodities, as the electric company and cable television companies signed up large numbers of new customers, and banks and retail chains opened favela branches under the protection of UPP police occupation (Freeman 2012).

## Conclusion

Capitalism has always depended on the displacement of human populations. Capital must grow and expand, but it regularly reaches limits to that expansion, requiring restructuring, spatial fixes, the colonisation of new territories, the privatisation of communally held assets, and the displacement of people. The built environment that facilitates capital accumulation during one period becomes a constraint on further growth and is abandoned by capital in the next (Harvey 1982). Working-class populations are left behind, benefiting from the use values of remaining housing and neighbourhoods, creating commons on the edge of processes of capital accumulation until capital is ready to flow back in, reintegrating derelict urban territories into processes of capital accumulation, and displacing those populations (Blomley 2008).

In this chapter I have identified three global urban displacement regimes. The first is Haussmannisation, which restructured central city areas to accommodate nineteenth-century industrial capitalism, imposing new techniques of social control and shifting working-class populations to the edge of the city. In Rio, Haussmannisation absorbed capital that was released at the end of the slave regime, made the city more efficient for the movement of goods and people, created new consumption spaces for an urban elite, and imposed a new spatial order on a class structure that was being remade.

The second displacement regime is urban renewal, the remaking of central working-class neighbourhoods in the context of post-World War II automobile suburbanisation in North America, displacing racialised populations to high-rise public housing or further marginalising them in inner-city neighbourhoods. Urban renewal in the Third World coincided with import substitution industrialisation, the industrialisation of agriculture, and another major wave of rural–urban migration. Third World governments carried out authoritarian evictions of centrally located slums, rehousing some of the displaced people in government-built housing on the urban periphery. In Rio, the slum removals of 1962–1975 were the second major displacement of the poor from central neighbourhoods in the history of that city, abandoning hundreds of thousands on the urban periphery.

The third regime is the neoliberal gentrification displacement regime, where central city areas that were devalued during the urban renewal phase are ideologically and economically revalued. The state collaborates with private capital to change the social character of central neighbourhoods. In the Third World, housing has come to be considered a human right and *in situ* slum upgrading has been declared best practice. And yet capital flows into Third World and First World cities alike, looking for a spatial fix, and ready to dispossess in order to accumulate. City governments are focused on quality of life, business climate, city branding, and accommodating every real estate development project they can. Central city slums are once again in the way of capital accumulation and Third World governance coalitions are finding ways around the anti-eviction consensus.

These circumstances gave us the 2007–2016 mega-event displacement regime in Rio de Janeiro, which is only the latest round of urban accumulation by dispossession. A focus on urban design, mega-projects, and mega-events as part of a programme to rebrand the city came into conflict with a visible and unruly favela population, who threated the aspirational brand. Despite the hard-won legal regime and social consensus against 1960s-style favela removals, the governance coalition used a series of institutions and programmes as pretexts for removal. The displacement of tens of thousands of poor Rio residents facilitated billions of dollars in broadcasting rights, sponsorship agreements, construction contracts, real estate deals, and commodity transactions.

## Notes

1. Lacerda was essentially the mayor of Rio de Janeiro.
2. These firms were implicated in the Car Wash corruption investigation and several key players have been jailed. Sergio Cabral, Rio's governor at the time, reportedly took a 25 per cent kickback on all infrastructure projects and is also serving a jail sentence.

## References

Abreu, M. d. A. (1997). *Evolução urbana do Rio de Janeiro* (3rd ed.). Rio de Janeiro, Brazil: Iplanrio.

AlSayyad, N. (2011). *Cairo: Histories of a city* (1st ed.). Cambridge, MA: Belknap Press of Harvard University Press.

Berman, M. (1982). *All that is solid melts into air: The experience of modernity.* New York, NY: Simon and Schuster.

Betancur, J. J. (2014). Gentrification in Latin America: Overview and critical analysis. *Urban Studies Research, 2014*, 1–14. https://doi.org/10.1155/2014/986961.

Blomley, N. (2008). Enclosure, common right and the property of the poor. *Social & Legal Studies, 17*(3), 311–331. https://doi.org/10.1177/0964663908093966.

Brakarz, J., & Aduan, W. E. (2004). *Favela-Bairro – Scaled-up urban development in Brazil.* Washington, DC: http://web.worldbank.org/archive/website00819C/WEB/PDF/BRAZIL_F.PDF.

Burgos, M. B. (1998). Dos parques proletários ao Favela-Bairro: as políticas públicas nas faveals do Rio de Janeiro. In A. Zaluar & M. Alvito (Eds.), *Um século de favela* (pp. 25–60). Rio de Janeiro, Brazil: Editora Fundação Getulio Vargas.

Caldeira, T., & Holston, J. (2014). Participatory urban planning in Brazil. *Urban Studies, 52*(11), 2001–2017. https://doi.org/10.1177/0042098014524461.

Caro, R. A. (1974). *The power broker: Robert Moses and the fall of New York* (1st ed.). New York, NY: Knopf.

Centre on Housing Rights and Evictions. (2008). *One world, whose dream? Housing rights violations and the Beijing Olympic Games.* Geneva: Switzerland.

Chalhoub, S. (1996). *Cidade febril: Cortiços e epidemias na corte imperial.* São Paulo, Brazil: Companhia das Letras.

Chronopoulos, T. (2013). Robert Moses and the visual dimension of physical disorder. *Journal of Planning History, 13*(3), 207–233. https://doi.org/10.1177/1538513213487149.

Clark, T. J. (1985). *The painting of modern life: Paris in the art of Manet and his followers* (1st ed.). New York, NY: Knopf.

Collins, W. J., & Shester, K. L. (2013). Slum clearance and urban renewal in the United States. *American Economic Journal: Applied Economics, 5*(1), 239–273. https://doi.org/10.1257/app.5.1.239.

Comitê Popular da Copa e Olimpíadas do Rio de Janeiro. (2015). *Olimpíada Rio 2016, os Jogos da exclusão. Megaeventos e Violações dos Direitos Humanos no Rio de Janeiro, Dossiê do Comitê Popular da Copa e Olimpíadas do Rio de Janeiro*. Rio de Janeiro, Brazil.

Cunha, J. B. (2011). O pac e a upp no 'complexo pavão-pavãozinho-cantagalo': Processo de implementação de políticas públicas em uma favela da zona sul da cidade do rio de janeiro. *XI Congresso Luso Afro Brasileiro de Ciências Sociais, Diversidades e* (Des).

Dias Simpson, M. (2013). *Urbanising favelas, overlooking people: Regressive housing policies in Rio de Janeiro's progressive slum upgrading initiatives* (DPU Working Paper, 155). https://www.ucl.ac.uk/bartlett/development/sites/bartlett/files/migrated-files/WP155_0.pdf.

Engels, F. (1968). *The condition of the working class in England.* Stanford, CA: Stanford University Press.

Faulhaber, L., & Azevedo, L. (2015). *SMH 2016: Remoções no Rio de Janeiro Olímpico.* Rio de Janeiro, Brazil: Mórula.

Freeman, J. (2012). Neoliberal accumulation strategies and the visible hand of police pacification in Rio de Janeiro. *REU, Sorocaba, SP, 38*(1), 95–126.

Freeman, J. (2018). *The rise and fall of Rio de Janeiro's Police Pacification Unit program.* Paper presented at the Annual Congress of the Latin American Studies Association, Barcelona, Spain.

Freeman, J., & Burgos, M. (2016). Accumulation by forced removal: The thinning of Rio de Janeiro's favelas in preparation for the games. *Journal of Latin American Studies, 40*(3), 1–29.

Fullilove, M. T., & Wallace, R. (2011). Serial forced displacement in American cities, 1916–2010. *Journal of Urban Health, 88*(3), 381–389. https://doi.org/10.1007/s11524-011-9585-2.

Gilbert, A., & Gugler, J. (1992). *Cities, poverty, and development: Urbanization in the Third World* (2nd ed.). Oxford, England; New York, NY: Oxford University Press.

Goldman, M. (2005). *Imperial nature: The World Bank and struggles for social justice in the age of globalization.* New Haven, CN; London: Yale University Press.

Gomes, M. d. F. C. M., & Motta, T. S. d. (2013). Empresariamento Urbano e Direito à Cidade: Considerações sobre os programas Favela-Bairro e Morar Carioca no Morro da Providência. *Revista LIbertas, 13,* 1–21.

Gonçalves, R. S. (2013). *Favelas do Rio de Janeiro: História e direito.* Rio de Janeiro, Brazil: Pallas.

Hall, P. (1996). *Cities of tomorrow: An intellectual history of urban planning and design in the twentieth century* (Updated ed.). Oxford, England; Cambridge, MA: Blackwell Publishers.

Harvey, D. (1982). *The limits to capital.* Chicago, IL: University of Chicago Press.

Harvey, D. (1985). *Consciousness and the urban experience: Studies in the history and theory of capitalist urbanization*. Baltimore, MD: John Hopkins University Press.

Harvey, D. (1989). From managerialism to entrepreneurialism: The transformation in urban governance in late capitalism. *Geografiska Annaler, Series B: Human Geography, 71*(1), 3–17.

Harvey, D. (2003). The 'new' imperialism: Accumulation by dispossession. *The Socialist Register, 2004*, 63–87.

Harvey, D. (2005). *A brief history of neoliberalism*. Oxford, England; New York, NY: Oxford University Press.

Harvey, D. (2006). *Paris, capital of modernity (1st Routledge paperback ed.)*. New York: Routledge.

Harvey, D. (2008). The right to the city. *New Left Review, 53*, 23–40.

Hobsbawm, E. J. (1996). *The age of revolution 1789–1848 (1st Vintage Books ed.)*. New York, NY: Vintage Books.

Housing and Land Rights Network. (2011). *Planned dispossession: Forced evictions and the 2010 Commonwealth Games* (Vol. 14). New Delhi, India.

Inzulza-Contardo, J. (2016). Contemporary Latin American gentrification? Young urban professionals discovering historic neighbourhoods. *Urban Geography, 37*(8), 195–1214. https://doi.org/10.1080/02723638.2016.1147754.

Jacobs, J. (1961). *The death and life of great American cities*. New York: Random House.

Janoschka, M., Sequera, J., & Salinas, L. (2014). Gentrification in Spain and Latin America – a critical dialogue. *International Journal of Urban and Regional Research, 38*(4), 1234–1265. https://doi.org/10.1111/1468-2427.12030.

Jeffrey, A., McFarlane, C., & Vasudevan, A. (2012). Rethinking enclosure: Space, subjectivity and the commons. *Antipode, 44*(4), 1247–1267. https://doi.org/10.1111/j.1467-8330.2011.00954.x.

Kirkland, S. (2013). *Paris reborn: Napoléon III, Baron Haussmann, and the quest to build a modern city* (1st ed.). New York, NY: St. Martin's Press.

Lees, L., Shin, H. B., & López Morales, E. (2015). *Global gentrifications: Uneven development and displacement*. Bristol, England: Policy Press.

Lewis, O. (1959). *Five families; Mexican case studies in the culture of poverty*. New York, NY: Basic Books.

Ley, D., & Teo, S. Y. (2014). Gentrification in Hong Kong? Epistemology vs. ontology. *International Journal of Urban and Regional Research, 38*, 1286–1303.

López-Morales, E., Shin, H. B., & Lees, L. (2016). Latin American gentrifications. *Urban Geography, 37*(8), 1091–1108. https://doi.org/10.1080/02723638.2016.1200335.

Macedo, R. d. (1955). *Barata Ribeiro: Administração do primeiro prefeito do Distrito Federal*. Rio de Janeiro, Brazil: DASP.

Mann, C. C. (2012). *149: Uncovering the new world Columbus created (1st Vintage books ed.)*. New York, NY: Vintage Books.

Marx, K. (1976). *Capital: A critique of political economy* (Vol. I, B. Fowkes, Trans.). New York, NY: Vintage Books.

McCann, B. (2014). *Hard times in the marvelous city: From dictatorship to democracy in the favelas of Rio de Janeiro*. Durham, NC: Duke University Press.

Moraes, J., Mariano, S. R. H., & Franco, A. M. d. S. (2015). Unidades de Polícia Pacificadora (UPPs) no Rio de Janeiro: Uma história a partir das percepções e reflexões do gestor responsável por sua implantação. *Revista de Administração Pública, 49*(2), 493–518. https://doi.org/10.1590/0034-7612121746.

Müller, M.-M. (2018). Policing as pacification: Postcolonial legacies, transnational connections and the militarization of urban security in democratic Brazil. In M. Bonner, M. Kempa, M. R. Kubal, & G. Seri (Eds.), *Police abuse in contemporary democracies* (pp. 221–247). Basingstoke, England: Palgrave Macmillan.

Needell, J. D. (1995). Rio de Janeiro and Buenos Aires: Public space and public consciousness in Fin-de-Siècle Latin America. *Comparative Studies in Society and History, 37*(3), 519–540.

O'Donnell, G. A. (1973). *Modernization and Bureaucratic-authoritarianism: Studies in South American politics, 9*. Berkeley, CA: Institute of International Studies, University of California.

Oliveira, N. G. d. (2015). *O poder dos jogos e os jogos de poder: Interesses em campo na produção da cidade para o espetáculo esportivo*. Rio de Janeiro, Brazil: UFRJ.

Peck, J. (2005). Struggling with the creative class. *International Journal of Urban and Regional Research, 29*(4), 740–770.

Perlman, J. E. (1976). *The myth of marginality: Urban poverty and politics in Rio de Janeiro*. Berkeley, CA: University of California Press.

Perlman, J. E. (2010). *Favela: Four decades of living on the edge in Rio de Janeiro*. Oxford, England; New York, NY: Oxford University Press.

Platt, H. L. (2015). *Building the urban environment: Visions of the organic city in the United States, Europe, and Latin America*. Philadelphia, PA: Temple University Press.

Pugh, C. (2001). The theory and practice of housing sector development for developing countries, 1950–99. *Housing Studies, 16*(4), 399–423.

Ribeiro, L. C. d. Q. (1997). *Dos cortiços aos condimínios fechados: As formas de produção da moradia na cidade do Rio de Janeiro*. Rio de Janeiro, Brazil: Civilização Brasileira: IPPUR, UFRJ: FASE.

Roberts, B. R. (1978). *Cities of peasants: The political economy of urbanization in the Third World*. London: E. Arnold.

Rolnik, R. (2011). Democracy on the edge: Limits and possibilities in the implementation of an urban reform agenda in Brazil. *International Journal of Urban and Regional Research, 35*(2), 239–255. https://doi.org/10.1111/j.1468-2427.2010.01036.x.

Rolnik, R. (2015). *Guerra dos lugares: A colonização da terra e da moradia na era das finanças (1ª edição. ed.)*. São Paulo, Brazil: Boitempo.

Sassen, S. (1991). *The global city: New York, London, Tokyo*. Princeton, NJ: Princeton University Press.

Saxenian, A. (1994). *Regional advantage: Culture and competition in Silicon Valley and Route 128*. Cambridge, MA: Harvard University Press.

Scott, J. C. (1998). *Seeing like a state: How certain schemes to improve the human condition have failed (Yale Agrarian Studies; The Yale ISPS series)*. New Haven, CT: Yale University Press.

Shin, H. B. (2009). Property-based redevelopment and gentrification: The case of Seoul, South Korea. *Geoforum, 40*(5), 906–917.

Smith, N. (1987). Gentrification and the rent gap. *Annals of the Association of American Geographers, 77*(3), 462–465.

Smith, N. (1996). *The new urban frontier: Gentrification and the revanchist city*. London; New York, NY: Routledge.

Speck, M. (1982). *The industrialization of Rio de Janeiro and Sao Paulo, 1870–1920*. Stanford, CA: Center for Latin American Studies, Stanford University.

Steiker-Ginzberg, K. (2014, 10 September). Morar Carioca: The dismantling of a dream favela upgrading program. *RioOnWatch*. http://www.rioonwatch.org/?p=17687.

Storper, M., & Christopherson, S. (1987). Flexible specialisation and regional industrial agglomerations: The case of the U.S. motion picture industry. *Annals of the Association of American Geographers, 77*(1), 104–117.

Teixeira, M. C. (1994). A habitação popular no século XIX—característica morfológicas, a transmissão de modelos: as ilhas do Porto e os cortiços do Rio de Janeiro. *Análise Social, 29*(127), 555–579.

Tenorio, M. (1996). 1910 Mexico City: Space and nation in the city of the centenario. *Journal of Latin American Studies, 28*(1), 75–104.

Tochterman, B. (2012). Theorizing Neoliberal Urban Development: A Genealogy from Richard Florida to Jane Jacobs. *Radical History Review, 2012*(112), 65–87. https://doi.org/10.1215/01636545-1416169.

UN-Habitat. (2003). *The challenge of slums: Global report on human settlements*. London and Sterling, VA: Earthscan.

Vainer, C. (2011). Cidade de exceção: relexões a partir do Rio de Janeiro. http://br.boell.org/web/51-1266.html.

Vaz, L. F. (1994). Dos cortiços às favelas e aos edifícios de apartamentos—a modernização da moradia no Rio de Janeiro. *Análise Social, 29*(127), 581–597.

Wakely, P. (1988). The development of housing through the withdrawal from construction. *Habitat International, 12*(3), 121–131.

Walker, R. (1981). A theory of suburbanization: Capitalism and the construction of urban space in the United States. In M. J. Dear & A. J. Scott (Eds.), *Urbanization and urban planning in capitalist society* (pp. 383–429). London; New York: Methuen.

Williams, E. E. (1994). *Capitalism & slavery*. Chapel Hill, NC: University of North Carolina Press.

Winkler, T. (2009). Prolonging the global age of gentrification: Johannesburg's regeneration policies. *Planning Theory, 8*(4), 362–381. https://doi.org/10.1177/1473095209102231.

# Part III

## Section Three: Journeys of Displacement

# 20

# Intervention: Women's Narratives from Refugee Camps in the Kurdistan Region of Iraq

Nazand Begikhani

Our Last Day in the Orchard
Author of the story: Anonymous with the help of tutors Ms Shara Ali & Mr Awat Muhamad
Translated from Arabic by Ara Raza Edited by Nazand Begikhani and Luke Hodgkin

**Fig. 20.1** Untitled, painting by a Yazidi ISIS survivor, Sharya Refugee Camp, Kurdistan (supplied by Begikhani)

I do not know where to begin. My tale is a long one. I am 17 years old, and I have spent 4 of them living in a camp. Before life in the camp I was like any other child. I would go to school with my friends. We had a beautiful house and a large orchard, where my sisters and I would play together. Our tribe, the *Bohashma*, was a very large and prominent tribe. Our relatives and family members would visit often, and we led a happy life. I have many memories of that time, but my fondest memories are those of our orchard, where I grew up among the pomegranate and orange trees. All that happiness is now over, and my entire life changed from the day our orchard was bombarded.

On that fateful autumn day, my mother, my three sisters, and I were in our orchard, that place we loved, where we made our fondest memories working, playing, and being merry. And although the cruelties of war and fighting had already spread to our town of Yathrib, we still continued with our sweet little life, unruffled, tending to our vegetables and our eight cows. That evening, after we were finished picking the ripe oranges and pomegranates, me and my mom sat under our large fig tree to rest. There was a breeze, and birds circled round the trees, landing on a twig occasionally looking for food. My three sisters had walked to the banks of the Tigris, which encircled our town of Yathrib, and were busy taking pictures. I suddenly heard the sound of a helicopter soaring above, and as soon as I looked up, the helicopter began to drop bombs on our heads. I screamed and threw myself in mother's lap, I became hysterical with dread. My sisters also ran toward us, screaming with fear. My mother kept instructing us to hide, shouting, "Hide girls, hide by the cows!" In our orchard we found nowhere to take shelter except to lay under our cows, and so we did, for more than an hour, in dread, not daring to move. In an instant, our joy disappeared, the sweetness of the river and the shade of our trees forgotten, our orchard had been turned into a living hell.

After an hour of bombardment of our orchard and neighboring orchards, and the town of Yathrib, my mother began to weep and said, *"We must return home, your brothers are there all alone and God know what has become of them."* We had left our brothers Safe and Muhamad at home with my father. We ran as fast as we could to our house, which was about fifteen minutes away from our orchard but on that day, felt like an eternity that would not end. There was shelling everywhere, and on the road were strewn the bodies of the killed and injured, people in disarray running here and there. The sun was beginning to set. When we arrived home, we saw my two brothers curled up in a corner, terrified but my father was nowhere to be found.

---

N. Begikhani (✉)
Rue Victor Hugo, Luynes, France
e-mail: nazand.begikhani@sciencespo.fr

My mother immediately began to scream and to weep, as the town had been shelled extensively, and we thought surely anyone outside had been killed. The ISIS terrorists were fighting with the tribal Sahwa, the Awakening, forces on the ground as the Iraqi Army shelled constantly from the air. Thankfully, moments later my father returned, and we embraced and we wept, he had gone outside to look for us.

The city had been besieged by ISIS for six months, and destruction and decay had preceded them. They had cut off all entry and exit routes, isolating us from the other villages and towns. They had detonated the *Zlweya* bridge, which led into *Zlweya* province. On that day, and without warning, the Iraqi Army had begun shelling our town and the outlying villages in an operation aimed at purifying the area of ISIS presence. They bombarded the area using helicopters, mortar, and even ground missiles and we townsfolk were caught in the crossfire, being shelled by the Iraqi Army on the one hand and having our houses and buildings detonated and the Sahwa slaughtered mercilessly to the hands of ISIS on the other. The ISIS terrorists would not allow people to leave town and would vent their spleen against the townspeople. I was very scared of them, their black clothing, thick long beards, and cruel, harsh faces. That night we did not sleep, spent the night calling our cousins and relatives far and wide, checking on their safety, thanking the Lord for our own.

In the morning we were shocked to find out my uncle Abdulla's house had been surrounded, looted, and detonated, due to his being a member of the *Sahwa* forces, *Awakening*, and because he had killed three ISIS insurgents in skirmishes with the terrorists the day before. They brought out my uncle and murdered him in front of our eyes, right before his children and his wife. They threatened us all, that the same fate would meet any that dare cross them. After the crowds dispersed, his children picked up his body and buried it. It was a dreadful incident, one I shall never be able to forget for as long as I live.

After this tragic incident, another of my uncles who lived in Erbil called my father and said to him, *"Take your daughters, and leave that place."* My family then decided that we would escape to Kurdistan, as ISIS terrorists were bound to come back eventually and take our lives, if the shelling did not first. We spent the next few days before leaving town in complete fear, not only of being killed but because of the disgusting ISIS practice of taking girls as sexual slaves and concubines. They would pick out any girls that they fancied and would take them away to have their way with them, in broad daylight and in front of everyone's eyes. As for the widows, they would marry them off to their own chiefs by force. So many girls in our town were taken by force by ISIS terrorists and made into concubines and sex slaves, never to be heard of again. Those ISIS terrorists spoke a myriad of Arabic dialects, they were without conscience and barbaric.

We prepared for escape. My father had arranged for a lorry driver to smuggle us out of town. He transported watermelons out of town, and we hid under the watermelons out of sight, shivering from dread of being discovered. When we reached the checkpoint, the lorry was allowed passage without a search. We left the town and drove through the adjacent villages, a journey of hunger, fear, and cold. We reached *Hawija* district, which was also under the control of ISIS, who would not allow anyone to leave unless they showed identity cards and provided an official statement issued by them. We told them we had an uncle imprisoned in Susa prison, and so we managed to escape *Hawija* also.

We spent three months traveling between various villages and towns, entering various areas caught between the ISIS terrorists and the Kurdish *Peshmerga* forces. We were in a very difficult plight, we often had nothing to eat, and my father would cry in despair, helpless as he was, and it gave me unspeakable grief to see him thus weep, as I had never before seen him shed a single tear. It was heart rending to see him cry. Various times, we came close to dying of thirst, hunger, and cold, but we finally were able to reach the refugee camps in Arbat.

The first week, the Kurdistan Regional Government (KRG) had prepared an elementary school where the refugees were to stay until camps could be set up. We were then transported to the camp, where we were given a tent and cooking supplies, and the organizations provided food, medicine, and other supplies. It was very strange to go from living in a house with walls and a roof over your head to a tent without doors or a window. To be surrounded by tents, and paths covered in gravel, instead of the freedom of wide-open fields, grass, and beautiful orchards, limited only by the blue sky and the Tigris, unfamiliar with dire want. And then, all of a sudden, you find yourself hungry, powerless, your brothers incapable of finding work even for a lowly daily stipend.

You spend your days awaiting the arrival of aid, living in poverty and humiliation. In the camp, our neighbors were very different from us, spoke different languages, and had different traditions, but we were able to establish friendly relations with all of them. There were Shia', Sunni, Shabak, and Yezidi, and not all from our own area, but from many different places including Mosul, Sinjar, and the villages surrounding Samarra, yet there was no conflict in our camp, and we did not feel hatred toward one another but enjoyed friendly relations. We all face the same difficulties. Life here in camp is very hard and the education is not good enough, as my siblings and I have failed general examinations for three years in a row now, and so we left school. I am trying to adapt to my life here in camp, but I really miss my old life, my friends, my old school, and my neighbors.

But what I miss the most is our orchard that we left behind, where I would play and I would be merry with my siblings. I miss the scents, the sounds, songbirds, and the rustle of little critters in the grass, I miss our eight cows that saved my life and my sisters' lives that fateful day, I even miss our faithful guard dog. The orchard must be in ruins now, the trees withered. I do not know what has become of the cows, nor our house. All I want to know is what our fate will be and for how much longer we will live in this tent.

**Fig. 20.2** Untitled, by Sohaila, Yazidi ISIS survivor, Sharya Refugee Camp, Kurdistan (supplied by Begikhani)

## Background of the Story

'Our Last Day in the Orchard' is the life story of an internally displaced girl in Iraq, who fled the ISIS attack on Yathrib in Salah al-Deen, in 2014. This is a painful story, full of suffering and longing narrated by a teenage girl of only seventeen years, who experienced different forms of violence caused by war, terrorism, forced displacement, dislocation, and hardship in a refugee camp in Arbat, Kurdistan Region.

The story is part of a collection of narratives titled *Shattered Hopes*, produced by internally displaced women living in the camps in Kurdistan Region, which, along with the two images used here, emerged from a research project on 'Gender-Based Violence and Displacement,' led by the University of Bristol's Centre for Gender and Violence Research (CGVR) and funded by the ESRC and the Arts and Humanities Research Council. The Principal Investigator was Emma Williamson with myself (Nazand Begikhani) and Nadia Aghtaie as co-investigators (for more information please see https://www.nazandbegikhani.com/info/university-of-bristol-gender-based-violence-and-displacement-647).

The research team adopted the knowledge co-production method, which consisted of collaborative techniques between academics and refugee community partners, involving both women and men. They used art and creative writing as a way to produce knowledge based on the experience of refugees and displaced communities and also as a means for healing and therapy.

The creative writing workshop consisted of thirty participatory sessions which took two different forms: the general creative writing sessions which included all the fourteen participants and individual sessions with each participant. While the general sessions were about principles of writing, verbal, and narrative expressions of real life experiences, the individual sessions were adopted in order to provide a more reliable and confidential atmosphere between the facilitator and the participant. The majority of the storytellers chose pseudonyms to keep their privacy.

In the 'Our Last Day in the Orchard,' the author used her own real life with a touch of imagination, and she was helped by the facilitators to give the best expression to her experience in her native language, Arabic.

Through the process of this research, the author was confident in telling her story because 'for the first time' she felt that not only was she given the opportunity to talk about her personal suffering but that she was genuinely being listened to. The whole process, according to the women who participated in the project, helped them gain deeper understanding of their experiences and to rebuild their sense of self and social esteem.

The full collection of *Shattered Hopes* was expected to be published in 2020 in three languages: Arabic, English, and French. However, because of the COVID-19, the publication has been delayed.

# 21

# Constraints and Transgressions in Journeys of Displacement

Joris Schapendonk and Milena Belloni

## Introduction

Migration studies is an important academic field that has contributed to empirical and conceptual understandings of displacement. One of the severe binaries that have segregated debates in migration studies for a long time is that of voluntary versus forced migration (e.g. King 2002; Erdal and Oeppen 2018). Especially in policy and public discourses, voluntary migration is all too often framed as a process of active decision-making based on economic interests or lifestyle aspirations. Forced migration, on the other hand, is related to situations whereby people have been deprived from choices and are forcedly uprooted from their living places due to violent conflict, political prosecution, environmental distress, or as a consequence of large-scale development projects. The notion of displacement, then, is mostly associated with those involuntary moves.[1]

---

J. Schapendonk (✉)
Department of Human Geography, Nijmegen Centre for Border Research (NCBR), Radboud University, Nijmegen, Netherlands
e-mail: j.schapendonk@fm.ru.nl

M. Belloni
University of Antwerp / Flemish Research Foundation, Antwerp / Brussels, Belgium
e-mail: milena.belloni@uantwerpen.be

Based on empirical and analytical arguments, however, several migration scholars have explicitly problematised the use of the oppositional categorisation of forced versus voluntary migration (e.g. Crawley and Skleparis 2018; Erdal and Oeppen 2018). Instead, they position migration processes on a forced–voluntary spectrum (e.g. Schuster 2016). In line with these critical accounts, we argue that a focus on journeys and trajectories of displacement provides an insightful analytical lens to better understand the various ways in which elements of agency *and* constraints and choice *and* force constantly blend into each other. In so doing, we approach displacement not as a clear-cut and static situation bounded by a specific place—be it in a camp, hot spot, or city—but as a dynamic process of im/mobility across various places. A process that may include onward movements, periods of emplacement, detours, transit statuses, waiting, entrapment, serial decision-making, and continuous navigation (Schapendonk et al. 2018).

From this dynamic standpoint with a profound temporal-spatial stretch, we become sensitive to the ways migrants and refugees face very different degrees of freedom at different stages of their journeys of displacement and how specific 'migrant' characteristics and 'refugee characteristics' coexist in migrants' life stories (Crawley and Skleparis 2018). In other words, as Mainwaring and Brigden (2016, p. 244) argue, a focus on migration journeys helps us to transcend conceptual as well as physical borders (see also Hyndman 1997).

This chapter starts with a discussion of the literature on journeys of displacement. Although it is not meant to be an all-inclusive overview, it explores the variety of geographical orientations characterising contemporary migration journeys and identifies some of the most common biases and significant contributions. The second section highlights specific constraints and elements of force during mobility processes. As we illustrate, journeys can indeed be unsettling and, hence, they can be considered the defining feature of the exilic process for its connotation of 'expulsion, enforcement and uprooting' (BenEzer and Zetter 2015, p. 299). From there, however, we transcend conventional notions of displacement by examining dimensions of journeys and trajectories that are usually related to voluntary processes of migration, including aspirations, social networks, and navigation. With regard to the latter, we mostly rely on our knowledge regarding migration from West and East Africa.

## Journeys and Trajectories of Displacement: An Overview of an Expanding Field

In this section we provide an overview of the literature related to journeys and trajectories of displacement. In so doing, we discuss the geographical coverage of these studies, the sociopolitical framing, as well as the conceptual contributions involved. It is thereby important to note that, although migratory patterns have diversified in recent years, many journeys are historically shaped. Some journeys follow age-old trade routes across deserts, as in the case of trans-Saharan migration. Other routes run along the same pathways and travellers are guided by networks of previous generations of co-nationals. In some cases, there are striking similarities between journeys of the past and those of today.[2] While we acknowledge the richness of studies that focus on displacement across history—and we can include novels here (e.g. Seghers 2012)—here we focus on the recent rise of a subfield of migration studies that investigates the dynamics of journeying.

This emerging field covers a great variety of geographical areas, yet it is biased to movements from the so-called Global South to the Global North. These studies thereby reflect a societal and political fascination for undocumented or unauthorised migration. While many scholars emphasise the life-marking experiences for the people undertaking these journeys, many policy and media reports frame the same movements as a security threat for the receiving states in question. Geographically, we may distinguish studies that focus on displacement journeys from Central America towards the United States (Singer and Massey 1998; Vogt 2013), African movements towards Europe (Collyer 2007; Belloni 2019; Schapendonk and Steel 2014; Press 2017; Triulzi and Mckenzie 2013) and Israel (BenEzer 2002; van Reisen et al. 2014), Central Asian movements towards Europe (Khosravi 2011; Kuschminder 2017; Crawley et al. 2016),[3] and migratory movements from different Asian regions to Australia (Viviani 1984; Arfish and Olliff 2008; Munt 2012; Nardone and Correa-Velez 2015). In order to counterbalance this geographical bias, it is thus important to pay attention to those studies with a focus on journeys in a context of South–South migration. Among others we can mention studies on the Central African region (Lyytinen 2017; Wilson 2018), Somali displacement to South Africa (Steinberg 2015; Jinnah 2015), and studies that focus on migration trajectories from Afghanistan to Turkey (Kaytaz 2016), and those of Burmese workers to Malaysia (Franck et al. 2018) and journeys from Indonesia to Malaysia (Spaan and van Naerssen 2018; see Lui 2007 for an overview of displacement in Asia).

Most of the above studies are set in a geopolitical context of securitisation and fortified borders that 'produce' illegal migrants within and outside national borders (De Genova 2017). Journeys are in this context often fragmented processes full of geographical detours, unexpected transit situations, and successive moments of displacement and replacement (Schapendonk et al. 2018). Whereas these processes are sometimes framed as a cat-and-mouse game where powerful states chase after migrants, other scholars actually invert the logics of this power play between state authority and migrants. The 'autonomy of migration' literature, for instance, considers unauthorised migration as a social movement and approaches migrant itineraries as dynamic spatial configurations that always exceed the power of states and supra-states to manage and control them (Casas-Cortes et al. 2015, pp. 900–901; Papadopoulos and Tsianos 2013; De Genova 2017).

In the conceptual sense, terms like journeys, itineraries, and trajectories are often used as synonyms. Some scholars, however, explicitly distinguish journeys from trajectories. Journeys can be viewed as border-crossing movements between two ends (departure/arrival). These journeys can still be highly dynamic (Belloni 2016a), fragmented (Collyer 2007), and may include protracted transit situations (Missbach 2015), but they concentrate on the act of travelling towards a specific or imagined destination. As BenEzer and Zetter (2015, p. 299) write, the journey can then be seen as an in-between phase—a medium—that connects the beginning and end stage of the migration process as well as the causes and consequences of displacement.[4] The concept of trajectories, instead, goes explicitly beyond the idea of a movement between two places. Trajectories may include multiple journeys in various directions as well as multiple episodes of rest, emplacement, and settlement (Schapendonk et al. 2018). In some instances, trajectories are closely linked to people's biographies (Pascual-de-Sans 2004) and their vital conjunctures (Johnson-Hanks 2002; Carling 2017). Trajectories may include emplacement processes, even in situations of displacement (Lems 2018). In other words, displaced people also often make new place attachments and invest in home-making along their trajectories, even in ephemeral conditions such as informal camps (e.g. Calais Writers 2017, pp. 111–157; Brun and Fábos 2015). In addition, the broader analytical framework of trajectories may also include return movements towards a place of origin that actually may bring their own experiences of displacement (Kleist 2018; Lems 2018).

Another distinction in this field can be made on the extent to which mobility is seen as the ontological lens to understand migration. Some authors take the journey as a mere empirical unit of analysis. They, for instance, look at the violence and risks involved (e.g. van Reisen et al. 2014; Press 2017) or they investigate how displaced people use social networks and new mobile devices during the act of travelling (e.g. Zijlstra and van Liempt 2017). Other scholars, however, take journeys and trajectories as a conceptual starting point to rethink migration and displacement (Mainwaring and Brigden 2016; Schapendonk et al. 2018; Belloni 2019). The latter group of studies moves away from sedentarist thought in migration studies, which is particularly present on notions around refugees and displaced people (Malkki 1992; Khosravi 2011). In other words, by starting from mobility—as a source of empowerment as well as distress—research on journeys and trajectories inherently challenges us to rethink the 'national order of things' and territorialised notions of belonging (Malkki 1992; Crawley and Skleparis 2018; Schapendonk 2020). This preference of dynamics *over* statics also bleeds into other important concepts of migration, such as borders (that can be displaced too) (Andersson 2014; Massa 2018), social networks (Schapendonk 2015), and identity (Wilson 2018).

Taking journeys and trajectories as a focus to study displacement helps scholars to understand who can move and who remains stuck in different situations. This particular feature links processes of displacement to the politics of mobility (Adey 2006). Mobility is thus seen as an unequal resource, being stratified according to class (Van Hear 2006), availability of transnational social networks (Belloni 2016b), age, and physical fitness. Evidently, journeys of displacement are also highly gendered (Gerard and Pickering 2013; Soto 2016; Vogt 2016; see Gururaja 2000 for a general argument around gender and displacement). As argued by several scholars (Gerard and Pickering 2013; Kofman 2018; Belloni et al. 2018), today, displaced women continue to have lesser access to resettlement and limited possibilities to engage in onwards mobility as they tend to be more vulnerable to violence along the way, an issue that is articulated in the next section.

## Constraints in Journeys of Displacement: Limited Prospects, Risks, and Death

Whereas the reasons for displacement vary across social–political settings, it is crucial to understand the ways journeys themselves become sources of threat, risk, and trauma (BenEzer and Zetter 2015; Press 2017). For many people on the move, room to manoeuvre is rather limited due to racialised mobility regimes (Schwarz 2018) and the exploitative acts of fraudulent state actors and traffickers that occasionally cooperate with each other. In other occasions, other structural problems—like the inaccessibility of rights and severe poverty—result in people getting stuck on their journeys, leaving them with limited prospects for better futures.

The everyday violence involved in journeys of displacement is documented for all the geographical routes outlined above. In the context of Central American crossings through Mexico, for instance, Wendy Vogt (2013) documents how spaces of liminality (camps, borderlands, migrant enclaves) are not only sites where migrants are exposed to various vulnerabilities but how the journey itself has become a site of violence. She argues that the physical abuse, kidnapping, and exploitation of migrants—locally known as the *cachuco industry*[5]—have become 'far more systematic and inescapable in recent years' (Vogt 2013, p. 764). Another infamous case of violence related to displacement journeys is that of Rohingya refugees trying to find ways from Myanmar to Bangladesh. Rape, torture, and other brutalities are reported as being part of the journeys of these displaced people (Cairns 2017; see also Jones 2016). In the African context, and especially in Libya and across the Sinai desert road to Israel, sub-Saharan Africans are exposed to similar violent acts, including forced labour, captivity, and torture (Bialasiewicz 2012; van Reisen et al. 2014). Scholars explain this widespread violence as a result of the intersection of local and global economies that profit from mobility (Vogt 2013)—which includes the business of bordering states (Andersson 2014; Jones 2016). In these different contexts, gender relations shape for a large part the violent landscapes, whereby female bodies are often violently exploited by authority agents, smugglers, and other migrants. However, as Vogt (2016, p. 379) also notes, the gendered dynamics of displacement journeys are often more complex than the narrative of the female victim. In line with scholars that reposition women in processes of globalisation (Davids and Van Driel 2009), the work of Vogt illustrates that migrant women are strategic agents who navigate not only routes but also gendered relations during their journeys, resulting in reciprocal protection mechanisms and different forms of care.

Above all, the most bitter evidence of violent processes of displacement is the fact that many migrants find death as the final destination of their journeys. In this regard, the Missing Migrant project of the International Organization for Migration (IOM) indicates that the European–African border is indeed the deadliest border on earth (Ferrer-Gallardo and Van Houtum 2014). While many of these fatalities are framed as tragic incidents at sea, many could have been avoided with a bit more political will. This is clearly illustrated by the work of Charles Heller (2015) and colleagues who have mapped the non-response of different EU-based authorities in the case that became publicly known as the 'left-to-die-boat.'[6]

## The Other Side of Displacement Journeys: Transgressions, Aspirations, and Navigations

The literature on journeys and trajectories does not only discuss violence, precarity, and processes of subjectification. In fact, many studies articulate the active attitudes of displaced people and their skills in navigating borders, their capacities to re-ground in unknown places (Lems 2018), their resilience in searching for security (Innes 2016; Munt 2012), and the importance of trust relationships (Lyytinen 2017). This section therefore discusses this 'other' side of journeying. In this context, displacements—like other types of mobility—are regularly framed as socially embedded phenomena. Individual aspirations and social imaginaries (Salazar 2011) as well as family and community expectations vis-à-vis migration are thus of crucial importance to understand the directions and motivations involved in journeys.

The notion of aspiration has emerged as a key concept in the study of migration at large (Carling 2002; Carling and Collins 2018). In a world where involuntary immobility is the norm for most people—especially in the Global South—aspirations to a better life elsewhere and the related imaginaries of destinations help to explain why some migrants take high risks to leave their homes in times of economic and political crises (Bal and Willems 2014). These considerations are relevant no matter what label the migrant is assigned by asylum authorities. A focus on aspirations enables researchers to understand journeys beyond frameworks of mere victimisation. Even in case of violent conflicts, displaced people often take their life aspirations with them. In this respect, it is worth noting that people's aspirations are likely to change along the course of journeys and trajectories. If one has reached the safer grounds of a refugee camp in a conflict situation, for instance, there might be other motivations at stake that shape onward movements. Similarly, presumed points of transit may actually turn into destinations when people become socially and economically attached to them over time (Crawley and Skleparis 2018). The levels of enforcement and volition thus become highly complex and diffuse if one approaches displacement as a process instead of a one-off event.

The latter is also underlined by the work of Erdal and Oeppen (2018) who critically discuss the level of volition in processes of forced migration. By unpacking the journeys of Pakistani and Afghani migrants, they investigate how and when the distinction between forced and voluntary migration matters. In so doing, they combine migrants' own experiences with the categorisation of immigration authorities along the course of their journeys. They approach both departures of the migrants as 'forced'—as they lack meaningful life alternatives back home. However, the aspirations and pathways of the Afghani and Pakistani migrants in question were influenced by the prospects of institutional protection in Europe. One group (Afghani migrants) are usually framed by bureaucracies as forced migrants while the other (Pakistani migrants) are framed as voluntary migrants. This is not only an instance of how aspirations and structural possibilities interact in displacement journeys; it also points to the fact that—although a binary divide between voluntary and forced migrants may not reflect the realities of migrants' journeys—the labels that emerge from this division do impact individuals' prospects and pathways.

In our work on West African trajectories (Schapendonk) and East African journeys (Belloni) we create explicit room for concepts that transgress conventional notions of displacement (as uprooting and enforcement). By analysing the pathways of Eritrean asylum seekers, Belloni (2016a, 2019) embraces her informants' gambling-related terminology when they talk about bets, jackpots, luck, and multiple attempts. Interestingly, this notion of gambling suggests a proactive attitude among Eritrean asylum seekers while reflecting a feeling of entrapment. Due to emotional, economic, and social investments, these migrants feel entrapped in Italy, like gamblers who have lost too much already to then give up. Despite the fact that Dublin regulations restrict their movements in Europe, they feel they have no other choice but to repeatedly try to reach another destination. This last move, in their perspective, would allow them to fulfil their families' and personal expectations for which they embarked on the journey in the first place. Belloni illustrates that her informants' decision-making at different stages of their journeys cannot be understood without taking into account the previous steps they took.

Schapendonk (2015, 2020) builds on the notion of social navigation (Vigh 2006, 2009)[7] to study how migrants en route from Africa cope with obstacles and borders on their way to, and within, Europe. He emphasises how his informants negotiate their way across a fuzzy web of migration facilitation and migration control. He thereby articulates how migrants become experts in improvising tricks and informal tactics to relate with, or circumvent, specific actors of the migration industry, be it mobility facilitators, NGO workers, or detention officers. In similar ways, other scholars have highlighted how migrants are able to turn borders and specific categorical labels into resources of mobility (Massa 2018; Wilson 2018). In relation to these tactical processes, the researcher should always carefully navigate the information that she or he reveals since it might harm the migratory projects of her/his informants as well as that of migrants in the near future. As Khosravi (2018) highlights people en route do have a fundamental right to opacity. He argues that not everything of the journey 'should be seen, explained, understood and documented' (Khosravi 2018, p. 3).

## Conclusion: Journeys and Protection

This chapter has reviewed the emerging field of literature on (contemporary) migrant journeys and trajectories. This bundle of studies unpacks the dynamics behind the arrows that we see so often on maps to illustrate yet another 'migration crisis.' In contrast with these arrows, that indeed give the impression of a 'linear, uninterrupted flow of people' (Crawley and Skleparis 2018, p. 49), the literature on journeys stress non-linear movements, fragmentations, and complex border crossings. We have highlighted that—from the perspective of journeys and trajectories—conventional categories of displacement are contested. Migrants may turn into refugees, and vice versa, along their pathways. This, of course, also relates to the discursive power of policy labels. As Erdal and Oeppen (2018, p. 984) write:

> Following De Genova (2016), those labelled as refugees may be pitied as 'victims' of forced displacement whilst they reside in camps in neighbouring countries; yet, as soon as they show more 'entrepreneurial' agency by choosing to leave the camp and head farther afield, they become suspect, labelled as 'illegal migrants' or 'bogus asylum seekers.'

The scholarly question then is indeed how to question taken-for-granted categorical labels without harming the right to protection in this time where migrant rights are under severe pressures (Schuster 2016). We follow Crawley and Skleparis (2018) in their argument that there is an inherent danger in simply working with the same labels of policymakers and asylum bureaucracies in order to make our research policy relevant (see also Dahinden 2016). In line with other scholars (Zetter 2007; Erdal and Oeppen 2018), we argue that these categorical labels have *not* become redundant. In fact, their normative weights influence people's experiences of displacement profoundly. In other words, it does matter (in terms of protection, everyday experiences, social–political positions) how one is framed by migration regimes through discursive labels and policy categories. However, by showing how the categorical divide of voluntary versus forced movements blurs along journeys, we rather hope to contribute to an understanding of migration and displacement that does not flatten realities into static, over-victimised, and essentialist notions of deserving/'forced' refugees versus undeserving/'voluntary' migrants.

## Notes

1. It is important to note that some scholars are engaged in discussions on the dividing line between forced migration studies and refugee studies. Forced migration is then seen as a broader field that actually brings in different forms of migration and different motivations to understand refugee-like situations. Some see this broader field of forced migration studies as a potential risk regarding the protection of refugees (e.g. Hathaway 2007); others portray it as a productive ground to debate different connections between different forms of migration (Van Hear 2011). They plead for a decoupling of academic discussions from the legal definitions and policy categories (DeWind 2007).
2. For instance, by reading diaries and chronicles of Ethiopian and Eritrean refugees in the 1960s and 1970s (Getahun 2007), one is able to find many similarities with contemporary journeys: the negotiation with smugglers, the avoidance of authorities, the fear of spies, the hiding in order to cross borders, the obstacles to dealing with bureaucratic diplomatic services without having a passport.
3. The recent Syrian conflict, and the subsequent crisis of reception of refugees in Europe, has clearly boosted the interest in refugee journeys toward and within the EU.
4. Interestingly, later in their article the same authors question the clear beginning and end points of journeys of displacement (BenEzer and Zetter 2015).
5. *Cachuco* is a violent term that translates to 'dirty pig.'
6. The detailed reconstruction follows the series of non-responses regarding a boat in distress carrying 72 sub-Saharan migrants in 2011. It reveals a series of interactions between the passengers of the boat and authorities, including the military, NATO, and the Italian and Maltese maritime rescue centres. Despite existing legal obligations and despite the fact that the exact location was known by the authorities, none of the actors involved intervened and this resulted in the death of 63 of the 72 passengers.
7. Social navigation is introduced by the anthropologist Vigh (2006) in a West African context of social becoming, conflict, and uncertainty and has later been applied to the context of migration (Vigh 2009).

## References

Adey, P. (2006). If mobility is everything then it is nothing: Towards a relational politics of (im)mobilities. *Mobilities, 1*(1), 75–94.

Andersson, R. (2014). *Illegality Inc. clandestine migration and the business of bordering Europe*. Oakland, CA: University of California Press.

Arfish, H. M., & Olliff, L. (2008). 'It's difficult to stay, and it's hard to leave.' Stories of refugee journeys in Australia. *Australasian Review of African Studies, The, 29*(1/2), 104.

Bal, E., & Willems, R. (2014). Introduction: Aspiring migrants, local crises and the imagination of futures 'away from home'. *Identities, 21*(3), 249–258.

Belloni, M. (2016a). Refugees as gamblers: Eritreans seeking to migrate through Italy. *Journal of Immigrant & Refugee Studies, 14*(1), 104–119.

Belloni, M. (2016b). My uncle cannot say 'No' if I Reach Libya: Unpacking the social dynamics of border-crossing among Eritreans heading to Europe. *Human Geography, 9*(2), 47–56.

Belloni M. (2019). *The Big Gamble*: The Migration of Eritreans to Europe. Berkeley: Univ. of California Press.

Belloni, M., Pastore, F., & Timmermann, C. (2018). Women in Mediterranean asylum flows: Current scenario and ways forward. In C. Timmermann, M. L. Fonseca, L. Van Praag, & S. Pereira (Eds.), *Dynamic interplays between gender and migration*. Leuven, Belgium: University of Leuven Press.

BenEzer, G. (2002). *The Ethiopian Jewish exodus: Narratives of the journey*. London & New York, NY: Routledge.

BenEzer, G., & Zetter, R. (2015). Searching for directions: Conceptual and methodological challenges in researching refugee journeys. *Journal of Refugee Studies, 28*(3), 297–318.

Bialasiewicz, L. (2012). Off-shoring and out-sourcing the borders of Europe: Libya and EU border work in the Mediterranean. *Geopolitics, 17*(4), 843–866.

Brun, C., & Fábos, A. (2015). Making homes in limbo? A conceptual framework. *Refuge: Canada's Journal on Refugees, 31*(1), 5–17.

Cairns, E. (2017). 'I still don't feel safe to go home': Voices of Rohingya refugees. *Oxfam*. Retrieved from https://www.oxfamnovib.nl/Files/rapporten/2017/20171218%20bp-voices-rohingya-refugees-181217-embargo-en.pdf

Calais Writers. (2017). *Voices from the Jungle. Stories from the Calais Refugee Camp*. London: Pluto Press.

Carling, J. (2002). Migration in the age of involuntary immobility: Theoretical reflections and Cape Verdean experiences. *Journal of Ethnic and Migration Studies, 28*(1), 5–42.

Carling, J. (2017). On conjunctures in transnational lives: Linear time, relative mobility and individual experience. In E. Mavroudi, B. Page, & A. Christou (Eds.), *Timespace and international migration* (pp. 33–47). Cheltenham, England: Elgar Publishing.

Carling, J., & Collins, F. (2018). Aspiration, desire and drivers of migration. *Journal of Ethnic and Migration Studies, 44*(6), 909–926.

Casas-Cortes, M., Cobarrubias, S., & Pickles, J. (2015). Riding routes and itinerant borders: Autonomy of migration and border externalization. *Antipode, 47*(4), 894–914.

Collyer, M. (2007). In-between places: Trans-Saharan transit migrants in Morocco and the fragmented journey to Europe. *Antipode, 39*(4), 668–690.

Crawley, H., Düvell, F., Sigona, N., McMahon, S., & Jones, K. (2016). *Unpacking a rapidly changing scenario. Migration flows, routes and trajectories across the Mediterranean* (MEDMIG research brief 1). Retrieved from http://www.medmig.info/research-brief-01-unpacking-a-rapidly-changing-scenario/

Crawley, H., & Skleparis, D. (2018). Refugees, migrants, neither, both: Categorical fetishism and the politics of bounding in Europe's 'migration crisis'. *Journal of Ethnic and Migration Studies, 44*(1), 48–64.

Dahinden, J. (2016). A plea for the 'de-migranticization' of research on migration and integration. *Ethnic and Racial Studies, 39*(13), 2207–2225.

Davids, T., & Van Driel, F. (2009). The unhappy marriage between gender and globalisation. *Third World Quarterly, 30*(5), 905–920.

De Genova, N. (Ed.). (2017). *The borders of 'Europe': Autonomy of migration, tactics of bordering*. Durham, NC: Duke University Press.

DeWind, J. (2007). Response to Hathaway. *Journal of Refugee Studies, 20*(3), 381–385.

Erdal, M. B., & Oeppen, C. (2018). Forced to leave? The discursive and analytical significance of describing migration as forced and voluntary. *Journal of Ethnic and Migration Studies, 44*(6), 981–998.

Ferrer-Gallardo, X., & van Houtum, H. (2014). The deadly EU border control. *ACME, An International Journal for Critical Geographies, 13*(2), 295–304.

Franck, A. K., Brandström Arellano, E., & Trawicki Anderson, J. (2018). Navigating migrant trajectories through private actors. *European Journal of East Asian Studies, 17*(1), 55–82.

Gerard, A., & Pickering, S. (2013). Gender, securitization and transit: Refugee women and the journey to the EU. *Journal of Refugee Studies, 27*(3), 338–359.

Getahun, S. A. (2007). *The history of Ethiopian immigrants and refugees in America, 1900–2000: Patterns of migration, survival, and adjustment*. New York: LFB Scholarly Pub LLC..

Gururaja, S. (2000). Gender dimensions of displacement. *Forced Migration Review, 9*, 13–16.

Hathaway, J. (2007). Forced migration studies: Could we agree just to 'date'? *Journal of Refugee Studies, 20*(3), 349369.

Heller, C. (2015). *Liquid trajectories. Documenting illegalised migration and the violence of borders* (Doctoral dissertation). Goldsmiths, University of London, London.

Hyndman, J. (1997). BorderCrossings. *Antipode, 29*(2), 149–176.

Innes, A. (2016). In search of security: Migrant agency, narrative, and performativity. *Geopolitics, 21*(2), 263–283.

Jinnah, Z. (2015). Rational routes? Understanding Somali migration to South Africa. In M. van der Velde & T. van Naerssen (Eds.), *Mobility and migration Choices. Thresholds to crossing borders* (pp. 43–54). Farnham, England: Ashgate.

Johnson-Hanks, J. (2002). On the limits of life stages in ethnography: Toward a theory of vital conjunctures. *American anthropologist, 104*(3), 865–880.

Jones, R. (2016). *Violent borders: Refugees and the right to move.* London & New York, NY: Verso Books.

Kaytaz, E. (2016). Afghan journeys to Turkey: Narratives of immobility, travel and transformation. *Geopolitics, 21*(2), 284–302.

Khosravi, S. (2011). *'Illegal' traveller: An auto-ethnography of borders.* Basingstoke, England & New York, NY: Palgrave Macmillan.

Khosravi, S. (2018). Afterword. Experiences and stories along the way. *Geoforum.* https://doi.org/10.1016/j.geoforum.2018.05.021

King, R. (2002). Towards a new map of European migration. *International Journal of Population Geography, 8*(2), 89–106.

Kleist, N. (2018). Trajectories of involuntary return migration to Ghana: Forced relocation processes and post-return life. *Geoforum.* https://doi.org/10.1016/j.geoforum.2017.12.005

Kofman, E. (2018). Gendered mobilities and vulnerabilities: Refugee journeys to and in Europe. *Journal of Ethnic and Migration Studies, 45*(12), 2185–2199. https://doi.org/10.1080/1369183X.2018.1468330.

Kuschminder, K. (2017). Afghan refugee journeys: Onwards migration decision-making in Greece and Turkey. *Journal of Refugee Studies, 31*(4), 566–587. https://doi.org/10.1093/jrs/fex043.

Lems, A. (2018). *Being-here. Placemaking in a world of movement.* Oxford, England: Berghahn.

Lui, R. N. (2007). Such a long journey: Protracted refugee situations in Asia. *Global Change, Peace & Security, 19*(3), 185–203.

Lyytinen, E. (2017). Refugees' 'journeys of trust': Creating an analytical framework to examine refugees' exilic journeys with a focus on trust. *Journal of Refugee Studies, 30*(4), 489–510.

Mainwaring, Ċ., & Brigden, N. (2016). Beyond the border: Clandestine migration journeys. *Geopolitics, 21*(2), 243–262.

Malkki, L. (1992). National geographic: The rooting of peoples and the territorialization of national identity among scholars and refugees. *Cultural Anthropology, 7*(1), 24–44.

Massa, A. (2018). Borders and boundaries as resources for mobility. Multiple regimes of mobility and incoherent trajectories on the Ethiopian–Eritrean border. *Geoforum.* https://doi.org/10.1016/j.geoforum.2018.01.007

Missbach, A. (2015). *Troubled transit: Asylum seekers stuck in Indonesia.* Singapore: ISEAS-Yusof Ishak Institute.

Munt, S. R. (2012). Journeys of resilience: The emotional geographies of refugee women. *Gender, Place & Culture, 19*(5), 555–577.

Nardone, M., & Correa-Velez, I. (2015). Unpredictability, invisibility and vulnerability: Unaccompanied asylum-seeking minors' journeys to Australia. *Journal of Refugee Studies, 29*(3), 295–314.

Papadopoulos, D., & Tsianos, V. S. (2013). After citizenship: Autonomy of migration, organisational ontology and mobile commons. *Citizenship Studies, 17*(2), 178–196.

Pascual-de-Sans, À. (2004). Sense of place and migration histories idiotopy and idiotope. *Area, 36*(4), 348–357.

Press, R. (2017). Dangerous crossings: Voices from the African migration to Italy/Europe. *Africa Today, 64*(1), 2–26.

Salazar, N. B. (2011). The power of imagination in transnational mobilities. *Identities, 18*(6), 576–598.

Schapendonk, J. (2015). What if networks move? Dynamic social networking in the context of African migration to Europe. *Population, Space and Place, 21*(8), 809–819.

Schapendonk, J. (2018). Navigating the migration industry: Migrants moving through an African-European web of facilitation/control. *Journal of Ethnic and Migration Studies, 44*(4), 663–679.

Schapendonk, J. (2020). *Finding ways through Eurospace: West African movers reviewing Europe from the Inside*. Oxford: Berghahn Books.

Schapendonk, J., & Steel, G. (2014). Following migrant trajectories: The im/mobility of sub-Saharan Africans en route to the European Union. *Annals of the Association of American Geographers, 104*(2), 262–270.

Schapendonk, J., van Liempt, I., Schwarz, I., & Steel, G. (2018). Re-routing migration geographies: Migrants, trajectories and mobility regimes. *Geoforum*. https://doi.org/10.1016/j.geoforum.2018.06.007.

Schuster, L. (2016). Unmixing migrants and refugees. In A. Triandafyllidou (Ed.), *The Routledge handbook of immigration and refugee studies* (pp. 297–303). Abingdon, England: Routledge.

Schwarz, I. (2018). Migrants moving through mobility regimes: The trajectory approach as a tool to reveal migratory processes. *Geoforum*. https://doi.org/10.1016/j.geoforum.2018.03.007.

Seghers, A. (2012 [1944]). *Transit*. Amsterdam: Van Gennep

Singer, A., & Massey, D. S. (1998). The social process of undocumented border crossing among Mexican migrants. *International Migration Review, 32*(3), 561–592.

Soto, L. (2016). The telling moment: Pre-crossings of Mexican teenage girls and their journeys to the border. *Geopolitics, 21*(2), 325–344.

Spaan, E., & van Naerssen, T. (2018). Migration decision-making and migration industry in the Indonesia–Malaysia corridor. *Journal of Ethnic and Migration Studies, 44*(4), 680–695.

Steinberg, J. (2015). *A man of good hope. One man's extraordinary journey from Mogadishu to Tin Can Town*. London: Vintage.

Triulzi, A., & Mckenzie, R. (Eds.). (2013). *Long journeys. African migrants on the road.* Leiden, Netherlands: Brill.

Van Hear, N. (2006). 'I went as far as my money would take me.' Conflict, forced migration and class. In F. Crepeau et al. (Eds.), *Forced migration and global processes. A view from forced migration studies* (pp. 125–158). Lanham, MD: Lexington.

Van Hear, N. (2011). Forcing the issue: Migration crises and the uneasy dialogue between refugee research and policy. *Journal of Refugee Studies, 25*(1), 2–24.

van Reisen, M., Estefanos, M., Rijken, C., & Borgman, E. (2014). *The human trafficking cycle: Sinai and beyond.* Nijmegen, Netherlands: Wolf Legal Publishers.

Vigh, H. (2006). Social death and violent life chances. In C. Christiansen, M. Utas, & H. Vigh (Eds.), *Navigating youth generating adulthood: Social becoming in an African context* (pp. 31–60). Uppsala, Norway: Nordiska Afrikainstitutet.

Vigh, H. (2009). Motion squared. A Second look at the concept of social navigation. *Anthropological Theory, 9*(4), 419–438.

Viviani, N. (1984). *The long journey: Vietnamese migration and settlement in Australia.* Carlton, Australia: Melbourne University Press.

Vogt, W. (2013). Crossing Mexico: Structural violence and the commodification of undocumented Central American migrants. *American Ethnologist, 40*(4), 764–780.

Vogt, W. (2016). Stuck in the middle with you: The intimate labours of mobility and smuggling along Mexico's migrant route. *Geopolitics, 21*(2), 366–386.

Wilson, C. (2018). Spatial mobility and social becoming: The journeys of four Central African students in Congo-Kinshasa. *Geoforum.* https://doi.org/10.1016/j.geoforum.2018.05.018.

Zetter, R. (2007). More labels, fewer refugees: Remaking the refugee label in an era of globalization. *Journal of refugee studies, 20*(2), 172–192.

Zijlstra, J., & van Liempt, I. (2017). Smart (phone) travelling: Understanding the use and impact of mobile technology on irregular migration journeys. *International Journal of Migration and Border Studies, 3*(2–3), 174–191.

# 22

# Migrants' Displacements at the Internal Frontiers of Europe

Martina Tazzioli

## Introduction

The term 'displacement' is by now widely used both in academic literature and in policy documents; however, it is characterised by a sort of conceptual fuzziness. United Nations agencies—among them the UNHCR and the UNDP—constantly update reports with numbers, stories, and figures about displaced persons, warn about the unprecedented mass displacement across the world, and share 'good practices' and craft 'comprehensive responses' that involve states, international actors, and nongovernmental organisations. Yet, far less attention is paid to the heterogeneity of experiences, causes, and impacts of displacement, that in fact are difficult to contain in a single word which works as a unifying conceptual umbrella.[1] According to UNESCO, 'the displacement of people refers to the forced movement of people from their locality or environment and occupational activities. It is a form of social change caused by a number of factors.'[2] Thus, on the basis of such a definition, displacement is related to the non-voluntary character of movement and to the social transformations that are engendered by the spatial dislocation of people. More broadly, displacement is conceptualised as the effect of

M. Tazzioli (✉)
Department of Politics & International Relations, Goldsmiths, University of London, London, UK
e-mail: martina.tazzioli@gold.ac.uk

protracted crises or catastrophes, with no reference to the violent dispossessions that are connected to it, nor to the way in which states' politics contribute to triggering multiple forms of displacement. Relatedly, displaced persons are posited as subjects who are forced to escape, leaving undefined the directionality of such a move.

This chapter engages with modes of displacements that are implemented by states at the internal frontiers of Europe as a way for regaining control over unauthorised migrations. A focus on how migrants are governed at the internal frontiers of Europe enables highlighting that displacement is enacted also to govern migrants, by keeping them on the move. Importantly, such a gaze on displacement does not intend to replace other perspectives nor to provide an overwhelming analytical grid on displacement. Rather, it sheds light on how displacement is used as a 'political technology' (Foucault 2012) of migration governmentality, shifting the focus from a depoliticised view of displacement as a phenomenon triggered by crises and emergencies. Indeed, while commonly used definitions of displacement—such as UNESCO's one—do refer more or less explicitly to a dislocation in space and to protracted moments of strandedness, this chapter shows that displacement works also as a political technology deployed by states for regaining control over 'unruly' migrants, keeping them on the move. Indeed, by being constantly displaced and chased away, migrants are forced to undertake convoluted geographies. Hence, the chapter argues that *displacement is not only about moving but also about keeping on the move*. Related to that, displacement generates forced and convoluted hypermobility, while at the same time it also consists of dismantling migrants' infrastructures and material conditions for living. In so doing, bringing in temporality in the analysis is of fundamental importance for thinking of displacement beyond forced spatial dislocation from a point A to a point B. In order to grasp how displacement measures keep migrants on the move, we need to move beyond punctual moments and snapshots—for example, when migrants are moved or forced to flee, as well as when they are stranded in a camp—and take into account how temporality can play an active role in generating displacement. On this point a clarification is needed: by rethinking displacement in relation to mobility, this chapter does not speak of temporality as 'the continued state of being displaced that can be maintained over time' (Bakewell 2011, p. 23); rather, it draws attention to migrants being repeatedly pushed back or chased away from borderzones in Europe.

As long as forced hypermobility is analysed in terms of displacement, the very nexus between mobility and freedom is questioned. In fact, displacement does not refer exclusively to massive flights caused by exceptional phenomena or natural disasters; it is also a political technology used for disciplining unruly mobility and as a way to take legal and material terrain away from migrants. The chapter proceeds in three steps. In the first section, it takes stock of the main theoretical elaborations on displacement, foregrounding those analyses that politicise the notion by thinking of it in relation to processes of dispossession. Then, it moves on to consider modes of migration displacement at the internal frontiers of Europe, with a specific focus on the French–Italian border and on Calais. It shows displacement is often characterised not by forced movements from a point A to a point B nor by immobility; rather, it consists of forcing migrants to undertake convoluted geographies and to remain on the move. Building on research fieldwork conducted between 2015 and 2018, the chapter brings attention to two modes of migrant displacements that end up enhancing migrant mobility. The final section develops the theoretical implications that stem from thinking of displacement through mobility, further elaborating on the nexus between displacement and containment. Through such an analytical angle, the chapter brings a contribution to current debates on migration displacement by drawing attention to the nexus between displacements and hypermobility.

## Displacement and the Stealing of Space and Time

The notion of displacement is at the core of a flourishing debate that often sees collaboration between scholars, politicians, and policymakers. In refugee studies, displacement has been widely debated, mostly through policy-driven approaches which question the inadequacy of the definition of 'refugee' enshrined in Article 51 of the Geneva Convention for addressing situations of displacement caused by environmental disasters, conflicts, and crises (Betts 2013; Zetter and Morrissey 2014).[3] More broadly, over the last two decades 'displacement' has become a catchword in the scholarship on forced migration that debates who should be entitled of protection. In this sense, the notion of displacement provides an analytical lens for getting a sense of the legal and political transformations taking place in the field of refugee protection. In brief, migration and refugee studies scholarship has mainly dealt with the causes of displacements, the conditions of protracted displacements that many face, and the legal gaps in international law around the protection of displaced persons. However, such a perspective on displacement has also partly contributed to tracing normative and exclusionary boundaries between people 'in real need of protection' and other migrants. As part of a debate which engages in fitting migrants into exclusionary categories and profiles, displacement has been mainly tackled according to a 'seeing like a State' (Scott

1998) approach, that is by fully assuming the idea that migration should be governed, and governed in a better way. Or, to put it another way, analyses about displacements in the field of migration and refugee studies have tended to reiterate what Nicholas De Genova compellingly defined as 'the reification of migrants and migrations' which ultimately '(re)fetishizes and (re)naturalizes the epistemological stability attributed to the ("national") state' (De Genova 2013a, p. 253).

This chapter mobilises a different analytical angle, interrogating displacement from the standpoint of how migrants are constantly dispossessed and 'stolen' of their time and lives. It builds on critical geography literature on migration, drawing, in particular, on works that have reflected on the use of mobility as a political technology used for governing migrants (Darling 2011; Gill 2009a; Moran et al. 2012). For instance, introducing the notion of 'governmental mobility,' Nick Gill has highlighted that asylum seekers are being increasingly moved from one detention centre to another across the UK, showing in this way that mobility is used as a technology of government (Gill 2009a). Jonathan Darling has shed light on the UK refugee dispersal policy, which consists of moving asylum seekers from big urban centres to peripheral areas, and pointing to the effects of destitution that it generates on asylum seekers. As Darling explains, the production of 'enforced immobility of asylum seekers through dispersal' (Darling 2016, p. 236) can be traced back to the 1990s and it generates effects of destitution on the migrants. This literature helps in rethinking displacement beyond forced migration, shedding light on how migrants are moved and displaced by state authorities. This chapter draws on this scholarship while at the same time mobilises a slightly different angle by focusing on illegalised migrants who are not inside the asylum channels nor who are detained in reception centres. That is, it looks at how displacement is enacted by the states at the internal frontiers of Europe in order to regain control on what the EU calls 'secondary movements.'[4] However, an exclusive spatial approach is not sufficient: in order to politicise migrants' displacement, there is a need to supplement works about the spatial disciplining of migration with analyses that centre on how political technologies of government affect lives (Fassin 2009; Foucault 1982). In such a way, displacement should be read as part of modes of governing that, even beyond migration, generate effects of destitution and dispossession. Ayse Çaglar and Nina Glick Schiller have compellingly argued that we need to look at the 'interrelated processes of displacement, dispossession, accumulation, and emplacement through which urban life is constituted' (Çaglar and Schiller 2018, p. 3; see also Mezzadra and Neilson 2017).

## Displacing by Keeping on the Move

Since summer 2016, every week a bus leaves from Ventimiglia, the Italian city close to the French border along the coast, to take between forty and fifty migrants to the city of Taranto, which is located 1200 kilometres south, and where a hotspot[5] was opened that year. In the same way, migrants are taken by force to Taranto from Como, which is instead situated the Swiss–Italian frontier.[6] Every journey costs to the Italian state about 5000 euros. Ventimiglia and Como have become actual borderzones for the migrants in transit, although Ventimiglia has a longer history as a borderzone than Como. As far as Ventimiglia is concerned, in 2011 it was the main crossing point for the Tunisian migrants who landed in Lampedusa after the outbreak of the Tunisian revolution. As a response, France suspended Schengen, that is re-established border controls, for a few months and enforced border controls. In 2015, with the increasing of migrant arrivals from Libya, the French government decided to again suspend Schengen. Como started to be visible in the media as a border for the migrants, only quite recently, in 2016, when the Swiss authorities substantially enforced a border-closure policy.

The migrants who are taken to Taranto are usually those who are apprehended by the French and Swiss police and pushed back to Italy. Frequently, the Italian police also arbitrarily arrest them in the streets of the two cities and force them to board the buses. After being detained for few days inside the hotspot of Taranto, migrants are released with a decree of expulsion, and they start to make their journey back to Ventimiglia and Como, using trains and buses (Tazzioli and Garelli 2018; Tazzioli 2018). Most of them have undertaken the same journey eight or ten times: they do not give up and, despite the cost of going back by train, they are determined to reach the national border and make it to France or Switzerland.

How should we analyse these forced internal transfers as long as migrants go back again to the border and Italy spends a huge amount of money for every trip? This chapter suggests that forced internal transfers are part of modes of displacement that are enforced by state authorities to regain control over the migrants, not by stopping or deporting, but by forcing them to undertake convoluted geographies, to decelerate and to divert their routes. Such a forced restless condition of hypermobility wears out the migrants, although, as has been mentioned above, most of them do not give up (Ansems de Vries and Guild 2019). Thus, the frequent police operations of North–South internal displacement generated an overall increase of migrants' mobility across the national territory: migrants are *moved* through official and forced channels—the bus transfers to Taranto, and they also *move*—as they try to get back to the border with their own means, or they try other routes. Therefore, in this case a mix of state-led forced transfers and migrants' movements are at play.

We can speak of displacement as long as migrants are taken away from the border, while at the same time these modes of displacement are not intended to relocate migrants to Taranto on a permanent basis nor to detain them somewhere. Displacement in this case is not even about migrants fleeing a crisis or a conflict. Rather, in this context displacement is enforced by the state authorities and it consists of chasing migrants away while at the same time exhausting them, by pushing them to restart their journey multiple times. Asking about the political and economic rationale behind such a state-drive enforcement of migrants' hypermobility is not a question that can be answered in straightforward way. Indeed, there is not even an overwhelming rationale of governing nor a coherent logic behind it. Rather, migrants are repeatedly displaced, without being relocated permanently, for diverse reasons that coexist and sometimes also conflict with each other and that also usually involve different institutional levels. For instance, in the case of Ventimiglia the interests of the municipality in chasing migrants away have played a major role, while the national authorities oscillated over time between invisibilising migrants in the media—regardless of their actual presence—and staging the 'border spectacle' (De Genova 2013b) of the migration crisis. On the one hand, migrants' displacements have been a state's response to the complaints raised by citizens in Ventimiglia and in Como—showing them that the 'migrant problem' could be solved; and, on the other, it was a sort of punitive logic against the migrants mixed with tactics of deterrence. 'We do not want that Ventimiglia becomes the Italian Calais,'[7] the words of the Angelino Alfano, who was Minister of the Interior at that time (2016), encapsulates well the state's narrative on migrants' displacements. In fact, evacuation from the frontier was presented as a way to lighten migrants' pressures at the border.

These tactics of internal displacement, this chapter contends, are not a specificity of the Italian context nor of the French–Italian border (Fontanari 2018). Instead, the French–Italian border helps shed light on a spatial disciplining of unruly migration that is apparent at many borderzones and internal frontiers in Europe. Arguing this does not mean subsuming the heterogeneity of mechanisms of control through which migrants are obstructed and temporarily blocked in different contexts; rather, the point is to foreground recursive political technologies of government, that is modes for regaining control over migrants that are widely used at different frontiers in Europe. Drawing attention to these tactics of repeated displacement also contributes in shifting the focus from analyses that centre on—and take for granted—the so-called migration crisis towards an account of more daily practices of migration governmentality that have a tangible effect on migrant lives and geographies. Investigating how migrants are displaced by being kept on the move equips us with an analytics of displacement that is not about 'migration crisis' nor about disaster-led migrant flights and that, instead, comes to grips with more banal and frequent, but not less violent modes of *being moved*.

Along these lines, the other paradigmatic case study that it is worth mentioning is the constant displacement of migrants from Calais. As is well known, the city of Calais is a critical borderzone for the migrants, 'a site of persistent border struggle and harsh border enforcement.'[8] The fact that Calais, as a borderzone for migrants, has a quite long history of border enforcement as well as of violent police evictions is remarkable for retracing a continuity in modes of displacement as political technologies for governing migration. Indeed, tactics of migrant dispersal and displacement are not new there, as these have been enacted by the French police since the late 1990s, when migrants started to build temporary shelters and informal encampments (Bernardot 2008). Nor has displacement been enacted only at the time of the repeated evictions of the so-called Jungle. Since the last eviction of the Jungle in October 2016, the displacement of the migrants from the Calais area has become a systematic police tactic, aimed at targeting them not only individually but also as part of temporary groups and multiplicities: indeed, in local decrees enforced after the last eviction of the Jungle, the municipality of Calais has forbidden the gathering and occupations in the Calais area.[9] Hampering migrants from gathering is actually a recursive police tactic at many borderzones in Europe—such as Calais, Ventimiglia, and Como. Indeed, modes of displacements are enacted not only for disciplining individual migrants, or making their presence invisible to citizens, but also for preventing and disrupting the emergence of collective political subjects and of temporary migrant multiplicities (Tazzioli 2017a): as soon as migrants get organised to build common spaces of liveability, or if they lay political claims, they are divided and dispersed by the police.

As in the cases of Ventimiglia and Como, migrants are not moved away to be relocated somewhere nor to be detained; rather, they are temporarily chased away from the area, being sometimes transferred to detention centres, other times to the so-called Centres of Hosting and Orientation (CAO), while sometimes they are just dropped far away in the street. After being chased away, migrants tend to go back to Calais, in the span of a few days. In fact, engaging with the geographies of displacement involves exploring the non-linearity of migrants' trajectories and the convoluted movements that they are forced to undertake. An insight into modes of displacement that consist in keeping migrants on the move enables grasping the twisted and fragmentary nature of migrant journeys. What does displacement mean for the migrants who are temporarily stranded in Calais? A spatial approach, this chapter contends, is not fully adequate for grasping how modes of displacement impact on migrant lives. Indeed, as the Calais case clearly shows, migrants are not only moved, chased away from the area; at the same time, they are also subjected to police operations apt at dismantling temporary encampments and informal settlements, at hampering food distribution, and at chasing them away with pepper spray[10] (Davies et al. 2017). Therefore, modes of displacement in Calais consist in moving migrants away and, at the same time, depriving them of the infrastructures and material conditions for staying. As part of this twofold spatial displacement—being moved away from Calais and being removed of the essential infrastructures of liveability—migrants have been also disrupted in getting support from locals. Indeed, since 2017 the municipality of Calais has started to physically hamper and also criminalise migrant solidarity practices carried out by locals as well as by activists coming from abroad (Fekete 2018).

Hence, modes of displacements that keep migrants on the move cannot be critically analysed disjoined from police measures that target migrants by taking away and demolishing spaces of liveability—'*lieux de vie*,' as the migrants themselves called these in the Calais jungle. In fact, as Claudia Aradau has aptly pointed out, French authorities undo 'the very conditions of liveability' of the migrants through 'the destruction of conditions of collectivity' (Aradau 2017, p. 7). It follows that convoluted geographies of migrant displacement at the internal frontiers of Europe should be investigated in conjunction with an insight into the 'politics of life' (Fassin 2009) that is exercised upon the migrants, taken both individually and as part of collective formations. Both from Calais and from Ventimiglia and Como, migrants are displaced, without however being moved somewhere in particular, in order to stay there. And even when they are not moved by the state authorities, migrants are de facto forced to constantly displace themselves, that is to keep on the move, in order not to be apprehended by the police, or in order to find new routes to travel.

## Rearticulating Displacement, Containment, and Mobility

The above section has showed that displacement is used at the internal frontiers of Europe as a political technology for governing 'unruly' migrants by keeping them on the move. Building on that, the intertwining between displacement, containment, and mobility requires attentive examination. First, if displacement is enacted by keeping migrants on the move and forcing them to undertake tortuous routes, it becomes apparent that mobility constitutes not only an object of government but also a technology, a means, for disciplining migration. Ultimately, *governing through mobility* has been recently explored by scholars, in the field of migration and mobility studies (Moffette 2018; Tazzioli 2017b), critical geography (Gill 2009b; Mountz 2011), and politics (Pinkerton 2019). However, no particular attention has been paid to the effects of displacement that such a mode of governing generates, and therefore to the nexus between displacement and mobility as a political technology.

An insight into the articulation between mobility and displacement in fact enables challenging a liberal conceptualisation of mobility that associates mobility with freedom. Indeed, the movements of the migrants who try to go back to the border cannot be grasped through the lens of free mobility: in fact migrants are de facto forced to undertake such an expensive journey—paying for transport—and to conceal themselves in order not to be checked by the police. In this sense, migrants' hypermobility induced by forced transfers is not synonymous of autonomous movements: on the contrary, migrants are repeatedly hampered in their free movement and their temporality of mobility is likewise disrupted, as they are forced to make multiple attempts before crossing or they have to make many detours. In fact, as Elena Fontanari argued, migrants' displacements at the internal frontiers of Europe dispossess migrants: 'being in state of protracted transit renders those caught in it deprived of their lifetime' (Fontanari 2018).

Hence, these modes of displacement impact on migrant lives by constantly withdrawing terrain from them and disrupting autonomous mobility. Bringing to the fore displacement, mobility cannot be read as condition or expression of freedom, nor it can be grasped through the binary opposition between free and unfree (Adey 2010; Bigo 2010; Cresswell 2010). Nick Gill invites us to pay attention to the 'subjugating and disempowering effects of mobility' (Gill 2013, p. 31).

This does not mean that the movements of the migrants who are displaced from borderzones in Europe are totally unfree nor that migrants do not have agency and do not strategically deal with their condition of being on the move. On the contrary, this chapter contributes to critical migration scholarship that conceives of border controls and modes of migrant displacements as ways for taming and disciplining migration movements that constitutively exceed state policing (De Genova 2017; Mezzadra 2011). However, migrants' struggles and tactics of resistance against displacement take place within what William Walters and Barbara Luthi defined as 'cramped spaces,' taking this as a notion that 'registers degrees of deprivation, constriction and obstruction, but always and simultaneously a concern for the ways in which such limits operate to stimulate and incite movements of becoming' (Walters and Luthi 2016, p. 361). Developing such an analysis in relation to measures of displacement at Europe's borderzones, it can be argued that migrants are not fully blocked or stranded, nor they are deprived of leeway of action and reduced to bare life. Rather, they are constantly disrupted, both physically and on a legal level, in enacting autonomous movement and in staying—as in the case of Calais. Modes of displacement that are enforced by keeping migrants on the move do not only obstruct migrant journeys, but they also disrupt and 'choke' migrants' spaces of life. Therefore, the nexus between displacement and mobility becomes clearer once we engage with how migrants' lives are affected, taking stock of the 'cramped spaces' from within which they can struggle and resist. By being kept on the move, migrants are deprived of a space to stay and somewhere to build collectively; their autonomous movements are constantly disrupted and their 'time is stolen' (Khosravi 2018), that is migrants are robbed of the very possibility of thinking about their future. Therefore, recalling the above-mentioned analysis of Ayse Çaglar and Nina Glick Schiller, modes of displacement, and the effects that these engender, should be connected with tactics of dispossession and destitution. The second aspect to consider in relation to modes of displacement that are enacted by keeping migrants on the move is containment. Building on the analysis above about migrants being disrupted in their autonomous movements and divested of their future, it appears that modes of displacement are part of political technologies that 'take terrain away from them' (Tazzioli 2017b). In this sense, modes of displacement that force migrants to restart their journey multiple times and to divert their routes generate containment, even if not spatial confinement or detention (Fili 2018; Mountz et al. 2013). While speaking of containment could sound counterintuitive as long as migrants are kept on the move, the impact of modes of displacement on migrant lives results in forms of containment if we consider that disrupts their geographies and lifetime.

Therefore, the repeated displacement of migrants from critical borderzones is ultimately a mode for containing migrants' autonomous movements and spaces. This pushes us to reformulate containment beyond detention, widening containment to the disrupting of migrants' spaces of life and geographies.

## Conclusion

Drawing on Bridget Anderson's argument that 'the instability of the migrant [...] exposes the instability of who counts as a citizen, and indeed of citizenship itself' (Anderson 2017, p. 1535); it can be similarly argued that the instability and fuzziness of the 'displacement' lexicon (which also includes 'displaced persons') enables shedding light on the variable geometries of the state. In fact, this chapter has explored how displacement is used as a technology for governing and regaining control over 'unruly' migrations, what the EU calls 'secondary movements,' at the internal frontiers of Europe. In particular, displacement refers here to migrants being constantly transferred or chased away from critical borderzones, such as Ventimiglia and Calais, without however being relocated or detained somewhere. As a result of these forms of displacement, migrants' mobility is not reduced but enhanced: migrants' displacements at the internal frontiers of Europe are enacted by keeping migrants on the move, by forcing them to undertake convoluted geographies and to interrupt their journeys. Therefore, the chapter has shown that forced hypermobility turns out to be a way to disrupt migrants' autonomous movements and lifetimes and to take away the infrastructures of liveability. Not only individual migrants are targeted but also migrant multiplicities, that is emergent collective formations which could become political subjects.

Such an insight on displacement has enabled questioning the nexus between freedom and mobility that underpins most of migration literature, as well as political theory, showing how forced hypermobility becomes for the migrants a mode of containment, even beyond detention and spatial confinement. Firstly, while policy documents and migration literature have notably associated displacement with exceptional or critical events—people fleeing conflict or natural disasters, this angle on displacement foregrounds more ordinary forms of displacement that are not, however, less harmful for the migrants. Secondly, while displacement is usually presented as a consequence of crises and as something that migrants should be protected from (protecting migrants from displacement), this chapter has shown, it is actually also used as a political technology by states in order to regain control over migrants (enacting *displacement for* governing migration). However, in analysing how displacement is used as a political technology for governing migration we could conclude that migrants are merely bodies upon which heterogeneous forms of dispossession and control are exercised. On the contrary, keeping migrants on the move is a way to regain control over the 'incorrigibility' (De Genova 2010) of migrants' presence that constantly exceeds and dodges, dispositive of capture and extraction. In this respect, Michel Foucault's account in *The Punitive Society* (1972–1973) of the penalisation of unruly mobility echoes such a constitutive exceeding of migrant subjective drives: unauthorised and disordered practices of mobility deemed to be potentially harmful had been increasingly criminalised and subjected to fixation or forced displacement (Foucault 2016). In fact, displacement is today, together with the multiplication of detention sites and measures, one of the main ways through which 'unruly' migrants' presences and movements are tamed and deprived of the material and legal terrain to stay.

## Notes

1. See for instance the UN Declaration for Refugees and Migrants (2016), http://www.globalcrrf.org/wp-content/uploads/2018/05/The-New-York-Declaration-Full-text-19Sep2016.pdf
2. http://www.unesco.org/new/en/social-and-human-sciences/themes/international-migration/glossary/displaced-person-displacement
3. Indeed, the definition of 'refugee' enshrined in the Geneva Convention does not include people who have been displaced from their countries or from their areas due to environmental disasters, wars, and conflicts and it always refers to the country of origin. In fact, according to the 1951 Refugee Convention a refugee is 'someone who is unable or unwilling to return to their country of origin owing to a well-founded fear of being persecuted for reasons of race, religion, nationality, membership of a particular social group, or political opinion' (http://www.unhcr.org/3b66c2aa10.html).
4. The expression 'secondary movements' is used in EU documents to designate the convoluted and erratic routes that migrants undertake across Europe as an outcome of spatial restrictions imposed by the Dublin Regulation and rejected asylum applications. Such a designation is taken here in a critical way, since it repurposes an image of linear movements, from point A to point B, and subordinates these 'secondary movements' to supposedly primary or first movements.
5. Hotspots are identification centres for migrants that had been implemented by the EU in 2015 on Greek islands and in Italy for quickly identifying and selecting migrants. However, the term 'hotspot' also refers to the broader

hotspot approach launched by the EU on the same year, which is not limited to detention infrastructures and which rather includes a series of standardised identification procedures and mechanisms for managing migration after landing.

6. https://www.ecre.org/asylum-seekers-transferred-from-northern-italy-to-taranto-hotspot
7. http://www.rainews.it/dl/rainews/articoli/Alfano-Ventimiglia-no-sara-la-nostra-Calais-migranti-17eca50c-a3cf-4989-975d-d568dc044860.html; https://www.thelocal.it/20180616/in-calais-of-italy-tension-soars-over-migrant-crisis
8. http://cherish-de.uk/migrant-digitalities/#
9. http://www.gisti.org/IMG/pdf/arrete_2007-03-06_calais-maire.pdf
10. https://www.hrw.org/report/2017/07/26/living-hell/police-abuses-against-child-and-adult-migrants-calais

# References

Adey, P. (2010). *Mobility*. London: Routledge.

Anderson, B. (2017). Towards a new politics of migration? *Ethnic and Racial Studies, 40*(9), 1527–1537.

Ansems de Vries, L., & Guild, E. (2019). Seeking refuge in Europe: Spaces of transit and the violence of migration management. *Journal of Ethnic and Migration Studies, 45*(12), 2156–2166.

Aradau, C. (2017). Performative politics and international Relations. *New Perspectives, 25*(2), 2–7.

Bakewell, O. (2011). Conceptualising displacement and migration: Processes, conditions and categories. In K. Koser & S. Martin (Eds.), *The migration–displacement nexus: Patterns, processes and policies* (pp. 14–28). Oxford, England: Berghahn Books.

Bernardot, M. (2008). *Camps d'étrangers*. Bellecombe-en-Bauges, France: Croquant.

Betts, A. (2013). *Survival migration. Failed governance and the crisis of displacement*. Ithaca, NY: Cornell University Press.

Bigo, D. (2010). Freedom and speed in enlarged borderzones. In V. Squire (Ed.), *The contested politics of mobility* (pp. 51–70). Abingdon, England: Routledge.

Çaglar, A., & Schiller, N. G. (2018). *Migrants and city-making: Dispossession, displacement, and urban regeneration*. Duke University Press.

Cresswell, T. (2010). Towards a politics of mobility. *Environment and planning D: Society and space, 28*(1), 17–31.

Darling, J. (2011). Domopolitics, governmentality and the regulation of asylum accommodation. *Political Geography, 30*(5), 263–271.

Darling, J. (2016). Asylum in austere times: Instability, privatization and experimentation within the UK asylum dispersal system. *Journal of Refugee Studies, 29*(4), 483–505.

Davies, T., Isakjee, A., & Dhesi, S. (2017). Violent inaction: The necropolitical experience of refugees in Europe. *Antipode, 49*(5), 1263–1284.

De Genova, N. (2010). The queer politics of migration: Reflections on 'illegality' and incorrigibility. *Studies in Social Justice, 4*(2), 101–126.

De Genova, N. (2013a). We are of the connections': Migration, methodological nationalism, and 'militant research. *Postcolonial Studies, 16*(3), 250–258.

De Genova, N. (2013b). Spectacles of migrant 'illegality': The scene of exclusion, the obscene of inclusion. *Ethnic and Racial Studies, 36*(7), 1180–1198.

De Genova, N. (Ed.). (2017). *The borders of Europe. Autonomy of migration, tactics of bordering*. Durham, NC: Duke University Press.

Fassin, D. (2009). Another politics of life is possible. *Theory, Culture & Society, 26*(5), 44–60.

Fekete, L. (2018). Migrants, borders and the criminalisation of solidarity in the EU. *Race & Class, 59*(4), 65–83.

Fili, A. (2018). Containment practices of immobility in Greece. In M. Karakoulaki, L. Southgate, & J. Steiner (Eds.), *Critical perspectives on migration in the twenty-first century* (pp. 162–180). Bristol, England: E-International Relations Publishing.

Fontanari, E. (2018, May 29). Dispersing their lives: Refugees kept on the move between Italy and Germany. *Border Criminologies*. Retrieved from https://www.law.ox.ac.uk/research-subject-groups/centre-criminology/centreborder-criminologies/blog/2018/05/dispersing-their

Foucault, M. (1982). The subject and power. *Critical Inquiry, 8*(4), 777–795.

Foucault, M. (2012). *Discipline and punish: The birth of the prison*. New York, NY: Vintage.

Foucault, M. (2016). *The punitive society: Lectures at the Collège de France, 1972–1973*. Basingstoke, England & New York: Palgrave Macmillan.

Gill, N. (2009a). Governmental mobility: The power effects of the movement of detained asylum seekers around Britain's detention estate. *Political Geography, 28*(3), 186–196.

Gill, N. (2009b). Longing for stillness: The forced movement of asylum seekers. *M/C Journal, 12*(1). Retrieved October 19, 2018, from http://journal.media-culture.org.au/index.php/mcjournal/article/view/123.

Gill, N. (2013). Mobility versus liberty? The punitive uses of movement within and outside carceral environments. In D. Moran, N. Gill, & D. Conlon (Eds.), *Carceral spaces: Mobility and agency in imprisonment and migrant detention* (pp. 31–48). Bristol, England: Ashgate.

Khosravi, S. (2018, March 2). Stolen time. *Radical Philosophy*. Retrieved from https://www.radicalphilosophy.com/article/stolen-time

Mezzadra, S. (2011). The gaze of autonomy: Capitalism, migration and social struggles. In V. Squire (Ed.), *The contested politics of mobility* (pp. 141–162). Abingdon, England: Routledge.

Mezzadra, S., & Neilson, B. (2017). On the multiple frontiers of extraction: Excavating contemporary capitalism. *Cultural Studies, 31*(2–3), 185–204.

Moffette, D. (2018, June 15). Dispersal policing as borderwork in Barcelona. *Border Criminologies*. Retrieved October 22, 2018, from https://www.law.ox.ac.uk/research-subject-groups/centre-criminology/centreborder-criminologies/blog/2018/06/dispersal.

Moran, D., Piacentini, L., & Pallot, J. (2012). Disciplined mobility and carceral geography: Prisoner transport in Russia. *Transactions of the Institute of British Geographers, 37*(3), 446–460.

Mountz, A. (2011). Specters at the port of entry: Understanding state mobilities through an ontology of exclusion. *Mobilities, 6*(3), 317–334.

Mountz, A., Coddington, K., Catania, R. T., & Loyd, J. M. (2013). Conceptualizing detention: Mobility, containment, bordering, and exclusion. *Progress in Human Geography, 37*(4), 522–541.

Pinkerton, P. (2019). The biopolitics of the migration-development nexus: Governing migration in the UK. *Politics, 39*(4), 448–463.

Scott, J. C. (1998). *Seeing like a state: How certain schemes to improve the human condition have failed.* New Haven, CT: Yale University Press.

Tazzioli, M. (2017a). The government of migrant mobs: Temporary divisible multiplicities in border zones. *European Journal of Social Theory, 20*(4), 473–490.

Tazzioli, M. (2017b, July 20). Calais after the jungle: Migrant dispersal and the expulsions of humanitarianism. *OpenDemocracy*. Retrieved October 22, 2018, from https://www.opendemocracy.net/beyondslavery/martina-tazzioli/calais-after-jungle-migrant-dispersal-and-expulsion-of-humanitarianis.

Tazzioli, M. (2018). Containment through mobility: Migrants' spatial disobediences and the reshaping of control through the hotspot system. *Journal of Ethnic and Migration Studies, 14*(16), 2764–2779.

Tazzioli, M., & Garelli, G. (2018). Containment beyond detention: The hotspot system and disrupted migration movements across Europe. *Environment and Planning D: Society and Space.* https://doi.org/10.1177/0263775818759335.

Walters, W., & Luthi, B. (2016). The politics of cramped space: Dilemmas of action, containment and mobility. *International Journal of Politics, Culture & Society, 29*(4), 359–366.

Zetter, R., & Morrissey, J. (2014). The environment–mobility nexus: Reconceptualising the links between environmental stress, mobility, and power. In E. Fiddian-Qasmiyeh, G. Loescher, K. Long, & N. Sigona (Eds.), *The Oxford handbook of refugee and forced migration studies* (pp. 342–354). Oxford, England: Oxford University Press.

# 23

## Carceral Journeys

### Nick Gill and Oriane Simon

## Introduction

This chapter is concerned with carceral journeys, which refer to the transportation or transfer of anyone whose freedom is significantly and deliberately curtailed. An obvious example of such journeys is inter-prison transfers, which occur for a variety of reasons ranging from administrative convenience in the management of carceral populations across space (Moran et al. 2013) to the 'ghosting' of prison populations—that is, the movement of certain prisoners as a way to make them invisible to authorities and other inmates. Carceral journeys is a broader concept than simply prison-to-prison movement though. It also includes the historic transportation of convicts to the colonies of Western powers, the shifting of immigration detainees between detention centres, the within-prison transfer of inmates to different parts of the same

---

This chapter was supported by the European Research Council, grant number StG-2015_677917, acronym ASYFAIR.

---

N. Gill (✉)
Department of Human Geography, University of Exeter, Exeter, UK
e-mail: N.M.Gill@exeter.ac.uk

O. Simon
KPMG Wirtschaftsprüfungsgesellschaft AG, Mainz, Germany

© The Author(s) 2020
P. Adey et al. (eds.), *The Handbook of Displacement*,
https://doi.org/10.1007/978-3-030-47178-1_23

establishment, the well-trodden city block to prison circuit in the US, the extraordinary rendition of terror suspects, and the meandering routes of 'delinquent' individuals between different 'carceral' institutions like hospitals, schools, and factories.

Viewing carceral journeys as a coerced form of displacement allows us to focus on the interstitial spaces between, surrounding, and within these sorts of institutions. This connective tissue has traditionally been sidelined in prison studies, partly because *prison* studies is already conceptualised in institutional terms, and partly owing to the intellectual heritage of the discipline, which has tended to emphasise the prison's reified boundary via concepts such as Goffman's 'total institution' (Goffman 1961). Geographers have argued that it is important to focus not just on the institutions that make up the carceral landscape, but also on the connections between them for various reasons. First, in reality, carceral institutions are not isolated but crisscrossed by multiple practical and social networks (Moran 2015; Moran and Keinänen 2012). Second, the carceral system is showing signs of denser connectivity between disparate institutions. The *integration* of diverse carceral sites into a cohesive carceral system is part of the emergence of a profitable network of governing institutions that are increasingly coordinated in the management of populations that are seen as deviant or dangerous (Gill et al. 2018). Third, inter-institutional carceral spaces are in some ways less visible, even though they are outside the formal carceral boundary. The conditions of inter-institutional transfers of inmates, for example, are harder to monitor than prison conditions generally. Consequently, fourth, interstitial carceral spaces are sites of acute power asymmetry, which produce both a heightened sense of carcerality in the subject (the subject is often acutely aware of their incarceration during carceral journeys) and particular opportunities for the abuse of power by the carceral agent.

A focus on carceral journeys is a distinctly geographical approach that emphasises mobility, connectivity, and lived experience. In what follows we explore carceral journeys both conceptually and practically. Conceptually, we emphasise the importance of attending not only to static forms of incarceration and exclusion but also to forms that combine mobility and incarceration; we examine the enmeshment of carceral journeys into profit-making and exploitative systems; and we attend to the various visibilities and invisibilities that they entail. Practically, we reflect on the use of extraordinary rendition as a particularly controversial and hidden type of carceral journey that entails distinctive combinations of forced mobility and inertia. Through examining the case of extraordinary rendition, we are able to highlight further geographical aspects of carceral journeys, including their viscerality and the opportunities for resistance that persist within even the most violent journeys. We also draw attention to the infrastructure required to bring carceral journeys about, and the effect that carceral journeys have not just on subjects, but a wide network of supporters. Overall, we argue that carceral journeys, as a particular form of displacement, are an increasingly key element of evolving global carceral circuitry.

## Conceptualising Carceral Journeys

At first sight it might seem as though 'carceral journeys' is something of an oxymoron. Surely to be incarcerated is to be immobilised and inert? And surely to be free is to be able to move and journey at will? Certainly many forms of incarceration do entail imposed stasis, and many forms of freedom afford movement, but the correspondence is not perfect in either direction and, as geographers, it is important to be critical of the sometimes assumed homology between freedom and movement (Gill 2009). Many journeys, for example, are undertaken regretfully and grudgingly. The number of forced migrants fleeing the threat of violence in their home countries has increased markedly in recent years, for instance. As a result, many of these migrants, who often have to move multiple times and whose journeys can take many years, long for stillness as a form of freedom.

Furthermore, although contemporary culture makes travel sound glamorous and adventurous (the travel industry has a vested interest in achieving this), it is often burdensome, tiring, disruptive, disorientating, and frightening owing to its uncertainties and risks (Anderson 2015; Favell 2008). The horrors of the shipment of slaves from Africa to the colonies at the height of the slave trade indicate well the impositions of movement. Indeed, journeys have historically been systematically imposed on prisoners as a form of punishment. 'Transportation' was a common sentence in the seventeen hundreds in Britain and other world powers, for instance (Woodward 2006). The essence of the convicts' punishment when they were sentenced to transportation was precisely the ordeal of the journey itself: the difficult conditions, the dangers, the hard work, and the distance from loved ones and familiar surroundings.

In different ways, today's governments around the world also seek to enrol mobility in their attempts to control populations. The threat and reality of deportation of non-citizens from the sovereign territory of Western developed countries, for example, is reminiscent of the transportation of convicts in previous centuries. The difference is that deportation is not a legal punishment, but an administrative measure applied to those who are deemed 'out of place,' despite the fact that it is often experienced as a bitter form of punishment by both those who are deported and their loved ones left behind (Khosravi 2018). It's classification as administrative rather than legal in nature, though, can mean that deportation is not subject to the sort of judicial oversight that would ordinarily govern state-sanctioned punishment and imprisonment. In turn, this can mean that it is more arbitrarily used, as well as more poorly scrutinised and monitored. Another way states enrol mobility as a tool of social control is through the requirement to keep moving as part of their regulation of urban spaces. The exclusion of socially undesirable sections of the population like homeless people, drug users, sex workers, and the mentally ill from cities in the US, for example, has been achieved in part via the outlawing of prolonged immobility in city centres through laws prohibiting loitering as well as sleeping and camping in public spaces (Beckett and Herbert 2010). These render personal mobility a legal requirement and function as a pernicious form of socio-spatial regulation.

Another example of social control via movement concerns immigration detention (i.e. incarceration for the purposes of migration control). The transfer of immigration detainees from one detention centre to another is used in some countries as a means to reward 'good,' 'compliant' behaviour, by moving detainees to centres with better facilities like gyms, libraries, and IT access, for instance, and punish 'bad,' 'non-compliant' behaviour by moving detainees to centres with poorer facilities (Gill 2009).

Turning to mainstream prisons, while prison sentences take into account the length of time that inmates will serve, they rarely take into account the distance that such inmates will be from their everyday lives and loved ones whilst completing their sentences. It makes a huge difference to inmates, though, if they are imprisoned near their families and other networks of support (including their lawyers) because they are likely to get more visits if they are nearby and impose fewer hardships on their loved ones (Comfort 2008).

In short, carceral journeys are important tools for governments who are seeking to control and manage populations—especially ones perceived to be deviant, troublesome, threatening, or undesirable—and can be experienced as an acute imposition and a painful form of punishment. It is all the more concerning, then, that carceral journeys are also profitable. The profitability of carceral systems has been identified by geographers as a key cause of the rise in rates of incarceration, especially, but not exclusively, in the US (Bonds 2012; Gilmore 2007). Prison building has been viewed as a way to address high local rates of unemployment and there is evidence that justice systems themselves, including the likelihood that certain minor crimes will carry a custodial sentence, are calibrated to ensure that prisons are filled with inmates in order to prop up prison economies (Gilmore 2007). This approach to generating employment is flawed. Not only does it render justice contingent on factors other than the crime committed, but it eventually leads to either a higher unemployment rate, because imprisonment often effectively dispossesses incarcerated individuals of future employability, or to spiralling imprisonment rates as more prisons are built to contain the unemployable (Western and Beckett 1999). Unfortunately, though, the temptation to build prisons is heightened by the promise of a flexible, bored, exploitable labour force, available at low rates of pay to carry out work from electronics manufacture or recycling to call centre work from within prisons themselves (Nowakowski 2013).

The global prison population swelled by nearly 20 per cent between 2000 and 2015 leading to over ten million people living in prisons in 2016 (Rope and Sheahan 2018). Where there are more prisons, the connections between imprisonable populations and prison institutions are sustained by carceral journeys. Massaro (2015) describes the 'revolving door' of American prisons in which the circulation of poor, black African Americans between city blocks and prisons is so common that communities have adapted to the high likelihood of repeated incarceration by sharing the burden of visits and bail payments. Wacquant (2001) has suggested that prisons and ghettos developed a symbiotic relationship in American culture towards the end of the previous century: a relationship bound together by carceral journeys into, around, and out of carceral institutions.

Another reason why carceral journeys are becoming more common is because they are included by some prison authorities in rehabilitation programmes. As part of their sentences, inmates in some countries are expected to relearn how to play a 'useful' role in society, which often means travelling outside the prison to engage in education, work, community service, victim compensation schemes, or paid employment. Scholars have argued that this form of carceral journeying is related to a shift in the logics that underpin imprisonment in some countries (Mincke and Lemonne 2014). No longer are prisons intended to simply confiscate time—rather they are intended to make prisoners' time more productive and active as a way to reconstruct an engaged liberal subject (Rose 2000). Ironically though, inmates who refuse this form of conditioning and show no interest in participating in their own improvement are *also* likely to be moved around under carceral conditions. They typically find themselves diagnosed as needing special treatment by experts, especially psychologists (Mincke and Lemonne 2014). This results in what Martel (2006, p. 600) calls 'a perpetual loop-line' of passing on: from juvenile delinquent institutions, to psychiatric hospitals and psychiatric prison wings (Schliehe 2014), to local community parole officers.

Carceral journeys introduce particular forms of visibility and invisibility to the carceral landscape. In terms of visibility, parading prisoners serves not only to humiliate them but to instil fear in the audience, thereby promoting compliance among the wider population. For this reason, occupying colonial powers have historically been particularly fond of prisoner parades, but the practice is exclusive neither to the distant past nor to colonies. In Arizona, Maricopa county sheriff Joe Arpaio created a tent prison, which remained open for twenty years until 2016, which required inmates to wear a humiliating pink colour (Fernández 2017). Press were frequently shown around the facility as a way to broadcast the sheriff's 'tough-on-crime' political image, and prisoners also wore the colours outside the prison in chain gangs that had mostly been abandoned in the US since the 1950s. The pink colour made the prisoners look ridiculous and sexually inferior.

In terms of invisibility, in his detailed study of the 'dirty protests'[1] inside the notorious Maze prison in Northern Ireland, Feldman (1991) describes the tense and violent relationship between the prison guards and the inmate population in the context of the acute political tensions in the country during the late 1980s. Desperate to suppress the political will of the dissidents in prison, prison authorities moved prisoners between prison cells and prison wings routinely as a way to sever their connections to fellow inmates, isolate them, and break their resolve.

Frequently moving prisoners can also make them invisible to family and other supporters on the outside. Besides increasing the harm of immigration detention for the detainees, Mountz and Loyd (2014) demonstrate that the remoteness and frequent transfers of detained immigrants affect their families and legal advocates (see also Gill 2009). Mountz and Loyd (2014, p. 389) conclude that the 'remote geographic locations' of detention sites and the 'frequent transfers of detainees […] have proven [to be] significant issues of concern for anyone who has tried to track down a loved one.'

## Extraordinary Rendition as a Carceral Journey

Having set out some of the features of carceral journeys, we now turn to an examination of extraordinary rendition as a particularly obfuscated example of carceral journeys (Paglen 2011). We show how extraordinary rendition—the forced transfer of suspected terrorists without due legal process—highlights the viscerality of carceral journeys, the resistance that they encounter, the extent to which they need to be supported by a wide infrastructure, and the influence they have in wider society, beyond the immediate subject.

The empirical material presented in this section draws upon PhD research conducted by Oriane Simon: in particular archival work and interviews (Simon 2017). Ten in-depth interviews were conducted over two years (between the end of 2014 and end of 2016) with human rights lawyers, who defended victims or investigated extraordinary rendition, as well as human rights investigators supporting NGOs, such as Amnesty International, Human Rights Watch, and The Rendition Project.

The prisoners' transfers across state borders constitute the most obvious, large-scale journey that is involved in extraordinary rendition. The extraordinary rendition victim El-Masri suffered two lengthy carceral journeys across state borders, which he describes in detail. Firstly, he was transferred from Macedonia, where he was seized, to a secret prison in Afghanistan. Secondly, for his release, he was again transferred across state borders and flown first to Albania, where he was driven up and down the mountains for hours before eventually being set on a flight to his home in Germany (El-Masri 2005, 2006; Watt et al. 2008). As part of the preparations for transfer to the prison, El-Masri underwent a 'medical examination' at the hands of US agents (El-Masri 2006; European Court of Human Rights 2012; Watt et al. 2008).

> As I was led into this room, I felt two people violently grab my arms, one from the right side and the other from the left. They bent both my arms backwards. This violent motion caused me a lot of pain. I was beaten severely from all sides. … finally they stripped me completely naked and threw me to the ground. My assailants pulled my arms back and I felt a boot in the small of my back. I then felt a stick or some other hard object being forced in my anus. I realized I was being sodomized. […] [Later] I was dressed in a diaper, over which they fitted a dark blue sports suit with short sleeves and legs. I was once again blindfolded, my ears were plugged with cotton, and headphones were placed over my ears. A bag was placed over my head and a belt around my waist. My hands were

chained to the belt. They put something hard over my nose. Because of the bag, breathing was getting harder and harder for me. I struggled for breath and began to panic. (El-Masri 2006, pp. 9–10)

This quote illustrates the viscerality of certain carceral journeys and their punitive effect on the victim. In fact, the European Court of Human Rights (ECHR) asserts that the physical force and measures for this transfer were excessive, unjustified, and that they were used purposefully to punish, intimidate, and cause pain (European Court of Human Rights 2012). The visceral and punitive aspects are further evidenced in the research interviews Simon (2017) conducted and the centrality of mobility to the achievement of disorientation becomes clear. One human rights lawyer, for example, asserts that those practices are 'prolonging the shock of capture' and aimed at making victims feel 'as uncomfortable as possible' (Simon 2017, p. 202):

Moving them to different locations is just another way of disrupting the person's way of thinking, making sure that they are never ever comfortable, they never relax, that they are always questioning what is going to happen. So putting a sack over someone's head, making sure that they are sensory deprived, is all part of the process. […] They take them to different locations for a number of reasons, but one of them is to keep the disorientation up. (Simon 2017, p. 202)

Besides the carceral journeys across state borders that are undertaken as part of extraordinary rendition, there are manifold smaller scale transfers, which enrol individual mobility as a technique in mental and physical coercion. Amongst others, for instance, Rose (2004), a journalist, points out the abusive qualities of forced cell extractions, describing the manner in which one prisoner was 'pepper-sprayed' in the face, which led the prisoner to vomit 'five cupfuls [sic]' (Rose 2004, p. 71). Another example is given by the extraordinary rendition victim Errachidi (2013), who recalls how Guantánamo Bay functionaries devised techniques of pretending to release prisoners by giving the detainees new clothes and bringing them near the airport just to then re-interrogate them because of supposedly novel evidence, so playing with the prisoners' hopes (Errachidi 2013).

Another point that extraordinary rendition illustrates about carceral journeys is that prisoners' forced mobility sometimes leads to unexpected forms of resistance. As Pain (2009) argues, manipulations are not simply absorbed passively by individuals. For one, carceral journeys provide opportunities for communication and information exchange between prisoners (El-Masri 2006; Errachidi 2013). In the American Kandahar prison (Afghanistan), two prisoners were tasked to empty the toilet buckets of all the enclosures every morning. While this task was meant to be punitive, the prisoners soon turned it into a means to exchange information (Errachidi 2013). Similarly, the prisoners actively used the rotation of prisoners to different cell blocks within Guantánamo Bay to exchange information and coordinate resistance actions (Errachidi 2013; Hicks 2010). Another example is that, at times, Errachidi was not allowed to go to the toilet and so he soiled himself. While this humiliated him, it also implied that he was seen by others as a fellow prisoner rather than a US spy (Errachidi 2013). The guards' aim to punish prisoners by refusing them the opportunity to relieve themselves and forcing them to soil themselves was turned into a means to constitute solidarity amongst the prisoners.

Carceral journeys also require an infrastructure and multiple actors. Extraordinary rendition's cross-state carceral journeys rely on a large infrastructure spread across the globe, which necessitates various forms of direct and indirect support from different countries (Marty 2006; Paglen and Thompson 2006; Singh 2013). Singh's study of torture (2013) delineates forms of involvement of fifty-four countries beside the US and a recent analysis adds another fifteen countries (Cordell 2017). El-Masri's comparatively short extraordinary rendition involved the Macedonian officials who originally arrested, detained, and interrogated him for about two weeks before handing him over to the CIA (El-Masri 2006; Marty 2006); his transfer to the secret prison in Afghanistan, including a refill stop in Iraq (Marty 2006); and eventually his release in Albania (El-Masri 2006; Marty 2006). It is noteworthy that owning a plane, organising a flight, and flying are different activities, for which different actors are responsible (Blecher 2007; Paglen and Thompson 2006). El-Masri's extraordinary rendition thus highlights the wide-ranging and complex infrastructure and actors involved in carceral journeys. The involvement of foreign states and state agencies is a crucial means for avoiding responsibility. Carceral journeys are a major tool in involving various US agencies, foreign officials, countries, and private contractors (Paglen and Thompson 2006).

Finally, extraordinary rendition demonstrates that carceral journeys have wide-ranging effects that extend beyond the immediate victims to people commonly understood as either indirectly or tangentially involved, such as the functionaries, detainees' lawyers, and all these actors' families and friends. Prisoners' regular carceral journeys affect their bonding with functionaries like guards and other prison staff, for example. With regard to asylum seekers, Gill (2009, p. 193) outlines that as detainees are constantly moved they become a fleeting and ghostly presence and the lack of bonds with detention custody officers increases the likelihood of abuse (Gill 2009). Bonding or the lack thereof not only renders the detainees' life easier or harder, it also affects the functionaries' propensity 'to support legal appeals, […] act as character references, […] secure good legal representation […] and block the transfer of detained asylum seekers' (Gill 2009, p. 194). Gill concludes that, 'given the degree of influence wielded by management, the way in which detainees are presented to managers is of critical importance' (2009, p. 194).

Similarly, Rose (2004) stresses that the guards' rotations in Guantánamo Bay made it more difficult to get to know their prisoners: the guards were explicitly watched and prohibited from bonding with their wards. Building a relationship with prisoners was discouraged in that the guards 'assigned to each block were changed every day' (Rose 2004, p. 67). In addition, the former Guantánamo Bay guard Hickman (2013) recalls that Guantánamo Bay functionaries were further discouraged from bonding with other US officials outside their small units.[2]

Besides the remote location, all of the interviewees who participated in the research Simon conducted who had been on the island reported that Guantánamo Bay isolates functionaries in that there is a bad internet connection, very little cell phone coverage and only an expensive public phone (see also Mountz and Loyd 2014; Rose 2004). In other words, functionaries are affected by what they do and experience. The very geographical location itself isolates functionaries and shields them from public outcry. As another interviewee, a human rights lawyer, points out 'the people who did the torture, […] they suffer' (Simon 2017, p. 170). She cites a guard saying, 'I have PTSD [post-traumatic stress disorder], not for what happened to me but for what I did to people' (Simon 2017, p. 170). Their experiences in Guantánamo Bay in turn affect their families and their later work on the mainland (Corsetti 2013; Lakemacher 2010a, b; Walls 2015).

Travelling to faraway places takes a toll on the body and mind of the lawyers, activists, journalists, and investigators (Simon 2017). One interviewee, a lawyer, explains that at first it is exciting to go to Guantánamo Bay, but then it becomes 'Oh, I need to go to Gitmo; I'm going to be eating crap food, and it will be 90° [Fahrenheit]' (Simon 2017, p. 301). Guantánamo Bay also employs tedious regulations, as one of the investigators explains: 'When you arrive you are assigned a minder, who escorts you wherever you want to go during your time there, including to buy groceries, to get food, to go everywhere. And, you have to travel in groups' (Simon 2017, p. 301). In these and similar ways, the prisoners' carceral journeys also affect people who are less directly involved.

# Conclusion

This chapter has discussed the characteristics of carceral journeys as a particular form of displacement. It has argued that carceral journeys are a by-product of expanding carceral circuits, which integrate various types of institution into an overall carceral system. This circuitry is increasing in complexity as the number of prisoners and other types of detainees increases, driven in part by profitability. Carceral journeys affect the visibility and invisibility of carceral practices, exposing and displaying certain inmates, punishments, and conditions, whilst obfuscating, hiding, and isolating others.

The case of extraordinary rendition, as a particular type of carceral journey, highlights the physical and forceful viscerality of carceral journeys as well as the burdens they can impose, not just on detainees but on other actors tangentially involved in the carceral system. This said, carceral journeys can also expose carceral systems to forms of resistance that would not otherwise be possible: extraordinary rendition victims have documented the unexpected ways in which they and other prisoners have responded, which have defied efforts to suppress them.

A key advantage of thinking about and with carceral journeys is that the experience of the journeyer is foregrounded. Carceral systems are often dehumanising (think of prison numbers that replace names during custodial sentences). Paying attention to carceral journeys, on the other hand, facilitates a focus on the individuals concerned, as well as the interstitial, inter-institutional spaces that they occupy. As such, the lens of carceral journeys offers an important epistemological window on to the nature and effects of the recent development of carceral systems. For all these reasons, carceral journeys are a conceptually innovative way to approach and understand the intersection between displacement and carceral space. Carceral journeys can be experienced as an acute imposition and a painful form of punishment, and their proliferation is the logical result of the expansion and evolution of global carceral systems. Carceral journeys illustrate that displacement can be a form of social regulation and control. This implies that an awareness of carceral journeys is necessary in order to reckon with the complexity of today's many forms of state power.

## Notes

1. So named to refer to the practice adopted by Irish republican prisoners of daubing their own faeces on the walls of their cells in protest against their treatment.
2. Nonetheless bonding between prisoners and functionaries occurs. In the extraordinary rendition victim Al-Hajj's words 'I was always glad to see the familiar faces of our military escorts, who became like old friends over the months. They always greeted me with happy smiles or hugs. "Long time!" they'd joke sarcastically' (Khan 2008, p. 202).

## References

Anderson, J. (2015). Exploring the consequences of mobility: Reclaiming jet lag as the state of travel disorientation. *Mobilities, 10*(1), 1–16. https://doi.org/10.1080/17450101.2013.806392.

Beckett, K., & Herbert, S. (2010). Penal boundaries: Banishment and the expansion of punishment. *Law & Social Inquiry, 35*(1), 1–38.

Blecher, S. (2007). Declaration of Sean Belcher, Civil Action 5:07-Cv-02798 (JW). Retrieved from https://www.therenditionproject.org.uk/pdf/PDF%20239%20[Mohamed%20et%20al%20v.%20Jeppesen%20Dataplan,%20Inc.%20-%20Memo%20of%20Plaintiffs%20in%20Opp%20to%20Gvt%E2%80%99s%20Motion%20to%20Dismiss%20(Dec%202007)].pdf

Bonds, A. (2012). Building prisons, building poverty: Prison sitings, dispossession, and mass incarceration. In J. Loyd, M. Mitchelson, & A. Burridge (Eds.), *Beyond walls and cages: Prisons, borders and global crisis* (pp. 129–142). Athens, GA: University of Georgia Press.

Comfort, M. (2008). *Doing time together: Love and family in the shadow of the prison.* Chicago, IL: Chicago University Press.

Cordell, R. (2017). Measuring extraordinary rendition and international cooperation. *International Area Studies Review, 20*(2), 179–197. https://doi.org/10.1177/2233865916687922.

Corsetti, D. (2013). I'll be your bad guy. *Witness to Guantánamo*. Retrieved from http://witnesstoguantanamo.com/wp-content/uploads/2015/01/DamienCorsetti_IllBeYourBadGuy.pdf

El-Masri, K. (2005). Statement: Khaled El-Masri. *American Civil Liberties Union.* Retrieved from https://www.aclu.org/other/statement-khaled-el-masri?redirect=human-rights_national-security/statement-khaled-el-masri

El-Masri, K. (2006). Declaration of Khaled El-Masri, Civil Action 1:05cv1417-TSE-TRJ. Retrieved from https://www.aclu.org/sites/default/files/pdfs/safefree/elmasri_decl_exh.pdf

Errachidi, A. (2013). *The General: The ordinary man who became one of the bravest prisoners in Guantánamo*. London: Chatto and Windus.
European Court of Human Rights. (2012). Case of El-Masri v. the former Yugoslav Republic of Macedonia. Judgment (39630/09). Retrieved from https://www.opensocietyfoundations.org/sites/default/files/CASE_OF_EL-MASRI_v__THE_FORMER_YUGOSLAV_REPUBLIC_OF_MACEDONIA.pdf
Favell, A. (2008). *Eurostars and Eurocities: Free movement and mobility in an integrating Europe*. Malden, England: Wiley-Blackwell.
Feldman, A. (1991). *Formations of violence: The narrative of the body and political terror in Northern Ireland*. Chicago & London: University of Chicago Press.
Fernández, V. (2017, August 21). Arizona's 'concentration camp': Why was Tent City kept open for 24 years? *The Guardian*.
Gill, N. (2009). Governmental mobility: The power effects of the movement of detained asylum seekers around Britain's detention estate. *Political Geography, 28*(3), 186–196.
Gill, N., Conlon, D., Moran, D., & Burridge, A. (2018). Carceral circuitry. New directions in carceral geography. *Progress in Human Geography, 42*(2), 183–204.
Gilmore, R. W. (2007). *Golden gulag: Prisons, surplus, crisis, and opposition in globalizing California*. Berkeley, CA: University of California Press.
Goffman, E. (1961). *Asylums: Essays on the social situation of mental patients and other inmates*. Garden City, NY: Anchor Books.
Hickman, J. (2013). *Selling Guantanamo: Exploding the propaganda surrounding America's most notorious military prison*. Gainesville, FL: University Press of Florida.
Hicks, D. (2010). *Guantanamo: My journey*. Sydney, Australia: William Heinemann.
Khan, M. (2008). *My Guantánamo diary. The detainees and the stories they told me*. New York, NY: Public Affairs.
Khosravi, S. (Ed.). (2018). *After deportation: Ethnographic perspectives*. London: Palgrave Macmillan.
Lakemacher, D. (2010a). Coming to terms with what I did. *Witness to Guantánamo*. Retrieved from http://witnesstoguantanamo.com/videos/coming-to-terms-with-what-i-did/
Lakemacher, D. (2010b). Irony behind the wire. *Witness to Guantánamo*. Retrieved from http://witnesstoguantanamo.com/wp-content/uploads/2015/01/DanielLakemacher_IronyBehindtheWire.pdf
Martel, J. (2006). To be, one has to be somewhere: Spatiotemporality in prison segregation. *British Journal of Criminology, 46*(4), 587–612.
Marty, D. (2006). *Alleged secret detentions and unlawful inter-state transfers of detainees involving Council of Europe Member States* (Doc. 10957). Retrieved from http://assembly.coe.int/committeedocs/2006/20060606_ejdoc162006partii-final.pdf
Massaro, V. (2015). The intimate entrenchment of Philadelphia's drug war. *Territory, Politics, Governance, 3*(4), 369–386.
Mincke, C., & Lemonne, A. (2014). Prison and (im)mobility: What about Foucault? *Mobilities, 9*(4), 528–549.

Moran, D. (2015). *Carceral geography: Spaces and practices of incarceration*. Farnham, England: Ashgate.

Moran, D., Gill, N., & Conlon, D. (Eds.). (2013). *Carceral spaces: Mobility and agency in imprisonment and migrant detention*. Farnham, England: Ashgate.

Moran, D., & Keinänen, A. (2012). The 'inside' and 'outside' of prisons: Carceral geography and home visits for prisoners in Finland. *Fennia-International Journal of Geography, 190*(2), 62–76.

Mountz, A., & Loyd, J. (2014). Transnational productions of remoteness: Building onshore and offshore carceral regimes across borders. *Geographica Helvetica, 69*(5), 389–398. https://doi.org/10.5194/gh-69-389-2014.

Nowakowski, K. (2013). Landscapes of toxic exclusion: Inmate labour and electronics recycling in the United States. In D. Moran, N. Gill, & D. Conlon (Eds.), *Carceral spaces: Mobility and agency in imprisonment and migrant detention*. Farnham, England: Ashgate.

Paglen, T. (2011). *Blank spots on the map: The dark geography of the Pentagon's secret world*. New York, NY: Penguin.

Paglen, T., & Thompson, A. (2006). *Torture taxi. On the trail of the CIA's rendition flights*. London: Icon Books.

Pain, R. (2009). Globalized fear? Towards an emotional geopolitics. *Progress in Human Geography, 33*(4), 466–486. https://doi.org/10.1177/0309132508104994.

Rope, O., & Sheahan, F. (2018). *Global prison trends 2018*. London: Penal Reform International & Bangkok, Thailand: Thailand Institute of Justice. Retrieved from https://cdn.penalreform.org/wp-content/uploads/2018/04/PRI_Global-Prison-Trends-2018_EN_WEB.pdf

Rose, D. (2004). *Guantánamo. America's war on human rights*. London: Faber and Faber.

Rose, N. (2000). Government and control. *British Journal of Criminology, 40*, 321–339.

Schliehe, A. K. (2014). Inside 'the carceral': Girls and young women in the Scottish criminal justice system. *Scottish Geographical Journal, 130*(2), 71–85.

Simon, O. (2017). *Evaluating extraordinary rendition* (Doctoral dissertation). University of New South Wales, Canberra, Australia.

Singh, A. (2013). *Globalizing torture. CIA secret detention and extraordinary rendition*. New York, NY: Open Society Foundations. Retrieved from https://www.opensocietyfoundations.org/sites/default/files/globalizing-torture-20120205.pdf.

Wacquant, L. (2001). Deadly symbiosis: When ghetto and prison meet and mesh. *Punishment & Society, 3*(1), 95–133.

Walls, G. (2015). Failing our discharged vets. *Witness to Guantánamo*. Retrieved from http://witnesstoguantanamo.com/videos/failing-our-discharged-vets/

Watt, S., Dakwar, J., Turner, J., Goodman, M., & Wizner, B. (2008). Petition alleging violations of the human rights of Khaled El-Masri by the United States of

America with a request for an investigation and hearing on the merits. Retrieved from https://www.aclu.org/sites/default/files/field_document/elmasri_iachr_20080409.pdf

Western, B., & Beckett, K. (1999). How unregulated in the U.S. labor market?: The penal system as a labor market institution. *American Journal of Sociology, 104*(4), 1030–1060.

Woodward, N. (2006). Transportation convictions during the Great Irish Famine. *Journal of Interdisciplinary History, 37*(1), 59–87.

# 24

# Precarious Migrations and Maritime Displacement

Vicki Squire and Maurice Stierl

## Introduction

Maritime displacement is characterised by forms of migration in which the journey itself is highly precarious. This is no more evident than in the Mediterranean, where large numbers of people are reported to die or go missing annually. Regularly exceeding 3000 fatalities per year and peaking at a total of 5096 in 2016 alone (UNHCR 2018), the death and vulnerability of people on the move via the Mediterranean has become a norm—especially over recent years (Squire 2017; Stierl 2016). Nevertheless, maritime displacement is not a new phenomenon, nor is it experienced only in the Mediterranean region. Perhaps most notoriously, many Vietnamese escaped to South East Asia by boat from the mid-1970s to the 1990s, leading to 'thousands and thousands of agonized deaths at sea' (Vo 2006, p. 1). During the 1970s and 1980s, there was also an increase in 'boat people' travelling to the USA from Cuba and Haiti across the Atlantic and the Florida Straits (Masud-Piloto 1996; Mitchell 1994). The 2000s saw growing numbers of boat migrations in the Middle East and Africa across the Red Sea via Djibouti and the Arabian Sea via Puntland (REF 2017, p. 5), while more recently ethnic Rohingya have escaped Myanmar by crossing the Andaman Sea to Indonesia (IOM 2018).

---

V. Squire (✉) • M. Stierl
University of Warwick, Coventry, UK
e-mail: v.j.squire@warwick.ac.uk; maurice.stierl.1@warwick.ac.uk

These examples are not exhaustive, but indicative of the significance of precarious migratory journeys across the sea. Indeed, maritime displacement is a long-standing and widespread phenomenon, occurring across various regions and posing significant dangers for people on the move (Mann 2016).

This chapter reflects on the specificity of maritime displacement as a form of precarious migration, while charting the ways in which existing research addresses the experiences of people on the move who cross the sea by boat. It focuses specifically on the Mediterranean region, primarily due to increased migrations and the plethora of analyses that have emerged in this context over recent years. However, it does so in terms that reflect both on the broader significance of maritime displacement more generally and that recognises the need to address boat journeys across the Mediterranean as involving longer sea and land journeys across diverse regions (Crawley et al. 2017; Squire et al. 2017). The first section reflects on the symbolic significance of the sea as a gendered and racialised site of governing migration and highlights maritime displacement as a highly precarious form of migration that is constituted as such through governing practices. The way in which this is concretely manifest in the context of the Mediterranean is examined in the second section, which traces maritime migration over several decades to show shifting routes and governing practices as indicative of how contestations over migration play out at sea. The final section reflects on both the symbolic and material significance of practices of managing displacement through search and rescue (SAR) and draws attention both to the growing military–humanitarian nexus as well as to the criminalisation of humanitarian civil society SAR organisations in the Mediterranean context. This, it argues, reflects a broader struggle over migration across the Mediterranean in which nongovernmental groups have become a critical force.

## The Sea as a Site of Governing Migration

As the racial of nature, the Sea's other side, its underside, represents all the built up fears about those racially characterised from the not-here: monsters from elsewhere, shadows from places unknown and threatening in their unknowability. (Goldberg 2017, p. 105)

In his discussion of migration via the sea, David Theo Goldberg points to the racialised imaginaries that are bound up with people coming via the sea 'from elsewhere,' who are conceived as 'monsters' and as 'threatening in their unknowability.' The fears provoked by longer-standing histories of people getting lost at sea 'without a trace' (Evans 2017) is in this regard not a process devoid of politics but one that is deeply embedded in the processes of racialisation that are integral to contemporary practices of governing migration (Squire 2020). More than just a racialised entity, however, the sea is also gendered. As Terri Gordon-Zolov and Amy Sodaro argue, enduring oceanic metaphors of 'engulfment, submersion and dissolution' motivate 'the construction of an armoured male body' as a defence against 'fears of the feminine' (2017, p. 16). To put it another way, gendered and racialised imaginaries of the sea pull between fear and control, between victimhood and heroism, with masculinity defined by the successful mastery of its unruly forces (see also Squire 2017, pp. 69–70). It is in this context that the death of people at sea becomes a moment whereby people who provoke fear can suddenly switch into those who provoke pity. The mourning of border deaths following a fatal shipwreck off the coast of Lampedusa is just one instance in which the outpouring of pity has been documented as persisting alongside hostile and restrictive policies towards people on the move (Albahari 2016; Rygiel 2016; Stierl 2016; Squire 2017). When rendered powerless in death—when emasculated and de-racialised, one might say—people on the move become benign rather than threatening figures as they succumb to the forces of the sea. Yet when people arrive to the shores of the EU alive, they provoke a series of gendered and racialised fears that are caught up in the symbolism of the sea's unruly dimensions.

Despite perceptions of an unruly materiality, often leading to assumptions that the sea is essentially 'ungovernable,' Philip Steinberg (2001) has importantly documented the highly regulated nature of ocean sites. This forms a backdrop for scholars of migration who have focused attention on the complex legal and regulatory frameworks and interventions through which maritime displacement is governed (e.g. Heller and Pezzani 2016). The territorial division of the sea into distinct jurisdictional areas is enshrined in the United Nations Convention on the Law of the Sea (UN 1982), while SAR obligations are enshrined in the International Convention for the Safety of Life at Sea (IMO 1974) and in the International Convention of Maritime Search and Rescue (IMO 1979). These form the legal basis for SAR operations (Jumbert 2018) and thus for maritime cooperation between states in the assistance of people on the move who cross the Mediterranean by boat (Pugh 2000).

Yet despite the existence of these legal frameworks, researchers in the field have pointed to the ways in which SAR obligations and those for disembarkation have lacked both the legal clarity and broader solidarity required to function effectively (Den Hertog 2012; Trevisanut 2010). It is in this context that scholars such as Violeta Moreno-Lax have engaged sustained legal analyses in the attempt to clarify SAR obligations (2011) and in order to draw attention to the ways in which such obligations relate to those enshrined in international refugee law (2012).

Indeed, over recent years, research at the intersections of international maritime and international refugee law have become increasingly critical, in a situation whereby the interdiction of people on the move, preventing them from completing journeys across the sea, has become a frequent occurrence (Heller and Pezzani 2012). Scholarship in the field has shown how practices of interdiction have persisted over time (Mann 2016), while also intensifying in the Mediterranean as 'rescue-through-interdiction/rescue-without-protection' has been redefined as a lifesaving device (Moreno-Lax 2018). In order to contextualise further how the sea has become a site of governing migration, we will now map the shifting scene of maritime displacement across the Mediterranean before going on to consider the humanitarian–military practices through which 'rescue-without-protection' has taken centre stage.

## The Shifting Scene of Maritime Displacement

Migration across the Mediterranean to southern European states has evolved over several decades, with modalities of migration and compositions of migratory groups varying over time and with new routes opening up as others have closed down. While precarious maritime journeys are thus not a novel phenomenon, crossings to Europe have become an increasingly contested political issue over recent years. Peaking in 2015, unprecedented levels of arrivals propelled the issue to the forefront of political debate and was 'reflected not only in images of human misery and suffering that dominated newspapers, TV screens and social media feeds but also in growing public fears about the perceived economic, security and cultural threats of increased migration to Europe' (Baldwin-Edwards et al. 2018, p. 2). This section highlights the ways in which scholars have assessed the relationship between maritime displacement and practices of governing migration, first by exploring the historical evolution of Mediterranean migration and subsequently by examining the three main geographical routes of maritime migration to the EU. In so doing, the analysis emphasises the significant, though not totalising, role that practices of governing migration play in constituting maritime displacement as a highly precarious form of migration (cf. Squire 2015). As such, we define maritime displacement as a form of migration that is the result of 'intersecting drivers and conditions of flight' (Squire et al. 2017), individual and collective migration projects (Andrijasevic 2010), and attempts to govern migration by the authorities of various EU institutions, Member States, and third countries (Vaughan-Williams 2015).

Research on the history of Mediterranean migration has shown that attempts by European states to govern migration have been constitutive factors in shaping trans-Mediterranean crossings (Mountz and Hiemstra 2012). In order to curtail labour migration in light of the 1973 oil crisis and the hike in unemployment, several western European states imposed more stringent visa restrictions on people from North Africa and Turkey (Bakewell and De Haas 2007). As scholars have argued, these restrictions on legal migration did not put an end to migration to Europe but were rather productive of increased unauthorised crossings, including those via the sea (Samers 2010; Boswell and Geddes 2010). In the context of the gradual consolidation of freedom of mobility within the EU and expanded immigration restrictions for citizens of the Global South during the late 1980s and 1990s, maritime migration became an increasingly common, politicised, and fatal phenomenon (Weber and Pickering 2011). In the 2000s a range of measures were introduced in order to thwart maritime and other irregularised forms of migration to the EU, including new surveillance regimes, the creation of the EU border agency Frontex (Neal 2009), and, crucially, agreements between the EU and North African states (Klepp 2010, 2011). During this period, Morocco, Tunisia, and Algeria criminalised irregular exits and intensified persecutions of human smugglers (Jakob and Schlindwein 2018), while Italy and Libya concluded a treaty in 2008 with the prime objective of reducing northbound migrations (Bialasiewicz 2012). Although some of these measures and agreements contributed to a decrease in Mediterranean crossings in 2009 and 2010, from 2011 the Arab uprisings 'reopened' the North African migration corridor (Heller et al. 2019). With the fall of several North African regimes and thus the (temporary) disintegration of what Ataç et al. (2015, p. 3) have referred to as the '(post-colonial) wardens of the European border regime,' as well as the increase in violent conflict in Syria and elsewhere that prompted millions to flee, a new phase of Mediterranean migration began (Perkowski 2016). Between 2014 and the end of 2017, a staggering 1,765,217 sea arrivals to the EU were recorded as well as 15,486 fatalities at sea (UNHCR 2018).

In view of this history of Mediterranean migration, research has drawn attention to the ways in which the constitution of maritime displacement reflects a struggle *over* migration, resulting from the interplay between restrictions on legal migration to Europe, the securitisation of migratory routes, as well as the needs and desires of many to move (Stierl 2016). Given this complex interplay that involves elements of coercion and choice in the phenomenon of precarious migration, several scholars have questioned its binary categorisation into voluntary or forced migration (Chimni 2009; Mountz 2010). For example, Alison Mountz and Jenna Loyd (2014) have pointed out how practices of governing migration impact on maritime displacement and migrant vulnerability by redirecting people on the move on to routes that are often lengthier, costlier, and more dangerous than previous ones and by contributing in this way to the rise in diverse practices of human smuggling (see also Albahari 2018). In the Mediterranean context, these dynamics have been examined along three geographical routes: the western, the eastern (or Aegean), and the central Mediterranean routes. These routes have frequently shifted over time, indicating how the interplay between migratory dynamics and attempts to govern migration manifest dynamically and materially in the Mediterranean.

The dynamics of western Mediterranean routes have evolved in the context of this struggle and in relation to attempts by Spanish, EU, and third-country authorities to govern migration (Andersson 2014; El Qadim 2018). As research has pointed out, when sea arrivals in Spain increased from the 1990s onwards, intensified Spanish–Moroccan collaborations in border enforcement temporarily reduced crossings, especially across the Strait of Gibraltar (De Haas 2008). Nonetheless, in the mid-2000s, migratory movements via the more dangerous Atlantic route brought increasing numbers of people to the Canary Islands, in turn leading to the first large-scale operation of Frontex in 2006 (Carrera 2007). As scholars such as Nick Vaughan-Williams (2008) and Bernd Kasparek (2010) have observed, these attempts by the EU and Spain to pre-empt movements to the Spanish archipelago effectively externalised and outsourced EU border practices to third countries, both by intervening within West African waters and by providing incentives to governments in the region to interdict departures. As previously seen, the subsequent decrease in migrant crossings to Spain did not last. About ten years on, in 2018, the western routes were identified as the most frequented Mediterranean migration routes (UNHCR 2018).

Similar to the western routes, precarious journeys resulting from the interplay between migratory dynamics and attempts to govern migration have been observed across the Aegean Sea (Jones 2016). Since its accession to the EU in 1981, Greece has been an entry point for people on the move via Turkey, though mainly across the land border. As several scholars have noted, the securitisation of the land border did not halt but rather redirected migration via the sea, especially from 2014 onwards (Topak 2014; Pallister-Wilkins 2015). Between 2015 and spring 2016, constituting what has been referred to either as the 'migration crisis' (New Keywords Collective 2016) or 'the long summer of migration' (Kasparek and Speer 2015), over one million people reached the Greek islands via perilous maritime routes. Comparable to measures taken in the two other Mediterranean regions, collaborations between the EU and Turkey, particularly through the so-called EU–Turkey deal of March 2016, increased Turkish interdictions of departures and prompted the plummeting of sea arrivals in Greece, at least temporarily (Stierl 2019).

The central Mediterranean routes to Italy and Malta have often been the focal point of maritime migration research, largely due to the high number of deaths that occur along these routes annually as well as the multiple interventions by EU and North African states in this region (Albahari 2015). Arrivals via the Strait of Sicily have increased over recent years—from an estimated average number of arrivals of 23,000 per year between 1997 and 2010 (McMahon and Sigona 2016) to about 64,000 in 2011, and even further to an annual average of around 156,000 between 2014 and 2017 (UNHCR 2018). In light of this increase in crossings, North African authorities, primarily those in Libya, were tasked with the role of containing departures as they had done before the Arab uprisings (Squire et al. 2017). These efforts, in combination with EU border enforcement operations, led to a significant decrease in arrivals to approximately 23,000 in 2018 (UNHCR 2018). Scholars in the field have noted how maritime migration has prompted both a military and a humanitarian response in the context of the central Mediterranean route (Cuttitta 2017). As we emphasise in the next section, the emergent military–humanitarian nexus not only implicates state actors but also civil society organisations, which have intervened at sea in the struggle over precarious migration and maritime displacement. This has turned the Mediterranean into an ever-more 'contested borderzone' over recent years (Squire 2011).

## The Contested Field of Maritime Displacement

As the analysis so far has highlighted, the Mediterranean is not an ungovernable space but rather the site of a struggle over migration where a humanitarian concern to save lives has morphed with border security concerns. Recent scholarship consists of two broad strands in this regard. First, scholars of humanitarianism have pointed to the ways in which humanitarian commitments have been *co-opted* by processes of securitisation, as border security concerns have increasingly trumped SAR obligations. For example, Maria Jumbert (2018) points to the ways in which arguments justifying the use of surveillance technologies have used rescue as a legitimating factor when required, while humanitarian missions have been utilised to prevent the arrival of people to EU shores. From another angle, Daniel Ghezelbash et al. (2018) have emphasised how suggestions that SAR acts as a 'pull' factor driving migration across the Mediterranean (and elsewhere) have led authorities to claim that an increase in border security measures are in effect humanitarian. Yet while such analyses of the co-optation of humanitarianism by security concerns remain critical, a second strand of research has emerged which goes further to emphasise the *inherent* linkage of humanitarian and security concerns in the management of migration across the Mediterranean (Pallister-Wilkins 2015; Vaughan-Williams 2015; Stierl 2018). In the context of maritime displacement, such works have emphasised the significance of a nexus of military and humanitarian practices in the emergence of SAR operations across the Mediterranean, beginning with the Italian Mare Nostrum operation in 2013–2014 and continuing in the activities of the European border agency, Frontex, and the military intervention of EUNAVFOR Med (e.g. Cuttitta 2014; Garelli and Tazzioli, 2018; Garelli et al. 2018; Heller and Pezzani 2016; Tazzioli 2016).

Analyses of military–humanitarian SAR operations across the Mediterranean are critical in highlighting the ways in which practices of governing maritime displacement produce precariousness at sea. Emphasising such operations as engaging in a politics of and over life, Martina Tazzioli argues that 'humanitarian spaces of containment' have been forged at the EU's borders in terms that reduce humanitarianism to a politics of rescue and that capture and channel the movements of people crossing the Mediterranean (Tazzioli 2016, p. 3). Going further, Charles Heller and Lorenzo Pezzani argue that this politics of rescue not only involves practices of assistance but also of non-assistance, with rescue, at times, in effect deadly and even 'murderous' (2016, p. 16). Indeed, that SAR operations play a key role in producing precariousness at sea is perhaps unsurprising, given the ways in which border security concerns are integral to such operations. As Cuttitta argues, from the very beginning the Mare Nostrum operation had a 'security mission' to capture smugglers, despite it being 'praised and supported' in its humanitarian emphasis on the provision of SAR to people in distress (2014, pp. 26–27). Nevertheless, when Mare Nostrum ended, the number of deaths in the Mediterranean dramatically increased (Garelli et al. 2018). This is unsurprising, given that the Frontex operation Triton was focused primarily on border management and the military EUNAVFOR Sophia operation was focused on tackling smuggling networks (see Ghezelbash et al. 2018). Despite being faced with SAR obligations, the mission of these operations was not directly to undertake SAR. Rather, such practices of governing maritime displacement were embedded in a broader and longer-standing preventative EU agenda on migration (Squire et al. 2017).

Although SAR operations and authorities have shifted over time, there remains a consistency to the symbolic significance of sea rescue for institutional actors that are tasked with the mission of protecting EU territorial borders and managing migration. On the one hand, rescuing people at sea bears proof of the state's continued power over 'life and death' (Cuttitta 2017, p. 791), while on the other it provides a 'good border spectacle' by which authorities are able retain an image of benevolence in the face of increasing numbers of border deaths (Tazzioli 2015, p. 66). Nevertheless, the ability of institutional actors to retain power over these symbolic markers of rescue came under challenge in the Mediterranean from 2014, as humanitarian civil society organisations stepped in as a response to states' failures to effectively provide SAR (Cusumano 2017; Stierl 2018). While initially such groups acted alongside SAR authorities within the Mediterranean, over time such relationships became more strained. In the summer of 2017, civil society SAR missions were faced with a Code of Conduct that required them to have armed judicial police on board, and although not all groups accepted the terms of the code, some were nevertheless subsequently subject to criminalisation based on accusations of collusion with smugglers (Tazzioli 2018, pp. 5–6). Moreover, as the EU made efforts to shift responsibility for SAR to the so-called Libyan coastguard and as disputes within the EU over disembarkation spiralled out of control, many SAR vessels were impounded or refused permission to sail, including aerial surveillance vehicles tasked with the work of monitoring the situation at sea and coordinating rescue operations (Squire 2018). Such responses on the part of EU authorities are indicative of the ways in which civil society intrusions within SAR operations across the Mediterranean disrupt the power of states in both symbolic and material terms, thus becoming a critical force in the maritime struggle over migration.

Indeed, there is a growing body of critical research that has precisely paid attention to the ways in which maritime displacement across the Mediterranean has emerged as a site of struggle and contestation in which nongovernmental groups are intimately bound. While scholars have raised caution about humanitarian civil society SAR initiatives that perform rescues at sea only to reinsert those who are rescued back into the hands of governing authorities (see Tazzioli 2018, pp. 8–9), they have also pointed to the critical role such groups play in constituting the Mediterranean as a site of struggle over migration (Cuttitta 2018; Stierl 2018, 2019; Squire 2020). For example, Cuttitta (2018) draws attention to the ways in which some groups seek to politicise actions around maritime displacement, rather than enacting SAR as part of a politically 'neutral' form of humanitarianism. Going further, Maurice Stierl distinguishes groups that enact rescue as an act of compassion to those that 'support, in solidarity, unruly enactments of the freedom of movement' (2018, p. 721). Moreover, activist scholars Charles Heller et al. (2017) have written about the ways in which border struggles at the maritime frontier have turned militarist technology against itself through 'disobedient' forms of surveillance and mobile lines of communication that operate 'beyond the gaze of sovereign control.' This includes the WatchTheMed network, which is an online platform mapping border deaths and rights violations (WatchTheMed n.d.), as well as the Alarm Phone, an activist project running a hotline for people in distress at sea that has engaged in over 2500 maritime emergency situations since October 2014 and that has repeatedly denounced state authorities for failing to render assistance effectively (Alarm Phone n.d.; Stierl 2016). More than simply providing an analysis of precarious migration and maritime displacement in the Mediterranean, such scholarship thus enacts forms of knowledge production that are directly enmeshed in the constitution of the sea as a contested borderzone.

## Conclusion

The Mediterranean has been the site of maritime displacement for several decades. Although migration patterns and routes have changed over time, they have always been the consequence of several intersecting factors, including the motivation of people to cross, the different drivers and conditions of flight, and the increasingly restrictive approaches to governing migration. As this chapter and the scholarly research reviewed therein have shown, only when these factors are considered in their inter-relation and complexity can we understand precarious migration across the sea. Viewed as such, the Mediterranean does not constitute an empty or ungovernable space but a contested field of action implicating legal and regulatory regimes, technologies of surveillance, and a range of vessels that crisscross it—migrant boats, state, and nongovernmental assets, as well as commercial and other ships. What emerges is a rather busy space, in which border deaths are not a 'quasi-natural' phenomenon or a product of 'fate' but a consequence of the interplay of precarious migrations and the practices that attempt to govern them.

As highlighted throughout this chapter, a plethora of academic studies on maritime displacement have emerged over recent years that reveal the various modalities of precarious migration across the Mediterranean. Further research is however required concerning the racialised, classed, and gendered dimensions of maritime migration, since it is also at sea that 'experiences of migration and displacement differ significantly, depending on how people are positioned in hierarchies of gender, race, class, age, religion, and sexuality' (Carastathis et al. 2018, p. 6). While important works have begun the task of addressing the gendered (Pickering and Cochrane 2012) and racialised (Saucier and Woods 2014) aspects of maritime displacement, much remains to be explored about how hierarchies of mobility are concretely manifest through maritime displacement. Adding nuance to the analysis in this way will also further problematise binary accounts of voluntary and forced migration, which fail to do justice to the complexity of migrant mobilities and maritime displacement in this context. Moreover, further research is also required concerning migratory dynamics that, while connected, precede migrant mobilities across the sea. Often overshadowed by the rather spectacular nature of maritime displacement, these dynamics involve long journeys across diverse regions before reaching the Mediterranean yet tend to be underacknowledged in scholarly research.

In September 2018, about five years after a large-scale shipwreck near the Italian island of Lampedusa on 3 October 2013 instigated the state-led military–humanitarian campaign Mare Nostrum, a shipwreck occurred off the coast of Libya. Based on testimonies of survivors, Médecins Sans Frontières (2018) informed the public about the capsized boat. This time, however, the story of the death of about one hundred people on the move from Africa was little noticed by the world media. Many of the bodies could not be recovered and will only feature as a rough estimate in statistics on maritime fatalities, if at all. That such deaths will continue to occur across the Mediterranean regions is inevitable if the increasingly restrictive approach to governing migration remains. As the history of Mediterranean migration has shown, if one route closes another opens, and people continue to cross by costlier and more dangerous means. So long as EU states continue to focus on the externalisation and outsourcing of border practices to third countries, and so long as attempts to delegitimise and criminalise the work of humanitarian civil society SAR organisations are sustained, the plight of the people on the move will remain invisible and unacknowledged and the Mediterranean will likely remain the deadliest borderzone in the world.

# References

Alarm Phone. (n.d.). Home page. Retrieved August 31, 2018, from https://alarmphone.org/en/

Albahari, M. (2015). *Crimes of peace: Mediterranean migrations at the world's deadliest border*. Philadelphia, PA: University of Pennsylvania Press.

Albahari, M. (2016). After the shipwreck: Mourning and citizenship in the Mediterranean, our sea. *Social Research: An International Quarterly, 83*(2), 275–294.

Albahari, M. (2018). From right to permission: Asylum, Mediterranean migrations, and Europe's war on smuggling. *Journal on Migration and Human Security* [Online]. Retrieved August 13, 2018, from https://doi.org/10.1177/2311502418767088

Andersson, R. (2014). *Illegality Inc*. Oakland, CA: University of California Press.

Andrijasevic, R. (2010). *Migration, agency and citizenship in sex trafficking*. Basingstoke, UK: Palgrave Macmillan.

Ataç, I., Kron, S., Schilliger, S., Schwiertz, H., & Stierl, M. (2015). Struggles of migration as in-/visible politics. Introduction. *Movements. Journal für kritische Migrations- und Grenzregimeforschung, 1*(2), 1–18.

Bakewell, O., & De Haas, H. (2007). African migrations: Continuities, discontinuities and recent transformations. In L. de Haan, U. Engel, & P. Chabal (Eds.), *African alternatives* (pp. 95–118). Leiden, Netherlands: Brill.

Baldwin-Edwards, M., Blitz, B., & Crawley, H. (2018). The politics of evidence-based policy in Europe's 'migration crisis'. *Journal of Ethnic and Migration Studies* [Online]. Retrieved September 13, 2018, from https://doi.org/10.1080/1369183X.2018.1468307

Bialasiewicz, L. (2012). Off-shoring and out-sourcing the borders of EUrope: Libya and EU border work in the Mediterranean. *Geopolitics, 17*(4), 843–866.

Boswell, C., & Geddes, A. (2010). *Migration and mobility in the European Union*. Basingstoke, UK: Palgrave Macmillan.

Carastathis, A., Kouri-Towe, N., Mahrouse, G., & Whitley, L. (2018). Introduction. *Refuge, 34*(1), 3–15.

Carrera, S. (2007, March 22). *The EU border management strategy: FRONTEX and the challenges of irregular immigration in the Canary Islands* (CEPS Working Document, No. 261).

Chimni, B. S. (2009). The birth of a 'discipline': From refugee to forced migration studies. *Journal of Refugee Studies, 22*(1), 11–29.

Crawley, H., Duvell, F., Jones, K., McMahon, S., & Sigona, N. (2017). *Unravelling Europe's 'migration crisis'*. Bristol, UK: Policy Press.

Cusumano, E. (2017). Emptying the sea with a spoon? Non-governmental providers of migrants search and rescue in the Mediterranean. *Marine Policy, 75*, 91–98.

Cuttitta, P. (2014). From the Cap Anamur to Mare Nostrum: Humanitarianism and migration controls at the EU's maritime borders. In C. Matera & A. Taylor (Eds.),

*The common European asylum system and human rights: Enhancing protection in times of emergencies* (pp. 21–37). Den Haag, Netherlands: Asser Institute.

Cuttitta, P. (2017). Delocalization, humanitarianism and human rights: The Mediterranean border between exclusion and inclusion. *Antipode, 50*(3), 783–803.

Cuttitta, P. (2018). Repoliticisation through search and rescue? Humanitarian NGOs and migration management in the central Mediterranean. *Geopolitics, 23*(3), 632–660.

De Haas, H. (2008). The myth of invasion—The inconvenient realities of African migration to Europe. *Third World Quarterly, 29*(7), 1305–1322.

Den Hertog, L. (2012, July 20). *Two boats in the Mediterranean and their unfortunate encounters with Europe's policies towards people on the move* (CEPS Papers in Liberty and Security in Europe, No. 48). Retrieved August 30, 2018, from https://papers.ssrn.com/sol3/papers.cfm?abstract_id=2117749

El Qadim, N. (2018). The symbolic meaning of international mobility: EU–Morocco negotiations on visa facilitation. *Migration Studies, 6*(2), 279–305.

Evans, B. (2017). Dead in the waters. In A. Baldwin & G. Bettini (Eds.), *Life adrift: Climate change, migration, critique* (pp. 59–78). London: Rowman & Littlefield.

Garelli, G., Heller, C., Pezzani, L., & Tazzioli, M. (2018). Shifting bordering and rescue practices in the central Mediterranean Sea, October 2013–October 2015. *Antipode, 50*(3), 813–821.

Garelli, G., & Tazzioli, M. (2018). The humanitarian war against migrant smugglers at sea. *Antipode, 50*(3), 685–703.

Ghezelbash, D., Moreno-Lax, V., Klein, N., & Opeskin, B. (2018). Securitisation of search and rescue at sea: The response of boat migration in the Mediterranean and offshore Australia. *International and Comparative Law Quarterly, 67*(2), 315–351.

Goldberg, D. T. (2017). Parting waters: Seas of movement. In A. Baldwin & G. Bettini (Eds.), *Life adrift: Climate change, migration, critique* (pp. 99–114). London: Rowman & Littlefield.

Gordon-Zolov, T., & Sodaro, A. (2017). Introduction: At sea. *Women's Studies Quarterly, 45*(1 & 2), 12–26.

Heller, C., & Pezzani, L. (2012). Forensic oceanography: '*Left-to-die boat*' (Part of the European Research Council project 'Forensic Architecture'; Centre for Research Architecture, Goldsmiths, University of London). Retrieved January 30, 2019, from http://migrantsatsea.files.wordpress.com/2012/04/forensic-oceanography-report-11april20121.pdf

Heller, C., & Pezzani, L. (2016, March). *Ebbing and flowing: The EU's shifting practices of (non-)assistance and bordering in a time of crisis* (Near Futures Online 1 'Europe at a Crossroads'). Retrieved January 20, 2019, from http://nearfuturesonline.org/wp-content/uploads/2016/03/Heller_Pezzani_Ebbing_2016.pdf

Heller, C., Pezzani, L., & Stierl, M. (2017, June). Disobedient sensing and border struggles at the maritime frontier of EUrope. *Spheres, Journal for Digital Cultures*. Retrieved August 31, 2018, from http://spheres-journal.org/disobedient-sensing-and-border-struggles-at-the-maritime-frontier-of-europe/

Heller, C., Pezzani, L., & Stierl, M. (2019). Toward a politics of freedom of movement. In R. Jones (Ed.), *Open borders: In defense of free movement* (pp. 51–76). Athens, GA: University of Georgia Press.
IMO (International Maritime Organisation). (1974). International Convention for the Safety of Life at Sea (SOLAS). Summary. Retrieved August 30, 2018, from http://www.imo.org/en/About/Conventions/ListOfConventions/Pages/International-Convention-for-the-Safety-of-Life-at-Sea-(SOLAS),-1974.aspx
IMO (International Maritime Organisation). (1979). International Convention of maritime search and rescue. Summary. Retrieved August 30, 2018, from http://www.imo.org/en/About/Conventions/ListOfConventions/Pages/International-Convention-on-Maritime-Search-and-Rescue-(SAR).aspx
IOM (International Organisation for Migration). (2018, April 27). IOM steps in to aid new Rohingya boat arrivals in Indonesia. Retrieved August 28, 2018, from https://www.iom.int/news/iom-steps-aid-new-rohingya-boat-arrivals-indonesia
Jakob, C., & Schlindwein, S. (2018). *Diktatoren als Türsteher Europas*. Bonn: BPB.
Jones, R. (2016). *Violent borders*. London: Verso.
Jumbert, M. G. (2018). Control or rescue at sea? Aims and limits of border surveillance technologies in the Mediterranean Sea. *Disasters*, forthcoming. https://doi.org/10.1111/disa.12286
Kasparek, B. (2010). Borders and populations in flux: Frontex's place in the European Union's migration management. In M. Geiger & A. Pécoud (Eds.), *The politics of international migration management* (pp. 19–140). London: Palgrave Macmillan.
Kasparek, B., & Speer, M. (2015). *Of hope, Hungary and the long summer of migration*. Retrieved January 30, 2019, from http://bordermonitoring.eu/ungarn/2015/09/of-hope-en/
Klepp, S. (2010). A contested asylum system: The European Union between refugee protection and border control in the Mediterranean Sea. *European Journal of Migration and Law, 12*(1), 1–21.
Klepp, S. (2011). A double bind: Malta and the rescue of unwanted migrants at sea, a legal anthropological perspective on the humanitarian law of the sea. *International Journal of Refugee Law, 23*(3), 538–557.
Mann, I. (2016). *Humanity at sea: Maritime migration and the foundations of international law*. Cambridge, UK: Cambridge University Press.
Masud-Piloto, F. R. (1996). *From welcomed exiles to illegal immigrants: Cuban migration to the US, 1959–1995*. Lanham and Oxford, UK: Roman & Littlefield.
McMahon, S., & Sigona, N. (2016). Boat migration across the central Mediterranean: Drivers, experiences and responses (MEDMIG Research Brief No. 3). Retrieved from http://www.medmig.info/research-brief-03-Boat-migration-across-the-Central-Mediterranean
Médecins Sans Frontières. (2018, September 10). Refugee Libya shipwreck survivors condemned to drown at sea or face arbitrary detention. Retrieved September 20, 2018, from https://www.msf.org/refugee-libya-shipwreck-survivors-condemned-drown-sea-or-face-arbitrary-detention

Mitchell, C. (1994). US policy toward Haitian boat people, 1972–93. *The Annals of the American Academy of Political and Social Science, 534*(1), 69–80.

Moreno-Lax, V. (2011). Seeking asylum in the Mediterranean: Against a fragmenting reading of EU Member States' obligations accruing at sea. *International Journal of Refugee Law, 23*(2), 174–220.

Moreno-Lax, V. (2012). Hirsi v Italy or the Strasbourg court v extraterritorial migration control? *Human Rights Law Review, 12*(3), 574–598.

Moreno-Lax, V. (2018). The EU humanitarian border and the securitisation of human rights: The 'rescue-through-interdiction/rescue-without-protection' paradigm. *Journal of Common Market Studies, 56*(1), 119–140.

Mountz, A. (2010). *Seeking asylum: Human smuggling and bureaucracy at the border.* Minneapolis, MN: University of Minnesota Press.

Mountz, A., & Hiemstra, N. (2012). Spatial strategies for rebordering human migration at sea. In T. Wilson & D. Hastings (Eds.), *A companion to border studies* (pp. 455–472). Malden: Blackwell Publishing Ltd.

Mountz, A., & Loyd, J. (2014). Constructing the Mediterranean region. *ACME, 13*(2), 173–195.

Neal, A. W. (2009). Securitization and risk at the EU border: The origins of FRONTEX. *Journal of Common Market Studies, 47*(2), 333–356.

New Keywords Collective. (2016). *Europe/crisis: New keywords of 'the crisis' in and of 'Europe'.* New Futures Online, Zone Books.

Pallister-Wilkins, P. (2015). The humanitarian politics of European border policing: Frontex and border police in Evros. *International Political Sociology, 9*(1), 53–69.

Perkowski, N. (2016). Deaths, interventions, humanitarianism, and human rights in the Mediterranean 'migration crisis'. *Mediterranean Politics, 21*(2), 331–335.

Pickering, S., & Cochrane, B. (2012). Irregular border-crossing deaths and gender: Where, how and why women die crossing borders. *Theoretical Criminology, 17*(1), 27–48.

Pugh, M. (2000). *Europe's boat people: Maritime cooperation in the Mediterranean* (Chaillot Paper 41). Retrieved August 30, 2018, from https://www.peacepalacelibrary.nl/ebooks/files/chai41e.pdf

REF (Research and Evidence Facility). (2017). Migration between the Horn of Africa and Yemen: A study of Puntland, Djibouti and Yemen. London and Nairobi, Kenya: EU Trust Fund for Africa (Horn of Africa Window) Research and Evidence Facility. Retrieved August 28, 2018, from https://www.soas.ac.uk/ref-hornresearch/research-papers/file122639.pdf

Rygiel, K. (2016). Dying to live: Migrant deaths and citizenship politics along European borders: Transgressions, disruptions and mobilizations. *Citizenship Studies, 25*(5), 545–560.

Samers, M. (2010). *Migration.* London: Routledge.

Saucier, P. K., & Woods, T. P. (2014). Ex aqua: The Mediterranean basin, Africans on the move, and the politics of policing. *Theoria, 61*(141), 55–75.

Squire, V. (Ed.). (2011). *The contested politics of mobility: Borderzones and irregularity.* Abingdon, UK: Routledge.

Squire, V. (2015). *Post/humanitarian border politics between Mexico and the US: People, places, things.* Basingstoke, UK: Palgrave Macmillan.

Squire, V. (2017). Governing migration through death in Europe and the US: Identification, burial, and the crisis of modern humanism. *European Journal of International Relations, 23*(3), 513–532.

Squire, V. (2018, July 4). Crew of NGO ship grounded in Malta sound alarm as search and rescue co-operation founders in Mediterranean. *The Conversation.* Retrieved August 31, 2018, from https://theconversation.com/crew-of-ngo-ship-grounded-in-malta-sound-alarm-as-search-and-rescue-co-operation-founders-in-mediterranean-99308

Squire, V. (2020). *Europe's migration crisis: Border deaths and human dignity.* Cambridge, UK: Cambridge University Press.

Squire, V., Dimitriadi, A., Perkowski, N., Pisani, M., Stevens, D., & Vaughan-Williams, N. (2017). *Crossing the Mediterranean Sea by boat: Mapping and documenting migratory journeys and experiences* (Final project report). Retrieved August 28, 2018, from http://www.warwick.ac.uk/crossingthemed

Steinberg, P. (2001). *The social construction of the ocean.* Cambridge, UK: Cambridge University Press.

Stierl, M. (2016). A sea of struggle: Activist border interventions in the Mediterranean Sea. *Citizenship Studies, 20*(5), 561–578.

Stierl, M. (2018). A fleet of Mediterranean border humanitarians. *Antipode, 50*(3), 704–724.

Stierl, M. (2019). *Migrant resistance in contemporary Europe.* London: Routledge.

Tazzioli, M. (2015). The desultory politics of mobility and the humanitarian-military border in the Mediterranean: Mare Nostrum beyond the sea. *REMHU, 23*(44), 61–82.

Tazzioli, M. (2016). Border displacements: Challenging the politics of rescue between Mare Nostrum and Triton. *Migration Studies, 4*(1), 1–19.

Tazzioli, M. (2018). Crimes of solidarity: Migration and containment through rescue. *Radical Philosophy, 2.01.* Retrieved August 31, 2018, from https://www.radicalphilosophy.com/commentary/crimes-of-solidarity

Topak, Ö. (2014). The biopolitical border in practice: Surveillance and death at the Greece–Turkey borderzones. *Environment and Planning D: Society and Space, 32*(5), 815–833.

Trevisanut, S. (2010). Search and rescue operations in the Mediterranean: Factor of cooperation or conflict? *The International Journal of Marine and Coastal Law, 25*(4), 532–542.

UN (United Nations). (1982). United Nations Convention on the Law of the Sea. Retrieved August 20, 2018, from http://www.un.org/Depts/los/convention_agreements/texts/unclos/unclos_e.pdf

UNHCR (Office for the United Nations High Commissioner for Refugees). (2018). *Operational portal, refugee situations: Mediterranean situation.* Retrieved October 3, 2018, from http://data2.unhcr.org/en/situations/mediterranean

Vaughan-Williams, N. (2008). Borderwork beyond inside/outside? Frontex, the citizen–detective and the War on Terror. *Space and Polity, 12*(1), 63–79.

Vaughan-Williams, N. (2015). *Europe's border crisis: Biopolitical security and beyond.* Oxford, UK: Oxford University Press.

Vo, N. M. (2006). *The Vietnamese boat people, 1954 and 1975–1992.* Jefferson, NC and London: McFarland & Company.

WatchTheMed. (n.d.). Home page. Retrieved August 31, 2018, from http://watchthemed.net

Weber, L., & Pickering, S. (2011). *Globalization and borders, death at the global frontier.* Basingstoke, UK: Palgrave Macmillan.

# 25

# Maintaining Health on the Move: Access and Availability for Displaced People

## Jennifer Cole

Health is a fundamental human right, established through the UN International Covenant on Economic, Social and Cultural Rights (UNGA 1966) and through resolutions of the World Health Organization (WHO 2015 (no longer available, see WHO 2017 for updated version)). Framing health as a human right creates a legal obligation on states to ensure access to timely, acceptable, and affordable healthcare of appropriate quality and to provide the conditions that support the underlying determinants of health such as safe and potable water, sanitation, sufficient food, housing, health-related information and education, and gender equality. Displaced people have as much right to health as anyone but providing the necessary services, materials, and environment can be challenging, particularly when people are on the move. This chapter explores how the legal requirement to provide the right to health is challenged throughout journeys of displacement through three lenses: the legal position of the individual; the agencies who are legally obliged and legally able to provide healthcare; and the different stages within the journey of displacement, which may influence both of the other two factors. Health conditions that are a direct consequence of displacement will be considered as well as existing health conditions that need to be managed within the context of displacement.

J. Cole (✉)
Royal Holloway, University of London, Egham, UK
e-mail: jennifer.cole@rhul.ac.uk

## Legal Status and Health

The UN International Covenant on Economic, Social and Cultural Rights is codified in the domestic law and constitutions of most countries but is prone to breakdown during times of economic turmoil, natural disasters, or conflict and instability—the very conditions that drive displacement. This can leave displaced individuals extremely vulnerable. For example, the Constitution of the Syrian Arab Republic holds the state accountable for the provision and protection of health to its citizens but this has failed to protect large numbers of the population during the conflict arising from the Arab Spring (Akbarzada and Mackey 2017). During preflight, in-flight, and postflight phases of their journeys, displaced people often find their access to healthcare services, pharmaceuticals, and medical treatments severely disrupted or unavailable (Thomas and Thomas 2004; Cole 2014). As a result, mortality rates can be up to sixty times higher than expected (UNHCR), with 1–5-year-olds particularly vulnerable due to poor nutrition and greater vulnerability to infectious disease (Yip and Sharp 1993), especially where vaccination programmes have broken down.

The degree of vulnerability to health risks to which an individual or community is exposed is strongly influenced by the legal status afforded them throughout their journey. Five distinct legal categories make up displaced populations (Jaff 2018): refugees, asylum seekers, internally displaced people (IDPs), stateless people, and returnees. Individual status within and between these categories is often fluid and overlapping.

For IDPs and returnees, the state in which they hold citizenship remains responsible for their health and healthcare provision, though it may be unable to provide this in the aftermath of natural disaster, instability, or conflict. Governments may be unwilling to provide for a persecuted minority, particularly one that it is forcibly trying to displace. As a result, IDPs in particular often experience far greater morbidity and mortality rates compared with both the surrounding population and with refugees from the same context who have crossed a border (Toole 1990); returnees may return to a disrupted area ahead of healthcare infrastructure being fully restored.

People displaced across borders, but who do not fall into the definition of being legal refugees, including asylum seekers and stateless people, become the responsibility of the host nation but may be unable to easily access or afford the healthcare available. This is particularly true if they enter the host country illegally or without appropriate documentation (Sullivan and Rehm 2005; Magalhaes et al. 2010; Woodward et al. 2013).

Legally recognised refugees, on the other hand, are the responsibility of the United Nations High Commissioner for Refugees (UNHCR) and this can afford them greater access to healthcare services, particularly those provided by international NGOs.

The impact legal status can have on displaced people has led to discussions on strengthening the global governance needed to protect them. In January 2017, the World Health Organization Executive Board requested that its Secretariat develop a Framework of Priorities and Guiding Principles to Promote the Health of Refugees and Migrants (WHO 2018) in collaboration with the International Organization for Migration (IOM) and UNHCR. This was followed in May 2017, at the World Health Assembly, by a call from the WHO Director-General to identify country practices, experiences, and lessons learned on the health of refugees and migrants and to contribute to the development of a draft Global Action Plan (GAP), *Promoting the Health of Refugees and Migrants*, discussed at the 72nd World Health Assembly in May 2019 (WHO 2019). The GAP defines the term 'refugee' as per the 1951 Convention Relating to the Status of Refugees and its 1967 Protocol, and notes that whilst there is no universally accepted definition of 'migrant,' anyone identifying as such is entitled to healthcare under the targets for universal health coverage set out in Sustainable Development Goal 3: 'Ensure healthy lives and promote well-being for all at all ages.' How these terms are applied will have important implications for access to and entitlement to healthcare and healthcare services for such people, influencing how individuals are treated within national legal frameworks and contexts (WHO 2018), including how the health of displaced people is monitored and reported. The GAP makes no distinct or additional provision for those displaced by climate change, who are currently excluded from the legal definition of a refugee, though there are calls from the academic community for this to be addressed in future iterations (Turner et al. 2019).

At present, a number of international monitoring bodies, including several departments of the UN, report violations to the existing protocols and attempt to keep states accountable (Gable and Meier 2013); these include UN committees including the Committee on the Rights of the Child, the Committee on the Elimination of all forms of Discrimination Against Women; and NGOs including the Red Cross. Nonetheless, many people remain unacceptably vulnerable.

## Healthcare Providers

The legal status of displaced people also affects who is able to provide healthcare services and materials during the journey of displacement and in receiving communities—whether settlement is temporary or permanent. The landscape of healthcare providers is complex, including governments, the World Health Organization, the International Committee of the Red Cross, the office of the United Nations High Commissioner for Refugees, nongovernmental organisations such as Oxfam, Medicins Sans Frontiers, USAID, Catholic Relief Services, and many others. Such organisations provide medical staff, facilities, provisions, checklists, guidance, or a combination of these. WHO, led by the department of Health Action in Crisis, provides a variety of support including material to assist with environmental health issues such as safe water, sanitation, and waste disposal, as well as medicine. The UN designates safe zones in which temporary facilities can be built, providing field hospitals, staff, equipment, and medical provisions (Akbarzada and Mackey 2017). The United Nations Population Fund provides a Minimum Initial Services Package for reproductive health in emergency situations (Krause et al. 2015). WHO also provides pre-stocked kits containing a full set of supplies appropriate to certain situations, for example, emergency diarrhoeal disease kits that contain pre-prepared materials to treat 100 severe or 400 mild cholera cases, or 100 cases of dysentery (WHO 2016).

However, while UNHCR has the legal mandate for the protection of refugees, a role that logically includes their protection from preventable causes of morbidity and mortality, there is no agency charged specifically with the protection and assistance of IDPs. While the ICRC can come to the aid of IDPs, UNHCR cannot. IDPs may not be able to benefit from assistance offered by NGOs unless their government allows it (Owoaje et al. 2016), potentially putting them at a disadvantage in comparison with legally recognised refugees, particularly in conflict- and disaster-affected zones where populations may be heavily dependent on humanitarian aid. In Columbia, taking healthcare services to IDPs through mobile clinics, rather than expecting them to seek out such services, has seen some success (Ramirez et al. 2016).

Weak or corrupt governments may be unwilling or incapable of providing support to IDPs, resulting in high dependence on foreign aid and NGOs to run a functioning healthcare service (UNHCR 1997). Problems can be particularly acute if a government will not allow international aid agencies to enter the country, as was the case in Burma/Myanmar prior to the dissolution of the ruling military junta in 2011 (Morrison et al. 2013). In weak states, conditions in IDP camps run by NGOs can be better than in the surrounding regions, giving little incentive for people to leave and return home or move on.

Health assistance provided to IDPs from local authorities can be problematic if the officially recognised legitimate authorities lack resources and agency to act efficiently. In Syria, for example, the Assistance and Coordination Unit was formed by the National Coalition of Syrian Revolution and Opposition Forces to coordinate provision of humanitarian assistance, including polio vaccinations, to IDPs and all remaining Syrian citizens, and to monitor health conditions in camps (Akbarzada and Mackey 2017). It is recognised as the legitimacy authority but operates within a deeply divided and challenging geopolitical region. ISIS appears to have established its own healthcare system in the same region—the 'Islamic State Health Services'—with hospitals and healthcare infrastructure, though the extent to this which this is a genuine humanitarian service rather than part of a propaganda and recruitment campaign has been the subject of fierce debate (Gardham 2015; Miller 2016). IDPs within Syria, in the meantime, are left vulnerable and at the mercy of the factions vying to control whatever new state emerges.

When displaced people cross borders, the pressure this can place on the healthcare systems of receiving countries can be immense, even when legal refugee status is recognised. In Turkey, for example, which was hosting close to 3.5 million Syrian refugees in 2017, the Turkish government provides free access to shelter, education, and healthcare (including emergency care) via any public hospital (Assi 2018). NGOs are able to apply for permits to run clinics that employ Syrian personnel, providing government-run facilities tailored for Syrians and employment opportunities for displaced people, with the costs covered by a government agency, the Disaster and Management Presidency AFAD (in Turkish, *AfetveAcil Durum YönetimiBaşkanlığı, AFAD*), but there is concern that the strain this has put on the Turkish public sector may be unsustainable. A further concern is that there may be poor mental health provision in such facilities due to the lack of Syrian mental health practitioners. Detection and treatment of mental health issues among displaced children and youth in particular lacks robust measurement and has been highlighted as an area where more work is desperately needed, particularly due to its known impact on lifelong morbidity and mortality (Gadeberg et al. 2017).

## Stages of Displacement

How the individual experiences displacement, and the different stages they experience, is a further defining factor in their health status. Displacement is rarely a single journey. It can involve a number of moves, with stops along the way of various durations and equally inconsistent levels of safety and security, resulting in a parallel journey from health to poorer health. Displacement often moves people from one vulnerable situation into another or even into more vulnerable circumstances (Strong et al. 2015) in which they may become stigmatised and marginalised, seen as an 'other' who threatens the health of local populations through diseases they carry and the drain they may become on local public health resources (Grove and Zwi 2006).

During this journey, their legal status (or lack thereof) can have a strong influence on local and international opinion of them: undocumented individuals can be stigmatised and even criminalised, increasing actual and perceived barriers to healthcare access (Hacker et al. 2015). This section categorises three stages of a displacement journey—preflight, in-flight, and postflight—and highlights the key health challenges individuals are likely to encounter during each stage. Whilst issues specific to legal refugees, documented migrants, recognised IDPs, and undocumented individuals are drawn out where possible, it is also important to recognise that the health conditions to which all persons on a journey of displacement are vulnerable are often independent of their official status, which may itself be fluid depending on the immediate context. In such an analysis, it may be neither appropriate nor possible to separate people by legal labels. In such cases, the general term 'displaced people' is used.

## Preflight

Preflight, individuals and affected communities may already have seen disruptions to their local healthcare systems. Displacement is often symptomatic of wider conflict or instability within society. Communicable and non-communicable diseases, mental health issues, including depression, insomnia, fatigue, anxiety, uncertainty, and disorientation (Richards et al. 2011; van Loenen et al. 2017), as well as incidences of conflict-related trauma all tend to increase some time before large-scale displacements begin (Cousins 2014). Where the conditions that drive displacement result in hospitals closing or being destroyed, individuals—including healthcare staff in the facilities that remain—may hoard pharmaceuticals and equipment that is out-of-date and unsafe. The cost of medicines may increase as supplies run low (Strong et al. 2015), with unregulated or black markets emerging to satisfy demand.

Pre-conflict, Syria had one of the most efficient public healthcare systems in the Middle East (Abu Sa'Da and Serafini 2013) but this was significantly impacted by the conflict (Cole 2014; Akbarzada and Mackey 2017), leaving a healthcare system that lacked well-trained and experienced medical personnel, appropriate medical equipment, and necessary financing (Cousins 2014). More than 50 per cent of Syrians who were being treated for chronic illness before the conflict reported having treatment disrupted (Kherallah et al. 2012), often because hospitals give priority to injured and wounded individuals rather than to the chronically ill, suggesting that the care of those with longer-term conditions needs more attention.

Damaged infrastructure and reduced public services result in a lack of basic survival necessities such as food, shelter, and clean water. In general, the degradation of these systems pushes people to leave (Thomas and Thomas 2004), but for those with existing health conditions, such as diabetes, for example, the factors influencing decisions to stay or leave can be more complex. Displaced Syrians and Palestinians have reported the lack of medical supplies for heart disease and chronic pain as a reason for flight (Strong et al. 2015), whereas during Hurricane Katrina in the USA, decisions to evacuate the city were delayed by concerns that appropriate healthcare for existing conditions might not be available along the journey or at destinations (US GAO 2006); this was exacerbated by healthcare insurance being tied to local healthcare systems and not transferable to others outside the local area (Brodie et al. 2006).

Regions within or close to conflict and disaster zones may be unable to procure or enable a regular supply of essential medicines—of the right amount, at the right place, at the right time, for the right condition, and at the right cost—to meet basic health needs, including for sexual and reproductive health (IAWG 2010). Separation from one's medical records, combined with a reluctance to register due to security worries, particularly for those from persecuted minorities or with tenuous legal status, can make consistency of care difficult.

This makes the choice over whether to leave or remain neither easy nor straightforward for such people. Poverty, restrictions on movement, and concerns about what might lie at the other end of a journey can trap people with existing health conditions in troubled locations for a considerable time before they are able to leave (IASC 2008). The most vulnerable are often minorities; poorly educated and with low or insecure incomes (Žilić 2015), poor nutritional status, and fragile health even before flight begins (Ramirez et al. 2016). People who remain in conflict-ridden areas, in particular, have poorer access to healthcare systems than IDPs and cross-border refugees who leave the affected area (Cambanis 2015).

Populations who are dependent on NGOs and humanitarian aid to provide services may have to move to where such organisations are able to operate, leaving those who are unable to do so without any access to healthcare at all.

## Inflight

During flight, poor living conditions, barriers to accessing healthcare, insecurity, and inconsistency of supply—of pharmaceuticals, medical advice, healthcare professionals, and treatment for chronic conditions—as well as difficulty maintaining a nutritious and balanced diet (Strong et al. 2015) can impede health and prevent the early diagnosis and treatment of emerging conditions such as cancers or heart disease. In addition, a tendency to focus on the health of children and the elderly can result in the health of (non-pregnant) younger adults being neglected.

This results in high rates of mortality and morbidity (APHA 1992), including undernutrition, diarrhoea, pneumonia, and malaria—often caused or exacerbated by a lack of preventative measures including vaccination programmes, washing facilities, mosquito nets, and pesticides. In temporary camps in particular, lack of sanitation infrastructure, waste disposal services, and crowded conditions can lead to an increase in infectious disease (Siriwardhana and Wickramage 2014) including sexually transmitted diseases (Matthews and Ritsema 2004) and HIV/AIDS (Austin et al. 2008). Disrupted vaccination schedules can lead to the re-emergence of diseases that have previously been eradicated from the affected region, such as polio (Cole 2014). In Chad, vaccination rates of 18 per cent have been recorded in IDPs, compared with 76 per cent in the surrounding urban population (Guerrier et al. 2009) and refugee children arriving in Germany have been found to have MMR vaccination rates 20 per cent lower than that of adult refugees (Jablonka et al. 2016).

Refugees and IDPs in overcrowded camps with inadequate facilities (Parameswaran and Wijesinghe 2012) are vulnerable to increased incidences of diarrhoea (Owoaje et al. 2016), whilst the impact of poor nutritional status on the immune system also increases their susceptibility to all infectious diseases, including respiratory infections, malaria, measles (Connolly et al. 2004), tetanus (Jablonka et al. 2017), diphtheria (Janati and Taylor 2018), and cholera (Shultz et al. 2009; Lam et al. 2017). High incidence of colds, flu, respiratory infections, urogenital infections, eye infections, scabies, diarrhoea, viral gastroenteritis, vomiting and dehydration, and dental problems have been recorded in refugees and IDPs seeking asylum in Europe (van Loenen et al. 2017). High (perceived) incidence of infectious disease among arriving immigrants can be used to dehumanise displaced people and frame them as disease carriers who pose a threat to receiving communities (Esses et al. 2013).

Financial constraints also impose barriers: those inflight may forego paying for medications needed to treat chronic conditions in order to pay for other basic necessities, increasing the risk of severe complications and making their conditions more difficult to manage. Stroke, for example, may result from foregoing medicines for diabetes and hypertension (Strong et al. 2015). Conditions such as type 2 diabetes that need specialised diets can also become difficult to manage adequately on the move. The situation may be made worse by the stress of displacement increasing behaviours such as smoking, which increase or exacerbate the risk of non-communicable diseases including hypertension, chronic obstructive pulmonary disease, and cancers (Owoaje et al. 2016; Weaver and Roberts 2010; Ezard et al. 2011; Akbarzada and Mackey 2017), though evidence for an increase in damaging behaviours is weak and sometimes contradictory. Women appear to be less likely to resort to such behaviour than men (Žilić 2015).

The lack of training available to healthcare workers in temporary or transit facilities has also been identified as a challenge. Integrated, compassionate, and culturally competent healthcare needs to be provided (van den Muijsenbergh et al. 2013; van Loenen et al. 2017). In camps and transit centres, displaced people and healthcare workers alike experience lack of trust, time, and information; even refugees with legal status prefer to continue their journey as soon as possible over receiving care. Accessing a doctor at busy border crossings and in longer-term reception centres can be difficult (van Loenen et al. 2017), while those who are temporarily or insecurely settled can find it difficult to navigate their way through local healthcare systems. Administrative problems in particular can hamper accessibility, as can language barriers (Flores 2005). Continuity of care may be impeded by a lack of information on previous treatment (e.g., no personal health record, or health records only in their first language), difficulties in recording medication taken during the journey, and a lack of knowledge about what care might be available at subsequent destinations.

Overcrowded situations in hotspots and transit centres, the large numbers of people arriving in such locations who show signs of illness, and the lack of sufficient medical staff all emphasise a need for guidelines and instruments for screening and rapid health assessment of newly arriving refugees and other migrants (Pottie et al. 2011; van Loenen et al. 2017). Better mechanisms are needed for integrating those still in the process of their journey with the available healthcare services at touchpoints along the way.

## Postflight

Displaced people are likely to arrive at final destinations in a state of mental and physical health that is poorer than average for the receiving communities in which they settle. Linguistic, literacy, and cultural barriers can leave people feeling isolated from receiving communities and from the healthcare workers and services available to them. Pharmaceutical and health services may be more difficult to access than in camps and transit centres; or may not be in place at all in conflict- and disaster-damaged regions that are yet to fully recover. Conditions in the new location, or on return to home regions, may be just as challenging as those experienced preflight. In Sri Lanka, for example, health systems in the Northern Province were in dire need of being rebuilt when displaced populations returned at the end of the conflict in 2009 (Siriwardhana and Wickramage 2014). Uncleared conflict ordinance, such as landmines in Ethiopia, can pose severe threats to the health and well-being of returnees.

Postflight populations often face poverty, particularly while legal status to remain is being arranged. In Sri Lanka, scores for post-traumatic stress disorder (PSTD), anxiety, and depression all decreased after obtaining residence permits and self-rated general health improved (though there was no such change for chronic conditions) (Lamkaddem et al. 2015). Housing conditions, employment, and other social factors also improved after obtaining the permits, suggesting correlation rather than causation, but without a permit conditions conducive to improved health can be more difficult to achieve.

Low socioeconomic status can be particularly problematic in countries with a two-tier medical system of lower-quality public sector provision for the poor and higher-quality private healthcare for the more affluent (Ramirez et al. 2016). More than three-quarters of refugees and IDPs report financial difficulties affecting their ability to obtain such diverse items as mobility devices and sanitary products. Approximately 10 per cent report lack of knowledge about where to seek help and others report physical challenges in travelling to a healthcare facility. Only 1.5 per cent reported no difficulties in obtaining care when needed (Strong et al. 2015).

Financial challenges mean that displaced children may have to work to supplement family income; in Colombia, this has been shown to disrupt their education (Ramirez et al. 2016). As strong links between educational attainment, especially of girls and women (Sen and Östlin 2008), and long-term individual and family health are well established, ensuring access to education for such children may be a particularly effective preventative health measure. Health knowledge that is usually passed on by family members, such as reproductive education, can be disrupted where separation has occurred and may need to be provided formally.

Whilst it is generally considered to be better to settle refugees within existing settlements and communities than in temporary camps if possible (Toole 1990), this can lead to real or perceived pressures on the receiving communities' resources, resulting in local resentment and tensions. Protracted conflicts erode community resilience strategies and challenge meaningful recovery. Promoting connectedness in the aftermath can be a key enabler of health (Siriwardhana and Wickramage 2014) and should be aimed for wherever possible; legal status can be key to making the right connections and can be a vital gateway to social inclusion and societal acceptance.

## Conclusions

This chapter has shown that the health situation facing individuals and communities along their journeys of displacement is challenging, but increasingly humanitarian aid and support, in countries, in transit camps, temporary settlements, and receiving communities, is identifying the unique needs of displaced populations and acting to meet them. While this can be challenged by the legal definition of their status or not as a refugee, new guidelines are being developed that aim to ease these challenges.

Increasing attention is being given to nutritional requirements of people along their journeys, not only with regard to protein–energy provision but also the provision of essential micronutrients including vitamins A–C, iron, and folic acid. The distribution of vitamin tablets to refugees and displaced people can have a positive impact on health, as can the distribution of appropriate cooking utensils and clean fuel that ensure food can be cooked safely.

In terms of infectious diseases, which can be exacerbated by the poor-quality environments found in transitional spaces and in low-quality housing in more permanent destinations, malaria nets, sanitation materials, disinfectant, soap, and contraceptives can all help to prevent health problems, as will ensuring that vaccine schedules are not disrupted or neglected. More positive understandings of mental health impacts, cognitive and sensory impairments, the needs of menstruating and pregnant women, and the importance of culturally appropriate and sensitive healthcare have all, rightly, received more attention in recent years than had previously been the case.

Finally, it is important to stress that while refugees, cross-border migrants, and internally displaced people face many health challenges, they should not be considered to be, or framed as, disease carriers who threaten receiving populations (Burnett and Peel 2001; Harper and Raman 2008), nor should they be criminalised: it helps no-one for media and political pressure to be aimed more on health services that will 'protect' the receiving population than on tending to the needs of those arriving (Fassil 2000). Displaced people face complex and challenging health journeys and need support and understanding to ensure they arrive at a final destination of security and robust health.

## References

Abu Sa'Da, C., & Serafini, M. (2013). Humanitarian and medical challenges of assisting new refugees in Lebanon and Iraq. *Forced Migration Review, 44*, 70–73.

Akbarzada, S., & Mackey, T. K. (2017). The Syrian public health and humanitarian crisis: A 'displacement' in global governance? *Global Public Health, 13*(7), 914–930.

APHA (American Public Health Association). (1992). *The health of refugees and displaced persons: A public health priority*. American Public Health Association. Retrieved February 1, 2019, from https://www.apha.org/policies-and-advocacy/public-health-policy-statements/policy-database/2014/07/29/10/34/the-health-of-refugees-and-displaced-persons-a-public-health-priority

Assi, R. (2018). K-7 access to health and healthcare: The case of Syrian refugees in Turkey. *The European Journal of Public Health, 28*(Suppl. 1), cky044-007.

Austin, J., Guy, S., Lee-Jones, L., McGinn, T., & Schlecht, J. (2008). Reproductive health: A right for refugees and internally displaced persons. *Reproductive Health Matters, 16*(31), 10–21.

Brodie, M., Weltzien, E., Altman, D., Blendon, R. J., & Benson, J. M. (2006). Experiences of Hurricane Katrina evacuees in Houston shelters: Implications for future planning. *American Journal of Public Health, 96*(8), 1402–1408.

Burnett, A., & Peel, M. (2001). Asylum seekers and refugees in Britain: Health needs of asylum seekers and refugees. *BMJ: British Medical Journal, 322*(7285), 544–547.

Cambanis, T. (2015, November 27). In Syria, you can't go home again. *Foreign Policy*. Retrieved June 7, 2018, from https://foreignpolicy.com/2015/11/27/in-syria-you-cant-go-home-again/

Cole, J. (2014). Conflict, post-conflict and failed states: Challenges to healthcare. *RUSI Journal, 159*(5), 14–18.

Connolly, M. A., Gayer, M., Ryan, M. J., Salama, P., Spiegel, P., & Heymann, D. L. (2004). Communicable diseases in complex emergencies: Impact and challenges. *The Lancet, 364*(9449), 1974–1983.

Cousins, S. (2014). Experts sound alarm as Syrian crisis fuels spread of tuberculosis. *BMJ: British Medical Journal (Online), 349*. https://doi.org/10.1136/bmj.g7397.

Esses, V. M., Medianu, S., & Lawson, A. S. (2013). Uncertainty, threat, and the role of the media in promoting the dehumanization of immigrants and refugees. *Journal of Social Issues, 69*(3), 518–536.

Ezard, N., Oppenheimer, E., Burton, A., Schilperoord, M., Macdonald, D., Adelekan, M., … van Ommeren, M. (2011). Six rapid assessments of alcohol and other substance use in populations displaced by conflict. *Conflict and Health, 5*(1), 1.

Fassil, Y. (2000). Personal views: Looking after the health of refugees. *BMJ: British Medical Journal, 321*(7252), 59.

Flores, G. (2005). The impact of medical interpreter services on the quality of health care: A systematic review. *Medical Care Research and Review, 62*(3), 255–299.

Gable, L., & Meier, B. M. (2013). Global health rights: Employing human rights to develop and implement the Framework Convention on Global Health. *Health and Human Rights, 15*(1), 17–31.

Gadeberg, A. K., Montgomery, E., Frederiksen, H. W., & Norredam, M. (2017). Assessing trauma and mental health in refugee children and youth: A systematic review of validated screening and measurement tools. *European Journal of Public Health, 27*(3), 439–446.

Gardham, D. (2015). Islamic State creates jihadi health service. *BMJ: British Medical Journal (Online), 350*. https://doi.org/10.1136/bmj.h3487.

Grove, N. J., & Zwi, A. B. (2006). Our health and theirs: Forced migration, othering, and public health. *Social Science & Medicine, 62*(8), 1931–1942.

Guerrier, G., Zounoun, M., Delarosa, O., Defourny, I., Lacharite, M., Brown, V., & Pedalino, B. (2009). Malnutrition and mortality patterns among internally displaced and non-displaced population living in a camp, a village or a town in Eastern Chad. *PloS one, 4*(11), e8077.

Hacker, K., Anies, M., Folb, B. L., & Zallman, L. (2015). Barriers to health care for undocumented immigrants: A literature review. *Risk Management and Healthcare Policy, 8*, 175.

Harper, I., & Raman, P. (2008). Less than human? Diaspora, disease and the question of citizenship. *International Migration, 46*(5), 3–26.

IASC. (2008). *Humanitarian action and older persons: An essential brief for humanitarian actors*. New York, NY and Geneva, Switzerland: ISAC.

IAWG (Inter-agency Working Group on Reproductive Health in Crises). (2010). *Inter-agency field manual on reproductive health in humanitarian settings: 2010 revision for field review*. Geneva, Switzerland: World Health Organization.

Jablonka, A., Behrens, G. M. N., Stange, M., Dopfer, C., Grote, U., Hansen, G., … Happle, C. (2017). Tetanus and diphtheria immunity in refugees in Europe in 2015. *Infection, 45*(2), 157–164.

Jablonka, A., Happle, C., Grote, U., Schleenvoigt, B. T., Hampel, A., Dopfer, C., … Behrens, G. M. N. (2016). Measles, mumps, rubella, and varicella seroprevalence in refugees in Germany in 2015. *Infection, 44*(6), 781–787.

Jaff, D. (2018). Mental health needs in forcibly displaced populations: Critical reflections. *Medicine, Conflict and Survival, 34*(1), 10–12.

Janati, N., & Taylor, E. (2018, 22 January). *Yemen: Children bear the brunt of the worst diphtheria outbreak for a generation.* Save the Children Press Release.

Kherallah, M., Alahfez, T., Sahloul, Z., Dia Eddin, K., & Jamil, G. (2012). Health care in Syria before and during the crisis. *Avicenna Journal of Medicine, 2*(3), 51–53.

Krause, S., Williams, H., Onyango, M. O., Sami, S., Doedens, W., Giga, N., ... Tomczyk, B. (2015). Reproductive health services for Syrian refugees in Zaatri camp and Irbid city, Hashemite Kingdom of Jordan: An evaluation of the minimum initial services package. *Conflict and Health, 9*(S1), S4.

Lam, E., Al-Tamimi, W., Russell, S. P., Butt, M. O. I., Blanton, C., Musani, A. S., & Date, K. (2017). Oral cholera vaccine coverage during an outbreak and humanitarian crisis, Iraq, 2015. *Emerging Infectious Diseases, 23*(1), 38–45.

Lamkaddem, M., Essink-Bot, M.-L., Deville, W., Gerritsen, A., & Stronks, K. (2015). Health changes of refugees from Afghanistan, Iran and Somalia: The role of residence status and experienced living difficulties in the resettlement process. *The European Journal of Public Health, 25*(6), 917–922.

Magalhaes, L., Carrasco, C., & Gastaldo, D. (2010). Undocumented migrants in Canada: A scope literature review on health, access to services, and working conditions. *Journal of Immigrant and Minority Health, 12*(1), 132.

Matthews, J., & Ritsema, S. (2004). Addressing the reproductive health needs of conflict-affected young people. *Forced Migration Review, 19*(19), 6–8.

Miller, M. (2016). Syrians suffer from war on healthcare. *The Century Foundation.* Retrieved June 2, 2018, from https://tcf.org/content/commentary/syrians-suffer-war-health-care/

Morrison, J., Cullison, T., Hiebert, M., Summers, T., & Hammergren. L. (2013). *Rehabilitating health in the Myanmar transition. 'This is the time we have been dreaming of for decades'.* A Report of the CSIS Global Health Policy Center. Centre for Strategic and International Studies.

Owoaje, E. T., Uchendu, O. C., Ajayi, T. O., & Cadmus, E. O. (2016). A review of the health problems of the internally displaced persons in Africa. *Nigerian Postgraduate Medical Journal, 23*(4), 161.

Parameswaran, A., & Wijesinghe, P. R. (2012). Was there a disparity in age appropriate infant immunization uptake in the theatre of war in the north of Sri Lanka at the height of the hostilities?: A cross-sectional study in resettled areas in the Kilinochchi district. *BMC International Health and Human Rights, 12*(1), 26.

Pottie, K., Greenaway, C., Feightner, J., Welch, V., Swinkels, H., Rashid, M., ... Zlotkin, S. (2011). Evidence-based clinical guidelines for immigrants and refugees. *Canadian Medical Association Journal, 183*(12), E845–E925.

Ramirez, J., Barrera, A., & Franco, H. D. (2016). The effect of conflict and displacement on the health of internally displaced people: The Colombian crisis. *University of Ottawa Journal of Medicine, 6*(2), 26–29.

Richards, A., Ospina-Duque, J., Barrera-Valencia, M., Escobar-Rincón, J., Ardila-Gutiérres, M., Metzler, T., & Marmar, C. (2011). Post-traumatic stress disorder, anxiety and depression symptoms, and psychosocial treatment needs in Colombians internally displaced by armed conflict: A mixed-method evaluation. *Psychological Trauma: Theory, Research, Practice and Policy, 3*(4), 384–393.

Sen, G., & Östlin, P. (2008). Gender inequity in health: Why it exists and how we can change it. *Global Public Health, 3*(S1), 1–12.

Shultz, A., Omollo, J. O., Burke, H., Qassim, M., Ochieng, J. B., Weinberg, M., ... Breiman, R. F. (2009). Cholera outbreak in Kenyan refugee camp: Risk factors for illness and importance of sanitation. *The American Journal of Tropical Medicine and Hygiene, 80*(4), 640–645.

Siriwardhana, C., & Wickramage, K. (2014). Conflict, forced displacement and health in Sri Lanka: A review of the research landscape. *Conflict and Health, 8*(1), 22.

Strong, J., Varady, C., Chahda, N., Doocy, S., & Burnham, G. (2015). Health status and health needs of older refugees from Syria in Lebanon. *Conflict and Health, 9*(1), 12.

Sullivan, M. M., & Rehm, R. (2005). Mental health of undocumented Mexican immigrants: A review of the literature. *Advances in Nursing Science, 28*(3), 240–251.

Thomas, S. L., & Thomas, S. D. M. (2004). Displacement and health. *British Medical Bulletin, 69*(1), 115–127.

Toole, M. J. (1990). Refugee and displaced populations in developing countries. *JAMA, 263*, 3296–3302.

Turner, R., & PLOS Medicine editors. (2019). Migrants and refugees: Improving health and well-being in a world on the move. *PLoS Med, 16*(7), e1002876.

UNGA (United Nations General Assembly). (1966). International covenant on economic, social and cultural rights. *United Nations, Treaty Series, 993*(3).

UNHCR. (1997). *Refugee operations and environmental management: Selected lessons learned.* Engineering and Environmental Services Section. UNHCR. Retrieved June 7, 2018, from http://www.unhcr.org/protect/PROTECTION/3b03b2754.pdf

US GAO (Government Accountability Office). (2006). *Transportation-disadvantaged populations: Actions needed to clarify responsibilities and increase preparedness for evacuations* (GAO-07-44). Washington, DC: US Government Accountability Office.

van Loenen, T., van den Muijsenbergh, M., Hofmeester, M., Dowrick, C., van Ginneken, N., Mechili, E. A., ... Lionis, C. (2017). Primary care for refugees and newly arrived migrants in Europe: A qualitative study on health needs, barriers and wishes. *The European Journal of Public Health, 28*(1), 82–87.

van den Muijsenbergh, M., van Weel-Baumgarten, E., Burns, N., O'Donnell, C., Mair, F., Spiegel, W., ... MacFarlane, A. (2013). Communication in cross-cultural consultations in primary care in Europe: The case for improvement. The rationale for the RESTORE FP 7 project. *Primary Health Care Research and Development, 15*(2), 122–133.

Weaver, H., & Roberts, B. (2010). Drinking and displacement: A systematic review of the influence of forced displacement on harmful alcohol use. *Substance Use & Misuse, 45*(13), 2340–2355.

WHO (World Health Organization). (2016). *WHO donates emergency medical supplies for the response in Borno State, Nigeria.* World Health Organization. Retrieved June 2, 2018, from http://www.who.int/features/2016/emergency-supplies-nigeria/en/

WHO (World Health Organization). (2017). *Human rights and health.* World Health Organization. Retrieved January 31, 2020, from https://www.who.int/en/news-room/fact-sheets/detail/human-rights-and-health

WHO (World Health Organization). (2018). *Draft 4+ global action plan to promote the health of refugees and migrants.* World Health Organization. Retrieved December 19, 2018, from https://www.who.int/migrants/Global_Action_Plan_for_migration.pdf?ua=1

WHO (World Health Organization). (2019). *Promoting the health of refugees and migrants: Draft global action plan, 2019–2023* (No. A72/25). Retrieved October 31, 2019, from https://apps.who.int/gb/ebwha/pdf_files/WHA72/A72_25-en.pdf

Woodward, A., Howard, N., & Wolffers, I. (2013). Health and access to care for undocumented migrants living in the European Union: A scoping review. *Health Policy and Planning, 29*(7), 818–830.

Yip, R., & Sharp, T. W. (1993). Acute malnutrition and high childhood mortality related to diarrhea: Lessons from the 1991 Kurdish refugee crisis. *JAMA, 270*(5), 587–590.

Žilić, I. (2015). Effect of forced displacement on health. *Radnimaterijali EIZ-a, 3*, 1–41.

# Part IV

## Section Four: Traces of Displacement

# 26

## Intervention: Disasters and Displacement: When There Is No Time to Stop

### Ayesha Siddiqi

In the early hours of the morning of 1 April 2017, residents of the town of Mocoa, the capital city of the Colombian department of Putumayo on the border with Ecuador, awoke to the sights and sounds of slashing rain, howling winds, large boulders rolling down hills and into people's homes, and walls of muddy brown water sweeping away cars, trees, and even houses. It was later estimated that this deadly landslide in Mocoa resulted in the deaths of about 400 people, displaced entire neighbourhoods, and left immense destruction in its wake.

In October of the following year, I was doing fieldwork in some of the worst hit areas, lying desolate since the landslide. Amongst the abandoned houses and caution signs, asking people not to return to their homes, I also came across two single lost shoes. The body of each of these shoes was buried deep within the layers of mud and rocks brought down by the landslide but the soles were clearly visible. I was not quite sure why they were there until I interviewed a young woman who emphasised her most vivid memory from the day she fled her home.

---

A. Siddiqi (✉)
Department of Geography, University of Cambridge, Cambridge, UK
e-mail: as3017@cam.ac.uk

What do you remember from that day when you left your home?

**I was running so fast my shoes came off. There was no time to stop, not even to pick them up.**

# 27

# Antipodean Architectures of Displacement

## Anoma Pieris

Policies on the containment, accommodation, or detention of displaced persons are closely linked to a new, neoliberal politics of border security that seeks to manage population flows in a supposedly borderless world. In architecture these issues are taken up by a small coterie of scholars who look at the politics of offshore detention centres, refugee camp design, and border politics (Anderson and Ferng 2013; Pugliese 2008). By highlighting the kinds of spatial violence displaced populations are subjected to, they align themselves with a broader discourse on borders (Wastl-Walter 2011) and global conflicts, for example, in Sarajevo (Herscher 2010), the Middle East, and elsewhere (Franke and Weizman 2014; Kenzari 2011; Bevan 2016). Architecture has been identified as one of the biopolitical technologies applied in refugee camp environments revisited most recently by Katz et al. (2018).

Architecture's contribution to the theme of displacement crosses numerous mediums, most notably in international exhibitions like the Museum of Modern Art, New York (MoMA 2016–2017) exhibition *Insecurities* and the exhibition *Exit* by architects Diller+Scofidio and Renfro with Paul Virilio (2008–2009). Architectural practitioners have also been attracted by the humanitarian design responses anticipated in UNHCR refugee camp facilities and several innovative solutions have been tested, applied, critiqued (Kennedy

A. Pieris (✉)
Faculty of Architecture Building and Planning, The University of Melbourne, Parkville, VIC, Australia
e-mail: apieris@unimelb.edu.au

2008), and exhibited in prominent architectural biennales such as the 15th International Architecture Exhibition 2016, in Venice, curated by Alejandro Aravena and themed 'Reporting from the front.'

Despite these many ethical considerations of an ongoing crisis in refugee accommodation, such issues have been poorly historicised in the architecture disciplines. Canonical texts, predicated on static and materially resilient structures, look for stable populations and buildings. Advancements in this area are left to urban planners or policymakers who seek to integrate refugee communities in cities or communities, in approaches not necessarily mediated by aesthetics. Such preferences are surprising given that 'displacement' and the associated violence, whether of Indigenous-Aboriginal populations or of settler colonists, is the basis of settler colonialism; and émigré architects, fleeing Europe during the Second World War, are key figures in the modernist canon. Displacement seemingly underwrites architectural histories, particularly in settler societies where ontological security is historically tenuous or contrived. Not only does architecture assume an exaggerated value as a technology for emplacing settler populations, but it is tied to the expropriation and demarcation of property that normalises imported cultures in a substantially different environmental context from the one in which that culture was originally conceived.

As a settler society, continuously populated by immigrants, Australia offers unique insights into the physical traces of displacement across a specific geography over time. Its physical spaces and place-consciousness are constantly reconfigured by waves of displaced persons, internally due to colonial forces or by the passage to or from the continent, further intensified due to its small population when compared to comparable land masses in Europe or America. Consequently, from among several scholars who grapple with these issues globally, the work of a few Australians in the spatial disciplines stands out. Social difference has already been raised by geographers Fincher and Jacobs (1998), highlighting the complex social identities that are spatialised in global cities. Jane M. Jacobs' (1996) work on postcolonial and immigrant spatialities and Indigenous identities is an early contribution to this discourse. Stephen Cairns (2004) and Mirjana Lozanovska (2016, 2019) have theorised migration for the architecture discipline. The flawed progress of Australia's multicultural policy (Hage 2003; Gunew 2003) has been an incentive for much of this work. While their interest in and attention to seemingly marginal or minority spaces appears to fall outside Australia's modernist canon, these authors convey the fundamental instability of a settler colonial teleology. Narratives of displacement threaded through Australia's built environment offer a different set of insights into its history and historiography; insights that are buried in more populous settler societies where these forces have long been normalised. This chapter interprets 'displacement' architectures as including those 'total environments' in which people forcibly removed from secure ontological spaces are confined for a given period of time. It argues that 'displacement' as a social phenomenon undergirds settler institutional and architectural histories, particularly of those used as spaces of quarantine, and examines the case of Australia to illustrate this point. Making these connections is important because despite their ubiquity, architectures used to punish, process, and assimilate errant citizens, racially different Aboriginal people, or new non-Anglophone immigrants into Australia (or to the more affluent settler societies globally) have been poorly represented so far. Sites of incarceration and human displacement lie outside the progressive settler teleology of 'white' settlement as facilitating liberal democratic values fundamental to nation-building. Troubling practices of social and racial differentiation on which those values were contingent have been repressed in the construction of 'settler'—architectural historiographies.

A looser interpretation of displacement within architectural histories documents the two-fold process by which immigrant British settlers modelled urban centres and rural pastures after the faraway motherland and later, during the twentieth century, cultivated strains of modernism derived from endemic features of the precolonial landscape. Both these aesthetic processes, either of estrangement or engagement, often excluded or selectively included materialities perceived as morally or culturally subordinate to Anglo-Celtic cultural patrimony. For example, in constructing a hegemonic narrative for an identifiable Australian architecture, derived from European precedents but made distinctive through adaptation to local climate and geography, the material cultures of Indigenous populations were initially marginalised. Belated recognition of the influence of Indigenous cultures, usually as design orientations by elite Anglo-Australian architects often reduced or suppressed aspects of their complex cultural past, including a past of genocide, exclusion, and incarceration. Failure to acknowledge this racially different population within Australian architectural historiography mirrored attitudes to racial or social difference in institutional and political culture at the time. However, these exclusions had also much to do with the aesthetic preferences of the discipline, where authored or permanent works and perceptions of an elite practice repressed those problematic institutional legacies. Among them, the four spatial histories of segregation examined in this chapter, of the differential containment of certain categories of displaced human subjects, demonstrate how certain institutionalised physical environments act as socio-spatial counterpoints to the progressive settler teleology. These observations are applicable across settler environments, where forms of spatial violence imposed on socially or racially differentiated populations regarded by governments as potentially dissenting minorities is normalised. These groups face exceptional levels of social and spatial repression alongside the political and legal systems that constrain them.

Socio-spatial patterns of repression recur across several displacement histories in Australia, where their repetition exposes unresolved ontological tensions in the settler psyche. They appear as sites of institutionalisation, which would ordinarily attract architectural histories regarding their form, typology, and programme, adaptations of which continue to inform contemporary professional practice. Their elision from histories of institutional architecture can be understood in two ways. These institutional types seem more closely aligned with the nineteenth-century institutions critiqued by Michel Foucault, most notably the prison (1979); however, their replication in an Antipodean continent and away from populous urban centres emphasise their double displacement. They enter a field of political spaces whose opacity helps perpetuate the myth of everyday civility and cultural homogeneity. My examples include the convict prison, the Aboriginal mission, the prisoner of war and internment camp, and the immigrant reception centre, the latter being the precursor to Australia's more recent, notorious, offshore detention centres for 'illegal' asylum seekers. These remote spaces, which reduce and regulate their human captives, embed forms of structural exclusion and dehumanisation as a normative practice, traceable in the histories of many comparable settler colonies. Together they form a carceral taxonomy that undergirds social values predicated on the construction of a 'white' settler population conceived as a spatial extension of an exogenous parent state. This is true, for example, for the settler colonies of North America, Australia, and New Zealand but also to varying degrees for other colonial environments where Aboriginal genocide enabled impositions and erasures far greater than in populous indigenous societies. Anglophone Australian identity was preserved by the White Australia Policy from 1901–1973, an era after Federation when legislation restricted non-European immigrants' arrivals into Australia (Jupp 2002).[1] The end of these policies and investment in multiculturalism has not necessarily erased its damaging legacies.

If we focus on Australia as exemplifying these traces, we see the continuous serial production of spaces of quarantine. Historically, encounters with waves of displaced persons have heightened postcolonial socio-spatial anxieties. They travel across race distinctions, reinforcing settler categories of class, culture, professionalisation, and perceived national loyalty. Fresh generations of immigrants, even of diverse cultural backgrounds, while acknowledging

Aboriginal dispossession, appear to buy into these settler values as a condition of being emplaced (James 2012, pp. 259–260). In settler colonial environments, or in any colonial environment where alien norms and forms are imposed, emplacement appears as an exaggerated value. However, due to Australia's distance from its parent geography, its vastness, small population, and location in the Asia Pacific, these values manifest in very particular ways.

In this chapter, I interpret the defensive incarceration of displaced persons as a response to a form of agoraphobia, following the prominent Australian scholar Paul Carter's (2002) reflection on the term's historiography and poetics. Carter describes a sensibility based on the fear of open spaces and of crowds. He examines the interpretations of various philosophers and artists, highlighting etymological roots in the *agora*, the meeting place or assembly in ancient Greece. Through the work of Sigmund Freud, among several other authors, Carter tracks agoraphobia across individual psychosis and generalised estrangement leading to urban neuroses (2002, p. 35). Perhaps the most useful interpretation for this chapter is the alignment of space phobias with ontological insecurity (Laing 1965), which characterises settler societies. Displacement, whether of early British settlers to Australia or of the histories they repressed in staking out their historical territory, are rooted in displacement-related ontological crises. The fact that migrant populations who first arrived by boat have turned hostile to new generations of so-called 'boat people' is an underlying irony in Australian society and politics. Discomfort with displacement appears as a manifestation of insecure or increasingly destabilised settler ontologies.

Far from the fear of populous crowded spaces or of the frenetic pace of modern urban life, the Australian version of agoraphobia had to do with the continent's vastness and peripheral distribution of fertile land. Fear of the seeming alien-ness of the central Australian desert was accompanied by ignorance of its inhabitants. Numerous pathogenic beliefs regarding morality, race, and cultural difference were overlaid on that space. Underlying the construction of normative 'Australian' values was a eugenic schema for Anglophone citizenship secured by culturally homogeneous cities and landscapes. Psychosocial commitment to sedentarisation was achieved through exclusion or containment of unruly, racially different subjects, as deviants or dissenters. Australia's success as a colony and a nation has been due to segregation and subsequent assimilation of supposed 'deviants' into sedentary values. Spaces of segregation recur as a serial taxonomy circumscribing territories outside property ownership and excluded from settler privilege. Wage-earning home or property ownership in a suburban grid or a rural acreage appears as the generalised aspiration and reward. Their geometric rigour eliminates agoraphobia. The prison, the internment camp, or the mission station is their racially or socially differentiated and thereby opaque counterpart.

## Punishment

By including the convict heritage of Australia as a key exemplar of displacement architecture, we recognise that social repression is not limited to race. Between 1787 and 1868, numerous convict sites in Australia collectively received around 166,000 men, women, and children transported to build a new society (Brooke and Brandon 2005, p. 246). While the various systems of punishment, forced labour, and reform over an eighty-year period are too complex for this chapter, as is its extensive literature, we need to understand transportation as a strategy of imperial expansion, calculated on the forced displacement of convicts. The preeminent site among these, Port Arthur convict settlement (1830–1877), the most substantial from among several thousand across Australia's states and territories (Australian Convict Sites 2008), was a total institution designed as a small township on an isolated, deep seaport (see Brand 1978 for a detailed account).[2] Convicts employed in forestry, mining, and for sustaining the settlement were integral to its institutional history (Nicholas 1988; Tuffin 2018). Port Arthur was a major industrial complex evident in the inclusion of a Dockyard, Master Ship Wright's House, Clerk of Works' House, Lime Kiln, and Dairy among around thirty convict-built structures on the 136-hectare site (Australian Convict Sites, p. 32). Four institutions built on the hill overlooking the site during the mid-nineteenth century—the hospital, the pauper house, the asylum, and in particular the panopticon or separate prison—reflected global advancements in institutional segregation and psychological control. Despite the evident architectural significance, archaeological or heritage studies of the sites supersede any architectural interest; in fact very little architectural attention is afforded to Australia's convict sites (excepting Kerr and Kerr 1988). While they offer keen insights into the darker side of immigrant heritage, that heritage was also redeemed because convict emancipists and property owners' contribution to Australian society recentred this discourse elevating convict genealogies in this settler teleology of progress. The career of Sydney's town planner, the convict architect Francis Greenway (Broadbent and Hughes 1997) is a case in point. Previously repressed convict histories are increasingly drawn into a well-substantiated majoritarian past. Such redemption is not available to other racialised identity categories.

## Racial Exclusion

Parallel sites of incarceration of displaced Traditional Owner or First Nations populations are comparatively decentred and minoritised and therefore unlikely to attract comparable internal or international tourist interest. Their racial difference, cultural distance, and perceived civilisational backwardness (in the eyes of their colonisers) prompted structural exclusions that persisted until the mid-twentieth century. The elimination and containment of Traditional Owner populations made vast arable areas of Australia, particularly in the coveted south-eastern states, available to pastoralists (Broome 2005). Assuming Indigenous cultures were nomadic, based on their seasonal movement to different camping grounds, white settlers expropriated traditional lands and rationalised them in cadastral grids. These processes were facilitated by the ignorance of Indigenous land management (Gammage 2012) and the sanctimonious conviction of European superiority.

The architectures that housed displaced Indigenous populations have proven less tangible than those of convicts. In south-eastern states like Victoria, Indigenous peoples who survived expulsion, genocide, and European diseases were corralled into missions or reservations irrespective of their language or Country—the places to which they are culturally connected and from which they derive their identity (see Pieris et al. 2014, pp. 12–25).[3] Desocialisation, incarceration, and forced assimilation embedded displacement as an ontological reality (Broome 2005). The remote, mainly rural, carceral spaces that first mediated this violent history were operated by Christian institutions on civilising missions. These were replaced by the Aboriginal Protectorate during the mid-nineteenth century with stations established in various districts. Continuing hostilities between Indigenous and settler communities, fragmentation of Indigenous groups, and the inability (or reluctance) of the Protectorate to defend them saw further dissolution into smaller stations and reserves between 1858 and 1869 (Mission Voices n.d.).[4] These came under the Central Board for the Protection of Aborigines established in 1860. Various groups confined to these stations were compelled to form new identities derived from the mission rather than language groups or Country. New sensibilities of Indigenous identity were forged in these displacement environments, in lieu of lost traditional lands. Continuing hostilities and dispossession throughout the nineteenth century likewise brought Indigenous groups to settlements in the urban fringes, often regarded as 'slums'. These strategies of forced displacement were ratified under the Aborigines Protection Act of 1869.

We can imagine the mid-nineteenth-century landscape of south-eastern Australia as populated by these parallel convict sites and Aboriginal missions, both spaces for containing and to some extent criminalising the black and white populations of Australia. Both typologies combined institutional and religious architectures with camp-like temporary dwellings. Both would transform by the end of the nineteenth century: convict sites due to the ending of transportation (in 1868) and the missions due to the Aborigines Protection Law Amendment Act of 1886—known as the Half-caste Act. The Act justified the removal of Aboriginal and mixed-race children, breaking up family groups and undermining the sustainability of mission environments through depopulation (Museum of Australian Democracy n.d.; Yarra Healing n.d.). So-called 'half-castes' were institutionalised and trained as domestic workers or used as unpaid labour. Strategies of exclusion grew even more insidious during the twentieth century following the decision to remove mixed-race Indigenous children, so that they could be assimilated into the wider population. The policy continued from 1937 until the 1960s.

Whereas convict stories and sites became part of national memory, the mission created a new frame of reference for twentieth-century Indigenous activism wherever concentrated communities remained attached to mission sites. New place meanings and definitions were created for what was held sacred, and mission stories accumulated as extensions of the Country. The pain of displacement would thus become the basis for forging and politicising new Traditional Owner place-identities through First Nations calls for sovereignty. Group emplacement had some clear advantages for establishing custodianship of Indigenous sites or artefacts if under the management of resident First Nations groups. But the struggle to sustain these sites against numerous pressures from surrounding pastoralists was a constituent part of their persistence or failure. Groups discordant with the vision of colonial citizenship were excised and contained. As with the prison system, institutionalised moral education played an important role in mediating inclusion through their purported salvation or reform.

## Containing the Enemy

Although the third example of displacement is likewise physically constrained, its history is part of a larger global border conflict where military, penal, and immigrant histories intersect. Whereas Indigenous peoples were displaced internally within the Australian geography, from lands over which they had traditional ownership, 'enemy aliens' were segregated from among the

European settler population because of their 'hostile' nationalities. Indigenous populations were displaced to enable the expropriation of land, because they preceded European settlement, whereas non-Anglophone settlers in Australia were seen as newcomers with less legitimacy than the British. The history of their segregation, containment, and, in some cases, punitive incarceration occurred in repurposed or adapted military camp facilities, many of them as programmatically complex as small country towns. A pattern emerges of a series of total environments which gradually become more temporary, even as they increase in capacity, because the objective is to restrain a transient population. These are typically hutted encampments using timber-framed structures with corrugated metal or asbestos sheet cladding, designed to hold around 6000 persons in the largest such facilities, and used for training the almost one million recruits who fought in the Second World War ('Second Word War' n.d.).

Mid-twentieth-century Australia is populated by these camp environments. Military field camps, used during the First World War for enemy prisoners of war (POWs), were followed during the Second World War with purpose-built POW camps for captive soldiers held by Australia on behalf of Britain and the USA (National Archives Australia n.d.).[5] Camps were laid out as geometric figures: rhomboids, pentagons, and dodecagons, and surrounded by barbed wire fencing, guard towers, and search lights. Military training camps and camps for garrison battalions would be built adjacent as related facilities. While prisoners of war were enemy combatants, captured and sent to Australia from other theatres of conflict by Allied partners, Australia was more directly culpable for the incarceration of 'enemy aliens' and refugees: nationals of the major powers in the Axis partnership sent to or already resident in Australia (Weiss 2003, pp. 239–240; Report on Directorate of Prisoners of War and Internees 1939–1951).[6] Some 42,516 persons were held captive across the course of the war in approximately eighteen camps across Australia (Bevege 1993; Report on Directorate of Prisoners of War and Internees 1939–1951). Among the most troubling features of internment was the placement of German and Austrian Jewish refugees, who had fled Europe for Australia, alongside Nazi sympathisers, in adjacent compounds. Australia also held Japanese internees from northern Australia, Queensland, and the Pacific Islands (Nagata 1996).

Wartime sensitivities were omnipresent in the Australian population, given the large numbers who died in the overseas conflicts, and the fierce fighting against the Japanese in the Pacific. The remoteness of camps and their relative opacity protected the internee populations from Anglo-Australian prejudices fuelled by political rhetoric. This also informed the rapidity with which camps were decommissioned after the war and salvaged for much needed building materials. Once properties were returned to private owners, from whom they had been requisitioned earlier, all traces of the camps disappeared from the rural landscape. Their impermanence, lack of aesthetic significance, and pragmatic, modular construction retarded architectural interest.

## Migrant Assimilation

Australia's desire for rapid population growth and industrialisation during the postwar years prompted the repurposing and adaptation of the military camp typology for immigrant reception and accommodation from 1947 to the 1970s (Pennay 2007). The brief was no longer punitive. The Federal Department of Immigration was established in 1945 with an ambitious migration programme offering assisted passage for United Kingdom residents, for British and US ex-servicemen, and for 12,000 displaced persons per year from the refugee camps in Europe (Australian Government, Department of Home Affairs n.d.). Formal and informal agreements with a number of European countries were tagged to industrial labour creation. The Australian population, which at the end of the war stood at over seven million, reached ten million by 1960.

Non-Anglophone so-called 'new migrants' were differentiated from the many more British immigrant arrivals after the war. Very few came with independent sponsorship. New arrivals were processed and allocated jobs, and were assimilated into Australian language and ways in reception and training centres.[7] These facilities also acted as holding centres for families and an 'outside workers' hostel for their menfolk who returned to the camp between work assignments in remote or rural areas. Their co-location with military camp facilities or administration by the military (Pennay 2009, pp. 44–47, 49) continued a global practice whereby the underlying structures of refugee environments were militarised. Regimentation, routine, military presence, and the reuse of austere hutted troop accommodation blurred the lines between wartime and peacetime spatial experiences for many refugees.

Commonwealth-administered hostels were set up by the Australian government in 1951 as the next step in this process (Commonwealth Hostels Ltd 1968).[8] Hostels and guesthouses, typically in urban areas, provided temporary accommodation for newcomers transitioning into Australian society. At first these facilities were Nissan and Quonset huts or barrack-type architectures inherited from the military camp model, but gradually purpose-designed facilities domesticated the institutional brief. It seemed as though the model was finally being humanised as a suitable transition space for displaced persons, and was increasingly used to provide accommodation and services for new influxes of refugees. The largest group to enter during the 1970s–1980s were fleeing war in Cambodia and Vietnam. The White Australia Policy was ended by 1973; attitudes to migration shifted from assimilation (1947–1964) to integration (1964–1973) and to multiculturalism (1973 onwards) (Koleth 2010). However, systems of exclusion became more focused on securing the national border against the Communist threat from Asia, and controlling the already narrow pathways to citizenship.

## Detention

In the decades that followed the 1970s, with each consecutive wave of maritime refugee arrivals, Australia's immigration policy grew more punitive. By 1992, a policy of mandatory detention encompassed all non-citizens without a valid visa entering Australia (Parliament of Australia 2013) and remote detention centres were opened up.[9] Accordingly, a subject category of 'unlawful non-citizens,' including asylum seekers, were to be detained while their legitimacy was established. The *mandatory* detention of unauthorised asylum seekers (those who circumvent established procedures) and both onshore and offshore facilities that housed them became the subject of political conflict, human rights controversies, and untold social trauma. The poor quality of camp environments, overcrowding, the duration of stay, unrest, and mental health problems have created continued public division over the associated policies. The taxonomy of facilities created to accommodate displaced persons expanded to include Immigrant Reception and Processing Centres, Immigrant Detention Centres (IDCs), residential housing, and transit accommodation, as well as alternative places of detention. Of these the IDCs had established precedents in the examples recounted so far.[10] The creation and maintenance of environments like them for asylum seeker accommodation is due to the conviction expressed by successive Australian governments that punitive deterrence is effective.

## Conclusion

This paper has argued that forced displacement to or within Australia is evident across a series of remote and segregated facilities, largely absent from the architectural canon and its teleology of modernity and national progress.

While the case of Australia bears attributes applicable to many settler colonial societies that evolved as liberal democracies, its distance from Europe and consecutive histories of segregation linked to displacement make for a unique history. The examples of carceral spaces highlighted in this paper appear as punitive antecedents to a continuing citizenship practice of funnelling errant subjects through structures of confinement to be re-educated or picked out. One way of approaching this topic would be by highlighting spaces and practices by which social or racial difference is criminalised. But such an approach would fail to establish how displacement and incarceration become historically linked to citizenship creation, or to the boundaries that are drawn at various times around normative citizenship. Their displacement from the societies or circumstances that emplace and dignify them is what makes the various subjects that pass through these institutional spaces vulnerable to state violence and control. Drawing these environments and their questionable practices into a subfield of displacement architectures is a challenging but ethical move, intended at acknowledging all cultures of institutionalisation, including those that are socially damaging. While their ephemeral materialities and lack of exceptional aesthetic features has absented them from Australian architectural histories, recognition of these spaces as historically pervasive and distributed across Australia will contribute to politicizing the discipline.

## Notes

1. The Immigration Act No. 17 of 1901 placed restrictions on immigration and provided for the removal from the Commonwealth of prohibited immigrants.
2. Key convict sites in Australia include Kingston and Arthur Vale Historic Area, Norfolk Island; Old Government House and Domain in Parramatta, Hyde Park Barracks in Sydney, and the Old Great North Road and Cockatoo Island Convict Site, also in New South Wales; Brickendon Woolmers Estates, Cascades Female Factory, Coal Mines Historic Site, and Port Arthur Historic Site in Tasmania, and Fremantle Prison in Fremantle Western Australia.
3. 'Country' is an integrated conception of land, law, and culture.
4. For example, in Victoria, four Church Mission stations were established at Ebenezer at Lake Hindmarsh in the Wimmera (1858–1904), Ramahyuck at Lake Wellington in West Gippsland (1862–1908), Lake Tyers in East Gippsland (1861–present), and Lake Condah in the Western District (1867–1918 as a station and 1861–1951 as a reserve). Additionally, government stations were established at Framlingham reserve, near Warrnambool (1865–1890 as a station and 1861–present as a reserve), and Coranderrk

reserve, near Healesville (1863–1924). Cummeragunja near Maloga was on the New South Wales side of the Murray River (1889–1953 station and 1883–present reserve). In addition, there were several smaller stations and reserves.
5. At Cowra, Hay, Holsworthy, Bathhurst, Long Bay, Orange, in NSW; Tatura (including Murchison and Dhurringile) in Victoria; and Loveday in South Australia. There were internment camps at Tatura, Loveday, Harvey, Marrinup, and Hay, and several other smaller wood camps and work camps.
6. Some 7103 aliens, naturalised British subjects, and those born of enemy parents were apprehended and 1875 were released. Some 7862 overseas national aliens were shipped to Australia, of whom 6110 were POWs; at its peak in 1942, 10,731 local and overseas internees were held in Australia. Some 25,727 POWs were held in Australia.
7. Key migrant camp facilities were at Bonegilla, Benalla, Rushworth, and Somers in Victoria; Greta and Kapooka in New South Wales; Stuart and Wacol in Queensland; and Northam Holden, Graylands, and Cunderdin in Western Australia.
8. By 1967, there were 28 migrant hostels and 20 guesthouses and non-migrant hostels run by Commonwealth Hostels Ltd.
9. Migration Amendment Act 1992.
10. Key IDCs include those at Maribyrnong, Villawood, Perth—near the airport, Christmas Island, Berrimah, Curtin RAF Base, Scherger, Pontville, Wickham Point, Yongah Hill, and, formerly, Port Hedland, Woomera, and Baxter. Offshore Processing Centres are at Nauru and Papua New Guinea.

# References

Anderson, S., & Ferng, J. (2013). No boat: Christmas Island and the architecture of detention. *Architecture Theory Review, 18*(2), 212–226.
Australian convict sites, World Heritage Nomination. (2008). Canberra, ACT: Australian Government, Department of Environment, Water, Heritage and the Arts.
Australian Government, Department of Home Affairs. (n.d.). *Fact sheet: More than 65 years of post-war migration*. Retrieved from https://www.homeaffairs.gov.au/about/corporate/information/fact-sheets/04fifty.
Bevan, R. (2016). *The destruction of memory: Architecture at war*. London: Reaktion Books.
Bevege, M. (1993). *Behind barbed wire: Internment in Australia during world war II*. St Lucia, QLD: University of Queensland Press.
Brand, I. (1978). *Penal peninsula: Port Arthur and its outstations, 1827–1898*. West Moonah, TAS: Jason Publications.

Broadbent, J., & Hughes, J. (1997). *Francis greenway, architect.* Sydney, NSW: Historic Houses Trust.

Brooke, A., & Brandon, D. (2005). *Bound for Botany Bay: British convict voyages to Australia.* Kew: National Archives.

Broome, R. (2005). *Aboriginal Victorians: A history since 1800.* Sydney, NSW: Allen & Unwin.

Cairns, S. (Ed.). (2004). *Drifting: Architecture and migrancy.* Abingdon & New York, NY: Routledge.

Carter, P. (2002). *Repressed spaces: The poetics of agoraphobia.* London: Reaktion.

Commonwealth Hostels Ltd. (1968). *Facts about commonwealth hostels limited.* Milson's Point, NSW: Commonwealth Hostels Ltd.

Diller+Scofidio & Renfro. (2008–2009). *EXIT* [45-minute panoramic installation, Foundation Cartier pour l'art contemporain, Paris]. Retrieved from https://dsrny.com/project/exit.

Fincher, R., & Jacobs, J. M. (1998). *Cities of difference.* New York, NY: Guilford Press.

Foucault, M. (1979). *Discipline and punish: The birth of the prison.* New York, NY: Vintage Books.

Franke, A., & Weizman, E. (2014). *Forensis: The architecture of public truth.* Berlin: Sternberg Press.

Gammage, B. (2012). *The biggest estate on earth: How aborigines made Australia.* Crows Nest, NSW: Allen & Unwin.

Gunew, S. M. (2003). *Haunted nations: The colonial dimensions of multiculturalism.* New York, NY: Routledge.

Hage, G. (2003). *Against paranoid nationalism: Searching for hope in a shrinking society.* Annandale, VIC: Pluto Press.

Herscher, A. (2010). *Violence taking place: The architecture of the Kosovo conflict.* Stanford, CA: Stanford University Press.

Jacobs, J. M. (1996). *Edge of empire: Postcolonialism and the city.* London & New York, NY: Routledge.

James, S. W. (2012). Indigeneity and the intercultural city. *Postcolonial Studies, 15*(2), 249–265.

Jupp, J. (2002). *From white Australia to Woomera: The story of Australian immigration.* Cambridge: Cambridge University Press.

Katz, I., Martin, D., & Minca, C. (2018). *Camps revisited: Multifaceted spatialities of a modern political technology.* London: Rowman and Littlefield.

Kennedy, J. (2008). *Structures for the displaced: Service and identity in refugee settlements.* Doctoral dissertation, International Forum on Urbanism, TU, Delft.

Kenzari, B. (2011). *Architecture and violence.* Barcelona: Actar.

Kerr, J. S., & Introduction by Kerr, J. (1988). *Out of sight, out of mind: Australia's places of confinement, 1788–1988.* Sydney, NSW: S.H. Ervin Gallery, National Trust of Australia (N.S.W.).

Koleth, E. (2010). *Multiculturalism: A review of Australian policy statements and recent debates in Australia and overseas.* Parliament of Australia, Social Policy Section,

Research Paper no. 6 2010–11. Retrieved January 7, 2018, from https://www.aph. gov.au/About_Parliament/Parliamentary_Departments/Parliamentary_Library/pubs/rp/rp1011/11rp06#_Toc275248120.

Laing, R. D. (1965). *The divided self: An existential study in sanity and madness.* Harmondsworth: Penguin.

Lozanovska, M. (2016). *Ethno architecture and the politics of migration.* London & New York, NY: Routledge.

Lozanovska, M. (2019). *Migrant housing: Architecture dwelling and migration.* Abingdon & New York, NY: Routledge.

Mission Voices. (n.d.). ABC. Retrieved from https://web.archive.org/web/20040831012400/http://www.abc.net.au/missionvoices/.

MoMA (Museum of Modern Art, New York). (2016–2017). *Insecurities: Tracing displacement and shelter* [exhibition].

Museum of Australian Democracy. (n.d.). *Documenting democracy, Aboriginal Protection Act 1869.* Retrieved from http://foundingdocs.gov.au/item-did-86.html.

Nagata, Y. (1996). *Unwanted aliens: Japanese internment in Australia.* St Lucia, QLD: University of Queensland Press.

National Archives of Australia. (n.d.). *Wartime internment camps in Australia.* Retrieved from http://www.naa.gov.au/collection/snapshots/internment-camps/index.aspx.

Nicholas, S. (1988). *Convict workers: Reinterpreting Australia's past.* Cambridge & Melbourne, VIC: Cambridge University Press.

Parliament of Australia. (2013, March 20). *Immigration detention in Australia.* Retrieved from https://www.aph.gov.au/About_Parliament/Parliamentary_Departments/Parliamentary_Library/pubs/BN/2012-2013/Detention.

Pennay, B. (2007). *The Army at Bonegilla, 1940–71.* Albury & Wodonga, VIC: Parklands.

Pennay, B. (2009). Remembering Bonegilla: The construction of a public memory place at block 19. *Public History Review, 16,* 43–63.

Pieris, A., Tootell, N., Johnson, F., McGaw, J., & Berg, R. (2014). *Indigenous place: Contemporary buildings, landmarks and places of significance in south East Australia and beyond.* Parkville, VIC: Melbourne School of Design, Faculty of Architecture, Building and Planning, The University of Melbourne.

Pugliese, J. (2008). The tutelary architecture of immigration detention prisons and the spectacle of 'necessary suffering'. *Architecture Theory Review, 13*(2), 206–221.

Report on Directorate of Prisoners of War and Internees (Australian Army Headquarters, Directorate of Prisoners of War and Internees). (1939–1951). ORMF0024 (Official Record).

Second World War 1939–45. (n.d.). *Australian war memorial.* Retrieved from https://www.awm.gov.au/articles/second-world-war.

Tuffin, R. (2018). Convicts of the 'proper description': The appropriation and management of skilled convict labour. *Labour History: A Journal of Labour and Social History, 114,* 69–92.

Wastl-Walter, D. P. (2011). *The Ashgate research companion to border studies*. Farnham: Ashgate Publishing.

Weiss, J. P. (2003). *It wasn't really necessary: Internment in Australia with emphasis on the second world war*. Eden Hills, SA: J. P. Weiss.

Yarra Healing. (n.d.). *Aboriginal mission stations and reserves in Victoria*. Retrieved from http://www.yarrahealing.catholic.edu.au/teaching-learning/index.cfm?loadref=119.

# 28

# Spiritual Geographies of Displacement and Resilience

Julia Christensen and Veronica Madsen

## Introduction

Experiences of displacement extend well beyond the disruptive or disorienting effects of physical movement from one place to another. Displacement, whether it involves a physical move or occurs within one's ancestral homelands, is an experience laden with despair, longing, and placelessness. Yet displacement goes even deeper, and is even wider-reaching, than the emotional—it has profound spiritual implications as well. Though there is a dearth of literature on spiritual displacement specifically, the spiritual dimensions of displacement have been taken up in various ways in the literature, from psychology to geography, with an emphasis on forced migration (cf. Gozdziak and Shandy 2002) as well as Indigenous peoples in settler colonial states (cf. Thistle 2017). These literatures describe the forms of spiritual displacement that arise through experiences of involuntary migration as people are relocated from one place to another, leaving significant spiritual sites behind but carrying spiritual practices in tow. They also describe the forms of spiritual displacement that arise in-place for Indigenous peoples under settler colonialism, where Indigenous peoples experience historical and contemporary displacement from their homelands, spiritual practices, ontologies, and

---

J. Christensen (✉) • V. Madsen
Memorial University, St. John's, NL, Canada
e-mail: jchristensen@mun.ca; vlm867@mun.ca

epistemologies through settler colonial agents, structures, and policies. Despite rich and diverse engagements with the spiritual dimensions of displacement, these literatures remain conceptually unintegrated, with little crossover between bodies of work to better inform a broader conceptualisation of the spiritual in geographies of displacement. In this chapter, we assemble into dialogue three principal channels in the conceptualisation of spiritual forms of displacement: (1) the geographies of home as a place of spiritual or ontological belonging, including an examination of spiritual homelessness; (2) the spiritual dimensions of involuntary migration; and (3) the spiritual displacement of Indigenous peoples in settler colonial contexts. Together, these literatures scaffold the development of a conceptual understanding of spiritual displacement as material, ontological, and both spatially and temporally mobile. They also reveal myriad examples of the ways in which, despite physical displacement, spiritual connections with (home)lands are actively maintained and allow for forms of resilience through spiritual *em*placement.

## Home as a Place of Spiritual or Ontological Belonging

The concept of place as a site for the reproduction of culture and identity is well established in geography (cf. Aitken and Valentine 2014; Blunt 2005; Crang 2013), but what of spiritual places? The literature conveys a sense of the spiritual that is both embedded in places and transcendent of places; one that is both rooted and that travels. Spirituality transcends, or indeed lies beyond, geography: not only is it unchartable, it '[calls] into question what can be charted' (Bartolini et al. 2018, p. 1). Spiritual geography has been described as a place of religious tradition, a holy place, a cultural framework for being-in-the-world, and a set of spiritual or religious practices (Henderson 1993). Spirituality emerges through a human desire to 'organize reality to compensate for the disparity between the known and the unknown' (Henderson 1993, p. 470). In Bartolini et al. (2018), spiritual practices must be attended to as more than a set of beliefs, and in so doing the spaces of spirituality—of active relationship with the divine—become the site for geographical analysis. Colonial and capitalist efforts attempt to bring these transcendent spaces of spirituality under control through a simultaneous disciplining and domination of spiritual places and practices (cf. Wildcat et al. 2014). For example, the settler colonial project has centred the inextricable link between Indigenous spiritualities and land by dispossessing Indigenous peoples of their homelands in an effort to not just control and extract from those lands but also control, subdue, and extract from Indigenous peoples themselves (cf. Watson 1997). At the same time, spirituality is cited as a source of strength and resilience. For example, Hansen and Antsanen (2016) describe Indigenous Elders' strategies to share traditional spiritual teachings with youth in order to cope with the negative effects of residential school, while Corntassel (2012) illustrates how efforts to rebuild Indigenous spiritual connections with their homelands have been central to Indigenous self-determination.

In this chapter, we first consider the broad literature on the cultural geographies of home and homelessness. In particular, we attend to the conceptualisation of home as multi-scalar and both a lived, imagined, and indeed spiritual place of belonging (Blunt 2005; Blunt and Dowling 2006; Mallett 2004), and in turn homelessness as material and ontological (Somerville 1992). The predominant view of homelessness, following Somerville (1992), is a practical, or literal, one that understands homelessness as a matter of 'rooflessness.' Instead, Somerville suggests, homelessness can be conceptualised as 'rootlessness,' an ontological state of being that implies the absence of a sense of place or a sense of home. Blunt and Dowling (2006, p. 254) suggest that the meaning of home, then, arises through the relationship between material and imaginative realms and processes—home is lived as well as imagined, conveying a distinct spiritual element to the experience of home as a site of tension between 'the real and the ideal' (Bartolini et al. 2018, p. 1).

In the context of Indigenous peoples globally, homeland, territory, or country are profoundly cultural and political scalar concepts that are rooted in the same relations of belonging and identity embedded in the meaning of home. Writing from the New Zealand context, Kearns (2006) suggests that the ontological or spiritual significance of home is demonstrated by the experiences of Maori who live away from their homelands but whose connections to those places are nonetheless potent. Thornton (2008, p. 4) writes that place has been central to cultural (re)production in all societies and 'can be said to constitute a cultural system.' Though this relationship is one that is experienced across cultures, Thornton (2008) asserts that Indigenous peoples have a special relationship to place, maintained with the landscapes they have inhabited since time immemorial. Easthope (2004) also suggests that when external forces threaten one's sense of place, the possibility for rootedness diminishes. A profound sense of spiritual rootlessness can therefore result when a relationship to place, both collectively and individually formed, becomes fragmented or fractured (Christensen 2013).

Spiritual homelessness situates experiences of homelessness within the cultural significance of place and the consequences of the diverse forms of displacement experienced by Indigenous people in settler societies. George and Geisler (2006) call the myriad colonial interventions and subsequent displacements experienced by Indigenous people the 'other homelessness,' that is, homelessness that is spiritual rather than simply the absence of shelter. Meanwhile, spiritual displacements lie at the heart of what Keys Young et al. (1998) term 'spiritual homelessness,' a concept used to explain the broader significance of homelessness in the Australian Aborigine and Torres Strait Islander context. Keys Young et al. (1998) found that Indigenous homelessness was perceived by many Indigenous people as being commonly experienced at the level of family and community, rather than solely by individuals. Through their findings, they argue that Australian Indigenous peoples' experiences of homelessness cannot be extracted from the historical and ongoing colonial experiences of Indigenous people, and in this way are qualitatively distinct. Thus, spiritual homelessness serves to articulate Indigenous experiences of homelessness as multilayered and multi-scalar, and brings together 'a broad range of situations and experiences, including physical, spiritual and cultural dimensions' (Keys Young et al. 1998, p. 25).

The elements that have most profoundly shaped Indigenous homelessness are those that fracture and delegitimise both material and imaginative elements of home (see Blunt and Dowling 2006; Marston 2004), disrupting a sense of belonging and connection to 'place,' and represented through detachments from family, the land, and independence (Christensen 2017). Equally, the transformation of land and cultural modes of spatiality through settlement and housing, the introduction of new social categories of difference, and economic transitions have reoriented many Indigenous peoples around colonial modes of sociocultural organisation and paved an uneven material landscape. The colonial past and present in settler colonial societies reproduce a social and material landscape of 'collective homelessness with enduring consequences' (George and Geisler 2006, p. 4). Conceptualising these forms of displacement as 'spiritual homelessness' (Keys Young et al. 1998) advances an explanation of the sociocultural dimensions of Indigenous homelessness as well as the profound irony of being displaced on one's homeland (Christensen 2017). Nevertheless, Christensen (2016) also describes the ways in which family, the land, and cultural teachings are actively engaged in the lives of Indigenous peoples experiencing homelessness as a form of resilience and home-journeying.

There has been considerable postcolonial geographical work on home, nation, and empire to explore the material and imaginative spaces of home, the contested politics of identity and belonging, and the material and political implications of different modes of belonging, place, and identity shaped by the long and continued processes of migration, displacement, settlement, dispossession, and the growing recognition of the rights of Indigenous people (Blunt 2005). In settler colonial contexts, colonial forms and processes manifest in the present through modern systems of law and government which silence or severely restrict the ability of Indigenous peoples to contest their dispossession (cf. Gelder and Jacobs 1998). Challenges to loss of land, for instance, must be articulated through the language and value system of the state. Such legal and governance frameworks serve to perpetuate the spiritual displacement of Indigenous peoples through the undermining of Indigenous knowledge frameworks in order to justify the continued dispossession of homelands. As Gibson (1999) articulates, government policies in the name of Indigenous self-determination can be strategies for fragmenting, containing, and absorbing Indigenous challenges to the hegemony of the capitalist/colonial state, leaving colonial systems of knowledge and structures of authority relatively undisturbed.

The spiritual displacement of Indigenous peoples under settler colonialism also takes place discursively. In their analysis of claims for Indigenous sacredness in Australia, Gelder and Jacobs (1998) highlight the contrast between tradition, authority, and possession on the one hand, and 'modernness,' loss of authority, and dispossession on the other. This dichotomy 'sets up a structure that simply does not speak to the modes of empowerment many modern Aboriginal people are actually experiencing' (Gelder and Jacobs 1998, p. 51). In contemporary Australia, efforts to bring sacred Indigenous sites into national protection can have the contradictory effect of dispossessing sacred spaces from Indigenous peoples: the delimiting of sacred space simultaneously preserves and compromises the spiritual emplacement and self-determination of Indigenous peoples.

## The Spiritual Dimensions of Involuntary Migration

There are significant conceptual parallels to be drawn between work on the spiritual dimensions of migration and Indigenous experiences of displacement and resettlement. The literature on forced migration attends to the spiritual in its engagement with displacement, though as Gozdziak and Shandy (2002) argue, this attention is limited to the role of religious persecution in conflict settings and the politicisation of religious identity. Hagan and Ebaugh (2003) also suggest that the role of religion in the stages of the migration process has been overlooked by both immigration and sociology of religion scholars. Meanwhile, there is a wide diversity of religious and spiritual beliefs and practices that sustain many refugees and involuntary migrants in their processes of displacement, migration, and integration into a host society, most of which are neglected by researchers and policymakers (cf. Gozdziak and Shandy 2002).

Another emergent theme within the study of spirituality and refugees is the role of religious practice in coping with trauma. The importance of spiritual beliefs in coping with physical displacement and relocation extends beyond the refugee camp to integration into the country of settlement. Current research has revealed the important role of spiritual development as a positive human right in refugee children's lives, which takes place within the ecological system of spatial environments, communities, and the individual child. At its core, the right to spiritual development recognises that children continually seek to make sense of their experiences and to understand their place in the world, and that this cognitive-cultural process is implicated in children's development of self-understanding, emotion, and value systems. In their research, Ojalehto and Wang (2008) have examined the role of the spiritual in the forced migration experiences of children. They argue that as children begin to organise their social worlds, the concept of home becomes an important social anchor, helping to inform their identities, sense of continuity, and perception of the world as a stable place (cf. Garbarino and Kostelny 1996). The spiritual significance of displacement, therefore, is high amongst children, for whom place serves as an important object of spiritual discourse (Ojalehto and Wang 2008). In this vein, Ojalehto and Wang (2008) assert that there is significant therapeutic potential in efforts to 'spiritualise' the spatial environments of children who have been geographically displaced through the mobilisation of cultural narratives of past, present, and future.

Spiritual displacement is not just enacted, experienced, and embodied spatially. Just as there are timescapes to spirituality, so are there too to spiritual displacement. The temporal dimensions of spiritual displacement are illustrated in Kraly and Flowers' (2016) work on the spiritual geographies of Indigenous children who were removed from their families and communities to attend residential schools in Australia. The role of time in displacement is evident in this work in two ways: firstly, the ways in which time was employed to impose order and control over Indigenous children in the schools through the use of the clock and rigid daily and weekly schedules. These modes of institutionalisation served to disrupt the temporal rhythms of life in the children's home communities. Secondly, the temporal nature of spiritual displacement was also made apparent through the artwork of Indigenous children.

The children's paintings and drawings that Kraly and Flowers (2016) examine depict the spiritual significance of both place and time as they illustrate not only spiritually significant places but also spiritually significant times of day—in particular, sunset, which for the Noongar people is an observed time for storytelling and learning from kinfolk. The resilience of these children was made evident in the power of their imaginations as they continued to depict the spirituality of family and community despite being displaced from land and kin. Similarly, Noguera (2013) found that refugee children who had escaped armed conflict in their home country used spiritual practice as a source of comfort and strength. In particular, the practice of prayer travelled with children, even amidst violence and war zones, as they prayed for peace, strength, and the realisation of dreams.

## The Spiritual Displacement of Indigenous Peoples in Settler Colonial Contexts

Finally, we examine the spiritual elements in the displacement of Indigenous peoples in (settler) colonial contexts, primarily through the dispossession of homelands and thus the very foundation for Indigenous spiritual life. In particular, we address the ways in which the spiritual frames Indigenous experiences of housing insecurity, homelessness, and homesickness, not only drawing on the spiritual homelessness literature but also literature examining Indigenous conceptualisations of home and place (Havemann 2005). The spiritual and the land are inextricable, compelling us to not only consider the materiality of the spiritual and the wide-ranging consequences of separation or disconnection from homelands in the settler colonial context (Christensen 2013, 2017; Kake 2016; Memmott 2014; Thistle 2017) but also the ways in the which the land—and in turn, the spiritual—have been the deliberate site of historical and contemporary colonial attack (Simpson 2017). In attending to the concerted focus on the spiritual within the colonial project, we shed light upon the effects that spiritual displacement and cultural disconnections have on Indigenous health, and identify directions towards spiritually oriented, culturally safe interventions in the decolonisation of Indigenous health and home (Christensen 2016).

Spiritual displacement as an experience that is embodied, and that changes and accumulates and reproduces itself across time and space, is something one understands from the literature on intergenerational trauma (Brokenleg 2012). In writings of intergenerational trauma among Indigenous peoples, the dispossession of Indigenous homelands, the separation of children from their families, and the assimilative aims of residential schools and colonial social policy have all worked to disrupt and disconnect Indigenous spirituality and spiritual practices. Intergenerational trauma is thus necessarily spiritual, and the experience of these forms of pervasive spiritual displacement is passed down from generation to generation. Spiritual displacement is a wound that deepens across generations, and in this sense, as Brokenleg (2012, p. 10) reflects: 'trauma became the carrier for cultural genocide.' We must, however, attend to the ways in which, in spite of spiritual displacements, spirituality and spiritual practice nevertheless strengthen amidst displacement. Spiritual displacement and spiritual resilience can be experienced simultaneously, and in the case of Indigenous peoples surviving and thriving in spite of settler colonialism, the spiritual remains a source of strength and resilience in spite of those agents and structures who work hard to displace. Indeed, in their work on the teachings of Dene Elders in northern Saskatchewan, Canada, Hansen and Antsanen (2016, p. 14) consider the ways in which the Land-as-Spirit and respect for land are active and mobile teachings; teachings that 'enabled the people to live well in the past and … promote Indigenous resilience and wellbeing in the present day.' In a similar vein, Andrade Vinueza (2017) found that the mobility or portability of spirituality was incredibly important to refugee children in Central America, for it was something that, even whilst experiencing displacement from home and family, could not be taken from them.

This sense of intergenerational spiritual displacement is evident in Griffin-Pierce's (1997) writing on Navajo sacred geographies. Indigenous homelands are deeply imbued with spiritual meaning. As a result, Griffin-Pierce writes that an Indigenous sense of place is necessarily spiritual. In her work, even Navajo who are raised off their traditional territories describe a sense of longing for the land of their homelands. Significantly, homelands have spiritual meaning as an assemblage of places and journeys, but also as the site for the reproduction and sharing of Indigenous spiritual knowledge. Moreover, Navajo sacred teachings, like those of other Indigenous nations, are nested in the Land and call upon its beneficiaries to care for it. Griffin-Pierce writes that the land acts as a visual text which encodes the core perceptions of life. Griffin-Pierce (1997) suggests that a perhaps unconscious stress results for Navajo who are far from their homelands and therefore cannot uphold the sacred teachings. Many make regular pilgrimages to their homelands in order to renew their relationships with homelands.

The land cannot be extricated from Indigenous knowledge, and Indigenous knowledge cannot be extricated from Indigenous spiritualties and cultures.

Dispossession of Indigenous lands was/is Indigenous spiritual displacement. These connections underlie the Indigenous Resurgence movement in North America, one that promotes the kind of action and provides the spiritual and ethical bases for a transformative movement that has the potential to liberate both Indigenous peoples and settlers from colonialism (Alfred 2005). Integral to Indigenous Resurgence is the centring of the Land in Indigenous pedagogy. In its disembodied epistemology and praxis, settler pedagogy enforces and reproduces the spiritual displacement of Indigenous peoples (Wilson and Restoule 2010). Indigenous Resurgence requires Indigenous peoples to grow up in relationship with homelands, and immersed in Indigenous languages and spiritualities (Simpson 2014), which stands in radical contrast to settler pedagogical systems and institutions.

The localness of relationship, the context-dependence (the subjectivity) of relationship and its significance to Indigenous ways of being in the world means that it has been undermined in relation to the God's-eye view (cf. Haraway 1988) of Western scientific knowledge and therefore Western forms of spirituality and knowledge (which are inextricable from one another). Greenwood (2013, p. 99) writes that 'despite the utter centrality of relationships to Indigenous ways of being in the world, relationship remains a mutable concept, always geographically and contextually expressed and as diverse as the environments in which it, and associated knowledge(s), is rooted.' The embracing of diversity, subjectivity, and the power of place in Indigenous spirituality and knowledge has been muted in contrast to the universalism in Western knowledge's self-image, and in the colonising technologies of Christianity. While Indigenous spiritualities are enmeshed in the Land, the symbols and spiritual places of Christianity derive strength from their rootlessness; they travel and colonise. Christianity, settler colonialism, and capitalism have conspired in the spiritual displacement of Indigenous peoples (Greenwood 2013). In fact, as Heather McDonald (2001) notes, in conjunction with the physical extension of the British Empire, missionary practice was busy trying to produce a parallel empire of its own—the 'Empire of the Spirit.' In this way, Eurocentrism is not just the cognitive legacy of colonisation (Battiste and Youngblood Henderson 2000) but the spiritual legacy as well. Meanwhile, Indigenous peoples believe that all existence is connected. Indigenous ontology *is* spirituality. Therefore, disconnection from relationships with other things in the world results in spiritual displacement (Wilson 2008). Ontology, then, is the act of being in relationship (Greenwood 2013). There is an inherent politics of knowledge to the spiritual displacement of Indigenous peoples under settler colonialism.

There are also health consequences to spiritual displacement, for spiritual connection is a central element of Indigenous health (Reading and Wien 2009). The colonisation process in Australia rendered many Indigenous peoples 'spiritually and/or physically placeless. ... fundamentally [undermining] their ontology, which is based on spiritual and physical connectedness to their land and sea country' (Havemann 2005, p. 68). Data on the effects of the settler colonial state's continued displacement of Indigenous peoples provides a statistical vocabulary for the loss of connection with place, cultural identity, and spirituality, as well as the intergenerational transmission of profound ontological insecurity (Havemann 2005). Conversely, Indigenous spirituality and efforts to reclaim Indigenous spirituality have been shown to have positive benefits on health and wellness at the individual and community scale (Fleming and Ledogar 2008; Hatala 2008: Robbins and Dewar 2011).

## Conclusion

In this chapter, we have attempted to move beyond an understanding of spiritual displacement that is narrowly rooted; one that is just about the effects of moving/being moved away from spiritually significant places or sites. Rather, it is also, perhaps even more so, about experiencing ruptures in spiritual connection through violence, be it physical, structural, social, environmental, or cultural. This is how spiritual displacement in its more profound and damaging forms can be experienced without being physically displaced. Part of this oversight in our understanding of the spiritual comes from a privileging of Judeo-Christian or Western understandings of spirituality. In many non-Western cultures, including diverse Indigenous cultures, the spiritual underpins and is totally inextricable from all other areas of human life. In this way, we can understand how seemingly un-spiritual practices and policies (e.g., modern housing programmes, residential schooling, climate change, and extractive industries) drive spiritual displacements in profound ways. Similarly, we can see how experiences of involuntary migration also encounter displacing elements through institutionalised spaces—including spaces of care—that do not promote or support, or in various ways even discourage, integral spiritually enriching elements like access to family, engagement with important cultural narratives and traditions, artistic or creative expression, culturally embedded forms of healthcare, and so on. This includes relationships with time and timescapes, as the spiritual is not only placed but rather has complex and mutually constitutive spatial, temporal, social, and cultural dimensions. At the same time, spiritual practices travel in the sense that bodily practices that activate and build human connections to the sacred and divine can be, and are, taken up in a variety of ways in an effort to build resilience amongst people as well as lands. They provide protection and comfort to child migrants; they provide pathways to a renewed sense of place for Indigenous peoples; they support health, well-being, and self-determination. Our intention in this chapter was to not only explore what is means to be spiritually displaced, but to also understand what spiritual emplacement or re-placement can—and does—offer in terms of capacity for resilience and self-determination. For, in spite of all of the ways in which violence results in the marginalisation and displacement of people around the world, or within our own countries, spiritual spaces, practices, and connections remain, and indeed strengthen, even in the face of profound adversity.

# References

Aitken, S. C., & Valentine, G. (Eds.). (2014). *Approaches to human geography: Philosophies, theories, people and practices*. Thousand Oaks, CA: Sage.

Alfred, T. (2005). *Wasase: Indigenous pathways of action and freedom*. Toronto, ON: University of Toronto Press.

Andrade Vinueza, M. A. (2017). The role of spirituality in building up the resilience of migrant children in Central America: Bridging the gap between needs and responses. *International Journal of Children's Spirituality, 22*(1), 84–101.

Bartolini, N., Mackian, S., & Pile, S. (2018). Spaces of spirituality: An introduction. In N. Bartolini, S. Mackian, & S. Pile (Eds.), *Spaces of spirituality* (pp. 15–34). Abingdon: Routledge.

Battiste, M., & Youngblood Henderson, J. (2000). *Protecting indigenous knowledge and heritage: A global challenge*. Saskatoon, SK: Purich.

Blunt, A. (2005). Cultural geography: Cultural geographies of home. *Progress in Human Geography, 29*(4), 505–515.

Blunt, A., & Dowling, R. (2006). *Home*. Abingdon: Routledge.

Brokenleg, M. (2012). Transforming cultural trauma into resilience. *Reclaiming Children and Youth, 21*(3), 9–13.

Christensen, J. (2013). 'Our home, our way of life': Spiritual homelessness and the sociocultural dimensions of indigenous homelessness in the Northwest Territories (NWT), Canada. *Social & Cultural Geography, 14*(7), 804–828.

Christensen, J. (2016). Indigenous housing and health in the Canadian north: Revisiting cultural safety. *Health & Place, 40*, 83–90.

Christensen, J. (2017). *No home in a homeland: Indigenous peoples and homelessness in the Canadian north*. Vancouver, BC: UBC Press.

Corntassel, J. (2012). Re-envisioning resurgence: Indigenous pathways to decolonization and sustainable self-determination. *Decolonization: Indigeneity, Education & Society, 1*(1), 86–101.

Crang, M. (2013). *Cultural geography*. Abingdon: Routledge.

Easthope, H. (2004). A place called home. *Housing, Theory and Society, 21*(3), 128–138.

Fleming, J., & Ledogar, R. J. (2008). Resilience and indigenous spirituality: A literature review. *Pimatisiwin, 6*(2), 47.

Garbarino, J., & Kostelny, K. (1996). The effects of political violence on Palestinian children's behavior problems: A risk accumulation model. *Child Development, 67*(1), 33–45.

Gelder, K., & Jacobs, J. M. (1998). *Uncanny Australia: Sacredness and identity in a postcolonial nation*. Melbourne, VIC: Melbourne University Publishers.

George, L., & Geisler, C. (2006). Homeless in the heartland: American dreams and nightmares in Indian country. In P. Milbourne & P. Cloke (Eds.), *International perspectives on rural homelessness* (pp. 37–56). Abingdon: Routledge.

Gibson, C. (1999). Cartographies of the colonial/capitalist state: A geopolitics of indigenous self-determination in Australia. *Antipode, 31*(1), 45–79.

Gozdziak, E. M., & Shandy, D. J. (2002). Editorial introduction: Religion and spirituality in forced migration. *Journal of Refugee Studies, 15*, 129–135.

Greenwood, M. (2013). Being Indigenous. *Human Development, 56*(2), 98–105.

Griffin-Pierce, T. (1997). 'When I am lonely the mountains call me': The impact of sacred geography on Navajo psychological wellbeing. *American Indian and Alaska Native Mental Health Research, 7*(3), 1–10.

Hagan, J., & Ebaugh, H. R. (2003). Calling upon the sacred: Migrants' use of religion in the migration process. *International Migration Review, 37*(4), 1145–1162.

Hansen, J. G., & Antsanen, R. (2016). Elders' teachings about resilience and its implications for education in Dene and Cree communities. *The International Indigenous Policy Journal, 7*(1), 2.

Haraway, D. (1988). Situated knowledges: The science question in feminism and the privilege of partial perspective. *Feminist Studies, 14*(3), 575–599.

Hatala, R. A. (2008). Spirituality and aboriginal mental health: An examination of the relationship between aboriginal spirituality and mental health. *Advances in Mind Body Medicine, 23*(1), 6–12.

Havemann, P. (2005). Denial, modernity and exclusion: Indigenous placelessness in Australia. *Macquarie Law Journal, 5*, 57.

Henderson, M. L. (1993). What is spiritual geography? *Geographical Review, 83*(4), 469–472.

Kake, J. (2016). Why are our people overrepresented amongst te pani me te rawakore?: Reflections on the root causes of Maori urban homelessness. *Parity, 29*(8), 8.

Kearns, R. (2006). Places to stand but not necessarily to dwell: The paradox of rural homelessness in New Zealand. In P. Milbourne & P. Cloke (Eds.), *International perspectives on rural homelessness* (pp. 247–260). New York, NY: Routledge.

Keys Young (Firm), Australia. Dept. of Health, Aged Care, Australia. Dept. of Family, & Community Services. (1998). *Homelessness in the aboriginal and Torres Strait islander context and its possible implications for the supported accommodation assistance program (SAAP)*. Canberra, ACT: Commonwealth Dept. of Health & Aged Care.

Kraly, F. P., & Flowers, E. (2016). Mobilizing a 'spiritual geography': The art and child artists of the Carrolup native school and settlement, Western Australia. *Transfers, 6*(1), 103–109.

Mallett, S. (2004). Understanding home: A critical review of the literature. *The Sociological Review, 52*(1), 62–89.

Marston, S. (2004). A long way from home: Domesticating the social production of scale. In E. Sheppard & R. B. McMaster (Eds.), *Scale and geographic inquiry: Nature, society and method* (pp. 170–191). Oxford: Blackwell.

McDonald, H. (2001). *Blood, bones and spirit: Aboriginal Christianity in an East Kimberley town*. Melbourne, VIC: Melbourne University Publishers.

Memmott, P. C. (2014). 'Spiritual homelessness' amongst Australian indigenous people—What does it mean? *Homelessness Australia*, 35–37.

Noguera, R. T. (2013). The narratives of children in armed conflict: An inference to spirituality and implication to psychological intervention. *International Journal of Children's Spirituality, 18*(2), 162–172.

Ojalehto, B., & Wang, Q. (2008). Children's spiritual development in forced displacement: A human rights perspective. *International Journal of Children's Spirituality, 13*(2), 129–143.

Reading, C. L., & Wien, F. (2009). *Health inequalities and the social determinants of aboriginal peoples' health*. Prince George, BC: National Collaborating Centre for Aboriginal Health.

Robbins, J. A., & Dewar, J. (2011). Traditional indigenous approaches to healing and the modern welfare of traditional knowledge, spirituality and lands: A critical reflection on practices and policies taken from the Canadian indigenous example. *The International Indigenous Policy Journal, 2*(4), 2.

Simpson, L. B. (2014). Land as pedagogy: Nishnaabeg intelligence and rebellious transformation. *Decolonization: Indigeneity, Education & Society, 3*(3), 1–25.

Simpson, L. B. (2017). *As we have always done: Indigenous freedom through radical resistance*. Winnipeg, MB: University of Minnesota Press.

Somerville, P. (1992). Homelessness and the meaning of home: Rooflessness or rootlessness? *International Journal of Urban and Regional Research, 16*(4), 529–539.

Thistle, J. (2017). *Definition of indigenous homelessness in Canada*. Toronto, ON: Canadian Observatory on Homelessness Press.

Thornton, T. F. (2008). *Being and place among the Tlingit*. Seattle, WA: University of Washington Press.

Watson, I. (1997). Indigenous peoples' law-ways: Survival against the colonial state. *Australian Feminist Law Journal, 8*(1), 39–58.

Wildcat, M., McDonald, M., Irlbacher-Fox, S., & Coulthard, G. (2014). Learning from the land: Indigenous land based pedagogy and decolonization. *Decolonization: Indigeneity, Education & Society, 3*(3), 86–101.

Wilson, S. (2008). *Research is ceremony: Indigenous research methods*. Halifax, NS: Fernwood Publishing.

Wilson, D. D., & Restoule, J. P. (2010). Tobacco ties: The relationship of the sacred to research. *Canadian Journal of Native Education, 33*(1), 29–45.

# 29

# Mapping Trajectories of Displacement

## Nishat Awan

This chapter gives an overview of recent research into migrant journeys with an emphasis on the methods used to trace the displacement of people across international borders. I begin by addressing the ways in which displacement has been usefully conceptualised in this context through a focus on the nature of movement. This includes notions of truncated journeys, clandestine routes, and migrant trajectories understood as phases of movement and stasis (Collyer 2010; Hage 2009; Khosravi 2014; Mainwaring and Brigden 2016; Schapendonk 2012). Whilst a lot of work in this area is inherently spatial, there is also a recent emphasis on describing migrant journeys through alternative temporalities, such as body chronologies or in relation to certain pivotal life events (Collins 2018; Shubin 2015). This work aims to understand the specificities of the experience of crossing borders without documentation, including the ways in which the border regime has made such journeys increasingly difficult and too often fatal. It also focuses on the role of a wide variety of actors, including people smugglers and border officials (Garelli and Tazzioli 2018; Icli et al. 2015; Macías-Rojas 2018). Yet, there is an inherent contradiction and ambiguity in the use of methods such as tagging, tracking, and tracing of individuals within research, since these are all key to how the border regime itself functions. In such a context, it is impossible for research to be either apolitical or neutral. Whether understood through the

---

N. Awan (✉)
Goldsmiths, University of London, London, UK
e-mail: N.Awan@gold.ac.uk

© The Author(s) 2020
P. Adey et al. (eds.), *The Handbook of Displacement*,
https://doi.org/10.1007/978-3-030-47178-1_29

bureaucratised language of university ethical committees as the imperative to protect vulnerable participants, or understood as a personal ethical commitment towards those in precarious and often highly difficult situations, many researchers choose to take a stance that actively helps undocumented migrants. But this is of course not always the case; many research projects exist that are designed to help states govern their borders, whether in the form of new technologies of securitisation or through providing expert evidence to aid deportation (Hatton 2018; iBorderCtrl 2016).[1]

In this chapter, the term 'activist research' is used to denote those who align themselves consciously with undocumented migrants, often using the techniques of border control against itself, in what Pezzani and Heller have called 'a disobedient gaze' (2013). Official narratives can thus be challenged, for example monitoring the number of people making a particular journey can be used to account for the number of lives lost as a result of border securitisation, or statistics can be interrogated in order to turn the logic of the border around by focusing on the number of people who *started* a particular journey rather than the number that arrived (Tazzioli 2014). Despite such possibilities, it is important to emphasise that in using similar techniques to those of border control, researchers face a whole set of ethical issues that must be dealt with care and a critical perspective in order to not inadvertently harm the very people we attempt to advocate for. An ubiquitous example of such a technique is the use of GPS for locational positioning, which as many have noted is based on military technologies of surveillance and control (Holmes 2006; Kaplan 2006; Schuurman and Pratt, 2002). In thinking about the ethical implications of GPS, Petrescu points to the difference between tracking and tracing, using the example of the psychoanalyst Deligny's practice of tracing the movements of severely autistic children through hand-drawn maps (2007). In making these maps there was a desire to spend time with the children, to get to know them through their movements and gestures. This kind of tracing is the precise opposite of the tracking of individuals made available through GPS, which allows knowledge of a person's location without sharing a presence with them; extreme care is therefore required in the way that this knowledge is mobilised.

## Migrant Journeys and Temporalities *en route*

It is now commonly understood that in many journeys of displacement, and especially those made by forced migrants, there is no direct route that can be traced from A to B (Ahsan Ullah 2013; Hassan and Biörklund 2016; Shaffer et al. 2018). Instead, the journey is emphasised as a social process that is an ongoing event in people's lives, as well as shaping the places through which people travel. There may be periods where migrants remain static, but that does not necessarily signal the end of a journey, which can also encompass settling in, finding a home, deciding to move on, or being deported. At the same time, deportation regimes and militarised borders keep people moving, what Khosravi terms circulation: 'a controlled movement of people sent back and forth between undocumentedness and deportability: between countries, between laws, between institutions' (2018). Any mapping of such journeys requires an ethical position in relation to a securitisation regime that derives surplus value from the very conditions of precarity that people are forced into. This means that journeys that are regarded as illegal and therefore assumed to be hidden are not necessarily so. Both the demonisation of so-called human smugglers, who are often undocumented migrants themselves, and the pretence of a hidden dimension to border crossing are merely ways for actors such as the EU, who are pushing for evermore policed borders, to evade responsibility for the deadly effects of their policies (Achilli 2018; Garelli and Tazzioli 2018). As the Windrush scandal in the UK and the changing fate of the Dreamers in the US have shown, the confluence of securitised borders and legal regimes produces illegality within racialised subjects.[2] Coutin refers to this aspect of migrant lives as 'clandestinity,' which she defines 'as a hidden, yet known, dimension of social reality' and further observes that when migrants are clandestine they embody 'both law and illegality' (2005, pp. 195, 198). In this sense, migrants are of course illegalised but their presence is accepted, wanted even, for the neoliberal economies of the north to draw value from working lives that can be exploited due to their illegal status and can easily be cut short through deportation. In Europe and the US, undocumented migrants are included in our economic and other systems, we rely on their labour for our cities and our rural economies to function, for example through providing domestic labour or for fruit picking, and yet they remain illegal, not having the correct documentation and the right to remain in the country, as well as not being able to access services. As Mezzadra and Neilsen have argued, borders become a method for the production of exploitable and precarious labour (2013).

This contradictory status of being known yet hidden is one of the most important elements of contemporary forced migration and a quality that must be appreciated when attempting to trace and tag such journeys. For whom is the tracing and tagging taking place? What purpose does it serve? And crucially, how does it intervene within the complex politics of (in)visibility that many migrants embody? In some of the examples I relate below, this tracing takes the form of a single story or event that is then used to understand and comment on wider processes, whether it is the bodily experience of such journeys or the failings of the border regime. But as Mainwaring and Brigden note: 'Human smuggling and migration routes are sustained practices, leaving visible traces in the physical, social, economic and political landscape of transit communities' (2016, p. 246). Examples in this chapter show how we can find these traces not only in the physical landscape but also in social relations or in the virtual realm of digital signatures. For many migrants the journey becomes a near permanent condition that is perpetuated through deportation regimes. What type of time does such a life embody and how does it affect an understanding of displacement? Often in cases of deportation, time is used as a mechanism of control. Being 'sent back'—sometimes to a place you might not even know—can, of course, also be part of the journey. As Griffiths explains, time in such instances can slow down whilst people wait months and years for a decision on their asylum application (2014). For many, time becomes a bodily experience that weighs heavy in waiting. In other cases, time is accelerated with deportation decisions being made in days, or even hours, making the threat of having to leave always imminent. The time of deportation is therefore contradictory, both fast and slow, disrupting life rhythms.[3] In traumatic and difficult journeys, the body can also act as a timekeeper. In the dark, in the back of a lorry, time might be understood as rhythms, but rhythms that are interrupted and broken. Conversely, over longer timeframes the body offers its own chronologies based in events; the birth of a child, an illness, or a reunion. To understand the temporalities of such journeys, the linear, concatenated time of modernity needs to be replaced by time that is embodied and relational (Grosz 2005). In the examples related below, time acts in different ways. It is used to verify places and locations in order to assign responsibility for journeys that ended in death, but in other examples time-based media is used to disorient, to reflect the experience of travelling clandestinely (Awan 2016). In other examples, alternative temporalities are described that remove people from what is considered to be the 'normal' flow of time.

## Mapping Stories and Accounts

The use of maps within the normative discourse on migration, whether in policy reports, academic research, or journalism, has often been critiqued for showing only a particular perspective that is aligned to the interests of border securitisation and migration control. In the European context, for example, maps are used to show the movement of people in one direction only—towards Europe—ignoring the large numbers of people moving within the African continent, or the circulations caused by the deportation regime described above. Such representations have been described as 'static invasion maps' that 'not only represent moral panics, they also co-construct them' (van Houtum 2012, p. 410). More recently, the use of mapping by policy-led organisations has become more sophisticated, so that rather than focusing on the start and end of the journey, as the arrows on earlier maps did, the focus shifts to the routes that people take, pinpointing important nodes and places of passage that should be targeted in the quest to reduce migration towards Europe and North America. Casas-Cortes et al. (2017) describe the International Centre for Migration Policy Development (ICMPD) as exemplary of this attempt to map the shifting routes that people take in order to evade border control. Their i-Map is interactive and regularly updated to reflect the changing nature of the routes. The aesthetic of the map is relational, rather than using arrows to map flows in one direction only, the map has a networked logic that encompasses some of the circulations of undocumented people. As Casas-Cortes et al. have noted, the map attempts to capture a shifting subject in order to 'facilitate control and not movement' (2017, p. 13). The different thicknesses of lines are used to denote major and minor routes, as well as distinguishing between what they call 'migration hubs' and 'migration route cities.'[4]

In contrast to the i-Map, there are others who use maps to understand the experience of crossing borders without documentation. In the Mapping Journey Project, the artist Bouchra Khalili asks migrants to trace the routes they have taken on a map of the world (Khalili 2008). Their stories speak of difficult journeys across international borders: a thick line drawn in black permanent marker shows the circuitous routes taken to evade border control. In my own work, I have also used maps in a similar way, but instead of a geopolitical map people were given a blank piece of paper on which to draw a story of their journey (Awan 2016). This meant that the maps people drew were necessarily distorted with the scale changing according to the details of passage. Often parts of the journey that took longer and were more difficult were shown in greater detail, taking up more space on the page than areas that were easier to pass through. Both in Khalili's work and my own, videos focused on the hands that drew the maps and the voices that spoke of their own experiences. The difference between these maps and the approach of the i-Map is clear in both the methods and the politics of representation. The i-Map is a matrix, a grid over territory, that is approached as a growing repository and a database. Techniques of crowdsourcing are used to collect and collate information from lots of different sources. The maps of Khalili and the ones I use in my own work are different, they are hand-drawn by migrants themselves, and they tell stories of journeys with an ethnographic focus on how people account for their own experiences. They could be understood as a way of doing trajectory ethnography, asking people where they went, why, and how, in order to understand their motivations and desires (Fig. 29.1).

What hopefully becomes clear from these three examples is that the way in which displacement is tagged and tracked is very much related to the political motivations of those doing the work. In the i-Map the subject is removed, and we are left with de-territorialised and de-subjectified lines that attempt to hide their politics, whilst also being largely ahistorical in their approach. In the use of hand-drawn maps by migrants, the politics behind the work is acknowledged even if the faces of the people we interviewed are hidden. There is a desire to protect individual identities whilst not erasing the subject, so that the focus on hands and being able to see the pattern and fabric of a sleeve give a glimpse of the person. Combined with their voice, the map is embodied through stories that speak of what it felt like to move clandestinely, the material conditions of such journeys, and the lived experiences that lead to time being measured through the movement of bodies in space.

**Fig. 29.1** 'Migrant Narratives of Citizenship' exhibition (Awan 2016); interview with Afghan woman in Odessa, Ukraine. (Image: Cressida Kocienski)

## Mapping Signals and Traces

The ambiguity of tracking people's movement through space means that the way in which these routes and trajectories are understood is key, as are the motivations for doing such work. Whereas the examples above understood the movement of people through their own accounts, the practices described below aim to track displacement through uncovering the signals and traces that are left behind. These can be digital signals embedded in different kinds of archives, from satellite images to web traffic, or they can be material traces left within the physical environment. To detect and make sense of such traces requires careful investigative work that is able to deploy a set of interdisciplinary methods, an approach that is exemplified in many of the examples related below. At the same time, it requires an appreciation of what Weizman calls 'weak sensors,' material or otherwise; these are traces that can only ever be faint, remote, and far from objective in the normative sense of the word, being always 'suggestive rather than conclusive' (Weizman 2014, p. 29). The role of such weak sensors can be appreciated in an investigation carried out by the Forensic Architecture research agency, where a blurry line in video footage of a Palestinian demonstration was the key to building a case around the culpability of the Israeli army in the death of a demonstrator (Forensic Architecture 2014, pp. 83–95). The faint trace in the video prompted the reconstruction of events in a three-dimensional model of the scene populated with information from several videos, which was then presented in a court of law. On its own, the faint line in the video footage would not amount to much, but buttressed with other accounts and presented within the constructed space of the model it became decisive. Such 'weak sensors' therefore demand a political and aesthetic project alongside investigative or academic concerns.

The Undocumented Migration Project (De León 2010) addresses the lived reality of people crossing the US–Mexico border. Through combining approaches from ethnography, archaeology, and forensic science, the project documents and analyses the material traces people leave behind. There is in this work an explicit desire to not interfere with what is a clandestine process. Instead fieldwork in the Sonoran Desert takes place in the hottest part of the day, since an exchange with human smugglers could put migrant lives at risk. The project meticulously records material left behind by migrants that is usually considered rubbish, approaching these as an archive that gives a glimpse into the harsh realities of the desert crossing. Drawing on the practices of museum archives, everyday objects such as discarded clothing, shoes, and water bottles are given a status and an importance within a carefully curated collection and in this way become part of the historical record. The interdisciplinary nature of this work gives it a potency, by mixing archaeological fieldwork with an ethnographic approach that produces powerful accounts of the realities of lives lost in the desert. Through a process of tagging found objects and by placing emphasis on the materialities of the journey, the project highlights techniques of US border control, which are designed to enrol nature itself in its work; the environmentality of the desert, animal, and other non-human agents are all put to use to form a lethal barrier. As de León writes, it is 'a killing machine that simultaneously uses and hides behind the viciousness of the Sonoran Desert' (De León 2015, pp. 3–4).

The investigative approach described above is also deployed by the Forensic Oceanography project, which centres on the deaths of migrants at sea by attempting to piece together the traces and faint signals that are left behind after a devastating event (Forensic Architecture 2014; Heller et al. 2012). Centred on the 'left-to-die boat'[5] in the Mediterranean, their research traced its path after it ran out of fuel and drifted at sea. The aim of doing this work was to prove that a number of actors were aware of the boat in distress but chose not to launch a rescue. The report and maps produced by the group were used to assist a legal case brought against the European Union (EU), Frontex, and NATO by 'Groupe d'information et de soutien des immigrés' (GISTI). Forensic Oceanography's methodology most explicitly uses the border regime against itself by mobilising the very technologies that are used to control borders, such as satellite imagery, remote sensing, and visual analysis. They combine this data-driven approach with the testimony of survivors to uncover the precise details of events as they have unfolded. In an interview with a survivor they used aide-memoires, such as photographs of vessels and planes that were in the area, to verify and cross-reference the locational details found through the sensing data. Unlike the interviews related above that focused on the experiential aspects of border crossing, these were designed to uncover factual details that could stand as evidence to support a legal case. Here the dense practice of tagging and tracking journeys at sea through vessel-tracking technologies, thermal imaging, radar, and satellites that are part of the Eurosur surveillance programme used to police the borders of the EU is put to a very different use—to prove the culpability of those who claim jurisdiction over the seas.

The Forensic Oceanography project consciously uses digital technologies of border control against the grain within the very specific context of a legal case, but without such a framework the practice of tracking and tagging journeys can become ethically problematic. In parallel to Eurosur, the EU's Eurodac programme is designed to aid border control through collecting biometric data on migrants, which is shared with European law enforcement agencies. At the same time, a number of European governments as well as the US are using smartphone data, 'proactive monitoring of social media,' and in some cases have the ability to seize laptops and smartphones from asylum seekers without identification documents (Latonero and Kift 2018, p. 7). In such a context, to use social media analysis of migrants' networks requires a very strong rationale that has the capacity to overturn a surveillance gaze.

The recent project, Mapping Refugee Media Journeys: Smartphones and Social Networks, aimed to understand refugees' use of social media through a mixed methods study. The project report provides very useful insights into the ways migrants have used social media, particularly WhatsApp and other potentially private communication channels, such as Telegram and Viber (Gillespie et al. 2016, p. 64), based on extensive interviews with refugees themselves. But the project also aimed to uncover the behaviour of refugees online through an attempt at tracking their use of social media using 'computational social network analysis of social media communications networks' (Gillespie et al. 2018, p. 3). This was achieved by extracting and sorting information on particular users from data sets provided by Facebook and Twitter through their proprietary application programming interfaces (APIs). The data made available through each of these APIs is limited in different ways, from the type of data available to its size. What is surprising in the write-up is that the researchers seemed to lament their own restricted access to information, since they were only able to use data from public groups on Facebook and public tweets, rather than having access to private areas of the platforms. They couched this as 'technical problems … that restricted our unfettered access to refugee information on social media' (Gillespie et al. 2016, p. 59). At the same time, they seemed to be unaware that the ability to track people online was not restricted simply because people chose to hide their identity, but that through making connections with the accounts that they were engaging with online, identities can often be revealed (Priedhorsky et al. 2014; Sharma et al. 2012). Therefore, the project's attempt to use Twitter data to follow people's 'likes' and follower networks raises serious methodological and ethical questions in the context of a group for whom surveillance can be an issue of survival. What seems to have curtailed this attempt is a lack of technical know-how rather than a critical perspective that would allow what is essentially a technology of surveillance in the name of marketing to be overturned in a countermove.

The report was published before the recent Cambridge Analytica scandal, where the company acquired the data of millions of people from Facebook without their consent, using it to micro-target adverts. It is now widely acknowledged that social media influenced the results of a number of votes, including the 2016 US presidential election and the UK's Brexit referendum (Cadwalladr 2017; Russia 'Meddled in All Big Social Media' 2018). The public outcry resulted in Facebook and other social media platforms restricting the amount of data they make available through their APIs. Yet many academics, and particularly those in the field of digital journalism, complained that the actions of one academic and a disreputable company had lost them access to extremely valuable social data (Bastos and Walker 2018; Bruns 2018). This goes to the heart of the problem in the majority of academic research involving such data, where the standard epistemological questions somehow fall by the wayside. As with any other research method, it is not only the information that is acquired but the ways in which it has been gathered and the purpose for which it is collected that is of importance. The data provided by social media APIs has been curated and collected by these platforms for the explicit purpose of marketing (Lanchester 2017). As Venturini and Rogers state: 'They have purposely and relentlessly built the *self-fulfilling prophecy* of "computational marketing" and, to do so, created a new type of data devoted to support it' (2019). To use such data uncritically is to go along with the assumptions behind it, including the fact that such platforms are designed to undermine public discourse by encouraging certain types of behaviour based on clicks and likes. The above example highlights the difficulties faced by academics when engaging with technologised ways of doing research. It seems that while the research team was very well represented by people who had contacts within the refugee community and wider related networks, the technical aspects were not as well represented or thought through. This would suggest the need for stronger ethical guidelines concerning research using social media, something that is underway in many universities. But more importantly, it requires researchers to not only think of adding new digital methods to the mix to make their research 'cutting edge' but to take seriously the implications of using such methodologies in the research design itself. In particular, it is important to think through the lived ethical implications of such work, which go far beyond simplified concerns around surveillance or the fact that digital methods only reach limited types of audience, although these are clearly also important.

## The Politics and Possibilities of Looking Askance

The examples discussed above all attempt to unpack the experience of displacement through finding ways to trace what are often difficult journeys. The most successful of these are able to mobilise a series of interdisciplinary methods within a practice-based research designed to interrogate the technologies and techniques of the border regime. There is a desire in this work to move beyond analysis to affecting the situation on the ground, often in collaboration with migrants themselves, whether in the service of bringing people and institutions to account for crimes that have been committed, or in order to make others empathise with the lived experience of displacement, or simply to help people along their way. This work, therefore, is unapologetically political; it takes a stance that does not fit in the neutral space of the objective researcher (Haraway 1988). In this, the research is not only concerned with producing academic or investigative work, but, through bringing a collage of different perspectives and techniques, it also takes part in an aesthetic project that is able to reveal some of the contradictions inherent within the mainstream discourse on migration and the displacement of people.

There is a concern across the different examples of beginning to account not only for the spaces of displacement but also for the times of it. The use of stories and narratives speak for an embodied time that allows for moments of empathy and exposes the facile nature of a modernist understanding of linear time. When difficult journeys are recalled, they are not related chronologically but follow their own embodied logic. But the mobilisation of time as timeline is both common and highly destructive within asylum practices, where it is used to track and verify someone's story. Time is often used to deny asylum by questioning specific details and chronologies. In this context, it is interesting that the Forensic Oceanography project consciously uses time understood as linear to question instead the narrative of the border regime. The passing of time is also dangerous for the way in which it can easily erase precious traces of lives lost in remote locations or cover the traces of faint signals, which gives the work of uncovering traces an added urgency.

The ethical question of how to approach a vulnerable subject underlies all work discussed in this chapter, and for most the answer lies in choosing an indirect approach. In the context of visual representation, for example, the subject is present but identities are not completely revealed by choosing to only show people's hands and through hearing their voices. In the Undocumented Migration Project, De León is adamant that direct participant observation of the harsh desert crossing will always remain a problematic choice; it not only serves to obscure the migrant subject by foregrounding the experience of the researcher, it also puts migrant lives at risk (De León 2015, pp. 12–14). Rather than addressing this difficult and clandestine process directly, he deploys a range of other anthropological methods that indirectly trace the route of migrants. There is much to be said for such indirect approaches when dealing with vulnerable people and difficult situations. When military and corporate technologies of surveillance and control are brought to bear upon such research, whether in the use of GPS tracking or social network analysis, the ethical obligation lies with researchers to fully understand the implications of such technologies.

I started this chapter by asking, for whom is the tagging and tracking taking place and what purpose does it serve? These remain central questions. While there are varied impulses for doing this work, they all seem to share a common political project of revealing the violence of the border. This could be achieved through supporting the movement of people, allowing those on the move to share their own realities, or by using the border *dispositif* against itself. Both the potential and danger of this work to track and tag migrant journeys, is that, whether used by agencies such as Frontex or activist groups such as the No Borders Network, there is a similar conceptualisation of space that underlies this work. It is an understanding of space as relational and the border regime as fluid and dynamic. The possibilities of intervening within such a space, or of finding cracks in the grid, means gaining a close understanding of this topological space of surveillance and control. It might act as a grid but is not necessarily homogeneous; it is the technologised biopolitical manifestation of the dispersed border that is shifting yet ever present. To find a path through such a space requires an intimate knowledge of this thick terrain, but also of the gaps within these contours that bodies can silently slip through. Slipping through the gaps, that is, finding a space that is neither tracked nor tagged, is both a necessity and a danger for those who are compelled to cross borders without documents. Research in this area would do well to follow the example of migrants themselves, by being aware of the dangers as well as the politics and possibilities of looking askance and of finding the cracks.

## Notes

1. Considering the EU's investment in fortifying its borders, it is perhaps unsurprising that a large number of technologically oriented EU-funded projects exist that focus on border security, all of whom justify their research through a

discourse of security and convenience (EFFISEC (Efficient Integrated SeCurity Checkpoints) 2013; 'MobilePass, a secure, modular and distributed mobile border control solution for European land border crossing,' MobilePass 2014; Kyriazanos 2018).
2. The Windrush scandal concerned people from mostly Caribbean backgrounds who were wrongly detained and deported by the UK Home Office. Many of these were British subjects that arrived in the UK as citizens before the Immigration Act of 1971 that restricted the citizenship rights of those from the former colonies. For more information see El-Enany 2019; 'Windrush scandal' 2019. The Dreamers refers to undocumented migrants within the US that arrived as children. They were provided temporary rights under the Obama administration but their fate has become a political bargaining tool under the Trump administration. For more information see 'DREAM Act' 2019; Walters 2017.
3. Griffiths identifies 'four experiential temporalities,' including 'sticky time,' 'suspended time,' 'frenzied time,' and 'temporal ruptures' that reveal the contradictions of the way time is experienced in detention and deportation (2014, p. 1994).
4. The map used to be available to view freely, but since the controversy surrounding it, the website now requires a password: http://www.imap-migration.org/index.php?id=4. For a detailed account of the i-Map and its complicities, and a conceptual distinction between routes and trajectories, see Casas-Cortes et al. (2017). A graphic derived from the interactive i-Map can be viewed on the Reuters website ('Europe's Migration Crisis' 2016).
5. A boat that set sail from Tripoli on 27 March 2011 and spent 14 days drifting in the open sea with no rescue despite a number of ships and fishing boats present in the area. It resulted in the deaths of 63 people.

# References

Achilli, L. (2018). The 'good' smuggler: The ethics and morals of human smuggling among Syrians. *The Annals of the American Academy of Political and Social Science, 676*(1), 77–96. https://doi.org/10.1177/0002716217746641.

Ahsan Ullah, A. (2013). Theoretical rhetoric about migration networks: A case of a journey of Bangladeshi workers to Malaysia. *International Migration, 51*(3), 151–168. https://doi.org/10.1111/j.1468-2435.2009.00579.x.

Awan, N. (2016). *Migrant narratives of citizenship: A topological atlas of European belonging* [Exhibition, Yorkshire Sculpture Park]. Retrieved from http://www.ysp.co.uk/exhibitions/nishat-awan.

Bastos, M., & Walker, S. T. (2018, April 11). Facebook's data lockdown is a disaster for academic researchers. *The Conversation*. Retrieved July 12, 2019, from https://

theconversation.com/facebooks-data-lockdown-is-a-disaster-for-academic-researchers-94533.

Bruns, A. (2018, April 25). Facebook shuts the gate after the horse has bolted, and hurts real research in the process. *Internet Policy Review*. Retrieved July 12, 2019, from https://policyreview.info/articles/news/facebook-shuts-gate-after-horse-has-bolted-and-hurts-real-research-process/786.

Cadwalladr, C. (2017, May 7). The great British Brexit robbery: How our democracy was hijacked. *The Guardian*. Retrieved from https://www.theguardian.com/technology/2017/may/07/the-great-british-brexit-robbery-hijacked-democracy.

Casas-Cortes, M., Cobarrubias, S., Heller, C., & Pezzani, L. (2017). Clashing cartographies, migrating maps: The politics of mobility at the external borders of E.U.rope. *ACME: An International Journal for Critical Geographies, 16*(1), 1–33.

Collins, F. L. (2018). Desire as a theory for migration studies: Temporality, assemblage and becoming in the narratives of migrants. *Journal of Ethnic and Migration Studies, 44*(6), 964–980. https://doi.org/10.1080/1369183X.2017.1384147.

Collyer, M. (2010). Stranded migrants and the fragmented journey. *Journal of Refugee Studies, 23*(3), 273–293. https://doi.org/10.1093/jrs/feq026.

Coutin, S. B. (2005). Being en route. *American Anthropologist, 107*(2), 195–206. https://doi.org/10.1525/aa.2005.107.2.195.

De León, J. (2010). *The undocumented migration project*. Retrieved October 28, 2018, from http://undocumentedmigrationproject.com/home/about/.

De León, J. (2015). *The land of open graves: Living and dying on the migrant trail*. Oakland, CA: University of California Press.

DREAM Act. (2019). *Wikipedia*. Retrieved from https://en.wikipedia.org/w/index.php?title=DREAM_Act&oldid=881916989.

EFFISEC (Efficient Integrated SeCurity Checkpoints). (2013). EFFISEC website. Retrieved February 27, 2019, from http://www.effisec.reading.ac.uk/project.htm.

El-Enany, N. (2019). *(B)ordering Britain*. Retrieved from https://www.bloomsbury-professional.com/uk/bordering-britain-9781509917792/.

Europe's migration crisis. (2016, January 5). Reuters website. Retrieved March 5, 2019, from http://graphics.thomsonreuters.com/15/migrants/index.html.

Forensic Architecture (Ed.). (2014). *Forensis: The architecture of public truth*. Berlin: Sternberg Press.

Garelli, G., & Tazzioli, M. (2018). The humanitarian war against migrant smugglers at sea. *Antipode, 50*(3), 685–703. https://doi.org/10.1111/anti.12375.

Gillespie, M., Ampofo, L., Cheesman, M., Faith, B., Iliadou, E., Osseiran, S., et al. (2016). *Mapping refugee media journeys: Smartphones and social media networks*. London: The Open University/France Médias Monde.

Gillespie, M., Osseiran, S., & Cheesman, M. (2018). Syrian refugees and the digital passage to Europe: Smartphone infrastructures and affordances. *Social Media + Society, 4*, 1–12.

Griffiths, M. B. E. (2014). Out of time: The temporal uncertainties of refused asylum seekers and immigration detainees. *Journal of Ethnic and Migration Studies, 40*(12), 1991–2009. https://doi.org/10.1080/1369183X.2014.907737.

Grosz, E. (2005). *Time travels: Feminism, nature, power* (Annotated ed.). Durham, NC: Duke University Press.

Hage, G. (2009). Waiting out the crisis: On stuckedness and governmentality. In G. Hage (Ed.), *Waiting* (pp. 97–106). Carlton, VIC: Melbourne University Press.

Haraway, D. (1988). Situated knowledges: The science question in feminism and the privilege of partial perspective. *Feminist Studies, 3*, 575–599.

Hassan, A., & Biörklund, L. (2016). The journey to dreamland never ends: A refugee's journey from Somalia to Sweden. *Refugee Survey Quarterly, 35*(2), 116–136. https://doi.org/10.1093/rsq/hdw007.

Hatton, J. (2018). MARS attacks!—A cautionary tale from the UK on the relation between migration and refugee studies (MARS) and migration control. *Movements. Journal for Critical Migration and Border Regime Studies, 4*(1), 103–129.

Heller, C., Pezzani, L., & SITU Research. (2012). *The left-to-die boat.* Retrieved November 2, 2018, from https://www.forensic-architecture.org/case/left-die-boat/.

Holmes, B. (2006). Counter cartographies. In J. Abrams & P. Hall (Eds.), *Else/where: Mapping new cartographies of networks and territories.* Minneapolis, MN: University of Minnesota Design Institute.

iBorderCtrl. (2016). iBorderCtrl website. Retrieved November 4, 2018, from https://www.iborderctrl.eu/.

Icli, T. G., Sever, H., & Sever, M. (2015). A survey study on the profile of human smugglers in Turkey. *Advances in Applied Sociology, 5*, 1–12.

Kaplan, C. (2006). Precision targets: GPS and the militarization of U.S. consumer identity. *American Quarterly, 58*(3), 693–714. https://doi.org/10.1353/aq.2006.0061.

Khalili, B. (2008). *The mapping journey project* [Video Installation]. Retrieved from http://www.bouchrakhalili.com/the-mapping-journey-project/.

Khosravi, S. (2014). Waiting. In B. Anderson & M. Keith (Eds.), *Migration: The COMPAS anthology.* Oxford: Centre on Migration, Policy and Society.

Khosravi, S. (2018, December). Stolen time. *Radical Philosophy, 2*(3). Retrieved from https://www.radicalphilosophy.com/article/stolen-time.

Kyriazanos, D. (2018). *robusT Risk basEd Screening and alert System for PASSengers and luggage.* Tresspass website. Retrieved February 27, 2019, from https://www.tresspass.eu/.

Lanchester, J. (2017, August 17). You are the product. *London Review of Books*, pp. 3–10.

Latonero, M., & Kift, P. (2018). On digital passages and borders: Refugees and the new infrastructure for movement and control. *Social Media + Society, 4*(1). https://doi.org/10.1177/2056305118764432.

Macías-Rojas, P. (2018). The prison and the border: An ethnography of shifting border security logics. *Qualitative Sociology, 41*(2), 221–242. https://doi.org/10.1007/s11133-018-9382-2.

Mainwaring, C., & Brigden, N. (2016). Beyond the border: Clandestine migration journeys. *Geopolitics, 21*(2), 243–262. https://doi.org/10.1080/14650045.2016.1165575.

Mezzadra, S., & Neilson, B. (2013). *Border as method, or, the multiplication of labor*. Durham, NC: Duke University Press.

MobilePass. (2014). *MobilePass, a secure, modular and distributed mobile border control solution for European land border crossing*. Retrieved February 27, 2019, from http://mobilepass-project.eu/.

Petrescu, D. (2007). The indeterminate mapping of the common. *Field, 1*(1), 88–96.

Pezzani, L., & Heller, C. (2013). A disobedient gaze: Strategic interventions in the knowledge(s) of maritime borders. *Postcolonial Studies, 16*(3), 289–298. https://doi.org/10.1080/13688790.2013.850047.

Priedhorsky, R., Culotta, A., & Del Valle, S. Y. (2014). Inferring the origin locations of Tweets with quantitative confidence. In *CSCW: Proceedings of the conference on computer-supported cooperative work* (pp. 1523–1536). https://doi.org/10.1145/2531602.2531607.

Russia 'meddled in all big social media'. (2018, December 17). *BBC News*. Retrieved from https://www.bbc.com/news/technology-46590890.

Schapendonk, J. (2012). Turbulent trajectories: African migrants on their way to the European Union. *Societies, 2*(2), 27–41. https://doi.org/10.3390/soc2020027.

Schuurman, N., & Pratt, G. (2002). Care of the subject: Feminism and critiques of GIS. *Gender, Place & Culture, 9*(3), 291–299.

Shaffer, M., Ferrato, G., & Jinnah, Z. (2018). Routes, locations, and social imaginary: A comparative study of the on-going production of geographies in Somali forced migration. *African Geographical Review, 37*(2), 159–171. https://doi.org/10.1080/19376812.2017.1354308.

Sharma, N. K., Ghosh, J., Benevenuto, F., Ganguly, N., & Gummadi, K. (2012). Inferring who-is-who in the twitter social network. *SIGCOMM Computer Communication Review, 42*(4), 533–538. https://doi.org/10.1145/2377677.2377782.

Shubin, S. (2015). Migration timespaces: A Heideggerian approach to understanding the mobile being of eastern Europeans in Scotland. *Transactions of the Institute of British Geographers, 40*(3), 350–361. https://doi.org/10.1111/tran.12078.

Tazzioli, M. (2014). *Spaces of governmentality: Autonomous migration and the Arab uprisings*. London & New York: Rowman & Littlefield International.

van Houtum, H. (2012). Remapping borders. In T. M. Wilson & H. Donnan (Eds.), *A companion to border studies* (pp. 404–418). Chichester: John Wiley & Sons.

Venturini, T., & Rogers, R. (2019). 'API-based research' or how can digital sociology and journalism studies learn from the Cambridge Analytica affair. *Digital Journalism*. Retrieved from https://hal.archives-ouvertes.fr/hal-02003925.

Walters, J. (2017, September 14). What is Daca and who are the Dreamers? *The Guardian*. Retrieved from https://www.theguardian.com/us-news/2017/sep/04/donald-trump-what-is-daca-dreamers.

Weizman, E. (2014). Introduction: Forensis. In F. Architecture (Ed.), *Forensis: The architecture of public truth* (pp. 9–32). Berlin: Sternberg Press.

Windrush scandal. (2019). *Wikipedia*. Retrieved from https://cn.wikipedia.org/w/index.php?title=Windrush_scandal&oldid=885962573.

# 30

# Uncovering Internally Displaced People in the Global North Through Administrative Data: Case Studies of Residential Displacement in the UK

Janet C. Bowstead, Stuart Hodkinson, and Andy Turner

## Introduction

At its simplest, displacement means the involuntary movement of individuals and households from their established or expected places of residence due to a change in conditions beyond their reasonable ability to control (Zuk et al. 2018). In the contemporary nation-state era, displacement has been synonymous with the movement of refugees across national borders, fleeing various forms of conflict, violence, human rights' violations, or persecution because of their race, religion, ethnicity, nationality, social group, or political opinion. But displacement also takes place *within* national borders, and it is this form of forced migration that has exploded since the early 1990s to the extent that many more people are now internally displaced within nations than are displaced internationally. According to the Geneva-based Internal Displacement Monitoring Centre (IDMC), an estimated 28 million people were internally displaced during 2018 (IDMC n.d.). Many of those displaced may be on

---

J. C. Bowstead (✉)
Department of Geography, Royal Holloway, University of London, Egham, UK
e-mail: janet.bowstead@cantab.net

S. Hodkinson • A. Turner
School of Geography, University of Leeds, Leeds, UK
e-mail: S.N.Hodkinson@leeds.ac.uk; A.G.D.Turner@leeds.ac.uk

their way to cross a national border but never make it; as a result, these 'refugees in all but name' are not entitled to the same legal protections afforded to refugees under the United Nations' 1951 Refugee Convention.

Currently it is estimated that 95 per cent of internal displacement takes place in countries of the Global South, home to the world's major conflict and disaster zones such as Colombia, Syria, and Yemen. Yet counts of internally displaced people (IDP) are also high in countries of the Global North, where between 2008 and 2018, over 12.8 million people were recognised as IDP, the majority displaced by increasingly frequent and severe events such as hurricanes and wildfires, with over half of these living in the USA (IDMC n.d.). Moreover, the true scale of internal displacement in the Global North is far higher than is currently being recorded as many residential displacements are not included in official IDP counts. Take for example the collateral human damage of residential displacement as a consequence of the collapse of housing markets in several countries before, during, and after the 2008 global financial crisis. Bank repossessions of homes after home owners defaulted on their mortgage payments resulted in at least 10 million people being evicted in the USA (Gottesdiener 2013), and an estimated 380,000 household evictions in Spain (Gutiérrez and Domènech 2017). In the UK, many householders were also displaced as a result of the demolition and gentrification of public housing estates (Watt and Smets 2017); evictions following cuts to welfare support resulting in households struggling to keep up with the rent (UNHCR 2018); and fleeing domestic violence—mostly women and children (Bowstead 2015a, 2017).

The reason these residential displacements are not included in official IDP counts is due to the UN's guidelines on what constitutes internal displacement remaining wedded to the classical figure of refugees fleeing persecution. Similarly, while the UN regards 'forced evictions' as a violation of the international human right to adequate housing (UN General Assembly 2013), this is clearly imagined in a Global South context of landless and often indigenous communities being forcibly removed from urban slums or ancestral or communal lands over which the state or a private actor claims ownership. In contrast, the normalised, individual-level evictions of tenants by landlords or struggling home owners by banks, and even the collective displacement of residential communities by public and private landowners in a process of so-called urban regeneration, are excluded from this official definitional framework. This is despite such movements being beyond any person's reasonable ability to control, generating displacement on a large scale, and causing immense human suffering from disruption to schooling, employment, and family life, the loss of vital social networks, and the development of post-traumatic stress disorders (Elliott-Cooper et al. 2019; Watt 2018). Instead, Global North governments frame such involuntary moves as part of the continuum of internal migration *choices* made by individuals, households, and families for personal, financial, or employment reasons (Fielding 2012).

The politics of what counts as internal displacement, and what does not, is thus highly relevant. Official international recognition that increasingly draconian government policies towards housing and welfare generate displacement would imply duties on the state to address the material needs of those displaced and act to prevent further and prolonged displacement (UN Office for the Coordination of Humanitarian Affairs 2004; OHCHR 2016). Given the often complex and changing geographies of state administration in different national contexts, with different degrees of devolved responsibility to national, regional, and local authority areas, there are real financial and legal incentives for public or voluntary bodies to either not prevent or actively facilitate the forced movement of people beyond their own administrative borders whilst failing to record such movements to evade liability for service or welfare provision (Clarke and Cochrane 2013; Bowstead 2015b). More fundamentally, acknowledging that evictions by property owners constitute internal displacement would bring the entire neoliberal direction of capitalist economies into question and politicise the violent population upheavals wrought by urbanisation and mega-developments in the Global South.

It is vital, therefore, that academics and others concerned about the injustices of hidden displacement continue to evidence the causes and consequences, in order to pressure governments and relevant intergovernmental organisations like the UN to mitigate (Huisman 2013). However, such efforts must overcome long-standing methodological challenges for displacement studies, such as the largely unpredictable and hidden nature of the displacement itself, and the absence of purposively collected official data on residential mobility and causality. In this chapter, we draw on our own research to focus on the potential of administrative data to overcome some of the methodological hurdles to displacement research. By administrative data we mean information stored and used by or on behalf of public bodies to administer public policy. Given the reluctance to acknowledge and evidence the existence of other forms of residential displacement beyond disaster migration, it might seem illogical to suggest that administrative data has the potential for advancing displacement research. But below the surface of recent media headlines on mass displacement from the housing crisis we discover that the underlying evidence often comes from information held and used by or on behalf of public bodies.

The chapter continues with a brief review of the broad challenges of identifying displacement within existing data sources. It then sets out the tantalising potential of administrative data for identifying displacement and presents two case studies of their use in action from our UK-based research into regeneration and welfare changes (Hodkinson et al. 2016) and people (mostly women and children) escaping domestic violence (Bowstead 2015a, 2019). A final section offers some critical reflections on the promise of administrative data for research in general.

## The Methodological Challenges of Tracing Residential Displacement

Residential displacement in the Global North has a long history, from the violent land enclosures that accompanied the spread of capitalist property relations over many centuries (Hodkinson 2012) to the common nineteenth-century working-class experience of eviction and spatial segregation under private landlordism, to the mass displacements under governments' twentieth-century slum clearance programmes. There has been renewed political and academic interest in residential displacement in the UK from the rising numbers of households being evicted or pressured to leave urban areas as the overheating housing market meets stagnant or falling household incomes, exacerbated by austerity cuts to housing benefit and other welfare support since 2010 (Hodkinson and Robbins 2013; Paton and Cooper 2016; Watt 2018). However, our knowledge about residential displacement continues to lag behind.

As outlined in the introduction, one major reason behind the methodological challenges of tracing residential displacement is the failure of governments and intergovernmental bodies like the UN to collect monitoring data about different forms of internal displacement. Another challenge lies in the shadowy nature of forced home moves that makes displacement so hard to quantitatively track and qualitatively understand: residential displacement tends to be an indiscernible process involving an unknown number of people in unperceivable locations, who have usually relocated before they can be discovered, leaving few indicators behind (Atkinson 1998). This epistemological problem has been compounded by the lack of accurate or inclusive population movement research more generally. Instead of an official population register that is updated continually—such as in Sweden through its tax registration system—the UK is reliant on *estimates* of migration from modelling the ten-yearly census figures in combination with other population data sets such as birth and death registers and health and higher education records for internal migration (changes of residence), and survey data and National Insurance number applications for international migration (Office for National Statistics 2018). Such estimates are often disputed by local authorities, especially when they are used to allocate resources (County Councils Network 2018). Disputes arise not least because social surveys tend to only sample from the settled population and often do not cover anyone in temporary accommodation, such as hostels or staying informally with others.

These methodological limitations have both shaped and limited existing displacement scholarship in the UK. For example, quantitative researchers looking at macro-scale forced movements of households from gentrification have relied on proxies of displacement—such as changing incidences of occupational status or benefit claimants in longitudinal data sets (Atkinson 2000; Fenton 2016). While this has shown plausible causal correlations between the net growth of high-income households and the net loss of lower-income households in the rapidly gentrifying spaces of inner London, such quantitative studies are fraught with problems, not least because they cannot prove that any displacement has actually taken place. Various research has thus pursued more localised case studies of residential displacement in regeneration areas with an emphasis on the human experience (Hardy and Gillespie 2016; Hodkinson and Essen 2015; Lees et al. 2014; Watt 2018). And through painstaking work these studies have arguably produced more accurate statistics and richer experiential data of displacement, but usually only at very localised scales at given moments in time with no capacity to continue to track or update the data sets in the longer term.

Similarly, with the responsibility for providing domestic violence services being devolved to local government, as well as homelessness duties and the provision of social housing, there is no incentive for a local authority to provide any continuity of service support beyond its boundaries. In fact, it is generally legally prevented from doing so and will typically also limit the voluntary sector organisations it funds to provide services to individuals who live, work, or study within its boundaries. Not only does this cause problems for organisations trying to provide seamless and holistic support to those (usually women and children) affected by abuse, whether or not they cross an administrative boundary, it causes discontinuity of administrative records with data collection stopping at the boundaries. This can cause difficulties for individuals having to start again in terms of schooling, accessing health services to provide for health needs, claiming benefits, and organising housing. It also fragments the administrative record. As a result, there has been a failure by authorities and service providers to recognise the complex, multistage displacement journeys caused by domestic violence, with each service access, report, or relocation generally being treated instead as an individual incident (Bowstead 2017).

## The Tantalising Promise of Administrative Data

We know from official data that 85,000 households in England and Wales were evicted by their landlords over 2014 and 2015 from the publication of Ministry of Justice records of court decisions (MOJ n.d.). It has also been reported that 50,000 homeless families were moved out of their London borough between 2011 and 2015, with identified flows from inner to outer London, and from London to other regions ('Over 50,000 Families Shipped Out of London' 2015). Some academics and journalists are using Freedom of Information law to access unpublished data. For example, since 1997, Lees (2018) has collected official planning data on actual and planned demolitions of social housing estates to produce an estimated displacement of over 135,000 London council tenants and leaseholders.

In scalar terms, administrative data can cover the whole of the UK as well as the (changing) administrative relationship with the European Union, different nations and territories under devolution (England, Wales, Scotland, and Northern Ireland), regions, metropolitan and local authorities, combined authorities, and sub-local and cross-local geographies through the operational reach of actors such as social landlords. There is a mass of administrative data generated to monitor statutory and non-statutory service provision for the purposes of official statistics and evaluation. These service monitoring data normally contain counts and demographic profiles of the individuals accessing a service, measures of the service provided, and some details about outcomes. Such administrative data thus comprise an incredibly varied mixture of historic and contemporary information about people. To further explore their potential, this chapter will now turn to two different case studies of our own research into displacement in the UK using administrative data.

## Case Study 1: Mapping the Displacement Effects of Welfare Reform Using Local Authority Housing Benefit Administrative Data

In 2014, Hodkinson and Turner were funded by the Engineering and Physical Sciences Research Council (EPSRC) to explore the implications for welfare administrators, advisors, and claimants of the Coalition Government's (2010–2015) rollout of a fully digitalised welfare claiming system. The project focused on the Leeds Local Authority District and worked in partnership with Leeds City Council and voluntary advice organisations (see Hodkinson et al. 2016 for more details). One aim of the research was to identify what kinds of administrative data were held by public and voluntary bodies, and understand whether they could evidence involuntary changes to residential location as a consequence of welfare policy changes and benefit cuts. The research focused on cuts to housing benefit,[1] specifically the 'Removal of the Spare Room Subsidy' policy, more commonly known as the 'Bedroom Tax' that was introduced in April 2013 in England, Scotland, and Wales, and from November 2017 in Northern Ireland.

The UK government argued that reforms to housing benefit were needed to address the chronic shortage of social rented housing—the most affordable and secure form of housing in the UK with the lowest average rental levels and lifetime tenancies being the norm. It claimed that one of the main causes of this shortage was the significant under-occupation of social housing by its existing residents (Wilson 2019). Official data collected as part of the 2011 Census classified 1.6 million households living in social housing in England and Wales—39.4 per cent of the total—as having one or more 'spare bedrooms' (Office for National Statistics 2014). The UK government asserted that under-occupation of social housing had been permitted to grow over time due to the poor design of the housing benefit system, which had hitherto not linked the amount of housing benefit households could receive to the size of the property they were renting. Significantly, the census also found 2.1 million households renting in the private sector (49.5 per cent) and 12.4 million owner occupying households (82.7 per cent) were also 'under-occupying' their homes, but the government did not propose any measures to address this. It is argued elsewhere that the Bedroom Tax is better understood as part of a concerted neoliberal attack on social housing (Hodkinson and Robbins 2013).

The Bedroom Tax targeted all social housing tenants of working age who were claiming housing benefit and deemed to occupy a home with more bedrooms than their family size warranted. Official data estimates that 547,341 households were affected in May 2013, after the first month of the policy (DWP n.d.). Households deemed to have one surplus bedroom have seen their weekly housing benefit award cut by 14 per cent of their weekly asking rent level, or 24 per cent for two or more surplus bedrooms. So, an asking rent of £100 per week saw a housing benefit shortfall of £14 or £25 every week. The official rationale of the policy was to incentivise social housing tenants to either come off housing benefit altogether—by getting a job or increasing their hours of work—or avoid having to find the extra weekly benefit shortfall (with the risk of going into rent arrears and facing eviction) by *voluntarily* moving to a smaller home so they were less impacted financially. This would free up the home they were under-occupying for other households on the social housing waiting list. But moving home under such circumstances was hardly voluntary; and for those tenants that were evicted for rent arrears and made homeless, local authorities had no statutory obligation to rehouse under homeless law because the tenants would be regarded as having made themselves 'intentionally homeless' by not paying the rent. In other words, the Bedroom Tax in practice could foreseeably cause residential displacement.

To compound the sense of injustice, the government's calculation of 'surplus bedrooms to need' was not based simply on comparing the number of occupants with the number of bedrooms; instead, the age and sex of any children and the partnership status of adults were also considered. Two children under ten, irrespective of their sex, were expected to share a room, as were two children under sixteen of the same sex. This meant that in many cases, homes with three bedrooms occupied by two adults and two children were deemed to be 'under-occupying' and their housing benefit was cut. Linking to our second case study, the government also did not allow women victims of domestic violence to have a secure panic room excluded from the bedroom calculation; instead the government increased funding to local authorities 'discretionary housing payments' schemes so that domestic violence victims and others with specific bedroom needs such as the disabled could apply for temporary financial assistance to meet the costs of the Bedroom Tax.

## The Data

UK local authorities began to take over the administration of housing benefit claims from central government after 1982 until this became a universal requirement in 1989. As part of this role, local authorities were required to send the Department for Work and Pensions (DWP) a monthly data set known as the Single Housing Benefit Extract (SHBE) for the purpose of generating national statistics. These monthly SHBE files contain quantitative information on every current housing benefit and council tax benefit claimant and every member of that claimant's household in the local authority area. This is collected in approximately 290 fields of personal information with attributes including: name, date of birth, National Insurance number, tenancy type, full postal address, income, benefits claimed, household size and characteristics, weekly rent level, weekly housing benefit award, disability, stated ethnicity of main claimant, and landlord's name and address.

Leeds City Council provided a de-identified and less detailed version of SHBE dating back to April 2008 and rolling forwards on a monthly basis from January 2013 to October 2015. All identifying personal information was redacted, except for the full postcode, for approximately 90,000 claimant households each month. These extracts were generated from a live client management database that contains even more information about past and current housing benefit claimants, including qualitative records of verbal and written correspondence between a claimant and the local authority benefit team. Leeds City Council also provided a de-identified data set for all social housing tenancies affected by the Bedroom Tax. This data set was also collated on a monthly basis and had ten fields of information, including: number of bedrooms required according to the government regulation, number of bedrooms in the property, the age and sex of any dependent children, and the number of non-dependents. Rent arrears data were also included for council-only tenancies as the local authority did not have access to this information for those renting from housing associations.

Bespoke open-source Java software was developed to automate the processing: cleaning, linking, generalising, and visualising of these two types of data provided on a monthly basis. Automation allowed for outputs to be generated from the source input data, requiring only a small amount of manual intervention, producing processed data that could be easily reproduced. The analysis focused on identifying unique claimants across the data sets and tracking their claims over time during the period of January 2013 to October 2015. This enabled individuals to be spatially located each month, their residential movements tracked around the local authority area, as well as identifying any changes to both their tenancy type (e.g. council, housing association, private landlord, and temporary accommodation) and to the amount of housing benefit they received (usually associated with changes in employment status—households with sufficient income are not eligible for housing benefit).

## Findings

The administrative data provided a number of important and original insights. First, a far larger number of households were found to be affected by the Bedroom Tax policy than official figures suggested. Between April 2013 and October 2015, the official monthly head count of those in the Bedroom Tax in Leeds fell gradually from 8780 to 6332 social tenancies. However, unique benefit claim identifiers pointed to a cumulative total of 13,738 social tenancies (11,089 council and 2712 housing association) and 26,207 individuals—including a minimum of 2663 children under ten—affected at some point by the Bedroom Tax in the same period. If extrapolated to the UK scale, then over the same period an estimated 856,420 households could have been affected. This demonstrates that looking at the administrative data that produce official statistics in a different way can show a very different reality. Here the difference was between monthly snapshots—which are used in media and policy narratives—and cumulative totals, which neither central government nor local authorities were counting and reporting statistically. The consequence has been a recognition that official statistics significantly undercount the total numbers of tenancies and people affected by the policy.

A second set of findings related to identifying which households had been affected, for how long, and with what consequences. In Leeds, 6822 households either initially affected in April 2013 or subsequently affected were no longer subjected to the Bedroom Tax at the end of October 2015. Of these, just under half—3166 households—had escaped because their children's age or household size had changed, or the number of bedrooms in their home had been officially reclassified. Another 2621 households had left the housing benefit caseload altogether. Yet, 4642 households were effectively trapped in the Bedroom Tax either continuously or predominantly throughout April 2013 to October 2015. The data showed that the majority of these households were either single adults or adult couples, and thus to comply with the government's Bedroom Tax policy, they would have to move into one-bedroom social housing, which happens to be the scarcest type of social housing. Consequently, these particular households were stuck in the Bedroom Tax and could only downsize by being forced to give up their secure (lifetime), lower-cost social rental tenancies and instead move into the insecure and higher-cost private rental sector where short-term tenancies with limited rights are the norm. The data also demonstrated a significant rise in rent arrears for affected council tenancies; 84.7 per cent of council tenants still affected by the Bedroom Tax in October 2015 were in rent arrears, which may have prevented them from bidding for smaller social rented homes due to local authority housing allocation rules.

Third, and most relevant to the theme of this chapter, the administrative data revealed a significant amount of previously hidden displacement. Despite the much larger population of social tenancies affected and the growth in rent arrears, official evictions linked to the Bedroom Tax in Leeds were at the time negligible. However, we found at least 1565 households had been displaced by the Bedroom Tax in the first two and half years of the policy being introduced in Leeds with some households in fact moving more than once: 1395 households had moved into smaller accommodation within Leeds in order to escape the Bedroom Tax including 288 households who moved home and lost their secure council or housing association tenancies by entering the private rental sector, four of which went on to become homeless; and a further 170 households who downsized and yet actually remained trapped in the Bedroom Tax during this period. It is also important to add a cautionary note about the 2621 households who had left the housing benefit caseload altogether. While some may well have improved their financial situation so as to no longer be eligible for housing benefit, it is equally possible that some were also displaced, moving in with family or friends, or relocating to another city altogether, most likely as tenants in the private sector. Once again, if extrapolated to the UK scale, we are looking at between 62,351 and 166,776 households displaced from their homes by the Bedroom Tax alone. When we consider that the Bedroom Tax is just one of dozens of welfare policy changes that have resulted in drastic cuts to the incomes of the poorest members of society, the potential scale of policy-induced residential displacement since 2010 and the onset of austerity is much larger. Yet none of these forced moves are recognised as such by the UK government.

## Case Study 2: Tracing Forced Migration Due to Domestic Violence in the Supporting People Services Monitoring Data

Bowstead's research focuses on women's domestic violence journeys in England (Bowstead 2015a, 2017). Such journeys are generally kept secret, because of the ongoing risk from an abusive partner who knows your contacts and likely destinations. Women may access services as part of their journeys and therefore leave some trace in the administrative record, but other stages of their displacement journeys remain hidden from all records. However, the administrative data discussed here enabled research into aspects of these journeys. The key methods of analysis that had not previously been possible were the mapping of patterns of displacement, of rates of displacement per local authority, and the association of such patterns and journey distances with demographic characteristics of the individuals concerned, and characteristics of place such as rurality and deprivation. Importantly, it was also safe and ethical analysis on what are often hidden and dangerous journeys of escape.

Neither specialist domestic violence services, such as women's refuges, nor other kinds of temporary accommodation and support that are used by women and children escaping abuse are statutory services. As a result, provision is not based on any formula per population (or any other criteria) and varies enormously across the country; and services open and close over time (Towers and Walby 2012). This unstable pattern of services, accessed by individuals at risk and in need of support, does not generate any consistent data on the scope or scale of the displacement. However, through a specific funding programme and its associated monitoring requirements, data were generated which crossed administrative boundaries, and therefore enabled research on these domestic violence journeys.

## The Data

The Supporting People programme was launched in April 2003 by the UK government to help vulnerable people in England and Wales live independently and included funding for a wide range of housing-related support services. Two data collection systems were developed by government[2] to record standard information about clients (service users) at the start and end of receiving these services (ODPM 2002). The Client Record system (i.e. intake) ran from April 2003 and the Short Term Outcomes system (i.e. exit) from May 2007, both ending with the termination of the programme in March 2011. Data monitoring was carried out by each service provider and submitted to the Client Record Office at the Centre for Housing Research (CHR) in St Andrews for data collection, processing, and preliminary statistical analysis. Data were kept as discrete annual datasets and used in Annual Reports (CHR 2012). Supporting People services provided support to a range of 'Client Groups' and the Client Record therefore identifies both Primary and up to three Secondary Client Groups for each case.

In total, there were around 210,000 cases of service access per year: a total of nearly 1.7 million over the eight years. The key methodological opportunities of using Supporting People data are the safety of using de-identified data which are at the individual level but not disclosive when used under the terms of the Special Licence.[3] The sample is large, enabling statistical analysis with confidence in the significance of findings, and comprehensive in terms of these types of service across England. Other funding programmes may collect detailed monitoring data but be restricted to specific local authorities, or major cities—such as London—or over a shorter period of time. They may remain simply held by the funder for the time of the funding programme and only used for service monitoring and evaluation. Unusually, the Supporting People data were archived and made available to researchers under Special Licence from the UK Data Service (DCLG and University of St Andrews, Centre for Housing Research 2012).

The intake data sets include basic demographic variables on age, sex, ethnic origin, age of accompanying children, and some information on disability and additional needs; and the exit data sets included some of the same variables. Crucially, geolocation data were present at the local authority level or could be derived from other variables. Variables recorded the local authority area in which households lived before accessing housing-related support, whether they remained in the same accommodation or moved to access the services, their previous location if they moved, and their household location after leaving support. This research focused on cases where the Primary Client Group code was 'Women at risk of domestic violence,' 'People at risk of domestic violence,' and the data sets, when processed and combined, therefore enabled displacement from domestic violence to be both quantified and mapped.

Variables were generated on the frequency of women relocating within the same local authority area (residential mobility), the frequency of women leaving their local authority area (internal migration), and the frequency of women arriving in that local authority area to access a service, as well as rates per population. These rates were used to generate choropleth maps of rates of leaving, arriving, residential mobility, and net leaving with location variables processed as the centroids of the local authority area. This enabled the generation of flow maps and the measurement of straight-line distances for migration journeys across administrative boundaries. Because the location data were only at the local authority level, no distance measurement was possible on the journeys of residential mobility. Totals of origin and service location data were also used to generate annual data sets with the 354 English local authorities (2001 boundaries) as the individual cases, plus Scotland, Wales, Northern Ireland, and outside the UK as additional origins.

## Findings

The Supporting People data set allowed, for the first time, the scale of domestic violence displacement to be researched, both the geographical scale and that it involves tens of thousands of women and children a year—plus those that did not access services. Forced migration due to domestic violence was identified where domestic violence was the primary reason for accessing services, and where the individual had changed accommodation at the point of accessing the service. This gave a total of approximately 18,000 cases per year, with around 9500 women (over half with children) migrating across local authority boundaries to access services and around 8500 relocating within their local authority (i.e. residential mobility). Despite the similarity in the total numbers of cases each year over the eight years, the journeys to access services were found to be very varied with regard to distance moved, demographic characteristics of the cases, and characteristics of places. That there are such varied and individual journeys was evidenced by the data.

The pattern is of spatial churn, with women and children leaving everywhere and going to anywhere which provided a service. The geographical scale is particularly striking, raising important issues for service provision, with the mass of individual-level journeys not aggregating into any major flows; so that most local authorities have a net rate of around zero. Until these patterns were revealed in the England-wide data, many local authorities presumed that they experienced net arriving, because in their own data they were only aware of women accessing their services (either locally or from elsewhere) and had no idea about the scale of women leaving their area due to domestic violence. As discussed previously, it may suit local authorities for women to leave—to be displaced away—and be beyond their responsibility. Whilst many women need to relocate across boundaries, in terms of the abuse they are trying to escape, it is also clear from other research the incentives for local authority officers to advise women to leave to a refuge elsewhere—thus ending their statutory duties towards them.

This service monitoring data, because of its inclusion of location variables, its aggregation at the England-wide scale, and its archiving with the UK Data Service to provide licenced access for research, thereby enabled new insights on forced internal migration in the UK. Insights that are no longer possible because of the end of the Supporting People programme and the reversion to local funding and diverse and local service monitoring data collection (CHR 2015).

## Concluding Reflections

This chapter has demonstrated that it is both conceptually desirable and methodologically possible to broaden existing international understandings of internal displacement to incorporate involuntary residential moves in the Global North caused by the deliberate design or unintended consequences of government welfare and housing policies. In particular, we have shown that administrative data in two different areas of social policy—housing benefit reform and access to support for women and children experiencing domestic violence—can help to overcome long-standing methodological challenges for displacement studies, such as the largely unpredictable and invisible nature of displacement itself, and the absence of purposively collected official data on residential mobility and causality. Our analysis of two longitudinal data sets maintained and used for different forms of welfare support identified tens of thousands of people being displaced through changes to, or conditions of, accessing state support.

The potential of administrative data to enrich displacement studies in the UK and other countries with similar welfare and public administrative systems is enormous. There is a plethora of other existing live or historical administrative data sets that capture various forms of household moves and thus potential displacement. These include the allocation and management of social housing tenancies, local authorities discharge of homelessness duty, adjudications of eligibility and payments of social security and welfare claims, health and social care, education, advice-giving, the policing and justice system, state pension, tax, and electoral representation. This likely goldmine of displacement data is further enhanced by continual advances in computational architecture that enable researchers to stitch together and systematically map, analyse, and generalise large volumes of individual-level digital data from disparate sources covering variegated scales (Kitchin 2014). The data can be de-identified before being made available to researchers, enabling safer analysis, especially for issues such as domestic violence research about what are often secret and dangerous journeys of escape. Institutional self-interest to share data more openly is also growing as public spending cuts and rising service demand encourage local authorities and voluntary bodies to work with universities to better utilise their own administrative data for organisational planning and the wider social good. It shows that instead of relying on heavily redacted and overly generalised versions of data, academic collaborations with public bodies can unlock rich and complex internal administrative data that exist to provide far more detail on displacement flows and causality than ever before.

At the same time, several important caveats need to be made about the future use of administrative data for displacement research and research more widely. In general terms, these are data sets that were designed primarily to monitor and facilitate the administration of public funds and services provision. They can be highly complex, voluminous, and detailed; or simple, limited, and threadbare. Some administrative data are collected and managed in a systematic, consistent, and continuous way, while others can run on a more temporary and ad hoc basis, with varying degrees of updating. In addition, the data are generated by frontline staff in hundreds of services and, despite best efforts to clean the data, will always include some data errors and omissions. Some errors can make it seem like there was displacement when in reality there wasn't and vice versa.

This leads to the rather obvious limitation that administrative data are created for administrative purposes and therefore do not necessarily provide ideal variables for research of any kind. There is no displacement variable in these data, and displacement must be inferred through analysis. The quality of administrative data also suffers from the complex and changing geography of state administration in the UK, including different degrees of devolution from central government to Scotland, Northern Ireland, Wales, and England, and then a complex and changing patchwork of institutional boundaries within a nation. England, for example, has both unitary authorities and two-tier authorities (county and borough), as well as some combined authorities, directly elected mayors; and metropolitan authorities such as the Greater London Authority. In this mosaic of administrative responsibilities and boundaries, and multiple layers of administrative geographies, real challenges exist to information sharing and consistency of policy and practice responses, creating both gaps and cliff edges for particular people and data to fall off and effectively (administratively) disappear.

Finally, there is the problem of access, which despite a growing institutional self-interest to share data and work with academics remains a huge problem. Administrative data held about individuals are rarely de-identified or archived and are not usually made available for research at an individual level. Such data sets may also be regarded as commercially sensitive in a competitive tendering environment. Both the housing benefit and the Supporting People data were made available on special licences; however, the latter archive is no longer being updated because of the end of the Supporting People programme and the reversion to local funding and diverse and local service monitoring data collection (CHR 2015). More practically, sharing data, especially sensitive data of this kind, poses major resource, legal, and reputational risks to public and voluntary bodies. Local authorities in particular are wary of releasing data that under analysis could show them in a bad light, and the political toxicity of displacement creates a strong predisposition against sharing. This poses a new challenge of cultivating trusted and productive data-sharing partnerships between social scientists and external bodies in ways that still protect academic freedom (Administrative Data Taskforce 2012).

## Notes

1. Housing benefit is a means-tested social security payment for low-income households to meet housing costs for rented accommodation provided by social and private landlords.
2. Initially developed under the Office of the Deputy Prime Minister (ODPM), subsequent reorganisation brought it under the Department of Communities and Local Government (DCLG).
3. The licence restricts use of the data to a named researcher and strict conditions of data storage and access.

## References

Administrative Data Taskforce. (2012). *The UK Administrative research network: Improving access for research and policy*. UK Administrative Data Research Network website: Retrieved from https://esrc.ukri.org/files/research/administrative-data-taskforce-adt/improving-access-for-research-and-policy/.

Atkinson, R. (1998). Displacement through gentrification: How big a problem? *Radical Statistics*, 69. Retrieved from http://www.radstats.org.uk/no069/article2.htm.

Atkinson, R. (2000). Measuring gentrification and displacement in Greater London. *Urban Studies, 37*(1), 149–165. https://doi.org/10.1080/0042098002339.

Bowstead, J. C. (2015a). Forced migration in the United Kingdom: Women's journeys to escape domestic violence. *Transactions of the Institute of British Geographers, 40*(3), 307–320. https://doi.org/10.1111/tran.12085.

Bowstead, J. C. (2015b). Why women's domestic violence refuges are not local services. *Critical Social Policy, 35*(3), 327–349. https://doi.org/10.1177/0261018315588894.

Bowstead, J. C. (2017). Segmented journeys, fragmented lives: Women's forced migration to escape domestic violence. *Journal of Gender-Based Violence, 1*(1), 43–58. https://doi.org/10.1332/239868017X14912933953340.

Bowstead, J. C. (2019). Women on the move: Administrative data as a safe way to research hidden domestic violence journeys. *Journal of Gender-Based Violence, 3*(2), 233–248. https://doi.org/10.1332/239868019X15538575149704.

CHR (Centre for Housing Research). (2012). *Client records and outcomes (housing-related support): Annual report 2011–2012.* St Andrews, Scotland: Centre for Housing Research, University of St Andrews.

CHR (Centre for Housing Research). (2015). *Supporting People client records & outcomes: CHR final report 2003–2015.* Centre for Housing Research, University of St Andrews website. Retrieved from http://kimmckee.co.uk/wp-content/uploads/2011/11/SP-Final_Report-03-15.pdf.

Clarke, N., & Cochrane, A. (2013). Geographies and politics of localism: The localism of the United Kingdom's coalition government. *Political Geography, 34*, 10–23. https://doi.org/10.1016/j.polgeo.2013.03.003.

County Councils Network. (2018). *Counties welcome fair funding proposals but argue new formula must be able to fund social care pressures.* County Councils Network website. Retrieved from https://www.countycouncilsnetwork.org.uk/counties-welcome-fair-funding-proposals-argue-new-formula-must-able-fund-social-care-pressures/.

DCLG (Department for Communities and Local Government) and University of St Andrews, Centre for Housing Research. (2012). *Supporting people client records and outcomes, 2003/04–2010/11: Special Licence access [computer file] (No. SN: 7020).* Retrieved from https://doi.org/10.5255/UKDA-SN-7020-1.

DWP (Department for Work and Pensions). (n.d.). *Stat-Xplore.* Department for Work and Pensions website. Retrieved from https://stat-xplore.dwp.gov.uk/.

Elliott-Cooper, A., Hubbard, P., & Lees, L. (2019). *Moving beyond.* Marcuse: Gentrification, displacement and the violence of un-homing.. *Progress in Human Geography.* https://doi.org/10.1177/0309132519830511.

Fenton, A. (2016). *Gentrification in London: A progress report, 2001–2013* (CASE, No. 195). CASE, London School of Economics and Social Sciences. Retrieved from http://sticerd.lse.ac.uk/dps/case/cp/casepaper195.pdf.

Fielding, T. (2012). *Migration in Britain: Paradoxes of the present, prospects for the future.* Cheltenham: Edward Elgar.

Gottesdiener, L. (2013, August 1). The great eviction: The landscape of Wall Street's creative destruction. *The Nation.* Retrieved from https://www.thenation.com/article/great-eviction/.

Gutiérrez, A., & Domènech, A. (2017). Spanish mortgage crisis and accumulation of foreclosed housing by SAREB: A geographical approach. *Journal of Maps, 13*(1), 130–137. https://doi.org/10.1080/17445647.2017.1407271.

Hardy, K., & Gillespie, T. (2016). *Homelessness, health and housing: Participatory action research in East London.* University of Leeds/Feminist Trust website. Retrieved from http://www.e15report.org.uk/.

Hodkinson, S. (2012). The new urban enclosures. *City, 16*(5), 500–518. https://doi.org/10.1080/13604813.2012.709403.

Hodkinson, S., & Essen, C. (2015). Grounding accumulation by dispossession in everyday life: The unjust geographies of urban regeneration under the Private Finance Initiative. *International Journal of Law in the Built Environment, 7*(1), 72–91. https://doi.org/10.1108/IJLBE-01-2014-0007.

Hodkinson, S., & Robbins, G. (2013). The return of class war conservatism? Housing under the UK Coalition Government. *Critical Social Policy, 33*(1), 57–77. https://doi.org/10.1177/0261018312457871.

Hodkinson, S., Turner, A., & Essen, C. (2016). Exploring the impacts and implications of a changing UK welfare state under digitalisation and austerity: The case of Leeds. *University of Leeds*. https://doi.org/10.5518/wp/3.

Huisman, C. (2013). *Resourceful researchers: Tackling the elusive problem of displacement by going further upstream*. Presented at the RC21-Berlin, Berlin, Germany. Retrieved from http://www.rc21.org/conferences/berlin2013/RC21-Berlin-Papers/08-1-Huisman.pdf.

IDMC (Internal Displacement Monitoring Centre). (n.d.). *Global internal displacement database (GIDD)*. Internal Displacement Monitoring Centre website. Retrieved from http://www.internal-displacement.org/database.

Kitchin, R. (2014). *The data revolution: Big data, open data, data infrastructures and their consequences*. London: Sage.

Lees, L. (2018, March 16). *Challenging the gentrification of council estates in London*. Urban Transformations blog website. Retrieved from https://www.urbantransformations.ox.ac.uk/blog/2018/challenging-the-gentrification-of-council-estates-in-london/.

Lees, L., London Tenants Federation, Just Space, & SNAG. (2014). *Staying put: An anti-gentrification handbook for council estates in London*. Southwark Notes Archive Group website. Retrieved from https://southwarknotes.files.wordpress.com/2014/06/staying-put-web-version-low.pdf.

MOJ (Ministry of Justice). (n.d.). *Mortgage and landlord possession statistics*. Ministry of Justice website. Retrieved from https://www.gov.uk/government/collections/mortgage-and-landlord-possession-statistics.

ODPM. (2002). *Supporting people: Guide to accommodation and support options for households experiencing domestic violence*. Retrieved from https://webarchive.nationalarchives.gov.uk/20091211213042/http://www.spkweb.org.uk/NR/rdonlyres/C5A1BAEC-2F69-4BB3-8338-3F6F623DB852/1771/Guide_to_Accomodation_and_Support_Options_for_Hous.pdf.

Office for National Statistics. (2014). *Census suggests 1.1 million households in England and Wales were overcrowded*. Office for National Statistics website. Retrieved from https://webarchive.nationalarchives.gov.uk/20160105214004/http://www.ons.gov.uk/ons/rel/census/2011-census-analysis/overcrowding-and-under-occupation-in-england-and-wales/sty-household-occupancy-and-overcrowding.html.

Office for National Statistics. (2018). *Methodology guide for mid-2017 UK population estimates (England and Wales): June 2018*. Retrieved from https://www.ons.gov.uk/peoplepopulationandcommunity/populationandmigration/

populationestimates/methodologies/methodologyguideformid2015ukpopulationestimatesenglandandwalesjune2016.

OHCHR. (2016). *Mandate of the Special Rapporteur on the human rights of internally displaced persons: Resolution adopted by the Human Rights Council.* UN Office for the Coordination of Humanitarian Affairs website. Retrieved from http://www.ohchr.org/Documents/HRBodies/SP/CallApplications/HRC33/A_HRC_RES_32_11_en.docx.

Over 50,000 families shipped out of London boroughs in the past three years due to welfare cuts and soaring rents. (2015, April 29). *The Independent.* Retrieved from https://www.independent.co.uk/news/uk/home-news/over-50000-families-shipped-out-of-london-in-the-past-three-years-due-to-welfare-cuts-and-soaring-10213854.html.

Paton, K., & Cooper, V. (2016). It's the state, stupid: 21st gentrification and state-led evictions. *Sociological Research Online, 21*(3), 1–7. https://doi.org/10.5153/sro.4064.

Towers, J., & Walby, S. (2012). *Measuring the impact of cuts in public expenditure on the provision of services to prevent violence against women and girls.* Trust for London & Northern Rock Foundation website. Retrieved from https://www.trustforlondon.org.uk/publications/measuring-impact-cuts-public-expenditure-provision-services-prevent-violence-against-women-and-girls/.

UN General Assembly. (2013). *Report of the Special Rapporteur on adequate housing as a component of the right to an adequate standard of living, and on the right to non-discrimination in this context, Raquel Rolnik: Addendum – Mission to the United Kingdom of Great Britain and Northern Ireland* (No. A/HRC/25/54. Add. 2). Retrieved from https://www.ohchr.org/EN/HRBodies/HRC/RegularSessions/Session25/Documents/A_HRC_25_54_Add.2_ENG.DOC.

UN Office for the Coordination of Humanitarian Affairs. (2004). *Guiding principles on internal displacement* (2nd ed.). UN Office for the Coordination of Humanitarian Affairs website. Retrieved from https://www.brookings.edu/wp-content/uploads/2016/07/GPEnglish.pdf.

UNHCR. (2018). *Global trends: Forced displacement in 2017.* UNHCR website. Retrieved from https://www.unhcr.org/5b27be547.pdf.

Watt, P. (2018). 'This pain of moving, moving, moving': Evictions, displacement and logics of expulsion in London. *L'Annee Sociologique, 68*(1), 67–100.

Watt, P., & Smets, P. (2017). *Social housing and urban renewal: A cross-national perspective.* Bingley: Emerald Publishing.

Wilson, W. (2019). *Under-occupying social housing: Housing Benefit entitlement* (House of Commons Library Briefing Paper No. 06272). Retrieved from https://researchbriefings.files.parliament.uk/documents/SN06272/SN06272.pdf.

Zuk, M., Bierbaum, A. H., Chapple, K., Gorska, K., & Loukaitou-Sideris, A. (2018). Gentrification, displacement, and the role of public investment. *Journal of Planning Literature, 33*(1), 31–44. https://doi.org/10.1177/0885412217716439.

# Part V

Section Five: Governing Displacement

# 31

# Intervention: Forensic Oceanography—Tracing Violence Within and Against the Mediterranean Frontier's Aesthetic Regime

Charles Heller and Lorenzo Pezzani

Illegalised migration across the Mediterranean and fatalities at sea have been structural and highly politicised phenomena only as of the end of the 1980s, when, in conjunction with the consolidation of the freedom of movement within the EU, migrants from the Global South were increasingly excluded from accessing European territory. This however only resulted in their movement operating through clandestine strategies, in particular by crossing the sea on overcrowded vessels. In turn, European states have deployed across the Mediterranean frontier a vast array of bordering practices and techniques to contain and channel migrants' movements. The dialectical process between control and escape has had a harrowing human cost: more than 30,000 migrants have perished at sea since the end of 1980.[1]

Most migrants' deaths across the Mediterranean frontier have not only occurred *at* sea but *through* the sea, which has been turned into a deadly liquid as a result of the EU's exclusionary policies. While the *liquid violence* of the maritime frontier is thus *mediated* by water,[2] it is also mediated by images and a constantly shifting *aesthetic regime*. We use the term 'aesthetic' here in the sense underlined by Jacques Rancière as what presents itself to sensory experience (Rancière 2004, p. 13). Distinct conditions of (in)visibility and (in)audibility are imposed on to the maritime frontier by states' restrictive

C. Heller (✉) • L. Pezzani
Goldsmiths, University of London, London, UK
e-mail: l.pezzani@gold.ac.uk

policies but also shaped, transformed, and contested by the multiple other actors. While illegalised migrants seek to cross borders undetected, agencies aiming to control migration try to shed light on acts of unauthorised border crossing through surveillance means in order to make the phenomenon of migration more knowable, predictable, and governable. However, as Nicholas De Genova (2013) has incisively analysed, the *border spectacle* is highly ambivalent and hides as much as it reveals. For their part, migrants in distress may do everything they can to be seen so as to be saved from drowning. Conversely, border agents not only attempt to deliberately hide the structural violence inherent to practices of policing maritime migration, they may also choose not to see migrants in certain instances, considering that rescuing them at sea entails the responsibility for disembarking them and processing their asylum claims and/or deporting them. This has led to repeated cases of migrants who have been left abandoned to drift at sea.

It is in the aim of contesting the EU's liquid violence that we initiated the Forensic Oceanography project in 2011, within the wider Forensic Architecture agency. To contest the violence of borders, we also had to contest the boundaries of what could be seen and heard at the maritime frontier and exercise what we have called a *disobedient gaze*: revealing what state actors have sought to conceal, and not revealing that which they seek to shed light upon.

Our project was sparked by a 2011 incident that came to be known as the 'left-to-die boat' case.[3] In wake of the Arab uprisings that led to renewed crossings, and at the height of the NATO-led military intervention in Libya, during which more than thirty-eight warships were deployed off the coast, seventy-two migrants fleeing the war zone were left to drift in the central Mediterranean Sea for fourteen days. Sixty-three human lives were lost, despite the survivors calling Father Zerai (an Eritrean priest based in Rome) via satellite phone, despite distress signals sent out to vessels navigating in this area, and despite several encounters with military aircraft and a warship. While survivor testimonies indicated increasing instances of non-assistance, during this period the Mediterranean appeared as a 'black box' for civilian actors, in which the capacity to see and document the events occurring at sea was nearly entirely in the hands of state actors. The challenge we faced as we embarked on our investigation in support of the nine survivors and a coalition of NGOs was precisely in wresting the capacity to sense the sea away from state actors, so as to make the violence of abandonment visible and breach the impunity in which it was being perpetrated.

Images could be of only limited assistance in the process. While several photographs were taken at different moments during these tragic events by military personnel as well as the passengers themselves, only one of

**Fig. 31.1** Reconnaissance picture of the 'left-to-die boat' taken by a French patrol aircraft on 27 March 2011. (Credit: Council of Europe)

them—taken by a French surveillance aircraft during the first day of the migrants' journey (Fig. 31.1)—was released in response to a parallel investigation by the Council of Europe (PACE 2012). In the absence of revelatory images documenting these events, our investigation had to rely on the 'weak signals' that underpin truth production practices in the field that Thomas Keenan (2014), after Allan Sekula, has called 'counter-forensics.' By corroborating survivors' testimonies with information provided by the vast apparatus of remote sensing technologies that have transformed the contemporary ocean into a digital archive, we assembled a *composite image* of the events. The expertise of an oceanographer allowed us to model and reconstruct the drifting boat's trajectory, and satellite imagery analysis to detect the presence of a large number of vessels in the vicinity of the drifting migrant boat that did not heed their calls for help (see Fig. 31.2). While these technologies are often used for

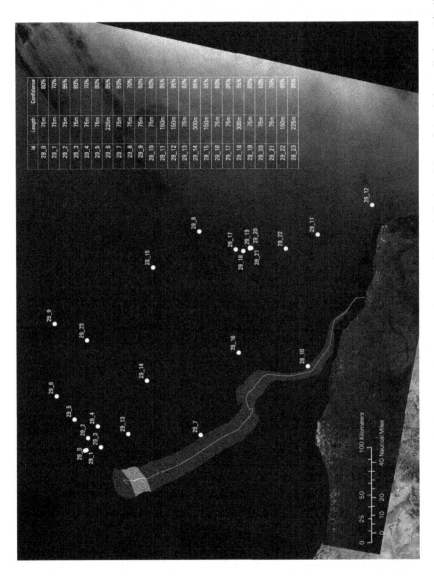

**Fig. 31.2** Analysis of the 29 March 2011 Envisat satellite image showing the modelled position of the 'left-to-die boat' (yellow diagonal hatch) and the nearby presence of several military vessels who did not intervene to rescue the migrants. (Credit: Forensic Oceanography and SITU Research, *Report on the Left-to-Die Boat Case*)

the purpose of policing and detecting illegalised migration as well as other 'threats,' we repurposed them to find evidence of the failure to render assistance. Not only did our reconstruction of the migrants' drift allow us to demonstrate that the migrants had remained within NATO's maritime surveillance area during their fourteen days of deadly drift but, by identifying many ships in the vicinity of the migrants' boat (see Fig. 31.2), our report allowed the NGO coalition we collaborated with to file several legal cases against the different states—including France, Spain, Italy, and Belgium—whose assets had taken part in the NATO-led operation and who shared a degree of responsibility for the death of the sixty-three passengers.[4]

## Notes

1. See the list of migrant deaths at the European borders established by UNITED for Intercultural Action: http://unitedagainstrefugeedeaths.eu/about-the-campaign/about-the-united-list-of-deaths/.
2. Here we draw on Sean Cubitt's expanded understanding of mediation, which, beyond technologically mediated communication processes between humans, he defines as 'the material processes connecting human and nonhuman events (…). Mediation is the primal connectivity shared by human and nonhuman worlds' (Cubitt 2017, p. 3). The way in which, in another text, he talks about sunlight as that which 'mediates the sun and the earth' (Cubitt 2014) further points to the understanding of mediation that inspires us here.
3. For our reconstruction of these events, see our report: https://content.forensic-architecture.org/wp-content/uploads/2019/06/FO-report.pdf (Heller et al. 2012). Our video animation Liquid Traces summarises our findings: https://vimeo.com/128919244.
4. https://www.fidh.org/La-Federation-internationale-des-ligues-des-droits-de-l-homme/droits-des-migrants/63-migrants-morts-en-mediterranee-des-survivants-poursuivent-leur-13483.

## References

Cubitt, S. (2014, August 13). How to connect everyone with everything [Web log post]. Retrieved from http://seancubitt.blogspot.com/2014/08/how-to-connect-everyone-with-everything.html.
Cubitt, S. (2017). *Finite media: Environmental implications of digital technologies*. Durham, NC: Duke University Press.

De Genova, N. (2013). Spectacles of migrant 'illegality': The scene of exclusion, the obscene of inclusion. *Ethnic and Racial Studies, 36*(7), 1180–1198.

Heller, C., Pezzani, L., & Situ Research. (2012). Report on the left-to-die boat. London: Forensic Architecture. Retrieved December 10, 2013, from https://content.forensic-architecture.org/wp-content/uploads/2019/06/FO-report.pdf.

Keenan, T. (2014). Getting the dead to tell me what happened: Justice, prosopopoeia, and forensic afterlives. In Forensic Architecture (Ed.), *Forensic: The architecture of public truth* (pp. 35–55). Berlin, Germany: Sternberg Press.

PACE (Parliamentary Assembly of the Council of Europe). (2012). Lives lost in the Mediterranean Sea: Who is responsible? PACE. Retrieved from http://assembly.coe.int/CommitteeDocs/2012/20120329_mig_RPT.EN.pdf

Rancière, J. (2004). *The politics of aesthetics: The distribution of the sensible*. London and New York, NY: Continuum.

# 32

# Governing the Displaced: Contradictory Constellations of Actors, Ideas, and Strategies

Lama Tawakkol, Ali Bhagat, and Sarah E. Sharma

## Introduction

Displaced populations face uneven and discriminatory discursive and institutional governance techniques as a result of labels and classification, geographic location, and other indicators of identity such as class, gender, and race. This chapter examines how states, non-state organisations, and the private sector unequally govern displaced populations in the context of contemporary neoliberalism, particularly in light of recurring financial crises and resulting waves of austerity at international, national, and municipal levels. It argues that there are contradictions in the discursive and the material, indicating tensions between official narratives framing policy debates on displaced groups and the actual practices of capitalist governance, paired with societal unwillingness to accept and accommodate economic, environmental, and conflict-based refugees. The most prominent contradictions prevalent in the neoliberal management of displaced populations discussed here include: first, differing realities for displaced populations due to how they are labelled or where they are (re)settled; second, clashing commitments from different scales of governance

---

L. Tawakkol (✉) • S. E. Sharma
Department of Political Studies, Queen's University, Kingston, ON, Canada
e-mail: l.tawakkol@queensu.ca; sarah.sharma@queensu.ca

A. Bhagat
Department of Politics, University of Manchester, Manchester, UK
e-mail: ali.bhagat@manchester.ac.uk

and the devolution of responsibility to municipal states and individuals; and, third, how the increasing role of private actors in governing displaced groups results in further disenfranchisement. To highlight how these contradictions manifest themselves, and the tensions that arise when attempting to resolve them, the chapter surveys the narratives and discourses surrounding different displaced populations, how refugees are sheltered, and how private solutions are introduced to support social reproduction.

Forced displacement is defined at the international level as 'the involuntary movement, individually or collectively, of persons from their country or community, notably for reasons of armed conflict, civil unrest, or natural or man-made catastrophes' (IOM 2011, p. 39). For the purposes of this chapter, we use UNHCR's broad definition of 'displaced populations,' including refugees, asylum seekers, internally displaced populations (IDPs), and stateless persons, who are displaced due to conflict, environmental, and economic conditions. While the boundaries and labelling of such categories are blurry and constantly in flux, depending on policy changes, we echo Ehrkamp (2017) in noting that these groups all forcibly migrate and should not be seen merely as labour in transit (De Genova 2002; Milner 2011). With these changing categorisations, however, international rights attributed to displaced populations vary and an increasing number of actors become involved in governing fragmented populations across various geographies and scales. We focus on how this is done particularly through neoliberal governance tactics, and define neoliberalism as complex and contradictory modes of ideological and material force surrounding market-based governance (Peck and Tickell 2002; Roberts and Soederberg 2014). With reductions in overall official development assistance, private actors are being courted to fund programmes for displaced groups, both within and outside refugee camps, with tactics like market-based housing and refugee microfinance. For willing corporate actors, displaced groups are presented as an *opportunity* as well as a challenge, so long as proper (i.e. market-based and credit-driven) development solutions are undertaken in comparison to previously failed and inefficient solutions (public spending). We counter this narrative by exploring the tensions inherent within the governance structures of the forcibly displaced and the processes of neoliberalism, both at local and at global scales.

## Governing Labels: Narratives and Categorisations of Displaced Populations

Forcibly displaced populations are governed, at the most basic level, by the labels they are given. In outlining their legal status and rights, these labels, in turn, have major consequences on determining who is forcibly displaced, what they are entitled to, and who is excluded from these protections and services (Crawley and Skleparis 2018, p. 59). Typically, forcibly displaced populations are recognised through the criteria provided by international law, UNHCR, and other international organisations, all of which reflect a lack of choice by the displaced. The 1951 Geneva Convention on the Status of Refugees defines a refugee as someone who 'owing to a well-rounded fear of persecution for reasons of race, religion, nationality, membership of a particular social group or political opinions, is outside the country of his nationality and is unable or, owing to such fear, is unwilling to avail himself of the protection of that country' (IOM 2011, p. 79). Similarly, an asylum seeker is 'a person who seeks safety from persecution or serious harm in a country other than his or her own and awaits a decision on the application for refugee status under relevant international and national instruments' (IOM 2011, p. 12). These definitions, predominantly based on a notion of extenuating circumstances and coercion, distinguish the forcibly displaced from other migrant groups. The 'migrant' label, on the other hand, signifies groups who do not necessarily have much in common; it is usually understood as including 'all cases where the decision to migrate was taken freely by the individual concerned for reasons of "personal convenience" and without intervention of an external compelling factor' (IOM 2011, p. 61). While creating neat and seemingly objective categories, such distinctions create discrepancies in the ways displaced peoples are governed, at both the popular and policy levels, and in the discourses that surround them. Groups officially recognised as 'genuine' refugees receive sympathy and aid, and are prioritised over 'migrants,' who are often viewed with suspicion and antagonism as 'opportunistic' and potentially security threats (Jacobsen 2006, p. 274; Sajjad 2018, p. 46).

In terms of formal governance, officially recognised refugees and IDPs are under the charge of their host countries. These countries have a responsibility under international law to provide them with the safety and services they need, even though the latter are not awarded formal refugee legal status (Jacobsen 2006, p. 276; Williams 2008, p. 511). Nevertheless, it is the host governments who decide how their responsibilities are to be fulfilled and create the policies governing the displaced. Even in the cases where the responsibility for the displaced does not directly lie with the host countries, as in the cases of UN assistance or private sponsorships, host governments are the ones who lay out the rules in which these entities operate (Jacobsen 2006, p. 278; Kumin 2015). They determine eligibility for the assistance and delineate the boundaries within which it is provided. Accordingly, more often than not, refugees' guaranteed rights are granted within strict boundaries. In some cases, forms of employment for refugees are restricted (Crawley and Skleparis 2018, p. 56). As will be outlined in more detail in the coming sections, refugees might only be able to find low-level employment with minimal benefits. This does not have to be for official reasons but could also be because of structural barriers, such as language skills (Jacobsen 2006, p. 282; Lewis et al. 2015, p. 592). In other cases, refugee movement can be constrained by not being allowed into certain areas (Crawley and Skleparis 2018, p. 56).

Matters become even more complicated for refugees who deviate from these structures in any way. In countries where refugees are mandated to live in camps, they are only offered aid and protection in those areas (Jacobsen 2006, p. 278). Those living outside camps, for any reason, are left to fend for themselves, regardless of their legal rights. Refugees in urban areas are often harassed in various ways; they are abused, discriminated against, and even arrested. They also have less access to the services and assistance provided in refugee camps by both the government and other entities (Jacobsen 2006, p. 277). In more extreme cases, states have provided alternatives to refugee status and pressured forcibly displaced populations to accept them in an effort to get around their own legal obligations (Fawaz et al. 2018, p. 12). In many ways, the governance of refugees is also considerably predicated on their geographical location, since Global North countries generally have more resources and are able to fulfil their legal obligations towards these populations. On the other hand, Global South countries, which lack the capacity to adequately protect or provide refugees with services, will usually either leave their obligations unfulfilled or delegate the responsibility of governing displaced populations to UNHCR and other national and international organisations (Jacobsen 2006, p. 281). Nevertheless, despite the obstacles they face, refugees are in relatively more secure positions than other displaced populations.

Even though people are forced to flee their homes for many reasons, the formal criteria for seeking refuge do not recognise all of them or guarantee them the same rights. Labels other than refugee do not offer legal status and, thus, these displaced populations lack formal protection and opportunities (World Bank 2017, p. 113). Asylum seekers are often not even allowed to work and are sometimes denied other rights by their host governments, including a choice in housing (Jacobsen 2006, p. 281; Lewis et al. 2015,

p. 591). Persons forced to flee for reasons that are not listed in the 1951 Geneva Convention, such as environmental and climate refugees, are not considered refugees in the legal sense of the term. When environmental/climate refugees are considered, it is only as IDPs; their transnational displacement is yet to be recognised as a legitimate cause for formal refugee status (Williams 2008, p. 513). Because they are only rhetorically, not officially, refugees, environmental refugees are not guaranteed any rights by international law or their host states; rather, they are generally regarded as (voluntary) migrants (Duong 2010, p. 1249). Migrants, deemed to have had a choice in their displacement, are subject to immigration laws and not guaranteed any protection or services. In fact, they are sometimes seen as 'undeserving' of that protection (Crawley and Skleparis 2018, p. 49).

The discrepancies in recognition and rights are even more pronounced in the cases of irregular, or 'illegal,' migrants, or asylum seekers who are denied refugee status but continue to live in their host countries. These groups lack even proper access to employment and services (Jacobsen 2006, p. 281) and are in worse situations than refugee populations, insofar as they are more vulnerable to exploitative and coercive work conditions (Lewis et al. 2015, p. 591). Because they are seen as having had a choice in leaving their home countries (officially regarded as safe) or having illegally entered or stayed in the country, or both, (irregular) migrants are subject to stricter policies, procedures, and even detention (Crawley and Skleparis 2018, p. 52). They are often also blamed for social problems, by the public and policymakers, even when their host countries had been struggling well before they arrived (World Bank 2017, p. 7). These same discourses apply to refugees when they seek to resettle in other areas (especially in the Global North). In those cases, they become viewed as 'migrants,' seeking better economic prospects, rather than 'genuine' refugees (Crawley and Skleparis 2018, p. 58). It becomes clear, then, how the inconsistencies in how forcibly displaced populations are viewed and governed are inextricably linked to the labels they are given.

Because of their grave impact on governance, the labels defining and distinguishing the forcibly displaced are increasingly proving problematic. They do more than demarcate rights and responsibilities; rather than being objective markers of status, these labels create and shape social realities (Sajjad 2018, p. 41). As detailed above, they determine who gets what. As they stand today, these labels, and the hierarchy they reflect, ignore how there are various other legitimate drivers of forced displacement than the ones officially recognised (Crawley and Skleparis 2018, p. 50). They do not take into account new or intensified factors forcing people to flee both within and beyond their state borders. As neoliberal policies continue to expand across the globe, people are driven away from home and displaced by nature-related causes as well as for socioeconomic reasons, including a lack of opportunities and prospects, poverty, or debt (Lewis et al. 2015, p. 593). Instead, these labels remain heavily embedded in a neoliberal and individualistic understanding and serve to predominantly exclude and marginalise people fleeing 'natural disaster, war or broadly based political and economic turmoil' rather than persecution in the sense of the Geneva Convention (Duong 2010, p. 1250). This also contains a geographic element, since the largest numbers of refugees and forcibly displaced populations reside in the Global South; the World Bank (2017, p. 10) estimates that 88 per cent of refugees and 94 per cent of IDPs live in countries that are economically below the global average. The rest of the chapter builds on the discrepancies of these labels to expand on the various other contradictions inherent to the structures governing displaced populations, showcasing them with particular focus on the housing and employment of refugees (arguably the most secure of the forcibly displaced).

## Governing Housing: Displaced Groups and Precarious Shelter

Another important indicator of governance when it comes to forcibly displaced groups is housing. This section focuses on refugee housing in the context of global neoliberal austerity, which results in a lack of adequate and affordable shelter for refugees both within and outside refugee camps. It reveals that the resettlement of forcibly displaced populations must be situated against the unequal relations of market-based housing provision which individuals are expected to provide for themselves, if not through existing resources then through forms of often coercive (in)formal credit. Although the politics and geographies of refugee camps continue to be a large focus in academic literature (Ehrkamp 2017), a growing body of scholarship documents the experience of displaced populations in the urban context (Darling 2016; Soederberg 2019). Three themes in the literature emerge: (1) within refugee camps, displaced populations are expected to rely on themselves for governance, despite contradictory institutional arrangements across space and place; (2) at the urban level, refugees must compete with other marginalised populations for scarce affordable housing in the context of neoliberal austerity, and (3) both within and outside of camps, the social reproduction of displaced groups is constricted by discriminatory features of governance and capitalism, which play out unevenly according to time, geography, and other social indicators such as class, gender, and race.

Refugee camps, far from being temporary safe havens, can house displaced populations over decades, with examples of Palestinians in Lebanon and Somalis in Kenya (Hyndman and Giles 2016). Within refugee camps, Easton-Calabria and Omata (2018) write that economic 'self-reliance' has become the key framework for refugees to sustain themselves. UNHCR (2005, p. 1) defines self-reliance as, 'the social and economic ability of an individual, a household or a community to meet essential needs in a sustainable manner.' Although historical narratives of self-reliance suggest environments where refugees can actually have the rights, administrative leeway, and space for employment and income-generating activities, refugees housed in camps have long been denied opportunities that can facilitate social reproduction (i.e., 'life's work,' or the labour required to both meet day-to-day needs and biological reproduction) (Mitchell et al. 2003). In fact, refugee camps, rather than presenting opportunities for marginalised groups to become self-reliant, have become environments that discipline vulnerable populations through daily processes of regulation and control (Conlon 2010; Darling 2011; Mountz 2011). Further academic scholarship explores these sites as states of exception where the law is suspended (Agamben 1998; Darling 2009), leading to various levels of maltreatment based on differentiated bureaucratic labels of deservedness. Contesting a heterogeneous narrative, however, are analyses that stress the political agency of refugees in camps: for instance, the development of housing alongside representative political structures in Palestinian refugee camps in Lebanon (Tuastad 2017). Despite these debates, poor housing conditions in refugee camps can be understood in terms of neoliberal development paradigms resulting in declining public funding spread thin across a global landscape of growing displaced populations (Alnsour and Meaton 2012; Zamore 2018). As a result, housing issues such as overcrowding, inadequate protection, and expensive rent reflect poor general outcomes in social, health, environmental, and educational indicators in refugee camps.

While the majority of refugee camps are located in the Global South, there has been renewed attention to camps in the Global North since the 2015 European refugee crisis, particularly to the poor conditions of those in Greece and Italy (UNHCR 2017; Pinelli 2018). This complements the ongoing existence of detention centres in European and North American countries, which have come under increasing scrutiny in the United States and England (Gill 2009). While camps have historically been the solution for protracted displacement, urban areas are increasingly housing the various forms of burgeoning displaced populations in both the Global North and South: for instance,

60 per cent of refugees and 60–80 per cent of IDPs live in cities (Landau 2006; UNHCR 2016; iDMC 2018). Scholarship examining urban displaced populations has explored how these groups in the Global South are more likely to move to informal settlements (Sanyal 2012) and how poor refugees are more vulnerable to eviction and harassment in comparison to richer refugees who can pay for personal protection (Campbell 2006). Further work should be undertaken to theorise and understand how the housing of internally displaced populations and climate refugees overlaps with the above phenomena, particularly in countries that face multiple forms of displacement at the urban level.

In the Global North, a growing body of scholarship examines displaced populations that have arrived in European cities since the 2015 refugee crisis. The majority of these populations are Syrian, Afghani, Iraqi, and Libyan, with other displaced populations migrating from the Middle East and Africa. Soederberg (2019) argues that the resettlement and housing of refugees must be situated against the backdrop of austerity urbanism, or the intensification of existing neoliberal practices that have resulted in entrenching pro-market solutions to housing provisioning alongside the state's fiscal retrenchment (Peck 2012). In this scenario, refugees compete with pre-existing marginalised populations for scarce and expensive housing in free market systems. Further, as countries implement reforms where fiscal responsibilities are devolved to the municipal level, cities grapple with housing crises and growing urban populations. As a result, rates of evictions and homelessness have risen for refugee populations across European cities that have hosted these groups since 2015.

Following the above call, further scholarship could also analyse processes of uneven displaced experiences in urban centres. This is increasingly significant as refugee camps become agglomerated within cities and heightened numbers of displaced groups migrate to cities both internally and externally for various reasons (Darling 2017), and in light of their overlap with local processes of displacement marked by informality, evictions, and homelessness. With discourses of scarce state resources and heightened levels of indebtedness, narratives of self-reliance are not only peddled to refugees but also espoused as disciplinary modes of governance towards a growing number of poor and dispossessed globally. With the further retrenchment of public funding and the growing interest of private actors, questions pertain to how the forcibly displaced will be governed in the future in terms of housing, and social reproduction, more generally. The following section considers their experiences as labour and in finding employment.

## Governing Employment: Entrepreneurship and Financialised Philanthropy

The third prong of refugee survivability concerns the contradictions surrounding livelihood and work—or lack thereof—in camps and urban settlements. Labour is managed on multiple and interrelated scales including the international, national, and urban. Everyday dimensions of survival deserve analytical attention as many refugees face various barriers to access upon relocation. Labour is also contextualised by neoliberalism in the various contexts of migration. With this in mind, refugees are framed as marginalised populations unwanted by state and society. Despite general xenophobia towards refugees, global finance vis-à-vis microcredit loans and other forms of cash transfers are increasingly significant in reshaping refugee survival in physical and virtual space. The penetration of global finance feeds into wider trends of neoliberal governance, which focus on self-reliance through advancements in financial technologies (fintech) that shrink time, space, and scale for the purposes of capital accumulation. Gabor and Brooks (2017) refer to this as fintech-philanthropy-development (FPD) where poverty is a new frontier for profit-making (Soederberg 2014).

UNHCR's (2005) refugee self-reliance policy was first aimed at finding durable solutions to protracted refugee situations through the Millennium Development Goals (MDGs). Three main strategies come out of this policy framework: cash-for-work, food-for-work, and labour-based infrastructure reconstruction. Refugees are paid directly in cash for small-scale labour efforts or work for food (UNHCR 2005, pp. 68–70). Austerity underpins these initiatives as do the overarching aims of local economic development. These initiatives aim to reduce aid flows and turn refugees from passive recipients of aid to hard-working populations deserving of humanitarian assistance. The Kenyan case illustrates decades of governance strategies surrounding the world's largest camps in Dadaab and Kakuma and is used in this section as a way to empirically ground these international regulatory movements. In recent years, the Kenyan state has threatened to close these camps in favour of voluntary repatriation for Somalis in Dadaab—an inherently racialised process fraught with ethnic and sociopolitical issues. The threatened closures are in response to ongoing national austerity which led to refugee camp alternatives in Kakuma such as the Kalobeyei settlement which focuses instead on repatriation. The central purpose of the Kalobeyei project is to integrate refugees and residents of Turkana county where the settlement is located. Within the context of diminished funding, the Kalobeyei integrated settlement moves away from the 'aid model' and instead focuses on refugee self-reliance (UNHCR Kenya 2017).

Policies of putting refugees to work that emerged in the early to mid-2000s were gradually accompanied by policies surrounding microfinance loans framing refugee settlements as a new opportunity for private development. UNHCR states, 'microfinance is the provision of financial services in a sustainable way to micro-entrepreneurs or any people with low incomes who do not have access to commercial financial services … microfinance is banking with the poor' (Azorbo 2011, p. 2). These microfinance initiatives were created for both refugee camps and urban environments. The built-in contradiction of microfinance, following the model of the Grameen Bank, involves the exclusion of the poorest strata of society and only engages with clients deemed moderately risky, that is, those able to repay their loans or at least the interest. The creation of Kalobeyei marks a concentrated and intentional shift in model that further demonstrates a shift from 'aid' to 'self-reliance' in the refugee policy regime. FPD in Kakuma and Kalobeyei has allowed private capital to penetrate these settlements, co-opting informal and customary financial terrains, connecting refugees to donors and private financial companies in the Global North. For example, Mastercard's Executive Vice President of Public-Private Partnerships sees the building of these financial landscapes from Global North to South as the best way to effect development and humanitarian outcomes while also providing rich data for private sector finance-led accumulation (Utley 2017). In addition, the World Bank (2018), in collaboration with Mastercard and Western Union, released *Kakuma as a Marketplace*—a report focusing on self-reliance-based entrepreneurship and credit access as a way out of poverty. Missing in these conceptions are the inherent power relations of debt that result in deepening poverty when refugees cannot pay back their loans.

Furthermore, FPD as refugee assistance is less regulated at the urban scale as many refugees lack the basic citizenship rights to live in major cities. This is true in Western Europe as much as it is in parts of Africa and other long-term conflict zones. For example, while France and Germany both have institutionalised policies of social housing, many refugees are barred from accessing shelter in major cities, like Paris and Berlin, and are forced out of these cities (Bhagat and Soederberg 2019). In attempting to survive, refugees work flexible, low-paying jobs but must also rely on microcredit loans to set up businesses as a tool for survival. Entrepreneurship as a solution for poverty traverses North–South relations as it requires little state involvement in terms of language training or any form of welfare. It also avoids recent far right politics in Europe and other parts of the world that have framed refugees as job-stealing migrants. The Council of Europe Development Bank (CEB), for example, has linked with national microcredit organisations, such as MicroBank Spain and PerMicro Italy, to relieve states from 'large and unpredictable inflows of migrants and refugees' (Glisovic 2015, p. 5). European refugee integration is thus reliant on microfinance loans in the face of systemic debt and housing crises.

Alternatively, in the Global South, the global non-profit organisation (NPO) KIVA supports many of the world's poor in over eighty countries in sub-Saharan Africa, Latin America, the Middle East, and South East Asia. KIVA relies on field partners to receive and transfer loans to poor entrepreneurs through financial inclusion. While Kiva prides itself on charging no interest on their loan payments and relies on donors to tip them for operating costs, its field partners charge much higher interest rates to refugees in order to fund their projects and remain afloat within the neoliberal topography of refugee assistance (MacFarquhar 2010). Loans in urban Nairobi, for example, are transmitted through the M-Pesa mobile money transfer system, further generating profit for the Vodafone-owned telecom company. In this way, fintech plays a key role in the transformation of refugees as financial subjects and facilitates capital accumulation in the following ways: first, fintech generates profit for itself by acting as a technological provider; second, it generates capital for Kiva regardless of its zero interest policies; and, third, it funds NGO service programmes through interest and service delivery.

While fintech-backed microfinance loans for both housing and livelihoods have barely made a dent in the overall assistance of refugees, the UNHCR has lauded it as the future mode of refugee assistance. In Kenya, KivaZip is considered a resounding success despite only serving forty-five refugees between 2012 and 2014—an inadequate sample size for the perceived successes (UNHCR 2015). Additionally, UNHCR suggests that fintech will vastly improve financial access for refugees through easy digital payments and loan transfers, remittances, and increased competition and innovation in refugee camps through its interlinking of mobile wallets and micro-entrepreneurs to financial providers (Pistelli 2018). This reflects how UNHCR and other international policies have transformed the aid industry and the nature of refugee labour from cash-for-work to FPD-led entrepreneurship. Despite their claims about the benefits of these programmes, however, the poorest populations are excluded from loans, while the various issues of debt and profit remain unaccounted for.

## Conclusion

In this chapter, we have outlined the patterns and mechanisms of governing forcibly displaced populations, and the discrepancies and contradictions inherent therein. Examining the various labels offered to forcibly displaced populations showed how the categorisation process is itself problematic and can lead to vastly different sets of rights and experiences. Despite that, the succeeding sections demonstrated that even refugees, arguably the most secure of the forcibly displaced, face considerable challenges on a daily basis because of how they are governed. This becomes particularly evident when we look at two of the most basic requirements for survival and social reproduction: housing and employment. By situating our analysis within the context of neoliberalism, we showed how the governance of refugees' housing and labour is deeply embedded in its austerity policies and emphasised on private solutions. We illustrated how, against the backdrop of neoliberal policies, more actors, especially private entities, become involved in the governance of forcibly displaced populations and their interests are favoured over the livelihoods and survival of these vulnerable groups. In highlighting these contradictions and tensions, we have also identified future areas of research, both scholarly and policy-oriented, in relation to the governance of displaced populations.

## References

Agamben, G. (1998). *Homo Sacer: Sovereign Power and Bare Life*. Stanford: Stanford University Press.

Alnsour, J., & Meaton, J. (2012). Housing conditions in Palestinian refugee camps, Jordan. *Cities, 36*, 65–73.

Azorbo, M. (2011). *Microfinance and refugees: Lessons learned from UNHCR's experience*. Geneva: UNHCR.

Bhagat, A., & Soederberg, S. (2019). Forced migration in authoritarian neoliberalism. *South Atlantic Quarterly* [Special Issue – Jamie Peck and Nik Theodore (Eds.)], *118*(2), 421–438.

Campbell, E. H. (2006). Urban refugees in Nairobi: Problems of protection, mechanisms of survival, and possibilities for integration. *Journal of Refugee Studies, 19*(3), 396–413.

Conlon, D. (2010). Ties that bind: Governmentality, the state, and asylum in contemporary Ireland. *Environment and Planning D: Society and Space, 28*(1), 95–111.

Crawley, H., & Skleparis, D. (2018). Refugees, migrants, neither, both: Categorical fetishism and the politics of bounding in Europe's 'migration crisis'. *Journal of Ethnic and Migration Studies, 44*(1), 48–64.

Darling, J. (2009). Becoming bare life: Asylum, hospitality, and the politics of encampment. *Environment and Planning D: Society and Space, 27*(4), 183–189.

Darling, J. (2011). Domopolitics, governmentality and the regulation of asylum accommodation. *Political Geography, 30*(5), 263–271.

Darling, J. (2016). Asylum in austere times: Instability, privatization and experimentation within the UK asylum dispersal system. *Journal of Refugee Studies, 29*(4), 483–505.

Darling, J. (2017). Forced migration and the city: Irregularity, informality, and the politics of presence. *Progress in Human Geography, 41*(2), 178–198.

De Genova, N. P. (2002). Migrant 'illegality' and deportability in everyday life. *Annual Review of Anthropology, 31*(1), 419–447.

Duong, T. T. V. (2010). When islands drown: The plight of 'climate change refugees' and recourse to international human rights law. *University of Pennsylvania Journal of International Law, 31*(4), 1239–1266.

Easton-Calabria, E., & Omata, N. (2018). Panacea for the refugee crisis? Rethinking the promotion of 'self-reliance' for refugees. *Third World Quarterly, 39*(8), 1458–1474.

Ehrkamp, P. (2017). Geographies of migration I: Refugees. *Progress in Human Geography, 41*(6), 813–822.

Fawaz, M., Gharbieh, A., Harb, M., & Salamé, D. (2018). *Refugees as city-makers*. Beirut, Lebanon: Issam Fares Institute for Public Policy and International Affairs, American University in Beirut.

Gabor, D., & Brooks, S. (2017). The digital revolution in financial inclusion: International development in the fintech era. *New Political Economy, 22*(4), 423–436.

Gill, N. (2009). Governmental mobility: The power effects of the movement of detained asylum seekers around Britain's detention estate. *Political Geography, 28*(3), 186–196.

Glisovic, J. (2015). *Microfinance response to the refugee crisis*. Luxembourg: Council of Europe Development Bank (CEB).

Hyndman, J., & Giles, W. (2016). *Refugees in extended exile: Living on the edge*. London: Routledge.

iDMC. (2018). *Unsettlement: Urban displacement in the 21st century*. Geneva: International Displacement Monitoring Centre.

IOM (International Organization for Migration). (2011). *Glossary on migration* (2nd ed.). Geneva: IOM.

Jacobsen, K. (2006). Refugees and asylum seekers in urban areas: A livelihoods perspective. *Journal of Refugee Studies, 19*(3), 273–286.

Kumin, J. (2015). *Welcoming engagement: How private sponsorship can strengthen refugee resettlement in the European Union*. Brussels, Belgium: Migration Policy Institute (MPI) Europe.

Landau, L. B. (2006). Protection and dignity in Johannesburg: Shortcomings of South Africa's urban refugee policy. *Journal of Refugee Studies, 19*(3), 308–327.

Lewis, H., Dwyer, P., Hodkinson, S., & Waite, L. (2015). Hyper-precarious lives: Migrants, work and forced labour in the Global North. *Progress in Human Geography, 39*(5), 580–600.

MacFarquhar, N. (2010, April 14). Banks making big profits from tiny loans. *New York Times*, p. 3.

Milner, N. R. (2011). From 'refugee' to 'migrant' in Calais solidarity activism: Re-staging undocumented migration for a future politics of asylum. *Political Geography, 30*(6), 320–328.

Mitchell, K., Marston, S. A., & Katz, C. (2003). Introduction: Life's work: Introduction, review and critique. *Antipode, 35*(3), 415–442.

Mountz, A. (2011). The enforcement archipelago: Detention, haunting, and asylum on islands. *Political Geography, 30*(3), 118–128.

Peck, J. (2012). Austerity urbanism: American cities under extreme economy. *City, 16*(6), 626–655.

Peck, J., & Tickell, A. (2002). Neoliberalizing space. *Antipode, 34*(3), 380–404.

Pinelli, B. (2018). Control and abandonment: The power of surveillance on refugees in Italy, during and after the Mare Nostrum operation. *Antipode, 50*(3), 725–747.

Pistelli, M. (2018). 5 Things you should know about financial services for refugees: High repayment rates, technology solutions, and new research tell us that it's time for the financial industry to think about refugees [Web log post]. *FinDev Gateway.* Retrieved from https://www.findevgateway.org/blog/2018/jun/5-things-you-should-know-about-financial-services-refugees.

Roberts, A., & Soederberg, S. (2014). Politicizing debt and denaturalizing the 'new normal'. *Critical Sociology, 40*(5), 657–668.

Sajjad, T. (2018). What's in a name? 'Refugees,' 'migrants' and the politics of labelling. *Race & Class, 60*(2), 40–62.

Sanyal, R. (2012). Refugees and the city: An urban discussion. *Geography Compass, 6*(11), 633–644.

Soederberg, S. (2014). *Debtfare states and the poverty industry: Money, discipline and the surplus population*. London: Routledge.

Soederberg, S. (2019). Governing global displacement in austerity urbanism: The case of Berlin's refugee housing crisis. *Development and Change, 50*(4), 923–947.

Tuastad, D. (2017). 'State of exception' or 'state in exile'? The fallacy of appropriating Agamben on Palestinian refugee camps. *Third World Quarterly, 38*(9), 2159–2170.

UNHCR (UN High Commissioner for Refugees). (2005). *Handbook for self-reliance*. Geneva: UNHCR.

UNHCR (UN High Commissioner for Refugees). (2015). *Microfinance programmes in UNHCR operations – Innovative microlending in Kenya – Kiva Zip & RefugePoint*. Retrieved from https://www.unhcr.org/54edf4479.pdf.

UNHCR (UN High Commissioner for Refugees). (2016). *Global trends: Forced displacement in 2015*. Geneva: UNHCR.

UNHCR (UN High Commissioner for Refugees). (2017). *Situation on Greek Islands still grim despite speeded transfers.* UNHCR (UN High Commissioner for Refugees)

Kenya (2017). Kalobeyei Settlement. Retrieved September 14, 2018, from http://www.unhcr.org/ke/kalobeyei-settlement.

Utley, T. (2017, August 30). How Mastercard is bringing digital innovation to refugees around the globe. *Forbes*. Retrieved September 20, 2018, from https://www.forbes.com/sites/toriutley/2017/08/30/how-mastercard-is-bringing-digital-innovation-to-refugees-around-the-globe/#595ee1033ef7.

Williams, A. (2008). Turning the tide: Recognizing climate change refugees in international law. *Law & Policy, 30*(4), 502–529.

World Bank. (2017). *Forcibly displaced: Toward a development approach supporting refugees, the internally displaced, and their hosts.* Washington, DC: The World Bank.

World Bank. (2018). *Kakuma as a marketplace: A consumer and market study of a refugee camp and town in Northwest Kenya.* Washington, DC: International Finance Corporation, World Bank.

Zamore, L. (2018). Refugees, development, debt, austerity: A selected history. *Journal on Migration and Human Security, 6*(1), 26–60.

# 33

## Bureaucracies of Displacement: From Immigrants' Social and Physical Exclusion to Their Judicial Removal

### Cecilia Menjívar and Andrea Gómez Cervantes

Immigration laws and policies have created a coercive bureaucracy (Rodriguez and Paredes 2014) that has gained power and legitimacy by implementing strategies that displace immigrants socially, economically, and geographically, internally and externally. However, this bureaucracy targets certain groups of immigrants more than others, so it is useful to attend to this differentiation, which emerges when examined through the lens of race, class, and gender. In this chapter we shed light on the two components of the immigration regime that contribute to displace Latino/a immigrants socially, economically, and spatially, including physical separations. We focus first on the legislative side of the immigration bureaucracy as it increasingly narrows formal paths to admission and expands insecure, temporary legal statuses, both strategies contributing to keeping an increasing number of immigrants on the legal margins. This displacement through law has social and economic consequences for the migrants, their families, and their communities, both within the US and beyond borders. We then turn to discussing the enforcement side of the bureaucracy to highlight the internal displacement of immigrants (through

C. Menjívar (✉)
Department of Sociology, University of California, Los Angeles, CA, USA
e-mail: menjivar@soc.ucla.edu

A. Gómez Cervantes
Wake Forest University, Winston-Salem, NC, USA

© The Author(s) 2020
P. Adey et al. (eds.), *The Handbook of Displacement*,
https://doi.org/10.1007/978-3-030-47178-1_33

detention) and their displacement across borders (through deportation). And in the third section we look at the judicial displacement of immigrants by focusing on the case of those who apply for asylum but whose lives remain suspended in limbo. Although we focus our examination on the US case, we would like to note that bureaucracies of displacement have increased and spread around the world as more immigrant-receiving countries rely on narrowing paths to legal status, detention, and deportation to manage their migration flows. For instance, immigrant detention and apprehension have risen in Australia, New Zeeland, and Switzerland (Bosworth et al. 2018; Vogl and Methven 2017), and deportations are now taking place in 'transit' countries such as Mexico and Indonesia (Mountz 2015; Vélez Santiago and Fernandez Sanabria 2016) and wealth democracies are actively repelling asylum seekers even before they arrive at their borders to apply for asylum (FitzGerald 2019).

## The Legislative Side of the Immigration Bureaucracy

Immigration law dictates which immigrants are allowed into the country, as well as the rights and protections extended to some groups while denied to others. As a result, throughout US history, immigration policies and legislative mechanisms have contributed to the social and economic exclusion of certain immigrant groups and their displacement from institutions and communities. Thus, while some immigrants are admitted to the United States to fulfil certain needs (political, economic), their admission does not guarantee them full membership. This tends to be the case with labour migrants who are formally admitted to fill a labour shortage need but through policies of admission are constrained from fully participating in society (De Genova 2004; Espiritu 2003). This is also the case with certain groups who are admitted because they serve a national security purpose and who tend to be classified as refugees (Menjívar 2000; Pedraza-Bailey 1985). In any case, our point is that formal admission into the country does not translate automatically into inclusion; just like the case of domestic minorities, immigrant groups can be excluded and displaced through law even if they are admitted formally.

Furthermore, historically and contemporaneously, displacement begins before entry, as some groups have not been allowed in and new categories of admission have made entry difficult or impossible for others (see FitzGerald 2019). Race, or a combination of race and law, has always played a pivotal role in determining who is allowed in and who is not. For instance, the 1882 Chinese Exclusion Act barred Chinese labourers from entering the United States for a period of ten years, and the 1908 Gentlemen's Agreement sought to bar Japanese labourers from entry by getting Japan to only give passports to non-labourers (Boyd 1971). These two policies followed increases in Chinese and Japanese migration flows entering through Hawaii and California as the US government and railroad companies encouraged these workers to migrate. However, after the railroad was completed and recession loomed over the country, nativist sentiment grew to displace Chinese and Japanese workers, leading to anti-Chinese and anti-Japanese immigration policies (Calavita 2000).

During the past century or so, in an effort to fulfil labour shortages, migration from Mexico to the US was lightly regulated (Massey et al. 2002). It was common for Mexican farmworkers to travel back and forth during agricultural seasons without formal visas to enter the country. With increased labour shortages during wartime, the United States created programmes that formally recruited Mexican nationals as workers, such as the Bracero Program (1942–1964), which institutionalised the flow of seasonal workers from Mexico to the United States, making US agricultural growers dependent on Mexican labour, and labourers dependent on seasonal employment in the United States (Calavita 2010). The 1952 and amendments made in 1965 to the Immigration and Nationality Act (INA) ended the national quota system, in place since 1924, which limited migration from the Eastern Hemisphere to 170,000 total visas and from the Western Hemisphere to 120,000 visas, with no country exceeding 7 per cent of all visas (Chishti et al. 2015). Although the INA opened the legislative door to immigrants from Asia and Latin America, the new limit affected Mexico negatively. Cuts to visas for Western Hemisphere countries and especially for Mexico significantly shifted the nature of migratory flows as the demand for agricultural labour continued and, coupled with an unstable Mexican economy, Mexican workers continued to migrate to the United States. The biggest transformation, however, was that these workers migrate without legal status (Calavita 2010; Massey et al. 2002), which means marginalised lives, especially as the consequences of an undocumented status have increasingly worsened over the years.

The 1965 INA established new entry categories, in particular new family-based and employment-based visas (Enchautegui and Menjívar 2015). However, the formula used to distribute visas worldwide, where no country can exceed 7 per cent of visas in any one year, has created large discrepancies across national groups. We know that there are social, political, economic, and historical forces that shape migratory flows and thus not all countries send the same number of migrants to the United States at the same time.

Allocating visas equally across all countries without regard for structural and social forces in origin countries creates major imbalances in who can travel to the US with a visa. On the one hand, European countries that used to send the bulk of immigrants no longer do so and their visa allocations go unused. On the other hand, for some countries (e.g., China, India, Mexico, and the Philippines) the backlogs for certain family-based visa categories are so long (decades long in some cases) that in practice it is nearly impossible for these immigrants to sponsor the migration of a family member. This gets more complicated when powerful family ties, including extended kin, are meaningful for the immigrants but are not recognised in the immigration bureaucracy (Menjívar et al. 2016). However, immigrants from the largest sending countries who seek to reunite with their families will do so, regardless of whether they receive a visa, meaning they will likely live 'in the shadows of the law' (Chavez 1991). Thus, immigration law contributes to the social displacement of immigrants, first, by separating families (and keeping them apart), and, second, by excluding them from basic rights and protections (Enchautegui 2014; Enchautegui and Menjívar 2015).

Two pieces of legislation signed into law in 1996 have contributed to reinforcing the social and economic displacement of undocumented and temporary legal immigrants: the Personal Responsibility and Work Opportunity Reconciliation Act (PRWORA) and the Illegal Immigration Reform and Immigrant Responsibility Act (IIRIRA). These laws have restricted lawfully admitted immigrants who arrived after 1996 from accessing federally funded welfare programmes for their first five years in the United States; undocumented immigrants were banned indefinitely (Hagan et al. 2003; Viladich 2012). Additionally, inadmissible immigrants who use public services can become ineligible for lawful permanent residence and even face deportation if they are unable to replay the services that they use (Hagan et al. 2003, p. 447). The Public Charge rule, which went into effect on 24 February 2020, consider receiving public benefits for more than twelve months grounds for inadmissibility for lawful permanent residence (USCIS 2020a). The Public Charge rule makes an exception for screening, testing, treatment and receiving vaccines for COVID-19 (USCIS 2020b).

Thus, for decades, immigration policies have systematically excluded certain groups of immigrants from rights and protections, contributing to their social and physical displacement. A panoply of laws consistently reduces opportunities for legalisation and at the same time curtails access to social services and opportunities based on legal status. Such actions have direct and serious consequences for the social and economic incorporation of immigrants; research has demonstrated that legal status affects the jobs and wages immigrants obtain (Massey and Gentsch 2014); housing options (Hall and Greenman 2013); access to higher education (Abrego 2006; Gonzales 2016) and healthcare (Marrow and Joseph 2015; Philbin et al. 2017); and friendships and family relations (Das Gupta 2014; Dreby 2015; Enriquez 2020; Gomberg-Muñoz 2015). As such, legal status has become an important axis of stratification that intersects with other social positions (Abrego 2014; Cebulko 2018; Gonzales and Burciaga 2018; Menjívar et al. 2016) to produce important social cleavages along legal statuses. Legal status, and the multiplication of temporary statuses, shapes access to goods and services, rights and duties. As such, it pushes certain immigrants to the margins of society and permits discrimination against them, because in contrast to other social markers such as gender, race, or age, it is not against the law to discriminate on the basis of legal status; in fact, the law demands such discriminatory treatment (Menjívar et al. 2016). Latinas/os bear the brunt of such practices in the United States today.

# The Enforcement Side of the Immigration Bureaucracy

On the enforcement side of the immigration bureaucracy, the Immigration Reform and Control Act (IRCA) of 1986 contributed to amplifying the militarisation of the Mexico–US border, which started in earnest in the late 1970s with policies to clamp down on drug trafficking. The militarisation of the border has continued to expand over the decades, reaching the hyper-militarisation we see today, with drones, armed vehicles, and agents in military equipment, including uniform and military arms, and the overall institutional arrangement of US Customs and Border Protection (see Slack et al. 2016). The militarisation of the border has even engaged the National Guard, as President Obama and now President Trump have sent this military body to patrol the border, even though apprehensions at the US southern border have steadily decreased since 2008 and in 2017 the US Border Patrol reported the lowest number of apprehensions since 1971 (Border Patrol 2017).

Since their inception, these military tactics mirrored the doctrine of 'low-intensity conflict' that the United States implemented to subdue the opposition in Central American civil wars (and in other countries where it has intervened militarily), with the objective of controlling specific civilian populations but resulting in massive human rights violations (Dunn 1996). Such militaristic strategies at the border have contributed to controlling certain immigrants (e.g., those crossing the US southern border) through physically and geographically displacing them across dangerous terrain as they attempt to enter the United States (Dowling and Inda 2013; Massey et al. 2002). Increased border policing pushed immigrants from traditional pathways to more dangerous topography, such as desserts and rivers (Cornelius 2001; Eschbach et al. 1999; Massey et al. 2002). Utilising a 'security' narrative, the United States has poured millions of dollars into the militarisation of the Mexico–US border, increasing the budgets of the Border Patrol at unprecedented rates and expanding surveillance technologies, which has resulted in increased violence that today immigrants and non-immigrant communities experience both at the border and in migration corridors beyond the US border (Menjívar and Abrego 2012). In 2020 the Border Patrol budget was $18.2 billion (DHS 2020), compared to $10,049 million in 2010 (DHS 2020). This militarisation trend has led to an increase in migrants' deaths at the Mexico–US border (Cornelius 2001; Slack et al. 2016), among other harmful consequences.

Although immigration enforcement falls under federal authority, certain provisions in the IIRIRA 1996 have permitted other levels of government to engage in immigration enforcement. Indeed, with Section 286(g) of IIRIRA, the enforcement of immigration has engaged collaboration from local jurisdictions (Coleman 2012; Menjívar 2014) by providing local police agencies the tools to investigate immigrants' citizenship status and assist in their detention. In addition, other programmes to enhance cooperation in immigration enforcement among various levels of government have been created, including the Secure Communities Program in 2008, which extends local policing agencies' economic incentives to share with immigration enforcement the legal status of detained immigrants (Coleman 2012). Thus, the Secure Communities Program incentivised police officers to detain suspected undocumented immigrants, instead of focusing on combatting crime. Scholars and advocacy groups found such programmes, especially the 287(g), to promote racial profiling amongst Latinos/as (Fischer 2013). Thus, the Obama administration phased out the programme in 2012 and replaced it with the Priority Enforcement Program, which supposedly prioritised the detention and apprehension of immigrants who had committed severe crimes. However, most recently, the Trump administration has resurrected these programmes with the goal of enhancing their implementation in all states and regions of the country and has done away with prioritising, placing every undocumented immigrant in the country at risk of detention and deportation (Alvord et al. 2018). These initiatives have significantly enhanced immigration enforcement in the interior of the country (as well as at border, with calls for building 'the wall') and bringing back workplace raids, increasing detentions and deportations of mostly Latino immigrants (Menjívar et al. 2018; Provine and Doty 2011), thus displacing them from society and from increasingly scant economic opportunities.

In detention, immigrants are regularly transferred from one facility to another, often in the middle of the night, to another state, without informing relatives or even the detainee's legal representatives. These transfers pose challenges for families to remain in touch with their detained relatives and undermine efforts at legal representation (Dow 2004) as detention facilities tend to be located in remote locations (Ryo and Peacock 2018). Such isolation is exacerbated because phone calls from those facilities are expensive (e.g., over one dollar a minute at the Kernes facility) (Shepherd 2018). Detention facilities thus contribute powerfully to physically displacing immigrants from their families, communities, and legal channels, while also amplifying feelings of isolation.

The detention and deportation of Latinas/os is further sustained through the privatisation of immigration detention, or what some scholars have called 'the immigration-industrial complex,' as it functions in parallel fashion to the 'prison industrial complex' (Ackerman and Furman 2013; Doty and Wheatley 2013; Douglas and Saenz 2013; Welch 2000). For instance, the private companies CoreCivic (formerly Corrections Corporation of America [CCA]) and The Geo Group Inc. (GEO) have multi-year, multimillion-dollar contracts with the Department of Homeland Security (DHS) and Immigration and Customs Enforcement (ICE) to collaborate in various aspects of immigration enforcement, including detention, apprehension, and surveillance (Gómez Cervantes et al. 2018). Seventy-one per cent of detained immigrants were held in a privately owned or managed institution in 2017 (Tidwell Cullen 2018). The removal of these immigrants, mostly Latino men, from their families and communities has significant short- and long-term effects, as women are left financially (and socially) responsible for their families (Dreby 2015).

Deportations physically remove immigrants from the country, cementing their displacement from US society as bars to readmission (between 3 and 10 years, but up to 20 years in some cases) are activated with a deportation, bars which apply to immigrants who entered without inspection (likely surreptitiously through the border), but not to those who overstay their visas (and more likely to be non-Latino arriving from other parts of the world). Between 2003 and 2018, a total of 4,617,463 immigrants were deported from the United States, the overwhelming majority were from Mexico (65 per cent), Guatemala (11 per cent), Honduras (9 per cent), and El Salvador (6 per cent) (TRAC Immigration 2019). And given that an estimated 4.1 million

US citizen children under the age of eighteen live with at least one undocumented parent, mass deportation contributes to the physical, social, and economic displacement of certain immigrant families, separating them indefinitely (American Immigration Council 2017) and making it hard to survive financially (in the United States or in the countries that deportees are returned to). Between 2011 and 2013 an estimated half a million children were directly affected by the deportation of a parent (American Immigration Council 2017).

Latino/a, but also Muslim, immigrants have felt most directly the costs of the bureaucracy of enforcement (Menjívar and Kanstroom 2014; Rivera 2014) in various forms of displacement. Through mechanisms that permit formal discrimination in institutions, primarily against Latino immigrants, and the 'war on terror' that targets Muslim immigrants, both groups face systematic exclusion and social displacement, reproduced in different areas of life and sustained by immigration policies (Bosworth et al. 2018; Maghbouleh 2017; Rivera 2014). Immigration enforcement does not occur in a vacuum; institutionalised racism is at the core of policing and control strategies (Bosworth et al. 2018). For instance, 89 per cent of the detained population come from Mexico, Guatemala, El Salvador, and Honduras, and 79 per cent are men (Ryo and Peacock 2018); over 90 per cent of all deportees are men of Mexican or Central American origin, making Latinos and their communities the preeminent targets of the enforcement bureaucracy (TRAC Immigration 2019). And only days after the USA PATRIOT Act of 2001 was signed in the aftermath of the 9/11 attacks, hundreds of Muslims and Arab immigrants in the United States were held in detention for extensive periods of time without charges (see Akram and Johnson 2002; Sinnar 2003). The racialised mechanisms of exclusion are perhaps most noticeable in the 2017 Executive Orders (EOs), accompanied by various anti-immigrant initiatives (Alvord et al. 2018; Menjívar et al. 2018). The EOs have expanded the mechanisms that facilitate the mass detention and deportation of Latina/o immigrants, while attempting to negate Muslim immigrant entry to the United States and deny Mexican and Central American asylum seekers even the chance to present their cases to US authorities. These EOs have further facilitated the systematic targeting and marginalisation of Latinos (Alvord et al. 2018; Menjívar et al. 2018) and the construction of Muslims as a threat to the nation, which justifies their exclusion. It is Latinos/as, however, who are overwhelmingly detained and deported, experiencing internal displacement similar to what is experienced in African American communities, as approximately one-third of their male members are displaced from their communities through incarceration (Alexander 2012; Comfort 2007).

Immigration enforcement (e.g., apprehension, detention, and deportation) is perhaps the most violent form of displacement and exclusion that immigrants and their families experience. When immigrants who have lived for years in the United States—for example, about one-third of undocumented immigrants have lived in the United States for ten years or longer (see Passel and Cohn 2016)—are apprehended and detained, they are indefinitely separated from families and communities. Given the time these immigrants have spent in the United States, they likely live in 'mixed-status' families, that is, they have children and/or spouses who are US born or have more secure legal statuses. Thus, such separations have significant short- and long-term consequences for the displaced immigrant but also for their family members, including the US born. These effects are immediate, such as emotional distress, financial insecurity, family reconfiguration, depression, and anxiety (see Brabeck et al. 2015; Brabeck and Xu 2010), but also long term, as some of these effects endure and may even be irreversible. Thus, the exclusion that the enforcement side of the immigration bureaucracy makes possible contributes to displace immigrants geographically, physically, socially, and economically, immediately and in the future.

## Judicial Displacement

Starting in 2014, media and political attention has focused on Central American 'unaccompanied undocumented minors' and families who, travelling from Honduras, El Salvador, and Guatemala, upon reaching the Mexico–US border turn themselves in to Border Patrol agents and ask for asylum. These migrants are fleeing gender-based and gang violence, as well as severe poverty and deep inequality resulting from neoliberal, US-guided policies, and exorbitant international debt (McMichael 2000; UNHCR 2015).

Since 2000, the Trafficking Victims Protection Act requires immigration enforcement to screen unaccompanied youth as 'potential victims of human trafficking' and promotes access to immigration lawyers for court proceedings[1] (Menjívar and Perreira 2017). Additionally, per requirements to file for asylum, immigrants conduct a 'credible fear' interview where immigration officers assess whether immigrants' claims to asylum are 'credible.' In 2016 alone, DHS officials conducted 46,961 'credible fear' interviews with Central American asylum seekers, 36,324 of which were considered 'credible' (DHS 2016). While the majority of asylum-seeking youth and women migrate to the United States because they have relatives there to receive them, these asylum seekers are placed in immigration detention facilities run by the DHS and private corporations. The court backlog in fiscal year 2020 reached over 1 million (TRAC 2020), which translates into long waits in detention or alternative to detention programmes and living in uncertainty (see also Gómez Cervantes et al. 2018).

As of this writing, the law distinguishes between youth asylum seekers coming from countries that border the United States and those coming from countries that do not share a contiguous US border; the former are returned immediately, while the latter remain and are processed through the system. Thus, if DHS determines the youth to be 'unaccompanied' and come from a non-adjacent country, they are placed with Health and Human Services, Office of Refugee Resettlement (ORR); however, if the youth are Mexican, they can be removed immediately (Menjívar and Perreira 2017). The majority of the apprehended youth (mostly Central Americans) are then placed in shelters, while others go to 'transition homes' or foster care (Menjívar and Perreira 2017). Thus, immigrant youth are displaced throughout the country even though most of these 'unaccompanied minors' do not travel alone and have relatives in the United States. They are placed in detention facilities and separated from their traveling companions, oftentimes close family members or friends (Shepherd 2018; Thompson et al. 2019).

Central American immigrants fleeing violence face the consequences of displacement in acute fashion. In 2018, the Trump administration purposefully separated children, as young as three years old and in one case a four-months-old baby, from their parents in efforts to supposedly deter Central American immigrants from arriving in the United States to seek asylum (Burnett 2018). As a result of Trump administration's policy changes starting in 2017 throughout 2018, DHS separated approximately 2654 children, most from Central America, from their parents as they entered the US–Mexico border and turned themselves in to Border Patrol requesting protection (ACLU 2019). Children were displaced to 121 different detention facilities or 'care' centres in 17 states across the United States with a median detention length of 154 days, while some children were in detention up to a year (ACLU 2019). While the majority of children have been reunited with their parents, many children continue to be held in detention facilities, violating laws against the indefinite detention of minors. Some parents were deported to their countries of origin while their children were detained, other parents were deemed 'unfit' to care for their children although they have no criminal record, and other parents remain in immigration custody, continuing the separations (ACLU 2019). Additionally, partners, siblings, and friends are oftentimes separated during the asylum processing. The actual volume of family separation at the border, however, is unknown given the lack of information that DHS shares publicly and their lack of accurate data collection-practices (see ACLU 2019).

Displacement is also visible in the 2019 'Asylum Ban 2.0' which bars any person from asylum eligibility if they have transited through another country without applying for asylum in transit countries (DHS 2019). In the rule co-sponsored by the Department of Homeland Security and Department of Justice, migrants who are seeking asylum at the Mexico–US border are sent to countries that have entered into agreements with the United States including Mexico, Guatemala, Honduras, and El Salvador to await their asylum hearings. However, these countries are the same unsafe places that migrants are fleeing in the first place. In 2020 over 57,000 asylum seekers, over 80 per cent who have been victims of violence, awaited their asylum court hearings in make-shift camps in Mexican cities that have long histories of cartel violence along the US southern border (Agren 2020). Thus, immigration policies displace people seeking refuge from violence, sending them to dangerous and life-threatening spaces.

## Discussion/Conclusion

The three components of the immigration bureaucracy outlined in this chapter create various forms of exclusion and social, economic, physical, and geographical displacement of immigrants in the interior of the country and beyond its borders. (1) The legislative side includes laws and policies that create (reduce and eliminate) categories of admission that regulate immigration flows and restrict rights and protections to some immigrant groups while denying these to others. (2) The enforcement side criminalises Latinos/as and Muslims while promoting their surveillance and incarceration and physically displaces these immigrants to detention facilities, as well as geographically, to their origin countries through deportation. And (3) the judicial component separates immigrants from their families and communities through the asylum-seeking processes as well as through forceful removal and deportation proceedings. Such forms of bureaucratic displacement affect the individuals who apply for visas, or who are detained or deported, but also their families, in the immediate future but also long term. And although we focus on the US case, our examination is relevant to other immigrant-receiving countries today as they closely follow the US model in managing immigrant populations.

## Note

1. Since 2014, several attempts have been introduced in the legislation to speed up the processing of unaccompanied youth and deny their claims to asylum, thus, also speeding their deportation. See S.2666 in 2014, or H.R.495 in 2017 (Protection of Children Act 2017; Protect Children and Families Through the Rule of Law Act 2014).

## References

Abrego, L. J. (2006). 'I can't go to college because I don't have papers': Incorporation patterns of Latino undocumented youth. *Latino Studies, 4*, 212–231.

Abrego, L. J. (2014). *Sacrificing families: Navigating laws, labor, and love across borders*. Stanford, CA: Stanford University Press.

Ackerman, A. R., & Furman, R. (2013). The criminalization of immigration and the privatization of the immigration detention: Implications for justice. *Contemporary Justice Review, 16*(2), 251–263.

ACLU (American Civil Liberties Union). (2019). Family separation by the numbers. Retrieved from https://www.aclu.org/issues/immigrants-rights/immigrants-rights-and-detention/family-separation.

Agren, D. (2020, January 29). Remain in Mexico: 80% of migrants in Trump policy are victims of violence. *The Guardian*. Retrieved from https://www.theguardian.com/us-news/2020/jan/29/migrants-mexico-trump-policy-victims-of-violence.

Akram, S. M., & Johnson, K. R. (2002). Race, civil rights, and immigration law after September 11, 2001: The targeting of Arabs and Muslims. *NYU Annual Survey of American Law, 58*, 295–356.

Alexander, M. (2012). *The new Jim Crow: Mass incarceration in the age of colorblindness*. New York, NY: The New Press.

Alvord, D. R., Menjívar, C., & Gómez Cervantes, A. (2018). The legal violence in the 2017 executive orders: The expansion of immigrant criminalization in Kansas. *Social Currents, 5*(5), 411–420. https://doi.org/10.1177/2329496518762001.

American Immigration Council. (2017). *U.S. citizen children impacted by immigration enforcement*. Washington, DC. Retrieved from https://www.americanimmigrationcouncil.org/sites/default/files/research/us_citizen_children_impacted_by_immigration_enforcement.pdf.

Border Patrol. (2017). *United States Border Patrol nationwide illegal alien apprehensions fiscal years 1925–2017*. Washington, DC: Border Patrol.

Bosworth, M., Parmar, A., & Vazquez, Y. (Eds.). (2018). *Race, criminal justice, and migration control: Enforcing the boundaries of belonging*. Oxford: Oxford University Press.

Boyd, M. (1971). Oriental immigration: The experience of the Chinese, Japanese, and Filipino populations in the United States. *The International Migration Review, 5*(1), 48–61. https://doi.org/10.2307/3002046.

Brabeck, K., & Xu, Q. (2010). The impact of detention and deportation on Latino immigrant children and families: A quantitative exploration. *Hispanic Journal of Behavioral Sciences, 32*(3), 341–361.

Brabeck, K., Lykes, M. B., & Hunter, C. (2015). The psychosocial impact of detention and deportation on U.S. migrant children and families. *American Journal of Orthopsychiatry, 84*(5), 496–505.

Burnett, J. (2018, February 27). To curb illegal immigration, DHS separating families at the border. *Morning Edition*. Retrieved from https://www.npr.org/2018/02/27/589079243/activists-outraged-that-u-s-border-agents-separate-immigrant-families.

Calavita, K. (2000). The paradoxes of race, class, identity, and 'passing': Enforcing the Chinese Exclusion Acts, 1882–1910. *Law & Social Inquiry, 25*(1), 1–40.

Calavita, K. (2010). *Inside the state: The Bracero Program, immigration, and the I.N.S.* New York, NY: Routledge.

Cebulko, K. (2018). Privilege without papers: Intersecting inequalities among 1.5-generation Brazilians in Massachusetts. *Ethnicities, 18*(2), 225–241.

Cecilia Menjívar, Andrea Gómez Cervantes and Daniel Alvord. 2018. ""The Expansion of "Crimmigration," Mass Detention, and Deportation." Sociology Compass 12 (4): e12573

Chavez, L. (1991). *Shadowed lives: Undocumented immigrants in American society*. San Diego, CA: Harcourt Brace Jovanovich College Publishers.

Chishti, M., Hipsman, F., & Ball, I. (2015). *Fifty years on, the 1965 immigration and nationality act continues to reshape the United States*. Washington, DC: Migration Policy Institute. Retrieved from https://www.migrationpolicy.org/article/fifty-years-1965-immigration-and-nationality-act-continues-reshape-united-states.

Coleman, M. (2012). The local migration state: The site-specific devolution of immigration enforcement in the US South. *Law & Policy, 34*, 159–190.

Comfort, M. (2007). *Doing time together: Love and family in the shadow of the prison*. Chicago, IL: The University of Chicago Press.

Cornelius, W. (2001). Death at the border: Efficacy and unintended consequences of US immigration control policy. *Population and Development Review, 27*(4), 661–685.

Das Gupta, M. (2014). 'Don't deport our daddies': Gendering state deportation practices and immigrant organizing. *Gender & Society, 28*(1), 83–109.

De Genova, N. (2004). The legal production of Mexican/migrant 'illegality'. *Latino Studies, 2*, 160–185.

DHS (Department of Homeland Security). (2016). *Credible fear workload report summary FY 2016 total caseload*. Washington, DC. Retrieved from https://www.

uscis.gov/sites/default/files/USCIS/Outreach/Upcoming%20National%20 Engagements/CredibleFearReasonableFearStatisticsNationalityReports.pdf.

DHS (Department of Homeland Security). (2019). *DHS and DOJ issue third-country asylum rule*. Washington, DC. Retrieved from https://www.dhs.gov/news/2019/07/15/dhs-and-doj-issue-third-country-asylum-rule.

DHS (Department of Homeland Security). (2020). *FY 2020 budget in brief*. Washington, DC Retrieved from https://www.dhs.gov/sites/default/files/publications/19_0318_MGMT_FY-2020-Budget-In-Brief.pdf.

Doty, R. L., & Wheatley, E. S. (2013). Private Detention and the Immigration Industrial Complex. *International Sociology*, 7(4), 426–443.

Douglas, K. M., & Saenz, R. (2013). The criminalization of immigrants & the immigration-industrial complex. *Daedalus, 142*(3), 199–227.

Dow, M. (2004). *American gulag: Inside U.S. immigration prisons*. Berkeley, CA: University of California Press.

Dowling, J. A., & Inda, J. X. (2013). *Governing immigration through crime: A reader*. Sanford, CA: Stanford University Press.

Dreby, J. (2015). *Everyday illegal: When policies undermine immigrant families*. Berkeley, CA: University of California Press.

Dunn, T. J. (1996). *The militarization of the U.S.-Mexico border, 1978–1992: Low intensity conflict doctrine comes home*. Austin, TX: CMAS Books, Center for Mexican American Studies, University of Texas.

Enchautegui, M. (2014). Legalization programs and the integration of unauthorized immigrants: A comparison of S. 744 and IRCA. *Journal on Migration and Human Security, 2*(1), 1–13.

Enchautegui, M., & Menjívar, C. (2015). Paradoxes of family immigration policy: Separation, reorganization, and reunification of families under current immigration laws. *Law & Policy, 37*(1–2), 32–60.

Enriquez, Laura. (2020). *Of Love and Papers: How immigration policy affects Romance and Family*. University of California Press (Open Access).

Eschbach, K., Hagan, J., Rodriguez, N., Hernandez-Leon, R., & Bailey, S. (1999). Death at the border. *The International Migration Review, 33*(2), 430–454. https://doi.org/10.2307/2547703.

Espiritu, Y. L. (2003). *Home bound Filipino American lives across cultures, communities, and countries*. Berkeley, CA: University of California Press.

Fischer, A. (2013). Secure communities, racial profiling, & suppression law in removal proceedings. *Texas Hispanic Journal of Law and Policy, 19*, 63–94.

FitzGerald, David. (2019). *Refuge beyond reach: How rich democracies repel asylum seekers*. Oxford University Press.

Golash-Boza, T. (2009). The immigration industrial complex: Why we enforce immigration policies destined to fail. *Sociology Compass, 3*(2), 295–309.

Gomberg-Muñoz, R. (2015). The punishment/el castigo: Undocumented Latinos and US immigration processing. *Journal of Ethnic and Migration Studies, 41*(14), 2235–2252.

Gómez Cervantes, A., Menjívar, C., & Staples, W. G. (2018). 'Humane' immigration enforcement and Latina immigrants in the detention complex. *Feminist Criminology, 12*(3), 269–292.

Gonzales, R. (2016). *Lives in limbo: Undocumented and coming of age in America*. Berkeley, CA: University of California Press.

Gonzales, R. G., & Burciaga, E. M. (2018). Segmented pathways of illegality: Reconciling the coexistence of master and auxiliary statuses in the experiences of 1.5-generation undocumented young adults. *Ethnicities, 18*(2), 178–191.

Hagan, J., Rodriguez, N., Capps, R., & Kabiri, N. (2003). The effects of recent welfare and immigration reforms on immigrants' access to health care. *International Migration Review, 37*(2), 444–463.

Hall, M., & Greenman, E. (2013). Housing and neighborhood quality among undocumented Mexican and Central American immigrants. *Social Science Research, 42*(6), 1712–1725.

Maghbouleh, N. (2017). *The limits of whiteness: Iranian Americans and the everyday politics of race*. Stanford, CA: Stanford University Press.

Marrow, H. B., & Joseph, T. (2015). Excluded and frozen out: Unauthorized immigrants' (non)access to care after US health care reform. *Journal of Ethnic and Migration Studies, 41*(14), 2253–2273.

Massey, D. S., & Gentsch, K. (2014). Undocumented migration and the wages of Mexican immigrants. *The International Migration Review, 48*(2), 482–499. https://doi.org/10.1111/imre.12065.

Massey, D. S., Durand, J., & Malone, N. J. (2002). *Beyond smoke and mirrors: Mexican immigration in an era of economic integration*. New York, NY: Russell Sage Foundation.

McMichael, P. (2000). *Development and social change*. Thousand Oaks, CA: Pine Forge Press.

Menjívar, C. (2000). *Fragmented ties: Salvadorian immigrant networks in America*. Berkeley, CA: University of California Press.

Menjívar, C. (2014). The 'poli-migra': Multilayered legislation, enforcement practices, and what we can learn about and from today's approaches. *American Behavioral Scientist, 58*(13), 1805–1819.

Menjívar, C., & Abrego, L. J. (2012). Legal violence: Immigration law and the lives of Central American immigrants. *American Journal of Sociology, 117*(5), 1380–1421. https://doi.org/10.1086/663575.

Menjívar, C., & Kanstroom, D. (2014). *Constructing immigrant 'illegality': Critiques, experiences, and responses [eBook]*. New York, NY: Cambridge University Press.

Menjívar, C., & Perreira, K. M. (2017). Undocumented and unaccompanied: Children of migration in the European Union and the United States. *Journal of Ethnic and Migration Studies, 45*, 197–217. https://doi.org/10.1080/1369183X.2017.1404255.

Menjívar, C., Abrego, L. J., & Schmalzbauer, L. (2016). *Immigrant families*. Cambridge: Polity Press.

Menjívar, C., Gómez Cervantes, A., & Alvord, D. (2018). The expansion of 'crimmigration,' mass detention, and deportation. *Sociology Compass, 12*(4), e12573. https://doi.org/10.1111/soc4.12573.

Mountz, A. (2015). In/visibility and the securitization of migration: Shaping publics through border enforcement on islands. *Cultural Politics, 11*(2), 184–200. https://doi.org/10.1215/17432197-2895747.

Passel, J. S., & Cohn, D. (2016). *Overall number of U.S. unauthorized immigrants holds steady since 2009*. Washington, DC: Pew Hispanic Center.

Pedraza-Bailey, S. (1985). Cuba's exiles: Portrait of a refugee migration. *The International Migration Review, 19*(1), 4–34. https://doi.org/10.2307/2545654.

Philbin, M. M., Flake, M., Hatzenbuehler, M. L., & Hirsch, J. S. (2017). State-level immigration and immigrant-focused policies as drivers of Latino health disparities in the United States. *Social Science & Medicine*, 1–10. https://doi.org/10.1016/j.socscimed.2017.04.007.

Protect Children and Families Through the Rule of Law Act, Pub. L. No. S.2666. (2014). Retrieved from https://www.congress.gov/bill/113th-congress/senate-bill/2666.

Protection of Children Act of 2017, Pub. L. No. H.R.495. (2017). Retrieved from https://www.congress.gov/bill/115th-congress/house-bill/495.

Provine, D. M., & Doty, R. L. (2011). The criminalization of immigrants as a racial project. *Journal of Contemporary Criminal Justice, 27*(3), 261–277.

Rivera, C. (2014). The brown threat: Post-9/11 conflations of Latina/os and Middle Eastern Muslims in the US American imagination. *Latino Studies, 12*(1), 44–64.

Rodriguez, N., & Paredes, C. (2014). Coercive immigration enforcement and bureaucratic ideology. In C. Menjívar & D. Kanstroom (Eds.), *Constructing immigrant 'illegality': Critiques, experiences, and responses* (pp. 63–83). New York, NY: Cambridge University Press.

Ryo, E., & Peacock, I. (2018). *The landscape of immigration detention in the United States*. Washington, DC: American Immigration Council. Retrieved from https://www.americanimmigrationcouncil.org/sites/default/files/research/the_landscape_of_immigration_detention_in_the_united_states.pdf.

Shepherd, K. (2018). *Immigrants rights group sues U.S. Government over family separation at the border: Immigration impact*. American Immigration Council. Retrieved from http://immigrationimpact.com/2018/03/02/sues-family-separation-border/.

Sinnar, S. (2003). Patriotic or unconstitutional? The mandatory detention of aliens under the USA Patriot Act. *Stanford Law Review, 55*(4), 1419–1456.

Slack, J., Martínez, D. E., Lee, A. E., & Whiteford, S. (2016). The geography of border militarization: Violence, death and health in Mexico and the United States. *Journal of Latin American Geography, 15*(1), 7–32.

Thompson, A., Torres, R. M., Swanson, K., Blue, S. A., & Hernández, Ó. M. H. (2019). Re-conceptualising agency in migrant children from Central America and Mexico.

*Journal of Ethnic and Migration Studies, 45*(2), 235–252. https://doi.org/10.1080/1369183X.2017.1404258.

Tidwell Cullen, T. (2018, March 13). ICE released its most comprehensive immigration detention data yet. It's alarming. Retrieved March 17, 2018, from https://immigrantjustice.org/staff/blog/ice-released-its-most-comprehensive-immigration-detention-data-yet.

TRAC Immigration. (2019). *Tracking immigration and customs enforcement removals.* Retrieved from http://trac.syr.edu/phptools/immigration/remove/.

TRAC Immigration. (2020). *Immigration court backlog tool.* Retrieved from https://trac.syr.edu/phptools/immigration/court_backlog/.

UNHCR (United Nations High Commissioner for Refugees). (2015). *Children on the run: Unaccompanied children leaving Central America and Mexico and the need for international protection.* DC: Washington. Retrieved from http://www.unhcr.org/56fc266f4.html.

USCIS (Department of Citizenship and Immigration Services). (2020a). *Public charge.* DC: Washington. Retrieved from https://www.uscis.gov/greencard/public-charge.

USCIS (2020b). *Public Charge.* https://www.uscis.gov/green-card/green-card-processes-and-procedures/public-charge Accessed July 2, 2020.

Vélez Santiago, P., & Fernandez Sanabria, A. (2016, July 15). México levanta un muro invisible: Deporta a 9 de cada 10 centroamericanos que van a EEUU. *Univision Noticias.* Retrieved from http://www.univision.com/noticias/indocumentados/mexico-levanta-un-muro-invisible-deporta-a-9-de-cada-10-centroamericanos-que-van-a-eeuu.

Viladich, A. (2012). Beyond welfare reform: Reframing undocumented immigrants' entitlement to health care in the United State. *Social Science and Medicine, 74*(6), 822–829.

Vogl, A., & Methven, E. (2017, January 30). In a global trend towards crimmigration, Australia has led the world. *The Guardian.* Retrieved from https://www.theguardian.com/commentisfree/2017/jan/31/in-a-global-trend-towards-crimmigration-australia-has-led-the-world.

Welch, M. (2000). The role of the immigration and naturalization service in the prison-industrial complex. *Social Justice, 27*(3), 73–89.

# 34

## Police, Bailiffs, and Hired Hands: Researching the Distribution and Dissolution of Eviction Enforcement

### Alexander G. Baker

Forced evictions remain an iconic symbol of the crisis of displacement in the early twenty-first century. Media covering the collapse of American mortgage markets in 2008 peppered screens with images of broken furniture and families on sidewalks. Yet for all the discussions of eviction, less seems to be made of those who enforce eviction, and while urban sociology and geography are preoccupied with worthwhile questions in the conceptual register—from how to define gentrification to what an 'expanded meaning' of dispossession might look like—some of the more material processes and practices facilitating and making evictions have grown, unnoticed.

The problem of eviction enforcement continues to elude us. A large literature exists on the relation between urban securitisation patterns, policing, and displacement. Scholars in wealthier nations such as Wacquant (2008), Dikeç (2011), and Beckett and Herbert (2009) emphasise the systems of control and exclusion that emerge from the conquest of urban space by private interests, and the domination of urban policy by frameworks of stigma and exclusion. Gretchen Purser (2014) has argued that these scholars, while providing powerful narratives of the role of neoliberal policy in producing systems of spatial exclusion, elide or ignore the act of forced eviction itself. Narratives of counterinsurgency and postcolonial policing draw a much greater attention to the

A. G. Baker (✉)
University of Sheffield, Sheffield, UK
e-mail: alex.baker@sheffield.ac.uk

role of acts of 'domicide' and housing demolition, often emphasising the ways in which mass displacement forms a core part of the process of population management during phases of settler-colonialism and which are thereafter institutionalised in the cultures of policing and racialised logics of the former colony and global patterns of counterinsurgency (Khalili 2010; McMichael 2015). Through subsuming eviction into a wider agenda of racialised policing, these studies can sometimes occlude the specific properties of eviction.

So, what is this term, 'eviction enforcement'? And how might we use it? Researchers in the field of forced evictions who do not subscribe to the dissipation of eviction into wider categories of dispossession and displacement, or accept a specialisation towards the home implied in terms such as 'domicide,' tend to rest on definitions provided by NGOs and human rights law. Brickell, Vasudevan, and Fernandez Arrigoitia start from Amnesty's definition of forced eviction: 'when people are forced out of their homes and off their land against their will, with little notice or none at all, often with the threat or use of violence' (Brickell et al. 2017, p. 1). This draws on UN-HABITAT's (2014) definition which adds caveats around the lack of legal protections. For our purposes, it might be sufficient for now to recognise eviction enforcement as the 'threat or use of violence' at work in eviction itself, and the development of specialised or focused institutions and techniques for its enactment. It is therefore a structural, and deeply integral, aspect of eviction as recognised under international protocols.

In this chapter I want to outline four key points of potential intervention for those who are undertaking to study such institutions and techniques:

1. Eviction enforcement comes in various forms whose contexts are linked, through agencies and institutions enforcing eviction. It is situated within a larger relation of spaces and practices of eviction.
2. Eviction enforcement is not simply about the ways in which people are physically removed from space but goes beyond into the management of structured and weaponised threat, an emotional and affective form of discipline at work which is capable of producing complicities among the evicted.
3. Wherever there is resistance, there is power: Eviction enforcement develops alongside and connects all the forces that resist eviction and displacement, from social movements to 'hidden transcripts.' It is the shadow of spatial struggles.
4. Eviction enforcement demands a strong ethical orientation from those of us who study it, towards the dismantling of its institutions. That is to say, an ethical orientation towards *abolition*.

## A Global Power

Eviction enforcement is a global phenomenon. Its institutions, practices, tactics, and strategies are interlinked and historically intertwined through their implication in structural systems of exploitation and coercion. The present moment in which they are situated is one of a securitised and carceral urban agenda. Alongside work on the exclusionary frameworks of neoliberal policing, social scientists and planners have emphasised the growing militarisation and partition of urban space, which is grounded in circuits of colonial exchange, and technological systems of dominance with global markets and reach (Weizman 2006; Graham 2011), as well as global systems of logistical (Cowen 2014) and infrastructural (Easterling 2014) space which enact large-scale forms of violence that compromises a 'savage sorting' of the global population (Sassen 2014, p. 5).

Eviction enforcement also occurs within a planetary framework for the sale and development of land and housing. Dallas Rogers (2017) has emphasised that real estate markets are a planetary assemblage of material and symbolic infrastructures and political actors that reshape state and international relations, from legal technologies of land to software packages that allow for easy commodity trading across borders. This is dependent on a Eurocentric liberal property regime which is explicitly racialised and hegemonic across contexts and which occludes alternative articulations through its deployment of displacement (Blomley 2004, 2005). Eviction enforcement is by demand a planetary process and practice, as comparable legal frameworks create a need to demarcate and boundary property wherever it appears. When looking at the eviction enforcement system, we have to situate it within this planetary system of private property.

Yet remarkable diversities and forms of eviction enforcement occur. There are private individuals who operate at the edge of legitimate and formal economies. Lynette Ong (2018) has studied these groups (whom she describes as 'hired thugs') in operation in China, where the state chooses to outsource its violence to third parties on an informal basis. The conditions produced are similar to those described by Achille Mbembe (2001, p. 66) in his work on 'private indirect government,' where the state itself entangles coercive functions within private interests. Elsewhere formal state actors are the dominant form, as in some US states where the work of eviction enforcement is monopolised by the sheriffs' offices (Desmond 2016); although this practice is flipped in states such as New York where appointed private marshals do the work of eviction, to say nothing of enacted and proposed legislation in states such as

Illinois that would further expand the private market (Illinois General Assembly 2015). In some wealthier nations, such private markets are growing, and a burgeoning 'eviction industry' is now active in Britain (Paton and Cooper 2016). In many European states, these grew from a continuity of the feudal bailiff system, especially in the UK, where the bailiff as an institution often claims a heritage back to the end of the Anglo-Saxon period and the Norman Conquest of 1066, and the county courts retain salaried bailiffs focused on the work of debts and repossession. The more extreme end of state power is enacted through militaries in land clearances, as found in Cambodian villages (Schoenberger and Beban 2018), or housing demolitions by Israeli forces in the Palestinian West Bank (Khalili 2010), where the state explicitly uses military logic and rhetoric to justify displacements. Eviction enforcement takes a multitude of forms according to local conditions, including the objectives and histories of the local state and land economy.

One of the aspects of these diverse forms that may be most intriguing to research agendas is the links between these contexts. Foremost among these are networks of the colonial past and present, as evidenced by the wide reach of the British Empire and its racialised legacies. Hong Kong's squatter removal department of the police force emerged out of the Empire's policing agenda (Smart 2002). South Africa's anti-land invasion police units are specialised agencies with riot and crowd control training, and the urban pacification agendas they support have a deeper history in the apartheid era and the persisting racism found in regimes of South African urbanism (McMichael 2015). Colonialism has lent its form to more than one eviction enforcement practice and British domestic enforcement policy around evictions may itself have been heavily influenced by decisions to defend the impartiality of the police in Ireland in the 1830s, by keeping them at a distance to land struggles and leaving the legal enforcement of evictions to bailiffs (Palmer 1985, 321–322). One contemporary British 'eviction specialist' who advised on the eviction of prominent protests has a background in the army, private security in Iraq, and shipping security in the Indian Ocean (Carter 2013), suggesting a 'boomerang circuit' in which disciplinary knowledge and skills are transferred from the colonial periphery to the metropolitan core (Graham 2011, p. 12).

There are also more formal regimes of knowledge sharing. In the UK, prison services and others liaise with bailiffs and eviction enforcers to develop best practice (Baker 2017a). The *Union Internationale de Hussiers de Justice* allows for exchange between debt and civil enforcement agencies around the world and within Europe. And police forces also share information and practice. In a report for *Le Monde Diplomatique* the journalist Rene Vázquez Díaz

(2007) argued that the eviction of the Ungdomshuset squat in Copenhagen had explicitly been used by the police as a 'laboratory' for the exploration and demonstration of new eviction tactics to observers from a range of European police forces. This forms part of what the American security analyst and academic Mary Manjikian (2013) has identified as a growing focus on the securitisation of property squatting: Manjikian herself has taught in US armed forces colleges, and as I shall mention below, researchers and scholars are not immune from becoming entangled and deployed by these networks themselves.

This brief sketch of just some of the networks and contexts in which eviction enforcement takes place highlights the diversity, generality, and specificity of eviction enforcement. Existing research shows well how states and private industries enact general regimes of displacement and dispossession, yet they also specialise in specific actions and sustain these structural relations of violence through the use of infrastructural relationships. We see here what one US military strategist has termed the 'seam' between the operation of civilian and military activity (Cowen 2014, p. 82). Eviction is part of existing and emerging security agendas and cuts across research agendas on geopolitics, housing, and postcolonial legacies.

## Power and Process

Despite the way eviction is often perceived as a specific moment of rupture in the life of spaces, eviction enforcement is a process, not an event. Eviction enforcement is an affective and affecting process, one which involves prolonged action upon bodies to remove them from space. Eviction enforcement not only involves enforcing agencies but also enforcing techniques: technologies, strategies, and tactics. For a practical useful division, we may choose to apply some specific meanings to these terms. 'Technology' refers to what Foucault conceives of as part of a matrix of practical reason—the technologies of production, sign systems, power, and the self (1998, p. 18). The differentiation of 'tactics' and 'strategies' derives from the military theorist Clausewitz (1968, p. 86) who distinguishes between the use of forces in combat and the theory of the use of combats collected together. But I also build here on de Certeau's division between the strategic spatial reading of a formal rationality, and the tactical nature of momentary action (1984, p. xix). We have a durational distinction here, between the broad, planned, and plotted world of strategy, and the momentary seizing of opportunity we understand as tactics. Of course it should be noted that this distinction is the product of a Eurocentric model of action, yet it also reflects the fact that the phenomena we are dealing with represent an aspect of the globalisation of a liberal property model and real estate frameworks (Blomley 2005; Rogers 2017).

One productive way to think of evictions is as a genealogical process, without clear origins but increasing intensities through a process of displacement. In this understanding evictions do not have an origin specifically, but instead sets of actors which design specific trajectories through which bodies and land are processed. This becomes more visibly practical when we turn to examples used by scholars who call for expanded understandings of regimes of dispossession. The longitudinal and durational experience of eviction is most visible in technologies for the strategic management of housing finance. Financial instruments and rent extraction practices used in the management of rental housing markets in New York, for instance, shows how the regime of accumulation in rental yields creates feelings of uncertainty, indignity, and anxiety as properties are left to decay (Fields 2017a). Software packages that support tenant management in European and North American nations often rely on the production of digital shadows of tenants and a 'gamified' regime of rental recovery (Fields 2017b). Yet eviction also structures and involves less formal models of dispossession which involve complex regimes of negotiation between landed capitalists and tenants or residents. Leitner and Shepard (2018) argue that the process of accumulation through land displacements in Jakarta reveals the need for an expanded understanding of categories of accumulation by dispossession due to the ways in which new forms of communality, new complicities, and new strategies of survival and homemaking permeate negotiation and resistance to displacement. Similar negotiations, durations, and frustrations to eviction can be seen in prolonged and longitudinal case studies of evictions from squats to rented homes (Lancione 2017; Desmond 2016). Evictions and displacements also persist long after the effects of physical removal, through the effects of 'root shock' in which the trauma of displacement persists long after the loss of the home and neighbourhood (Fullilove 2016; Pain 2019). Understanding eviction as a duration helps us to understand both the material technologies which manage it and the necessary infrastructures of feeling, threat, care, and other forms of emotion that sustain it.

Building on this, the challenge is to articulate the ways in which evictions are an affecting and affective process. A practical theoretical tool to use here is the definition of affect posited by Felix Guattari (1996, p. 159) who describes 'a process of existential appropriation through the continual creation of heterogeneous durations of being.' The work of evicting people from their homes and/or land is both physical and immaterial or emotional, and involves continual work upon the evicted to produce new affects and effects. This means that the tactics used 'on the doorstep' when dealing with the evicted operate in multiple registers. In my own work (Baker 2017b) I have argued that English County Court Bailiffs use a form of emotional labour to constantly anticipate and incentivise occupants of homes which are being repossessed. Forms of 'bodywatching' focused on monitoring the self, and learned, routinised kinds of engagement mean experienced bailiffs understand their work as a complete process of managing the body. Observing the bodies of tenants and adjusting to forms of momentary response specialise and train the bailiff to manage the evictee. In their work on Cambodian land grabs Schoenberger and Beban (2018) point to the ongoing politics of fear and anxiety which structure the experience of displacement and are willingly produced by the evicting forces of the state. Matthew Desmond's (2016) detailed study also emphasises the lasting and prolonged affects of eviction before and after the event, while Michele Lancione's work in Romania (2017) has explored the way eviction resistance can be frustrated by feelings of 'inertia.' When we think about how eviction and evicting agencies act upon those persons facing eviction we must consider the ways in which the focused production of affective states, and appropriations of everyday life and being, structure the way the eviction plays out. Analysis of emotional labour is a key aspect of understanding eviction enforcement.

None of which is to deny the violence of forced evictions but to emphasise the pace and 'speed' of such violence can range from fast to what Rob Nixon has termed 'slow' violence (Pain 2019). This should rightly cause us to question some of the categories of 'wilfulness' at work in the established definition above, as eviction focuses not only on negating the will but also often on bending it, and disposing it towards complicity and collaboration with evicting agencies. For those of us looking for the smoking gun of a faster mode of violence, we are not short of case studies where physical encounters have occurred, but it is also necessary to consider the way more 'docile' responses are forged. This requires us to also look at what has been negated or opposed, and the response to resistance that eviction enforcement demands.

## Resistance

'The final word on power' wrote Gilles Deleuze 'is that resistance comes first, to the extent that power relations operate completely within the diagram, while resistances necessarily operate in a direct relation with the outside from which the diagrams emerge' (Deleuze 2006, p. 74). In research into the dynamics of eviction we must build on this inversion to articulate the ways in which *resistance*, not power, is the driving force in the growth of institutions and practices. This is to recognise that the tactics and strategies focused on evicting people are constantly responding to the new formulations of refusal that are produced by those resisting. When we think about resistance there are multiple potential axes, but, for ease, to illustrate these dynamics we might think of resistance as happening on two scales: the first of these is 'formal' forms of resistance—those used by social movements and organised groups; the second is 'informal' forms of everyday opposition or those articulated as a 'hidden transcript,' as described by James C. Scott (2008).

Major social movements have a tendency to provoke a response from the state, but what is interesting about eviction enforcement is the specialism that has emerged in some contexts. In the UK, the High Court Enforcement industry has grown alongside ecological and squatters' direct action movements, with organisations offering specialist squatter removal services. The National Eviction Team (2018), a private agency that is part of the High Court Enforcement Group firm, boasts a long list of protest evictions and removals that displays its origins in the Roads Protest movement of the mid-1990s. Writing near to the end of that movement, the scholar Brian Doherty (1999) described the 'siege warfare' of escalating tactics between evictors and evictees as tree squats, tunnels, and protest spaces used by activists who demanded increasingly specialised tools and personnel, illustrating how the practices of companies such as these and the strategies and tactics of eviction are intertwined. Police repression of squatters' movements in Europe has also seen an ongoing set of responses to forms of resistance and an increased specialisation, from the testing of experimental new technologies at significant evictions in front of observers (Vasudevan 2017, p. 3), through to incorporating those studying social movements into police research (CEPOL 2006). State agencies are often grounded in the experiences of historical struggles between the state and insurgent social movements. As mentioned, the South African enforcement system grew out of the histories of colonial policing of landless peoples' movements (McMichael 2015), while military language permeates the organisation of evicting agencies in Shanghai, deriving from the imaginary of the local state grounded in the revolutionary history of Chinese politics (Zhang 2017). Formal and explicit forms of political resistance to displacement and histories of social change are understandably incorporated into the development of eviction enforcement agencies. Tactics are retained as part of the institutional memories of these agencies and their skill sets.

However, it is at the level of informal forms of resistance that the most productive research might be conducted. The experience of frontline and street-level enforcement has shaped how eviction enforcement has developed, and often occurs through experience and unmodified and informal forms of knowledge. Eviction enforcement is a job subject to pressures, and to routines. UK County Court Bailiffs have reported that they learn 'on the job' through encountering new situations and forms of non-cooperation from the tenants they encounter, and develop the emotional skills they need in accordance with their own perceptions of increasing risks (Baker 2017b). Bailiffs are assigned areas and routes which they regularly shift around, and in doing so learn to respond to the uneven geographies of class in the spaces they work in; they get to 'know the areas.' Central to this process is the formation of intuition, which Lauren Berlant (2008), following Henri Bergson, understands as the reshaping of cadences, rhythms, and routines. The role of routine and intuition has been emphasised in policing ethnography, such as work by Didier Fassin (2013) who analyses the boredom of policing in the Parisian *banlieue* and the habituated responses that develop in response to violence. In the case of the bailiffs, the development of new habits and intuitions signifies the incorporation of resistance into the repertoire of removal. Enforcement is being reworked and reshaped through constant encounters with disruptions by resistant evictees.

Of course this is in practice a false division of resistance, and traversing the lines of formality and informality is the 'encroachment of the everyday,' described by Anne-Maria Makhulu (2015) in her study of informal settlement, rights, and the state in South Africa during and after Apartheid. Makhulu emphasises that the role of small, situated practices stimulate both cause for protest and response from the state (2015, p. 128). These kinds of small-scale formal and informal acts also stimulate larger, historical, and political responses from government, and development of new patterns of planning and government. This work provides an interesting glimpse into how a focused and dedicated eviction enforcement agency and agenda might emerge from the confluence of the everyday and unspoken, and the formal act of claiming space.

When researching eviction enforcement, we must consider the ways in which enforcement can be considered the shadow of spatial struggles, one which matches, mirrors, and follows resistances, and incorporates them into official and embodied memories. Organisations such as the National Eviction Team are the product of specific formal movements, and their practices, skills, and abilities are the record of the hitherto preceding tactics and strategies of resistance they have encountered. This is no less true of the habits, intuitions, and affective dimensions of forced eviction outlined above. While there are, of course, other agendas and practices at work in shaping institutions, where there is resistance, there is power, and the critique of the evicting agencies demands that we start from resistance. This also demands that we consider what it means to resist the formation of eviction enforcement, and the possibilities for what comes after eviction.

## The Trap of Research and the Possibility of Abolition

If eviction is a relation between resistance and power, then we must also consider what our intervention into this relationship serves. There are significant ethical problematics that frame our approach to researching eviction enforcement. Eviction researchers are increasingly beginning to explore the challenge of researching eviction under conditions in which research is geared towards the reproduction and mitigation of phenomena. The recent critique by producers of the Anti-Eviction Mapping Project (Aiello et al. 2018), of the Princeton-based Eviction Lab (run by Matthew Desmond), has argued that there are multiple ways in which data sourcing, mapping, and the presentation of research can involve the exclusion of the evicted from narratives, and fund or support aspects of the eviction industry. Among other things, the critics accuse the Eviction Lab of purchasing data from private companies that work in the rent recovery sector. While this points us to the perils of 'big data' at the expense of involving those affected by evictions, this is not a problem that limits itself to data analytics alone. Gretchen Purser (2014), one of the authors of the critique of the Eviction Lab, conducted ethnographic research published in 2014 which involved assisting in a number of evictions in poorer African-American neighbourhoods, as part of a study of day labourers. There are also issues around dissemination, as mentioned above in relation to squatters and public order. European squatting research has been invoked in events informing police practice and understanding of how to engage with protest groups and squatters during evictions (CEPOL 2006), where policing agencies take an active interest in the findings of academic research to pursue more effective strategies for managing public disorder during eviction. In an era where public policy impact is often a condition of research funding, it is important to consider where to avoid imbrication within making eviction 'better.' These are not criticisms that are raised to condemn or damn previous research to the dustbin of ethics. Instead we must recognise just how studying the process of eviction can enclose researchers within existing parameters that facilitate further eviction. Can we resolve the issues we confront as researchers while also informing the agents of eviction? Is it possible to study the functioning of institutions without falling into the politics of 'best practice'? And what other framework should we adopt?

In response to these questions, police research and anti-carceral black activism in the US has already provided us with an answer: abolition. In an interview with the researchers Jordan Camp and Christina Heatherton, the activist Asha Ransby-Sporn (2016) of the group We Charge Genocide argues that abolition moves organisers away from making improvements 'right now' towards asking bigger questions like, 'What do we want instead?' Eviction agencies are the perpetrators of both 'slow' and 'fast' forms of violence, from the destruction of homes through direct military action, to facilitating affects of anxiety and loss. When asking what we want instead, we are also made to consider what we would do to eliminate these forms of violence, and what systems of ownership, property, land, and power we should institute instead.

Of course, there remain a number of problems; this approach closes down closer proximity research to the institutions researchers intend to critique. Moreover, it transplants a local but highly relevant context of American prison abolition and black struggles, into the study of a global phenomenon, a problem which authors such as Ananya Roy (2017) and AbdouMaliq Simone (2018) have explored recently in their own work. Yet this is not the case of simply hyper-extending an abolitionist perspective, but, as Roy (2017) argues, recognising the linked and racialised systems of power at work in regimes of '(dis)possessive' capitalism. That eviction enforcement agencies in so many contexts, from Hong Kong, South Africa, and Ireland, share pasts and presents reflects the real vitality of the need for research into eviction enforcement oriented towards abolition. Ultimately, this is a utopian demand, one which emerges from the social movements that eviction enforcers often negate, and one which points to the 'not yet' of homes and land.

## References

Aiello, D., Bates, L., Graziani, T., Herring, C., Maharawal, M., McElroy, E., … Purser, G. (2018). Eviction Lab misses the mark. *ShelterForce*. Retrieved November 2, 2018, from https://shelterforce.org/2018/08/22/eviction-lab-misses-the-mark/.

Baker, A. (2017a). *The machinery of eviction*. Doctoral dissertation, Newcastle University. Retrieved from http://hdl.handle.net/10443/3640.

Baker, A. (2017b). Bailiffs at the door: Work, power, and resistance in eviction enforcement. In K. Brickell, M. F. Arrigoitia, & A. Vasudevan (Eds.), *Geographies of forced eviction* (pp. 145–166). London: Palgrave Macmillan UK.

Beckett, K., & Herbert, S. (2009). *Banished: The new social control in urban America*. Oxford: Oxford University Press.

Berlant, L. (2008). Thinking about feeling historical. *Emotion, Space and Society, 1*(1), 4–9.

Blomley, N. (2004). *Unsettling the city: Urban land and the politics of property*. London: Routledge.

Blomley, N. (2005). Remember property? *Progress in Human Geography, 29*(2), 125–127.

Brickell, K., Fernandez, M., & Vasudevan, A. (2017). Geographies of forced eviction: Dispossession, violence, resistance. In K. Brickell, M. Fernandez, & A. Vasudevan (Eds.), *Geographies of Forced Eviction* (pp. 1–23). London: Palgrave Macmillan UK.

Carter, D. (2013). Ken Somerville joins The Sheriffs Office as Head of Eviction and Security. Archived December 8, 2013. Retrieved 2015, from https://web.archive.org/web/20131208031653/http://thesheriffsoffice.com/articles/ken-somerville-joins-the-sheriffs-office-as-head-of-eviction-security.

CEPOL. (2006). Can urban violence be understood? [Web log post]. Retrieved February 1, 2018, from https://www.cepol.europa.eu/media/blog/can-urban-violence-be-understood.

de Certeau, M. (1984). *The practice of everyday life* (S. Randall, Trans.). Los Angeles, CA: University of California Press.

Clausewitz, C. (1968). *On war* (J. J. Graham, Trans.). London: Routledge and Kegan Paul.

Cowen, D. (2014). *The deadly life of logistics: Mapping violence in global trade*. Minneapolis, MN: University of Minnesota Press.

Deleuze, G. (2006). *Foucault* (S. Hand, Trans.). London: Continuum.

Desmond, M. (2016). *Evicted*. New York, NY: Penguin.

Dikeç, M. (2011). *Badlands of the republic: Space, politics and urban policy*. Chichester: John Wiley & Sons.

Doherty, B. (1999). Manufactured vulnerability: Eco-activist tactics in Britain. *Mobilization: An International Quarterly, 4*(1), 75–89.

Easterling, K. (2014). *Extrastatecraft: The power of infrastructure space*. London: Verso Books.

Fassin, D. (2013). *Enforcing order: An ethnography of urban policing*. Cambridge: Polity Press.

Fields, D. (2017a). Unwilling subjects of financialization. *International Journal of Urban and Regional Research, 41*(4), 588–603. https://doi.org/10.1111/1468-2427.12519.

Fields, D. (2017b). Rent, datafication, and the automated landlord. In J. Shaw & N. Graham (Eds.), *Our digital rights to the city* (pp. 16–19). Oxford: Meatspace Press.

Foucault, M. (1998). *The history of sexuality, Vol. 1* (R. Hurley, Trans.). New York, NY: Penguin.

Fullilove, M. T. (2016). *Root shock: How tearing up city neighborhoods hurts America, and what we can do about it*. New York, NY: NewVillage Press.

Graham, S. (2011). *Cities under siege: The new military urbanism*. London: Verso Books.

Guattari, F. (1996). Ritornellos and existential affects. In G. Genosko (Ed.), *The Guattari reader* (pp. 158–172). Oxford: Blackwell Publishers.

Illinois General Assembly. (2015). Full text of SB0871. Retrieved 2015, from http://ilga.gov/legislation/fulltext.asp?DocName=&SessionId=88&GA=99&DocTypeId=SB&DocNum=871&GAID=13&LegID=86295&SpecSess=&Session=.

Khalili, L. (2010). The location of Palestine in global counterinsurgencies. *International Journal of Middle East Studies, 42*(3), 413–433.

Lancione, M. (2017). Revitalising the uncanny: Challenging inertia in the struggle against forced evictions. *Environment and Planning D: Society and Space, 35*(6), 1012–1032. https://doi.org/10.1177/0263775817701731.

Leitner, H., & Shepard, E. (2018). From kampungs to condos? Contested accumulations through displacement in Jakarta. *Environment and Planning A, 50*(2), 437–456.

Makhulu, A. M. (2015). *Making freedom: Apartheid, squatter politics, and the struggle for home*. Durham, NC: Duke University Press.

Manjikian, M. (2013). *Securitization of property squatting in Europe*. New York, NY: Routledge.

Mbembe, A. (2001). *On the postcolony*. Berkeley, CA: University of California Press.

McMichael, C. (2015). Urban pacification and 'blitzes' in contemporary Johannesburg. *Antipode, 47*(5), 1261–1278.

National Eviction Team. (2018). Client list. Retrieved June 2, 2018, from https://nationalevictionteam.co.uk/clients/client-list.

Ong, L. H. (2018). Thugs and outsourcing of state repression in China. *The China Journal, 80*(1), 94–110.

Pain, R. (2019). Chronic urban trauma and the slow violence of housing dispossession. *Urban Studies, 56*(2), 385–400. Retrieved from http://www.ncl.ac.uk/library/linkit?sv=e&d=250354.

Palmer, S. H. (1985). *Police and protest in England and Ireland 1750–1825*. Cambridge: Cambridge University Press.

Paton, K., & Cooper, V. (2016). It's the state, stupid: 21st gentrification and state-led evictions. *Sociological Research Online, 21*(3), 134–140.

Purser, G. (2014). The circle of dispossession: Evicting the urban poor in Baltimore. *Critical Sociology, 42*(3), 393–415.

Ransby-Sporn, A. (2016). We charge Genocide: An interview with Breanna Champion, Page May, and Asha Rosa Ransby-Sporn. In J. T. Camp & C. Heatherton (Eds.), *Policing the planet: Why the policing crisis led to Black Lives Matter*. London: Verso [Ebook].

Rogers, D. (2017). *The geopolitics of real estate: Reconfiguring property, capital and rights*. London: Rowman and Littlefield.

Roy, A. (2017). Dis/possessive collectivism: Property and personhood at city's end. *Geoforum, 80*, A1–A11.

Sassen, S. (2014). *Expulsions*. Cambridge, MA: Harvard University Press.

Schoenberger, L., & Beban, A. (2018). 'They turn us into criminals': Embodiments of fear in Cambodian land grabbing. *Annals of the American Association of Geographers, 108*(5), 1–16.

Scott, J. C. (2008). *Weapons of the weak: Everyday forms of peasant resistance*. New Haven, CT: Yale University Press.

Simone, A. (2018). The urban majority and provisional recompositions in Yangon: The 2016 Antipode RGS-IBG Lecture. *Antipode, 50*(1), 23–40.

Smart, A. (2002). Agents of eviction: The squatter control and clearance division of Hong Kong's Housing Department. *Singapore Journal of Tropical Geography, 23*(3), 333–347.

UN-HABITAT. (2014). Forced evictions. Fact Sheet 25/Rev. 1. New York, NY and Geneva: United Nations.

Vasudevan, A. (2017). *The autonomous city: A history of urban squatting*. London: Verso.

Vázquez Díaz, R. (2007). Répression pour l'exemple à Copenhague. *Le Monde Diplomatique*. Retrieved April 2, 2018, from https://www.monde-diplomatique.fr/2007/04/DIAZ/14648.

Wacquant, L. (2008). *Urban outcasts: A comparative sociology of advanced marginality*. Cambridge: Polity.

Weizman, E. (2006). *Hollow land*. London: Verso.

Zhang, Y. (2017). 'It felt like you were at war': State of exception and wounded life in the Shanghai Expo-induced domicide. In K. Brickell, M. Fernandez, & A. Vasudevan (Eds.), *Geographies of forced eviction* (pp. 97–118). London: Palgrave Macmillan UK.

# 35

# Governing the Unwanted: Measuring European Migration Enforcement at Street Level

Lisa Marie Borrelli

## Introduction

Generally, entering Europe, legally or 'illegally,' involves complex procedures in order to find out whether or not the person is eligible to remain. In case of arrivals at the European Union's external borders, there are several national and international organisations, state agencies, and private actors who are in charge of the reception, housing, healthcare, asylum procedure, and legal information, and later on—if a negative decision on the stay or asylum application is taken—for the detention and deportation of migrant subjects. Hence, there are various (non)governmental bodies which try to regulate, categorise, and manage the arrival, stay, and eventual removal of migrants.

The following chapter focuses on government practices directed towards irregularised migrants and the underlying biopolitics of the used policies and practices. Irregularised individuals with precarious legal status are subjected to deportation orders, detention, and forcible removals—practices which in themselves are highly disruptive and violent. Thus, as argued in the following, they are governed and controlled through several mechanisms, including detention, or weekly obligations to register for those who have been detected by authorities, as well as being regularly searched for, apprehended, and

---

L. M. Borrelli (✉)
HES-SO, University of Applied Sciences and Arts Western Switzerland Valais, School of Social, Sierre, Switzerland
e-mail: lisa.borrelli@soz.unibe.ch

questioned for those who try to abscond and live clandestinely. The observed forms of governance are based on biopolitics that aim to exclude migrant individuals, attempting to manage human life and exerting biopower over the governed subjects.

This chapter proposes a closer look at the street-level practices of agencies and institutions tasked to govern migrants—here more specifically migrants with precarious legal status and with little chance of legally remaining in Europe. Not only can such practices give a detailed account on how policies are implemented, but they also tell us about the otherwise, often hidden, governing tools and procedures attempting to control migration. Using ethnographical material, including participant observation and semi-structured interviews with migration offices (Switzerland, Latvia, Italy), border police or border guard services (Latvia, Lithuania, Sweden), and local police units (Switzerland, Germany), collected between 2015 and 2017, this chapter aims for a twofold analysis. After a short theoretical discussion on the concepts of displacement, biopolitics, and biopower, the analytical section examines the infrastructure of governments created and used to manage irregularised migrants. It is argued that mechanisms and technologies of the presented infrastructure are used in order to cautiously 'displace' migrants, to control their mobility, and to eventually circumscribe their movements, as well as to remove them forcibly from the territory. Hence, displacement happens not only within the country of origin, or away from it. Rather, practices of deportation and detention have the same character of displacement, forcing individuals to move against their will, which not only has geopolitical but also biopolitical reasons.

Next, this work will highlight the street-level perspective of the abovementioned practices and examine biopolitical measures 'for achieving the subjugation of bodies and the control of populations' (Foucault 1990, p. 140) through the eyes of the street-level bureaucrats who exercise the biopower of states against the migrant subject. By studying the agencies dealing with the implementation of deportation orders, confinement, and territorial exclusion or restriction, we cannot only grasp how geographies of vulnerability are enacted but also map linkages between state agents and the migrant subject. Indeed, through the detailed examination of everyday practices and their critical discussion, the vulnerable body is brought forward, which otherwise remains silenced by the overshadowing professional discourse (Das 1996; cf. Philo 2005) and the discourse on securitisation (Huysmans 2014). Both analytical frames will be brought together in a brief conclusion.

## Who Governs Irregularised Migrants in the European Migration Regime?

Migrants with precarious legal status (or irregularised migrants) include—among others—rejected asylum seekers, who have refused to leave voluntarily; Dublin cases[1]; as well as undocumented and irregular individuals, who have not applied for asylum, but are in the territory without legal permission (e.g. visa-overstayers, as well as migrants who have crossed the borders 'illegally'). Their stay is deemed irregular or illegal and thus measures are taken to identify, detect, detain, and deport these individuals. While on the EU level both migrant trafficking and smuggling are tackled as assumed sources of irregular migration, thus prioritising external border controls via the European Border and Coast Guard Agency (Frontex) and the prevention of exploitation of migrants through the European Agenda on Migration, as well as the European Agenda on Security; the everyday work of street-level bureaucrats tasked with the detection, detention, and deportation of migrants with precarious legal status has other priorities. Border guard services (Lithuania and Latvia) and border police (Sweden, Germany, and Switzerland) focus on individuals already living in the respective countries, trying to detect them on trains, at their workplaces, as well as in their homes. Often, certain units are tasked with the detection, while others are given the responsibility to plan or even implement the deportation, once the sought-after individual is apprehended (Borrelli 2015).

More importantly, besides the European and national policies, it is the street level which has to implement ever-changing laws and who—according to previous research on street-level bureaucracy (Bouchard and Carroll 2002; Dubois 2016; Eule 2014; Eule et al. 2019; Heyman 2009; cf. Lipsky 2010; Maynard-Moody and Musheno 2000; Satzewich 2014; Zacka 2017)—have a certain range of discretion in their everyday decision-making (see also Butler 2004, p. 6 on the 'petty sovereign'). These bureaucrats are the ones on the 'street,' case workers and police officers, who are either in direct contact with the respective migrant or process their cases on the ground, take first decisions, read claims, and prepare deportation orders. Following their work shows the various ways how governments attempt to manage the removal.

## Displacement in Context of Irregularised Migration in Europe

Because of this forced departure, this chapter argues that the displacement of individuals, which according to the *Oxford Dictionary* (Displacement n.d.) is defined by 'the enforced departure of people from their homes, typically because of war, persecution, or natural disaster,' needs to be broadened in our understanding. UNESCO adds 'a form of social change caused by a number of factors' to the definition of displacement, highlighting that it includes individuals who are 'forced to leave their home region to which they are attached and for which they have the knowledge to make a living' (UNESCO n.d.), causing the displaced to become impoverished. This displacement is often caused by arbitrary policies and laws, causing status insecurities and increasing precarity even for long-residing migrant subjects (cf. Tuckett 2015, 2018).

Already before 'becoming irregular,' the asylum seeker is taken to a reception facility—for example, in Germany the distribution is carried out according to the *Königsteiner Schlüssel*, a statistical tool allocating asylum seekers proportionally to each Federal State. In relocation programmes, it is the authorities who choose the country to which asylum seekers are relocated, eventually considering language and family links. Waiting at times several years before the asylum application is rejected, these individuals are forced to move away from their known spheres yet again. Swedish border police officers explain that at times they have trouble understanding why families, whose children have learned Swedish, go to school, and even started apprenticeships, are still forced to leave. Similarly, two Latvian border guards express their unease in chasing after a young Ukrainian man who came to Latvia at the age of one, who is supposed to be deported because he could not pay the cost of renewing his non-citizen passport (Borrelli and Lindberg, 2018; OCMA n.d.). Hence, displacement applies also to individuals who are not residing in their country of origin.

While these procedures imply a high factor of uncertainty, this is maintained and at times purposefully reinforced by deterrence measures. Rejected asylum seekers are picked up at their housing facilities, or homes, to be detained or eventually deported, sometimes on special flights.[2] However, displacement also happens more mundanely. In Switzerland, territorial restriction and exclusion orders are used in order to control spatial mobility (Art. 74 of Foreign Nationals and Integration Act, FNIA 2005). Certain groups of migrants are not allowed to either enter or leave a certain territory—often cities, for example, due to previous criminal offences but also as pure deterrence strategies directed against rejected asylum seekers (cf. Swiss Federal Tribunal 2017).

The effects of such displacements on the body and mind of the migration subject are severe. Besides the spatial constrains and change of locations, the temporal aspect of displacement is crucial (Bosworth 2014; Gill 2016, p. 118). Bosworth (2014) describes how detainees are subjected to frequent transfers between detention centres, happening ad hoc and caused largely by efforts to discipline the subject. Overall, detention functions as (temporal) immobility, adding physical constraints to waiting in an uncertain limbo, where the outcome of their cases remains highly ambivalent (Griffiths 2014). Indeed, the Executive Committee on the High Commissioner's Programme (2004, p. 1) also defines this protracted situation as limbo, in which 'their lives may not be at risk, but their basic rights and essential economic, social and psychological needs remain unfulfilled after years in exile.'

Making use of the term displacement in the context of removing individuals from European territory challenges the common understanding of belonging—spatially and temporally. Not only does it shift the focus from the displaced individual—the suffering and vulnerable—it brings the role of government agencies within Europe to the forefront, highlighting their role in displacement. While the latter is often connected to a deep familiarity of the inhabited space, it remains related to the country of origin in most cases (e.g. regarding internal displacement). Applying it to European contexts allows us to critically examine deportation and detention practices, highlighting the forcefulness of those procedures and questioning their rightfulness regarding cases where individuals have been on the territory for many years. Breaking with the common understanding of displacement being caused by natural disasters, war, or persecution allows thus for a critical view on potentially 'displacing' effects European migration policies have. Migrants can be forced into immobility and mobility by migration control, as such the presented broader definition highlights the continuity of displacement and thus suffering and upheaval for migrants who have fled their home country.

## Enacted Biopower: Spatial (Im)mobility Within European Migration Enforcement

The removal of irregularised migrants is further deeply rooted in biopolitics. Much like Foucault's understanding of biopolitics and biopower as the rationalisation of 'problems posed to governmental practice by phenomena characteristic of a set of living beings forming a population: health, hygiene, birthrate, life expectancy, race' (Foucault 2008, p. 317), the management of migration is assumed to be key to biopolitical management. In this concept, power is 'not directly exercised on people but by inducing, facilitating, hindering, limiting, and preventing their actions' (Estévez 2016, p. 248). So-called Fortress Europe (Follis 2012) determines who is allowed in and who is not. Immigration restrictions are a classical tool to assert biopower on to bodies, causing their (im)mobility; deciding to let live and let die (Fassin 2009; Foucault 2003). Irregularised migrants are not able to seek general healthcare. Their entitlements to social services are low and much literature has argued for the Janus-faced state (Das and Poole 2004)—which rejects the unwanted, while at the same time embracing their workforce—being the underbelly of economic markets (Engbersen and Broeders 2009; Jahn and Straubhaar 1998; Papadopoulou-Kourkoula 2008). Most importantly, here disciplinary power is included in the state, which erects policies to regulate migration flows and punish those who are deemed ineligible.

Politics over life come into being through the creation of migration policies, which aim for the protection of the population 'inside,' defending it against the one 'outside' (Esposito 2005 in Estévez 2016, p. 249). The relevance of biopolitics and the exercise of biopower become especially visible in the implementation of deportation and detention orders. For example, mobile border police units in Sweden conducted their work controls in search of informal labour and irregularised migrants in certain areas, including suburbs with higher percentages of migrants and in places of low-cost labour, such as workshops, car washes, hairdressers, and smaller, local supermarkets (Borrelli 2017a). Industrial areas were preferred places where their 'targets' could be met. By mapping the city and the surrounding area according to potential 'hits' (finding irregularised individuals), a cartography of vulnerabilities can be created. The frequent visits do not only remind migrants with precarious legal status 'where they belong' but also create a certain tension, increasing uncertainty and exclusion.

At the same time, the bodies of the detained are managed in a certain way. Migrants, who are detained, cannot be held with 'regular prisoners,' and while the spatial separation is legally necessary, the actual living conditions are at times very similar. Their 'non-criminal' status does not necessarily entail greater rights and while state agencies are obliged to create entire sets of new governing principles, new categorisation, and systems to detain, practices vary greatly in the studied countries. Switzerland places prisoners and migrants (detained for administrative reasons) in the same facilities—prisons. Both groups have the same amount of walking time in the court, and have the same hours for food, for personal hygiene, and to receive visitors. However, detained migrants are not allowed to work (Borrelli 2017a).

Looking at the means of street-level bureaucrats to govern 'the unwanted,' and at biopolitical measures, allows for tracing the highly unequal power relation between migrant and state. The displaced bodies, after having reached the frontiers and territories of Europe, will be assessed.

Linking back to the relocation aspect—these programmes favour vulnerable individuals with special needs. Planning deportations, European migration regimes not only distinguish between those migrants with precarious legal status for whose deportation force and violence are needed, but also between bodies who are able to travel. Pregnant women, for example, are seen to be able to travel until their seventh month of pregnancy (Borrelli 2016a), whereas after that the deportation needs to be postponed. While these procedures are routine, followed by street-level bureaucrats, an ironic tone can be grasped when they recall individuals suddenly reappearing after having been deported, returning by flight, despite their advanced state of pregnancy. Hence, on the street level we find tendencies where the biopolitical management is inherited, lived, and reproduced.

Distinctions between how to biopolitically govern irregularised migrants' lives appear when studying detention and aspects of gender. Women and children are very rarely placed in detention in Switzerland. Since it is mostly prisons where individuals are held, there are fewer places for women. Also, if a family is 'deportable,' the general rule is to place the husband in detention, assuming that the wife and children will not abscond, but comply with a smooth deportation procedure (Borrelli 2016b, 2017b). In Sweden, similarly to Lithuania and Latvia, detention facilities are specialised and thus do not mix the housing of different confined groups. In Germany, detention is practiced differently in each *Bundesland* (Federal State) and at times not at all, due to a lack of nearby facilities and costs. Instead, unannounced deportations are practiced—often targeting families and adults with children, due to their less mobile lives.

However, there are moments and cases where the biopolitical power of the state is questioned and eventually contested. Migrants, as well as street-level bureaucrats, retain certain strategies of resistance, which can reduce biopower exerted upon them. At times, respective deportees cannot be sent back—due to the refusal to leave voluntarily or to support the authorities with information. Yet, the administration of life goes deep and instead of becoming enabled due to resistance, migrants might end up in limbo—without the rights to stay or work, but tolerated until a solution is found. Also, biopower can be contested by laws and policies themselves (Borrelli 2018 on the state's inherent limiting factors regarding successful policy implementation; cf. Castles 2004). Individuals with medical issues might not be able to be deported. Street-level bureaucrats are expected to enquire on the psychological and physical situation of the respective deportee—much to their dismay. In the case of medical issues, respective deportees can be assessed by medical doctors regarding their 'transportability.' The categorisation of bodies between 'able to fly' and not clearly reflects biopolitical control, sorting bodies into more capable and less. At the same time, it suggests that 'some bodies were [more] stubborn' (McWhorter 2017). A second example for inherent policy limitation is the Dublin Regulation, which represents a modern dilemma of biopower. While responsibility for asylum applications should be regulated by the Dublin Regulation, migrants end up being pushed back and forth between states who try to exercise power over them, while also reducing their responsibility vis-à-vis other Member States.

However, biopower is also exercised through weekly, or more frequent, obligations to register at the local police station or migration authority, as well as through the control of border police units in certain parts of the city or countryside. Mobility is an aspect of biopolitics, which decides who is eligible to move and who is not. Like Foucault's (2008) examples of the leper epidemic and the pest, we find political interest in separating and segmenting. Indeed, we can also find 'political dreams' (2008, p. 4), which Foucault associated with both epidemics. Biopolitics today aims for the exclusion of a broad group of individuals (see the leper) interested in marking and keeping the undifferentiated masses outside. At the same time biopolitics addresses similar points that Foucault associates with the plague. Once inside the territory, the individual needs to be assessed, analysed, and distributed. Not only does each asylum claim need to be assessed individually, but also the follow-up on each case, on each rejected asylum seeker, or on each Dublin deportee is different. Here it is the disciplined society, which excludes the individual (Foucault 2008).

Thus, exclusion can have different faces and is not always visible. One more mundane form is the differentiation of documents. Italian citizens have different passports and information on their citizen cards to asylum seekers and migrants with residence permits. In Switzerland, residence permits are categorised by capital letters, determining the (temporally limited) stay. It is through paperwork that mobility is governed and circumscribed—spatially and temporally (Borrelli and Andreetta 2019).

In a final step, the results of biopolitics become translated into what Mbembe (2003) has termed 'necropolitics.' Compared to biopolitics, which is also referred to as the politics of life, necropolitics (or the politics of death) takes an interest in the study of regimes or apparatuses where the state of exception (Agamben 2005) is permanent. It is no longer a temporally constricted state, but exceptional (and often restrictive) policies and procedures become the general rule—life is permanently subordinated to state power. 'Necropolitical studies of migration examine how people subjected to criminal and legal violence […] are left to die in their home countries or when they try to cross increasingly dangerous and lethal borders as a result of criminalization of undocumented migration, the creation of obstacles to asylum, and arbitrary deportations' (Estévez 2016, p. 250). The criminalisation of attempts to enter Europe is one example of necropolitics. The decision to send irregularised migrants back to their home countries implies that an evaluation regarding the safety and potential risk is conducted (see Country of Information units). Hence, it remains in the hands of the receiving countries to decide how and if migrants are allowed to live and ultimately even how they are supposed to die. Sending back individuals to highly violent, militarised zones or sending them back to 'developing' countries willing to take back individuals deemed not worthy of protection (see the EU-Turkey Agreement) does underline the power of Western states to decide on the bodies and thus lives of migrants. Here it is 'the ultimate expression of sovereignty' (Mbembe 2003), which comes into being.

## Conclusion

This chapter tries to challenge the common definition of displacement, linking it to the removal and restriction of mobility of irregularised migrants in the European Union (and Switzerland). It does so by deconstructing the understanding of displacement and by opening up the discussion with examples collected during several months of fieldwork in different government agencies tasked with migration enforcement. Using displacement in context of European migration policies opens up the field for a reconceptualisation of displacement, as well as for a critical examination of removal practices and their legitimacy. Further, it links the detention and deportation of migrants with precarious legal status to the discussion of biopolitics and the materialisation of biopower. The management of migration is heavily intertwined with biopolitics, examining how social and political control is exerted over people's lives in general.

This power to control is a very traditional attempt of the state to manage the population. In its disaggregatedness, neither being a single entity nor a graspable construct, the state is not only able to exercise a strong power to subordinate, categorise, and exclude, but it is also able to cast a great shadow of uncertainty on the bodies, who are subjected to its control (see Das 2004 on the state's illegibility; Eule et al. 2019). This again has a strong effect on the individuals, causing psychological distress, trauma, and a constant fear of being detected and potentially removed.

Bringing together both sections, we are able to map spatial and temporal vulnerabilities of migrants, marginalised and excluded due to migration policies. The exclusion and deportation of people follows certain patterns, disclosing certain categorisations (migrants as the excluded, non-belonging, other, the resistant, the ineligible).

At the same time these individuals should not be seen as mere victims of a 'heartless "system"' (Herzfeld 1992, p. 90) of public administration. Instead, they maintain certain ranges of agency (Eule et al. 2019), which need to be discussed and brought into the analysis. Here the autonomy of migration approach seems helpful (Casas-Cortes et al. 2015; De Genova 2017; Papadopoulos and Tsianos 2013). Further, to grasp smaller resistance practices against governing bodies, Scott's (1990) concept of 'hidden transcripts' is of help, which include at times less openly opposing strategies, such as foot-dragging (both from the street-level bureaucrat and the migrant subject; see Borrelli and Lindberg 2018).

In conclusion, the mapping of vulnerabilities in and beyond the migration regime is of crucial interest to broaden our understanding of street-level practices, the ambiguous engagement of various (non)governmental actors, including institutions as well as individuals, but also to further a critical perspective of the European migration regime, which is part of a biopolitical system of displacement.

## Notes

1. The Dublin III Regulation (Regulation 604/2013) defines which EU Member State (also including Denmark, Norway, Iceland, and Switzerland) is responsible for the evaluation of an asylum claim of third-country nationals, who arrived in Europe. Generally, the regulation states that third-country nationals have to apply in the first country they entered and where they were identified by state authorities.
2. These are specifically scheduled flights in which individuals who are deemed difficult to remove and assumed to cause trouble are deported forcibly, with entire teams escorting them.

# References

Agamben, G. (2005). *State of exception*. Chicago, IL: University of Chicago Press.

Borrelli, L. M. (2015). [Field notes]. Unpublished raw data.

Borrelli, L. M. (2016a). [Field notes, Swiss Cantonal Migration Office]. Unpublished raw data.

Borrelli, L. M. (2016b). [Field notes, Switzerland]. Unpublished raw data.

Borrelli, L. M. (2017a). [Field notes]. Unpublished raw data.

Borrelli, L. M. (2017b). [Field notes, Sweden]. Unpublished raw data.

Borrelli, L. M. (2018). Whisper down, up and between the lanes—Exclusionary policies and their limits of control in times of irregular migration. *Public Administration, 96*(4), 803–816. https://doi.org/10.1111/padm.12528.

Borrelli, L. M., & Andreetta, S. (2019). Introduction. Governing migration through paperwork. *Journal of Legal Anthropology, 3*(2), 1–9. https://doi.org/10.3167/jla.2019.030201.

Borrelli, L. M., & Lindberg, A. (2018). The creativity of coping: Alternative tales of moral dilemmas among migration control officers. *International Journal of Migration and Border Studies, 4*(3), 163–178. https://doi.org/10.1504/IJMBS.2018.10013558.

Bosworth, M. (2014). *Inside immigration detention*. Oxford and New York, NY: Oxford University Press.

Bouchard, G., & Carroll, B. W. (2002). Policy-making and administrative discretion: The case of immigration in Canada. *Canadian Public Administration, 45*(2), 239–257. https://doi.org/10.1111/j.1754-7121.2002.tb01082.x.

Butler, J. (2004). *Precarious life: The powers of mourning and violence*. New York, NY: Verso.

Casas-Cortes, M., Cobarrubias, S., & Pickles, J. (2015). Riding routes and itinerant borders: Autonomy of migration and border externalization. *Antipode, 47*(4), 894–914.

Castles, S. (2004). The factors that make and unmake migration policies. *International Migration Review, 38*(3), 852–884. https://doi.org/10.1111/j.1747-7379.2004.tb00222.x.

Das, V. (1996). *Critical events: An anthropological perspective on contemporary India*. Oxford: Oxford University Press.

Das, V. (2004). The signature of the state: The paradox of illegibility. In V. Das & D. Poole (Eds.), *Anthropology in the margins of the state* (pp. 225–252). Oxford: Oxford University Press.

Das, V., & Poole, D. (2004). *Anthropology in the margins of the state*. Oxford: Oxford University Press.

De Genova, N. (Ed.). (2017). *The borders of 'Europe': Autonomy of migration, tactics of bordering*. Durham, NC: Duke University Press Books.

Displacement. (n.d.).In *Oxford Dictionaries | English*.Retrieved from https://en.oxforddictionaries.com/.

Dubois, V. (2016). *The bureaucrat and the poor: Encounters in French welfare offices*. London and New York, NY: Routledge.

Engbersen, G., & Broeders, D. (2009). The state versus the alien: Immigration control and strategies of irregular immigrants. *West European Politics, 32*(5), 867–885. https://doi.org/10.1080/01402380903064713.

Estévez, A. (2016). The biopolitics of asylum law in the United States. In S. Prozorov & S. Rentea (Eds.), *The Routledge handbook of biopolitics* (pp. 247–259). London: Routledge.

Eule, T. G. (2014). *Inside immigration law: Migration management and policy application in Germany*. Farnham: Ashgate.

Eule, T. G., Borrelli, L. M., Lindberg, A., & Wyss, A. (2019). *Migrants before the law: Contested migration control in Europe*. Cham: Palgrave Macmillan.

Executive Committee of the High Commissioner's Programme. (2004). *Protracted refugee situations* (EC/54/SC/CRP.14 §). Retrieved from http://www.unhcr.org/40c982172.pdf.

Fassin, D. (2009). Another politics of life is possible. *Theory Culture Society, 26*(44), 44–60. https://doi.org/10.1177/0263276409106349.

Foreign Nationals and Integration Act (FNIA). (2005). Retrieved from https://www.admin.ch/opc/en/classified-compilation/20020232/index.html.

Follis, K. S. (2012). *Building Fortress Europe: The Polish-Ukrainian frontier*. Philadelphia, PA: University of Pennsylvania Press.

Foucault, M. (1990). *The history of sexuality, Vol. 1: An introduction* (Reissue edition). New York, NY: Vintage.

Foucault, M. (2003). *'Society must be defended'. Lectures at the Collège de France, 1975–76*. (M. Bertani & A. Fontana, Eds., D. Macey, Trans.) (1st ed.). New York, NY: Picador.

Foucault, M. (2008). Panopticism' from 'Discipline & punish: The birth of the prison. *Race/Ethnicity: Multidisciplinary Global Contexts, 2*(1), 1–12.

Gill, N. (2016). *Nothing personal?: Geographies of governing and activism in the British asylum system*. Chichester: Wiley Blackwell.

Griffiths, M. (2014). Out of time: The temporal uncertainties of refused asylum seekers and immigration detainees. *Journal of Ethnic and Migration Studies, 40*(12), 1999–2009.

Herzfeld, M. (1992). *The social production of indifference*. Chicago, IL: University of Chicago Press.

Heyman, J. M. (2009). Trust, privilege, and discretion in the governance of the US borderlands with Mexico. *Canadian Journal of Law and Society, 24*(3), 367–390.

Huysmans, J. (2014). *Security unbound. Enacting democratic limits*. Abingdon and New York, NY: Routledge.

Jahn, A., & Straubhaar, T. (1998). A survey of the economics of illegal migration. *South European Society and Politics, 3*(3), 16–42. https://doi.org/10.1080/13608740308539546.

Lipsky, M. (2010). *Street-level bureaucracy: The dilemmas of the individual in public service*. New York, NY: Russell Sage Foundation.

Maynard-Moody, S., & Musheno, M. (2000). State agent or citizen agent: Two narratives of discretion. *Journal of Public Administration Research and Theory, 10*(2), 329–358. https://doi.org/10.1093/oxfordjournals.jpart.a024272.

Mbembe, A. (2003). Necropolitics. *Public Culture, 15*(1), 11–40. https://doi.org/10.1215/08992363-15-1-11.

McWhorter, L. (2017). From scientific racism to neoliberal biopolitics. In N. Zack (Ed.), *The Oxford handbook of philosophy and race* (pp. 282–293). New York, NY: Oxford University Press. https://doi.org/10.1093/oxfordhb/9780190236953.013.22.

OCMA (The Office of Citizenship and Migration Affairs).(n.d.).Alien's (non-citizen's) passport. Retrieved October 30, 2018, fromhttps://www.pmlp.gov.lv/en/home/services/passports/alien%E2%80%99s-(non-citizens)-passport.html.

Papadopoulos, D., & Tsianos, V. (2013). After citizenship: Autonomy of migration, organisational ontology and mobile commons. *Citizenship Studies, 17*(2), 178–196.

Papadopoulou-Kourkoula, A. (2008). *Transit migration: The missing link between emigration and settlement*. New York, NY: Palgrave Macmillan.

Philo, C. (2005). The geographies that wound. *Population, Space and Place, 11*(6), 441–454. https://doi.org/10.1002/psp.400.

Satzewich, V. (2014). Visa officers as gatekeepers of a state's borders: The social determinants of discretion in spousal sponsorship cases in Canada. *Journal of Ethnic and Migration Studies, 40*(9), 1450–1469. https://doi.org/10.1080/1369183X.2013.854162.

Scott, J. C. (1990). *Domination and the arts of resistance: Hidden transcripts*. New Haven, CT and London: Yale University Press.

Swiss Federal Tribunal (Schweizerisches Bundesgericht) v A (2017) 2C 287/2017. (2017). Retrieved from https://www.bger.ch/ext/eurospider/live/de/php/aza/http/index.php?lang=de&type=highlight_simple_similar_documents&page=1&from_date=&to_date=&sort=relevance&insertion_date=&top_subcollection_aza=all&docid=aza%3A%2F%2F13-11-2017-2C_287-2017&rank=1&azaclir=aza&highlight_docid=aza%3A%2F%2F13-11-2017-2C_287-2017&number_of_ranks=2294.

Tuckett, A. (2015). Strategies of navigation: Migrants' everyday encounters with Italian immigration bureaucracy. *The Cambridge Journal of Anthropology, 33*(1). Retrieved from http://berghahnjournals.com/view/journals/cja/33/1/ca330109.xml.

Tuckett, A. (2018). *Rules, paper, status: Migrants and precarious bureaucracy in contemporary Italy*. Stanford, CA: Stanford University Press.

UNESCO (United Nations Educational, Scientific and Cultural Organization. (n.d.).Displaced Person/Displacement.Retrieved October 30, 2018, from http://www.unesco.org/new/en/social-and-human-sciences/themes/international-migration/glossary/displaced-person-displacement/.

Zacka, B. (2017). *When the state meets the street. Public service and moral agency*. Cambridge, MA: Harvard University Press.

# 36

## A Forced Displacement and Atrocity Crime Nexus: Displacement as Transfer, Annihilation, and Homogenisation

Andrew R. Basso

Forced displacement is often weaponised during processes of political violence. There is a vast array of scholarship which examines atrocity crimes[1] (genocide, crimes against humanity, and war crimes) in individual and comparative contexts, but only a small subset of this literature begins to conceptualise the relationships between forced displacement and atrocity crime perpetration. Human rights violations and geography are linked in many ways. Most important to this relationship is understanding how political geographies are made sites of violence. Different types of geographies do not necessarily perfectly determine what sorts of crimes occur: deserts can just as easily be used as sites of extermination as sites of guerrilla warfare (Gewald 1999; Hillebrecht 2010).[2] Densely populated areas can become sites of possible resistance, but they can also be used as ghettos, transit sites, and places of mass murder (Friedländer 2007; Akçam 2013; Hiebert 2017).[3] Geography is not necessarily a deterministic variable in understanding atrocities, but there can be no doubt that perpetrators have used a wide variety of spaces as central aspects of perpetrating the atrocities they are committing.

What follows is an introduction to a forced displacement and atrocity crime nexus. This chapter begins by assessing relevant atrocity crime and human rights laws and norms; positions forced displacement as transfer,

A. R. Basso (✉)
Western University, London, ON, Canada
e-mail: abasso2@uwo.ca

annihilation, and homogenisation; and concludes with a call to action for scholars and students to more rigorously focus conceptual and empirical studies on instances of forced displacement and how they relate to atrocity crime perpetration.

## Atrocities, Human Rights, and Forced Displacement

Forced displacement violates—or at best undermines—almost all internationally recognised humanitarian principles. Below is a brief examination of atrocity crime and human rights laws and norms which are violated by forced displacement. While limited, this discussion provides a brief introduction to how displacement intersects human rights.

### Atrocity Crimes

Forced displacement is central to the perpetration of many atrocity crimes. The 1948 Convention on the Prevention and Punishment of the Crime of Genocide did not specifically name forced displacement as a constitutive crime of genocide but Article 2, paragraphs b, c, and d, can be interpreted as having links with displacement. These articles deal with causing serious harm (mental or physical) on members of groups, inflicting conditions of life designed to destroy members of groups, and preventing births within groups (United Nations 1948a; Schabas 2009). Forced displacement can be used as a primary tool to annihilate populations (discussed below), almost always does extreme harm to both the mental and physical states of members of groups, and can cause infertility due to stresses of movement. Forced displacement itself is not necessarily genocide but it is often a tool during genocidal processes and—when paired with systemic deprivations of vital daily needs—can become in fact a genocidal process. Article 7, paragraph 1, section d, of the Rome Statute of the International Criminal Court states that 'deportation or forcible transfer of population' is a crime against humanity (United Nations 1998). It should be noted that displacement rarely occurs in a vacuum and is also an interrelated practice to other crimes against humanity: murder, extermination, torture, sexual violence, enforced disappearance, and inhumanely causing great suffering (United Nations 1998). Finally, forced displacement can be a war crime. It is illegitimate for parties to an armed conflict to deport or transfer populations in occupied territories elsewhere or transfer their own populations into occupied territories (United Nations 1998). Related practices like wilful killing, torture, causing great suffering, destruction of property, and the direction of attacks against civilians, inter alia, are also outlawed. While genocide and crimes against humanity can occur in armed conflict and in peace, war crimes can only be perpetrated during international or non-international armed conflicts (Clapham and Gaeta 2014; Paust et al. 2007; Sivakumaran 2012).[4]

## Human Rights

Forced displacement undermines and/or violates a plethora of rights delineated in non-binding and codified human rights documents (United Nations 1948b, 1966a, b). An analysis of all relevant human rights and how they are affected/violated by forced displacement is impossible in this limited space, but there are three major rights which can be highlighted. First, when individuals are forcibly uprooted from their home communities there are stronger structural opportunities for violence against them, posing serious threats to the right to life (United Nations 1948b, 1966a). Practices like human trafficking, slavery, sexual violence, domicide, ethnic cleansing, and atrocities against vulnerable displaced populations are widespread. Without established support and community networks, individuals are at a much higher risk of violence due to the lack of humanitarian relief possibilities and social stability. This means that displacement itself is typically a politically violent act designed to intentionally inflict pain on displaced persons, clearly a violation to the right to life. Second, the right to an adequate standard of living can be interpreted to include the right to adequate water, food, housing, and sanitation (United Nations 1948b, 1966b). The best place for individuals to enjoy this right to an adequate standard of living is their home; forced displacement denies individuals this opportunity by placing them in precarious material positions, threatening their access to provisions necessary for life. Forced displacement, whether as flight from violence, as a part of violent processes, or as a by-product of war and atrocity, is a threat to rights. Put plainly: forced displacement is the absence of an adequate standard of living. Third, the right to self-determination is violated during forced displacement processes (United Nations 1948b, 1966a, b). When groups are not allowed to live as a community in the ways that they see fit (so long as they are not violating human rights themselves) and are displaced, this is a grave breach of the right to self-determination; clearly, displacement disrupts these abilities. Perpetrators of systemic forced displacement rarely target only one or a few individuals.

Rather, perpetrators target groups and collectively punish or target groups for uprooting, disruption, and potentially annihilation.

## Displacement and Rights

There are established (but not robust) international legal principles which define forced displacement as an atrocity and human rights violation and outlaw the practice.[5] However, displacement has been, and will continue to be, a serious sociopolitical problem which is recognised and generally banned by international law. It is important to note that forced displacement is often an intentional, central tool of political violence, not only a by-product of war and atrocity—and these actions violate human rights and are often criminal.

# Displacement as Transfer: Direct and Indirect Killing

A common feature of many processes of annihilation is some type of forced displacement to concentrated killing sites. Perpetrators force targeted populations away from their homes to eliminate possibilities for resistance and to transport them to killing centres. These displacements allow perpetrators to engage in direct and indirect killing programmes away from population centres, where they can have total hegemony over targeted populations and the machineries of annihilation can be constructed in concentrated sites.

## Biological Direct Killing

Direct killing operations are perpetrated using physical violence to annihilate populations. Examples of this include the use of gas chambers, execution squads, blade strikes, gunshot wounds, and blunt force trauma, inter alia. Direct killing requires perpetrators to kill their targets using force typically with constructed materials (hand-held weapons, sites of mass murder like an extermination camp, and blades, inter alia) (Basso 2016, 2019). Displacement typically features prominently with these types of killing methods. The Nazi extermination camps are a perfect example of this: Jews, Slavs, Roma/Sinti, sexual minorities, and the disabled were sent to extermination sites to be killed quickly (Friedländer 2007; Kershaw 2009). Typically, trains were used to transport targeted populations to sites of mass murder far enough away from major cities so extermination could be focused and easier to implement, not dispersed and difficult to manage (not to mention somewhat hidden from onlookers) (Hilberg 2003). One of the other killing methods the Nazis deployed was the *Einsatzgruppen* killing squads in the east. These Schutzstaffel (SS) personnel often escorted targeted populations away from their communities and engaged in mass executions at various predetermined sites (Hilberg 2003; Megargee 2007). Once again, this allowed for focused, punctuated killing to occur.

In Rwanda, the act of displacement took many forms. At the end of the genocide, Hutus (with many Hutu extremists) fled Rwanda to Zaire en masse with the help of the French. Displacement also took the form of 'hunts' for Tutsis—derogatorily referred to as *inyenzi* ('cockroaches' in Kinyarwanda) by perpetrators (Dallaire 2003). Tutsis took flight from their homes out of sheer terror to avoid genocidal violence and hid in marshlands and whatever places they could find. Every day during the 100 days of genocide, Hutu extremists hunted Tutsis in killer bands, led by the *Interahamwe* ('those who kill together'), and turned the 'land of a thousand hills' into hundreds of sites of genocide (Hatzfeld 2006). While displacement was a tactic to flee from violence for the Tutsis, it was a tactic of hunting for perpetrators. Displacement plays many roles in biological killing programmes.

## Biological Indirect Killing

Helen Fein (1997) placed a new focus on indirect killing with her 'genocide by attrition' concept. She argued for a deeper understanding of Article 2(c) of the Genocide Convention—the creation of systemic conditions to destroy a group which are generally imposed at stationary killing centres (Fein 1997). Atrocities perpetrated by indirect killing methods are largely predicated on denying targets their vital daily needs (e.g. food, water, clothing, shelter, and medical care). By denying victims these necessities of life, the human body degrades and ultimately succumbs to the conditions of life imposed upon it (Totten 2012; Rosenberg and Silina 2013; Basso 2016). This can also include the use of intentional poisoning and forced labour—both designed to rapidly destroy the body and kill individuals. Killing centres can be as geographically small as a Nazi ghetto system, as large as the Cambodian killing fields, and as dispersed as Sudanese villages, towns, and cities (Fein 1997; Totten 2012; Hiebert 2017). Victims are displaced to these centres, deprived of their rights, and deprived of their vital daily needs causing mass death.

The modern concentration camp, where targets are deported to, incarcerated at, and killed en masse, was first implemented by Lord Kitchener during the Second Boer War (1899–1902), utilised by the Germans during the Herero Genocide in German South West Africa (GSWA) and by the Nazis during the Holocaust, and utilised in a plethora of other atrocities throughout the twentieth and early twenty-first centuries.[6] The premise of a concentration camp is simple: displace targeted populations to one single area, deprive them of vital daily needs, compel them into forced labour, conduct medical experiments upon them, and destroy through deprivation. Similar things could be said of the Jewish ghetto system during the Holocaust. Jews were concentrated and deprived of vital daily needs (largely while awaiting transfer to extermination camps) (Friedländer 2007; Fein 1997). In Cambodia, major cities were forcibly emptied by the Khmer Rouge and entire communities were sent to 'the killing fields' (Hiebert 2017). In these places, Cambodians were ideologically indoctrinated, tortured, executed, worked to death through compelled labour, and killed through intentional deprivation of vital daily needs.

## Cultural Killing

Raphael Lemkin (2013), the man who created the word 'genocide'—a portmanteau of the Greek *genos* (a people) and the Latin *cide* (to kill)—was prolific in identifying processes of group destruction. He made an important link between biological and cultural destruction. While killing programmes based on biological destruction kill members of a group, cultural destruction aims to destroy *what it means to be a member of a group*. This latter, cultural ontological destruction, for Lemkin has the equivalency in severity and destructiveness as biological destruction. A perfect representation of this is Canada's former Indian Residential School (IRS) system (late 1800s–1996). Approximately 150,000 Indigenous children were forcibly separated from their parents and the Canadian state, with the help of religious organisations, sent these children to 139 schools which operated across Canada (TRC 2015). Children were taught that their cultures were dirty and savage, their spiritual beliefs were backward, and their overall identities had to be left in history; Canada committed a genocidal assimilation of Indigenous children and it was made possible only through the separation of families and transfer of children great distances away from their homes (Woolford 2015). This distance attempted to sever all links Indigenous children had with their parents and communities, creating what could be called a 'broken circle' (Fontaine 2010). Forced displacement to these genocidal schools underpinned the entire system.

## Displacement as Transfer

The transfer of targeted populations to stationary killing centres is an important intermediate step for perpetrators of atrocity.[7] Stationary centres allow perpetrators to build infrastructures of annihilation in specific areas, allowing them to concentrate their resources to inflict maximum amounts of destruction. Displacement is central to these sorts of plans as it brings targeted populations to the perpetrators, expediting the killing process.

## Displacement as Annihilation: Displacement Atrocities

Displacement Atrocity (DA) crimes are processes which weaponise the act of displacement against targeted populations in order to destroy them in whole or in part (Basso 2016). This type of criminal activity can be defined as follows:

> A Displacement Atrocity is a type of killing process employed against a targeted population which uniquely fuses forced population displacement with primarily indirect deaths resulting from the dislocation and systemic deprivations of vital human needs (i.e., food, water, clothing, shelter, and medical care). The killing processes exploit various geographies to annihilate populations in whole or in part with genocidal or non-genocidal intent. Land is utilized in terms of area squared (kettling) or linear distance (escorting). (Basso 2018, p.3, 2019, p.7)

In DA crimes, targets are uprooted from their communities and intentionally kept moving to accelerate their deaths using indirect methods (Basso 2018). By denying the necessities of life, perpetrators create potent killing systems which eventually destroy the human body via exhaustion and deprivation.

## Kettling DA Crimes

Perpetrators of *kettling DA crimes* utilise large land areas to displace populations into, deprive them of vital daily needs, continually force them to move, and deny them the ability to escape from cordons of annihilation (Basso

2016, 2019). Perpetrators keep populations on the move by continually harassing them using physical force and limited violent engagements; direct killing enforces the cordon/blockade around the zone of death (Basso 2019). Victims may search for escape routes from zones of death but are usually unsuccessful and the quest for escape contributes to the destruction of the group gradually: if targets remain stationary they will have no chance of survival or escape, but if they continue to be displaced then the physical and mental exhaustion from movement will accelerate biological destruction. Perpetrators deny targets any resources by poisoning water sources and destroying all other materials which could be used as humanitarian relief. In short, perpetrators of kettling DA crimes use direct force to continually displace populations, kettle them into a zone of death, keep them there to ensure annihilation, and create a deadly cocktail of indirect killing enforced by direct killing (Basso 2016).

From 1904 to 1908 in GSWA, German colonial troops (and colonials) annihilated approximately 75 per cent of the Herero peoples through DA crimes and a widespread concentration camp system in one of the many colonial genocides of the twentieth century (Drechsler 1980). From the end of 1903 through August 1904, the Hereros revolted against German colonial rule and initially fought well until the Germans were reinforced with fresh troops under the command of General Adrian Dietrich Lothar von Trotha. Soon after his arrival in GSWA, he fought the Hereros into a concentric battle at the Waterberg (11 August 1904) and planned to exterminate them there (Drechsler 1980; Hull 2006). When his plan failed and the Hereros escaped into the Kalahari (Omaheke) Desert, he exploited the territory and forcibly kept the Hereros there for months in the fall of 1904. Approximately 30,000–45,000 were killed through DA crime policies alone as the Germans created a 250-kilometre cordon around the Omaheke, poisoned water wells, shot at Hereros driving them further into the desert, and did not allow them rest or vital daily needs. The remaining 15,000–30,000 Hereros were killed in concentration camps (Drechsler 1980; Gewald 1999; Basso 2019). The Herero Genocide is a perfect, crucial example of a kettling DA crime.

## Escorted DA Crimes

Perpetrators of *escorted DA crimes* exploit linear distance. Perpetrators uproot targeted peoples and force march them to their deaths over hundreds of kilometres. Victims are 'escorted' to their demise by armed perpetrators who ensure that victims continually march through limited uses of direct force and threats of the use of force (Basso 2016, 2018, 2019). If victims do not continue to move they are killed to induce other members to continue to move in the hopes they might survive the death march. This limited use of direct violence also asserts the hegemonic authority of the perpetrator group. When coupled with systemic deprivations of vital daily needs, forced displacement over hundreds of kilometres can become a deadly process (Basso 2016).

At the fall of the Ottoman Empire, from 1914 to 1925, approximately 2.5 million Christians (1.5 million Armenians, 750,000 Greeks, and 250,000 Assyrians) were annihilated through various killing methods, though the use of escorted DA crimes featured prominently (Balakian 2003; Hofmann 2011; Gaunt 2011). First the Young Turk regime implemented policies of annihilation against Christian minorities from 1914 to 1918 and then later the Kemalists (led by Mustafa Kemal—'Atatürk') from 1919 to 1925 (Akçam 2013; Hofmann 2011). The Young Turks used the Great War (1914–1918) as a cover and a justification for their atrocities against Christians. Christian men were forced to serve in labour battalions and worked to death at rates approaching 90 per cent (Stavridis 2011; Akçam 2013). The remaining populations were gathered and forced marched into the heartland of Anatolia (Asia Minor) towards Der Zor. Christians in these displacement caravans marched between approximately 600 and 1600 kilometres and succumbed to the death marches at rates of nearly 80–90 per cent (Basso 2019; Hofmann 2011). Displacement was a potent killing method linked with systemic deprivations of vital daily needs. The Kemalists, while first denouncing the crimes of the Young Turk regime, then engaged in these crimes after the failures of the Paris Peace Conference. These events resulted in a cumulative genocide of Christian minorities in the Ottoman Empire (Halo 2017; Hofmann 2011), the homogenisation of Asia Minor, and the formation of a new country: modern-day Turkey.

## Displacement as Annihilation

Displacement can be weaponised as a process of annihilation. The act of displacement itself is already extremely violent but perpetrators can additionally co-opt various geographies to implement programmes of annihilation based on movement and systemic deprivations. Rather than construct machineries of annihilation as seen in the Holocaust, perpetrators exploit and weaponise the geographies they inhabit to destroy the populations they target.

# Displacement as Homogenisation: Ethnic Cleansing

In contrast to DA crimes which are based on displacement as annihilation and biological and cultural killing programmes which use displacement as transfer to killing centres, perpetrators of 'ethnic cleansing' use displacement as a process of homogenisation (Naimark 2002). Violence in ethnic cleansing processes is typically not meant to destroy groups, but to forcibly displace them to demographically reengineer regions through identity homogenisation (typically in favour of the perpetrator group). Perpetrators of ethnic cleansing will use primarily direct killing methods to induce displacement from areas. In response to extremely violent acts, targeted populations may voluntarily take flight from their homes and communities, or they are actually transported by the perpetrator group to another region.

## The Ethnic Cleansing Concept

Ethnic cleansing as a concept became popularised during the wars of Yugoslav succession (1991–2001), particularly to explain the atrocities committed by Serbian regular and irregular military units against Bosniak Muslims in Bosnia (Naimark 2002). Benjamin Lieberman (2013) offers a cogent definition of the ethnic cleansing as the:

> removal of a group from a particular area. It is a means for forced remaking of human landscape. Definitions of ethnic cleansing do not specify the type of area from which a targeted group is to be removed, but in practice ethnic cleansing often targets groups living in border areas with mixed populations. (p.44)

Lieberman (2013) continues: '[ethnic cleansing] can refer to the forced removal not only of ethnic groups but also of similar related groups' (p.44).

Perpetrators of ethnic cleansing programmes largely utilise direct violence to compel targeted populations into states of disarray and displacement. Perpetrators will typically descend upon a town or a city in a region they wish to homogenise and commit extraordinary violence upon targeted populations (Lieberman 2013). Gendering understandings of ethnic cleansing is important as it is common for perpetrators to separate and kill men and inflict sexual violence upon remaining populations (especially women and girls) (Skjelsbaek 2012; Naimark 2002). This gendered action has a dual-track purpose: first, it deprives targeted groups of the typical demographic group which offers armed resistance (fighting-age men), and second, it instils fear in the remaining population which has been subjected to unspeakable horrors.

These actions destroy links individuals have with their homes and, sometimes, their communities. In this sense, violence serves a purpose other than annihilation alone. Violence is designed to make targeted populations feel unwelcome in their own homes and communities, to sever the ties that targets have with their traditional lands and cultures, and to induce displacement away from their homes. Where the victims go afterwards is a matter of debate, though perpetrators typically only plan to displace targets into other areas and do not necessarily care so long as they are removed from the place where they lived and that region is now homogenised.

Perhaps one of the most profound confusions about ethnic cleansing is when it is compared to genocide, crimes against humanity, and war crimes. Stated plainly: 'ethnic cleansing' is not a crime in international law and individual perpetrators of this crime cannot be indicted for their actions under any 'ethnic cleansing' crime (Schabas 2003). However, the individual acts which comprise ethnic cleansing can be prosecuted under the Genocide Convention and Rome Statute (United Nations, 1948a, 1998). Criminal actions which are part-and-parcel of ethnic cleansing campaigns include, but are not limited to, forced displacement/deportation, murder, extermination, sexual violence, and torture, inter alia, which can be considered war crimes, crimes against humanity, and even genocide (Lieberman 2013; Woodward 1995; Naimark 2002).

## Ethnic Cleansing Cases

These processes are best represented by the Bosnian Genocide (1992–1995). The United Nations established safe areas around specific enclaves in Bosnia during the Yugoslav Wars and meant to defend these enclaves with massive shows of force if challenged. However, Srebrenica was too vital a strategic goal for Serb forces and was a lynchpin to their ethnic cleansing operations which meant the enclave had to be attacked. In the summer of 1995, Serb forces tested the Dutch peacekeepers' resolve to defend the enclave and the peacekeepers proved to be an ineffective deterrent (Woodward 1995). What happened next shocked the world: Serb forces descended upon the enclave, separated Bosniak Muslims along gendered lines, and began their processes of ethnic cleansing. In what was later recognised as the Srebrenica Genocide, approximately 7000 to 8000 Bosniak Muslim men and boys were tortured and executed (Naimark 2002). Women and girls were systematically raped as sexual violence was a key cog in the atrocity machinery of Serb forces. This gendered tool of atrocity was used in broader Serbian ethnic cleansing campaigns (Naimark 2002). The populations not executed (but also subjected to violence) were then forcibly displaced on busses and on foot to Bosniak-controlled areas. Direct violence here was utilised in a ferociously short period of time in order to sever the links targets had with their homes, communities, and themselves and to break them in order to compel them to move.

Similar processes are taking place in Myanmar (2016–present) with genocidal ethnic cleansing of Rohingya Muslims. While the stripping of citizenship and rights of Rohingya Muslims has been a long-standing practice since 2008 (Uddin 2015), the violence against Rohingyas intensified in 2016. Myanmar state forces and non-state actors have been intentionally burning down Rohingya communities and committing acts of murder, extermination, and sexual violence upon Rohingyas themselves. These actions are designed to ethnically cleanse Myanmar of its Rohingya population—something the United Nations has recently labelled as genocide (Darusman et al. 2018).

## Displacement as Homogenisation

Finally, displacement has often—and will likely continue to be—used as a tool of homogenising political geographies. Perpetrators will use violence to induce displacement to homogenise a region along identity-based divisions. Displacement in this sense is the desired end goal of atrocities, as well as an atrocity in itself.

# Conclusions

Forced displacement plays a significant role in the perpetration of atrocities and flagrant human rights violations. Displacement may often be a by-product of political violence, especially in terms of flight from violence and as a result of instability, but it is also a central weapon used against targeted populations during times of violence. This chapter only introduces readers to some of the past and present weaponisations of forced displacement and geography against targeted populations. Displacement as a concept should be viewed as an event (transfer to stationary killing centres) and a process (displacement to annihilate and to homogenise).

While there are preliminary links being made between displacement and atrocity there is an urgent need for more conceptualisation and theorising. The current number of displaced persons in the world—71.44 million as of 2018—is already straining humanitarian institutions (UNHCR 2018). Possible displacements of individuals and communities due to climate change will likely push this number well into the hundreds of millions (Biermann and Boas 2010). The future structural opportunities for violence against climate refugees will likely be potent and permissive (Basso 2018). To combat this climate change displacement and politically violent displacement in general, it is important to identify past and present forms of atrocities which link displacement and violence to punish and prevent them now and in the future.

## Notes

1. While it is important to refer to specific atrocities—genocide, crimes against humanity, or war crimes—in reference to specific situations, these three crimes can be grouped together when discussing general processes of atrocity to simply 'atrocity crimes' for ease of communication (Scheffer 2006, 2007).
2. The Germans used the Kalahari Desert as a killing tool in the Herero Genocide (1904–1908) but during the Nama Genocide (1905–1908), the Namas used the desert as a site of resistance to the Germans (though that resistance was eventually broken).
3. As is seen during the Holocaust: the forced ghettoisation of Jews in the east and the heroic Warsaw ghetto uprising. In the Ottoman Genocide of Christian Minorities (1914–1925), Turks displaced Armenians, Greeks, and Assyrians towards Der Zor using cities as predetermined transit sites to create large death caravans. In Cambodia, the countryside was turned into 'the killing fields' after the cities were emptied.
4. Please refer to Common Articles 2 and 3 to the Geneva Conventions for more information on the differences between international and non-international armed conflicts.
5. Protection regimes for internally displaced persons are woefully inadequate compared to refugees. Please refer to Weiss (2003).
6. Previous renditions of the concentration camp have been used by the Spanish and other imperial powers in the Americas.
7. This occurs after discrimination and targeting and before annihilation and denial. Please refer to Stanton (1996) and Feierstein (2012).

# References

Akçam, T. (2013). *The Young Turks' crime against humanity: The Armenian Genocide and ethnic cleansing in the Ottoman Empire*. Princeton, NJ: Princeton University Press.

Balakian, P. (2003). *The burning Tigris: The Armenian Genocide and America's response*. New York, NY: Perennial.

Basso, A. R. (2016). Towards a typology of displacement atrocities: The Cherokee Trail of Tears, the Herero Genocide, and the Pontic Greek Genocide. *Genocide Studies and Prevention, 10*(1), 5–29.

Basso, A. R. (2018). Displacement atrocities and climate refugees: Exploring future structural opportunities for mass political violence against the displaced. In *Canadian Political Science Association Annual Conference*. Regina, Saskatchewan, Canada, May 30 to June 1.

Basso, A. R. (2019). *'All four seasons and I will die': A Typology of displacement atrocities*. Doctoral dissertation, University of Calgary, Canada.

Biermann, F., & Boas, I. (2010). Preparing for a warmer world: Towards a global governance system to protect climate refugees. *Global Environmental Politics, 10*(1), 60–88.

Clapham, A., & Gaeta, P. (Eds.). (2014). *The Oxford handbook of international law in armed conflict*. New York, NY: Oxford University Press.

Dallaire, R. (2003). *Shake hands with the devil: The failure of humanity in Rwanda*. Toronto: Random House.

Darusman, M., Coomaraswamy, T., &Sidoti, C. (2018, September 18). *Report of the detailed findings of the Independent International Fact-Finding Mission on Myanmar* (Human Rights Council, A/HRC/39/64). Retrieved from https://www.ohchr.org/Documents/HRBodies/HRCouncil/FFM-Myanmar/A_HRC_39_CRP.2.pdf.

Drechsler, H. (1980). *'Let us die fighting': The struggle of the Herero and Nama against German Imperialism (1884–1915)*. London: Zed Books.

Feierstein, D. (2012). The concept of 'genocidal social practices'. In A. Jones (Ed.), *New directions in genocide research* (pp. 18–36). New York, NY: Routledge.

Fein, H. (1997). Genocide by attrition 1939–1993: The Warsaw ghetto, Cambodia, and Sudan: Links between human rights, health, and mass death. *Health and Human Rights, 2*(2), 10–45.

Fontaine, T. (2010). *Broken circle: The dark legacy of Indian Residential Schools, A memoir*. Victoria, BC: Heritage House Publishing Company Ltd..

Friedländer, S. (2007). *The years of extermination: Nazi Germany and the Jews, 1939–1945*. New York, NY: HarperCollins.

Gaunt, D. (2011). The Ottoman treatment of the Assyrians. In R. G. Suny, F. M. Göçek, & N. M. Naimark (Eds.), *A question of genocide: Armenians and Turks at the end of the Ottoman Empire* (pp. 244–259). New York, NY: Oxford University Press.

Gewald, J. B. (1999). *Herero heroes: A socio-political history of the Herero of Namibia 1890–1923*. Athens, OH: Ohio University Press.

Halo, T. (2017). The genocide of the Ottoman Greeks 1913–1923: Myths and facts. In G. Shirinian (Ed.), *Genocide in the Ottoman Empire: Armenians, Assyrians, and Greeks, 1913–1923* (pp. 300–323). New York, NY: Berghahn Books.

Hatzfeld, J. (2006). *Machete season: The killers in Rwanda speak*. London: Picador.

Hiebert, M. S. (2017). *Constructing genocide and mass violence: Society, crisis, identity*. New York, NY: Routledge.

Hilberg, R. (2003). *The destruction of the European Jews*. New Haven, CT: Yale University Press.

Hillebrecht, W. (2010). The name and the war in the south. In J. Zimmerer & J. Zeller (Eds.), *Genocide in German South-West Africa: The Colonial War of 1904–1908 and its aftermath* (pp. 143–158). Pontypool, Wales: The Merlin Press.

Hofmann, T. (2011). Cumulative genocide: The massacres and deportations of the Greek population of the Ottoman Empire (1912–1923). In T. Hofmann, M. Bjørnlund, & V. Meichanetsidis (Eds.), *The genocide of the Ottoman Greeks: Studies on the state-sponsored campaign of extermination of the Christians of Asia Minor (1912–1922) and its aftermath: History, law, memory* (pp. 39–110). Athens: Melissa International Ltd..

Hull, I. V. (2006). *Absolute destruction: Military culture and the practices of war in imperial Germany*. Ithaca, NY: Cornell University Press.

Kershaw, I. (2009). *Hitler, the Germans, and the Final Solution*. New Haven, CT: Yale University Press.

Lemkin, R. (2013). *Totally unofficial: The autobiography of Raphael Lemkin* (D.-L. Frieze, Ed.). New Haven, CT: Yale University Press.

Lieberman, B. (2013). 'Ethnic cleansing' versus genocide? In D. Bloxham & A. D. Moses (Eds.), *The Oxford handbook of genocide studies* (pp. 42–60). New York, NY: Oxford University Press.

Megargee, G. P. (2007). *War of annihilation: Combat and genocide on the Eastern Front, 1941*. New York, NY: Rowman & Littlefield.

Naimark, N. M. (2002). *Fires of hatred: Ethnic cleansing in twentieth-century Europe*. Cambridge, MA: Harvard University Press.

Paust, J. J., Cherif Bassiouni, M., Scharf, M. P., Gurule, J., Sadat, L., & Zagaris, B. (2007). *International criminal law: Cases and materials*. Durham, NC: Carolina Academic Press.

Rosenberg, S. P., & Silina, E. (2013). Genocide by attrition: Silent and efficient. In J. Apsel & E. Verdeja (Eds.), *Genocide matters: Ongoing issues and emerging perspectives* (pp. 106–126). New York, NY: Routledge.

Schabas, W. A. (2003). 'Ethnic cleansing' and genocide: Similarities and distinctions. *European Yearbook of Minority Issues, 3*(1), 109–128.

Schabas, W. A. (2009). *Genocide in international law* (2nd ed.). New York, NY: Cambridge University Press.

Scheffer, D. (2006). Genocide and atrocity crimes. *Genocide Studies and Prevention, 1*(3), 229–250.

Scheffer, D. (2007). The merits of unifying terms: 'Atrocity crimes' and 'atrocity law'. *Genocide Studies and Prevention, 2*(1), 91–96.

Sivakumaran, S. (2012). *The law of non-international armed conflict*. New York, NY: Oxford University Press.

Skjelsbaek, I. (2012). *The political psychology of war rape: Studies from Bosnia and Herzegovina*. New York, NY: Routledge.

Stanton, G. (1996). The ten stages of genocide. *Genocide Watch*. Retrieved from http://www.genocidewatch.org/genocide/tenstagesofgenocide.html.

Stavridis, S. T. (2011). International Red Cross: A mission to nowhere. In T. Hofmann, M. Bjørnlund, & V. Meichanetsidis (Eds.), *The genocide of the Ottoman Greeks: Studies on the state-sponsored campaign of extermination of the Christians of Asia Minor (1912–1922) and its aftermath: History, law, memory* (pp. 277–296). Athens: Melissa International Ltd..

Totten, S. (2012). *Genocide by attrition: The Nuba Mountains of Sudan*. New Brunswick, NJ: Transaction Publishers.

TRC (Truth and Reconciliation Commission of Canada). (2015). *Honouring the truth, reconciling for the future: Summary of the final report of the Truth and Reconciliation Commission of Canada*. Winnipeg, MB. Retrieved from http://www.trc.ca/websites/trcinstitution/File/2015/Findings/Exec_Summary_2015_05_31_web_o.pdf.

Uddin, N. (2015). State of stateless people: The plight of Rohingya refugees in Bangladesh. In R. E. Howard-Hassmann & M. Walton-Roberts (Eds.), *The human right to citizenship: A slippery concept* (pp. 62–77). Philadelphia, PA: University of Pennsylvania Press.

UNHCR (United Nations High Commissioner for Refugees). (2018). *UNHCR statistics: The world in numbers*. Retrieved from http://popstats.unhcr.org/en/overview.

United Nations. (1948a). *Convention on the prevention and punishment of the crime of genocide*, General Assembly, New York, December 9, no. 1021.

United Nations. (1948b). *Universal declaration of human rights*, December 10, 217 A (III).

United Nations. (1966a). *International covenant on civil and political rights*, December 16, Treaty Series vol. 999, p. 171.

United Nations. (1966b). *International covenant on economic, social and cultural rights*, December 16, Treaty Series vol. 993, p. 3.

United Nations. (1998). *Rome statute of the international criminal court (last amended 2010)*, July 17. ISBN No. 92-9227-227-7.

Weiss, T. G. (2003). Internal exiles: What next for internally displaced persons. *Third World Quarterly, 24*(3), 429–447.

Woodward, S. L. (1995). *Balkan tragedy: Chaos and dissolution after the Cold War*. Washington, DC: Brookings Institution Press.

Woolford, A. (2015). *This benevolent experiment: Indigenous boarding schools, genocide, and redress in Canada and the United States*. Winnipeg, MB. University of Manitoba Press.

# Part VI

## Section Six: More-Than-Human Displacements

# 37

## Intervention: Flower Power—Khmer Women's Protests Against Displacement in Cambodia and the United States

### Katherine Brickell

Flowers have been used by women in collective action around the world in response to a range of social and political issues. One of the most iconic images of the 1960s and anti-Vietnam war protests was a photograph of a girl holding a flower facing the National Guard and their bayonets.[1] Later, in the 1980s, the Greenham Common Women's Peace Camp was established to protest the siting of a nuclear weapon at the Royal Air Force Greenham Common in the United Kingdom. Thousands of women occupied its perimeter fence, using wire to weave flowers, poems, ribbons, and embroidery. More recently, FEMEN, a Ukrainian, now global, feminist organisation fighting sexism and homophobia, have become well-known for carrying flower bouquets and wearing floral wreaths on their heads during protests (Betlemidze 2015). Just as the 'flower power' of the 1960s 'thrived on a juxtaposition of peaceful protests and armed police—flowers on (or in) the one hand and guns on (or in) the other' (Marder 2012, p. 26), women today are still drawing on the 'power' of flowers to make their peaceful calls for social justice and redress in contexts of displacement.

In this short intervention, four photographs explore Khmer women's engagement with flowers to aid their message to stop displacement, first, in relation to forced eviction in Cambodia and, second, against the deportation of their menfolk from the United States. Both forced eviction and

K. Brickell (✉)
Department of Geography, Royal Holloway, University of London, Egham, UK
e-mail: katherine.brickell@rhul.ac.uk

deportation are forms of 'root shock' that threaten to bring about a traumatic reaction to one's emotional ecosystem (Fullilove 2016). The use of culturally resonant flowers, namely the lotus and carnation, reflects women's lead roles in resisting displacement in both instances and speaks to their desire to display 'positive' and productive dissent against governmental actions in non-violent ways.

## Contesting Forced Eviction in Cambodia

Located in the northern heart of Phnom Penh, Cambodia's capital city, Boeung Kak Lake (BKL) was once home to an estimated 4000 families. In February 2007, without public consultation, 133 hectares of BKL and the surrounding area were leased for 99 years from the municipality of Phnom Penh to Shukaku Inc, a private Cambodian–Chinese corporate venture. Setting futuristic city plans in motion in August 2008, the lake started to be filled with sand, displacing water and flooding homes so they became permanently uninhabitable. Others in the vicinity endured the emotional harm of being forced to destroy their own home with a sledgehammer so they could not return. Only then would the miniscule compensation be paid. It was the women of BKL who took the lead in contesting these evictions and those to come. For nearly a decade, they have engaged in interventions to exert their rights to defend their homes. Women drew on the associational value of Khmer women with peace to do this, the lotus flower habitually conscripted in this battle (Fig. 37.1).

Followers of Buddhism use and admire flowers in their rituals and everyday lives, and the lotus is archetypal in this regard. The lotus is often stated to represent enlightenment and is the symbol of knowledge and the Buddha. As Buchmann (2015, p. 112) writes, the 'lotus is thought to represent in Buddhism four human virtues: scent, purity, softness and beauty.' Held at protests and offered to police, the lotus-wielding activists aimed to soften divisions across the barricade line. As one woman explained to me:

> we would like to say that we all are Cambodians. We follow Khmer ethics, cultures, traditions, and Buddhism. Furthermore, we as women follow the non-violent protest principle. We have no weapons but lotuses. We pray and beg for sympathy from those governmental officers … we hold the lotuses for we would like to express that we are brave enough to solve the issue and dig for solutions. Also, the lotus flower is for worship. Thus, when we confront the police force, we hold up the lotus in front of them; it seems that they are also aware of what

**Fig. 37.1** Activist offering a lotus flower across the barricade line, Phnom Penh, Cambodia, 2012 (© Erika Piñeros)

we mean. It is the chance we have to explain to those police to learn about their own religion and righteous acts.

The lotus is used, then, as a moral reminder to those who (may) attack them. A sacred Buddhist emblem, it legitimates women's public presence and evokes shared cultural and religious identities.

One outcome of BKL women's activism has been their imprisonment, their non-violent protest criminalised (see Brickell 2020 for the timeline of incarcerations between 2012 and 2018). In May 2012, during a peaceful protest, BKL women were arrested and quickly imprisoned for illegal occupancy of public property and obstructing public officials with aggravating circumstances. They were sentenced to two and a half years each. A campaign called 'Free the 15' was launched, again with the sacred lotus as its representational heart (Fig. 37.2). The NGO video 'Flowers of Freedom' by LICADHO shows community members praying for their release, lotuses in hand.[2] They were released at the end of June that year.

**Fig. 37.2** Free the 15 'Stop the Violence' T-shirt, Phnom Penh, Cambodia, 2012 (© Erika Piñeros)

## Fighting Deportation from the United States

Immigration and Customs Enforcement (ICE) are targeting long-time 'Khmerican' residents with criminal convictions, most of whom have lived in the United States since fleeing the Khmer Rouge genocide (1975–1979) as children. Few have ever lived in Cambodia and/or have little memory of it.

It is estimated that over 700 residents, mainly men, have been deported to Cambodia since 2002. Under the Trump administration, Cambodian deportations have heightened, with a record year in 2018 when over 100 Cambodian community members were deported. Sixty-eight per cent of South East Asian Americans with orders of removal have children (NAPAWF and SEARAC 2018). The toll on women has been profound as the National Asian Pacific American Women's Forum (NAPAWF) and Southeast Asia Resource Action Center (SEARAC) (2018, p. 4) explain:

> While battling the legal and administrative barriers of fighting their family member's detention, women struggled financially with one less income and the increased responsibility of caring for children and family members alone. As a result, women and their families sometimes faced severe trauma and depression

that often lasted long after their loved one's release from detention. Suddenly, their family member's immigration status rocked every aspect of their lives and their ability to thrive.

Women are at the front lines fighting to reunite with their loved ones separated by ICE.

The lotus flower is also emblematic for South East Asian Americans of the struggle against displacement, its stylised image emblazoned (much like BKL) on their T-shirts (Fig. 37.3). As Suely Saro of the Asian Pacific Islander American Professional Network of Long Beach noted on Cambodian Genocide Remembrance Day in April 2019, 'The lotus flower is very important in Cambodian culture … it is a symbol of revival. It blooms beautifully from the deepest and thickest mud. And, like a lotus flower, Cambodians have and will continue to rise above our deep, dark history and blossom beautifully into our future' (City News Service 2019). The lotus thus suggests hope and resilience in still challenging times. The women meeting on 10 May 2019, *Mother's Day*, outside the Sacramento Capitol building, sung to stop the deportation of Hay Hov, Roeun Pich, Amlorn Siratana, and Kang Hen. As 'Mothers Against Deportation,' they called on Governor Gavin Newsom and Santa Clara District Attorney Jeff Rosen to use their power to issue clean-slate decisions that would provide relief from deportation (Fig. 37.4).

**Fig. 37.3** Protest outside Sacramento Capitol building, 2019. (Photo courtesy of Asian Law Caucus)

**Fig. 37.4** Pink paper carnation, Sacramento, 2019. (Photo courtesy of Kevin Lo)

The pink paper carnations were the idea of an undocumented woman named Maria Hu Wu.[3] With pink symbolising deep love, the Khmer mothers and grandmothers wore the carnations around their wrists, standing together and defying ICE attempts to break familial bonds. As Kevin Lo, staff attorney at the Asian Law Caucus, commented, 'We tied carnations onto their wrists to honour their strength after losing so much to genocide, incarceration and deportation' (Lo 2019). One of the mothers presented a petition to Gavin Newsom with 47,000 signatures and wearing a single pink carnation as she did. Two of the men have since been pardoned, and two have found relief from deportation through other means.

## Conclusion

Flowers hold symbolic weight with their resultant political potentialities not unbeknownst to women contesting displacement. Flowers are caught up in biopolitics, the intimate management of living (Sandilands 2016), their potency drawn upon to refuse violences of displacement and familial separation. In Cambodia and the United States flowers have been symbolically, materially, and affectively mobilised through women's bodily performances. In both cases, 'flower power' has been brought about through the holding and gesturing of flowers for political affect and has been represented on T-shirts as a medium for demonstrating and feeling shared solidarities in group protest. While different flowers share different moods and sentiments, the lotus and carnation have both been rallied to call for political action to address the familial intimacies denied in forced eviction and deportation cases.

## Notes

1. See https://www.magnumphotos.com/newsroom/politics/behind-the-image-protesting-vietnam-war-with-flower/.
2. See http://www.licadho-cambodia.org/video.php?perm=35.
3. For more information about Maria Hu Wu's story and her creative work with undocumented immigrants, see https://www.pri.org/stories/2016-11-07/how-art-can-help-us-better-understand-fastest-growing-group-undocumented.

## References

Betlemidze, M. (2015). Mediatized controversies of feminist protest: FEMEN and bodies as affective events. *Women's Studies in Communication, 38*(4), 374–379.

Brickell, K. (2020). *Home SOS: Gender, violence and survival in crisis ordinary Cambodia*. Oxford, UK: Wiley.

Buchmann, S. (2015). *The reason for flowers: Their history, culture, biology, and how they change*. New York, NY: Scribner.

City News Service. (2019, April 16). Cambodian Genocide Remembrance Day honors millions killed. *NBC*. Retrieved from https://www.nbclosangeles.com/news/local/LA-County-Declares-Cambodian-Genocide-Remembrance-Day-2019-508674421.html.

Fullilove, M. (2016). *Root shock: How tearing up city neighborhoods hurts America, and what we can do about it*. New York, NY: New York University Press.

Lo, K. (2019, 12 May). Post on Twitter. Retrieved from http://pic.twitter.com/FBQmDINGdX.

Marder, M. (2012). Resist like a plant! On the Vegetal Life of Political Movements. *Peace Studies Journal, 5*(1), 24–32.

NAPAWF (The National Asian Pacific American Women's Forum) & SEARAC (Southeast Asia Resource Action Center).(2018). *Dreams detained, in her words: The effects of detention and deportation on Southeast Asian American women and families*. Retrieved May 24, 2019, from https://www.searac.org/wp-content/uploads/2018/09/dreams_detained_in_her_words_report-2.pdf.

Sandilands, C. (2016). Floral sensations: Plant biopolitics. In T. Gabrielson, C. Hall, J. M. Meyer, & D. Schlosberg (Eds.), *The Oxford handbook of environmental political theory* (pp. 226–237). Oxford, UK: Oxford University Press.

# 38

## Animals, People, and Places in Displacement

Benjamin Thomas White

## Introduction

Living as an exile in Paris in the late 1920s, the Russian writer Teffi published a memoir of her flight south from Moscow during the civil war. In one scene, she describes sitting on a sealskin coat in the freight car of a train, just after she and her companions have survived a terrifying stay in the frontier town of Klintsy and crossed into German-occupied Ukraine in autumn 1918. 'It's not for nothing that I just mentioned my sealskin coat,' she writes (Teffi 2016, p. 104), 'A woman's sealskin coat represents an entire epoch in her life as a refugee.'

Bought in Moscow or St. Petersburg before the revolution, the sealskin coat is taken on the journey south despite the summer weather because it is warm and valuable and no one knows how long the journey will last. In freight cars or on the decks of steamers, it serves as mattress and blanket. As the refugee woman's circumstances decline, so does the coat: luxurious on departure, shiny-elbowed in Odessa, patched with cheaper foreign fur in Novorossiysk, and shortened to the knee in Constantinople. By the time the refugee settles in Paris in 1922–1924, the sealskin will be no more than scraps sewn into an ordinary woollen coat, eventually 'obliterated by an invading horde of dyed cats.' 'Seals,' writes Teffi (2016, p. 105), 'are remarkable beasts. They can

B. T. White (✉)
University of Glasgow, Glasgow, UK
e-mail: BenjaminThomas.White@glasgow.ac.uk

© The Author(s) 2020
P. Adey et al. (eds.), *The Handbook of Displacement*,
https://doi.org/10.1007/978-3-030-47178-1_38

endure more than most horses.' And again, a little later (p. 106): 'Dear, gentle beast, comfort and defence in difficult times, banner of our lives as refugee women: A whole epic could be written about you. I remember you and salute you.'

It is no accident that this reflection occurs immediately after Teffi crosses the military frontier into Ukraine: it is at this point in the narrative, in temporary and relative safety, that she becomes a 'refugee.' The seal she apostrophises connects the exiled writer not just to Moscow and St. Petersburg and her lost position of privilege but to every mode of transport and stopping place on her journey as a refugee and every mental and material state she has passed through on the way. Though Teffi does not mention it, the coat also connected her to a Russian imperial economy spanning thousands of miles: the seal was probably hunted in the Commander Islands, east of Kamchatka in the Bering Strait.[1] The dyed (and dead) cats that replace the seal, by contrast, met their end in Paris alleyways.

Animals play other roles in Teffi's narrative. On the train south from Moscow, some peasant women threaten her group: their hostility is aroused by the Pekingese lapdog sitting in the lap of one of her companions (Teffi 2016, p. 51). This woman can't understand why the peasants hate her dog so much, but the reader recognises the unthinking privilege that goes with keeping a non-working animal fed in a period of wartime food shortages. The animal's vulnerability indicates that Teffi's companion has been displaced from her position of privilege: the peasant women know it, even if she still does not. At Klintsy, meanwhile, Teffi glimpses a dog dragging a human arm through the dirt down by the railway track. This sight, above all, establishes the town as a place of fearful danger.

Teffi's sealskin coat makes a good starting point for a chapter on animals in displacement, in a section on 'the stuff of displacement.' The coat may be 'stuff,' but the seal is not, and the animals in Teffi's narrative flag up three interlocking issues to explore. The first is the way animals figure in representations of displacement, whether of displaced people—as Teffi's 'dear, gentle beast,' which can 'endure more than most horses' (2016, p. 105), comes to represent the resilience of the refugee woman—or of the spaces they move through, as the dyed cats represent her place of exile. This connects to the second theme, the role of animals in the 'emplacement' of displaced human populations: that is, their experience of place in displacement, not just the sense of being out of place (the dog at Klintsy) but also their establishment of connections—social, economic, psychological, ecological—with new places. And the third is the agency, in all this, of animals themselves. A stray dog is making its own choices when it decides to gnaw the arm from a human corpse, and its agency can't be reduced to that of a human owner.

This chapter starts by using camps as a site for thinking about the relationship between animals, people, and places in displacement: not because most displaced people live in camps (they do not), but because camps make it easier to 'see' how animals influence the lives of displaced people. By 'camps,' I mean purpose-built encampments constructed by others, whether states, international organisations, or humanitarian agencies, to contain displaced people, especially refugees—that is, people displaced across a border (Black 1998; Harrell-Bond 1998; Bakewell 2014). The practice of putting displaced people in camps, rather than allowing them to self-settle, became generalised after the First World War. For historical depth, this chapter looks as far back as 1918, informed by research on archival and published sources. For contemporary breadth, it draws on a range of 'grey' literature (reports, policy documents) produced by humanitarian agencies, collaborations with contemporary humanitarian practitioners, and news media sources.[2] The chapter discusses domesticated animals, including livestock, working animals, and pets, as well as wild animals, from migrating birds to elephants. From an initial focus on displacements caused by war and persecution, it broadens out to consider the relationship between human and animal displacements in the twenty-first century, in situations of increasing environmental stress. The conclusion suggests directions, and methods, for future research.

## Animals and Displacement: Representations, Emplacement, and Agency

In November 2015, the *Daily Mail* published a notorious racist cartoon which represented migrants entering Europe as sinister figures in face veils and turbans mingled with gleeful rats (Mac 2015). It is not unusual for representations of displaced people by hostile observers to play on the association between displaced people and animals, whether pictorially or in the language of 'swarms.' But people who are not refugees do this, too, in ways that are intended to garner sympathy—though perhaps problematically, as when news reporters try to humanise refugees by focusing on their companion animals. One notable recent example is Kunkush the cat, who left Iraq with his refugee family and was separated from them on Lesvos in November 2015, but was reunited with them in Norway the following year: the story was widely reported and later turned into not one but two picture books ('The Epic

Journey,' 2016; Kuntz and Shrodes 2017; Ventura 2018). Other observers use animals to represent spaces of displacement—stressing proximity to rats as an indicator of the squalor of a camp, say (BBC News 2017). Displaced people, of course, can speak for themselves. 'I fear we and our children will become like animals if we have to keep living this refugee life,' Noor Ilyas, a Rohingya refugee at Jamtoli refugee camp in southern Bangladesh, wrote recently, and it is common for people living in camps to say that they are being treated like animals (Ilyas 2018; Neuman and Torre 2017).

In camps we can also see how animals very directly shape the human experiences of displacement and emplacement. With their twin functions of 'care and control' (Malkki 1995, p. 498), camps are heavily surveilled and bounded spaces: surveilled so that displaced people's humanitarian needs can be met, and bounded so that their movements can be monitored and controlled. And when humans are displaced, they often take animals with them, especially livestock and working animals. When the organisations running a camp take an interest in these animals, the same regimes of surveillance register their presence, too, creating a record of how they shape displaced people's experiences in both contemporary and historic cases. This helps explain why the existing literature on animals in displacement, mostly produced by practitioner organisations, focuses particularly on livestock and working animals (UNHCR/IUCN 2005; LEGS 2013, 2014; Beirne and Kelty-Huber 2015; FAO 2016; World Animal Protection n.d.; Veterinarians Without Borders 2015): these animals contribute to the livelihoods of displaced people. Indeed, often the animals *are* the people's livelihoods, as well as their means of transport into displacement. If humanitarian assistance does not extend to the animals, the people's experience of displacement will worsen. For example, in 2003, some 14,000 donkeys carried internally displaced people to the Abu Shawk camp in Darfur, Sudan—but 18 months later only 2300 survived. Nearly 85 per cent had died, mostly for lack of fodder. This left the people with no means of transport, nor of earning a living: they became much more dependent on humanitarian assistance in the camp and much less able to return home and resume their normal lives (Sprayson 2006, p. 50). The example illustrates both how animals' presence in camps is registered in humanitarian records and how important they are in the lives of displaced people. But it is not only livestock and working animals that are important, nor is it only in transport and livelihoods that animals make a difference.

Animals shaped these spaces of confinement from the start: for example, in the camp created in 1918 by British occupying forces in Ottoman Mesopotamia—now Iraq—at Baquba on the river Diyala about 30 miles north-east of Baghdad. The site of this camp was chosen because it was defensible and readily accessible (and therefore easily supplied) by road and rail, with water for drinking, washing, and sanitation provided by the river and two canals. This made it suitable for housing not only almost 50,000 people, Assyrian and Armenian refugees from eastern Anatolia and the Caucasus, but also the thousands of animals that they had brought with them. The British commander of the camp, H. H. Austin, estimated that in the autumn of 1918 there were seven or eight thousand sheep and goats, as well as about six thousand large animals: horses, ponies, donkeys, mules, camels, and cattle. Another eight hundred ponies arrived when the 'Assyrian contingent' of armed refugees, set up under British command to protect the parties travelling south, settled at Baquba in mid-November (Austin 1920, pp. 21–22).[3]

An annotated blueprint from the British War Office archives shows some of the place afforded to animals in the camp. At its south-westernmost corner was an animal enclosure a few hundred metres across, bisected by a canal and downwind of the rest of the camp according to an arrow showing the direction of the prevailing wind. (It is too small to be the 300-acre forage farm mentioned in other sources.) A butchery is shown nearby. In the centre of the camp was a bazaar, set up early on to stimulate economic life, in which animals and animal produce played an important role. Close to it were isolation and disinfection areas: these were intended to limit access to the camp for one specific type of animal, lice. Typhus, for which lice are the vectors, killed tens of thousands across the Middle East during and after the First World War, and this was one of the main reasons why death rates in the camp's early days were appallingly high. Isolation and fumigation eliminated lice before refugees were allowed to move around the camp freely. A note made by the camp's senior medical officer when the site was being laid out also mentioned that mosquito nets were to be supplied.[4] The veterinary regimes for animals, meanwhile, mirrored the medical regimes for people, with segregation paddocks in use to eliminate mange and glanders, a nasty contagious disease affecting horses (Austin 1920, p. 22). (Sick animals, unlike sick people, could also be destroyed to control the spread of disease.) The Assyrian contingent and its ponies were housed in a large enclosure across the river from the main camp. Textual sources indicate that animals shaped the spaces of the camp in other ways, not visible on the plan: for example, it soon included a number of poultry farms and a piggery.[5]

Baquba is an early example of a modern refugee camp. In contemporary spaces of displacement there are many similar instances. Manuel Herz's study of the long-established camps in south-western Algeria where Sahrawi refugees have lived since the 1970s notes that 'goat barns'—circular enclosures cheaply constructed from wire and scrap metal—are a prominent spatial feature (Herz 2011, pp. 340–347). Camel butcheries are important shops (pp. 310–311), while an animal market is located outside one camp 'so that the noise and smell of the animals do not cause discomfort to the inhabitants' (p. 302). Goats and camels are socially, economically, and culturally important animals in Sahrawi society, and this is reflected in the space of the camps even as they gradually become permanent urban settlements. We could multiply other contemporary examples, like the camel economy at the Dadaab complex in northern Kenya (Rawlence 2016), where the Dagahaley, Ifo, and Hagadera camps each have slaughter slabs and marketplaces.[6] In different ways, and across different cases separated by a hundred years and thousands of kilometres, animals have shaped spaces for displaced people in camps.

Animals at Baquba also shaped refugees' encounters with spaces beyond the camp, in ways that find many parallels in more recent cases. They became the means of connection between people in the camp and the people and landscape around them, partly because they were crucial to British plans to stimulate the camp's economic life, which would be recognisable to modern humanitarian practitioners concerned with refugee livelihoods. Dairy products from the refugees' own animals (sheep, goats, and presumably cattle too) were sold for cash as well as exchanged for barter. On the one hand, this took refugee herders out beyond the western edges of the camp to graze their animals. On the other, it took them to surrounding villages: refugees in groups of two or three would set out each morning, 'laden with curds and other goods for which they would buy eggs and similar articles from the Arabs, returning to camp at dusk' (Austin 1920, p. 90). The piggery and poultry farms provided food (meat and eggs) but perhaps also feathers: a large quilt-making enterprise was run by the American Persian Relief Commission, presumably using feathers and yarn produced by the camp's animals (this is not clear from British sources), which employed thousands of women and children for cash wages. Meanwhile, men were sent to do contract labour outside the camp, and this involved animals too. Examples from March 1920 included a small party of men and oxen for a private farm and a very large one—'2500 persons and 1000 animals'—sent north to work for the occupation administration's Labour Directorate near Mosul.[7]

Animals, then, were a means of providing a livelihood in displacement and not only a subsistence livelihood: they were also a means of integration into the life of the host society. But this integration was defined in primarily economistic terms, which often predominate in contemporary refugee integration initiatives, too. Since 2015, for example, UNHCR has provided significant support to the animals of Malian refugees in Burkina Faso. The programme, funded by the IKEA Foundation, is intended to 'help more than 6,000 people earn a sustainable income through small-scale dairy farming,' as well as improving the nutrition of refugee children (IKEA Foundation 2015, n.p.).[8] It provided support for animals, including veterinary care, because UNHCR recognised that they were a means of livelihood and integration for the people—defined economistically, in terms of giving the refugees a means of access to a market economy for dairy produce.

But this is an anthropocentric conception, and it doesn't say much about *place*: landscape, natural environment, local ecology. If we go beyond the economic 'self-reliance' that has been a problematic structuring concept for refugee settlement for over a century (Easton-Calabria and Omata 2018), and think instead in terms of *emplacement*—the development of a new sense of place—as essential to integration, then we need to recognise ecological as well as economic factors (Kindon 2018).[9] As Assyrian herdsmen took their flocks of mountain sheep and goats out west of the Uthmaniyya canal each day, familiarising themselves with the landscape, the vegetation, the predators, and the local human inhabitants of the lowlands and listening to the sound of jackals in the darkness, they were developing a sense of place beyond a simple awareness of the local 'market potentialities' (Austin 1920, p. 90) of their produce. The connections they developed with their new place were economic and material, but they were also ecological, emotional, and psychological.

In a historical case as far back as Baquba, the absence of refugee voices from institutional archives can make it hard to recover the refugees' sense of place. But in more recent cases, sources *are* available for us to develop a textured understanding of emplacement, and the roles animals play in it, informed by work in disciplines from archaeology to anthropology (Cummings and Harris 2011; Gooch 2008).[10] This does not just cover livestock and working animals and refugees who work with them: even at Baquba, most refugees were not pastoralists, and some were not agriculturalists at all. And it is not only through livelihoods, as another modern example illustrates. Animals are largely excluded from the sterile, hypermodern, profoundly bleak refugee camp at Azraq in Jordan. Caged birds are an exception, and people seem willing to pay a high price for them: in 2016, 30–40 Jordanian dinars each (roughly €35–50), when 'incentive workers'—refugees employed by humanitarian agencies for work within the camp—were paid 10 dinars (€12) per day.[11] Why? Partly, perhaps, because looking after a pet offers some sense of agency and control to people who lack it in other areas of their lives in displacement. And partly because a companion animal can make even a metal and plastic shelter into a home: making a home in a place is a step towards making that place home, however provisionally.

But working animals and livestock as well as pets can play a role in the emotional lives of displaced people. We can see this in Austin's lyrical description of the flocks returning to the Baquba camp each night and being lavished with affection by the refugee men, women, and children, even as the female animals were milked (Austin 1920, pp. 88–89)—a depiction that returns us to the role animals play in representations of displacement.[12] Like other British sources on the camp at Baquba, Austin's book overwhelmingly concentrates on one of the three groups of refugees living there: the Hakkari Assyrians, a population of seminomadic pastoralists from the Hakkari mountains now on the border between Turkey and Iraq. Not coincidentally, these were the people to whom most of the camp's animals belonged. One reason for this attention lies in nineteenth- and early twentieth-century European visions of the Middle East, which commonly juxtaposed a depiction of the region's urban life as corrupted by modernity with an idealised depiction of the 'authentic' way of life of its pastoralists, whether the Bedouin of the desert or the Assyrians of the mountains, who were viewed as more amenable to imperial rule. Austin's account uses the Hakkari Assyrians' closeness to animals as a literary means of representing them as a timeless and apolitical pastoral population deserving sympathy and support[13]: greeting their flocks with affection and 'a holy calm' and giving up their own army-issue blankets to keep the lambs and kids warm (Austin 1920, p. 89), in scenes that can be read as an implicit evocation of the shepherds and stable of the nativity story. Elsewhere, Austin stressed the close companionship of Assyrian men and their horses.

Humanitarians often represent refugees as, in Liisa Malkki's famous words, 'speechless emissaries' (1996): here, animals are made to speak for them. Like the humanitarian representations of refugees as passive and needy victims, the British representation of the displaced Hakkari Assyrians as an amenable client population also had real effects on their lives. It led to the incorporation of Assyrian men *and their steeds* into the coercive forces of the occupation, in ways that permanently damaged the refugees' relationship with their host society (White 2019). Assyrians were not passive in this: they made individual and collective choices to participate in the occupation. But we also need to recognise the agency of the horses that carried them in these mounted military units.

In *Vibrant Matter: A Political Ecology of Things*, Jane Bennett (2010, chapter 7) offers an account of non-human agency that starts with Charles Darwin's observations of the 'small agencies' of earthworms, whose constant digestive work produces the topsoil in which human history is rooted. Bennett borrows the term *actant* from Bruno Latour to describe their position—and that of other non-humans—in an ecological assemblage 'in which agency has no single locus, no mastermind, but is distributed' among humans and non-humans alike (Bennett 2010, p. 96). 'A certain nonhuman agency,' she says, is 'the condition of possibility of human agency.' To explore the political implications of this, Bennett expands on John Dewey's theory of conjoint action, whereby political publics emerge—are continually emerging and re-emerging—when bodies are 'pulled together not so much by choice … as by a shared experience of harm that, over time, coalesces into a "problem."' This is an understanding, Bennett observes, of a political system as 'a kind of ecosystem' (2010, p. 100), and it

> paves the way for a theory of action that more explicitly accepts nonhuman bodies as members of a public, more explicitly attends to how they, too, participate in conjoint action, and more clearly discerns instances of harm to the (affective) bodies of animals, vegetables, minerals, and their ecocultures. (Bennett 2010, p. 103)

Bennett's account of non-human agency shows that we can never have the whole view without a sense of how the animal or object helps shape what we might otherwise think of as a human story. It requires us to consider the agency of horses and ponies in British imperial plans for the Hakkari Assyrians, or of microbes, lice, and mosquitoes and nearby stagnant water in making the camp at Baquba a place of human precarity. And it helps us to make sense of the tensions that arose around Baquba in the spring and summer of 1919, as the Hakkari Assyrians and their animals took their place in the agricultural landscape of Diyala province, and the flocks began to grow. Sheep and goats from the camp, the refugees' livelihoods, would break into fields on neighbouring farms and munch the tender, new crops that were the farmers' livelihoods. The herders may not have minded too much about this damage: they may even have encouraged it. But goats are independent-minded animals (Gooch 2008, p. 70), and to understand the political problem that arose here we need to give their agency its due. Bennett's book offers a framework for understanding how the sheep and goats, the new crops, and any fences that were broken figured alongside Hakkari Assyrian herders and Arab farmers as part of the emergent political 'public' around Baquba.

Such complex situations continue to arise when humans and animals are displaced today. For example, in 2011–2012 a military offensive by the Sudanese government drove nearly 125,000 people and hundreds of thousands of animals from Blue Nile state across the new border into South Sudan, where they settled in Maban county (and where their animal numbers rapidly halved, to about 50,000 cattle and 80,000 sheep and goats, stressed by the journey and the wetter conditions on arrival). They considerably outnumbered the host community. Over the next two years, animals played a complex political role in the relationship between refugees, local residents, and a third group, Mbororo nomads, also pastoralists, who regularly pass through Maban county as part of their seasonal migrations. Sources of tension included crop damage—which triggered fights causing up to 20 human deaths—as well as degradation of common grazing and competition for water. Local protocols to resolve conflicts were developed with input from the South Sudanese government, humanitarian agencies, and Veterinarians Without Borders, taking animal as well as human agency into account. They included moving refugee animals to sparsely populated grazing zones as far as 60 km from the camps where the refugee people lived, scheduling access to watering points, and rerouting the Mbororo migration somewhat to the west, further from the camps (Hoots 2018).

It is harder to reroute migrations when these involve wild animals rather than humans and livestock. After renewed persecution by the Myanmar military began to drive hundreds of thousands of Rohingya refugees into Bangladesh in late 2017, points of conflict arose between refugees and wild elephants, especially along the western boundary of what quickly became the largest refugee camp accommodating Rohingya, at Kutupalong. The camp blocks a migration route between forest ranges in Bangladesh and Myanmar for a critically endangered population of Asian elephants, which 'always follow their traditional routes and corridors for regular movement': the resulting conflicts killed nine people in and around the camp between September 2017 and January 2018 (IUCN Bangladesh 2018, p. 4). The problem implicates the elephants, the refugees in the camp they are trying to pass through, and the host community, who may also be affected as elephants move around the edges of the camp. It is representative of a larger set of issues caused by the camp's emergence, which has rapidly degraded local forests as refugees seek to 'fulfill their basic needs, like food, shelter, and income generation' (IUCN Bangladesh 2018, p. 4). The acute humanitarian emergency at Kutupalong, in other words, also highlights the intersection of human and animal displacements under conditions of increasing ecological stress. The final section of this chapter zooms out to think about these at a global scale.

## Human and Animal Displacements in the Twenty-First Century

In our time, the intersection of human forced migrations with animal migrations is increasingly recognised. The prolonged Syrian drought in the years up to 2010, one important factor in the political crisis that ignited in 2011, is an example. Worsened in its intensity and duration by anthropogenic climate change, it had already pushed over a million rural Syrians off the land shortly before the war began (Kelley et al. 2015). Those people were pushed into worse poverty than would otherwise have been the case because of the liberalisation of the Syrian economy over the previous decade and the state's retreat from providing basic services, among other reasons: ecological factors are not separate from political factors. But they were important, and their impact has been felt by animals as well as humans, in complex interlocking ways. Syria's total livestock fell by as much as a third because of the drought (ACSAD 2011), while in the Jordanian desert, the water table is falling—the impact of the regional drought exacerbated by the arrival of hundreds of thousands of Syrian refugees in what was already one of the most water-scarce countries in the world. The falling water table means that oases are shrinking, and bird species that migrate across the desert are losing the important staging posts they provided. The Syrian war highlights the intersection of human and animal displacements under environmental stress, and we are likely to see more such conflagrations.

You could argue, indeed, that a 'great displacement' is going on all around us right now: the displacement of all other species within the biosphere to make way for human beings and the things we like to eat. Vertebrate animal populations are estimated to have fallen, across the board, by around 60 per cent between 1970 and 2014 (WWF 2018, p. 70), from a baseline that had already shifted a long way since the start of the Holocene. The steep decline in invertebrate populations is, if anything, even more alarming (Dirzo et al. 2014).

We tend to conceptualise this displacement as a displacement *of* other living things *by* humans. But the processes that have driven this destruction of biodiversity have also displaced humans. In the archaeological record the relationship between the spread of settled agriculture and the shift away from hunter-gatherer societies is uncertain (Leary and Kador 2016; Cummings and Harris 2011), but in recorded history the geographic marginalisation of hunter-gatherers is clear (Scott 2009). If Britain has one of the most depleted natural environments in the world, it is not just because of early industrialisation and urbanisation but also the related processes of the early commercialisation of agriculture, which pushed people off the land through enclosures and clearances, and its early mechanisation. Translated to colonial settings, commercial agriculture displaced Indigenous peoples; in settler-colonial settings these displacements reached genocidal intensity. Plantation economies also displaced people for labour: enslaved Africans or, later, indentured labourers. The spread and intensification of industrialised agriculture, the relentless march of urbanisation, and the pharaonic routines of resource extraction all continue to displace and re-place human populations today, at the same time as they diminish animal habitats. One of the great drivers of biodiversity destruction is anthropogenic climate change, which literally displaces animal species out of existence. It is also already contributing to human displacement, as humanitarian NGOs and journalists have signalled ('Climate Change,' 2018; 'A Warming World,' 2018), and this is only likely to increase. The people who will be worst affected by climate change are those who have contributed least to it.

It is not, then, simply a case of displacement of other animals by humans: the processes that drive the 'great displacement' are processes of human displacement, too. Our understanding of human experiences of displacement deepens when we recognise the roles that animals play in them, and the reverse is also true: we will understand the destruction of biodiversity, the displacement of other species, better when we think of them not (only) as a human impact on a natural environment but also a displacement and destruction of human lives and livelihoods.

## Conclusion

How should we research human and animal displacements? As the processes causing them interlock in increasingly clear but increasingly complex ways, we need to bring them into a single conceptual frame, capable of making sense of them together, not as separate phenomena studied on the one hand by natural scientists and on the other by social scientists and a smattering of humanities scholars. Interdisciplinary animal studies already offers the methodological toolkit and cross-disciplinary approach that are required: it is more a question of defining the object of study.

The scope of this handbook shows how broadly, yet coherently, displacement can be defined, beyond the war- and persecution-related displacements that this chapter began with. In urban displacements from informal or 'slum' districts, animals—'vermin'—can figure in the construction of the squalor to be eliminated, as well as in representations of the populations to be relocated. In disaster evacuations, the importance of providing for companion animals is increasingly recognised, both to persuade people to leave and to sustain their mental health later. Catastrophic recent floods in North Carolina (2018) and Queensland (2019) have highlighted the impact of natural disasters on livestock and wild animals, in the Global North as well as the Global South. The extreme weather events that caused them are becoming more frequent as a result of anthropogenic climate change; their impacts are exacerbated by human practices of land use and patterns of settlement. Such moments of acute crisis also bring us back to human and animal displacement under chronic, though intensifying, ecological pressures: the Queensland floods followed a five-year drought that had put cattle herds as well as wild animals under stress.

The scale of such studies, meanwhile, could run from the micro to the macro. At the microscale, the case of Jamal, the 'Bird Man of Red Road,' shows how a single human/companion animal relationship can bring together different forms of displacement and illustrate the themes explored in this chapter (Leslie 2016, pp. 11, 18). An asylum seeker from Iraq, Jamal was, with his two canaries, the last resident of the Red Road tower blocks in Glasgow—an immense housing complex born of 1960s slum clearances and demolished in 2015. At the macroscale are global changes displacing animal populations and the humans that depend on them.

As for methods, I would like to suggest that this work should start *with* humans in displacement. The Scottish wildlife artist Derek Robertson, who also has hands-on experience of participating in scientific studies of bird migrations, has recently been exploring this intersection. His series Migrations—'a field study of diversity'—follows the parallel routes of human and bird migration from the Middle East across the Mediterranean, through Europe, and as far as Scotland.[14] One image (Fig. 38.1) illustrates the environmental pressure of the falling water table in the Jordanian desert: it shows a desert bird (a horned lark) that can tolerate extremely dry conditions, and a group of children gathered around a migratory bird (a bee-eater) that had come to ground by an abandoned blue bus, perhaps mistaking it for water, and been unable to fly on. In the Middle East and in the Calais 'Jungle,' Robertson worked with refugees and exiles to capture a sense of place in spaces of displacement, including the place within them of migrating birds. Naomi Press and other clinical psychotherapists working at Calais with a small British NGO, ArtRefuge, found that exiles represented their own displacement by modelling or drawing animals.

**Fig. 38.1** Derek Robertson, *The Desert Is Full of Promises* (by permission of the artist)

At Kutupalong, meanwhile, Bangladeshi artist Kamruzzaman Shadhin recently produced an *Elephant in the Room*: an installation commissioned by IUCN and UNHCR, taking the shape of a life-size pair of papier-mâché elephants coated in discarded clothes (Fig. 38.2).[15] Shadhin worked with refugees to produce this artwork, with the process as much as the finished product aiming not just to warn refugees about elephants but to create mutual understanding. Taking Jane Bennett's (2010) approach, the artwork could be seen as a focus for articulating an emerging human/non-human public, recognising the elephants' agency, and trying to make it comprehensible to the human inhabitants of Kutupalong. It could also be seen as a way of establishing interspecies solidarities (Coulter 2016) among what Donna Haraway (2003, 2007) has termed companion species.

In these examples, scientific practice informed the artwork that was produced, but the art in turn informs (natural and social) scientific practice. More important, in all of them, displaced humans were active participants in the production of artworks that exist at least as much to benefit them as to benefit the other, undisplaced humans who were involved. These artistic

**Fig. 38.2** Kamruzzaman Shadhin, from *Elephant in the Room* (by permission of the artist)

examples can all inform scientific research on human and animal displacements—but they also provide a model for *doing* ethical research at the points, and in the spaces, where those displacements intersect. We need such ethical research urgently.

## Notes

1. I would like to thank Bathsheba Demuth for this information.
2. My work with humanitarian practitioners was supported by a Wellcome Trust Seed Award in Humanities and Social Science, 'Humans and animals in refugee camps' [award reference 205708/Z/16/Z].
3. There were certainly other animals at Baquba: rats, cockroaches, and other 'pests' and wild animals including birds and perhaps jackals. But they are hardly visible in the British archival record. Perhaps surprisingly, nor are dogs, though the herders surely had some.
4. The National Archives (UK), War Office records WO 95/5238: War diary of Assistant Director Medical Services, 29 August 1918.
5. The National Archives (UK), Foreign Office records FO 371/6359, folios 80–87: 'Armenian and Assyrian Refugees in Irak' (cover 22 December 1921).

6. Plans for each sub-camp are available on the UNHCR data portal: https://data2.unhcr.org/en/documents/details/31535 (Hagadera); https://data2.unhcr.org/en/documents/download/31534 (Dagahaley); https://data2.unhcr.org/en/documents/download/31533 (Ifo). All accessed on 12 December 2018.
7. British Library, India Office Records, IOR/L/PS/10/775: 'Refugee Camp, Baquba. Monthly report. For the month of March 1920,' 31 March 1920.
8. In late 2015, according to UNHCR, there were around 34,000 Malian refugees in Burkina Faso (https://data2.unhcr.org/en/documents/download/49009); they had brought about 47,000 animals with them (https://twitter.com/Refugees/status/1033752179025227776).
9. This section draws on the excellent unpublished research of the geographer Sara Kindon.
10. Gooch (2008, p. 73), 'Successful pastoralism demands a strong feeling of understanding between herders and the animals they herd, tantamount to a shared world-view, whereby the world can be perceived through the senses of the animals in question.'
11. Personal communication from Ann-Christin Wagner, observed during fieldwork for Wagner (2019).
12. Animals mattered in Austin's own 'emplacement' at Baquba, too. The only reference in his book to his own leisure time, in what must have been a stressful and dangerous job, is a mention of long, solitary morning rides on horseback in the countryside around the camp (1920, p. 90).
13. It goes without saying that the Hakkari Assyrians were *not* apolitical.
14. Paintings from the series have been exhibited in the UK and internationally. They can be viewed at https://www.creativepastures.com/migrations.
15. These details come from the Facebook page of UNHCR Bangladesh (saved to the Internet Archive, since Facebook content is highly dynamic): https://web.archive.org/web/20181212233809/https://www.facebook.com/UNHCRBangladesh/posts/art-used-as-a-way-to-promote-to-co-existence-between-wild-elephants-and-rohingya/513555982401033/. Accessed 12 December 2018.

# References

ACSAD (Arab Center for the Studies of Arid Lands and Dry Zones). (2011). *Drought vulnerability in the Arab region. Case study—Drought in Syria. Ten years of scarce water (2000–2010)*. Damascus, Syria: Arab Center for the Studies of Arid Lands and Dry Zones. Retrieved December 12, 2018, from https://www.unisdr.org/files/23905_droughtsyriasmall.pdf.

Austin, B.-G. H. H. (1920). *The Baqubah refugee camp: An account of work on behalf of the persecuted Assyrian Christians*. London & Manchester, England: The Faith Press.

Bakewell, O. (2014). Encampment and self-settlement. In E. Fiddian-Qasmiyeh, G. Loescher, K. Long, & N. Sigona (Eds.), *The Oxford handbook of refugee and forced migration studies* (pp. 127–138). Oxford, England: Oxford University Press.

BBC News. (2017). *Syria refugee children 'bitten by rats' in camps* [Video report]. Retrieved December 12, 2018, from https://www.bbc.co.uk/news/av/world-middle-east-39698249/syria-refugee-children-bitten-by-rats-in-camps

Beirne, P., & Kelty-Huber, C. (2015). Animals and forced migration. *Forced Migration Review*, (49), 97–98. Retrieved December 12, 2018, from http://www.fmreview.org/climatechange-disasters/beirne-keltyhuber.html.

Bennett, J. (2010). *Vibrant matter: A political ecology of things*. Durham, NC: Duke University Press.

Black, R. (1998). Putting refugees in camps. *Forced Migration Review*, (2), 4–7. Retrieved December 12, 2018, from https://www.fmreview.org/camps/black.

Climate change is exacerbating world conflicts, says Red Cross president. (2018). *The Guardian*, October 21. Retrieved December 12, 2018, from https://www.theguardian.com/world/2018/oct/21/climate-change-is-exacerbating-world-conflicts-says-red-cross-president.

Coulter, K. (2016). *Animals, work, and the promise of interspecies solidarity*. New York, NY: Palgrave Macmillan.

Cummings, V., & Harris, O. (2011). Animals, people and places: The continuity of hunting and gathering practices across the Mesolithic–Neolithic transition in Britain. *European Journal of Archaeology*, *14*(3), 361–393. https://doi.org/10.1179/146195711798356700.

Dirzo, R., Young, H. S., Galetti, M., Ceballos, G., Isaac, N. J. B., & Colle, B. (2014). Defaunation in the Anthropocene. *Science*, *345*(6195), 401–406. https://doi.org/10.1126/science.1251817.

Easton-Calabria, E., & Omata, N. (2018). Panacea for the refugee crisis? Rethinking the promotion of 'self-reliance' for refugees. *Third World Quarterly*, *39*(8), 1458–1474. https://doi.org/10.1080/01436597.2018.1458301.

The epic journey of a refugee cat to find its family [video report]. (2016). *The Guardian*, February 19. Retrieved December 12, 2018, from https://www.theguardian.com/world/video/2016/feb/19/refugee-family-who-fled-iraq-are-reunited-with-cat-video.

FAO (Food and Agriculture Organization of the United Nations). (2016). *Livestock-related interventions during emergencies—The how-to-do-it manual*. Rome, Italy: Food and Agriculture Organization of the United Nations. Retrieved December 12, 2018, from www.fao.org/3/a-i5904e.pdf.

Gooch, P. (2008). Feet following hooves. In T. Ingold & J. L. Vergunst (Eds.), *Ways of walking: Ethnography and practice on foot* (pp. 67–80). Farnham, England: Ashgate.

Haraway, D. J. (2003). *The companion species manifesto: Dogs, people, and significant otherness*. Chicago, IL: Prickly Paradigm Press.

Haraway, D. J. (2007). *When species meet*. Minneapolis, MN: University of Minnesota Press.

Harrell-Bond, B. (1998). Camps: Literature review. *Forced Migration Review*, (2), 22–23. Retrieved December 12, 2018, from https://www.fmreview.org/camps/harrellbond.

Herz, M. (Ed.). (2011). *From camp to city: Refugee camps of the Western Sahara*. Zürich, Switzerland: Lars Müller Publishers.

Hoots, C. (2018). The role of livestock in refugee–host community relations. *Forced Migration Review*, (58), 71–74. Retrieved December 12, 2018, fromhttps://www.fmreview.org/economies/hoots.

IKEA Foundation. (2015). *Circles of prosperity. Annual review 2015*. Leiden: IKEA Foundation. Retrieved December 12, 2018, from https://ikeafoundation.org/about/documents/.

Ilyas, N. (2018, August 27). I am a Rohingya refugee: We will become like animals if we stay in these camps. *The Guardian*. Retrieved December 12, 2018, from https://www.theguardian.com/world/2018/aug/27/i-am-a-rohingya-refugee-we-will-become-like-animals-if-we-stay-in-these-camps.

IUCN (International Union for the Conservation of Nature) Bangladesh Country Office. (2018). *Survey report on elephant movement, human–elephant conflict situation, and possible intervention sites in and around Kutupalong camp, Cox's Bazar*. Dhaka, Bangladesh: IUCN Bangladesh Country Office. Retrieved December 12, 2018, from https://www.unhcr.org/protection/environment/5a9946a34/survey-report-elephant-movement-human-elephant-conflict-situation-possible.html.

Kelley, C. P., Mohtadi, S., Cane, M. A., Seager, R., & Kushnir, Y. (2015). Climate change and the recent Syrian drought. *Proceedings of the National Academy of Sciences, 112*(11), 3241–3246. https://doi.org/10.1073/pnas.1421533112.

Kindon, S. (2018, 26 September). *Rethinking the place of 'place' in refugee resettlement: Aotearoa New Zealand*. Unpublished paper presented at the Glasgow Refugee, Asylum, and Migration Network seminar series, Glasgow, Scotland.

Kuntz, D., & Shrodes, A. (2017). *Lost and found cat: The true story of Kunkush's incredible journey* [Illustrated by S. Cornelison]. New York, NY: Crown Books for Young Readers.

Leary, J., & Kador, T. (2016). Movement and mobility in the Neolithic. In J. Leary & T. Kador (Eds.), *Moving on in Neolithic studies: Understanding mobile lives* (pp. 1–13). Oxford, England: Oxbow.

LEGS (Livestock Emergency Guidelines and Standards). (2013). *Livestock interventions in camps* (briefing paper). Retrieved December 12, 2018, from http://www.livestock-emergency.net/wp-content/uploads/2013/11/LEGS-Livestock-in-Camps-Briefing-Paper-2013.pdf.

LEGS (Livestock Emergency Guidelines and Standards). (2014). *Livestock emergency guidelines and standards* [2nd edition]. Bourton on Dunsmore, England: Practical

Action Publishing. Retrieved December 12, 2018, from http://www.livestock-emergency.net/resources/download-legs/.

Leslie, C. (2016). *Disappearing Glasgow: A photographic journey*. Glasgow, Scotland: Freight Books.

Mac [Stanley McMurtry]. (2015, November 17). Europe's open borders (cartoon). *Daily Mail*. Retrieved December 12, 2018, from http://www.dailymail.co.uk/news/article-3321431/MAC-Europe-s-open-borders.html.

Malkki, L. (1995). Refugees and exile: From 'refugee studies' to the national order of things. *Annual Review of Anthropology, 24*, 495–523.

Malkki, L. (1996). Speechless emissaries: Refugees, humanitarianism, and dehistoricization. *Cultural Anthropology, 11*(3), 377–404.

Neuman, M., & Torre, C. (2017, July 14). Médecins Sans Frontières: 'Calais has become a cage in a jungle' [Web log post]. *Border Criminologies*. Retrieved December 12, 2018, from https://www.law.ox.ac.uk/research-subject-groups/centre-criminology/centreborder-criminologies/blog/2017/07/medecins-sans.

Rawlence, B. (2016). *City of thorns: Nine lives in the world's largest refugee camp*. London: Portobello Books.

Scott, J. C. (2009). *The art of not being governed: An anarchist history of upland Southeast Asia*. New Haven, CT: Yale University Press.

Sprayson, T. (2006). Taking the lead: Veterinary intervention in disaster relief. *In Practice, 28*(1): 48–51. Retrieved December 12, 2018, from https://www.bva.co.uk/uploadedFiles/Content/About_BVA/Association/Overseas_Group/sprayson_disaster_relief.pdf.

Teffi. (2016). *Memories: From Moscow to the Black Sea* (Trans. Robert Chandler, Elizabeth Chandler, Anne Marie Jackson, & Irina Steinberg; introduction by Edythe Haber). London: Pushkin Press.

UNHCR/IUCN (United Nations High Commission for Refugees/International Union for the Conservation of Nature). (2005). *Livestock-keeping and animal husbandry in refugee and returnee situations: A practical handbook for improved management*. Geneva, Switzerland: UNHCR/IUCN. Retrieved December 12, 2018, from http://www.unhcr.org/uk/protection/environment/4385e3432/practical-handbook-improved-management-livestock-keeping-animal-husbandry.html.

Ventura, M. (2018). *Kunkush: The true story of a refugee cat* [Illustrated by B. Guo]. North Mankato, MN: Capstone Press.

Veterinarians Without Borders. (2015). Helping people and animals survive conflict and forced displacement. In *Focus—Maban County, South Sudan*. Ottawa, Canada: Veterinarians Without Borders. Retrieved December 12, 2018, from https://www.vetswithoutborders.ca/images/pdfs/VWBVSF_South_Sudan_Maban_Project_Feb_2015.pdf.

Wagner, A.-C. (2019). *Transnational mobilities during the Syrian war—An ethnography of rural refugees and Evangelical humanitarians in Mafraq, Jordan* (Doctoral dissertation). University of Edinburgh, Edinburgh, Scotland.

A warming world creates desperate people. (2018). *The New York Times,* June 29. Retrieved December 12, 2018, from https://www.nytimes.com/2018/06/29/opinion/sunday/immigration-climate-change-trump.html.

White, B. T. (2019). Humans and animals in a refugee camp: Baquba, Iraq, 1918–20. *Journal of Refugee Studies, 32*(2), 216–236. https://doi.org/10.1093/jrs/fey024.

World Animal Protection. (n.d., but post-2014). *Livestock in refugee camps: The case for improved management.* No place of publication given. Retrieved from http://www.worldanimalprotection.ca/sites/default/files/ca_-_en_files/casestudy-ethiopia-livestockinrefugeecamps.pdf

WWF (World Wide Fund for Nature). (2018). *Living planet report 2018: Aiming higher.* Gland, Switzerland: WWF/Zoological Society of London. Retrieved December 12, 2018, from https://wwf.panda.org/knowledge_hub/all_publications/living_planet_report_2018/.

# 39

# Energy on the Move: Displaced Objects in Knowledge and Practice

Jamie Cross, Craig Martin, and G. Arno Verhoeven

## Introduction: Energising Migrant Materialities

Over a decade ago, calls for studies of transnational migration to address material culture and take 'migrant materialities' seriously (Basu and Coleman 2008) gave new impetus to a critical engagement with the artefacts and architectures of forced displacement in sub-Saharan Africa. For some scholars this engagement was motivated by concerns with citizenship and the question of how conditions of bare life (Agamben 1998) are enshrined in bureaucratic materials, from identity papers, passports, certificates, and other documents (e.g. Long 2012). For others, new attention to the materiality of forced displacement opened up news understandings of forms of knowledge and practices that allowed displaced people living in exile to transform 'spaces' into 'places' (Kaiser 2008).

Our contribution to this edited collection builds on this tradition to argue that migrant materialities are also constituted through 'energetic objects': in particular the thermal technologies through which people meet their everyday needs for cold drinking water and hot fires. These energetic objects—from a homemade water cooler to a pair of bellows—are often invisible in discussions of displacement, yet they convey meaning and significance, act as vessels of knowledge and practice, and are grounds for improvisation and design.

---

J. Cross (✉) • C. Martin • G. A. Verhoeven
University of Edinburgh, Edinburgh, UK
e-mail: Jamie.Cross@ed.ac.uk; Craig.Martin@ed.ac.uk; a.verhoeven@ed.ac.uk

© The Author(s) 2020
P. Adey et al. (eds.), *The Handbook of Displacement*,
https://doi.org/10.1007/978-3-030-47178-1_39

To date the social study of energy in contexts of displacement has had very little to say about the material texture of everyday life in camps for displaced people across sub-Saharan Africa. Research by humanitarian practitioners, for example, has been concerned primarily with auditing what people use rather than analyse how people use things or interpret what materials and patterns of use might mean. Yet here—like anywhere (Shove and Walker 2014)— energy demand and use is a social practice that can be excavated and understood through an ethnographic engagement (Cross et al. 2019).

As the introduction to this collection argues, journeys of displacement involve 'myriad adjustments to unfamiliar ecologies,' in relation to a range of 'speeds, circumstances, mobilities, and actions.' In this chapter, we set out two points of departure for the exploration of the relationships between migrant worlds and energy cultures. Our approach involves a recalibration of focus around material artefacts, drawing on methods from social anthropology and design.

First, we map the mobilities of displacement by following two 'energetic objects' that are prominent in the lives of people living in Goudoubo refugee camp, northern Burkina Faso: a goatskin water cooler and a pair of bellows. This approach formed the basis of a larger research project into the material culture of displaced people living in two refugee camps—Kakuma camp in Kenya and Goudoubo camp in Burkina Faso. The fieldwork research on which this chapter is based took place in Goudoubo refugee camp, northern Burkina Faso, during eight trips between March and October 2017 by trained teams of Kenyan and Burkinabe researchers, supported by experienced social scientists, designers, and humanitarian energy specialists. During these trips our research team developed detailed studies of seemingly ordinary, mundane, even banal, energy objects and examined their histories, meaning, and active roles in people's social and economic lives. In each location they created inventories of things and mapped their connections to other things, places, and people, using these as a starting point for interviews. People were invited to discuss how they came to have them, used them, stored them, and kept them working, and then we asked them to show us how they used them, kept them, and maintained them.

Secondly, this methodological approach moves beyond an audit of what people have, to an understanding of how things are built, maintained and used, what they mean, and how they materialise people's knowledge of the world. It focused considerable attention on processes of ad hoc design, innovation, repurposing, modification, maintenance, and repair: the forms of material knowledge and skilled practice that people use and deploy to mend, rebuild, and redesign technologies. Such a focus transfers our analysis of the volume's key works. Our approach reframes 'adjustment' as the practical adjustment of knowledge and skill to unfamiliar materials, contexts, and challenges and reframes speed in its energetic sense, that is, in terms of the efficiency of a fire or a cooling device, how fast something gains heat/temperature, or how quickly something cools.

As we show, the everyday use of a water cooler and a set of bellows—two objects in the lives of the predominantly Tuareg people housed in Goudoubo—invokes histories and memories of mobility and movement. They are part of a material assemblage, with each deceptively simple device intimately connected to other materials that were either brought across the border from Mali or acquired in the camp. These particular things play a significant role in 'place making'—as people make the camp a home by re-establishing minimum forms of comfort or deploying labour-saving devices. But they are also significant as vessels of knowledge and skilled practice.

## The Bellows

Early in our study, the research team in Goudoubo was invited to take tea with a Tuareg man and his family who were living in the camp. Enjoying and participating in this man's hospitality was important in order to establish trust and good relations with members of the displaced community with whom we were working. The tea was prepared on a small, coal-fired stove, and our Tuareg host was keeping the coals hot with a set of bellows, a rather antiquated piece of equipment, but something made locally, and of which he was extremely proud. The whole event was elevated from a simple gesture to one involving status and ceremony.

The bellows are not an object normally identified as those which are fundamental to energy requirements amongst displaced populations, particularly when considering domestic circumstances of heating and cooking. Bellows provide air to help organic-based fires burn hot and clean, and this is particularly useful in craft practices involving metalwork and blacksmithing. Without a constant supply of air to the coals of the fire, temperatures do not reach appropriate levels with which to forge and shape metal into useful products and tools. Bellows are a simple technology, with more contemporary blacksmiths using electric fans and blowers to stoke their fires.

The presence of these bellows in the tea ceremony with our Tuareg host was intriguing. These bellows were not conventional equipment one might find in displaced households. It was not clear where they came from, but great attention to detail was evident in the use of red leather, brass tacks, and carved and decorated wood. It became clear that these bellows were not necessarily provided to residents within the camp, but somehow arrived based on someone's detailed understanding of construction and knowledge of operation of such devices.

During a later visit to the camp, the research team uncovered what seemed to be the source of understanding about bellows construction and design: a double-bag bellows, with two skins attached to dual windpipes joined through a central block of wood (Fig. 39.1). These specialist bellows were being used by a 19-year-old Tuareg blacksmith apprentice, who was learning the craft from his father. The bellows formed part of a makeshift forge, where red hot coals were able to make metal pliable and allow for the production and repair of a range of useful tools used within the camp.

The makeshift forge was constructed in the sand by creating a small, shallow pit at the end of the bellows where some cold coals are deposited. The blacksmith first covers the coals with a handful of wood shavings, followed by hot coals from the tea stove which are placed overtop the shavings. Once the

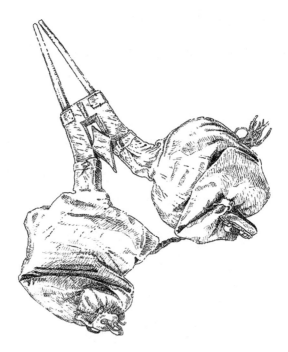

**Fig. 39.1** Illustration of double-bag bellows employed by Tuareg blacksmith apprentice (Illustration by Ann-Kathrin Müller 2018)

forge is primed, the bellows operations begins, with hands alternating in cyclic fashion to rise up and draw air into one of the bellows' skins. As one hand rises to draw air in, the opposite hand descends to push air out, squeezing the skin and igniting the coals. Like clockwork, the blacksmith's arms move up and down, one skin blowing on the fire while its counterpart takes a breath, getting ready to heat the coals.

Seconds later, smoke begins to rise and then disappears as the wood shavings burst into flames. Fire quickly consumes the wood shavings as the coals take the heat with each breath of air from the bellows. In minutes, the blacksmith moved coals at the top of the pile to one side with a large machete, uncovering a white ember coal fire. Leaving the machete in the coals, the blacksmith turned his attention to his improvised anvil, what can only be described as an oversized railroad spike, held firmly in position by a section of timber. With the blade of the metal machete reaching the optimum temperature, the blacksmith's father takes hold of a file and begins shaping the edge of the hot machete and tampering the pliable edge against the anvil with his hammer.

These blacksmiths bellows were significantly larger than those encountered during the tea ceremony with our Tuareg host. With two bags employed, they are classified as a dual-piston bellows and appear frequently in the historical records of communities of blacksmithing craft practice from across Asia, Europe, and Africa, demonstrating a significant trajectory in practice and knowledge, shared across generations and time.

The research team began asking detailed questions about the particular bellows before them, which allowed them to uncover personal details about the blacksmith's life, and ways in which he had acquired his specialist knowledge. These bellows, it turns out, were made by his father. A pair of double-bellows, we were told, was the best way to get the forge to a temperature hot enough to work metal, significantly outperforming simpler approaches like fanning the flames with sheets of cardboard.

The elder blacksmith explained that these bellows were constructed from goatskin, outlining that sheepskin is not strong enough. Handling the animal hide requires a considered approach, he continued, since the goat needs to be skinned from the rear. The skin around the mouth, throat, and windpipe of the animal needs to remain intact, where a connection is made to a wooden block which holds the pipes leading to the fire. The skin at the throat, he explained, needs to be strong since a reduction in the pathway leading to the tubes results in increased air pressure, allowing for increased flow of air through the pipes to the fire, requiring a strong connection in order to remain intact with repeated use.

Stoking the forge is but one use for a bellows, as any blacksmith knows. Our blacksmith explains that he sometimes lets his wife make use of the bellows at home, to get a good cooking fire started, especially during rainy season when firewood is inevitably collected, but too wet to light effectively.

Historians and social anthropologists working in West Africa offer some insights into the social and cultural significance of the blacksmith and the nature of the double-bag bellows at work (McNaughton 1988). Such accounts move beyond purely technical and practical dimensions to highlight the spiritual and emotional dimensions of blacksmith practices and role of the blacksmith in community life.

Getting a fire hot, to be able to forge, strike, and mould steel over an anvil in order to shape and temper tools, ornaments, farm implements, and often weaponry, has been revered across West Africa's Mande ethnic language groups since at least the early fifteenth century. Mande blacksmiths have been described by historians and social scientists as a caste, a group that has sought to retain specialist occupational knowledge and bloodlines through marriage within the community. Blacksmiths have also been described as figures of awe and fear, associated with the possession and control of *nyama*, a life force that, in many Mande traditions, is associated with power, knowledge, and creation. The blacksmith's alchemical ability to shape hard metal ores into forms useful for cultivating life and material culture offers, we might say, a quintessential display of *nyama*. Such ideas present Ousmane and Oumar in fresh light. The bellows are more than a technology for making fire efficient; they are also a medium through which a practical knowledge of fire is exchanged between master and apprentice, father and son.

In the context of forced displacement, such practices are subject to rapid transformation. In Goudoubo, for example, the UNHCR made blowtorches available to metalworkers in the artisan centre, in the expectation that this would catalyse activity. The blowtorch is a standard piece of equipment in many metalworking industries: highly efficient, clean, and portable, the blowtorch is an ideal tool for heating metal quickly and easily. But what the blowtorch gains in ease of use it loses as a vehicle for the transmission of practical mastery, craft traditions, and culture. Rather than simply providing him with the tools to forge a livelihood in the camp, the blowtorch presented Ousmane with a new challenge: how to pass on and showcase his skilled knowledge with dignity.

Firewood and charcoal remain the fuels of choice for most humanitarian agencies and their distribution remains the norm in many humanitarian settlements. As the humanitarian energy community pays closer attention to the impact of burning biomass on air quality, however, these fuels are increasingly presented as a 'dirty,' 'primitive,' and 'inferior' to alternatives like liquid petroleum gas or solar.

Framing biomass as dirty may be a necessary means of securing resources and new initiatives aimed at enhancing the provision of energy services in complex humanitarian environments. But this is also a narrow frame that erases our understanding of what makes fuel dirty. 'Dirtiness' is not an intrinsic quality of biomass; it is also an effect of materials, practices, and systems. Dirtiness is an effect of burning poor-quality materials, like wet wood. Dirtiness is an effect of inefficient combustion, with poor air circulation in and around a fire heightening pollution. Dirtiness is also an effect of failure in infrastructures of fuel distribution. When dry wood or 'clean' alternatives, like gas, are unavailable, people seek out and make do with the fuels they can find. Such an analysis does not imply that we need to be content with fuels that have negative impacts. Rather, it suggests that we might extend the range of possible interventions around humanitarian energy by examining what people do when they seek to improve the efficiency of fire.

The example of the bellows demonstrates how a deceptively simple technology reveals the heterogeneity of practices that take place around a fire. Bellows are an artefact of cottage industry and a tool of ritual and ceremony. They are important vehicles for the intergenerational transmission of artisanal skill. They are also vital technologies for the controlled circulation of air in a fire and thus temperature control and efficiency.

As this example suggests, humanitarian energy practitioners have much to gain from approaching material cultures around fire as a source of knowledge, practice, and innovation, rather than as impediments to change (Khandelwal and Lain 2018).

## The Water Carrier

When Goudoubo refugee camp in northern Burkina Faso was established in 2012, one of the first steps in its construction was the creation of an infrastructure for drinking water, connecting taps and hand pumps to ensure a continued supply of water for the camp's 7576 inhabitants. In 2017, the UNHCR's ubiquitous, standard-issue 10-litre plastic water containers can be found in and around Goudoubo camp. Lines of empty, yellow and white containers are lined up at water collection points, waiting to be filled. These unelaborated plastic water containers were used for general storage purposes, to collect water from taps and standpipes and carry it to the home. People used water directly from the container for bathing, cleaning, and washing up and in the preparation of cooked food, from millets and rice to porridge.

But few if any of the camp's inhabitant's—largely Tuareg and Fulani people originally from neighbouring Mali—actually drank this water directly. Why not? The answer was simply that it was widely perceived to be too hot to drink. Instead, before water from these standpipes was used for drinking, it was transferred to another vessel for cooling. These water-cooling devices tell us much about skilled knowledge and practice, as well as the lived experience of mobility and energy.

The cheapest and most commonly used solutions to the challenge of refrigerating water in and around Goudoubo involve the construction of what engineers and designers call 'an evaporative cooling technology.' Such technologies or devices may vary in terms of materials, but operate in a similar way: water is poured over the outer layers during the day; the water slowly heats up and evaporates, keeping the inner contents at a lower temperature—much as sweat cools the human body. Evaporative cooling devices can be made entirely from organic materials—animal skin, leather, and cloth—as well as from plastics and cardboard. They can be used to cool water as well as other fluids, like milk and juice. Similar evaporative cooling devices can be found across sub-Saharan Africa, from Burkina Faso to Kenya, Ethiopia, and Tanzania, and beyond. Yet while they are an important part of West Africa's Indigenous technological heritage, they have often gone unnoticed by the humanitarian community. Indeed, within the context of humanitarian assistance programmes around water and energy, these micro-technologies of heat and refrigeration have gone entirely overlooked.

One of the most prized evaporative cooling devices used inside Goudoubo refugee camp is the simple earthenware pot. Made from clay mixed with cattle dung, these can be manufactured and fired in and around Goudoubo camp. At home, they can be placed atop a plate on the ground, with the area around it kept damp throughout the day.

Less celebrated—but more ubiquitous in Goudoubo camp, where clay was in short supply—was the homemade or upcycled cooler, made by wrapping layers of insulating fabric around plastic water containers like those distributed by humanitarian agencies or sold by shopkeepers and traders from the nearby town of Dori. These homemade water coolers were often elaborate and often highly personalised objects: the fabric and the stitching were often a careful composition of colour and pattern. Yet these home-made plastic coolers were perceived as little more than functional devices by comparison with the goatskin bag, a leather sack that can be carried over the shoulder.

# 39 Energy on the Move: Displaced Objects in Knowledge and Practice

In northern Burkina Faso and southern Mali, goatskin water bags are prized objects. Most closely associated with the material culture and traditions of ethnic Tuareg people, they are valued by their Fulani and Bella neighbours not just for their functionality and effectiveness but also for the quality of the liquid that comes out of them. 'When you open it, the water tastes like it has come from a fridge,' one Fulani man from a village neighbouring the Goudoubo refugee camp told us. 'These things cool water so fast, it can become too cool: like ice!'

Across Goudoubo camp these goatskin vessels—called *Soumaley* in Fulani, *Agadoud* in Tamasheq, and *Guerba* in Arabic (Smith 2005)—could be found strung up on wooden sticks inside the shelter of homes (Fig. 39.2). They were widely held to be better at keeping water cool than either plastic jerry cans or clay pots. 'The water you drink from one of these is cooler than the water who can drink from any pot,' one Tuareg man told us in Goudoubo. 'If you have one of these and put it in your home, you'll never drink water from anything else again.'

In Goudoubo camp the goatskin water bag was an important vessel of cultural heritage and tradition. A goatskin carrier was often amongst the few

**Fig. 39.2** Illustration of goatskin water carrier (Illustration by Ann-Kathrin Müller 2018)

personal items that people carried with them when they fled or left Mali in the early 2010s, and almost a decade later, the taste of water from it could still evoke the sense of a life left behind.

'I once lived in Gao, at the edge of the waterhole of N'Tillit,' one Tamasheq-speaking Tuareg woman in Goudoubo said.

> The weather was colder than here. There were trees and there was shade. We used these goatskins to quench our thirst and to cool our water. But here in the camp, it's the desert. There are no trees. There is no shade, we have plastic containers and they can store our water but they cannot keep it cool. The water that comes out of the taps here is too hot. So we have to keep using these goatskins to keep our water cool.

Of course, not everybody romanticised the goatskin bag in this way and for other people it was also a reminder of former hardships. It takes a minimum of two weeks to make a *Soumaley* and the production process involves a gendered division of labour that has been well-documented by West African anthropologists (e.g. Hill 1972). Men kill the animal, cut the skin from the throat to the chest, and hang the carcass, before removing the hooves and skin. Women treat the skin and tan it, soaking the skin in water with a mixture of ash and ground seedpods (*Acacia arabica*, *bagaruwa* or *gabaruwa* in Hausa and *gaodey* in Fulani). The skins are soaked for several days before they are dried and sewn together with a leather thread. The animal's legs are tied together so that it can be hung in the shade and a small hole is left in the throat that can be pulled open or closed each time it is filled with water.

For some people, life in Goudoubo camp made access to drinking water significantly easier that it had been in the homes they left behind. For one young Tuareg couple, a married man and woman employed in the camp's school canteen, the goatskin bag carried strong associations of work and labour. 'We only used the goatskin bag because we didn't have lack of water,' the wife explained. 'Now, the tap is just right here, and you can buy pots in the market to keep your water in.' Her husband added, 'People don't need to make goatskin water bags anymore, when water is always available, close by, next to the place you live.'

It takes a minimum of two weeks to make a *Soumaley*. Depending on the intended contents, different kinds of manufacturing techniques and leather treatments are involved. Liquids like milk are considered to be 'thin' or 'transparent,' allowing any animal residue to be quickly skimmed off before drinking. The goatskins that are used to make water or milk carriers are often not reversed or turned inside out. 'Thicker' liquids like curd and porridge are seen as more difficult to filter and demand a cleaner carrier, and goatskins are turned inside out after being treated.

Plastic jerry cans are widely available in the camp but goatskin vessels are widely held to be better technologically and more efficient water coolers. With care and maintenance they can last for up to three years. Before they can be used, a dry and rigid skin must be carefully wetted, to prevent it breaking as it fills with water. If the skin is damaged or pierced, they can be repaired. A little cooked rice applied on to the surface of the skin works as an adhesive, and a new square of leather can be stuck on top and stitched in place. One of the risks is presented by the camp water itself. For some refugees in Goudoubo, the pumped water in the camp is so hot that it risks damaging these highly regarded devices. 'The tap water here is not good,' people said, 'it is so hot it damages the leather.' The leather water cooler is also associated with cleanliness and hygiene. In contrast to water stored in an open, earthenware pot, it is seen to reduce potential risks from water-borne disease by preventing children from touching the water with dirty hands.

In Goudoubo camp leather water vessels are desirable objects and the people who make them are acutely aware of their value on the market or as objects to exchange. During 2016, for example, one skilled craftswoman made two goatskin water coolers which she bartered for other goods, exchanging one for a hijab and one for a blanket. Yet such examples of local leather production are increasingly rare within Goudoubo camp. For many of the camp's inhabitants, goats are more valuable alive than dead. Each week, refugees can be found at the nearby cattle market of Seydou exchanging goats for cash to buy rice, oil, and other basic provisions. At a time of scarcity, leather is a relative luxury and the raw materials for a goatskin water cooler are less widely available.

## Heat and Water in Knowledge and Practice

The forms of knowledge and practice revealed in the bellows and water cooler—with their traces of improvisation (Malkki 1995; Brigden 2016, 2018)—mark journeys from home to spaces of displacement. They reveal the experiential or phenomenological dimensions of material life as well as the transactional dimensions, that is, the exchange-relationships through which people secure materials, components, and parts.

The reconfiguration of existing synthetic materials such as pipes and plastic canisters—their incorporation with natural materials such as fibres and skins—may seem ad hoc but it demonstrates a sophisticated understanding of material potentiality, in conjunction with high-level skills of designing, making, and construction. Engaging with the materiality of energy in spaces of displacement—spaces where fuel, electricity, and energy technologies may neither be readily available nor easily sourced—opens up new dimensions for our understanding of and approach to displaced people. Approaching displacement through the material offers new insight into the ways that objects and knowledges of displacement—from hand tools to practical skills—maintain links with home communities and establish new socioeconomic relations.

These forms of knowledge and practice outlined in this chapter demand to be understood as more than forms of 'Indigenous knowledge' manifested amongst displaced people. Rather, they constitute bodies of practical, technical knowhow that have been in themselves displaced and are here harnessed or adapted to meet new, localised needs. Such practices of improvisation and resourcefulness reveal people, as Clapperton Mavhunga has put it, 'deeply engaged in intellection, firmly anchored in their own philosophies, and alert to the world around and beyond them as a source of things that they render technological' (Mavhunga 2017, p. 8). Focusing afresh on these activities and material practices allows us to uncover the pathways and trajectories that connect displaced communities across places and through time, beyond the geopolitical space of the camp.

**Acknowledgements** The research data used here was generated collaboratively by the University of Edinburgh and Practical Action, with funding from the Economic and Social Research Council, as part of the research grant 'Energy and Forced Displacement: A Qualitative Approach to Light, Heat and Power in Refugee Camps', 2016–2018, Project Reference, ES/P005047/1. Information and data for this project was generated by the Displaced Energy Team: Jamie Cross, Craig Martin, Arno Verhoeven, Charlotte Ray, Megan Douglas (University of Edinburgh); Sarah Rosenberg-Jansen, Anna Okello, Elizabeth Njoki, Achille Lebongo, Adolpe Yemtim (Practical Action).

# References

Agamben, G. (1998). *Homo sacer: Sovereign power and bare life*. Stanford, CA: Stanford University Press.
Basu, P., & Coleman, S. (2008). Introduction: Migrant worlds, material cultures. *Mobilities, 3*(3), 313–330. https://doi.org/10.1080/17450100802376753.
Brigden, N. K. (2016). 'A focus on the everyday processes of identification, misidentification, and misdirection that accompany illusory national categories'.

Improvised transnationalism: Clandestine migration at the border of anthropology and international relations. *International Studies Quarterly, 60*(2), 343–354.

Brigden, N. K. (2018). Gender mobility: Survival plays and performing Central American migration in passage. *Mobilities, 13*(1), 111–125.

Cross, J., Douglas, M., Grafham, O., Lahn, G., Martin, C., Ray, C., & Verhoeven, A. (2019). *Energy and displacement in eight objects: Insights from sub-Saharan Africa*. London: Chatham House.

Hill, P. (1972). *Rural Hausa: A village and a setting*. Cambridge, England: Cambridge University Press.

Kaiser, T. (2008). Social and ritual activity in and out of place: The 'negotiation of locality' in a Sudanese refugee settlement. *Mobilities, 3*(3), 375–395.

Khandelwal, M., & Lain, K. (2018). The humble cookstove. *Limn*, (9). Retrieved from https://limn.it/articles/the-humble-cookstove/

Long, K. (2012). Rwanda's first refugees: Tutsi exile and international response 1959–64. *Journal of Eastern African Studies, 6*(2), 211–229.

Malkki, L. H. (1995). Refugees and exile: From 'refugee studies' to the national order of things. *Annual review of anthropology, 24*(1), 495–523.

Mavhunga, C. (2017). Introduction: What do science, technology, and innovation mean from Africa? In C. Mavhunga (Ed.), *What do science, technology, and innovation mean from Africa?* Cambridge, MA: MIT Press.

McNaughton, P. (1988). *The Mande blacksmiths: Knowledge, power and art in West Africa*. Bloomington, IN: Indiana University Press.

Shove, E., & Walker, G. (2014). What is energy for? Social practice and energy demand. *Theory, Culture & Society, 31*(5), 41–58. https://doi.org/10.1177/0263276414536746.

Smith, A. B. (2005). *African herders: Emergence of pastoral traditions*. Walnut Creek, CA: Altamira Press.

# 40

# Smartphones: Digital Infrastructures of the Displaced

Koen Leurs and Jeffrey Patterson

## Introduction

Smartphones are material, portable, embodied, and affective artefacts. In this chapter, the various meanings, roles, and usages smartphones play in the lives of displaced migrants will be assessed by unpacking smartphones as infrastructures. We provide an overview of research on transnationally displaced, digitally connected migrants. Our focus is particularly on studies on and with refugees fleeing from armed conflict or in fear of prosecution based on race, gender, sexual orientation, or religion—but we are also cognisant of the unjust and detrimental 'categorical fetishism' of labelling mobile populations (Crawley and Skleparis 2018, p. 49). During the so-called European refugee crisis in 2015, when newspapers featured photos of refugees taking selfies upon safe arrival on European shores, discussions erupted among politicians and social media circuits about whether smartphone-owning migrants were bogus asylum seekers. However, over time many came to realise that for

---

K. Leurs (✉) • J. Patterson
Graduate Gender Programme, Department of Media and Culture Studies, Utrecht University, Utrecht, The Netherlands
e-mail: K.H.A.Leurs@uu.nl

© The Author(s) 2020
P. Adey et al. (eds.), *The Handbook of Displacement*,
https://doi.org/10.1007/978-3-030-47178-1_40

displaced migrants smartphones are not luxury items, but bare essentials for survival and transnational communication. In their *Connecting Refugees* report, the United Nations High Commissioner for Refugees estimates that 68 per cent of refugees living in urban areas have an Internet-enabled phone (versus 22 per cent in rural areas) (UNHCR 2016). Through notions including 'connected migrants' (Diminescu 2020), the 'smart refugee' (Dekker et al. 2018) and the 'connected refugee' (Smets 2018), scholars are exploring the roles of smartphones in the lives of the displaced (e.g. Mancini et al. 2019). Displacement results in changing media routines and 'media-related needs' (Bellardi et al. 2018, p. 25). In this chapter, we map the conceptual contours of these emerging discussions in the fields of media, communication, migration studies, cultural anthropology, geography, and human rights. Scholarship on the role of smartphones is but one example of the emergent interdisciplinary research focus on 'digital migration studies' (Leurs and Smets 2018) that studies migration in and through smartphones, datafication, and digitally networked technologies.

By addressing smartphones as a component of the wider migration infrastructure, our aim is to move beyond fetishising the device and chart scholarship across the nexus of migrant-centric and non-migrant-centric scholarship and media-centric and non-media-centric approaches (see also Alencar 2020. By taking infrastructures as an analytic lens, scholars have begun to open up the 'black box of migration' (Lindquist et al. 2012) over the course of the last decade. This approach addresses migration as a constellation of non-migrants and migrants and human and non-human actors. The increased attention for infrastructures can be understood as part of a broader 'moment when a variety of scholars across a range of disciplines have brought the topic of infrastructures out of the expert seclusion of policy, engineering and technology development into the more open daylight of anthropology, sociology, political science, and urban studies' (Appadurai 2015, p.xii). Migration infrastructure, as argued by Biao Xiang and Johan Lindquist, refers to 'the systematically interlinked technologies, institutions, and actors that facilitate and condition mobility' (2014, p. 122). In particular, we build on Marie Gillespie et al. (2018) who took up an 'infrastructural lens' to research the 'digital passage' of Syrian refugees to Europe. Through this lens, they 'capture the dialectical dynamics of opportunity and vulnerability, and the forms of resilience and solidarity, that arise as forced migration and digital connectivity coincide' (Gillespie et al. 2018, p. 1).

This chapter focuses on three distinct ways in which smartphones are studied in the context of displaced migration, as part of infrastructures of (1) survival and surveillance, (2) transnational communication and emotion management, and (3) digital self-representation. Although we seek to provide an overview of scholarship on smartphones and displacement from across the world, our own fieldwork research has mostly concentrated on young 'connected migrants' residing in Europe. It includes work with young Syrian refugees, understanding their smartphones as personal pocket archives (Leurs 2017), as well as transnational and local digital connections maintained by gay displaced migrants and expatriates living in Amsterdam, the Netherlands (Patterson and Leurs 2019). Throughout the three sections below, we aim to be attentive to the dialectic of structure and agency, subordination and empowerment.

## Survival and Surveillance

In this section we discuss the usage of smartphones to assist in mapping journeys, maintaining transnational connectivity, as well as to access information during pre-migration, in-transit, and post-migration journeys. Developing the notion of 'smartphone travelling,' Judith Zijlstra and van Ilse Liempt (2017) focus on the role of smartphones in the journeys of irregular migrants from Turkey towards various destinations in Europe. They find 'mobile technology shapes and facilitate parts of the journey—like, for example, decisions on routes and modes of travel, final destinations and the financing of irregular migration' (Zijlstra and van Liempt 2017, p. 174). Zijlstra and van Liempt (2017) also demonstrate that smartphones, besides a relevant research topic, offer new means to conduct research; in their case they incorporated smartphones as a tool in their trajectory ethnography as an additional means of following their participants across online and offline spaces. During their journeys, smartphones enable migrants to facilitate their routes through GPS navigation, to maintain contact with smugglers, to send and access remittances and to place distress phone calls in case of emergencies. In his fieldwork with Congolese migrants and trans-Saharan migration, Max Schaub notes that although phones may facilitate migration as they allow migrants 'to tie together novel, geographically expansive networks,' an expanded 'communication infrastructure' (i.e. mobile phone coverage) is only one of the factors that turn a region into a 'transitable' one (2012, p. 126).

In Europe and the United States, refugee-led and oriented Facebook and WhatsApp groups have been popularised as a 'TripAdvisor for refugees' (Latonero and Kift 2018), but everyday realities are more complex. Displaced migrants have been found to struggle in navigating their 'information needs' (Maitland 2018, p. 6) in a complex information landscape, a continuum which researchers have described as ranging from situations of harsh 'information precarity' (Wall et al. 2017) to the abundance of the 'mobile commons' shaped by 'migrant digitalities' (Trimikliniotis et al. 2015). While information precarity concerns 'the condition of instability that refugees experience in accessing news and personal information, potentially leaving them vulnerable to misinformation, stereotyping, and rumors that can affect their economic and social capital' (Wall et al. 2017, p. 240), Nicos Trimikliniotis et al. (2015) define the mobile commons as follows:

> The invisible *knowledge of mobility* circulates between the people on the move (knowledge about border crossings, routes, shelters, hubs, escape routes, resting places; knowledge about policing and surveillance, ways to defy control, strategies against bio- surveillance, etc.), but also between trans-migrants attempting to settle in a place (knowledge about existing communities, social support, educational resources, access to health, ethnic economies, micro-banks, etc.). (p. 53)

As such, displaced populations face the challenge to locate, assess, and verify information about routes, procedures, and rights. For example, Rianne Dekker et al. argue, 'Syrian asylum migrants prefer social media information that originates from existing social ties and information that is based on personal experiences' (2018, p. 1). The emphasis on 'smart' journeys and 'smart' refugees centres the technology. Exploring 'the digital force in forced migration,' Saskia Witteborn's (2018) in-depth case study with two female asylum seekers in Germany suggests moving beyond a 'technocentric' focus by considering how 'imagined affordances' (p. 21) interact with institutional expectations and norms such as gender relations.

While humanitarian discourse and scholarly research on refugee experiences celebrate smartphones as empowering tools, greater attention is needed for the ways in which becoming a digital refugee through the use of digital devices and smartphones also means being imbricated in infrastructures of surveillance. In the words of Maria Jumbert et al., 'smartphones have gone from being seen as a vital resource for refugees, to becoming a tool of surveillance regarding their background and entitlement to international protection' (2018, p. 2). Marka Latonero and Paula Kift similarly summarise that refugees through their 'digital passage' to Europe, which is mediated by smartphones, social media, and networked technologies, co-shape an 'infrastructure for movement' which simultaneously also operates as an infrastructure of 'control' (2018, p. 1). The European Commission, for example, describes how Europe's border patrol agency (Frontex) and the European Police Office (Europol) increasingly tap into social media for migration monitoring, predictive analytics, and investigation:

Frontex primarily focuses on social media monitoring for preventive risk analysis purposes (e.g. performing analyses on irregular migration routes, to inform Member States who can then tailor responses to new phenomena). Europol on the other hand is involved in both the prevention and investigation aspects. (European Commission, European Migration Network 2016)

On the basis of locational data shared through social media platforms, in-transit displaced migrant populations in specific geographic regions can also be micro-targeted. For example, Facebook has been noted to collaborate with the Danish government to support their deterrence campaign. Although still relatively small in scale, smartphone and social media data are increasingly used to verify travel routes and asylum claims. In Germany, after extracting data from asylum seekers' phones was legalised in 2017, over 8000 phones were searched in a period of six months (Brenner and Frouws 2019). In response to these practices and rumours, asylum seekers have been noted to adapt and restrict their digital practices (Leurs 2017). In sum, the smartphone provides a materially grounded entry point to address the workings of the proliferating 'ban-opticon' which separates the 'kinetic elite' who have the right to travel and cross borders with 'the majority' from those who do not (Morley 2017, p. 80)

# Transnational Communication and Emotion Management

Migrants manage daily life in a new cultural context as they navigate their way through new language demands, cultural ideas, customs, and societal expectations that may not necessarily mirror those from where they migrated. Migration scholars have researched at length how human mobility is an intense transformative experience for both the migrant and their loved ones back 'home' (Stephan-Emmrich and Schröder 2018). The smartphone and social media are almost universal communication tools (Taipale 2019) that act as a 'lifeline' (Alencar et al. 2019; Maitland 2018) to help maintain pre-existing family bonds across physical geographical borders. The study of the smartphone and social media as tools for digital (emotional) resilience among migrants is not a newly researched phenomenon. There is a depth of literature that investigates the role of the smartphone and social media as emotional resources among transnational families, the 'networked family' (Kennedy et al. 2008), the 'digital family' (Taipale 2019), or the 'networked household' (Kennedy and Wellman 2007); it helps geographically displaced family members maintain the bi-directional exchange of emotional support (Baldassar et al. 2016). While it is crucial to acknowledge how transnational families are able to exchange reciprocal feelings of connectedness and togetherness (Patterson and Leurs 2019) through co-presence (Diminescu 2020; Baldassar et al. 2016), a multi-belonging and virtual feeling of being 'here and there,' also deserving of consideration is the examination of how the smartphone and social media can cause emotional disruption within transnational families. This section touches upon how the smartphone and social media contribute to the sense of social coherence and the positive emotional well-being of displaced migrants and their families back 'home,' while also considering the emotional labour of 'permanent connectivity' (Serrano-Puche 2015) and colliding family expectations (Wise and Velayutham 2017) and the extent to which these have obverse effects.

Smartphones and social media can be seen as valued tools that offer satisfying emotional gratification for transnational families. A breadth of literature highlights how the smartphone and social media positively impact displaced migrants' emotional well-being. The multiple modalities (text, voice, photo, and video messages) of many social media apps, such as WhatsApp and Skype, may increase feelings of belonging and emotional well-being among displaced migrants and their families (Stephan-Emmrich and Schröder 2018; Taipale 2019). The digital interaction through video is a useful tool to help transnational families cope and manage with everyday stressors and anxiety as a result of their disrupted lives (Díaz Andrade and Doolin 2019; Serrano-Puche 2015). For many digital families, FaceTime and Skype are 'total communication tools' because of their low/no cost and the (almost) face-to-face interaction it affords; it can feel as if they are in each other's physical space (Nedelcu and Wyss 2016). Mirjam Twigt describes how Iraqi refugee households in Amman, Jordan, draw on these 'affective affordances' in negotiating waiting as a form of 'mediation of hope' (2018, p. 1). In working with young Somalis stranded in Addis Ababa, Ethiopia, awaiting family identification, Koen Leurs charted means of transnational communication with parents who arrived in Europe and elsewhere through the notion of 'transnational affective capital' (2014, p. 87). Virtual visual connectedness, the ability to see the facial expressions and body movements of those geographically distanced, can often make the expressions and interpretations of emotions easier. Although the many modalities of smartphones can positively contribute to the emotional well-being of the members within transnational families, transnational co-presence can have obverse effects (Wise and Velayutham 2017).

The everyday lived experiences of displaced migrants and the expectations and beliefs of their families back 'home' can collide, which may eventually create tension leading to the deterioration of communication (Belloni 2019). Sakari Taipale (2019) highlights how the smartphone and social media can make the displaced migrant feel transnationally connected yet emotionally drained. The intensities of guilt, shame, worry, and frustration relating to the safety, health, and emotional well-being of themselves and their families can implicate transnational connectedness (Nedelcu and Wyss 2016), especially among those who fled from armed conflict including civil war (Patterson and Leurs 2019). Constant connectivity requires a set of negotiations—what is vocalised and what is left unsaid (to avoid upsetting the other), the regularity of communication, and sometimes the (unrealistic) expectations of the refugee from their family back 'home' (Wise and Velayutham 2017). Milena Belloni (2019) addressed the issue of avoidance in her study, suggesting it is the contrasting expectations between the refugee and their family that sever transnational family connections. Transnational co-presence can often leave refugees feeling 'pressure to communicate daily' and to send money, everyday material objects, and gifts (Serrano-Puche 2015; Wise and Velayutham 2017) back to their homeland; this can be experienced as an emotional burden (Witteborn 2015), an emotional labour (Hochschild 2008), and mentally taxing for both the refugee and their family. Consequently, the migrant chooses to limit or avoid their use of smartphones and social media to maintain transnational co-presence, which paradoxically is important for, and may arguably positively affect, the displaced migrants' emotional well-being.

## Digital Self-Representation

Through selfies, videos, text messages, and audio recordings, smartphones are portable archives charged with memories of past, present, and future lives—living archives of atrocities and injustices. This section explores how these self-representations contribute to a new archive of cultural heritage of diverse life upon arrival and whether it may sustain for an expansion of the commonly narrow representational stereotyping of displaced migrants. Leurs developed the notion of the smartphone as a pocket archive as a way to 'take serious young refugees' own digital archives as important sites of alternative knowledge production' (2017, p. 686). In co-creating knowledge with young Syrians by discussing photos, videos, texts, and social media posts stored on their smartphones, rich insights were established. Patterns range from identities performed for local and transnational audiences, human right violations, discrimination, and 'frustrations of not being able to control one's life course, having to make do with harsh external circumstances, lengthy procedures and seemingly arbitrary decisions' (Leurs 2017, p. 686). In her work on 'undocumented storytellers,' Sarah Bishop (2019) describes how undocumented immigrants growing up in New York City engage in narrative activism online and offline. She argues that the desire to express their voice digitally weighs heavier than risks which include hateful and xenophobic responses: 'Digital reclaimant narratives serve more than a single purpose, and the narrators demonstrate how the act of story sharing online may serve as a path to self-actualization, help to mitigate one's fear and uncertainty, offer a means for communal coping, or satisfy a sense of responsibility' (2019, p. 108). Analysing Meskhetian Turks who have dispersed over nine countries, Nurhayat Bilge describes how this refugee community mobilise social media for 'cultural identity preservation' (2018, p. 1). Maria Rae et al. studied digital practices of refugees kept in Australian-managed offshore detention centres on Manus and Nauru and show how smartphones and social media 'enable detained asylum seekers to conduct an unmediated form of self-represented witnessing that exposes human rights abuses and documents justice claims' (2018, p. 479). Behrouz Boochani's non-fiction book *No Friend But the Mountains: Writing from Manus Prison* (2018) also demonstrates how displaced populations may mobilise smartphones and digital networks to make human rights claims. Boochani, an Iranian Kurd, wrote his book on smartphones that were smuggled into the offshore processing centre. He sent poems and full chapters through text and voice messages to his translator Omid Tofighian through WhatsApp:

> A war waged with numbers /
> A numbers war /
> The frisking hands of the Papus /
> The imposing stares of the Australian officers /
> The prisoners trapped in a tunnel of tension /
> A huge feature of everyday life for the prisoners /
> Day to day…/
> A monstrous part of life /
> This is what life has become, after all…/
> This is one model constructed for human life /
> Killing time by leveraging the queue as a technology /
> Killing time through manipulating and exploiting the body /
> The body left vulnerable /
> The body an object to be searched /
> Examined by the hands of others /
> The body susceptible to the gaze of others /
> A program for pissing all over life. (Boochani 2018, p. 228)

## 40  Smartphones: Digital Infrastructures of the Displaced

Attending to refugee acts of contestation and activism is imperative, according to Ludek Stavinoha to 'avoid reifying the figure of the mute refugee so deeply embedded in the humanitarian imaginary' (2019, p. 1212). Building on fieldwork on the Greek island of Chios, he 'explores how refugees assert themselves as political subjects through communicative acts of citizenship—everyday forms of resistance against the border regime enacted in and through diverse media networks' (Stavinoha 2019, p. 1212).

The recent #SaveRahaf social media campaign is another example that demonstrates how refugees may strategically deploy smartphones and social media to make human rights claims. Rahaf Mohammed Alqunun, an 18-year-old Saudi woman, fled Saudi Arabia to seek asylum in Austria. She was stopped and detailed by Thai authorities while in transit. Her passport was taken by officials, and she was put in an airport hotel room awaiting deportation. She was able to keep her smartphone, allowing Alqunun to communicate her pleas through tweets and videos. She tweeted, 'based on the 1951 Convention and the 1967 Protocol, I'm rahafmohmed, formally seeking a refugee status to any country that would protect me from getting harmed or killed due to leaving my religion and torture from my family' 8.04 PM—6 Jan 2019, Rahaf Mohammed رهف محمد, (Mohammed 2019). Her plea went viral which was amplified by human rights organisations and activists. This led to UNHCR putting in an official request to Canada, which resulted in her being airlifted and granted asylum within a week from starting her campaign (Brenner and Frouws 2019). The question arises: how mainstream news media and host societies respond to these voices being inserted in the mediated arena?

At the height of the so-called European refugee crisis, the smartphone and selfie-taking refugees became a point of concern in mainstream news media coverage: smartphone-carrying displaced migrants did not fit with the image of the destitute refugee. There is a longer history of projection of fears over migrants' media and technology use as symbol of segregation and threat: the satellite dish, the Internet café, and now most notably the smartphone. Focusing on Western news media coverage of refugee smartphone practices, Lilie Chouliaraki scrutinises how the remediation of migrant-related selfies (photos of refugees taking selfies) operates as a site of 'symbolic bordering,' as their 'digital testimonies' are commonly marginalised (2017, p. 78) (see also Pitt, this volume). In contrast with photos of refugees taking photos, Roopika Risam in her quantitative analysis of US and UK news media found that the selfies refugees take themselves function as a form of 'self-representation that produces agency, creates communities, and resists the inscription of refugees as objects of knowledge' (2018, p. 58). Similarly, Kaarina Nikunen focuses on the 'selfie-activism' in the 'once I was a refugee' social media campaign in Finland and found that it 'expanded the "space of appearance" and introduced new voice and visuality to the public debate,' but also emphasised the difficulties of decoupling from expected performances of 'deservingness' (2019, p. 154). Scholars have also begun to address to what extent these new digital voices elicit responses. Natalia Sánchez-Querubín and Richard Roger's digital methods study of tourist responses on TripAdvisor to the mediated presence of refugees and 'media journeys' reveals fearful responses that mark an 'interrupted tourist route' (2018, p. 1). On a larger scale, Jeanine Guidry et al. (2018) compare #refugee hashtag posts on Instagram and Pinterest to locate expressions of hostility and solidarity; they find an overemphasis on security concerns. Finally, focusing on media witnessing, Zakaria Sajir and Miriyam Aouragh (2019) reflect on the impact of the heavy mediatisation of images of Alan Kurdi, a three-year-old whose dead body washed ashore in Bodrum, Turkey, in 2015, and Omran Daqneesh, a five-year-old whose photo was taken when he was sitting covered in blood and shell-shocked in the back of an ambulance, right after a bomb attack in Syria. Their study shows 'that although shocking images can awaken compassion toward the oppressed, they do not necessarily translate into movements of solidarity, but can rather degenerate into ineffective forms of pity' (Sajir and Aouragh 2019, p. 550). In sum, digital self-representations on the micro-political and intersubjective level offer forms of agency, but their impact on the overall structure of exclusionary and stereotypical mediation remains partial and paradoxical.

## Conclusions

In this chapter, we addressed the smartphone from the perspective of infrastructure. Infrastructure scholars raise awareness that besides being commonly rendered invisible, infrastructures are furthermore not singular, fixed, or stable entities that can be simply isolated or demarcated; rather they are commonly interpreted as 'not just as a "thing," a "system" or an "output," but as a complex social and technological process that enables—or disables—particular kinds of action' (Graham and McFarlane 2015, p. 1). Exploring the contours of emerging discussions on the smartphone from fields including media, communication, migration, and information studies as well as anthropology and geography, we discussed three dominant themes: (1) survival and surveillance, (2) transnational communication and emotion management, and (3) digital self-representation. While the chapter focused on smartphones as an object of research, smartphones can also be seen as important tools for research. Championing mobile methods for media and migration research, Katja Kaufmann argues the smartphone offers two broad ways to 'engage with migrants in inventive, more meaningful ways, to co-produce knowledge': (1) as a means of 'accompanying migrants in their mobile digital spaces,' it allows researchers to be digitally co-present in the digital practices and journeying of migrants and (2) to study automatically and user recorded data through 'reconstructing meanings and practices in co-productive data elicitation' (2020, p. 169). The chapter demonstrates that current scholarship has mostly been oriented towards questions around transnational communication, surveillance, and self-representation, while attention to the role of the smartphone and digital technologies in local processes of settlement, acculturation, and integration remains understudied (for a notable exception, see Alencar et al. (2019)). In keeping with Dana Diminescu's understanding of the possibilities for 'co-presence' of 'connected migrants' (2020, p. 74) in being able to maintain ties with their home country, as well as being able to forge new ties with their host society, greater awareness is needed for their interplay, as local and transnational connections are two sides of the same coin.

# References

Alencar, A., Kondova, K., & Ribbens, W. (2019). The smartphone as a lifeline: An exploration of refugees' use of mobile communication technologies during their flight. *Media, Culture & Society, 41*(6), 828–844. https://doi.org/10.1177/0163443718813486.

Alencar, A. (2020). Mobile communication and refugees: An analytical review of academic literature. *Sociology Compass* Online first. https://doi.org/10.1111/soc4.12802.

Appadurai, A. (2015). Foreword. In S. Graham & C. McFarlane (Eds.), *Infrastructural lives. Urban infrastructure in context* (pp. xii–xiii). New York, NY: Routledge.

Baldassar, L., Nedelcu, M., Merla, L., & Wilding, R. (2016). ICT-based co-presence in transnational families and communities. *Global Networks, 16*(2), 133–144.

Bellardi, N., Busch, B., Hassemer, J., Peissl, H., & Scifo, S. (2018). *Spaces of inclusion* (Council of Europe report DGI(2018)01). Retrieved from https://rm.coe.int/dgi-2018-01-spaces-of-inclusion/168078c4b4

Belloni, M. (2019). When the phone stops ringing: On the meanings and causes of disruptions in communication between Eritrean refugees and their families back home. *Global Networks.* https://doi.org/10.1111/glob.12230.

Bilge, N. (2018). Cultural identity preservation through social media: Refugees and community. In N. Bilge & M. I. Marino (Eds.), *Reconceptualizing new media and intercultural communication in a networked society* (pp. 1–17). Hershey, PA: IGI Global.

Bishop, S. (2019). *Undocumented storytellers: Narrating the immigrant rights movement.* New York, NY: Oxford University Press.

Boochani, B. (2018). *No friend but the mountains: Writings from Manus prison.* Sydney, NSW: Pan Macmillan Australia.

Brenner, Y., & Frouws, B. (2019). *Hype or hope? Evidence on use of smartphones & social media in mixed migration.* Mixed Migration Centre. Retrieved from http://www.mixedmigration.org/articles/hype-or-hope-new-evidence-on-the-use-of-smartphones-and-social-media-in-mixed-migration/

Chouliaraki, L. (2017). Symbolic bordering: The self-representation of migrants and refugees in digital news. *Popular Communication, 15*(2), 78–94.

Crawley, H., & Skleparis, D. (2018). Refugees, migrants, neither, both. *Journal of Ethnic and Migration Studies, 44,* 48–64.

Dekker, R., Engbersen, G., Klaver, J., & Vonk, H. (2018). Smart refugees: How Syrian asylum migrants use social media information in migration decision-making. *Social Media+ Society, 4*(1), 1–11.

Díaz Andrade, A., & Doolin, B. (2019). Temporal enactment of resettled refugees' ICT-mediated information practices. *Information Systems Journal, 29,* 145–174.

Diminescu, D. (2020). Researching the connected migrant. In K. Smets, K. Leurs, M. Georgiou, S. Witteborn, & R. Gajjala (Eds.), *The SAGE handbook of media and migration* (pp. 74–78). London: Sage.

European Commission, European Migration Network. (2016). *The use of social media in the fight against migrant smuggling.* Retrieved from http://www.emn.lv/wp-content/uploads/emn-informs-00_emn_inform_on_social_media_in_migrant_smuggling.pdf

Gillespie, M., Osseiran, S., & Cheesman, M. (2018). Syrian refugees and the digital passage to Europe: Smartphone infrastructures and affordances. *Social Media + Society, 4*(1), 1–12.

Graham, S., & McFarlane, C. (2015). Introduction. In S. Graham & C. McFarlane (Eds.), *Infrastructural lives. Urban infrastructure in context* (pp. 1–14). New York, NY: Routledge.

Guidry, J. P., Austin, L. L., & Carlyle, K. E. (2018). Welcome or not: Comparing #refugee posts on Instagram and Pinterest. *American Behavioural Scientist, 62*(4), 512–531.

Hochschild, A. (2008). Feeling around the world. *Contexts, 7*(8), 80.

Jumbert, M. G., Bellanova, R., & Gellert, R. (2018). *Smart phones for refugees: Tools for survival, or surveillance?* (Peace Research Institute Oslo Policy Brief). Retrieved from https://reliefweb.int/report/world/smart-phones-refugees-tools-survival-or-surveillance

Kaufmann, K. (2020). Mobile methods: Doing migration research with the help of smartphones. In K. Smets, K. Leurs, M. Georgiou, S. Witteborn, & R. Gajjala (Eds.), *The SAGE handbook of media and migration* (pp. 167–179). London: Sage.

Kennedy, T., & Wellman, B. (2007). The networked household. *Information, Communication & Society, 10*(5), 645–670.

Kennedy, T. L. M., Smith, A., Wells, A. T., & Wellman, B. (2008). *Networked families*. Washington, DC: Pew Internet & American Life Project.

Latonero, M., & Kift, P. (2018). On digital passages and borders: Refugees and the new infrastructure for movement and control. *Social Media + Society, 4*(1), 1–11. https://doi.org/10.1177/2056305118764432.

Leurs, K. (2014). The politics of transnational affective capital: Digital connectivity among young Somalis stranded in Ethiopia. *Crossings: Journal of Migration & Culture, 5*(1), 87–104.

Leurs, K. (2017). Communication rights from the margins: Politicising young refugees' smartphone pocket archives. *The International Communication Gazette, 79*, 674–698.

Leurs, K., & Smets, K. (2018). Five questions for digital migration studies: Learning from digital connectivity and forced migration in Europe. *Social Media + Society, 4*(1), 1–16.

Lindquist, J., Xiang, B., & Yeoh, B. (2012). Opening the black box of migration: Brokers, the organization of transnational mobility and the changing political economy in Asia. *Pacific Affairs, 85*(1), 7–19.

Maitland, C. (2018). Introduction. In C. Maitland (Ed.), *Digital lifeline: ICTS for refugees and displaced persons* (pp. 1–14). Cambridge, MA: MIT Press.

Mancini, T., Sibilla, F., Argiropoulos, D., Rossi, M., & Everri, M. (2019). The opportunities and risks of mobile phones for refugees' experience: A scoping review. *PLoS ONE, 14*(12). https://doi.org/10.1371/journal.pone.0225684.

Mohammed, R. (2019). [Post on Twitter]. Retrieved from https://twitter.com/rahaf84427714/status/1082005121016320001

Morley, D. (2017). *Communications and mobility: The migrant, the mobile phone, and the container box*. Hoboken, NY: Wiley Blackwell.

Nedelcu, M., & Wyss, M. (2016). 'Doing family' through ICT-mediated *ordinary co-presence*: Transnational communication practices of Romanian migrants in Switzerland. *Global Networks, 16*(2), 202–218.

Nikunen, K. (2019). Once a refugee: Selfie activism, visualized citizenship and the space of appearance. *Popular Communication, 17*(2), 154–170.

Patterson, J., & Leurs, K. (2019). We live here, and we are queer! Young gay connected migrants' transnational ties and integration in the Netherlands. *Media & Communication, 7*(1), 90–101.

Rae, M., Holman, R., & Nethery, A. (2018). Self-represented witnessing: The use of social media by asylum seekers in Australia's offshore immigration detention centres. *Media, Culture & Society, 40*(4), 479–495.

Risam, R. (2018). Now you see them: Self-representation and the refugee selfie. *Popular Communication, 16*(1), 58–71.

Sajir, Z., & Aouragh, M. (2019). Solidarity, social media, and the 'refugee crisis': Engagement beyond affect. *International Journal of Communication, 13*, 550–577.

Sánchez-Querubín, N., & Rogers, R. (2018). Connected routes: Migration studies with digital devices and platforms. *Social Media + Society, 4*(1), 1–13.

Schaub, M. X. (2012). Lines across the desert: Mobile phone use and mobility in the context of trans-Saharan migration. *Information Technology for Development, 18*(2), 126–144.

Serrano-Puche, J. (2015). *Emotions and digital technologies: Mapping the field of research in media studies*. London, England: London School of Economics and Political Science.

Smets, K. (2018). The way Syrian refugees in Turkey use media: Understanding 'connected refugees' through a non-media-centric and local approach. *Communications, 43*(1), 113–123.

Stavinoha, L. (2019). Communicative acts of citizenship: Contesting Europe's border in and through the media. *International Journal of Communication, 13*, 1212–1230.

Stephan-Emmrich, M., & Schröder, P. (2018). *Mobilities, boundaries, and travelling ideas: Rethinking translocality beyond Central Asia and the Caucasus*. Cambridge, England: Open Book Publishers.

Taipale, S. (2019). *Intergenerational connections in digital families*. Basel, Switzerland: Springer Nature.

Trimikliniotis, N., Parsanoglou, D., & Tsianos, V. (2015). *Mobile commons, migrant digitalities and the right to the city*. Basingstoke, England: Palgrave Macmillan.

Twigt, M. (2018). The mediation of hope: Digital technologies and affective affordances within Iraqi refugee households in Jordan. *Social Media + Society, 4*(1), 1–14.

UNHCR (UN Refugee Agency). (2016). *Connecting refugees*. Geneva, Switzerland: UN Refugee Agency. Retrieved from https://www.unhcr.org/5770d43c4.pdf.

Wall, M., Campbell, M. O., & Janbek, D. (2017). Syrian refugees and information precarity. *New Media & Society, 19*, 240–254.

Wise, A., & Velayutham, S. (2017). Transnational affect and emotion in migration research. *International Journal of Sociology, 47*(2), 116–130.

Witteborn, S. (2015). Becoming (im)perceptible. Forced migrants and virtual practice. *Journal of Refugee Studies, 28*(3), 350–367.

Witteborn, S. (2018). The digital force in forced migration: Imagined affordances and gendered practices. *Popular Communication, 16*(1), 21–31.

Xiang, B., & Lindquist, J. (2014). Migration infrastructure. *International Migration Review, 48*(1), 122–148.

Zijlstra, J., & Van Liempt, I. (2017). Smart(phone) travelling: Understanding the use and impact of mobile technology on irregular migration journeys. *International Journal of Migration and Border Studies, 3*(2–3), 174–191.

# 41

# Family Photographs in Displacement

Penelope Pitt

## Introduction

'I grabbed the dog, some photos, and we got out,' an evacuee of a bushfire in the Australian town of Tathra is reported to have told a journalist (Kraus 2018). People displaced from their homes commonly describe family photographs as among the personal items they gather up and take with them. Family photographs also get left behind; photographs can be destroyed or lost. People may produce photographs whilst en route to another place and after arriving. Photographs may be shared between family members living in displaced circumstances in different locales. These photographs take digital and hardcopy forms and they can, for example, be arranged on USBs and hard drives, in photograph albums and in frames within homes, and shared via emails, text messages, and social media. Family photographs (also known as 'domestic' and 'personal' photographs) are therefore of central importance to people experiencing displacement; they matter. In this chapter, I consider some of the ways in which they matter. I draw on existing research to present some key contemporary understandings about the relations between people and their family photographs in displacement.

P. Pitt (✉)
School of Education, Deakin University, Melbourne, VIC, Australia
e-mail: penelope.p@deakin.edu.au

© The Author(s) 2020
P. Adey et al. (eds.), *The Handbook of Displacement*,
https://doi.org/10.1007/978-3-030-47178-1_41

Research studies on the family photography of displaced people have been reported in diverse literatures including anthropology, sociology, art history, visual studies, psychology, and cultural studies. Studies in the broader field of photography and displacement include participatory visual methods projects, family photography studies, community art projects, and oral history studies. Due to the vastness of the broader literature, this chapter is limited to a discussion of research based on the study of 'found photos' (Tinkler 2013, p. xv). Found photos in this context are photographs produced *by* people living in displacement *for* themselves and/or their family and friends, and not photographs produced in response to the requests of researchers.[1] This chapter considers research studies in which family members were interviewed (in most cases) about photographs produced in a variety of circumstances of displacement.

Displacement is defined by UNESCO as 'the forced movement of people from their locality or environment and occupational activities' (2019, n.p.). Displacement indicates both the movement of people away from a locale and the *forced* nature of this movement. Displacements of people can be internal or across national borders and may be the result of factors including, but not limited to, poverty or economic change, natural and environmental disasters, persecution, violence (including domestic), and armed conflicts. Many of the studies discussed in this chapter are about the family photographs of displaced people as per this understanding of displacement and include the family photographs of refugees, exiles, prisoners of internment camps, and internally displaced persons.

Who should count as 'displaced' is not clear-cut, however. In recognition of this, Yarris and Castaneda (2015) have argued in favour of viewing 'contemporary displacement as always existing on a continuum between "force" and "will"' and paying attention to this continuum in each specific individual situation (p. 64). Questions of whether, and the extent to which, a movement is forced is particularly complicated with regard to dependant family members of migrants. It can be difficult to know the degree of agency and choice enjoyed by the *dependants* accompanying migrants not considered to be forced migrants. For example, some partners and children of migrants who have participated in family photography studies could be regarded as having had little choice but to leave their homes, to move away from a locality, in order to continue to be financially supported by their parent or partner. For some dependant partners and children, this experience may amount to a displacement.

For the purposes of this chapter, I regard people who have moved away from their home or locale in conditions of markedly limited agency as living 'in displacement.' Some examples from the more established body of work on migrant and diasporic family photographs are included in this chapter. I include these examples both in recognition of the complexity of the dynamics of force and will in circumstances of displacement and, also, in order to provide a holistic sense of the relations between people and their family photographs in situations of separation from a home.

Researchers have applied a range of key theoretical approaches to the relationship between people and their family photographs. Photographs have been conceptualised as representations (Gold 2013; Hall 1991), a medium for the performance of selves (Holland 1991, 2009; Kuhn 1995), visual narratives (Bach 2001; Harrison 2002; Tinkler 2013), material objects that people do things with (Edwards 2002; Edwards and Hart 2004; Rose 2010; Rose and Tolia-Kelly 2012), and multisensory (Edwards 2002; Mitchell 2005; Pink 2011). Less often put to use in research on family photographs in displacement are new materialist or other post-humanist theoretical approaches in which photographs may be considered as occurring within practices in which agency may be spread across humans, photographs, and other entities.

This chapter is divided into three main sections that consider the kinds of spatialities, temporalities, and emotions experienced and produced in the interactions of people and their family photographs in displacement. Whilst for readability these three aspects are divided into three separate sections, matters of spatiality, temporality, and emotions bleed into and produce each other. Hence there are significant overlaps and synergies between the sections below. The final section of this chapter pulls together key threads to suggest potential future directions for research on family photographs in displacement.

## Spatialities

A wide range of studies have reported that both newly arrived asylum seekers and migrants produce photographs to show family members back home that they have both arrived and survived their journey. On this point, it is common for researchers to draw on Barthes' (1980/2000) notion of a photograph being 'a certificate of presence' (p. 87), evidence that shows someone was in a particular place at a particular time. Chouliaraki (2017, p. 85) writes of 'celebration selfies,' produced by asylum seekers on their smartphones in recent years when they reach the beaches of Europe by boat. Chouliaraki (2017) suggests these particular selfies, produced on the shores of the Mediterranean and texted immediately to family left behind, are 'visual proofs' of both 'I am *here*' and 'I *am* here,' of 'arrival-as-survival' (p. 85).

Whilst the immediacy of the communication of this evidence of arriving and surviving and the 'selfie' nature of the photographs is relatively new, the practice more generally is many decades old. For instance, Hall (1991), in his study of photographs of post-World War II black settlement in Britain, describes the 'formal "high street" photo portraits' of newly arrived West Indians in postwar Britain as photos that 'you sent home as "evidence" that you had arrived safely, landed on your feet, were getting somewhere, surviving, doing alright' (p. 156). Migrants more generally are known to use family photography to produce visual 'evidence' of themselves as prospering and climbing the social ranks in a new state or country, which may then be shared with family left behind (Chambers 2003; Hall 1991; Smith 1999; Thomson 2011a, b).

Family photography practice has been theorised as producing the 'integration' and 'unity' of family (Bourdieu 1965/1996, p. 19), the 'connectedness' of family members (Sontag 1977/2008, p. 8), and family 'togetherness' (Hirsch 1997/2012, p. 7; Rose 2004, p. 549). Women and mothers in particular have been understood to take an active role in producing the closeness of family members through their production, curation, sharing, and display of family photographs. In a significant research study on middle-class mothers and their family photographs in recent-day England which included a few participants who were migrants, Rose (2003) found that it was through what mothers did with their photographs (e.g. the way they displayed and shared photos) that the women produced an 'extended domestic space' (p. 5) which stretched across large physical distances to encompass distant spaces. Mothers corporeally enacted family 'togetherness' across distance through their interaction with their photographs, 'bring[ing] near those far away' (Rose 2003, p. 12). In what could be considered a diasporic example of this production of togetherness, Solanke (1991), a child of Nigerian migrants to Britain, shares that when she first visited the home of extended family in Nigeria, she saw a photograph of one of her sisters on the wall and this photograph was a duplicate of a photograph in Solanke's family album in London. For Solanke (1991) this experience 'reinforces the sense of fitting into a structure of life elsewhere' (p. 132). A potential direction for future displacement research could be to explore whether this stretching of domestic space is similarly produced through interactions between family photographs and family members displaced in different locales or families in which some members remain in place and other members are displaced.

There is some evidence in the transnational care research literature (Fedyuk 2012; Baldassar 2008; Kea 2017) to suggest that this production of family connection and togetherness may become more important and intensified for families with displaced members. In a study of the domestic photographs of Gambian parents in the UK who sent their children to be fostered by carers in the Gambia, Kea (2017) found that 'it is important to Gambian parents that photographs of their children [produced in the Gambia] convey the sense that their children feel at home in their new environment and are treasured and loved' (p. 59). This sense was conveyed through the formality of the photographs, the types of poses, and the settings. Kea (2017) suggests that through the foster carers sharing these photographs with the parents in the UK, 'intimacy' is generated between the child in the Gambia who is the subject of the photographs, the photographer, and the parent viewing the photographs in the UK (p. 59). Using photographs to produce family as connected or intimate, and to 'shrink the distance' between dispersed family members (Smith 1999, p. 86), may be more crucial in more emotionally charged displacements such as the transnational separation of a parent and child.

More generally, family photography is known to be a way for people to create homely spaces from new, unfamiliar spaces. In her famous book, *On Photography*, Sontag (1977/2008) argued that the act of taking 'photographs … help[s] people take possession of space in which they are insecure' (p. 9). Whilst Sontag was referring to tourism and personal photography in making this argument, Chambers (2003) takes up this notion in describing the role of family photography when as a girl she moved with her family from Australia to New York in the 1950s for her father's work. Chambers (2003) argues that in the context of a family taking up residence in a foreign nation, the family photograph album is not simply a record of family events but consists of 'visual memories of the ownership and domestication of unfamiliar, alien space' (p. 103). Whilst Chambers (2003) and her family were expatriates or transnationals and were not displaced persons, Chambers' extension of Sontag's idea to family photography is potentially relevant to the study of displacement. Kunimoto (2004) details a similar theme of people using photography to take possession of insecure spaces in her substantive study of the family photograph collections of four Japanese-Canadians involuntarily interned by the Canadian government in camps during World War II.

Drawing on Lefebvre's theory of space, Kunimoto (2004) argues that 'taking a picture … transforms physical space into representational space, and this transformation may grant the possessor of the image a sense of control over the representation, and by extension, over physical space' (p. 144). A striking example of representational space as analysed by Kunimoto is a page in one of the albums in which a woman has placed four photographs of herself pictured both alone and together with other people living in the internment camp. Around these photographs, linking them together, the woman has drawn the shape of a house. In doing this, Kunimoto notes, the woman brings about 'a sense of control in an insecure environment' (p. 143). Kunimoto (2004) found that these forcibly displaced people responded to living in 'a liminal, transitional and marginal space' by creating photograph albums that 'foster a sense of place' similar to the homes from which they were forcibly displaced (p. 134).[2] A future productive direction for research studies in this field could be to tease out how taking possession of unfamiliar and/or insecure places through family photography practices may take different shapes for people living in different conditions of displacement.

## Temporalities

Temporal gaps in the visual narratives of photograph albums and collections of displaced people have been found to sometimes be the result of periods of instability, difficult living conditions, and separation from loved ones (Spitzer 1999; Naguib 2008; West 2014; Kunimoto 2004). For example, in Spitzer's (1999) study of his own family's album constructed during his family's sea voyage from Austria to Bolivia as refugees during World War I, when Spitzer's mother was pregnant with him, he finds gaps in his family's album: there are no photographs of particular people or the more traumatic stages of the family's migration. People enduring displacement sometimes fill the gaps in their albums and collections in creative and novel ways. Naguib (2008) describes how an Armenian couple, who were each orphaned as children in Armenia, survived World War I and met each other in Cairo as displaced young adults and as a result they had very few photographs of themselves before reaching adulthood. Naguib (2008) found that the couple filled out their photograph album with found photographs of strangers and newspaper clippings of the Armenian genocide: 'the images of skeletal children, destruction, and heaps of dead bodies fill in the gaps of her story and explain why possible photographs from her family albums are absent' (p. 243). In this way, photograph albums have been shown to be used by displaced people to not only document and remember celebrated personal milestones (as described in the general family photography literature) but also to document and preserve sometimes traumatic collective histories.

Outside of the family photography-related research literatures, different temporalities are known to be common to different kinds of displacements.

For example, as Peteet (2007) argues, 'refugee status indicates an impermanency that may not be consistent with diaspora's connotations of temporal depth and integration' (p. 633). This is an understanding that could be taken up and further developed in future research on the temporalities involved in family photographs in displacement. There is a place for future comparative work on the family photography practices involved in generating subjectivities linked to particular conditions of displacement (and to the variety of migration experiences detailed in the established literature).

There has been detailed work undertaken on the concept of *memory* in relation to the family photography of migrant and diasporic people that looks beyond individual visual narratives. Whilst there is not the space to explore this in depth here, the most influential concepts are Hirsch's (1997/2012) 'postmemory' and Tolia-Kelly's (2004) 're-memory.' Postmemory refers to the 'second-generation memories of cultural or collective traumatic events and experiences,' different to memory in that postmemories are not direct recollections of a subject but rather a connection to the source of attention through 'imaginative investment and creation' (Hirsch 1997/2012, p. 22). Tolia-Kelly (2004) describes 're-memory' as 'a form of memory that is not an individual linear, biographical narrative' (p. 314). It is a conceptualisation of a social kind of memory that Tolia-Kelly understands to be present in the lives of British Asian diasporic people; to be 'stimulated through scents, sounds and textures'; and to include imagined or recalled memories of narratives described by others and not originally or directly experienced by oneself (p. 314). The relevance of these concepts to the family photography practices of people displaced in various circumstances is an area worthy of attention.

Outside of the work on memory, there has been little research looking at the complexity of temporalities involved in making the lives of people and their family photographs in displacement. As I have argued elsewhere (Pitt 2015), reports on studies of migrant family photographs do not tend to comment on the broader temporal structure of the photograph albums or collections, and there is often an implicit assumption that the albums follow a linear progressive temporal order. My own research (Pitt 2015) has demonstrated how this is not always the case. In research with a woman who had migrated to Australia with her husband and children, I drew on subjectivity theories of Julia Kristeva (1979/2008) and Homi Bhabha (1994/2004) to suggest that multiple temporalities were produced in the relation between the woman and her photograph albums, entangling milestones of children and family members, cyclical time, and definitive moments (Pitt 2015). It is important for research on the family photography of displaced people to take into account theories of temporality if we are to gain deeper understandings of the lived experiences of displaced people.

The complexity of temporal experience in the lives of displaced people and their family photographs has been largely considered from a human-centred perspective to date. Whilst the concepts of re-memory and postmemory productively push memory into social, communal, cross-generational, and multisensorial territory, these concepts remain human-centred. Within a more-than-human approach, a family photograph may be considered to share some agency as a body alongside multiple other bodies including humans. Approaching the spatio-temporalities involved in the lives of displaced persons and their family photographs from a more-than-human position may be a way to unfold some more of the complexity of these relations.

## Emotions

An often noted point about emotion in the general family photography literature is that photographs tend to be produced to picture happiness in families. Chalfen (1987), for example, suggests that family photographs 'visually represent events in life that make a difference … in the direction of success and happiness' (p. 99). Family photographs produced by displaced people have been shown to do this in ways which entangle various events of displacement (such as arriving) with happiness and even euphoria. In writing about the 'celebration selfies' taken by migrants when they reach Europe's shores at the end of a dangerous boat journey, Chouliaraki (2017) describes 'smiling faces and V-signs' as demonstrative of 'the euphoria of arrival' and of survival (p. 85).

Another recurring theme regarding emotions in the family photography literature is that family photos provide an outlet for people to experience intense emotions about their familial relationships in an organised, structured, and relatively safe way. Chambers (2003), for example, argues that a family photograph album is a selectively constructed 'visual narrative that structures and organizes feelings by fixing meanings of "family"' (p. 113). Kunimoto (2004), in her study of interned Japanese-Canadians, found that the easy-to-handle form of the photograph and the photo album enabled forcibly interned people to process difficult emotions: 'within the space of the album, home can be folded up and tucked away,' enabling fears associated with living in an unfamiliar environment to ease (p. 140). Rose (2003, 2004) has described this—in relation to British mothers and family photographs of their children—as the co-production of emotions and spatiality. Drawing on the work of Barthes (1980/2000), Rose (2003) argues that 'family photos carry a material trace of the person photographed' (p. 557). Rose (2004) suggests that a mother can interact with 'the trace of a child in a photo' in a way she cannot with her actual children, as children in photos are 'miniaturized and mute' (p. 561). This interaction with the trace of their young children in photographs enables mothers to safely experience and to 'steady the ambivalent feelings of love and hate' they feel towards their children (Rose 2004, p. 561). Future research could consider drawing on visual theory as Rose (2003, 2004) did, to look at whether and how emotions and spatiality are co-produced in situations of displacement and, if so, which emotions and to what effect.

Some studies of displacement and photography note a discrepancy between the emotions displayed on the pictured faces and bodies of displaced people in the photos and the arduous situation they are known to have been enduring at the time (Spitzer 1999; West 2014). Spitzer (1999) struggles to reconcile his personal knowledge of the terrible losses his family were suffering at the time the photographs were taken with their smiling faces in the photograph album. Spitzer discovered notes in his father's handwriting on the backs of these photographs that 'expose[d] the intense feelings of grief and sadness that he associated with his voyage' (p. 216). A recent chapter by Hinkson (2017) about visual culture, family photographs, and an Aboriginal woman exiled from the Central Australian desert to live in Adelaide due to a family feud also describes an experience of a 'conjunction' of extremes of emotion: trauma and pleasure (p. 93). The coexistence of emotional extremes within the relations between displaced people and family photographs could be an area of future study.

## Potential Future Directions

Family photographs matter to people living in circumstances of displacement. In precarious and dangerous circumstances, family photographs can provide evidence to loved ones of arrival and survival. Through interactions with their family photographs, people living in displacement can take possession of new spaces in which they are insecure and can produce connection and intimacy with family members across large physical distances. Some displaced people are known to engage in family photography practices that avoid picturing trauma, whilst others engage in practices that actively preserve collective histories of the trauma of displacement. Intense and ambivalent emotions embedded in experiences of displacement are understood to be processed by people through their production of photographs and interactions with their family photographs. Whether certain experiences of spatio-temporalities may be associated with particular emotions or affects in the relations between displaced people and their family photographs is an area worthy of future attention.

Migrants, who may not necessarily be displaced, have been described in the more established research literature as valuing and using family photography in making and remaking their lives in new places. There is some indication that in situations of displacement, people's practices with family photographs may be of even more acute significance in the production of social connections and sense of self. This could perhaps be particularly the case where people have experienced less agency in their movement, or in other words, if they sit closer to the 'force' pole on the displacement continuum between 'force' and 'will' (Yarris and Castaneda 2015, p. 64). Future comparative studies of people displaced in differing circumstances could potentially consider this.

The common current understandings about the kinds of spatialities, temporalities, and emotions generated in the relationship between displaced people and their family photographs are largely grounded in human-centred perspectives. Different kinds of subjects and identities and relationships between family members have been variously understood as represented, negotiated, contested, mediated, and produced through displaced people's interactions with their family photographs. There is the potential for future research on this topic to expand on these ideas through embracing a post-humanist perspective. In commenting on the relation between photography and mobility more generally, Lisle (2013) has suggested post-humanist theory as a useful direction as 'it does not assume that things or people are bounded entities with unbreacheable limits' (p. 539). A post-humanist approach to displacement and family photography enables the primary focus on the displaced human to shift to a consideration of the entanglements of the movements of the photographs, the family members, and other entities, and the effects this has in the world. This is a further possible future direction for research about family photographs in displacement.

## Notes

1. Reports not considered in this chapter are those of studies in which photographs picturing displaced people were produced by others (such as researchers or photojournalists), reports of studies in which the researcher prompted the participants to produce photographs for the purposes of the research, reports

of studies based solely on archival photographs, and reports or commentary on the public representation or use of private family photographs picturing displaced people. For readings on this latter category, see, for example, Mannik (2012), Rose (2010) and Bussard (2018).
2. Kunimoto (2004 p. 132) explains that although 'Japanese nationals and Japanese-Canadians were banned from owning cameras' under the 1942 War Measures Act, this was not always strictly enforced in the camps. Some interned people smuggled cameras into the camps and then religious ministers had the film developed for them outside the camps (p. 135).

# References

Bach, H. (2001). The place of the photograph in visual narrative research. *Afterimage, 29*(3), 7.
Baldassar, L. (2008). Missing kin and longing to be together: Emotions and the construction of co-presence in transnational relationships. *Journal of Intercultural Studies, 29*(3), 247–266.
Barthes, R. (1980/2000). *Camera lucida: Reflections on photography*. London: Vintage.
Bhabha, H. (1994/2004). *The location of culture*. London: Routledge.
Bourdieu, P. (1965/1996). *Photography: A middle-brow art* (Trans. S. Whiteside). Cambridge, England: Polity Press.
Bussard, K. A. (2018). Some thoughts on the role of family and photography in today's refugee crisis. *Exposure, 50*(1), 17–21.
Chalfen, R. (1987). *Snapshot versions of life*. Madison, WI: University of Wisconsin Press.
Chambers, D. (2003). Family as place: Family photograph albums and the domestication of public and private space. In J. M. Schwartz & J. R. Ryan (Eds.), *Picturing place: Photography and the geographical imagination* (pp. 96–114). London: IB Tauris.
Chouliaraki, L. (2017). Symbolic bordering: The self-representation of migrants and refugees in digital news. *Popular Communication, 15*(2), 78–94.
Edwards, E. (2002). Material beings: Objecthood and ethnographic photographs. *Visual Studies, 17*(1), 67–75.
Edwards, E., & Hart, J. (2004). Introduction: Photographs as objects. In E. Edwards & J. Hart (Eds.), *Photographs objects histories: On the materiality of images* (pp. 1–15). London: Routledge.
Fedyuk, O. (2012). Images of transnational motherhood: The role of photographs in measuring time and maintaining connections between Ukraine and Italy. *Journal of Ethnic and Migration Studies, 38*(2), 279–300.

Gold, S. J. (2013). Using photography in studies of international migration. In S. J. Gold & S. J. Nawyn (Eds.), *Routledge international handbook of migration studies* (pp. 530–542). Abingdon, England: Routledge.

Hall, S. (1991). Reconstruction work: Images of post-war black settlement. In J. Spence & P. Holland (Eds.), *Family snaps: The meanings of domestic photography* (pp. 152–164). London: Virago.

Harrison, B. (2002). Photographic visions and narrative inquiry. *Narrative Inquiry, 12*(1), 87–111.

Hinkson, M. (2017). At the edges of the visual culture of exile: A glimpse from South Australia. In E. Gomez Cruz, S. Sumartojo, & S. Pink (Eds.), *Refiguring techniques in digital visual research* (pp. 93–104). Cham, Switzerland: Palgrave Macmillan.

Hirsch, M. (1997/2012). *Family frames: Photography, narrative, and postmemory.* Cambridge, MA: Harvard University Press.

Holland, P. (1991). Introduction: History, memory and the family album. In J. Spence & P. Holland (Eds.), *Family snaps: The meanings of domestic photography* (pp. 1–13). London: Virago.

Holland, P. (2009). 'Sweet it is to scan …': Personal photographs and popular photography. In L. Wells (Ed.), *Photography: A critical introduction* (4th ed., pp. 117–166). London: Routledge.

Kea, P. (2017). Photography, care and the visual economy of Gambian transatlantic kinship relations. *Journal of Material Culture, 22*(1), 51–71.

Kraus, C. (2018, March 19). Tathra residents reel as bushfire engulfs homes: 'I grabbed the dog, some photos, and we got out'. *The Guardian* (Australia edition). Retrieved from https://www.theguardian.com/australia-news/.

Kristeva, J. (1979/2008). Women's time. In V. Hesford & L. Diedrich (Eds.), *Feminist time against nation time: Gender, politics, and the nation-state in an age of permanent war* (pp. 23–39). New York, NY: Lexington Books.

Kuhn, A. (1995). *Family secrets: Acts of memory and imagination.* London: Verso.

Kunimoto, N. (2004). Intimate archives: Japanese-Canadian family photography, 1939–1949. *Art History, 27*(1), 129–155.

Lisle, D. (2013). Photography. In P. Adey, D. Bissell, K. Hannam, P. Merriman, & M. Sheller (Eds.), *The Routledge handbook of mobilities* (pp. 534–541). London: Routledge.

Mannik, L. (2012). Public and private photographs of refugees: The problem of representation. *Visual Studies, 27*(3), 262–276.

Mitchell, W. J. T. (2005). *What do pictures want?: The lives and loves of images.* Chicago, IL: University of Chicago Press.

Naguib, N. (2008). Storytelling: Armenian family albums in the diaspora. *Visual Anthropology, 21*(3), 231–244.

Peteet, J. (2007). Problematizing a Palestinian diaspora. *International Journal of Middle East Studies, 39*, 627–646.

Pink, S. (2011). Sensory digital photography: Re-thinking 'moving' and the image. *Visual Studies, 26*(1), 4–13.

Pitt, P. (2015). Exploring subject positions and multiple temporalities through an Iranian migrant mother's family photo albums. *Gender, Place and Culture: A Journal of Feminist Geography, 22*(2), 201–221.

Rose, G. (2003). Family photographs and domestic spacings: A case study. *Transactions of the Institute of British Geographers, 28*(1), 5–18.

Rose, G. (2004). 'Everyone's cuddled up and it just looks really nice'. An emotional geography of some mums and their family photos. *Social and Cultural Geography, 5*(4), 549–564.

Rose, G. (2010). *Doing family photography: The domestic, the public and the politics of sentiment*. Farnham, England: Ashgate.

Rose, G., & Tolia-Kelly, D. (2012). Visuality/materiality: Introducing a manifesto for practice. In G. Rose & D. Tolia-Kelly (Eds.), *Visuality/materiality: Images, objects and practices* (pp. 1–11). Farnham, England: Ashgate.

Smith, V. (1999). Photography, narrative, and ideology in 'Suzanne Suzanne' and 'Finding Christa' by Camille Billops and James V. Hatch. In M. Hirsch (Ed.), *The familial gaze* (pp. 85–98). Hanover, NH: University Press of New England.

Solanke, A. (1991). Complex, not confused. In J. Spence & P. Holland (Eds.), *Family snaps: The meanings of domestic photography* (pp. 128–138). London: Virago.

Sontag, S. (1977/2008). *On photography*. London: Penguin.

Spitzer, L. (1999). The album and the crossing. In M. Hirsch (Ed.), *The familial gaze* (pp. 208–220). Hanover, NH: University Press of New England.

Thomson, A. (2011a). Family photographs and migrant memories: Representing women's lives. In A. Freund & A. Thomson (Eds.), *Oral history and photography* (pp. 169–185). New York, NY: Palgrave Macmillan.

Thomson, A. (2011b). *Moving stories: An intimate history of four women across two countries*. Manchester, England: Manchester University Press.

Tinkler, P. (2013). *Using photographs in social and historical research*. Los Angeles, CA: Sage.

Tolia-Kelly, D. (2004). Locating processes of identification: Studying the precipitates of re-memory through artefacts in the British Asian home. *Transactions of the Institute of British Geographers, 29*(3), 314–329.

UNESCO (United Nations Educational Scientific and Cultural Organization). (2019). Displaced person/Displacement. Retrieved from http://www.unesco.org/new/en/social-and-human-sciences/themes/international-migration/glossary/displaced-person-displacement/.

West, T. (2014). Remembering displacement: Photography and the interactive spaces of memory. *Memory Studies, 7*(2), 176–190.

Yarris, K., & Castaneda, H. (2015). Special issue: Discourses of displacement and deservingness: Interrogating distinctions between 'economic' and 'forced' migration. *International Migration, 53*(3), 64–69.

# 42

# Displaced Home-Objects in Homing Experiences

## Mastoureh Fathi

## Introduction

In an era where the world seems to be ever more connected through media, virtual spaces, and mobile lives, a lack of attachment to places and locations is seen as normal. The historic and unprecedented number of people crossing borders means many people change and reconstruct their 'homes' in a different geographical location from the one they were born into. As such, the idea of home has not disappeared from our lives even though its forms have changed through displacement processes. These transformations necessitate an understanding of what home constitutes in relation to mobility, new forms of belonging, attachments, and the processes of detachments.

## What Is Home?

Home is referred to a place encompassing multiple positive or negative experiences that could be reminiscent of feelings of warmth, closeness, family, as well as detachment, displacement, and sometimes violence. What is common to all these concepts is the element of domesticity in relation to space and intimacy among its inhabitants. Furthermore, increased mobility means that these notions encompass local, national, and transnational elements that

---

M. Fathi (✉)
University College Cork, Cork, Ireland
e-mail: mastoureh.fathi@ucc.ie

include a variety of social relations and networks embedded within political structures that shape mobile people's decisions and future plans.

A space that is called 'home' includes objects including furniture, personal items, food, those with instrumental purposes, and those with symbolic functions. What distinguishes a home from another space is the existence of a sense of belonging, emotional attachments, and feelings of familiarity that convey a sense of continuity and conviviality to inhabitants. As such, one can argue that objects connect people to spaces and express larger meanings located within social, historical, and temporal structures. Objects that convey the sense of 'feeling at home' are important in the construction of a sense of belonging to significant people and places (Ahmed 1999; Ahmed et al. 2003). The significance of objects in the process of migration becomes clearer as mobility urges individuals to be 'selective' in what to carry, send, and take to a new place. This chapter deals with how objects make sense to individuals in the migration process and what the social and structural implications of certain objects are to people on the move.

Much of literature in migration studies problematises 'home' in relation to a longing and belonging to a 'homeland' by elaborating on ties and connections to the past life and the sense of identification to the 'country of origin' after displacement (Flynn 2007). By focusing on 'displaced objects' (objects that are sent or moved across transnational borders), one can direct attention to a different aspect of migrants' lives and that is materiality and everyday experiences of home-making in migration. Displacement in this chapter is taken as a social process and experiences of 'leaving' a home, a space, and a territory of familiarity and going towards a new geographical land with hopes for making a new home. Displacement is often used in the context of forced migration. In this chapter, I would argue that any form of migration needs to be read and located within the power relations that inform decisions to leave a place. Displacement is a considerably less-used term when discussing certain groups of migrants: middle-class migrants (Fathi 2017), international students, or middling migrants (Luthra and Platt 2016) who are navigating the regimes of mobility with substantial human and network capital. The author acknowledges the importance of differentiating between forced and voluntary migration (Bakewell 2011). However, the focus of this chapter is on the 'importance of objects' in giving a homing meaning to any form of mobility experience (at any point on the spectrum of voluntary to forced migration). As such, the approach taken in this chapter is that objects that are moved transnationally (or even internally), regardless of the regimes that control mobility, are displaced objects. In this chapter, the ways in which displaced stuff (objects carried across borders and/or locations) contributes to the formation of a migrants' construction of home will be outlined, drawing on specific examples of research conducted by the author in this area. This chapter conceptualises home in migration as an intersectional space that is constructed through moving and using objects and the ways in which certain meanings attributed to these objects are shaped, understood and reconstructed intersectionally after migration.

The chapter starts by discussing how home-objects contribute to the notion of migrant home in terms of the meanings they attach to the atmosphere of a home, the intimate space, or the first world, to which Bachelard (1994/1958) refers to as the *Poetics of Space*. Material contribution of stuff to home-making in migration needs to be seen through different layers. The main argument of the chapter is that the relationship between gender, nation, and emotions reconfirms social ties that exist in space and time, creating *a sense of continuity* in migration processes that connect here and there, rather than a separation of countries of origin and settlement. In the following section, I elaborate on my argument by reviewing the theories on materiality and objects in migrants' home-making practices before turning to the importance of (1) gender, (2) national identity, and (3) emotions in the role of displaced home-objects in homing experiences.

## Materiality of Migrants' Homes

The literature on home in migration has expanded rapidly in recent years (e.g. Boccagni 2014, 2017; Jacob and Smith 2008; Miller 2001; Pink 2004). This section specifically investigates the meanings objects take in migration processes in the process of home-making by focusing on the different aspects of material contribution objects have in constructing a sense of belonging to home in migration.

Homes in transnationalism are built in relation to multiple places and the countries of transit and settlement (Salih 2002). As discussed in the literature on transnationalism and diaspora studies, home in migrants' cases is linked to both roots and routes (Al-Ali and Koser 2002; Levitt and Waters 2015). In the former, the emphasis is placed upon an original homeland, but in the latter, focus is on the mobility and multiplicity of home in migration. Whilst the former is normally situated with emotions attached to a place 'there,' the latter connotes the importance of making home through networks, mobility, and transnational connections. Brah (1996) in her seminal work, *Cartographies of Diaspora*, argues that diaspora 'places the discourse of "home" and "dispersion" in creative tension, inscribing a homing desire while simultaneously critiquing discourses of fixed origins' (pp. 192–193). With 'homing desire,' Brah (1996) refers to a distinct desire to feel at home that is not the same as a desire for a homeland. It is an important distinction that Brah makes between homing desire and developing the idea of return to a homeland. A notable example of homing desire and actively reproducing a homeland in diaspora is evident in Ghorashi's (2004) research on Iranian diaspora in California and their organisational attempts to recreate Iran in Los Angeles: a nation within a nation. In this approach, where homing is practised through concerted efforts rather than a puritan attachment to a homeland, the meaning of home is intersected with multiple places, networks, organisations, institutions, people, and objects which add to evermore fluidity of the notion of home (see also Graham and Khosravi 1997). Indeed, some have argued that in the process of migration and dislocation, the notion of home becomes 'ambiguous' and 'uncertain' (Western 1992). The long-term disposition and displacement of material objects in the migration processes remains a topic that has been missing in transnationalism literature.

It is the material culture within our small domestic milieus that links us to the outside world (Tolia-Kelly 2004a). Material aspects of transnational lives manifest themselves in the use of everyday objects and are influential in shaping experiences of migration and home-making. For example, Boccagni's ethnography among Ecuadorian migrants in the north of Italy shows that active material construction of home 'here' and 'there' still persists. He argues that material home-making happens through remittances sent to Ecuador to invest in 'better' housing arrangements in Ecuador compared to their housing conditions in Italy to 'make them [houses in Italy] resemble "home"' (Boccagni 2014, p. 2). The meanings and intentions that are attached to components of home-making are fundamental to homing practices. For example, sending remittances is done in order to make a better house in Ecuador or objects are kept in migrants' houses in Italy in order to make them (Italian houses) look like Ecuadorian homes. Boccagni (2014) argues that these practices are reproduced through people's interactions from the same or different communities, their sense of self, and identifications. Different elements related to objects such as materiality, applicability, or uselessness, and emotional attachments (to name a few) are what locate objects within our individual and collective lives.

We build meanings around objects in terms of sense of nostalgia, familiarity, and closeness to other people, as well as negative reminiscences in the process of building relations and identifications with the outside world. Woodward (2001, p. 132) argues that 'the aspiration people have for their home, and their ideal ways of presenting and talking about their home and the objects inside it, are just as important as how they might actually live in their home.' Placing emphasis on narratives around objects, Woodward (2001) argues that narratives about objects are constructed texts that should be read within discourses around taste, the home and family, and everyday components that make people's everyday life.

What is common to most studies on home within disciplines of geography, sociology, gender studies, and history is the focus on multidimensionality of home and its reconstruction through the interrelationship of people, place, space, scale, identity, and power as Blunt and Dowling (2006) put it. Home could also be a space of belonging as well as alienation, intimacy and violence, desire and fear (Blunt and Varley 2004). These complex associated meanings and emotions show how home can be perceived (idealised) and experienced (practised) differently and how it is tied to the social relations that exist in that place and in relation to objects. Blunt and Dowling argue that 'home is much more than house or household' (2006, p. 3) as it encapsulates meanings that include 'socio-spatial relations and emotions.' However, home-making practices that are acknowledged as important in the construction of home are dependent on the use of objects. In what follows in this chapter, three widely discussed aspects of displaced home-objects have been chosen for review.

## Displaced and Gendered Home-Objects

Blunt and Dowling in their characterisation of a critical geography of home present three important attributes to home: home as material and imaginative; connections between home, power relations, and identity; and home as multi-scalar (2006, p. 6). Here in this section, I am focusing on the intersection of the first and the second to highlight how gendered relations within the household have been important to the understanding of home-objects and their transportation in migration and use afterwards. In other words, this section responds to the question: what role does gender play in taking, carrying, and using objects across borders?

Feminist scholars have been actively writing about 'home' for a long time (Ahmed 1999; Brah 1996; Salih 2003; Werbner 2000). One of the most important discussions in feminist approach is that home is both private and public space where gendered, classed, and racialised relationships are enacted and experienced. The reason feminist thinking has flourished in home studies is because gender is crucial to lived experiences of women (and men) and their home-making practices, that is, maintaining female and male identities through housework. Gendered relations in defining repressive attributed duties conventionally known as 'woman's duties' are rooted in the *distinction* between domestic (private) and social (public) binaries; these have been contested in different feminist movements due to the *justification* of such duties even when accompanied by violence, alienation, lack of control, and power (Friedan 1963). Other feminist works on home do not follow this path. For example, Rose (1993) points out that characterisation of home in humanistic geography has been masculinist. Young (1997) writes how gender has been fundamental in ownership, claims, reconstruction of the boundaries of the house, land, and titles. Whilst men have benefitted from the privilege of being involved in constructing a house and ownership, women have been seen as 'cultivating' that space and preserving it through domestic practices (Young 1997). The gendered understanding of home has permeated into the discourses around nation-building which will be discussed in the next section. The separation of spaces (male and female) and addition of objects (womanly and manly) to fill those spaces were a gendered division of social relations and material possessions: a point that is at the centre of my attention in discussing objects, their selection, and gendered roles in migration. Attachments to objects and the acquisition of certain practices are gendered and defined in relation to roles. This is particularly relevant in the section below where food practices in migration are discussed.

In *Poetics of Space* (1994/1958), Gatson Bachelard wrote that a home, even though it does have material properties and can be described to an extent in relation to its physical properties, is an orientation to the fundamental values that are gathered into one value: a myriad of intimate values of inside space. Home-making practices in transnational home are gendered, as Walter (2001) argues about migrant Irish women's involvement in home-making and use of domestic symbols. Walter argues that these symbols represent resettlement in Britain and the USA, what she describes as the process of 'feminising diaspora' that revolves around narratives of displacement and emplacement. Whilst displacement is linked to a homeland, emplacement is linked to the diasporic experiences and homing desire within the migratory experience. Similarly to Walter (2001), in her research with South Asian women, Tolia-Kelly (2004a, b) shows how material cultures as well as memories shape diasporic homes. By discussing the objects that represent pre-migration home (such as photos, or objects that travelled with families and particularly with women), landscapes from 'other' places are imported into British lives and shift the whole notion of British identity and lifestyle (2004a, pp. 676–691). In earlier research that I conducted with Iranian women migrants who work in the UK as doctors and dentists (Fathi 2015), their sense of belonging was comprised of a series of consumption practices and material use of 'stuff' that represented migrant women's understanding of social class that was different to their husbands or to other migrants. Their consumption habits before and after the migratory process were located within the classed and gendered frameworks of Iran and Britain (e.g. what decisions should be made in relation to clothing as a professional woman?). Gendered differences were also noted in research that was conducted on food and gender in migration where women showed stronger association with food from their countries of origin compared to men (Beoku-Betts 1995). What can be seen from these examples is that the meanings that objects take should be read within wider frameworks that form gendered (and classed, racialised) identities. Gendered practices in relation to home-making in diaspora or migration are specifically chosen and acted in order to convey 'cultural' meanings that go beyond the dynamics of a house as a material space. Gendered practices in diaspora help to revitalise those elements of homing that are located within wider concepts of nation, community, and customs which are perceived and/or understood to be lost in displacement. The link between gendered and national practices in making a home becomes clearer in the next section.

## Displaced Home-Objects and National Identity

A home can be explained beyond the boundaries of a material and structural entity that at times is seen as *separate from society*. Such a view towards home, and distinguishing it as domestic, is rather problematic as it isolates the space where social order and structures are practised and learned from society to where these practices are reiterated. Blunt and Dowling (2006) show how intimate and personal spaces are interconnected with wider power relations. They explore the critical geographies of home, nation and empire in three contexts: (1) 'imperial homes and home-making; [2)] homeland, nation and nationalist politics; and [3)] the politics of indigeneity, home and belonging' (Blunt and Dowling 2006, p. 142). Their argument in these three different contexts critically analyses the meaning of home in its material and imaginative ways, how it is tied up to the notion of nation and empire and what are considered as foreign, alien, and unhomely places. The meaning of home here is formed in relation to national boundaries and what is considered unhomely is if it falls outside such boundaries. Why does this matter in the context of displacement? Such characterisations of home, in relation to materiality, gender, race, and sexuality, all construct domestic home within empire as is also discussed in McClintock's work (1995). The movements of the colonising subjects and then the colonised to the empire encompass elements that were discussed above but within the context of displacement. McClintock particularly focuses on the role of gender and argues that women's role in building the empire was functional: not only did they bear the next generation of the builders of the [British] Empire, but also their contribution to building the domestic sphere was paramount to the image of the British Empire and reproduction of their masculinity and their power within colonial houses. As such, home was produced materially, in an imagined way, that was bound to the structural power relations (nation and empire-building) and to the power relations reproduced within the household (domesticity and gendered power imbalance). This links Dolores Hayden's (2002/1984) differentiation between home, gender, and nation in locating gender in home-making nation-building processes. A substantial part of nation-building in this sense comes from imaginations and visualisations of what a family looks like. She refers to the American context in the ways in which imaginations about suburban lifestyle are important in the construction of Americans' ideal life that is embodied in the house or neighbourhood. The second point she makes is about gendering the ideas and materiality of home. Through cleaning and maintaining the house, women's home-making practices have been constantly unrecognised. Suburban housing arrangements (separate dwellings with individual bathrooms and kitchens) multiplied caring duties for women and never offered privacy or haven to women.

The ways in which gender and nation are linked to objects have been discussed in some literature (see Tolia-Kelly 2004a). Displaced home-objects gain new significance within a new context that may place emphasis on their role in home-making and conveying a sense of belonging to a past life after migration. Savas (2014), in her research with Turkish migrants in Vienna, argues that a certain diasporic sphere is formed through Turkish migrants' collective taste in the use of objects that would give them a sense of collective identity, what she calls 'taste diaspora' that is an intersection of national and local preferences. Transportation and use of certain objects also render a collective identity to a migrant, for example, taking Persian carpets to a diasporic house as a usual practice among Iranian migrants. The act of transporting a carpet not only connotes the importance of home-making that constitutes a national identity to that household, but it also connects the homemaker to a wider discourses around belonging to a nation, class, and gender groups (handwoven carpets are treated as an essential part of the dowry that is given to brides in Iran), as well as the functionality of a carpet in a migrant's home.

Another well-researched example of the construction of national identity and belonging in relation to objects is through food production and consumption in migration and displacement (see also Murphy this volume).

Mata-Codesal (2008) argues that research into food production, consumption, and transportation among migrants evokes a new enquiry into sensorial re-creations of home. Sending and receiving food as a migrant material practice has received some attention in recent years (Jackson 2011; Kershen 2002; Povrzanović Frykman 2018; Salih 2003; Scholliers 2001; Walsh 2006). Povrzanović Frykman (2018) draws attention to food and food-related objects that help create 'continuity in transnational lives' (p. 41) by using methodological individualism, an approach that does not distinguish between old and new homes but concentrates on individuals and the journey they go through when they migrate. She further argues that by investigating an individual's motivations and practices, we enable ways of belonging in social fields that can be complemented with explorations made into ways of being (Povrzanović Frykman 2018, p. 41). As such national identities that are attached to certain objects gain a symbolic meaning in migration contexts which could be utilised as a way of making a home but also to assert one's social status and sense of belonging to a diasporic group, a nation, or a particular community. This leads us to the final part of the chapter on displacement and home-objects: the importance of emotions in characterisation of stuff in a migrant's home.

## Emotional Home-Objects in Displacement

Perhaps the most important way through which home-objects connect us to wider frameworks and identities is how we feel about them. Home, after all, is an emotional place (Massey 1992) and a product of how we feel about a dwelling, a place where certain emotional practices take place. The role of feelings towards a place characterised as home is an indicator in the theorisation of 'homely' and 'unhomely' places. For example, refugee camps, airports, or student hostels may seem 'unhomely' as these are places of transition and temporariness, but even unhomely places can become homely and sometimes homely places can render unhomeliness (Blunt and Dowling 2006).

What characterises objects and home settings as homely and unhomely has been discussed by some scholars who have written on geographies of home. Comfort lies at the heart of homeliness including furniture and design of the house. In the late nineteenth century and early twentieth century, technology was also intersecting with (here contributing to) the concept of comfort, according to Rybczynski (1986, cited in Blunt and Dowling 2006). Displacement is an unsettling experience. It can have an effect on making a place (un)homely depending on the context within which mobility is taking place. Social relations are vital elements of this context of mobility and displacement. In this regard, (un)homeliness highlights the role of women, and women's places have been important where women's contribution to home affairs, domestic labour, mothering, cooking, and caring has been the image that creates a space as a 'home.' Displaced objects can act as emotional anchors to link experiences of past, present, and future. Such emotional attachments intersect with gendered relations and displacement experiences. But there are scholars who highlight the role of objects in displacement contexts in linking domestic to public spheres. For example, Pels et al. (2002) argue that the practicality of objects and the ways in which they are used embody people's inclusion in certain locations and groupings. Objects in Pels et al.'s theorisation help migrants overcome segregation or exclusion from various social groupings. This approach is comparable to Bourdieu's notion of *habitus* (1984) in relation to consumption and the reproduction of taste that connects individuals to groups of people in terms of their bodily dispositions. However, Pels et al.'s argument is based on the performative aspect of objects or their 'everyday' usage. Thinking about objects in transnational lives, performativity (everyday usage) acts as a connector between the migrant and the wider network of people, locations, and meanings in migratory contexts. This connection is what they believe conveys homeliness to the person. Homeliness of objects should be regarded in relation to how wider sociopolitical and cultural systems attach meanings to objects (from here and there). Povrzanović Frykman (2012, cited in Povrzanović Frykman and Humbracht (2013), p. 62) proposes four hypotheses to relate objects and material practices to migrants' transnational lives in order to offer a theory of objects:

1. The issue of material continuity and perceived normalcy of everyday life in transnational context
2. The issue of presence in another location
3. The feeling of incorporation
4. The issue of social connections and obligations

From these hypotheses, it is evident that taking into account displaced objects in migration processes and transnational lives necessitates including a temporal, national, place-sensitive approach to understand the ways in which migrants live their lives with and through objects that they carry with them.

As argued above, displaced objects connect migrants to nations (e.g. the Persian carpet connecting an Iranian migrant to the homeland but also being seen as an object that has certain gendered and classed characteristics). The usage of certain commodities reconnects transnational links to populations at a distance, in different historical times and geographical locations. The usage of objects creates social ties that are reiterated on an everyday basis. In other words, homeliness here is translated into emotions that the object would convey as well as the functionality of a certain object in the migrant's daily life. This performative aspect of consumption is what makes certain objects homely and usually (for migrants or mobile people) 'worthy of transportation'. For example, Woodward (2001) argues that even the most banal objects have a function and can transmit the notion of home through their everyday use. The importance of banal objects was highlighted in the transportation of smartphones and laptops amongst migrants in recent studies (Dekker et al. 2018; see also Leurs and Patterson, this volume).

In an analysis of transportable objects across borders, Povrzanović Frykman and Humbracht (2013, p. 56) also discuss the importance of objects, particularly food, in relation to 'the sense of familiarity' they create when moved transnationally. They show, through narratives of migrants living in Sweden, that certain acts like 'liking, craving, and carrying' are done in different directions, for various reasons, and for the sake of different people who exist within the myriad spaces where migrants live (p. 55). Their contention is that through transnational mobility, migrants actively deploy the use of material (kitchen utensils, ingredients, furniture) that remind them of a different part of their life to the one they live on a daily basis in Sweden. They further argue that:

> the palpable "borders" of a migrant's transnational space are defined by the extent to which significant others and objects can become emplaced in different contexts. The fluency with which they are contextualised in different locations can determine the level to which migrants can experience that their everyday practices are "normal" and occur without interruption. (Povrzanović Frykman and Humbracht 2013, p. 57)

As can be seen from the above studies, objects are emotionally charged. Emotions here are formed in relation to other contexts, spaces, people, and familiar practices and tastes (such as a mocha coffee maker that would render a familiar taste of coffee to an Italian migrant in Povrzanović Frykman and Humbracht's study).

Research into home-objects' evocation of emotional attachments revolves around family photos and the ways in which they have been used as a useful window into understanding home before and after migration. As well as showing the relationships between members of the family, the dynamics of home, representations of space and materiality, Chambers (2003) presents the analysis of family photos as a powerful method to discover home. Gillian

Rose's (2004) study of family photos shows how home is made through photos and Tolia-Kelly (2004a) portrays the geographies of the home and its extension beyond the house and the place of residence in Britain (see also Pitt, this volume). In her research, Tolia-Kelly encouraged her participants to paint home (2006). The objects that were used in everyday activities of home-making were gendered, demonstrating relationships between members of the family which also highlighted power positions. For example, Gillian Rose's (2004) use of family photo albums enabled women participants to depict home as it would make sense from a gendered perspective that stretched the home beyond what was 'homeliness.' Their depictions and imagined usage were interpreted by the use of objects and how they play a role in people's everyday lives. Making sense of what objects are used for and how they are utilised is what makes objects meaningful for individual lives whilst linking these individual lives to collective groupings as well as to their immediate families and network of familiar people. Povrzanović Frykman and Humbracht (2013) argue that interpretation is an important tool as to when objects can be used to convey a sense of familiarity and belonging, a topic that has also been taken up by others (Noble 2012; Nowicka 2007). Through *continuity* of emotions and *ruptures* in emotional attachments to objects, one can bridge between the present and past (Povrzanović Frykman and Humbracht 2013) and between here and there (usually pre-migration life) to other locations.

## Objects, Being, and Belonging: A Conclusion

To conclude, selection, transportation, and usage of objects are used to negotiate belonging and identity in displacement. These material practices should be considered in an intersectional way to allow an analysis that renders meanings to migrants' lives, for example, to understand these practices in relation to gender and class. Furthermore, meanings of places that are associated with the concept of home (e.g. neighbourhood, place of birth, country) are intertwined with temporality, social relations, and structural and social frameworks and are important in how we make sense of objects. In other words, migration and displacement should be understood as a process. They are not a one-off separation from a home and a one-off resettlement. If we focus on the processes as practices, we can better understand the role of materialities and practices in home-making, displacement, and emplacement. In these processes, we actively recreate the meanings that we attach to objects in order to connect our worlds together (such as linking India to Britain in order to identify with both contexts and forge a new postcolonial identity). Moreover, displaced objects help in the construction of the sense of continuity with home via their capacity of bridging worlds before and after displacement/migration. If one takes the approach that objects, social relations, and identities are shared across time and space, then exploring other aspects of materiality can contribute to the field of displacement and migration studies. For example, the relationship between objects (transported and acquired in the new context) and their location within the class structures in migration needs further exploration. One may ask the questions:

- To what extent do migrants make sense of class through usage of certain products?
- How do such processes help their sense of belonging and homing practices?
- Can these structures change due to the reattribution and reclaiming of new class identities among migrants and the appropriation of the objects in a different way?

This chapter showed how material practices (acquisition, transportation, and usage) of objects ranging from furniture to food can help in understanding *continuity* as well as *ruptures* in migrants' experiences of displacement. By reviewing various studies on intersection of space, gender, emotions and objects, I highlighted that material practices and consumption of particular objects help migrants to distinguish homely and unhomely spaces which in effect offers different meanings of home in relation to gender, race, class, national identity and displacement.

# References

Ahmed, S. (1999). Home and away: Narratives of migration and estrangement. *International Journal of Cultural Studies, 2*(3), 329–347.

Ahmed, S., Castaneda, C., Fortier, A., & Sheller, M. (2003). Introduction: Uprootings/regroundings: Questions of home and migration. In S. Ahmed, C. Castaneda, A. Fortier, & M. Sheller (Eds.), *Uprootings/reground of home and migration* (pp. 1–19). Oxford, England: Berghahn Books.

Al-Ali, N., & Koser, K. (2002). *New approaches to migration? Transnational communities and the transformation of home*. London: Routledge.

Bachelard, G. (1994/1958). *The poetics of space* (Trans. M. Jolas). Boston, MA: Beacon Press.

Bakewell, O. (2011). Conceptualising displacement and migration: Processes, conditions, and categories. In K. Koser & S. Martin (Eds.), *The migration–displacement nexus: Patterns, processes and policies* (pp. 14–28). Oxford, England: Berghahn Books.

Beoku-Betts, J. A. (1995). We got our way of co-preservation of cultural identity among the Gullah. *Gender and Society, 9*(5), 535–555.

Blunt, A., & Dowling, R. M. (2006). *Home*. London and New York, NY: Routledge.

Blunt, A., & Varley, A. (2004). Introduction: Geographies of home. *Cultural Geographies, 11*, 3–6.

Boccagni, P. (2014). What is in a (migrant) house? Changing domestic spaces, the negotiation of belonging and home-making in Ecuadorian migration. *Housing, Theory and Society, 31*(3), 277–293.

Boccagni, P. (2017). *Migration and the search for home: Mapping domestic space in migrants' everyday lives*. New York, NY: Palgrave Macmillan.

Bourdieu, P. (1984). *Distinction: A critique of the social judgement of taste* (Trans. R. Nice). London: Routledge and Kegan Paul.

Brah, A. (1996). *Cartographies of diaspora*. London: Routledge.

Chambers, D. (2003). Family as place: Family photograph albums and the domestication of public and private space. In J. M. Schawartz & J. R. Ryan (Eds.), *Picturing place: Photography and the geographical imagination* (pp. 96–114). London: IB Tauris.

Dekker, R., Engbersen, G., Klaver, J., & Vonk, H. (2018). Smart refugees: How Syrian asylum migrants use social media information in migration decision-making. *Social Media and Society, 4*(1), 1–11.

Fathi, M. (2015). I make here my soil. I make here my country. *Political Psychology, 36*(2), 151–164.

Fathi, M. (2017). *Intersectionality, class and migration: Narratives of Iranian women migrants in the UK*. New York, NY: Palgrave Macmillan.

Flynn, M. (2007). Reconstructing 'home/lands' in the Russian Federation: Migrant-centred perspectives of displacement and resettlement. *Journal of Ethnic and Migration Studies, 33*(3), 461–481.

Friedan, B. (1963). *The Feminine Mystique*. New York: W. W. Norton and Co.

Ghorashi, H. (2004). How dual is transnational identity? A debate on dual positioning diaspora organizations. *Culture and Organization, 10*(4), 329–340.

Graham, M., & Khosravi, S. (1997). Home is where you make it: Repatriation and diaspora culture among Iranians in Sweden. *Journal of Refugee Studies, 10*(2), 115–133.

Hayden, D. (2002/1984). *Redesigning the American dream: The future of housing, work and family life*. New York, NY: W. W. Norton.

Jackson, P. (2011). Families and food: Beyond the 'cultural turn'? *Social Geography, 6*(1), 63–71.

Jacob, J., & Smith, S. (2008). Living room: Rematerialising home. *Environment and Planning A, 40*, 515–519.

Kershen, A. J. (2002). *Food in the migrant experience*. Farnham, England: Ashgate.

Luthra, R., & Platt, L. (2016). Elite or middling? International students and migrant diversification. *Ethnicities, 16*(2), 316–344.

Massey, D. (1992). A place called home. *New Formations, 17*, 3–15.

Mata-Codesal, D. (2008). *Rice and coriander. Sensorial re-creations of home through food: Ecuadorians in a northern Spanish city* (Sussex Centre for Migration Research Working Paper No. 50). Brighton, England: University of Sussex.

McClintock, A. (1995). *Imperial leather: Race, gender and sexuality in the colonial contest.* New York, NY: Routledge.

Mellin-Olsen, T., & Wandel, M. (2005). 'Changes in food habits among Pakistani immigrant women in Oslo', Norway. *Ethnicity & Health, 10*(4), 311–339.

Miller, D. (2001). Introduction. In D. Miller (Ed.), *Home possessions: Material culture behind closed doors.* Oxford, England: Berghahn Books.

Noble, G. (2012). Home objects. In S. Smith (Ed.), *International encyclopedia of housing and home* (pp. 44–438). London: Elsevier.

Nowicka, M. (2007). Mobile locations: Construction of home in a group of mobile transnational professionals. *Global Networks, 7*(1), 69–86.

Pels, D., Hetherington, K., & Vandenberghe, F. (2002). The status of the object: Performances, medications, and techniques. *Theory, Culture and Society, 19*(5/6), 1–21.

Pink, S. (2004). *Home truths: Gender, domestic objects and everyday life.* New York, NY: Berg.

Povrzanović Frykman, M. (2018). Food as a matter of being: Experiential continuity in transnational lives. In D. Mata-Codesal & M. Abranches (Eds.), *Food parcels in international migration. Anthropology, change, and development* (pp. 25–46). Cham, Switzerland: Palgrave Macmillan.

Povrzanović Frykman, M., & Humbracht, M. (2013). Making palpable connections: Objects in migrants' transnational lives. *Ethnologia Scandinavica, 43*, 47–67.

Rose, G. (1993). *Feminism and Geography: The limits of geographical knowledge.* Cambridge: Polity Press.

Rose, G. (2004). 'Everybody's cuddled up and it just looks really nice': An emotional geography of some mums and their family photos. *Social and Cultural Geography, 5*(4), 549–564.

Salih, R. (2002). Shifting meaning of 'home': Consumption and identity in Moroccan women's transnational practices between Italy and Morocco. In A. Al-Ali & K. Koser (Eds.), *New approaches to migration? Transnational communities and the transformation of home* (pp. 51–67). London: Routledge.

Salih, R. (2003). *Gender in transnationalism: Home, longing and belonging among Moroccan migrant women.* New York, NY: Routledge.

Savas, Ö. (2014). Taste diaspora: The aesthetic and material practice of belonging. *Journal of Material Culture, 19*(2), 185–208.

Scholliers, P. (2001). Meals, food narratives, and sentiments of belonging in past and present. In P. Scholliers (Ed.), *Food, drink and identity. Cooking, eating and drinking in Europe since the Middle Ages* (pp. 3–22). Oxford, England: Berg.

Tolia-Kelly, D. (2004a). Materializing post-colonial geographies: Examining the textural landscapes of migration in the South Asian home. *Geoforum, 35*(6), 675–688.

Tolia-Kelly, D. (2004b). Locating processes of identification: Studying the precipitates of re-memory through artefacts in the British Asian home. *Transactions, 29*(3), 314–329.

Tolia-Kelly, D. (2006). Mobility/stability: British Asian cultures of 'landscape and Englishness'. *Environment and Planning A, 38*(2), 341–358.

Van der Horst, H. (2010). Dwellings in transnational lives: A biographical perspective on 'Turkish-Dutch' houses in Turkey. *Journal of Ethnic and Migration Studies, 36*(7), 1175–1192.

Walter, B. (2001). *Outsiders inside: Whiteness, place and Irish women*. London: Routledge.

Walsh, K. (2006). British expatriate belongings: Mobile homes and transnational homing. *Home Cultures, 3*, 123–144.

Werbner, P. (2000). Introduction: The materiality of diaspora—Between aesthetic and 'real' politics. *Diaspora: A Journal of Transnational Studies, 9*(1), 5–20.

Western, J. (1992). *A passage to England: Barbadian Londoners speak of home*. London: University College London Press.

Woodward, I. (2001). Domestic objects and the taste epiphany: A resource for consumption methodology. *Journal of Material Culture, 6*(2), 115–136.

Young, I. M. (1997). House and home: Feminist variations on a theme. In I. M. Young (Ed.), *Intersecting voices: Dilemmas of gender, political philosophy, and policy* (pp. 134–164). Princeton, NJ: Princeton University Press.

# 43

# The Role of Design in Displacement: Moving Beyond Quick-Fix Solutions in Rebuilding Housing After Disaster

Esther Charlesworth and John Fien

## Displacement and Shelter

This chapter focuses on the antithesis of displacement—the resettlement of communities after the events of natural and/or human-induced disasters. Specifically, we focus on how the design of temporary and long-term housing in towns destroyed by disaster can assist in building long-term community recovery and resilience. Through two case studies of housing recovery after natural disaster events in Sri Lanka and Haiti, we will argue the case for integrating architecture and wider design thinking into housing and settlement infrastructure programmes for communities displaced by disasters.

*Why does displacement matter?* The number of people now displaced around the world today now totals almost 100 million. This exceeds the total number of people displaced since World War II and, if counted as a nation, would represent the world's fifteenth largest country (World Population Review 2019). The intensity and protracted nature of conflict and natural disasters have doubled the number of displaced persons in the past two decades. In 2017, there were 30.6 million new displacements across 143 countries worldwide. This is 80,000 people a day or, on average, one person every second every day. Compounding the impacts of such heavy rates of displacement is the fact that most of the world's protracted conflicts and natural disasters,

E. Charlesworth (✉) • J. Fien
School of Architecture and Design, RMIT University, Melbourne, VIC, Australia
e-mail: esther.charlesworth@rmit.edu.au; john.fien@rmit.edu.au

from earthquakes in Mexico and Indonesia to devastating droughts and conflicts in sub-Saharan Africa and typhoons in Southeast Asia and the Pacific, tend to occur in the most ecologically and economically vulnerable parts of the world (Fig. 43.1). This is a reflection of the human dimensions of environmentally related disasters and the increasing interplay of conflict, climate change, and displacement in complex emergencies (see Parsons, this volume).

*Why does design matter?* Facilitating the provision of culturally appropriate and technically efficient housing is one of the most important tasks for those working with displaced people. It provides the normalcy required to begin addressing health and education needs and rebuild livelihoods, saving both money and lives (Davis and Alexander 2015). Thus, part and parcel of a successful housing programme is the provision of appropriate settlement and civic infrastructure. However, as Sanderson (2018) argues, 'for humanitarian organizations, governments and others providing the right support to achieve

**Fig. 43.1** Cyclone Aila washed away the southern-most island of Bangladesh and people became refugees. This picture showed how the cyclone smashed people's houses there, 2009—description by the photographer Mayeenul Islam. (https://commons.wikimedia.org/wiki/File:Cyclone_Aila_Climate_Change_Nijhum_Dwip_2009_Dec_Bangladesh.jpg—used under Creative Commons Attribution-Share Alike 3.0 Unported License https://creativecommons.org/licenses/by-sa/3.0/legalcode)

this is anything but simple' despite the billions of dollars annually spent by donors and governments (p. 2). Thus, Sanderson asks us to consider questions related to:

> What kind of shelter best meets the needs of these ... people? How long is it meant to be used for? Where should it be located? What are the materials, and who will build it? And there are more questions: what do we do when there is no land to build on (say in a dense city), or when people need shelter for years or decades (as in the case of refugees and other forced displaced people), or there is not enough money, or no political will? (2018, p. 2)

Thus, restoring hope, dignity, and security after disaster is not always about the lack of resources—although this remains a significant problem—but rather their allocation in ways that ensure resilience and sustainability. This involves seeking to maximise the multiple 'capitals' of sustainable human development: human, social, economic, environmental, and built. Reaching towards these integrated goals through effective shelter (housing) and settlement projects (infrastructure, such as hospitals, schools, and roads) is the hallmark by which successful recovery and reconstruction can be assessed. This chapter addresses the intersecting issues of displacement, design, and disaster through the examination of two housing case studies in Sri Lanka and in Haiti. Both case studies demonstrate the critical role that shelter and settlement projects play in economic and social recovery after disaster and how successful housing solutions can only be developed through the close involvement of the people and communities originally displaced by disasters.

## Shelter as Critical Intervention

> Following almost every disaster and crisis, shelter is regarded as a critical, life-saving need alongside, for instance, health and protection. But the problems, scope, practice and benefits of providing humanitarian shelter and settlement assistance are still, for those outside the sector, not well understood. This lack of understanding results in poorly integrated responses at best, and at worst in a significant gap in meeting an essential need of affected people, denying them a decent, healthy and suitable place to live. (Serdaroglu and Moore 2018, p. xi)

Since the seminal work of Davis (1978) on shelter after disaster, the literature speaks of 'housing as a process.' This is because site-specific cultural and environmental factors must be negotiated in shelter and settlement design, unlike many health programmes where, for example, a universal solution such as a vaccine for cholera is often possible. However, there is a practice–theory gap between the participatory, community-led processes needed for long-term recovery and the product-delivery culture that characterises all but a small number of shelter and settlement programmes (Fan 2013; Charlesworth and Ahmed 2015). In many shelter programmes responding to displacement, from Haiti to Syria, the field is mostly characterised by a 'one size fits all' approach, with insufficient attention paid to the aspirations of the people most affected and the infrastructure needed. For example, a review of housing projects in Sri Lanka after the Indian Ocean tsunami found that most shelter programmes not only failed to improve the well-being of affected communities, but they also failed to provide the basic infrastructure of water, sewerage, and electricity and failed to ensure that new settlements were built in appropriate locations (Charlesworth and Ahmed 2015).

At least four factors in the policy and operating conditions of shelter and settlement programmes are commonly seen as responsible for such results:

1. *Responding to crises is an immediate need and shelter agencies are better prepared for this task than they are for longer-term development.* For example, international disaster and refugee agencies have guidelines for meeting emergency shelter needs, but the extended time periods and costs of recovery mean that these are not appropriate to addressing long-term recovery. Similarly, national disaster agencies are chiefly responsible for pre-disaster risk reduction and emergency management and relief, and are not often equipped for long-term recovery and reconstruction (Olshansky et al. 2012).
2. *Mono-sectoral approaches dominate the humanitarian and disaster fields.* Disciplinary specialists, for example, in protection, gender, shelter, health, and education, tend to operate in isolation and have evaluation criteria related to each specialty rather than to integrated or collective outcomes. For example, tools for evaluating housing projects typically address the technical issues of numbers housed and construction materials, rather than the psychosocial and livelihood benefits that could be delivered through a housing programme (Nath et al. 2017). There is also a lack of understanding of the significance of shelter among many specialist humanitarian sectors.
3. *Many shelter and settlement programmes have failed to adequately engage with the displaced people for whom they are intended.* As part of 'housing as a process', the UN Disaster Relief Office argued that 'the key to success ultimately lies in the participation of the local community—the survivors—in

reconstruction' (UNDRO 1982). Principles for community engagement and their integration in pre-disaster risk reduction are well known (Awotona 2017), but it is very difficult for people who have been traumatised by grief from the loss of loved ones and their homes and livelihoods to engage in public participation processes in meaningful ways, especially in emergency situations where finding missing family members, medical attention, and fresh water is more of an immediate priority.

4. *The dynamics and tensions in the international humanitarian and development sectors often mean that bureaucratic and self-interested barriers get in the way of timely, comprehensive, and lasting actions to settle people displaced by conflict and disasters.* This problem has been described in a report from the UK Overseas Development Institute in terms of the sector being 'crippled by high levels of organizational insecurity, competitive pressures and financial uncertainty, as UN agencies and NGOs compete to raise money and secure donor contracts' (Overseas Development Institute 2016). Additionally, the international assistance required to meet global shelter requirements for displaced populations annually generally falls far short of the need, with Youssef (2015) identifying a financial deficit of 50 per cent annually.

The international humanitarian community has sought to address this problem, at least in part, through the Sphere Project (2011) which produced a Humanitarian Charter and a suite of Minimum Standards for Humanitarian Response, including standards for shelter after emergencies. A Global Shelter Cluster has also been formed, led by the United Nations High Commission for Refugees (UNHCR) (for conflict-displaced persons) and the International Federation of Red Cross and Red Crescent Societies (IFRC) (for disaster-displaced persons), to promote networking and coordination after an emergency. These have been assessed as working quite well, especially in the relief phases after displacement (e.g. see McCluskey 2017). However, such integrated and coordinated support rarely lasts into the recovery and reconstruction period after conflict and disasters. Indeed, the humanitarian sector was not designed for such long-term work and, often, humanitarian agencies and their staff have moved on to the 'next' crisis before planning for long-term recovery can even commence. The converse is sometimes true also: when the scale of needed emergency assistance is so great, funds previously earmarked for long-term recovery and development can be diverted to humanitarian assistance.

Thus, it is necessary to find ways of bridging the gap between emergency humanitarian aid and long-term development aid to help people survive disasters and get back on the path to self-reliance and dignity (Gabaudan 2012). A key challenge in this—and one which requires systems thinking to understand the many permutations of relationships across sectors and as conditions change over long periods of recovery—is to find ways of laying the foundations for settlement planning for future infrastructure and livelihood needs *within* the provision of shorter-term emergency relief and shelter services.

Significantly, this range of root causes of unsuccessful shelter and resettlement programmes seem to be *systemic* rather than related to individual crisis events (Ray-Bennett 2018, Chap. 2). The causes are multiple, complex, and compounding, so much so, in fact, that the seemingly simpler task of mitigating the impacts, rather than the root causes, of displacement unfortunately often becomes the main focus of planned solutions. For this reason, as well as those of speed and economy, the result is often a 'one-size-fits-all' approach, inspired by an ethos of 'universal solutions,' much like the use of cholera vaccinations in health programmes after disaster. However, in this universal approach to resettlement and rebuilding, insufficient attention can be paid to the aspirations of the people most affected and the infrastructure they specifically need, including the use of local technologies, materials, and labour and the implications of particular sociopolitical and site vulnerabilities. As a result, some have described the problematic response to displacement as the result of 'planned failure' rather than 'a failure of planning' (Cutter and Gall 2006, p. 1).

## The Role of Design in Recovery from Displacement

> It's also crucial that "humanitarian" practice not be seen as uniformly associated with conditions of emergency. The slow task of improving cities, settlements, institutions and infrastructure should also be at the core of the work of architects seeking to redress both sudden and long-term issues of inequality and scarcity. (Sorkin 2014, p. iii)

The challenge of providing housing and settlement services for displaced people has all the hallmarks of a 'wicked problem' (Rittel and Webber 1973). This requires complex design thinking rather than the oft narrowly focused, linear solutions just described (Vandenbroeck et al. 2016). Indeed, many housing and resettlement challenges in addressing displacement are strongly related to design issues (Keenan 2018). However, the skills of the key design thinkers in this area, such as architects, are seldom employed. While there is a growing number of studies on, for example, the design of IT systems for disaster management (see Hawryszkiewycz 2016), there is a critical lack of research on the role of design in providing housing and infrastructure solutions to displacement (Lee 2013).

Developing design solutions at housing and settlement scales is the core competency in architecture. Architecture is an ongoing, iterative process of creating and reforming two and three-dimensional space. It requires an understanding of the interdependence of human aspirations, the natural environment, and the social, political, and cultural systems in which designs are to be constructed. Thus, architecture provides 'both a mindset and a methodology' (Vandenbroeck 2012, p. 33) that can help to address many of the problems undermining resettlement and recovery programmes. This requires not only the contextual understanding and creative problem-solving of the designer but also an understanding of the relationships and processes that link all the elements/components, drivers, feedback loops, etc. of any problem or location into a holistic system (Pourdehnad et al. 2018). This is a process known as systemic design which integrates systems analysis with design thinking.

However, this expertise is not often applied in recovery and reconstruction and the number of architects equipped to respond in such situations is very low. As Brett Moore (Chief, Shelter Division, UNHCR) noted after Typhoon Yolanda in the Philippines: 'Despite the enormity of the disaster, it is almost impossible to get trained architects or planners for the complex task ahead of rebuilding the shelters and settlements' (in Charlesworth 2014). In similar vein, Saunders (2015) identified that 'not all major international non-governmental organisations which implement large-scale housing and reconstruction activities have in-house technical expertise with a background in the built environment. Indeed, several agencies that are household names lack even a single in-house technical advisor to inform agency decision-making' about shelter (p. x). The reason for this, he argues, is that the shelter sector is poorly understood in the humanitarian and development worlds. The corollary of this is that the profession of architecture has only recently begun engaging 'with the idea of a specialist humanitarian form of practice and has yet to integrate it within professional training or the main curriculum of architectural education' (Brogden 2019, p. 4; see also Charlesworth 2014).

This chapter thus seeks to amplify the case for integrating architecture and wider design thinking into housing and settlement infrastructure programmes for people displaced by disasters. This will be done via an analysis of two case studies of housing recovery after natural disaster events (preceded by long-term political conflict) in Sri Lanka and Haiti. A growing range of 'disaster architecture' examples could have been chosen, such as those detailed in

*Beyond Shelter: Architecture for Crisis* (Aquilino 2011) or *Humanitarian Architecture* (Charlesworth 2014) or even more recent prefabricated temporary housing solutions such as IKEA's 'Better Shelter' structure which has received much press. However, the aim of this chapter is to look at the critical relationship between the need for coordinated 'shelter' and 'settlement' projects, rather than just one-off universal housing solutions.

The two case studies presented here are based upon fieldwork conducted by one of the current authors and her colleague Iftekar Ahmed for *Sustainable Housing Reconstruction: Designing Resilient Housing After Disaster* (Charlesworth and Ahmed 2015). Individually and together, they offer insights into how sensitive design approaches to the rebuilding of housing and surrounding settlements, post disaster, can significantly improve the long-term prospects for a community's' economic and social recovery. The aim here is to understand why certain recovery projects worked and why others were less successful and thus offer a guide to future reconstruction of housing projects after disaster.

## Case Study 1. The Indian Ocean Tsunami: Reconstruction in Seenigama, Sri Lanka[1]

Second only to Indonesia, Sri Lanka suffered the most damaging impacts of the December 2004 tsunami. A deep-sea earthquake off Sumatra caused the tsunami but it catalysed waves that impacted landfalls as far away as East Africa. The Sri Lankan government estimates that 35,322 people died, a further 516,150 lost their homes, 65,275 homes were totally destroyed, and another 38,561 were partially destroyed. The tsunami waves impacted nearly 65 per cent of Sri Lanka's coastline; the local impacts varied considerably, depending on the presence or otherwise of natural features that could mitigate the impacts. This was the case of Seenigama in southern Sri Lanka, near Galle, where nearshore reefs had been partially destroyed by coral diving (and extraction) and which allowed the full force of the tsunami waves to rush onshore, overturning a train causing 3000 deaths and destroying most buildings.

The reconstruction of Seenigama was led by a local NGO, the Foundation of Goodness (FoG), which was established in 1999 by Kushil Gunasekara, the son of a local land-owning family. After a career in business and sports administration at national and international levels, he returned to the family estates and established the FoG to improve the social and economic well-being of local people. It developed social enterprise models for providing community services, such as English and computer training, sporting opportunities, a maternity clinic, and connecting the water and electricity supply of village houses to national service providers. The foundation was thus well situated to facilitate an integrated recovery programme after the tsunami due to its 'insider' status and the high levels of trust and social capital it had established in the district and through Gunasekara's international networks and management skills, as well as those developed within the community through previous training programmes. This is a relatively rare situation as most post-disaster reconstruction projects, not only in Sri Lanka, have implementing agencies that are often external to the country and/or based away from the project site.

FoG was thus well positioned to act as a conduit for international corporate donations and amalgamate and channel them into a coordinated programme aimed at rebuilding the devastated homes, civic infrastructure, and livelihoods in Seenigama. With support from a number of national and international donors, 625 new houses were built and 401 repaired. An example of such houses in the KPMG village is depicted in Fig. 43.2.

Among the most innovative aspects of the Seenigama programme was the decision to build four re-settlement 'villages' that integrated housing with supporting community infrastructure and services. Each one was funded by different donors and designed and built by different contractors, primarily local. They were of varying sizes and designs, with the first three built ahead of the fourth, thereby allowing lessons from the early designs and building processes to be integrated into the final village. The four villages are:

- *AVIVA Village*—comprising ten single-storey detached houses built around a community centre/meeting hall, computer and English training centre, library, maternity clinic, and playground (funded by insurance company AVIVA (UK) and WNS Customer Solutions (Sri Lanka)).
- *Perth Village*—a cluster of nine single-storey detached houses next to AVIVA Village, thus allowing residents to access the community facilities in AVIVA Village (funded by Perth City Council, Western Australia).
- *KPMG/LOLC Village*—comprising 50 single-storey detached houses set out around a community centre, water supply tower, library, and playground (funded 50 per cent each by KPMG (UK) and LOLC (Sri Lanka)).
- *Victoria Gardens*—a planned settlement of 84 two-storey duplexes set out around a central community area that included a community centre, water purification plant, sewage treatment plant, and playground. This project was funded by the State Government of Victoria, Australia, hence the name.

**Fig. 43.2** KPMG Housing Village, Seenigama, 2013. (Photo Credit: Foundation of Goodness, 2019)

These villages were constructed on newly acquired land for rehousing tsunami-affected households that had lost their coastal properties and were restricted from rebuilding there because of the government's post-tsunami 'buffer zone' policy which prevented resettlement near the coast. Interviews with residents revealed a continuing fear of future tsunamis and a preference for two-storey houses. Several house design options were offered to potential residents of Victoria Gardens and, after protracted discussions, the design selected consisted of two-storey, two-bedroom houses in a duplex arrangement. These were structurally much stronger than other new and repaired houses due to the adoption of a concrete slab and wall construction system, while the upper floors were above the height of the tsunami. However, it should be noted that the consultation was not as effective as it should have been as post-occupancy interviews reveal that the treated water is too expensive for many families and families prefer to cook in outside structures rather than use the Western-style indoor kitchens that were included in the plans.

The FoG stressed the need for rebuilding and repair work to provide employment opportunities for local people and the development of further training and employment initiatives. For example, Gunasekera's international sporting contacts enabled a cricket academy and a swimming institute to be constructed, while Nokia's offer to provide mobile telephone support was renegotiated to become funding for the design, construction, and operation of a training centre for commercial divers. The diving centre was a significant resilience initiative, not just for the families who benefit from the employment that the high-level training has opened up but also for obviating the need for local divers to mine offshore reefs for cement making. Protecting the remaining reefs may minimise the effects of onshore waves driven by typhoons and any future tsunamis.

Being embedded in the local community and long-term engagement with the community and donors were key to the success of reconstruction in Seenigama. This resulted in a successful owner-driven, rather than donor-driven, approach that was underpinned by the principles of systemic design. Indeed, many aspects of the FoG process for project planning and delivery integrated the twin dimensions of systemic design—systems thinking and design thinking. These are seen, for example, in the integration of social, environmental, and livelihood concerns in Seenigama's reconstruction projects, thus obtaining the mutually reinforcing benefits of recognising the indivisibility of social, environmental, and economic systems. They are also displayed in the emphasis on land use planning and the integration of residential dwellings and community infrastructure in the layout of the villages. Systemic design also resulted in an integrated community development approach, which, unlike many reconstruction projects, recognised that building houses alone was not sufficient. Thus, a key lesson of the Seenigama case is that housing has to be contextualised and supported within a settlement infrastructure programme that provides both physical (roads, water, electricity, sanitation, etc.) and social (education, livelihoods, sports, etc.).

## Case Study 2: Villa Rosa, Port-au-Prince, Haiti[2]

On 12 January 2010, an earthquake (magnitude 7.5 on the Richter scale) struck Haiti's West Province, less than 20 kilometres from the capital, Port-au-Prince. The impact was devastating, resulting in one of the worst humanitarian crises ever, with over 200,000 people killed and the homes of 1.5 million destroyed or severely damaged. A total of 105,000 buildings were destroyed and more than 208,000 severely damaged, including hospitals, medical centres, and schools as well as the presidential palace, parliamentary buildings, courthouses, and many government offices. A series of damaging aftershocks, while the displaced were living in the open air, only made the problems of destruction and displacement worse. Damage to buildings and infrastructure was later estimated to exceed $7.8 billion with the housing sector suffering the most with total damage estimated at $2.3 billion (World Bank 2016).

Eventually, 1280 camps would be established to cater for these internally displaced people. The design and construction of the shelters for the camps, transitional housing, and eventual permanent homes display some of the worst and some of the best examples of shelter and settlement responses to natural disaster. An understanding of the reasons for the scale of the damage and loss of life cannot be explained by earthquake forces alone—and is necessary for appreciating the need for systemic design rather than piecemeal approaches to recovery.

Disaster management experts explain the impact of a disaster as a result of the vulnerability of a region, with vulnerability being a function of the relationships across three variables: *exposure* to risk (i.e. the earthquake), the *sensitivity* to risk of a location and its communities (e.g. the quality of infrastructure, the strength of buildings and building codes, and the degrees of political stability and economic security, etc.), and its *adaptive capacity* in responding to risk. Adaptive capacity concerns a community's ability to prepare for, and adjust or adapt to, the impacts of exposure and sensitivity. This includes activities such as information, education, and capacity building programmes; reliable warning systems; support processes for vulnerable groups such as female-headed households, the elderly, and the disabled; promoting resilient building structures; and blocking new construction in risk-prone areas.

Unfortunately, a history of repressive colonialism followed by a century of outside interference (for exploitative economic reasons) in internal political processes meant inadequate preparation of the Haitian people for independence and self-sufficiency. Business and government corruption weakened civil society and undermined the quality of the public service and its ability to provide health, education, and welfare services. As a result, at the time of the 2010 earthquake, Haiti was the poorest country in the Americas and had the highest degree of economic inequality. While 5 per cent of the population owned 75 per cent of the country's arable land and more of its wealth, 75 per cent of the population live on less than $2 a day. Haiti's 'vulnerability' to disasters was created over time, with these political, social, and economic disasters increasing risk sensitivity and reducing the capacity to adapt to risk, thus exacerbating the 'un-natural disaster' of 2010.

At the time of the earthquake, Haiti lacked widespread public healthcare and was reliant on international food donations. Most houses and buildings were poorly built due to lack of education and skills and lax enforcement of construction codes, as weak as these were. As a result, for example, many buildings were constructed with low amounts of cement in the concrete mix and inadequate steel reinforcings. The earthquake not only destroyed and damaged homes but also 1300 schools and 50 health centres, as well as roads, bridges, water supply, energy systems, and telephone and IT services. All undermined local relief efforts, while the great damage to public buildings in the capital, Port-au-Prince, significantly reduced the capacity of national authorities to organise comprehensive and timely relief services, such as medical care, water and food security, shelter, public safety, and the beginning of rubble clearing to facilitate transport movement and future rebuilding. Dependence on international humanitarian assistance was well-nigh total.

How do national and international agencies and donors plan for and implement a recovery and reconstruction programme in such a situation? The Global Shelter Cluster—and its coordination clusters including Protection and WASH (Water, Sanitation and Hygiene)—is a critical Inter-agency Standing Committee (IASC) coordination mechanism that supports people affected by natural disasters and internally displaced people affected by conflict with the means to live in safe, dignified, and appropriate shelter.

However, there is a significant variation in capacities and approaches of the different clusters and weak inter-cluster coordination. Worse, many small organisations from Europe and the USA, especially, rushed to provide assistance but worked outside of the collaborative cluster system. Together, these factors contributed to a fragmentation in disaster response which was only intensified by the lack of a government platform to lead recovery coordination and planning. Also, the clusters are humanitarian mechanisms and have no mandate to move from emergency relief to recovery and reconstruction.

This was a particular problem in the shelter sector where recovery was compounded by both the scale of the shelter crisis and the many agencies acting outside of any agreed philosophical or operational approach. Lessons learnt from previous shelter emergencies indicate that addressing issues of land tenure and rights is vital; that a one-step move from transitional to permanent homes is needed after emergency shelter is no longer viable; that local materials, construction systems, and labour should be utilised; that local control of decision-making is a necessity; and that shelter responses need to be integrated with, and lay the foundation for, health, education, civil infrastructure and services, and the development of employment and livelihood opportunities. As a result, many houses were built on land for people without secure tenure; consultation was poorly executed and largely ineffective with agency-driven approaches overwhelming owner-driven ones; there was a lack of coordination of housing with wider infrastructure, service, and employment projects; and many inadequate shelter 'innovations' were constructed. The latter included many imported, pre-built structures that were climatically and culturally inappropriate, reflecting the uninformed design aspirations of people wanting to help but unwilling or unable to work with local people and shelter agencies.

However, there were also a wide range of cases where the shelter response was exemplary. Some of these were led by Caritas International and its national agencies, for example, Catholic Relief Services (CRS) (based in the USA) and Cordaid (based in the Netherlands). Unlike FoG, in Seenigama, which was a local community organisation, these are large international NGOs. However, they have extensive experience in shelter projects in many different countries and seem to be able to adapt their responses to local conditions.

The case study described here was led by Cordaid and its partners and concerns the reconstruction of Villa Rosa, an informal settlement of approximately 10,500 people, located near central Port-au-Prince. Growing spontaneously for more than 30 years before the earthquake, Santa Rosa was built on steep land without terraces, retaining walls or other methods of site stabilisation. Despite this, the most common construction method consisted of precariously built, three- to four-storey houses on small, narrow blocks, some as small as small as 10 m2. Poor-quality concrete block work on unstable, sloping sites increased the vulnerability of these, as did the lack of drainage, sanitation, and waste management services. So, when the earthquake hit in 2010, 60 per cent of Villa Rosa was devastated. Out of 1335 houses, 595 were completely destroyed and 260 were severely damaged and casualties were high.

Among the main implementing agencies, Cordaid played the lead role by providing funding (20 million Euros), coordinating the various local and agency stakeholders in an owner-driven approach to reconstruction, and selecting the beneficiaries in partnership with a local organisation, Casek. As an informal settlement before the earthquake, few householders held any form of legal tenure to the sites of their homes. Thus, Cordaid also worked with Casek to secure the tenure rights necessary to enable reconstruction. Cordaid also worked with Architecture for Humanity and Build Change to operationalise the works. Architecture for Humanity facilitated the physical planning and construction of community facilities through action planning, which provided a high degree of control to local people. Architecture for Humanity also prepared a catalogue of housing typologies and a construction manual for consultation and training purposes. The actual repair and reconstruction was led by Build Change, a US-based organisation specialising in earthquake-resistant design and construction. Build Change developed house designs through community consultations, raised awareness of earthquake-resistant construction among homeowners, trained local builders, and supervised construction. Beyond this project, Build Change also worked with the Ministry of Public Works, Transport and Communications (MTPTC) to raise wider public understanding of the need for earthquake-resistant design and construction and to enrich the skills of engineers.

The overall project design was based upon a community action plan developed with the people and community groups in Villa Rosa to guide the physical, social, and economic recovery. The three key elements of the plan were owner-driven reconstruction, community action planning, and an integrated one-step approach in moving from camps to resettlement. Together, these resulted in a community development approach aimed at boosting social, economic, and environmental resilience. This resulted in the following achievements, identified by Charlesworth and Ahmed (2015):

- *Resilient housing:* Instead of looking for a new 'green-field' site, away from local employment, family, and friendship networks, Villa Rosa was rebuilt in situ, allowing people to continue living where they always had. Altogether, the partners built 120 new houses and repaired 320 more, for a total of 440 houses—all reflecting earthquake-and hurricane-resistant construction and in the style of typical Haitian houses, which helped make them attractive to households. Taking account of the typically small plots and growing population of Villa Rosa, two-storey houses were built on the very small blocks while provision was made to add an extra storey to one-storey houses at a later time. Figure 43.3 shows how in the rebuilt houses in Villa Rosa the foundations, walls, and rooves were designed to be solid through steel reinforcement, bracings, and straps, while columns and beams in the frame were reinforced.
- *Community infrastructure:* Community infrastructure and utilities were provided in the form of paved walkways, underground drainage, solar-powered streetlights, landscaping of some public areas, and the provision of some sporting facilities.
- *WASH:* The International Organization for Migration (IOM) provided WASH training and services, including public and household toilets, new wells, and repairs to hand pumps. Provision of solid waste management services was also improved.

**Fig. 43.3** Rebuilt houses in Villa Rosa. (Photo credit: Koolhaus and Urhahn, 2016. https://favelapainting.com/HAITI-PAINTING-2015-2016)

- *Cash-for-work:* Canadian Hunger Foundation (CHF) ran a cash-for-work programme for rubble removal and storage, not only providing employment and income but also clearing the area of rubble so reconstruction work could begin.

The relative success of Villa Rosa's reconstruction reflects an owner-driven participatory approach to rebuilding not only the physical structure of the town but also social and economic structures. Reconstruction was also supported through a strong emphasis on capacity building, particularly in action planning, collaboration, and project management as well as skills in hazard-resistant construction. As a director of Build Change noted:

Build Change is a multi-faceted agency focusing on earthquake-resistant housing. Our approach is to help build permanent housing together with strong local capacity building.

In the reconstruction process here in Haiti, we have trained hundreds of local builders and construction workers, as well as homeowners. We expect the capacity to build disaster resilient buildings to remain with the community. Now we are reaching out through the government to train engineers in Haiti and to raise awareness of people about ways to build safer houses. (Interview by Iftekar Ahmed, 23 May 2013. In Charlesworth and Ahmed (2015), p. 54)

However, notwithstanding the involvement of the local group, Casek, the Haitian government was unable to be involved in the project and, despite the focus on consultation through action planning, the project was still led, financed, and implemented largely by external agencies despite efforts to build technical capacity to ensure project sustainability. Nevertheless, the Villa Rosa project offers a number of key lessons in delivering housing projects after displacement. Charlesworth and Ahmed (2015) describe them this way:

Despite the inherent challenge of coordinating a range of actors, this project demonstrates that a multi-stakeholder and multi-disciplinary partnership between a funding agency, Cordaid, and a technical agency, Build Change, with urban planning support from Architecture from Humanity was fruitful in addressing the complexity of the problem. Bringing on board a range of other international partners, IOM and CHF, and the local government authority, Casek, maximised the project's potential for success. (p. 66)

The Villa Rosa project in many ways then sums up the key factors needed in any successful housing reconstruction project, multi-stakeholder and multidisciplinary partnerships led by and for the beneficiary community.

## Conclusion: Moving Beyond Quick-Fix Solutions in Rebuilding Housing After Disaster

The two case studies discussed in this chapter—implemented in different geographic contexts by a range of agencies—demonstrate that a coordinated design and planning approach is required for the long-term recovery of communities displaced by disasters. The Seenigama and Villa Rosa projects demonstrate diverse approaches to building community resilience through housing when responding to environmental hazards such as bushfires, cyclones, earthquakes, and tsunamis. More than resilient building products, success here has resulted from linking 'physical products' to a wider set of social and institutional processes. While technically resilient design and construction methods have often been used in recovery after disasters, implementing them on the ground is often extremely challenging. It is the special effort towards addressing these challenges that sets these projects apart.

Thus, it is possible to conclude with a review of the valuable lessons for effective housing reconstruction beyond quick-fix solutions of, for example, importing prefabricated housing from a foreign country. While some of these lessons were context-specific, many had wider significance. These include:

- *Owner-driven approach:* The Seenigama and Villa Rosa projects demonstrated the value of the owner-driven approach—leading to high levels of satisfaction with the resettlement process. Owner-driven projects do have some drawbacks, such as longer implementation times and difficulty of quality control, but these can be overcome through the use of local designs, materials, and construction systems, with which local people are already skilled, as well as technical support and training.
- *Integration of housing with community infrastructure:* A key message that runs through the Seenigama and Villa Rosa projects is that post-disaster housing reconstruction must not be seen as only building new houses to replace those damaged, but rather an integrated approach where a range of other elements are provided, particularly community infrastructure and facilities such as roads, water, sanitation, electricity, schools, community buildings, and parks.
- *Support for livelihoods and the local economy:* The opportunity to earn a living, that is, a livelihood, is essential for those affected by disaster. Throughout discussions with disaster-affected communities during the original fieldwork for the case studies, the regeneration of livelihoods was emphasised as being as great a need as housing; in many of these communities, the house is also a workplace for home-based livelihoods. Most of the implementing agencies supported livelihoods as part of their housing reconstruction initiatives, through mechanisms such as skills training, provision of equipment, the necessary infrastructure for a livelihood, and start-up supplies or through cash-for-work. In addition, the reconstruction projects supported local economies through the creation of jobs and marketing opportunities for a range of local building product suppliers and producers.
- *Sustained engagement:* The challenges of replicability, upscaling, and long-term support can be addressed by implementing organisations, such as FoG in Sri Lanka and Cordaid in Haiti, adopting community action planning approaches, and continuing to work in the reconstructed community, enriching and diversifying local capabilities and expanding programmes into wider sustainable and resilient community development.

- *Multi-stakeholder engagement:* The complexity of post-disaster housing reconstruction is reflected in the involvement of a wide range of stakeholders and professionals in both case studies. This is a paradigm that can be expected to grow in significance over the future as disasters become more complex and the global forces of climate change and urbanisation continue to create unprecedented challenges.
- *Role of design and planning:* Within the multi-stakeholder engagement paradigm, design and planning professionals played a significant leadership role because the effective design and construction of housing and community infrastructure is central to post-disaster reconstruction. The facilitators of the Seenigama and Villa Rosa projects may not have used the label systemic design to describe their approaches. The term is relatively new and, indeed, it is the analysis of projects, such as the case studies in this chapter, which has helped to identify the attributes of systems thinking and design thinking, and their integration into a comprehensive approach to design, that has contributed to the theory and practice of systemic design.

It is hoped that this chapter can serve as a reflexive guide to agencies and academics working on the displacement of communities after protracted civil conflict as well as disasters. In a world beset by an increasing spectre of 'unnatural disasters', understanding the social context of disaster-affected communities, strengthening professional capacity, and integrating housing with a range of culturally appropriate and technically efficient design features, the reconstruction of housing can serve as a vehicle to protect people and property over the long term and help establish resilient and sustainable communities.

## Notes

1. Detailed accounts of the impacts of the Indian Ocean Tsunami in Sri Lanka, the Foundation of Goodness, and reconstruction in Seenigama may be found in Charlesworth and Ahmed (2015); Karan and Subbiah (2011), Chaps. 8 and 11; and Mulligan and Nadarajah (2012).
2. Detailed accounts of the impacts of the 2010 earthquake in Haiti, Cordaid, and reconstruction in Villa Rosa may be found in Charlesworth and Ahmed (2015), Build Change (2014), Bornstein et al. (2013), and L'Etang (2012).

# References

Aquilino, M. (2011). *Beyond shelter: Architecture and human dignity*. New York, NY: Metropolis Books.

Awotona, A. (2017). *Planning for community-based disaster resilience worldwide*. London: Routledge.

Bornstein, L., Lizarralde, G., Gould, K., & Davidson, K. (2013). Framing responses to post-earthquake Haiti: How representations of disasters, reconstruction and human settlements shape resilience. *International Journal of Disaster Resilience in the Built Environment, 4*(1), 43–57.

Brogden, L. (2019). Sustainability, design futuring and the process of shelter and settlement. In A. Asgary (Ed.), *Resettlement challenges for displaced populations and refugees* (pp. 1–16). Cham, Switzerland: Springer.

Build Change. (2014). *Homeowner-driven housing reconstruction and retrofitting in Haiti: Lessons learned, 4 years after the earthquake*. Retrieved from https://www.preventionweb.net/go/36290.

Charlesworth, E. (2014). *Humanitarian architecture: 15 stories of architects working after disasters*. London: Routledge.

Charlesworth, E., & Ahmed, A. (2015). *Sustainable housing reconstruction: Designing resilient housing after disaster*. London: Routledge.

Cutter, S., & Gall, M. (2006) *Hurricane Katrina*. Retrieved from http://citeseerx.ist.psu.edu/viewdoc/download?doi=10.1.1.565.3352&rep=rep1&type=pdf.

Davis, I. (1978). *Shelter after disaster*. Oxford, England: Oxford Polytechnic Press.

Davis, I., & Alexander, D. (2015). *Recovery from disaster*. London: Routledge.

Fan, L. (2013). *Disaster as opportunity? Building back better in Aceh, Myanmar and Haiti*. London: Overseas Development Institute.

Gabaudan, M. (2012, January 19) From emergency aid to development aid: Agencies are failing to connect. *The Guardian*. Retrieved from https://www.theguardian.com/global-development/povertymatters/2012/jan/19/humanitarian-aid-development-assistance-connect.

Hawryszkiewycz, I. (2016). Design in complex environments: Application in emergency management. In *Australasian Conference on Information Systems 2016, Wollongong*. Retrieved from http://ro.uow.edu.au/cgi/viewcontent.cgi?article=1009&context=acis2016.

Karan, P., & Subbiah, S. (Eds.). (2011). *The Indian Ocean Tsunami: The global response to a natural disaster*. Lexington, KY: University of Kentucky Press.

Keenan, J. (2018). Seeking an interoperability of disaster resilience and transformative adaptation in humanitarian design. *International Journal of Disaster Resilience in the Built Environment, 9*(2), 145–152.

Koolhaus, J., & Urhahn, D. (2016) Haiti painting 2015-2016. Retrieved from https://www.jeroenkoolhaas.com/following/jeroenkoolhaas.com/HAITI-PAINTING-2015-2016.

L'Etang, G. (2012, May 2). *Haiti after the earthquake: Camps, shanty towns and housing shortages* (Trans. O. Waine). *Metropolitiques*. Retrieved from https://www.metropolitiques.eu/Haiti-after-the-earthquake-camps.html.

Lee, A. (2013). Casting an architectural lens on disaster reconstruction. *Disaster Prevention and Management, 22*(5), 480–490.

McCluskey, J. (2017). *Evaluation of the Global Shelter Cluster Strategy (2013–2017)*. Retrieved from http://www.sheltercluster.org/global-strategic-advisory-group/documents/gsc-strategy-2013-2017-evaluation-report.

Mulligan, M., & Nadarajah, Y. (2012). *Rebuilding local communities in the wake of disaster: Social recovery in Sri Lanka and India*. New Delhi, India: Routledge India.

Nath, R., Shannon, H., Kabali, C., & Oremus, M. (2017). Investigating the key indicators for evaluating post-disaster shelter. *Disasters, 41*(3), 606–627.

Olshansky, R., Hopkins, L., & Johnson, L. (2012). Disaster and recovery: Processes compressed in time. *Natural Hazards Review, 13*(3), 173–178.

Overseas Development Institute (ODI). (2016). Time to let go: Remaking humanitarian action for the modern era. ODI. Retrieved from https://www.odi.org/sites/odi.org.uk/files/resource-documents/10422.pdf.

Pourdehnad, J., Wexler, E. R., & Wilson, D. V. (2018). Integrating systems thinking and design thinking. *The Systems Thinker, 22*. Retrieved from https://thesystemsthinker.com/tag/volume-22/.

Ray-Bennett, N. (2018). *Avoidable deaths: A systems failure approach to disaster risk management*. Cham, Switzerland: Springer Briefs.

Rittel, H., & Webber, M. (1973). Dilemmas in a general theory of planning. *Policy Sciences, 4*(2), 155–169.

Sanderson, D. (2018). Beyond the better shed. In Global Shelter Cluster (2018), *The state of humanitarian shelter and settlements 2018: Beyond the better shed—Prioritizing people* (pp. 2–7). Retrieved from https://www.sheltercluster.org/sites/default/files/The%20State%20of%20Humanitarian%20Shelter%20and%20Settlements%202018.pdf.

Saunders, G. (2015). Preface. In E. Charlesworth & A. Ahmed (Eds.), *Sustainable housing reconstruction: Designing resilient housing after disaster*. London: Routledge.

Serdaroglu, E., & Moore, B. (2018). Foreword. In Global Shelter Cluster (2018) *The state of humanitarian shelter and settlements 2018: Beyond the better shed—Prioritizing people* (pp. x–xi). Retrieved from https://www.sheltercluster.org/sites/default/files/The%20State%20of%20Humanitarian%20Shelter%20and%20Settlements%202018.pdf.

Sorkin, M. (2014). *Humanitarian architecture: 15 stories of architects working after disasters*. London: Routledge.

Sphere Project. (2011). *The sphere handbook 2011*. Rugby, England: Practical Action Publishing.

UNDRO (United Nations Disaster Relief Organization). (1982). *Shelter after disaster: Guidelines for assistance*. New York, NY: UNDRO.

Vandenbroeck, P. (2012). *Working with wicked problems*. Brussels, Belgium: King Baudouin Foundation.

Vandenbroeck, P., Van Ael, K., Thoelen, A., & Bertels, P. (2016, October 13–15). *Codifying systemic design: A toolkit*. In Relating Systems Thinking and Design Symposium (RSD), Toronto, Canada. Retrieved from http://openresearch.ocadu.ca/id/eprint/1918/.

World Bank. (2016). *What did we learn? The shelter response and housing recovery in the first two years after the 2010 Haiti earthquake*. Washington, DC: World Bank.

World Population Review. (2019). Total population by country 2019. Retrieved from http://worldpopulationreview.com/countries/.

Youssef, H. (2015). *Ambassador Hesham Youssef: 6 propositions for a paradigm shift to change the system* [Web log post]. Retrieved from http://blog.worldhumanitariansummit.org/entries/ambassador-hesham-youssef-6-propositions-for-a-paradigm-shift-to-change-the-system.

# Part VII

## Section Seven: Representing Displacement

# 44

## Intervention: Activism, Research and Film-Making—Fighting for the Right to Housing in Bucharest, Romania

### Michele Lancione

Nicoleta is one of 100 people evicted from their homes in a street called Vulturilor, near Bucharest's city centre, on 15 September 2014. For two years, through harsh winters and tropical summers, half of these families dwelt on the street in front of their old homes, in tents and self-made shacks. They stayed to fight against their unlawful displacement, enacting the longest and most visible action-protest for housing rights in the history of contemporary Romania.

Evictions like the one that affected Nicoleta and her community are not uncommon in Romania and elsewhere in the neoliberalised world. In Eastern Europe, they are closely interrelated to the post-1989 privatisation of assets (including housing) previously nationalised under former 'communist' regimes. In 1995, Romania gave tenants the opportunity to buy the publicly owned houses in which they were living. Only the richest could afford to purchase their dwellings, however, and a number of poorer tenants, including a large majority of Roma people, continued to pay rent to the state. Things changed again in 2001, when the government, pressured by the EU, passed the so-called legea retrocedarilor ('restitution law'), which allowed the former owners of a building that had been nationalised by the socialist regime in the 1950s to request the return of their property. The restitution takes place *in natură*, meaning that there is no financial compensation to the public purse

---

M. Lancione (✉)
Urban Institute, University of Sheffield, Sheffield, UK
e-mail: m.lancione@sheffield.ac.uk

© The Author(s) 2020
P. Adey et al. (eds.), *The Handbook of Displacement*,
https://doi.org/10.1007/978-3-030-47178-1_44

despite a full transfer of property rights to private hands. In other words, this is a way to transform social houses currently owned by the state—often located in prime urban locations—into tools for the private accumulation of capital. Such a privatisation process is strategically deployed by the local government under pressure from the EU, in a collusion of trans-local neoliberal interest on urban real estate. Many poor tenants, like the people of Vulturilor, are caught in these state-led and market-driven mechanisms of privatisation: once the house is 'restituted' to its 'original' owner, they cannot afford to pay rent at market costs and are evicted. Considering that only 1 per cent of housing stock in Romania is now socially owned, the choices for tenants are very limited: either living with relatives in overcrowded apartments or ending up in homeless shelters (Fig. 44.1).

Hundreds of recent cases, like that of Nicoleta and her family, occur within a complex web comprising historical racism against the Roma, paradoxical laws, corruption, neoliberalism, disempowerment, and exogenous neoliberal forces. In the case of the Vulturilor community, the only offers made by local authorities to the evictees consisted of six months of financial help to rent on the private market or relocation to shelters for homeless people (where families had to be separated: women and children in one place, men in another). The community of Vulturilor refused these 'options,' deeming them unacceptable for people who were entitled to social housing by Romanian law, and decided to dwell on the street to protest against their forced displacement. Their fight for their right to housing came as a surprise to many, including

**Fig. 44.1** (Source: *All pictures in the text are from the author*)

long-term activists and experienced social workers, as the general understanding was that in Romania, Roma people simply do not protest with such vigour and visibility: they are historically seen as marginalised, depoliticised, second-class citizens. The situation in Vulturilor, however, played out differently: the people decided to occupy the pavement in front of their old homes, creating self-built shacks in which to live and protest, making their voices and political subjectivity heard (or 'apparent,' as Judith Butler conceived this in her own work on bodily politics) (Fig. 44.2).[1]

Inspired by their strength, my work with this community over the past four years has endeavoured to narrate the full history of their harassment and displacement. It has also intervened to support the forms of resistance that they have brought to the fore locally, to raise awareness, strengthen multiple forms of solidarity, and fight back. These incremental interventions are part of a collective effort by a grassroots group called the Common Front for the Right to Housing (FCDL), of which I am a member. Through FCDL, we aim to sustain communities that are facing eviction or fighting for their right to housing in Romania and beyond. My personal involvement comprised joining solidarity actions in the Vulturilor camp; supporting the production of the community's online blog, which narrates the story of their resistance and everyday struggle for housing (http://www.jurnaldinvulturilor50.org); producing a full-length video documentary around forced evictions in Bucharest (http://www.ainceputploaia.com); and ongoing projects with a number of comrades on related issues (some of which are connected to the European Action

**Fig. 44.2** (Source: *All pictures in the text are from the author*)

Coalition for the Right to Housing and the City). Thanks to the Antipode Foundation, a number of us are now (2018–2019) working with Nicoleta to transform the community's online journal into a printed book—part diary, part guide to housing resistance—in order to inspire other communities to fight against eviction and housing injustice across Europe (Fig. 44.3).

The documentary, *A început ploaia/It Started Raining*, is the first to document the ongoing harassment of poor Roma communities in Bucharest and the role of historical marginalisation and the restitution law in their displacement. The 72-minute film follows the struggle of the Vulturilor community contextualising it through a number of interviews with activists, scholars, and politicians. What emerges is a picture not only of racial discrimination, homelessness, and eviction but also of positive grassroots practices of resistance and social change. The aim of the project is threefold. Firstly, it supports this particular struggle for the right to the city in Bucharest, making it visible, giving it voice, and situating it critically within a particular political and social landscape according to their protagonists' aims. Secondly, the film makes a methodological and stylistic contribution, mixing academic fieldwork, activist-oriented work, and visual methods to create space for empowering experiences and meaningful exchange. In particular, it deploys an activist and participatory approach to film-making and editing, so that the final artwork is guided by feedback obtained from the community and members of FCDL. Thirdly, and perhaps most importantly, the film is intended to intervene in debates and action on housing resistance. Dozens of screenings have

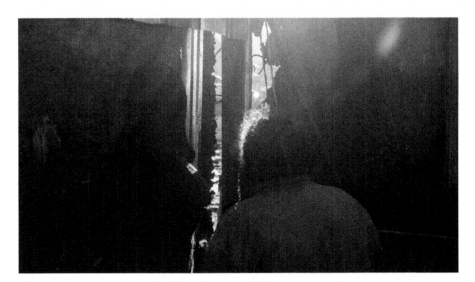

**Fig. 44.3** (Source: *All pictures in the text are from the author*)

taken place with communities that have been evicted or are facing eviction across Romania and elsewhere in Europe, including Hungary, Croatia, the Czech Republic, Sweden, Italy, Portugal, Germany, and beyond (Fig. 44.4).

Like the aforementioned book that we are creating from the community blog, the film is designed to be used as a tool for radical education and action. Anyone can see it, download it, and use it to organise meetings. It prompts discussion of eviction, housing rights, and modes of grassroots organising and

**Fig. 44.4** (Source: *All pictures in the text are from the author*)

is particularly useful in stimulating debate and action for the historically marginalised communities of Roma across Europe (access it at http://www.ainceputploaia.com). These kinds of projects, engagements, and collective productions complement my academic writings on marginality, diversity, and resistance. They open up possibilities for collective engagement that are not only fragile and difficult to maintain but also powerful and rewarding. I am thankful to Nicoleta, her community, and my comrades in FCDL for nurturing this sense of collectivity, action, and radical intellectual praxis.

*Author's selected writing on restitution, eviction and resistance:*

Lancione, M. (2017). Revitalising the uncanny: Challenging inertia in the struggle against forced evictions. *Environment and Planning D: Society and Space, 35*(6), 1012–1032.

Lancione, M. (2018). The politics of embodied urban precarity: Roma people and the fight for housing in Bucharest, Romania. *Geoforum.* https://doi.org/10.1016/j.geoforum.2018.09.008.

Lancione, M. (2019). Radical housing: On the politics of dwelling as difference. *International Journal of Housing Policy.* https://doi.org/10.1080/19491247.2019.1611121.

Lancione, M. (2019). Caring for the endurance of a collective struggle. *Dialogues in Human Geography.* https://doi.org/10.1177/2043820619850362.

## Note

1. Butler, J. (2011). Bodies in alliance and the politics of the street. *Transversal*, 1–15. Retrieved from https://transversal.at/transversal/1011/butler/en.

# 45

# How Not to Eat Human Stories: Ruts, Complicities, and Methods in Visual Representations of Refugees

Dominika Blachnicka-Ciacek

## Introduction

A surge of refugees arriving on European shores, which peaked in the summer of 2015, provoked a moral panic, as well as a renewed interest and sense of urgency in the politics and representation of displacement.[1] What the media referred to as the 'refugee crisis' of 2015 became a pan-European media spectacle that unfolded live on European screens for several months. For many in Europe, it was the first time that they learned about the long history of deaths in the Mediterranean Sea and the ways in which the EU has been a fortress preventing those without the 'right' documents from entering. Visual images have set the mood around the refugees and informed the ways in which European publics have imagined and perceived those arriving in Europe. How the media presented the 'refugee crisis' mattered not just for those consuming the images but, first and foremost, for the lives and livelihoods of those who were seeking asylum in Europe. Images from the infamous Hungarian Keleti train station, or rather the public outcry in response to those images, were said to inform Chancellor Merkel's decision to accept more refugees in Germany and her famous '*wir schaffen das* [we will manage]' speech (2015). The release

---

D. Blachnicka-Ciacek (✉)
SWPS University of Social Sciences and Humanities, Goldsmiths, University of London, London, UK
e-mail: dominika.blachnicka@gmail.com

© The Author(s) 2020
P. Adey et al. (eds.), *The Handbook of Displacement*,
https://doi.org/10.1007/978-3-030-47178-1_45

of the photographs of Alan Kurdi's body washed up on a Turkish beach in September 2015 caused a public outcry across the world, compelling a number of Western governments to adjust their policies. The UK and Canada announced modest resettlement schemes for Syrians, and Germany accelerated its 'open border' policy (Devichand 2016). Several months later the mood began to shift. The Paris bombings and the immediate suspicion that the attacks had been carried out by refugees changed the dynamics towards those arriving in Europe. Compassion and hospitality were overshadowed by anxiety and unease. It is, of course, not only the mainstream media that has strived to represent the plight of the asylum seekers. An abundance of artists and scholars have tried to challenge and unpack mainstream representations of the 'refugee crisis.' But even those with good intentions sometimes fall into the same traps, misrepresenting or abusing those they claim to speak for.

In this chapter, I explore the production of photographic and cinematic images of the displaced, focusing on the ways in which representations of refugees are constructed, canonised, and appropriated. I use two films, *Glimpse* (2016) by Artur Żmijewski and *Another News Story* (2017) by Orban Wallace, which have been created in response to the recent 'migration crisis' to explore the power relations involved in the production and consumption of images of displacement and the problematic nature of the 'gaze' and spectatorship. I close this chapter with a consideration of the academic use of audiovisual methods when representing displacement. Is it possible to engage with these issues in a way that does not turn into appropriation or pornography of the suffering?

## 'Going for the Refugees': Audiovisual Representations of Displacement

In the introduction to *After the Last Sky*, an essay by Edward Said with photographs by Jean Mohr, Said asserts that Palestinians are either associated with an image of a keffiyeh-wearing terrorist or that of a helpless refugee. He further comments that despite the abundance of textual and visual representations of Palestinians, their everyday lives and struggles under occupation remain virtually unknown to wider audiences (Said 1985, p. 9). In Said's view, the abundance of photographic and textual representation does not give insight into the experience of Palestinians in exile. Instead it obscures their situation.

Victor Burgin argues that images are decoded instantaneously, without the viewer being aware that the picture represents a 'point of view' and is implicated in reproduction of ideology (1982, p. 146). Terence Wright highlights that while visual representations of refugees play a crucial role in how the general public imagines them, there is little awareness of the politics of representation (2002, p. 53). Heather L. Johnson, talking about visual representations of refugees, reminds us that 'to construct a representation is an act of power' (2011, p. 1017). She says that 'how we imagine particular categories of people determines how we engage with them, who we accept as political actors and who is able to participate in the world' (Johnson 2011, p. 1017). These perceptions in turn inform the political climate and decisions about the treatment of the refugees.

The methods of representation follow certain canons. Nando Sigona reminds us that while the refugees' trajectories are diverse and represent multiple histories, contexts, and situations, this plurality does not translate into media, artistic, and academic practices (2014, p. 370). Whoever speaks about Palestinians, argues Said, never begins from scratch, but from the existing context of what has been said, written, and produced about Palestinians (1985, p. 4). Similar points can be made about the aesthetics of refugee representation in general in that they are entangled in and informed by cultural, historical, and political factors. As Malkki has famously argued, 'most of us have a strong visual sense of what "a refugee" looks like,' even though we might have never come across a refugee in our lives (1989, p. 9). Those representations follow certain tropes, and some argue that they come down to two main strategies of essentialisation (Johnson 2011; Wilmott 2017).

One of the most common audiovisual strategies for representing the theme of displacement is to present refugees as helpless victims (Malkki 1995; Johnson 2011; Sigona 2014). These depictions emphasise the refugees' passivity and powerlessness. The refugees are captured in exclusively humanitarian terms—as people in need without their own initiative, whose lives and well-being depend on others. Many critics recognise that these types of representations strip displaced people of their autonomy and political agency (Edkins and Pin-Fat 2005; Nyers 2005; Johnson 2011). Pupavac argues that, unlike political solidarity, which assumes a relationship between equals, the relationship with refuges is constructed on the foundation of their victimhood. It is therefore an asymmetrical relationship based on dependency (2008, p. 280). Peter Nyers asserts that the concept of a refugee is created through a series of 'ontological omissions' and observes that 'whatever is present to the political subject (i.e. citizen) is absent to the refugees' (2005, p. 3). In his view, while the citizen-subjects have access to 'visibility, agency, and rational speech,' the representations of refugees cast them as invisible, speechless, and, above all, non-political (Nyers 2005, p. 3).

There are number of aesthetic practices through which the refugee's state of victimhood is communicated. Wilmott (2017) explains how gender is one of the most important vehicles of victimhood. One semantic tactic used is the concept of 'women and children,' which becomes 'womenandchildren,' through which refugee women are relegated to the same category as children (2017, p. 1032). They are both depicted as vulnerable, helpless, and unable to decide for themselves. Wright (2002) demonstrates how the feminised representations of refugees derive from Christian imagery; they conform to the Christian ideal of passive, suffering, and yet surviving motherhood.

Nando Sigona reminds us that it is not just the media but also Western humanitarian organisations that resort to depicting displaced people as vulnerable in order to trigger a compassionate response (2014, p. 372). Some argue that the 'victimisation' mode of depicting refugees is especially important for NGOs and their campaigns to raise money for aid and advocacy activities (Wright 2002; Rajaram 2002; Sigona 2014). Blomfield and Lenette (2018) argue that in recent times the images of the displaced have increasingly been used by humanitarian agencies to evoke public sympathy and immediate action. They speak of 'humanitarian porn' indicating that the more traumatic the image, the more support they can get. Analysing the history of the UNHRC's engagement with refugees, Erin Baines explores how by the 1980s the bodies of a 'Third World' mother and child had become the emblematic image of the refugee in the UNHCR communication (2004, p. 36).

What happens when real life does not conform to these idealised depictions of feminised refugees? One piece of information which circulated in the media during the 'refugee crisis' was that most of the people crossing the Mediterranean Sea into Europe were men. This catalysed the suspicion of some right-wing politicians, tabloids, and Internet commentators about the authenticity of their claims to refugee status. They asked, 'Why are they not fighting in their homeland, but instead fleeing their countries, leaving their women and children behind?' (Rhodan 2015). On the reverse side of the victimisation strategy, there is a tendency to present refugees as a security threat (Bleiker et al. 2013; Wilmott 2017). To achieve this, different aesthetic techniques are used to trigger a sense of anxiety and fear rather than feelings of compassion and sorrow. Refugees are often presented in a crowd, as a sea of people or, as Rajaram evocatively writes, as a 'mute and faceless physical mass' (2002, p. 247). In these depictions, refugees are often presented as a racialised 'other.' Long-distance shots are used with no eye contact. Such depictions strip displaced people of their individuality and humanity. Wilmott (2017,

p. 67) sees such representations as a 'visual commutative act,' which aims to establish refugees as a threat. Once this is achieved, refugees can become subject to exceptional measures. If they are 'securitised,' then leaders get public acceptance to use extraordinary measures to combat the threat (Wilmott 2017, p. 68). 'Whether an audience sees issues such as immigration and terrorism as manageable—which can be dealt with through the usual process of law enforcement … or as security threats greatly depends on media framing' (Wilmott 2017, p. 68). Even 'do-gooders,' such as NGOs and aid groups, unintentionally contribute to these types of representations.

How our understanding of the refugee crisis is mediated and framed by the presence of cameras is the subject of Orban Wallace's film *Another News Story* (2017). Wallace's film tells the story of how the 'refugee crisis' of 2015 was produced as news. Rather than following the refugees, the film follows the caravans of international journalists as they travel across Europe following the refugees. The film gives insight into the ways in which the news about refugees is constructed and distilled. The boredom and mundane reality of everyday waiting and walking is replaced by a sense of action, drama, and urgency. The film is especially powerful in its aim to shed light on the ambiguous role of the journalists. On the one hand, some of the journalists claim to be on a mission to tell the 'truth' to the viewer. At the same time, they are often cynical and their good intentions are mixed with the need to 'get' the story quicker than the others. The desire to get the best angle often collides with what is happening on the ground. One of the most disturbing moments of the film takes place on Lesbos Island where journalists wait for refugee boats to arrive. When the boats appear, the crews jump into the water. But this sense of urgency is sparked not by the urge to help the refugees, but to film the drama. Another difficult moment happens at one of the train stations when crowds of people try to board a packed train to Austria. Here, right in the middle of what looks like extreme situation in which people are crammed together, children are screaming, and people are fighting to breathe, we see journalists throwing out questions: How do you feel? They are shooting a perfect picture of victimhood. As one of the journalists says, 'We are eating human stories' (45:09). Sometimes, Wallace's focus provokes irritation among his fellow journalists. 'If I was you I'd go for refugees not the journalists' (0:42), one journalist tells him. The film also portrays the decline of the interest in the refugee subject. The refugee crisis might be still there, but the media have moved on and the refugees are no longer the 'breaking news.'

## 'Eating Human Stories': On the Politics of Looking

And what about us, the onlookers? What is relationship between the public and images of suffering? In the complex matrix of representation, what is the role of the audience? How can we qualify this act of looking at the images of the displaced? Frank Moller positions the spectator as a 'participant witness' who is very much part of the situation and observes from a distance (2013, p. 22). In this construction, the act of 'spectatorship' carries a number of entanglements. Representations of refugees are used to trigger emotional responses: awareness, empathy, and action. There are, however, conflicting views as to the extent to which these images help people who seek asylum. Many critics argue that rather than evoking compassion and triggering action, photographs of the displaced have a neutralising effect on the viewer. Allen Feldman uses the term 'cultural anaesthesia,' to describe a process by which refugee experiences are reduced to sensationalised and visceral visualisations of anonymous corporeality (1994). Luc Boltanski (1999, p. xv) explores morally acceptable conditions that could justify 'the spectacle of distant suffering.' Perhaps one of the best-known articulations of the ambivalence involved in the act of 'looking' comes from Susan Sontag, who observes that while photographs remain the main channel through which Western audiences learn about atrocities, war, and the suffering of other people, the consumption of images may, in fact, have a 'neutralising' effect on the viewer (1979, p. 85). The sole effect of looking at the images 'clears our eyes,' creating a sense of catharsis for the viewer without alleviating the suffering (Sontag 1979, p. 85). She further argues that the photographs become reality in their own right and enable viewers to 'turn a blind eye' to the actual events unfolding; looking at a photograph, rather than the real scene, accentuates the passivity of the viewer whose only other option is not to look (1979, p. 132). In Sontag's view, the world is saturated with images that 'consume reality' (1979, p. 85). In this process of appropriating reality, they make it visible and then use it to play on our emotions to the extent that they make reality 'obsolete' (Sontag 1979, p. 85).

In her criticism of the 'onlookers,' Mieke Bal goes even further, arguing that the aesthetics of images of suffering give viewers a pleasure that is 'parasitical,' preying on the pain of others (2007, p. 21). She likens the act of looking at the photographic representations of people in pain as a 'theft of their subjectivity' and a secondary exploitation (Bal 2007, p. 21). Similarly, Mark Reinhardt (2007) argues in the same volume that by looking at the images of suffering we may, unintentionally, prolong the victims' shame. Exploring the case of the humiliating photographs from Abu Ghraib prison, he notes that the cameras were used to abuse and humiliate the prisoners, but the humiliation continued long after the cameras stopped working as onlookers gazed at the photographs after they had been released (2007, p. 17).

While recognising the problems embedded in representing suffering, a number of authors argue that such representations are, nevertheless, crucial for raising awareness of atrocities. In considering the role of photography, Frank Moller quotes Patrick Hagopian, a historian of the Vietnam War, who calls these representations a 'necessary violence of a photographic act' and explains that they might be necessary to trigger or at least demand a response (2013, p. 18). Moller also discusses Simon Norfolk, a photographer of conflict zones, who argues that the anesthetisation of violence is justifiable if it is used to get the message across (2013, p. 19).

Artur Żmijewski (2016) takes on the complexity of the power relations involved with the production of images of the displaced in his highly acclaimed film, *Glimpse* (2016), which was presented at several major art exhibitions including Documenta 14 in Athens in 2016. This short black-and-white film was shot on a 16-mm camera and combines documentary and more experimental work. At first the viewer is exposed to documentary shots portraying the everyday realities of people waiting for asylum in several places in Europe. We see Berlin's Tempelhof Airport, the site of a temporary refugee camp, and the ways in which the space is organised for the asylum seekers. Later, the camera takes us to Calais and the infamous 'Jungle Camp.' It is here that we see the first close-ups of asylum seekers. We notice an older man staring at the camera. The camera inspects his face and his body. It is here that for the first time that I feel uncomfortable with the intensity of this focus. The way the camera 'inspects' the man's body creates a sense of estrangement and alienation from the subject. It is at this point that the viewer might notice that this is not just an innocent film depicting the life of refugees in a temporary camp. There is a different dynamic at play. The film cuts to an image of a family: We see a man, a woman, and a little boy and girl. The kids smile at the camera. They are clearly excited by the presence of the film crew. But their father looks embarrassed and uneasy. One can tell he would prefer not to be part of this staged performance. The camera then moves to the intimate spaces of family's tent—we see the bedding, beds and sheets, and parts of underwear. There is a growing sense that we the camera should not be there, that I should not look. And then something unexpected happens. We see a close-up of a man in wet clothes and dirty pair of Crocs (shoes). An instant later the same man is given a brand-new pair of shoes. Leaving me puzzled, the film cuts to another man who is being given a dry coat, which he immediately puts on. We soon see that this invisible helping hand belongs to the artist but also that this act of giving is not innocent and that there is intention behind these gifts. The coat is given to serve as a canvas for the artist's intervention. The man is asked to turn around so the artist can paint a big X on the man's new coat. The film cuts to the streets of Paris and here we see a group of African asylum seekers, all male. One of them is given a broom and asked to clean the street. There is a bit of suspense. Neither I (the viewer) nor the protagonists know how to treat this bizarre request. Clearly confused, the man starts sweeping the streets and looks in anticipation at the camera. It looks like the camera plays on this anticipation of the protagonist's reactions. This sequence of handing a broom to a person and requesting they clean the street is repeated several times. The film ends with perhaps most disturbing image of all in which the artist moves the faces as if they were marionettes. The last image is of the face of a black man, which is painted with white paint. His face gradually becomes white.

Żmijewski's film is uncomfortable and perhaps this discomfort is the intention. The film systematically exposes the violent role of the camera, which is voyeuristic and intrusive. By offering these close-ups, we see the pretences of humanisation. But the methods of filming are dehumanising. This shows the thin line between engagement and intrusion. With these close-ups, we do not learn more. Instead they leave the viewers with a mix of emotions. And the 'performative actions' are unexpected, yet somehow subversively familiar. Is this act of giving a coat in exchange for a demand an evocative metaphor for the conditionality of our 'aid' to the refugees? Is the broom a telling object, illuminating the politics of settlement that place conditions on the type of behaviour and jobs asylum seekers can do? The process of 'subjectification' gradually becomes hard to bear and the references are telling. The act of painting an X and giving a broom is a literal reference to the ways in which European Jews were treated in Nazi-occupied Vienna in the 1930s. It is also a profound comment on the human implications of a dehumanising bureaucracy in which people are rendered as X's—cases in the register that fit certain categories. This final act of painting the protagonists white might speak to the racialised discourses that both point to 'race' and would like the refugees to be 'white.' The ambiguous role of the artist, a white man forcing people to do his bidding, exposes his privilege.

The film claims to be revealing the asymmetrical power relations between citizens and refugees (Żmijewski 2017). Indeed, we are sucked into this power game. And although we participate by looking, like Sontag writes, we remain passive. There is nothing we can do except for choose to look or not look. The film speaks to the structural forms of violence, racism, and exclusion that we are part of. Yet, Żmijewski's act of exposing is, in my view, subjectifying in itself. The film is made to make us feel uncomfortable, and indeed, I felt discomfort long after having watched it. Is this sense of discomfort worth the subject's humiliation? I miss having a 'counter-act,' something that would return to the people who participated in the project and break the expected cycle of exploitation. Even in this portrayal of dehumanisation and the objectifying character of the gaze, the refugees remain just objects. As such, in line with Reinhardt's observations, Żmijewski's performance becomes another instance of prolonging and, even accentuating, the victims' shame.

## Conclusion: Towards an Audiovisual Methodology of Displacement

Blomfield and Lenette (2018) question artists' claims that they challenge dehumanising media practices. They argue that some of these attempts to 'give voice' to human experiences of displacement fall into the trap of oversimplification and misrepresentation. Some representations might unintentionally contribute to 'humanitarian porn' by reinforcing 'othering, abjection, disempowerment and voiceless-ness' (Blomfield and Lenette 2018, p. 323). In the final part of this essay, I would like to explore the possibility of engaging with refugee experiences in an academic context in a way that minimises the shortcomings articulated by Blomfield and Lenette. As this chapter strives to demonstrate, there are an abundance of narratives and visual representations that shape our understanding of refugees and their plight. To be relevant, academics producing audiovisual representations of displacement need to establish their own space and reason for being. Below, I have drafted some of the initial considerations that such a critical audiovisual methodology for representing displacement should take into account.

### No Desire to Capture the Most Dramatic Part

Orban Wallace's film captures well the self-perpetuating cycle of 'breaking news' and the photojournalists that desire to get as close as possible to action. One of the things that academics can afford to do differently is to take their time. The production cycle of academic research means that researchers working on audiovisual representations of refugees do not need to take part in the race for 'breaking news.' As Blomfield and Lenette suggest, the role of academic research is to reply to the 'dominant tropes' and produce 'counter-narratives' (2018, p. 326). The privilege of taking time allows researchers to dig deeper and understand the nuance. In terms of refugee experiences, it might mean a shift of focus to what is happening at the margins of the journalistic field-of-vision and staying with the people after the cameras have been switched off. It employs a quiet presence, waiting, investing time in relationships, and accompanying people throughout their exilic trajectories.

## Think When and in What Ways the Camera Might Be Needed

Given the challenges and the risks of abuse and appropriation, the presence of a camera might not be needed or might not be welcome in all circumstances. Perhaps, there is no need for an image if it is just an illustration of a textual argument (Rose 2012). There are a number of reasons however where the presence of the camera might be needed: enabling researchers to go beyond the interview (Banks 2001), offering a non-verbal register (Knowles and Sweetman 2004), or offering access to context where words 'fail us' or become inadequate for expressing reality or accessing the context (Das 2007; Jackson 2002). Jackson (2002) observes that a recurring element in refugee stories is that the experiences of the dramatic past cease to be narratable. He writes: 'Not only is there a loss of the social context in which stories are told: the very unities of space, time, and character on which narrative coherence depends are broken' (Jackson 2002, p. 102). Through its ability to transcend time-spaces and reclaim the texture of the dispossession—the sounds, light, the landscape, and the relations between bodies and space—the camera, for both the refugees and the viewers, provides access to the contexts of people's experiences of displacement that are difficult to narrate or put into words.

## Shift Away from Faces and Focus on the Places, Spaces, and Structures of Displacement

One of the strategies for breaking away from the fixity of visual representations is turning to the spaces, places, and structures of dispossession. One of such attempts was an exhibition that I curated together with my colleague Przemysław Wielgosz, in 2016, in Poland called *Refugees. Present/Absent*. The aim of the exhibition was to engage with the horrifying paradox of refugee hypervisibility/invisibility in Poland and to offer a critical response to the ways in which the figure of 'the refugee' has been hijacked by the political right. We wanted to expose the dehumanising and demonising mechanisms of refugee representations in public discourse, in which refugees are reduced to a threat, the ultimate 'other.' Trying to avoid further appropriation and visual abuse of the images of refugees, which were circulating in Polish and European media during the 'refugee crisis,' we decided not to display the photographs of people who were fleeing to Europe. Instead, we turned to the spaces, places, and objects that had 'witnessed' people's flight across Europe. We invited two photographers Mikołaj Długosz and Kuba Czerwiński to present their photographs, which focus on the material traces of refugees fleeing across West Balkan routes. There were no close-ups, no photos of children, and no attempts to exoticise the displaced. Długosz and Czerwiński were interested in traces of refugees' everyday lives: rubbish, abandoned blankets, and wet clothes—the remains of refugees' material presence and the temporariness of their existence. The photos also captured the infrastructure of movement control: border crossings, registration centres, and bottleneck train terminals. This focus on the material traces of displacement enabled us to reclaim space for a critical analysis of political and economic entanglements over displacement and for an exploration of the ways in which the 'refugee crisis' is created as a media spectacle and used to legitimise a 'state of exception' on the national and European level.

## Refugees as Your Primary Audience

Blomfield and Lenette (2018) explore the conditions and circumstances of collaborative practices and argue that artists involved in such production must be aware of the broader political and societal preconceptions of their work. I would argue that developing an audiovisual representation of displacement is about treating refugees as one's primary audience, as Jean Rouch (1995, pp. 94–95), a French anthropologist and film-maker, argued was the potential role of ethnographic films. Rouch writes about the idea of 'communicating back' to the group that one is studying as a way of developing a more reflexive and critical audiovisual practice. He presents the footage first to the researched, which he calls an 'audio-visual counter-gift' that provides a stimulus for mutual understanding and feedback (Rouch 1995, pp. 94–95) In his words, 'film is the only method I have to show another just how I see him … In other words, for me, my prime audience is (after the pleasure of the 'cine-trance' during the filming and editing) the other person, the one I am filming' (Rouch 1995, p. 96). This practice of thinking about refugees as the primary audience can reduce manipulation and appropriation. It also offers a promise of developing relationships in which refugees are seen as fellow human beings rather than 'audiovisual material.' As part of my research into the memories of Palestinian dispossession, I collected oral histories of Palestinian refugees and their descendants and subsequently embarked on a journey with my camera to follow their routes of displacement back to Palestine and Israel using pre-agreed itineraries (Blachnicka-Ciacek 2016). We would agree where I should go and, in some cases, whom I should meet. I subsequently returned back from the fieldwork to visit the research participants in Poland and the UK again and share with them the material and affective artefacts from the journeys (i.e. stones from destroyed home village, mandarins from family trees, photographs, stories, and gossip). All those exchanges and encounters encouraged new conversations, reflections, and memories. But, most of all, they offered a way to develop a relationship with the research participants and make the research process, at least to some extent, multidirectional. The practices of involving research participants in the process of research and production do not necessarily alleviate the asymmetries and imbalances, but they may offer a way of reducing the objectification of the refugees. As in the case above, they compel the researcher to recentre his or her research participants. It also ensures that the integrity of the relationship with the research participants is the primary focus of the research.

These four areas of consideration for an audiovisual methodology of representing displacement are by no means exhaustive and do not automatically prevent one from falling into the traps I discussed earlier in the chapter. Rather, they are points that may be taken for consideration when planning, producing, or disseminating audiovisual knowledge on refugeehood that might offer a different way of thinking about the images of the displaced. They are part of an evolving set of ideas, approaches, techniques, and tools that seek to reposition refugees as agents in their own storytelling rather than as objects of the journalistic gaze.

## Note

1. In this article I do not use the world 'refugee' in a legal sense. I refer to people who have been forced to flee from their places of residence and seek safety and prosperity elsewhere.

## References

Baines, E. K. (2004). *Vulnerable bodies: Gender, the UN and the global refugee crisis* (1st ed.). Aldershot, Englandh, and Burlington, VT: Routledge.

Bal, M. (2007). The pain of images. In M. Reinhardt, H. Edwards, & E. Duganne (Eds.), *Beautiful suffering: Photography and the traffic in pain* (pp. 93–115). Williamstown, MA: Williams College Museum of Art & Chicago, IL: University of Chicago Press.

Banks, M. (2001). *Visual methods in social research*. London: SAGE.

Blachnicka-Ciacek, D. (2016). *The chronotopes of Palestine* [Video]. Retrieved from https://vimeo.com/186417555.

Bleiker, R., Campbell, D., Hutchison, E., & Nicholson, X. (2013). The visual dehumanisation of refugees. *Australian Journal of Political Science, 48*(4), 398–416. https://doi.org/10.1080/10361146.2013.840769.

Blomfield, I., & Lenette, C. (2018). Artistic representations of refugees: What is the role of the artist? *Journal of Intercultural Studies, 39*(3), 322nal. https://doi.org/10.1080/07256868.2018.1459517.

Boltanski, L. (1999). *Distant suffering: Morality, media and politics* (G. Burchell, Trans.). Cambridge and New York, NY:Cambridge University Press.

Burgin, V. (1982). *Thinking photography*. London: Macmillan.

Das, V. (2007). *Life and words: Violence and the descent into the ordinary*(1st Indian ed.). New Delhi and Oxford: Oxford University Press.

Devichand, M. (2016, September 2). Did Alan Kurdila death change anything? *BBC News*, sec. BBCTrending. Retrieved from https://www.bbc.com/news/blogs-trending-37257869.

Edkins, J., & Pin-Fat, V. (2005). Through the wire: Relations of power and relations of violence. *Millennium, 34*(1), 1–24. https://doi.org/10.1177/03058298050340010101.

Feldman, A. (1994). On cultural anesthesia: From Desert Storm to Rodney King. *American Ethnologist, 21*(2), 404.

Jackson, M. (2002). *The politics of storytelling: Violence, transgression and intersubjectivity*. Copenhagen, Denmarka, and Portland,OR: Museum Tusculanum Press.

Johnson, H. L. (2011). Click to donate: Visual images, constructing victims and imagining the female refugee. *Third World Quarterly, 32*(6), 1015–1037. https://doi.org/10.1080/01436597.2011.586235.

Knowles, C., & Sweetman, P. (2004). *Picturing the social landscape: Visual methods and the sociological imagination*. London: Routledge.

Malkki, L. H. (1989). *Purity and exile: Transformations in historical-national consciousness among Hutu refugees in Tanzania*.Doctoral dissertation,Harvard University, Cambridge, MA.

Malkki, L. H. (1995). Refugees and exile: From 'refugee studies' to the national order of things. *Annual Review of Anthropology, 24*, 495–523.

Merkel, A. (2015). Flrkel, A. (2015). Flüchtlingspolitik: 'Wir schaffen das' – Statement von Angela Merkel am 31.08.2015 [Video]. *Phoenix*. Retrieved from https://www.youtube.com/watch?v=kDQki0MMFh4.

Moller, F. (2013). *Visual peace: Images, spectatorship, and the politics of violence*. Basingstoke: Palgrave Macmillan.

Nyers, P. (2005). *Rethinking refugees beyond states of emergency* (1st ed.). New York, NY: Routledge.

Pupavac, V. (2008). Refugee advocacy, traumatic representations and political disenchantment. *Government and Opposition, 43*(2), 270–292. https://doi.org/10.1111/j.1477-7053.2008.00255.x.

Rajaram, P. K. (2002). Humanitarianism and representations of the refugee. *Journal of Refugee Studies, 15*(3), 247–264.

Reinhardt, M. (2007). Picturing violence: Aesthetics and the anxiety of critique. In M. Reinhardt, H. Edwards, & E. Duganne (Eds.), *Beautiful suffering: Photography and the traffic in pain* (pp. 13–36). Williamstown, MA/Chicago, IL: Williams College Museum of Art/University of Chicago Press.

Rhodan, M. (2015, November 20). Are the Syrian refugees all 'young, strong men'? *Time*. 2015. Retrieved from https://time.com/4122186/syrian-refugees-donald-trump-young-men/.

Rose, G. (2012). *Visual methodologies: An introduction to researching with visual materials* (3rd ed.). Los Angeles, CA and London: SAGE.

Rouch, J. (1995). The camera and man. In P. Hockings (Ed.), *Principles of visual anthropology* (pp. 79–98). Berlin & Boston, MA: De Gruyter. https://doi.org/10.1515/9783110290691.

Said, E. W. (1985). *After the last sky: Palestinian lives*. London: Faber.

Sigona, N. (2014). The politics of refugee voices. In E. Fiddian-Qasmiyeh, G. Loescher, K. Long, & N. Sigona (Eds.), *The Oxford handbook of refugee and forced migration studies* (pp. 369–382). Oxford: Oxford University Press. https://doi.org/10.1093/oxfordhb/9780199652433.013.0011.

Sontag, S. (1979). *On photography*. London: Penguin Books.

Wallace, O. [Director]. (2017). *Another news story*. IMDB. Retrieved from http://www.imdb.com/title/tt5345032/.

Wilmott, A. C. (2017). The politics of photography: Visual depictions of Syrian refugees in U.K. online media. *Visual Communication Quarterly, 24*(2), 67–82. https://doi.org/10.1080/15551393.2017.1307113.

Wright, T. (2002). Moving images: The media representation of refugees. *Visual Studies, 17*(1), 53–66. https://doi.org/10.1080/1472586022000005053.

Żmijewski, A. (2016). *Glimpse* [Video]. Retrieved from https://vimeo.com/220889854.

Żmijewski, A. (2017, June 3). About 'how we treat the others'. Conversation on 'Glimpse'. Artur Żmijewski in conversation with Michael Heitz. *Diaphanes*. Retrieved from http://www.diaphanes.fr/titel/conversation-on-glimpse-4706.

# 46

## Displacements of Experience: The Case of Immersion and Virtual Reality

### Emma Bond

From June 2017 to January 2018, acclaimed film director Alejandro G. Iñárritu ran an immersive virtual reality (VR) installation at the Fondazione Prada in Milan. Called *Carne y Arena (Flesh and Sand): Virtually Present, Physically Invisible*, it was a solo experience lasting six-and-a-half minutes, which aimed to allow visiting participants to experience a fragment of the journey made each year by hundreds of thousands of refugees in their attempt to cross the Mexican border and enter the United States. The installation incorporated material mined from true accounts that Iñárritu had collected during four years of personal interviews with refugees and had a specific artistic aim. In his own words:

> My intention was to experiment with VR technology to explore the human condition in an attempt to break the dictatorship of the frame, within which things are just observed, and claim the space to allow the visitor to go through a direct experience walking in the immigrants' feet, under their skin, and into their hearts. (Iñárritu 2017)

His emphasis here on the bodily experience of the participant adds critical flesh to Brenda Laurel's suggestion that VR technology offers users the rare

---

I would like to thank Dr Mona Bozdog and Dr Robin Sloan, both of Abertay University, for their immensely useful input into early drafts of this chapter.

---

E. Bond (✉)
School of Modern Languages, University of St Andrews, St Andrews, UK
e-mail: efb@st-andrews.ac.uk

© The Author(s) 2020
P. Adey et al. (eds.), *The Handbook of Displacement*,
https://doi.org/10.1007/978-3-030-47178-1_46

opportunity to take 'your body with you into worlds of imagination' (cited in Ryan 2001, p. 52). Once inside the installation, visitors to *Carne y Arena* would experience the vast, empty space one by one: they were asked to remove their shoes and to walk over the cold sand of the virtual desert, trying to avoid potentially dangerous encounters with border guards and agents. Throughout the immersive experience, they would come into virtual 'contact' with other migrants, whose hearts would beat red, visibly, through their bodies, if touched by the participants. The combination of the uncomfortable corporeal experience of the installation, as well as the virtual interaction with the corporeality of others, surely lies at the core of *Carne y Arena*'s overwhelming success. As Peter Aspden writes in his review for the *Financial Times*: 'We are of course observers; yet we feel like participants. [...] We feel the victims' plight through our toes, which in turn takes us outside our heads' (Aspden 2017). This seems to make *Carne y Arena* the perfect illustration of William Bricken's statement in his manifesto for VR, namely that 'our body is our interface' (cited in Ryan 2001, p. 52), and suggests that such bodily immersions can be particularly effective in inducing empathy for the real-life experiences of others. But who (and whose body) is capable of assessing the authenticity of such an empathetic experience? Can we really reproduce the experience of displacement through immersive or virtual means, and can anyone not directly affected themselves ever 'feel' what it means to be displaced?

This chapter maps out a contested territory of agency, empathy, and representation in relation to contemporary cultural representations of displaced peoples that employ new and innovative modes of audience engagement. Using a range of critical tools, including affect and haptic theory and the work of Jacques Rancière on the image and the spectator, it focuses its analysis on three recent 'immersive' representations in different forms of media that promise to communicate the refugee experience to their audiences: *Flight* (2017), *Limbo* (2017), and the soundscapes produced by the Refugee Hosts project (2016–). It also considers oppositional perspectives which push back against the notion that audiences can truly experience empathy through artistic media, and which therefore also question the ability of aesthetics to contribute real social and political change. Ultimately, it weighs up the potential risks and benefits in claiming to represent the experience of displacement through media that blur the lines between activism and entertainment and seeks in conclusion to determine how far experience itself can be 'displaced' through such immersive and virtual representations.

\* \* \*

The impetus to deploy new and sophisticated techniques in representations of the refugee experience comes most strongly from the feeling that displaced peoples are too often represented through image (for the most part photography), in a way that does not allow viewers to engage empathetically with their stories. This explains Iñárritu's reference to a 'dictatorship of the frame' above, since:

> Repeated exposure to images and stories of pain is not enough to stimulate identification with others, feed a sense of injustice, and promote political change. On one side, the dominant culture of spectatorship tends to [...] neutralize the moral force of photographs and visions of atrocities. Moreover, scenes of the hardships of others may even irritate, produce insecurity, and infuse the desire to withhold compassion. (Oliveri 2016, pp. 159–160)

In fact, as Rancière would have it, the underlying issue is not that we see too few or too many suffering bodies in contemporary media outlets, but that these bodies remain *nameless*, 'incapable of returning the gaze that we direct at them' and appear as 'an object of speech without themselves having a chance to speak': 'The problem is not counter-posing words to visual images. It is overturning the dominant logic that makes the visual the lot of multitudes and the verbal the privilege of the few' (Rancière 2009, p. 97).

It has been suggested that new techniques of immersion or VR might be able to return this privilege of agency to the multitude and to succeed in reversing the passivity of the spectator through enabling more profound connections between people, both in the sphere of the real and of the imagination. Chris Milk goes as far as to claim that such technologies will become 'the ultimate empathy machine' (cited in Bhutto 2018, p. 44). Could immersive and virtual techniques also then contribute to the formation of what Rancière calls an 'emancipated community': a community of actively engaged participants who by nature challenge the distinction between viewing and acting and who function as a '*community of narrators and translators?*' (Rancière 2009, p. 22, emphasis added). Michael Saler has written in similar terms about the importance of participatory virtual communities because of their ability to 'attune their inhabitants to be more responsive to others' (Saler 2012, p. 7) and claims that they foreground critical reason in the arena of empathetic imaginings. In fact, as Pierre Lévy suggests, virtual environments are inherently *creative* in and of themselves (1998).

> They provide safe and playful arenas for their inhabitants to reflect on the status of the real and to discuss prospects for effecting concrete personal and social

changes. They challenge their inhabitants to see the real world as being, to some degree, an imaginary construct amenable to revision. As a result of collectively inhabiting and elaborating virtual worlds, many become more adept at accepting difference, contingency, and pluralism: at envisioning life not in essentialist, 'just so' terms but rather in provisional, 'as if' perspectives. (Saler 2012, p. 7)

Yet, in opposition to these voices highlighting the potential for the virtual to engender positive change are standpoints that draw attention to the issues inherently present within those very same technologies. Immersion is, for example, not always seen as being conducive to critical thinking (see Ryan 2001, pp. 9–10), and whilst virtual worlds may be empirical, they are certainly not objective and are bound to 'embody their authors' biases, blind spots, ideologies, prejudices and opinions' (Kirschenbaum, cited in Risam 2017, p. 111). Given these reservations, can immersive and virtual technologies still be judged capable of generating empathy and allowing for an understanding of experiences of displacement, and if so, what tools can we use to assess their ability or potential to do so?

\* \* \*

The first piece I want to use to illustrate in my discussion, *Flight*, is the theatrical adaptation of Caroline Brothers' novel *Hinterland* (Brothers 2012) by Vox Motus, which premiered in 2017. In the stage version, the story is retold through miniaturised models that appear (albeit statically) in a revolving diorama. Audience members are greeted in the theatre lobby, as if by flight attendants, and then escorted individually to sit in private booths where they listen to pre-recorded actors' voices through headphones and watch the tale unfold through scenes that pass before them on a carousel. This is the description of the piece on the Vox Motus website:

> Two young orphaned brothers embark on a desperate odyssey to freedom and safety. With their small inheritance stitched into their clothes, they set off on an epic journey across Europe, in a *heart-wrenching* road story of terror, hope and survival. […] a *magical* experience that combines unsettling themes with *spellbinding* images. Flight brings you up close and intimate to this *heartbreaking* story in a unique, deeply individual experience […] draws you into its fragile miniature world and allows you to contemplate its *gripping* story of two children lost in dangerous lands. (Vox Motus 2017, emphasis added)

On first impact, the use of such emotive adjectives in this description may risk appearing callous or voyeuristic and figuring the retelling of a (true) story very much as a work of theatrical art risks relegating the audience members to a traditionally passive, receptive role. Indeed, some newspaper reviews criticised the makers of the piece for removing its political context: through—for example—not narrating the characters' lives prior to departure nor their reasons for fleeing their home country (see Billington 2017). I had the chance to informally interview the script adaptor, Oliver Emanuel, in May 2018, and he emphasised the production company's awareness that they were not making a story to be told back to refugees themselves but one that needed to be refashioned for privileged theatregoers instead. His main questions and motivations were how to draw this particular audience into narrative, especially with a visually led piece, in which words were less important to the overall meaning than a sense of immersion in the optical show.

In such a context, what is the audience role and capacity for empathy? Can they be effectively and actively engaged? Having been myself to see the piece, at the Edinburgh Festival in 2017, I can imagine productively framing it through a lens of haptic visuality, as theorised by Laura Marks and others. The attention to detail contained in the miniature models, the very close-up nature of the interaction (models pass so close to you that you can see every blade of grass), and the perception of texture, all recall the concept of haptic visuality as being a *tactile* art of seeing, which involves 'intimate, detailed images that invite a small, caressing gaze' (Marks 2000, p. 6). These are alternative economies of embodied looking, 'ways of seeing [...] in which the eye lingers over innumerable surface effects, instead of being pulled into centralized structures' (Marks 2000, p. 6). This haptic mode of perception grants spectatorship the chance to become more *active*: it 'allows us to experience in detail, but not to take a distance from experience in order to define it' (Marks 2002, p. 12). Haptic perception also manages in itself to mimic movement, something that takes on significance when trying to tell a story of physical displacement through the revolving appearance of static models:

> Haptic looking tends to rest on the surface of its object rather than to plunge into depth, not to distinguish form so much as to discern texture. It is a labile, plastic sort of look, more inclined to move than to focus. (Marks 2002, p. 8)

A sense of displacement is also echoed in the migratory movement of the piece itself—*Flight* has now travelled from Edinburgh to New York, on to Galway and Melbourne, and the complex diorama is itself transported each time by boat. The production company are also in discussions about taking the show to Mexico, which involves a debate around what to do about the multiple languages in the piece and how (and when) to translate them. Languages are currently used to convey a sense of strangeness in the piece, both for the subjects of the action and for the audience (Greek, e.g., is left untranslated in the English version), and so function as an effective mode of displacing a sense of disorientation from the action to the spectators themselves.

This evocation of a sense of disorientation is also the driving motivation behind *Limbo*, a *Guardian* VR piece that aims to allow viewers to experience 'first-hand' the period of time asylum seekers spend between their arrival in the United Kingdom and being granted a Home Office interview which will determine whether they can legally stay in the country or not. This description is from the *Guardian* website:

> What is it like to flee your home and start again in a new country? […]. While asylum seekers wait for their Home Office interview, they live on £5 a day and are unable to work or choose where they live. In Limbo you step into their shoes and experience their state of mind while you wait for the decision that will determine the rest of your life. ('Limbo' 2017)

The piece was made after conducting interviews with asylum seekers from twelve different countries as well as with immigration lawyers and barristers who have been involved in supporting asylum seekers through the legal process. The voices in the soundtrack are those of real refugees who guide the viewer through different segments of their own collective experience—their arrival in an unfamiliar city, their thoughts and concerns about loved ones they have left behind, their anxiety about not being allowed to work, and the long-awaited Home Office interview upon which much of their future safety depends.

*Limbo* was made in collaboration with the creative company ScanLAB Projects, which specialises in creating digital replicas of environments using innovative 3D-scanning technology. The resulting monochrome, sketch-like aesthetic has the quality of a dream (or indeed nightmare), where worlds are both transparent and fragmented. It is eerily effective in recreating the surreal sense of isolation and dislocation that asylum seekers report feeling, but in a mono-user experience such as this (and in *Flight* too), do virtual and immersive technologies still have the potential to produce the effect of a co-produced, affective experience? Without the participation of engaged others, can the audience member or user achieve what Sobchack calls 'volitional, deliberate vision' (Sobchack 1992, p. 93), a mode of viewing that works against the more usual 'characterization of the viewer as passive, vicarious, or projective' and replaces it 'with a model of viewer who participates in the production of the experience' (Marks 2002, p. 13)? For in an ideal state of affairs, as Marks says: 'the viewer has to work to constitute the image, to bring it forth from latency. [...] The act of viewing is one in which both I and the object of my vision constitute each other' (Marks 2002, p. 13).

Yet some argue that virtual technologies should be able to produce just such an emotional immersion, due to the capacity of the human mind to be emotionally affected by the contemplation of purely imaginary states of affairs (Carroll, cited in Ryan 2001, p. 157). Similarly, Frieze (2016) has argued for the capacity of immersive techniques to offer people the opportunity to reconnect and reaffirm bonds with each other, suggesting that they might hold a 'reparative' potential. He also hints at an educational dimension, citing Ryan's addition of the 'epistemic' to her previous categories of immersion: in this so-called forensic turn, participation becomes part of a search for evidence, or a pathway to determining 'truth' (Frieze 2016, p. 25). Yet the moral or ethical counterargument to using virtual and immersive technologies to represent the refugee experience is, of course, that they allow participants to 'parachute' into a story of trauma, but that they are then free to leave and return to their usual home comforts once the drama of the virtual experience has been terminated. In works such as the now famous National Theatre of Wales production *Bordergame* (2014), in which participants were asked to gather on the England/Wales border and try and prove their 'asylum seeker' credentials in order to gain entry to the Autonomous Republic of Cymru, the players were the protagonists but with limited freedom to act independently within the structure of the piece. Concerns about jeopardising the 'artistic' vision of the company might also discourage participants from diverting from the 'scripted' action of such pieces and assuming a degree of agency. But then again, might this sense of performed passivity not perhaps force us to look back on our real-life passivity towards such desperate situations and thereby gain a deeper understanding of the consequences of inactivity?

Aikaterini Antonopoulou has similarly argued against an uncomplicatedly positive conception of VR, warning that it may contribute to the 'gamification' of reality, where intense experiences can be had without participants suffering any of the real-life consequences (Antonopoulou 2017). VR, she says, prevents any sort of meaningful encounter in part because, by nature, it relies on the absence of substance. She takes issue with pieces such as *Clouds over Sidra* (2014), an immersive experience of the Za'atari refugee camp in Jordan as seen through the eyes of the twelve-year-old protagonist, Sidra. It was commissioned by the UN, sponsored by Samsung, and launched at Davos in 2015 with the aim of evoking 'humanitarian empathy' through immersion in the life of refugees ('UN uses virtual reality' 2016). Similarly, at Davos in 2017, participants in the World Economic Forum were invited to experience a new, tactile take on the refugee crisis: called *A Day in the Life of a Refugee*. This was a participatory, 'real-life' simulation, organised by Crossroads Foundation and designed by a team of refugees and NGO representatives. It was described as 'powerful' and 'moving' by participants, but—as Antonopoulou argues—a physical re-enactment cannot in any way come close to an experience of another's reality. In such 'games,' great attention is paid to the physical representation of the refugee camps, but the absence of pain, the ability to leave the simulation at any point, and most importantly, the absence of the fear of death emphasises the limits of any such virtualisation project.

These are issues that are also faced by historical re-enactments of displacement and disenfranchisement, such as those offered at the Conner Prairie Interactive History Park, in Indiana. As part of their experience here, visitors can pay additional $20 to play the role of a 'runaway slave' in an evening performance that sees park staff guiding a dozen people at a time through a two-hour 'immersive, real life encounter of what it was like to be a fugitive' enslaved person in Indiana in 1836 (see Bowman 2016). What Bowman found especially 'troubling' about her own visit there was the sense that the performance 're-enacted' social body types and behaviours, which she links to the 'compulsory visibility' of the Black body as 'seen/scene' (Bowman 2016, p. 64). Similarly, in his discussion of Wadi Khaled's 2012 theatre piece *66 minutes in Damascus*, Alston (2016) raises issues relating to the ethical compromises and gaps between the lived experience of the participating audience and the promised experience to be gained through immersion. We cannot access the experience being staged by having an emotional experience of our own that has been brought on by the immersion. By attempting to suggest otherwise, Alston argues that pieces like Khaled's erase the 'potential to transform the participant's perception in an empathetic way that preserves the necessary distance between one who empathises and one who is empathised with. Instead, what is invited is what Berthold Brecht might call "crude empathy": "a feeling for another based on the assimilation of the other's experience to the self"' (Alston 2016, p. 260).

So how else might new technologies work to create empathetic responses to experiences of displacement, and can we look for alternative 'immersive' representations that go beyond the visual or the performative? How can artistic representations work to effectively recreate aspects of the refugee experience, with what ethical aims, and for whom? How are audiences or participants expected to respond to such pieces? If the aim is to upturn expectations, assumptions, and preconceptions through creative engagement, interaction, and active participation, then two aspects we might identify as important are site-specific action, and self-advocacy or the opportunity for *self*-representation. One example of a site-specific representation can be seen in *Soundscapes* (2017), which is part of the AHRC-ESRC Global Challenges Research Fund Refugee Hosts project. The project aims to explore the multiple challenges of displacement, by carrying out interdisciplinary and participatory research in and with nine local communities in the Middle East. Research activity is focused on the roles played by local communities in 'hosting' and on refugee-led humanitarianism, thereby looking to disrupt conventional refugee/host binaries. The project features interventions by artists, writers, and other creative practitioners within a 'community of conversation,' which is made publically available via their website. Specifically, they focus on everyday representations and follow a policy of focusing their work on spaces and places, not faces.

The soundscapes function as truly immersive experiences, because—as Kendrick points out—hearing places us inside an event, giving us a spherical experience, whereas vision ordinarily requires a certain distance from a given object or event in order to gain perspective on it (Kendrick 2017, pp. 11–12). Building on elements of the haptic visuality that characterised *Flight* above, the soundscapes featured here become akin to an individual *affective* experience for the listener. 'It is nigh-on impossible to censor hearing. The ear has a more feral relationship to the outside world because we cannot see the act of hearing […] This is one reason why the auricular sense has evaded total commodification' (Kendrick 2017, p. 14). The Refugee Hosts soundscapes thus seek to answer the question: how is displacement *heard*, rather than seen? There are four locations of soundscapes on the website, recorded in Hamra (Beirut, Lebanon), Baddawi Refugee Camp (Lebanon), Beyoglu (Istanbul, Turkey), and Athens, Greece. This last piece is the only one to feature a brief accompanying photo essay, which frames the recordings as being a mode of 'anthropology in sound.' The three photos in the essay are generic landscape shots designed to help the listener 'situate themselves,' and there are no descriptions of what the sounds are—rather, the piece invites the listener to interpret the sounds as they wish, since, 'sound produces social relations, and sound can tell different stories about displacement and encounter, about space and representation' (Karimi et al. 2017).

In this way, a sense of agency is restored to the participant or listener, since they can engage with the material in a variety of different, and perhaps more empathetic ways. Yet it is important to note that the soundscapes project takes a *partial* approach to immersion, focusing on one 'sense.' Does this suggest that 'full body' immersion in the arena of representing displacement is too (ethically) problematic? This is especially true if one considers the body of VR player itself—who takes care of that, and how? Also, if we inhabit virtual bodies as a point of view, how will we know what these bodies look like and if we don't know, how will we be able to relate to them? (see Ryan 2001, p. 62). The 'partial' nature of the soundscapes thus circumvents the issue of whether one can fully 'represent' another person's reality and suggests another 'immersive' route towards empathy. It might be through following such new creative modes of immersion that virtual and other technologies of representation could be used to try and sensitise or raise awareness to issues such as displacement. And this might be less to do with *content*, than with the way they can influence players and 'mount claims through procedural rhetoric' (Bogost 2007, p. ix).

Using the example of video games, Bogost uses procedural rhetoric to show how they are fundamentally systems of rules that define how a process is run and that players understand a game and its permutations by interacting within this process. When that process simulates a real-world system of interactions, engaging with that process enables us to better understand the real-world scenario involved (so, e.g., by playing games that aim to explain the procedural factors within the refugee 'crisis,' we might better understand the context and actors involved). Gaming procedures can then be used *rhetorically* or designed in such a way that the interactive experience functions as an intellect-led argument to persuade players or users of a particular idea. In this understanding, there is theoretical backing to using games and related computer media (be that immersive, virtual, or augmented reality) to re-present challenging scenarios—doing so is in effect a form of interactive documentary. This could be seen as exploitative, but no more so than other artistic genres such as film or literature, and procedural rhetoric could be argued to be more effective than verbal rhetoric when representing complex, politically charged situations. Bogost's work might be used productively to encourage virtual reality and games developers to use their medium to investigate distressing scenarios through procedural rhetoric, focusing on how interaction enables the player to better understand the topic instead of aiming to trigger empathetic responses in users.[1]

\* \* \*

Yet I would like to conclude by evoking a particular empathetic response to the virtual immersion of Iñárritu's *Carne y Arena* piece, which I described in the opening to this chapter. Fatima Bhutto, an Afghan writer, who was born in Kabul, grew up in Syria, and now lives in Karachi, visited the installation in Milan and found that it triggered previously experienced bodily sensations of fear by allowing corporeally located memories of her own multiple displacements to resurface.

> I have had guns pointed at me before; in real life. I have been in a landscape, far from this, with dust and dirt and young men shouting with weapons raised and the threat of violence flowing between us and at this moment, I cannot separate those memories from this. Just as I can hear a child cry, I can feel my heart beat, I can hear it drumming against my body, and everything that I know. (Bhutto 2018, pp. 50–51)

We could say, then, that the experience of such immersive and virtual representations is real because of the real physical *impact* that it has on participants. 'Shifting the frame' (Lakoff 2004, pp. 33–34) of reference back to the bodily experience of haptic visuality, dreamlike disorientation, or unfiltered soundscapes—to reference the examples cited in this chapter—allows us to evade the dictatorship of the frame that troubles Iñárritu and which hampers engagement with and empathy for those affected by displacement.

There are so many complex ethical questions that arise when interaction, agency, intimacy, and responsibility come together, particularly in immersive environments where critical distance is arguably momentarily abandoned. I want to end this brief chapter with one further response to the trend for 'immersive' experiences of the refugee experience that speaks in a more urgent way to the necessity for self-representation and advocacy. A group of refugees in the Ritsona refugee camp in Greece, in summer 2016, made the ironic gesture of advertising their tent on Airbnb, inviting would-be holiday makers to share 'a real opportunity to experience life as a Syrian refugee' and promising scorpions, dehydration, and broken promises (Squires 2016). Although the photographs were quickly taken down from the website, such a gesture shows us how very far the kind of creative virtual and immersive pieces that I have discussed here are from ever being truly representative of the multiple complexities and hardships that are involved in experiencing displacement.

## Note

1. Bogost does, however, divert and expand his thinking in the later *How To Do Things With Videogames* (2011), in thinking through the short, vignette format of games that recreate personal experiences of complex historical crises, such as the 2003 Darfur genocide ('Darfur is Dying') and the Rwandan civil war of 1994 ('Hush'). The brevity of the play experience in both pieces, Bogost says, 'does not make an argument, but characterizes an experience' (2011, p. 23), allowing us to see how games can in fact inspire empathy in their players.

## References

Alston, A. (2016). The promise of experience: Immersive theatre in the experience economy. In J. Frieze (Ed.), *Reframing immersive theatre: The politics and pragmatics of participatory performance* (pp. 243–264). New York, NY: Palgrave Macmillan.

Antonopoulou, A. (2017, November 28). The virtual reality of the refugee experience. *Refugee Hosts*. Retrieved October 18, 2018, from https://refugeehosts.org/2017/11/28/virtual_reality_refugee_experience/.

Aspden, P. (2017, August 11). Carne y Arena: Alejandro Iñárritu's virtual reality project. *Financial Times*. Retrieved October 18, 2018, from https://www.ft.com/content/965826cc-7c63-11e7-ab01-a13271d1ee9c.

Bhutto, F. (2018). Flesh and sand. In V. Thanh Nguyen (Ed.), *The displaced: Refugee writers on refugee lives* (pp. 43–51). New York, NY: Abrams.

Billington, M. (2017, August 6). Flight review – miniature models tell epic refugee story. *The Guardian*. Retrieved October 21, 2018, from https://www.theguardian.com/stage/2017/aug/06/flight-review-edinburgh-festival-vox-motus-church-hill-theatre-refugees.

Bogost, I. (2007). *Persuasive games. The expressive power of videogames*. Cambridge, MA and London: MIT Press.

Bogost, I. (2011). *How to do things with videogames*. Minneapolis, MN and London: University of Minnesota Press.

Bowman, R. L. (2016). Troubling bodies in follow the North Star. In J. Frieze (Ed.), *Reframing immersive theatre: The politics and pragmatics of participatory performance* (pp. 63–76). New York, NY: Palgrave Macmillan.

Brothers, C. (2012). *Hinterland*. London and New York, NY: Bloomsbury.

Frieze, J. (2016). Reframing immersive theatre: The politics and pragmatics of participatory performance. In J. Frieze (Ed.), *Reframing immersive theatre: The politics and pragmatics of participatory performance* (pp. 1–25). New York, NY: Palgrave Macmillan.

Iñárritu, A. G. (2017). Carne y arena. *Fondazione Prada*. Retrieved October 18, 2018, from http://www.fondazioneprada.org/project/carne-y-arena/?lang=en.

Karimi, S. A., Kurdi, M. S., Sourmelis, G., Western, T., & Zafeiriou, S. (2017, October 5). Summer in Athens. A sound essay. *Refugee Hosts*. Retrieved October 22, 2018, from https://refugeehosts.org/2017/10/05/summer-in-athens-a-sound-essay/.

Kendrick, L. (2017). *Theatre aurality*. New York, NY: Palgrave Macmillan.

Lakoff, G. (2004). *Don't think of an elephant: Know your values and frame the debate – The essential guide for progressives.* New York, NY: Chelsea Green.

Lévy, P. (1998). *Becoming virtual: Reality in the digital age* (R. Bononno, Trans.). New York, NY: Plenum.

Limbo: A virtual experience of waiting for asylum. (2017, July 5). *The Guardian*. Retrieved October 18, 2018, from https://www.theguardian.com/technology/2017/jul/05/limbo-a-virtual-experience-of-waiting-for-asylum-360-video.

Marks, L. U. (2000). *The skin of the film. Intercultural cinema, embodiment, and the senses*. Durham, NC and London: Duke University Press.

Marks, L. U. (2002). *Touch. Sensuous theory and multisensory media*. Minneapolis, MN and London: University of Minnesota Press.

Oliveri, F. (2016). 'Where are our sons?': Tunisian families and the repoliticization of deadly migration across the Mediterranean Sea. In L. Mannik (Ed.), *Migration by boat* (pp. 154–177). New York, NY: Berghahn Books.

Rancière, J. (2009). *The emancipated spectator* (G. Elliott, Trans.). London: Verso.

Refugee Hosts. (2016). Local community experiences of displacement from Syria: Views from Lebanon, Jordan and Turkey. Retrieved October 18, 2018, from https://refugeehosts.org.

Risam, R. (2017). Postcolonial studies in the digital age: An introduction. In J. Ramone (Ed.), *The Bloomsbury introduction to postcolonial writing. New contexts, new narratives, new debates* (pp. 105–122). Bloomsbury: New York, NY and London.

Ryan, M.-L. (2001). *Narrative as virtual reality. Immersion and interactivity in literature and electronic media*. Baltimore, MD and London: Johns Hopkins University Press.

Saler, M. (2012). *As if: Modern enchantment and the literary prehistory of virtual reality*. Oxford and New York, NY: Oxford University Press.

Sobchack, V. (1992). *The address of the eye. A phenomenology of film experience*. Princeton, NJ: Princeton University Press.

Squires, N. (2016, June 23). Syrian refugees in Greece 'put their tent on Airbnb,' promising scorpions, dehydration and 'broken promises'. *The Telegraph*. Retrieved October 18, 2018, from https://www.telegraph.co.uk/news/2016/06/23/syrian-refugees-in-greece-put-their-tent-on-airbnb-promising-sco/.

UN uses virtual reality to inspire humanitarian empathy. (2016, May 19). *UN News*. Retrieved July 9, 2019, from https://news.un.org/en/story/2016/05/529752-feature-un-uses-virtual-reality-inspire-humanitarian-empathy.

Vox Motus. (2017). *Flight*. Retrieved October 21, 2018, from http://www.voxmotus.co.uk/flight/.

# 47

# Displacement in Contemporary Art

## John Potts

This chapter examines the representation of displacement in contemporary art. Displacement may involve persons displaced as refugees or asylum seekers, or due to climate change or other social disruption. The idea of displacement is treated in a range of ways by contemporary artists: at times obliquely, in other works more directly, confronting the social dislocation found in many parts of the contemporary world. I consider the theme of displacement in the context of the globalist aspect of contemporary art, itself a reflection of globalisation. The intensified movement of goods, information, capital, images—as well as tourists, refugees, and displaced persons—around the world provides the setting for contemporary artists' treatment of displacement.

The best international art transcends mere propaganda or the meagre illustration of theory. Artists are explorers within the networks of global images. In the word coined by art theorist and curator Nicolas Bourriaud, the artist is a 'semionaut,' navigating the virtual seas of global images, producing 'original pathways through signs' (Bourriaud 2002, p. 19). This is an updating of Marshall McLuhan's description in 1964 of artists as the antennae of their society, in that they pick up conceptual shifts and investigate them in creative works. McLuhan's formulation of the artist was itself borrowed from an

---

J. Potts (✉)
Macquarie University, Sydney, NSW, Australia
e-mail: john.potts@mq.edu.au

© The Author(s) 2020
P. Adey et al. (eds.), *The Handbook of Displacement*,
https://doi.org/10.1007/978-3-030-47178-1_47

assertion made in 1934 by the poet Ezra Pound that 'artists are the antennae of the race' (Pound 1934, p. 73). McLuhan added the metaphor of radar to that of the antenna, construing art as 'an early alarm system ... enabling us to discover social and psychic targets in lots of time to prepare to cope with them' (McLuhan 1964, p. xi). The most striking artworks probe the fault lines and fissures of contemporary meaning. They play with contradiction or expose the processes of meaning-making, national identity formation, wilful acts of cultural forgetting, or the consequences of migration or forced displacement. The best artworks evoke ambiguity of meaning while resonating with emotional power. They may be intellectual feats and visceral emotional statements at the same time.

Contemporary art is largely a global art, exhibiting works on an international scale at biennales, triennales, and major exhibitions, such as Documenta in Kassel, Germany, held every five years. Curators play a significant role in setting thematic parameters for commissioned works, often drawn from contemporary political and cultural theory. As an example, one of the themes of the 2007 Documenta exhibition was 'What is bare life?', a reference to the political theory of Giorgio Agamben. Agamben's theoretical writing has been applied to analysis of the social reality of displacement; his assertion in his 1998 book *Homo Sacer* that the camp is 'the new biopolitical *nomos* of the planet' (Agamben 1998, p. 176) has been related to the camps set up for refugees, asylum seekers, political prisoners, and displaced persons. Many recent international artworks have been concerned with the politically urgent issue of the displaced. The recent work of Chinese artist Ai Weiwei, in particular, has addressed the social dimensions of displacement in a series of compelling art installations. This chapter discusses these and other recent artworks.

## Global Art

Contemporary art has increasingly been conceived as global art. Art theorists and art historians searching for a successor term to modernism and postmodernism have proposed, among other terms, network culture and globalism, as cultural conditions reflected by contemporary art. The critic and theorist Rex Butler has suggested that the 'new style or movement of art that comes after postmodernism' should be called 'globalism' (Butler 2010, p. 58), incorporating both the impact of globalisation on the concerns and content of contemporary art and the international circuit of major art events at which the most recent artworks are showcased.

Nicholas Bourriaud proposes as the successor to postmodernism an 'altermodernity' comprising a 'translation-oriented modernity' (Bourriaud 2009, p. 43). For Bourriaud, such a conception of contemporary culture corresponds to the globalised world order, a modernity 'born of global and decentralized negotiations, of multiple discussions among participants from different cultures' (2009, p. 43). Such a culture must be 'polyglot,' because 'the immigrant, the exile, the tourist, and the urban wanderer are the dominant figures of contemporary culture' (Bourriaud 2009, p. 51). Altermodernity for Bourriaud embraces the styles and techniques of modernity in a 'globalised culture busy with new syntheses' (2009, p. 186). The global network becomes a space of exchange, of diverse representations of the world, in which translation of ideas and representations plays a crucial role. The nomadic function, celebrated by Baudelaire in early modernity in the figure of the *flâneur*, is intensified in the global age, as artists wander not just through cities but across continents. Bourriaud cites as exemplary artists in this regard Gabriel Orozco, Thomas Hirschhorn, Jason Rhoades, and Francis Alÿs, all of whom express the wandering aspect of modern urban life, effecting in their art a creative negotiation of geographical displacement.

The various instances of transfer and translation practised in contemporary art constitute for Bourriaud 'a practice of displacement,' involving 'the passage of signs from one format to another' (2009, p. 138). Bourriaud offers as examples of this practice the work of Korean artist Kim Soo-ja, combining a Tao-inspired vision with minimal art and ancestral motifs; the Thai artist Surasi Kusolwong fusing Thai folk and popular culture with minimal and conceptual art; and Indian artist Navin Rawanchakai installing the aesthetic of the Bollywood movie poster alongside Hollywood science fiction within a narrative epic in the style of conceptual art. These works have in common a focus on translation or transferal from one context to another. Displacement of cultural material, from one location to a new context, is for Bourriaud a central component of much contemporary global art.

## The Global Network of Contemporary Art

The art milieu has become, especially over the last two decades, a global network. The international biennales, triennales, and other recurring shows have established themselves as defining events of contemporary art. Of the world's art biennales, Venice, Sao Paolo, and Sydney are the oldest, while many other world cities have expanded the circuit of international art events by founding biennales or triennales. One feature of globalist art, as Butler has observed, is that it is 'defined not by artists and art works but by curators and exhibitions' (Butler 2010, p. 58). Large-scale exhibitions such as Documenta or the Venice Biennale are conceptualised by curators along loosely binding themes, with international artists chosen to respond to these themes with new works or with relevant existing works.

The curatorial concepts are often politically urgent and informed by critical theory, political theory, radical continental philosophy, and other academic disciplines including cultural geography and anthropology, as well as art history and theory. As an example, Documenta—the world's premier art event—was curated in 2007 by Roger Buergel and Ruth Noack, who posed three questions to which artists were invited to respond: Is Modernity our antiquity? What is bare life? What is to be done? The second two questions, inspired by Giorgio Agamben and Lenin, provoked responses in photographic and video works documenting refugees, detention camps, the aftermath of war, and social inequality. The 2012 Documenta, curated by Carolyn Christov-Bakargiev, eschewed an explicit theme but was organised around a set of ideas, crystallised by the curator as 'a significance given … to the act of being emplaced' (2012, p. 7). This notion of emplacement entailed 'the conditions in which artists and thinkers find themselves acting in the present—of being "on stage"/"under siege"/"in a state of hope"/"on retreat"' (Christov-Bakargiev 2012, p. 7).

Many of the recent artworks shown at biennales and other global art events favour digital photography and video, as well as other forms including installation and performance, in documenting social and political realities. The idea of place is often foregrounded in these works. Christov-Bakargiev articulated the vision of Documenta (13) in 2012 in these terms, whereby artists often engaged with the various sites of the exhibition in Kassel, Kabul, Alexandria–Cairo, and Banff, 'unfreezing the associations that are typically made with those places and stressing their continual shifting'(2012, p. 7). The 2017 Documenta, under artistic director Adam Szymczyk, effected a form of displacement in the very staging of the exhibition. Szymczyk decentralised Documenta away from Kassel, staging part of the exhibition in Athens, with a working title of 'Learning from Athens.' Many of the artworks within the exhibition examined a European perspective from the vantage point of Greece, even if the works were installed in Kassel.

## Images of Displacement

Because the contemporary artist-semionauts are operating within a globalised regime, many current works have a geopolitical aspect. Contemporary art should be understood in relation to the widespread perception that humans have altered their orientation in space–time. The previous orientation in space and time has been 'dis-placed,' it is widely assumed, by 'the great liberal utopia of globalisation, an ordering of the earth and its beings which claims to do away with our orientation' (Dean 2010, p. 20). The processes of globalisation—commonly understood as an intensified flow of capital, information, and people—function on many levels. Migration, within and across national borders, is a significant factor, not least with regard to the trauma of displacement. Digitised images are also part of this global flow, as is the capital invested in the global financial system and international art market.

One of the most thorough and engaging visualisations of the geopolitical reality of displacement has been effected in the audio-visual installation *Exit* (2015). *Exit* is a video installation by the American design studio Diller Scofidio + Renfro, with collaborators including the French urban theorist Paul Virilio. Originally exhibited in Paris in 2008, an updated version toured the world from 2015. *Exit* makes sophisticated use of data visualisation to show the flow of displaced persons due to globalisation and climate change. The installation comprises multiple large, curved screens, before which viewers sit within the installation. On the screens are projected maps and images of the globe, populated with imagery representing numbers of displaced persons. The screen-based installation creates an immersive effect for the viewer, who is surrounded by visualised and dynamic flows of data. Statistics are animated through data visualisation to convey the sheer volume of displaced individuals around the world. Thirty-six million people were displaced in 2008 alone; twenty-six million people have been displaced by environmental disasters since 2008. The likely impact of climate change on dispossession and displacement is modelled into the future and visualised in the data display of *Exit*.

As social theorist Mitchell Dean has observed, 'the rationality of globalisation portends an overcoming of space and time in a great frictionless circulation of things, information and people' (Dean 2010, p. 10). Much recent art has been concerned with refuting this liberal utopian vision, or at least with exposing some of the friction within the mechanism of world order. Objects and places are given visual representation and left to resonate within circuits of knowledge and patterns of local and global orientation.

Artists are often fascinated by the underside of globalisation—by the failures, the victims, the unseen casualties of the shining new networks of communication and trade. In the video work *Bade Area* (2005), Taiwanese artist Chen Chieh-jen explores an abandoned factory site. This derelict space is typical of industrial sites that once represented Taiwan's 'economic miracle,' now forgotten since the departure of industry to new territories offering cheaper production and labour costs. This ruthless economic logic, central to the globalised economy, has left the artist 'enraged' (Wang 2006, p. 96). For the *Bade Area* video work, he asked laid-off employees to revisit their former workplace—but in this context, the ex-workers are adrift like ghosts haunting their former abode. They are displaced in both time and space, 'like spectres in mourning, reminiscing their past life but lost in the present, not knowing where their future lies' (Wang 2006, p. 96).

## Displaced Persons: The Camp

In his book *Homo Sacer: Sovereign Power and Bare Life* (1998), Giorgio Agamben revived the distinction made in classical Greek thought between *zoē* (bare life) and *bios* (life of the *polis*—the political or good life). For Agamben, sovereign power, including that of the modern state, has always entailed the attempt to control and regulate bare life; this biopolitical impulse is nowhere more evident than in those 'states of exception' where sovereign power has been most apparent. In this context, Agamben regards the camp as the 'hidden paradigm of the political space of modernity' (1998, p. 123). The camp is 'the pure, absolute, and impassable biopolitical space' because it is 'founded solely on the state of exception'; it is 'the space that is opened when the state of exception begins to become the rule' (Agamben 1998, pp. 123, 168–169). This formulation is true not only of the concentration camps of World War II, but also of the contemporary camps set up for refugees and political prisoners. For Agamben, the camp is 'the hidden matrix and *nomos* of the political space in which we are still living' (1998, p. 166).

Agamben's treatise on bare life and the figure of the camp as 'biopolitical paradigm of the modern' (1998, p. 117) has been extremely influential in critical thought and curatorial practice. Agamben has built on the theorising of the biopolitical—the intersection of power and the body—developed by Michel Foucault, fusing it with a discourse on political theory and the state inaugurated by Carl Schmitt, Hannah Arendt, and other theorists. Agamben's conception of the camp and bare life has gained great purchase in the contemporary context of globalisation, in which the movements of refugees, asylum seekers, prisoners of war, and migrant workers are political realities.

The impact of Agamben's thought on contemporary art—as reflected in the enshrining of the question 'What is bare life?' as a central premise of the 2007 Documenta—may be attributed in part to his crystallisation of a complex argument in the concrete figure of the camp. Contemporary images of displaced persons held in prisoner of war or refugee camps are familiar to a mass audience due to their circulation in the media. It is readily apparent that sovereign states, including liberal democracies, operate spatial exclusion and enclosure as found in the detention camps for refugees or those for 'unlawful enemy combatants' held at Guantánamo Bay and other locations. As Mitchell Dean has remarked, these camps represent the 'spatialisation of the state of exception,' as a place where anything is possible, 'beyond the language of law and right, a kind of territorialisation beyond law, an orientation without order' (Dean 2010, p. 32).

Agamben has provided a critical reading of globalisation, focusing on the techniques 'designed to impose non-localised forms of order on specific locations in the world, as and when needed' (Murphie and Potts 2003, p. 171). The camp is for Agamben the emblem of this 'dislocating localization,' which embodies displacement and disjunction in its functions. The camp, and its contemporary 'metamorphoses' such as waiting areas at airports, constitutes 'the fourth, inseparable element that has now added itself to—and so broken—the old trinity composed of the state, the nation (birth), and land' (Agamben 1998, pp. 175–176).

The theme emanating from many contemporary artworks, including those exhibited at recent Documentas and Venice Biennales, is violence: of war, occupation, displacement, hatred. This violence is seen to emanate in and across states unable to control internal hostilities, unable to deal with the flow of refugees, immigrants, or victims of war across their borders. A related theme concerns the barriers erected in hostile and fearful times: at detention camps, refugee camps, within brutalised or paranoid states. Detention camps erected in liberal democratic states, troubled by the influx of refugees, also feature in contemporary art works.

Australian artist Rosemary Laing's large colour photograph of the Woomera detention centre in Australia is a stark testament to the force of barriers. The title of the work, *Welcome to Australia* (2004) is brutally ironic—but the physical reality of the detention centre in the desert is still more brutal. Laing has photographed the exterior of the camp as a severe construction of barbed wire fencing and search lights. It is a zone of exclusion mounted in the desert, with no sign of humanity or any living thing. These detention centres have been devised to accommodate refugees and asylum seekers while they are—to use the terminology of the official government rhetoric—'processed.' The detention centres exemplify the political state of exception constituted by the camp, as articulated by Agamben. The goal of these buildings, like the aim of the government directives that enabled them, is dehumanisation. The asylum seekers are to be locked up and kept out of view of the populace, while they are 'processed'—like objects or pieces of paper—by the government bureaucracy.

Laing has conveyed the iron will of the government's refugee policy through this stark image of its creation in the desert. This is the face of Fortress Australia. The forbidding boundary fence, the barbed wire, the searchlights, the prison-like buildings, the harsh surrounding landscape, all conspire to send the message: 'Keep out.' Conservative governments have exploited fears of refugee hordes and terrorists in sections of the electorate, but this political advantage can only be sustained by a constant repression. The faces of asylum seekers are resolutely kept from view, lest a glimmer of their humanity is glimpsed by the Australian people. Hence the camp is only visible to observers from the outside: it is protected and defined by its exclusionary walls. The dehumanisation of refugees is enforced behind these fortress walls, in this modern-day camp and state of exception, deep in the interior heartland of a wealthy liberal democratic nation.

Another Australian artist, Mike Parr, has confronted the political issue of refugees and camps in a series of performance art works. These works, including *Close the Concentration Camps* (2002), *Aussie, Aussie, Aussie, Oi, Oi, Oi (Democratic Torture)* (2003), and *Kingdom Come And/or Punch Holes in the Body Politic* (2005), involved physical ordeal and the endurance of pain on the part of the artist. In *Close the Concentration Camps* and *Aussie, Aussie, Aussie,* Parr's face was sewn with stitches through the skin and lips. In *Kingdom Come,* Parr, dressed in an orange suit resembling those worn by the inmates of Guantánamo Bay, received an electric shock every time a member of the audience ventured past a point marked by sensors.

These performance works function on several levels, beyond a simple political protest against government policies concerning asylum seekers and the 'war on terror.' As the performance theorist Edward Scheer has remarked, Parr's acts of physical ordeal, including violence enacted on his own body, were an attempt to embody the brutality of those policies condoning torture and the detention of displaced persons. This brutality had produced 'a similarly monstrous distortion in the image of Australian identity'; Parr's pain-riddled performances were 'a symbolic response to a reconfiguring of the symbolic order in Australia towards violence and paternalism' (Scheer 2009, p. 129).

One strategy adopted by artists making works concerned with displaced persons is to overcome the invisibility imposed on camp detainees. The South African artist Candice Breitz makes visible—and audible—the stories of six displaced individuals in her video installation *Love Story*, exhibited at the Museum of Fine Arts, Boston, in 2016. Video screens display detailed interviews with the six individuals, who have fled their homelands—Syria, Angola, Democratic Republic of Congo, India, Venezuela, Somalia—due to persecution or war, and seek asylum elsewhere. As viewers enter the installation, they first encounter an edited sequence of excerpts from the six interviews, performed by Hollywood actors Alec Baldwin and Julianne Moore. Breitz indicates, with this Hollywood version of displacement narratives, that many viewers will identify more with celebrity figures than with the displaced individuals of real life. The lengthy video interviews with the real displaced persons allow viewers to learn something of their circumstances and their plight.

When *Love Story* was exhibited at the NGV Triennial in Melbourne, in 2018, Breitz insisted on changing the title of the work to *Wilson Must Go*. This was the artist's protest against the National Gallery of Victoria (NGV) using security firm Wilson: this company had overseen the imprisonment of thousands of refugees and asylum seekers to Australia on Nauru and Manus Island in Papua New Guinea. In 2016, a set of documents was leaked from the Nauru detention centre, detailing more than 2000 incidents of self-harm and assault among detainees. These leaked documents motivated thirty-three Australian artists to contribute to an exhibition, *All We Can't See*, in Sydney, in 2018, focused on the conditions endured by those detained on Nauru. Artist Ben Quilty spoke of his motivation: 'What I think artists need to do now is try and humanise things … most of our recent governments go out of their way to dehumanise issues.' Fellow artist Luke Sciberras endorsed this view: 'The fundamental role of art is to make the invisible visible' (Petrovic 2018, p. 20).

Another means of rendering visible displaced persons otherwise kept from view is to deploy vision technologies, invented by the military for reconnaissance or surveillance purposes. The Irish artist Richard Mosse uses a long-range thermal imaging camera in his 2016 film *Incoming*, a work commissioned by the Barbican Art Gallery in London and the NGV in Melbourne. The subjects of this heat-seeking camera technology are arrivals by refugee boats in Lesbos, detainees at a refugee camp on the Greek–Macedonian border, as well as displaced persons in Turkey and Iraq. While this technology—which detects human activity at long distance and even at night—is used by the military for border surveillance and reconnaissance at a distance of up to fifty kilometres, in the hands of Mosse, it makes visible the everyday activities of detainees locked away in camps. Even at night, the refugees light up the video screen with heat signatures, denoting vitality and purpose within the camps.

## Colonial Displacement

The violence of displacement has been all too apparent, centuries before the current movements across the globe, to Indigenous peoples. The digital photographic work *Torch* (2006), by the Indigenous Canadian artist Rebecca Belmore, addresses the implications of colonialism as forced displacement of Indigenous peoples. The arm of Liberty, bound by the Stars and Stripes, is inverted and holds uprooted long black hair. This image connects liberty with discipline in the inverted bound arm and 'replaces the torch of enlightenment with the darkness of uprooted hair' (Dean 2010, p. 28). The Age of Reason oversaw the appropriation of land, the violent dispossession or uprooting of indigenes, and their 'binding within the nation-state and its very symbols of liberty' (Dean 2010, p. 28). Colonialism meant uprooting in many dimensions at once: physical, emotional, spiritual, political.

Tracey Moffatt, a member of the Aboriginal 'Stolen Generations'—when Australian Government policy was to remove certain Aboriginal children from their parents and place them in state care—has played out narratives of dislocation and identity within her art. In the short film *Night Cries* (1990), an Aboriginal woman cares for her dying invalid white mother. Moffatt, who was raised by white parents after being taken from her Aboriginal mother by state authorities, treats the relationship with a degree of ambivalence. Feelings of resentment and guilt are overlaid upon the mixed-race parental relationship. The heightened artifice of the studio setting, with its lurid colours and non-naturalistic background, invites a symbolic reading—or readings—of the unfolding narrative. The death throes of the mother provoke both grief and liberation for the distressed child, who has assumed the role of carer.

Yet Moffatt, who rejects the nomination of 'Aboriginal artist' because of the straitjacketing connotations and expectations imposed by such a term, produces work of rich complexity, art that resists simplistic postcolonial interpretation. The daughter in this film displays love and obligation as well as resentment towards her mother. A child of the Stolen Generations of necessity experiences a displaced emotional bond with substitute parents approved by the state. Yet, this displaced bond is also a genuine emotional relationship. Moffatt's series of digital images, shot, like *Night Cries*, to highlight artifice rather than naturalism, present complex layers of race, identity, emotional attachment, and shards of storytelling. The series *Something More* (1989), for example, depicts broken elements of an outback story involving mixed-race, sex, violence, road journey, and escape. The narrative suggested by these pictorial shards is enigmatic, yet clearly includes a dynamic element of movement across the land, alongside various racial and gender elements of identity, cross-fertilised in unspecified combinations.

## Ai Weiwei: Parables of Displacement

At the 2007 Documenta exhibition, Chinese artist Ai Weiwei offered an ingenious comment on the theme of mass migration in the globalised age. His work *Fairytale* involved the transport of 1001 Chinese citizens to Documenta in five stages. Awaiting their arrival were 1001 wooden chairs from the Quing dynasty, spread throughout the four main venues of Documenta. These antique chairs attested both to the Chinese past and to the contemporary Chinese citizens they would serve. The artwork to a large extent comprised the act of migration—the feat of physically moving 1001 Chinese citizens, who had not previously left China—to Kassel in Germany.

This conceptual work was at once a global fable—the title *Fairytale* referred to the Brothers Grimm, former residents of Kassel, and to the 1001 nights of the Scheherazade story—and a provocative act in the context of the Chinese Communist Party's sensitivity to travel and international relations. The art theorist Terry Smith considers *Fairytale* 'a delicious parody of the restrictions to travel that characterized China during the Mao years' (2011, p. 58). The welcoming chairs, arranged in clusters, fostered informal and lively discussions amongst the Chinese once they arrived, enabling expression of 'civil society in microcosmic form—exactly the kind of formation most threatening to Communist Party centralists' (Smith 2011, p. 58).

At this time, Ai Weiwei was able to exploit his standing as a major international artist while maintaining a dissident stance within China. *Fairytale* exposed the contradictions in the party's partial embrace of globalisation and the market while attempting to retain authoritarian control over its citizens. The 1001 Chinese participants in *Fairytale* experienced life outside China for the first time: for one month they were free to engage in the cultural and political pursuits available in a European city. Within China, however, such freedoms may be abruptly curtailed. Despite his international status, Ai's role as figurehead of dissent earned him severe attention from party officials: in

2011, he was detained and beaten by authorities for eighty-eight days, and his passport revoked.

Ai Weiwei's artworks focused on displacement with a new intensity from 2015, when he was allowed to leave China by party authorities. Now a political exile, his political perspective took on an international focus. His concern for displaced persons was triggered by his witnessing, in 2015, of exhausted and distressed refugees who had arrived at Lesbos by boat. This refugee vessel was later recreated, on enormous scale, by Ai in the work *Law of the Journey*, which was exhibited at the Sydney Biennale of 2018. The inflatable rubber boat of this work, made of the same material as the refugee boat on Lesbos, is 70-metres long, filled with 258 faceless refugee figures also made from rubber.

This monumental and striking work, the centrepiece of the Biennale, was described by its creator as 'a brutal and harsh work of art,' designed for a political purpose (Maddox 2018, p. 5) This political aim—the raising of awareness of the global refugee crisis—was furthered by Ai's documentary film *Human Flow*, which premiered at the Sydney Biennale before its international release in 2018. The gigantic artwork and the documentary generated great media attention for Ai, which he exploited by subjecting Australia's offshore detention policies to severe criticism in the press. The world's refugee crisis received significant media attention, at least for a short time, due to the artistic/political projects launched by Ai.

He has continued to take advantage of his status as one of the world's most high-profile contemporary artists to address concerns for the displaced. A series of large-scale installations, exhibited at major international venues or in biennales, have focused on refugees. *Laundromat* (2016), for example, installed at Deitch Projects in New York, assembled thousands of items—clothes, shoes, blankets—left behind by refugees from Syria, Afghanistan, and Iraq at Idomeni, a village on the Greek–Macedonian border. At one time, Idomeni accommodated up to 15,000 refugees in dire conditions. A similar public pronouncement was made by the artist in Berlin in 2016, when 14,000 orange life jackets, left behind by refugees arriving on Lesbos, were attached to columns of the Konzerthaus. Ai has also intervened directly in improving refugee camp conditions, distributing thousands of solar-powered lights to camps on the Greek islands.

Ai Weiwei's political call to action to address the crisis of the world's sixty-five million refugees is articulated in his artworks, his documentary, and in media interviews and articles. In a 2018 newspaper article in *The Guardian*, he described his experience visiting twenty-three nations and forty refugee camps while filming *Human Flow*. In this article, he challenged the leaders of

Western nations to bear their responsibility for the global crisis of displaced persons:

> the west—which has disproportionately benefited from globalisation—simply refuses to bear its responsibilities, even though the condition of many refugees is a direct result of the greed inherent in a global capitalist system. (Ai Weiwei 2018, p. 48)

In 2018, he encapsulated his own role as political artist:

> My role is very simple … I defend human rights, human dignity. I defend people who are voiceless or silent or will never even have a chance to speak from a platform. (Maddox 2018, p. 4)

Ai Weiwei, and many other international artists, continue to provide a platform for the millions of displaced persons around the world.

# References

Agamben, G. (1998). *Homo sacer: Sovereign power and bare life* (D. Heller-Roazen, Trans.). Stanford, CA: Stanford University Press.

Ai Weiwei. (2018, February 9). The west has profited from globalisation but refuses to bear its responsibilities toward displaced people. We have forsaken our belief in shared humanity. *The Guardian Weekly*, p. 48.

Bourriaud, N. (2002). *Postproduction: Culture as screenplay*. New York, NY: Lukas & Sternberg.

Bourriaud, N. (2009). *The radicant*. New York, NY: Lukas & Sternberg.

Butler, R. (2010). The world is not enough. In C. Merewether & J. Potts (Eds.), *After the event: New perspectives on art history* (pp. 57–67). Manchester: Manchester University Press.

Christov-Bakargiev, C. (2012). *Documenta (13): Catalog 111/3, the guidebook*. Kassel: Hatje Cantz.

Dean, M. (2010). Land and sea: 'In the beginning all the world was America'. In C. Merewether & J. Potts (Eds.), *After the event: New perspectives on art history* (pp. 20–37). Manchester: Manchester University Press.

Maddox, G. (2018, March 13). Taking refuge in the art of an endless struggle. *Sydney Morning Herald*, pp. 4–5.

McLuhan, M. (1964). *Understanding media*. London: Abacus.

Murphie, A., & Potts, J. (2003). *Culture and technology*. Basingstoke: Palgrave Macmillan.

Petrovic, S. (2018, January 31). Artists give a human face to official reports of misery. *Sydney Morning Herald*, p. 20.

Pound, E. (1934). *ABC of reading*. New York, NY: New Directions.

Scheer, E. (2009). *The infinity machine: Mike Parr's performance art 1971–2005*. Melbourne: Schwartz City.

Smith, T. (2011, May). Art of dissent. *The Monthly*, 58–59.

Wang, C. C. J. (2006). Chen Chieh-jen. In C. Merewether (Ed.), *Zones of contact: 2006 Biennale of Sydney catalogue* (pp. 96–97). Sydney: Biennale of Sydney.

# 48

# Reclaiming Safe Spaces: Arts-Based Research, Advocacy, and Social Justice

### Nelli Stavropoulou

## Introduction

In line with this section's theme, this chapter seeks to introduce arts-based research as a methodological approach that promotes self-expression and allows possibilities for challenging and reimagining experiences of displacement. This chapter aims to provide an introduction into key theoretical debates on arts-based research while also outlining the transformative potential of such methods in relation to self-representation, social justice, and advocacy. By introducing the mediated nature of refugee visualities, it outlines how the dominant visual culture frames displaced individuals in accordance with dominant narratives. The following sections move on to introducing the field of arts-based research and presenting exemplars of arts-based research that promote self-expression, participation, and accomplish social justice. The final section offers a critical interrogation of the potential of creative methods within the field of displacement and outlines a reflexive pathway for resistance.

N. Stavropoulou (✉)
Durham University, Durham, UK
e-mail: nelli.stavropoulou@durham.ac.uk

© The Author(s) 2020
P. Adey et al. (eds.), *The Handbook of Displacement*,
https://doi.org/10.1007/978-3-030-47178-1_48

## Media Politics of Representation

'What is it about displacement that detracts from one's ability to speak for oneself?' asks McNevin (2010, p. 144) when reflecting on processes of depoliticisation amongst displaced individuals as a consequence of their liminal legal status. In today's media-saturated visual culture, refugees and asylum seekers are increasingly represented through news media, popular culture, political manifestos, nongovernmental/humanitarian campaigns, art, and social media networks. Oscillating between invisibility and hypervisibility, between recognition and misrecognition, displaced individuals 'have an ambivalent relationship to the aesthetic forms that seek to represent them, one which touches on questions of communicability, visibility and ethics' (Woolley 2014, p. 3). As Haaken and O'Neill (2014, p. 82) observe, their stories and experiences are predominantly presented by intermediary others who have not directly experienced forced displacement themselves. Therefore, displaced individuals may lose their 'voice' as a result of losing their right to self-representation (Baker 1999).

Photographic images of individuals travelling across the Mediterranean on overcrowded boats became a major visual motif in documenting the 2015 story of Europe's 'refugee crisis'—the largest humanitarian crisis according to United Nations High Commissioner for Refugees (UNHCR) with an estimated record of 65.3 million people displaced worldwide as a result of conflict, persecution, and other human rights violations by the end of 2015 (UNHCR 2016). The European 'refugee' or 'migrant' crisis refers to the global mass movement of individuals in the European Union. The interchangeable use of 'migrant' or 'refugee' across media publications further highlights the constructed nature of representing displacement. Alongside the rising numbers of individuals crossing borders, we also witnessed the mass circulation of photographic images of individuals as they embarked on life-threatening journeys in search of safety (Stavropoulou 2019). Such proliferating visual documentation contributed towards fuelling a public imagination that conceptualises forced displacement as a humanitarian catastrophe, a threat to Europe's welfare, and an erosion of national borders and identities (Friese 2018, pp. 45–46).

The prevalence of refugees and asylum seekers in the press is not a new phenomenon. There is a rich body of academic work on visual representations of human mobility crises (King and Wood 2001; Wright 2002, 2004) and a growing literature on media representations of migration (Van Dijk 1991; Lenette and Cleland 2016; Chouliaraki and Stolic 2017). The visual culture surrounding communities affected by displacement has solidified the image of 'the refugee' as a universal 'special kind' of person (Malkki 1996; Mannik 2012) that challenges 'here' or 'there' divisions and instead occupies a liminal space in relation to citizenship, belonging, and recognition. Displaced individuals are increasingly represented in essentialist ways that follow the dominant narrative tropes of victimhood, loss, and suffering (Robertson et al. 2016; Chouliaraki and Stolic 2017), thereby perpetuating notions of a 'bare humanity' (Malkki 1996). According to Chouliaraki and Stolic (2017, p. 1164), the figure of the refugee is fundamentally ambivalent: 'on one hand, the refugee emerges as a victim of geo-political conflict in need of protection, yet on the other, she/he appears as a threat to the nation-based order and is to be excluded from "our" community.' Visual regimes of migration inform our understanding and our responses towards refugees and asylum seekers. As Bischoff (2018, p. 34) suggests, 'the naming and imaging of migration is ultimately also the foundation of political power.'

The political power of naming is particularly evident in understanding the difference between the categories of 'refugee' and 'asylum seeker,' as to be classified as a refugee presupposes the recognition of one's right to asylum in accordance with refugee law. Castles' (2003) concept of the 'asylum–migration nexus' is particularly useful in understanding the fluid relationship between migration and asylum, and in particular, the relationship between the political construction of displaced individuals in relation to distinctions between the 'deserving persecuted victim' and the 'opportunistic economic migrant.' Especially in the Global North, a dominant culture of disbelief underpins the contemporary asylum process, branding asylum seekers as liars or 'bogus' claimants (Sales 2002). Such understandings rest within a complex matrix of dominant narratives orchestrated by media, political, and social structures. As Nagarajan (2013, para. 1) suggests, 'politicians and the press are locked in a cycle of increasing anti-immigrant rhetoric, presented as "uncomfortable truth."' What is even clearer, however, according to O'Neill (2018) is:

> … that refugees and asylum seekers have become the folk devils of the twenty-first century and overall mainstream media representation of the asylum issue, the scapegoating of asylum seekers and tabloid headlines help to create fear and anxiety about the unwelcome "others" and to help set agendas that fuel racists discourses and practices. (2018, p. 76)

The media therefore play a central role in disseminating news, and in doing so, instruct particular 'ways of seeing' (Berger 1972) displaced individuals as a 'culturally inferior "other"' (Khan 2012, p. 54) with 'liminal social status'

(Lynn and Lea 2003, p. 446). Such a socially constructed 'other' is firmly placed within a set of contrasting binaries: deserving/undeserving, legal/illegal, and same/other (Sales 2002). The ways in which individuals seeking asylum are represented matters as such conceptualisations become engraved in the public imagination and support demarcating processes between 'us' and 'them.' In this context, the need for the telling of personal experiences of life in exile becomes imperative in order to raise awareness and to advocate for a perception of refugee lives which is more understanding of their situation (Stavropoulou 2019). While there is an extensive body of photographic work documenting the refugee experience, individuals seeking asylum remain on the other side of the camera retaining the role of the passive object of inquiry and therefore remaining excluded from making their worldview visible (see Robertson et al. 2016). This is where the use of participatory arts-based methodologies becomes political, by inviting participants who have experienced forced displacement to share their stories, and in doing so, become in charge of their self-representations.

## Defining Arts-Based Research

The term "arts-based research" (ABR) emerged between the 1970s and the 1990s as a methodological genre that incorporates creative practices in qualitative investigations (Leavy 2017). In recent decades, the field of arts-based research has increasingly gained momentum and institutional recognition as an epistemological process that challenges textual and linguistic constraints and sets in motion diverse processes of meaning-making and representation (Nunn 2017, p. 4; Pink 2001; Pentassuglia 2017). There is a rich body of literature concerned with arts-based research as a field of inquiry (Barone and Eisner 2011; Finley 2008) and a range of methodological approaches including a/r/tography, arts-informed research, and arts inquiry (Knowles and Cole 2008; McNiff 1998). Arts-based research incorporates creative means throughout different stages of the research process (e.g., data production, evaluation, dissemination, etc.) and is used across a vast array of disciplines including anthropology, sociology, psychology, and education (Knowles and Cole 2008). Furthermore, arts-based research involves a diverse range of artistic methods such as photography, filmmaking, digital storytelling, performance, creative writing, visual arts, and poetry, among others (Leavy 2017).

Arts-based approaches are closely associated with the assumption of 'giving voice' to research participants. Bogdan and Biklen (1998) define such a process as 'empowering people to be heard who might otherwise remain silent'

(1998, p. 204). Nevertheless, such a concept reveals multiple tensions as it presupposes that participants are lacking voice and may therefore reinforce hierarchical divisions between participants and researcher(s). Instead, a recognition of existing voices that become amplified through the research process might be more appropriate, as well as an acknowledgement that for a voice to heard, active and compassionate listening needs to be present (Vacchelli 2018). Another important ethical consideration of art-based research is the blurring of power relationships between participants and researcher within the investigative process (Foster 2015, p. 2). The importance of collaboration and the idea of 'handing over' creative control are captured by Mullen (2003) who suggests,

> We need to find ways not just to represent others creatively, but to enable them to represent themselves. The challenge is to go beyond insightful texts, and to move ourselves and others into action, with the effect of improving lives. (p. 117)

Nevertheless, it is important to recognise that despite the participatory nature of arts-based research, power hierarchies may not be completely eliminated. As Leavy (2017) argues, academic research is defined by an ethical responsibility to ensure that participants are at the forefront of the research process. The ability of arts-based research to reach decision-making audiences through creative emotive cultural texts makes it an ethically responsible research practice that allows knowledge dissemination and advocacy, while also fostering connections between wider social issues, thereby ensuring a socially engaged, integrated approach to research (Leavy 2009). Equally, attention to safeguarding participants' stories is essential as often creative outputs can become 'extracted and produced in ways (…) to conform to the aesthetics of powerful spectators' thus sacrificing the transformative potential that such creative approaches can offer (Haaken and O'Neill 2014, p. 84).

## Arts-Based Research in Practice

Arts-based research allows synergies between artistic expression and epistemological processes. In doing so, it reveals a creative way of engaging with individuals' emotional and symbolic worlds through inviting participants to reflect on their creations and reveal their interpretations (Pink 2001). Such creative methodologies support participation and inclusion in the production of knowledge, while also create new spaces for critical dialogue to emerge. More importantly, arts-based approaches can support enactments of 'cultural citizenship' (see Pakulski 1977), defined as 'the right to presence and visibility, not marginalization; the right to dignity and maintenance of lifestyle—not assimilation to the dominant culture; and the right to dignifying representation—not stigmatisation' (O'Neill 2018, p. 74). As this section will outline, an arts-based approach can be especially beneficial in engaging marginalised groups such as displaced individuals, as the arts can illuminate complex social worlds and reveal 'untold stories' (Cole and Knowles 2001, p. 211).

## New Vocabularies and Safe Spaces of Representation

As Tribe (2008, p. 941) argues, 'art extends our insights beyond the literal and more easily allows the symbolic, the impressionistic, the imaginative, the ironic and the surreal to challenge and extend our thinking.' According to Nunn (2017, p. 5), a principal strength of arts-based research is the 'range of vocabularies' it can offer for participants to engage with the complexities of displacement. Arts-based research can therefore support the creation of 'multivocal, dialogic texts' that invite new interpretations of experiences (O'Neill and Hubbard 2010, p. 47).

In her research with young asylum seekers, Bagnoli (2009) examined the use of visual arts-based and graphic elicitation methods within the context of an interview setting. Bagnoli observed how individuals departed from a verbal mode of thinking and instead shared non-standardised answers that illuminated the complexity of exilic experiences and encouraged a 'holistic narration of self' (2009, p. 566). As Bagnoli argues, lived experiences are 'made of a multiplicity of dimensions, which include the visual and the sensory, and which are worthy of investigation but cannot always be easily expressed in words' (2009, p. 547). Arts-based approaches are therefore particularly beneficial in imaginatively engaging with experiences of displacement through the combination of non-verbal means that allow individuals to share embodied, sensory, and tacit knowledge.

As part of their co-produced transdisciplinary arts and language research project *Migration and Home: Welcome in Utopia*, McKay and Bradley (2016) engaged newly arrived individuals and refugee support integration organisations in the United Kingdom in a series of arts-based workshops involving silk painting and songwriting. The project supported the creation of a safe place where participants could mix colours and personal experiences therefore creating a new mode of language, a practice of speech and collective dialogue (McKay and Bradley 2016). More importantly, the silk paintings provided opportunities for metacommentary (McKay and Bradley 2016; Rymes 2014) as individuals revealed hidden stories and creative motivations behind each piece. Recognition of one's individuality is a prerequisite for meaningful participation. Through their engagement with creative practices, individuals seeking asylum can 'make sense' of their experiences in ways that resist the homogenising effect of the 'refugee experience' as constructed by the media and legal discourses (https://welcomeutopia2016.wordpress.com).

Creative methods such as participatory photography invite participants to 'talk back' and to transgress dominant stereotypical representations through the creation of alternative images. Such a process was evident in Lokot's (2018) photovoice investigation with Syrian asylum seekers in Jordan as individuals were able to voice their feelings, viewpoints, and opinions on important social issues through the medium of photovoice. Photovoice is a community-based participatory visual method that involves giving cameras to participants and inviting them to identify core issues about their community in order to inform decision-making (Wang and Burris 1997). As Lokot (2018) observed, through their photographs, participants identified safe and precarious spaces and explored notions of hope, belonging/non-belonging, safety, and possible futures. Echoing Susan Sontag (1979), photography can be political as it can 'help people to take possession of space in which they are insecure' (Sontag 1979, p. 9).

## Empowerment, Transformation, and Social Justice

Due to its possibility to 'imagine what is possible' (hooks 1994, p. 281), art lends itself as a powerful tool to 'excavate the recurrent patterns of inequity and oppressions, as well as the acts of transformation and activism' (Villaverde 2008, p. 123). This is because 'art as inquiry has the power to evoke, to inspire, to spark the emotions, to awaken visions and imaginings, and to transport others to new worlds' (Thomas 2001, p. 274). As Keifer-Boyd suggests, 'a social justice approach to arts-based research involves continual critical reflexivity in response to injustice' (2011, p. 3). Such a reflexivity seeks to identify, highlight, and challenge structural and symbolic injustice.

The potential of arts-based research as a pathway to social justice has been explored by sociologist/criminologist Maggie O'Neill (2008, 2010, 2018; O'Neill et al. 2019) who argues that action-oriented participatory arts-based research invites the reimagining of lived experiences and the creation of praxis as purposeful knowledge. Throughout her thirty-year trajectory of participatory action-oriented research, O'Neill has been committed to conducting participatory research that invites participants to reflectively re-engage with their experiences through creative means. O'Neill developed the epistemological and methodological framework of ethno-mimesis in 1999 in order to capture the process of reflective ethnography and artistic production. For example, through the participatory project Sense of Belonging, O'Neill and Hubbard collaborated with different groups of asylum seekers, artists, and arts organisations across four different locations in the United Kingdom, in order to produce artwork that invited alternative understandings of exilic experiences. Through the project's guided walks and arts/research workshops, participants negotiated new spaces for their stories to emerge, which allowed them to accomplish social justice via a politics of recognition (O'Neill and Hubbard 2010) (Figs. 48.1 and 48.2).

Creative methodologies can therefore accomplish social justice by inviting individuals to imagine and reconstruct stories that allow them to challenge dominant narrative discourses. A similar transgression of narrative boundaries was accomplished through the interdisciplinary academic investigation *Refugee Hosts: Local Community Experiences of Displacement from Syria: Views*

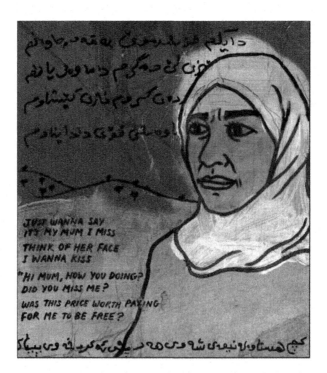

**Fig. 48.1** One of 48 panels painted as a result of interviews and workshops with the Dreamers Group of young accompanied refugees and asylum seekers in Loughborough. *I Had a Dream* by Paul Gent, Sense of Belonging—AHRC-funded project led by Maggie O'Neill and Phil Hubbard. (Courtesy of Prof. Maggie O'Neill and Paul Gent)

**Fig. 48.2** One of 48 panels painted as a result of interviews and workshops with the Dreamers Group of young accompanied refugees and asylum seekers in Loughborough. Paul Gent, Sense of Belonging—AHRC-funded project led by Maggie O'Neill and Phil Hubbard. (Courtesy of Prof. Maggie O'Neill and Paul Gent)

*from Lebanon, Jordan and Turkey.* Fiddian-Qasmiyeh, Ager, Rowlands, and Stonebridge examined experiences of local and displaced communities in the Middle East, focusing on the opportunities and challenges that arise when local communities become involved in activities that support displaced groups. An important aim of the project was to disrupt grand narratives and representational strategies that confine displaced individuals in discourses of victimhood, suffering, and the 'super refugee' (Fiddian-Qasmiyeh 2017). The project offered a critical space for individuals to document and trace their experiences of displacement but also to become politically engaged as they challenged and offered their responses to issues of representation (https://refugeehosts.org/).

Another example of a creative synergy between socially engaged research and creative approaches was accomplished through the Arts for Advocacy interdisciplinary research project, which explored the use of innovative participatory arts-based methods to facilitate critical engagement and advocacy in Morocco (https://artsforadvocacy.org). Jeffery, Palladino, and Bachelet engaged with a diverse network of displaced communities, artists, activists, researchers, migrants' representatives, and practitioners in Morocco to critically interrogate experiences of displacement and resettlement. Through working with artists, the project supported meaning-making and co-creation that interrogated notions of mobility/immobility and inclusion/exclusion through filmmaking and photography. Participants were invited to take photographs and produce short films as well as contribute towards the project's public exhibitions. Additionally, the project included arts-based intercultural workshops for members of the 'host' and 'forcibly displaced' communities, as well as supported knowledge exchange through a series of creative symposiums, workshops, public exhibitions, and conferences in both Morocco and the United Kingdom, culminating in a transferrable 'Arts for Advocacy' digital toolkit (https://artsforadvocacy.org/toolkit/toolkit-en.pdf) (Fig. 48.3).

Despite evidence of the transformative power of art, it is important to note that creative approaches should build upon participants' experiences, interests, and skills, as the expectation of 'creating art' can also serve as a barrier to access. Adaptability, modification, and on-going dialogue are necessary in ensuring an ethical and empowering research experience that invites participants to own the creative process as well as the final artefacts.

**Fig. 48.3** *Missed Call*. Photograph by Amine Oulmakki. Arts for Advocacy 2017. (Courtesy of Prof. Laura Jeffery, Dr. Mariangela Palladino, and Dr. Sébastien Bachelet)

## The Transformative Potential of Arts-Based Research in Examining Experiences of Displacement

As established throughout this chapter, the arts have a vital role in processes of social justice. In recent decades, we have witnessed a convergence between artistic production and participatory arts-based research. Writers including Bishop (2006) and Bourriaud (2002/1998) have enriched our understanding of participatory, collaborative, and relational art and have highlighted the potential of dialogue as a prerequisite for social engagement (Stavropoulou 2018). In *Conversation Pieces*, Kester (2005) argues that 'dialogical' art requires a shift in our understanding of what constitutes art. Building on Bakhtin's (1990) notion of conversation as a performative act, he suggests a departure from the visual and sensory and instead supports an understanding of art as 'a kind of conversation; a locus of differing meanings, interpretations and points of view' (Kester 2005, p. 79). Such a recognition of art as a location for negotiation and reinterpretation of subjectivities is integral to arts-based research, as such approaches can 'enable a diversity of experiences to be communicated in ways that disrupt "common sense" understandings' (Foster 2015, p. 1).

Scattered People is another example of a participatory arts-based project that re-dignified experiences of seeking asylum through music production and songwriting. The arts–health collaboration between community musicians, individuals seeking asylum, and researchers explored music participation as an embodied activity that can improve well-being as well as allow an 'unobtrusive' (Kellehear 1993) engagement with lived experiences of individuals seeking asylum in Australia (Lenette 2019, p. 178). The project established a communal creative space that initiated storytelling, self-expression, and knowledge exchange. Participants sang about their experiences of fleeing their home countries and trying to find a new home in a new place. As captured through the project's mini-documentary *Scattered People*, a sense of agency was gained through involvement in singing, songwriting, dancing, and music production as participants experienced forms of dignity that often become eroded due to their identification as 'illegals,' 'boat people,' and 'queue-jumpers' (Lenette et al. 2016, p. 137) (Fig. 48.4).

Through her international participatory arts-based project Dispersed Belongings, Nunn (2018) explored refugee-background young peoples' experiences of belonging in Gateshead, north-east England, and Bendigo, Central Victoria, Australia. In the United Kingdom, the project engaged newly arrived Syrian young people in a series of research interviews and creative workshops

**Fig. 48.4** Screenshot of *Scattered People* mini-documentary. (Available on YouTube http://www.youtube.com/watch?v=GV_US5RyIgM. Courtesy of Assoc. Prof. Caroline Lenette)

including photography, poetry, and songwriting. Working with local artists, the project supported the production of photographs and songs about participants' stories of displacement and resettlement. The project directly embedded participants in both the creative- and knowledge-production processes, as they were invited to contribute towards public exhibitions as well as conference presentations. According to Nunn (2018, p. 4), the project offered opportunities for young people 'to gain recognition as active agents—as artists, critical thinkers, representatives of their communities, and leaders' (Fig. 48.5).

Through stories individuals construct meaning, review their life trajectories, command attention, persuade, and evoke reactions while also revealing aspects of themselves to the world and across different audiences (Eastmond 2007; Riessman 1993). The notion of 'telling stories' is particularly relevant in the context of forced displacement, as individuals seeking asylum often rely on their storytelling skills in order to provide a personal account that is intelligible and which responds to legal expectations of the 'deserving refugee' (Sales 2002; Woolley 2014). Such a demand for a coherent refugee story not only constrains individual expression but most importantly (dis)places individuals in particular spaces of representation. In the asylum process, language becomes a 'point of tension, with ramifications for the asylum process' and were one can be seen to speak the 'wrong language' (McKay and Bradley 2016, p. 37).

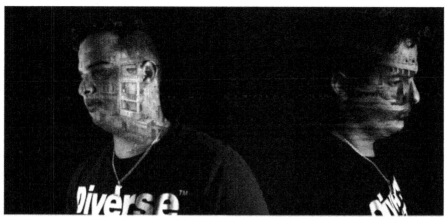

**Fig. 48.5** *My Country is Unforgettable* Mohamad El Hamood. Dispersed Belongings 2018. (Courtesy of Dr. Caitlin Nunn and Mohamad El Hamood)

Through arts-based research individuals can share stories that challenge mainstream conceptualisations and affirm their views and place within their changing social realities. However, as Greatrick and Fiddian-Qasmiyeh (2017) suggest, a meaningful and ethical engagement with experiences and stories of displacement involves not only supporting self-expression but also creating a space in which participants can resist and subvert different forms of externally imposed expectations, allowing them to step away from the narrative expectations of the 'vulnerable/violent/bogus/grateful refugee—, but also the very expectation that participants will (or should) be performing in an "authentic" way and "revealing" their "true self" during workshops' (2017, para. 14). Reflecting on their photovoice study with female asylum seekers in the northeast of England, Haaken and O'Neill (2014, p. 84) remind us that 'if a picture is thought to tell a 1000 words, it also masks a multitude of stories.' We cannot talk about a 'refugee experience'—instead, we should recognise the plurality and diversity of refugee experiences that become determined by individuals' identity markers, biographies, and past histories.

## Conclusion

As demonstrated in this chapter, art serves as a powerful tool for resisting displacement through its inherently political potential for resisting invisibility. Arts-based research is increasingly recognised and adopted across the academic community as a reflexive methodology that provides opportunities for self-expression and defamiliarizes the ordinary (Greene 1995, p. 4), while also serving as a reminder of what can be done differently. Arts-based research can transform displaced individuals' realities through facilitating cultural citizenship—understood as one's right to resist marginalisation, stigmatisation, and misrecognition. Nevertheless, it is important to remain critical as 'arts as a vehicle for social change when loosely bandied about remains simply an empty signifier; an empty slogan' (Counterpoints Arts 2017, para. 9). An ethical and reflexive approach to arts-based research involves continuously interrogating the promised transformative potential and ensuring that participants' realities become enriched through their active involvement in the production of knowledge about their lives.

## References

Bagnoli, A. (2009). Beyond the standard interview: The use of graphic elicitation and arts-based methods. *Qualitative Research, 9*(5), 547–570.

Baker, B. (1999). What is voice? Issues of identity and representation in the framing of reviews. *Review of Educational Research, 69*(4), 365–383.

Bakhtin, M. (1990). Author and hero in aesthetic activity. In M. Holquist & V. Liapunov (Eds.), *Art and answerability: Early philosophical essays by M.M. Bakhtin [Trans. V. Liapunov]* (pp. 4–256). Austin, TX: University of Texas Press.

Barone, T., & Eisner, E. W. (2011). *Arts based research*. Thousand Oaks, CA: Routledge.

Berger, J. (1972). *Ways of seeing*. Harmondsworth: Penguin.

Bischoff, C. (2018). Migration and the regime of the gaze: A critical perspective on concepts and practices of visibility and visualization. In D. Bachmann-Medick & J. Kugele (Eds.), *Migration: Changing concepts, critical approaches* (pp. 45–63). Berlin and Boston, MA: De Gruyter. Retrieved from https://www.degruyter.com/view/books/9783110600483/9783110600483-003/9783110600483-003.xml.

Bishop, C. (2006). The social turn: Collaboration and its discontents. In H. Francis, J. Jansen, & T. O'Connor (Eds.), *Rediscovering aesthetics: Transdisciplinary voices from art history, philosophy and art practice*. Stanford, CA: Stanford University Press.

Bogdan, R., & Biklen, S. K. (1998). *Qualitative research for education: An introduction to theory and method*. Boston, MA: Allyn & Bacon.

Bourriaud, N. (2002/1998). *Relational aesthetics*. Dijon: Presses du réel.

Castles, S. (2003). Towards a sociology of forced migration and social transformation. *Sociology, 37*(1), 13–34.

Chouliaraki, A., & Stolic, T. (2017). Rethinking media responsibility in the refugee 'crisis': A visual typology of European news. *Media, Culture & Society, 39*(8), 1162–1177. https://doi.org/10.1177/0163443717726163.

Cole, A. L., & Knowles, J. G. (2001). *Lives in context: The art of life history research.* Walnut Creek, CA: AltaMira Press.

Counterpoints Arts. (2017, May 17) *No boundaries 2017 – A symposium on the role of arts and culture.* Retrieved from http://counterpointsarts.org.uk/reflections-from-boundaries-2017-a-symposium-on-the-role-of-arts-and-culture/.

Eastmond, M. (2007). Stories as lived experience: Narratives in forced migration research. *Journal of Refugee Studies, 20*(1), 248–264.

Fiddian-Qasmiyeh, E. (2017). Disrupting humanitarian narratives? Representations of Displacement series. *Refugee Hosts.* Retrieved from https://refugeehosts.org/representations-of-displacement-series/.

Finley, S. (2008). Arts-based research. In J. G. Knowles & A. L. Cole (Eds.), *Handbook of the arts in qualitative research: Perspectives, methodologies, examples, and issues* (pp. 71–81). Thousand Oaks, CA: Sage Publications.

Foster, V. (2015). *Collaborative arts-based research for social justice.* London: Routledge.

Friese, H. (2018). Framing mobility: Refugees and the social imagination. In D. Bachmann-Medick & J. Kugele (Eds.), *Migration: Changing concepts, critical approaches* (pp. 45–62). Berlin, Germany, & Boston, MA: De Gruyter.

Greatrick, A., & Fiddian-Qasmiyeh, E. (2017, March 1). The roles of performance and creative writing workshops in refugee-related research. *Refugee Hosts.* Representations of Displacement Series. Retrieved from https://refugeehosts.org/2017/03/01/the-roles-of-performance-and-creative-writing-workshops-in-refugee-related-research/.

Greene, M. (1995). *Releasing the imagination: Essays on education, the arts, and social change.* San Francisco, CA: Jossey-Bass.

Haaken, J., & O'Neill, M. (2014). Moving images: Psychoanalytically-informed visual methods in documenting the lives of women migrants and asylum-seekers. *Journal of Health Psychology, 19*(1), 79–89.

hooks, b. (1994). *Outlaw culture: Resisting representations.* New York, NY: Routledge.

Keifer-Boyd, K. (2011). Arts-based research as social justice activism insight, inquiry, imagination, embodiment, relationality. *International Review of Qualitative Research, 4*(1), 3–19. Retrieved from https://www.jstor.org/stable/10.1525/irqr.2011.4.1.3.

Kellehear, A. (1993). *The unobtrusive researcher: A guide to methods.* St Leonards, NSW: Allen & Unwin.

Kester, G. (2005). Conversation pieces: The role of dialogue in socially-engaged art. In Z. Kucor & S. Leung (Eds.), *Theory in contemporary art since 1985.* Oxford: Blackwell.

Khan, A. (2012). *UK media's pathology of the asylum seeker and the (mis)representation of asylum as a humanitarian issue.* Glasgow: Esharpy.

King, R., & Wood, N. (2001). *Media and migration: Constructions of mobility and difference*. London: Routledge.

Knowles, J. G., & Cole, A. L. (2008). *Handbook of the arts in qualitative research: Perspectives, methodologies, examples, and issues*. Los Angeles, CA: Sage.

Leavy, P. (2009). *Method meets art: Arts-based research practice*. New York, NY: Guilford Press.

Leavy, P. (Ed.). (2017). *Handbook of arts-based research*. New York, NY: Guilford Press.

Lenette, C. (2019). *Arts-based methods in refugee research: Creating sanctuary*. Singapore: Springer.

Lenette, C., & Cleland, S. (2016). Changing faces: Visual representations of asylum seekers in times of crisis. *Creative Approaches to Research, 9*(1), 68–83.

Lenette, C., Weston, D., Wise, P., Sunderland, N., & Bristed, H. (2016). Where words fail, music speaks: The impact of participatory music on the mental health and wellbeing of asylum seekers. *Arts & Health: An International Journal for Research, Policy and Practice, 8*(2), 125–139.

Lokot, M. (2018, May 4). Mobility, hope and the 'appropriation' of space: Reflections from a photovoice project. *Refugee Hosts*. Retrieved from https://refugeehosts.org/2018/05/04/mobility-hope-and-the-appropriation-of-space-reflections-from-a-photovoice-project/.

Lynn, N., & Lea, S. (2003). 'A phantom menace and the new apartheid': The social construction of asylum-seekers in the United Kingdom. *Discourse and Society, 14*(4), 425–452.

Malkki, L. H. (1996). Speechless emissaries: Refugees, humanitarianism, and dehistoricization. *Cultural Anthropology, 11*(3), 377–404.

Mannik, L. (2012). Public and private photographs of refugees: The problem of representation. *Visual Studies, 27*(3), 262–276.

McKay, S., & Bradley, J. (2016). How does arts practice engage with narratives of migration from refugees? Lessons from 'utopia'. *Journal of Arts & Communities, 8*(1–2), 31–46.

McNevin, A. (2010). Becoming political: Asylum seeker activism through community theatre. *Local-Global: Identity, Security, Community, 8*, 142–159.

McNiff, S. (1998). *Arts-based research*. London: Jessica Kingsley Publishers.

Mullen, C. (2003). A self-fashioned gallery of aesthetic practice. *Qualitative Inquiry, 9*(2), 165–182.

Nagarajan, C. (2013, September 20). Politicians and the press are locked in a cycle of increasing antiimmigrant rhetoric, presented as 'uncomfortable truth.' *Open Democracy*. HYPERLINK "https://www.opendemocracy.net/transformation/chitra-nagarajan/how-politicians-and-media-made-us-hate-immigrants" https://www.opendemocracy.net/transformation/chitra-nagarajan/how-politicians-and-media-made-us-hate-immigrants

Nunn, C. (2017). Translations-generations: Representing and producing migration generations through arts-based research. *Journal of Intercultural Studies, 38*(1), 1–17.

Nunn, C. (2018). *Dispersed belongings: A participatory arts-based study of experiences of resettled refugee young people in regional cities in Australia and the United Kingdom. A report for project partners*. Manchester Centre for Youth Studies. HYPERLINK "http://e-space.mmu.ac.uk/621965/" http://e-space.mmu.ac.uk/621965/

O'Neill, M. (2008). Transnational refugees: The transformative role of art? *Forum Qualitative Sozialforschung/Forum: Qualitative Social Research, 9*(2), Art. 59. Retrieved from http://www.qualitative-research.net/index.php/fqs/article/view/403/873.

O'Neill, M. (2010). *Asylum, migration and community*. Bristol: Policy Press.

O'Neill, M. (2018). Walking, well-being and community: Racialized mothers building cultural citizenship using participatory arts and participatory action research. *Ethnic and Racial Studies, 41*(1), 73–97.

O'Neill, M., Erel, U., Kaptani, E., & Reynolds, T. (2019). Borders, risk and belonging: Challenges for arts-based research in understanding the lives of women asylum seekers and migrants 'at the borders of humanity'. *Crossings: Journal of Migration & Culture, 10*(1), 129–147.

O'Neill, M., & Hubbard, P. (2010). Walking, sensing, belonging: Ethno-mimesis as performative praxis. *Visual Studies, 25*(1), 46–58.

Pakulski, J. (1977). Cultural citizenship. *Citizenship Studies, 1*, 73–86.

Pentassuglia, M. (2017). 'The art(ist) is present': Arts-based research perspective in educational research. *Cogent Education, 4*(1). https://doi.org/10.1080/2331186X.2017.1301011.

Pink, S. (2001). *Doing visual ethnography: Images, media and representations in research*. London: Sage Publications.

Riessman, C. K. (1993). *Narrative analysis*. Newbury Park, CA: Sage.

Robertson, Z., Gifford, S., McMichael, C., & Correa-Velez, I. (2016). Through their eyes: Seeing experiences of settlement in photographs taken by refugee background youth in Melbourne, Australia. *Visual Studies, 31*(1), 34–49.

Rymes, B. (2014). *Communicating beyond language: Everyday encounters with diversity*. New York, NY: Routledge.

Sales, R. (2002). The deserving and the undeserving? Refugees, asylum seekers and welfare in Britain. *Critical Social Policy, 22*(3), 456–478.

Sontag, S. (1979). *On photography*. London: Penguin Books.

Stavropoulou, N. (2018, February 17). Active interpretation: At the meeting place of research and creative practice. *Open Democracy* [Special issue: Who are 'we' in a moving world?]. HYPERLINK "https://www.opendemocracy.net/en/active-interpretation-at-meeting-place-of-research-and-creative-practice/" https://www.opendemocracy.net/en/active-interpretation-at-meeting-place-of-research-and-creative-practice/

Stavropoulou, N. (2019). Understanding the 'bigger picture': Lessons learned from participatory visual arts-based research with individuals seeking asylum in the United Kingdom. *Crossings: Journal of Migration & Culture, 10*(1), 93–116.

Thomas, S. (2001). Reimagining inquiry, envisioning form. In L. Neilsen, A. Cole, & J. G. Knowles (Eds.), *The art of writing inquiry* (pp. 273–282). Halifax, NS: Backalong Books and Centre for Arts-Informed Research.

Tribe, J. (2008). The art of tourism. *Annals of Tourism Research, 35*(4), 924–944.

UNHCR (UN High Commissioner for Refugees). (2016). *Global trends: Forced displacement in 2015*. Retrieved from https://www.refworld.org/docid/57678f3d4.html.

Vacchelli, E. (2018). *Embodied research in migration studies: Using creative and participatory approaches*. Bristol: Policy Press.

Van Dijk, T. A. (1991). *Racism and the press*. London: Routledge.

Villaverde, L. E. (2008). *Feminist theories and education*. New York, NY: Peter Lang.

Wang, C., & Burris, M. A. (1997). Photovoice: Concept, methodology, and use for participatory needs assessment. *Health Education and Behavior, 24*(3), 369–387.

Woolley, A. (2014). *Contemporary asylum narratives: Representing refugees in the twenty-first century*. London: Palgrave Macmillan.

Wright, T. (2002). Moving images: The media representation of refugees. *Visual Studies, 17*(1), 53–66.

Wright, T. (2004). Collateral coverage: Media images of Afghan refugees, 2001. *Visual Studies, 19*(1), 97–112.

# Part VIII

## Section Eight: Resisting Displacement

# 49

# Intervention: An Interview with Anna Minton

## Vandana Desai

(1) What does 'displacement' mean to you as a concept?

For me, this means individuals and families being forced to move from their homes against their will. The context I've studied displacement in is policy driven, with regards to the consequences of estate regeneration programs in London, where more than 100 estates have been demolished and replaced with luxury apartment blocks with very little genuinely affordable housing. For example, in Elephant & Castle, the Heygate Estate has been replaced by Elephant Park, a luxury development where the entire first phase was sold to foreign investors. Of the 2704 new homes at Elephant Park, only 82 are for social rent and while 25 per cent are classed as 'affordable housing,' since the Conservative-led coalition redefined affordable housing in 2010 to mean up to 80 per cent of market rent, this is far from affordable for many Londoners.

In *Big Capital*, I also discuss the more diffuse but hugely important impact of the influx of global capital on the city, which has seen London change out of all recognition as a result of Quantitative Easing and foreign investment, raising property prices and rents and creating a 'trickle-down' effect of unaffordability throughout the city. At the same time, austerity policies and the capping of housing benefit have led to a situation where it is now policy to house homeless families in other cities.

---

V. Desai (✉)
Department of Geography, Royal Holloway, University of London, Egham, UK
e-mail: V.Desai@rhul.ac.uk

(2) What is the role of academics in resisting displacement in projects of regeneration and gentrification?

There are a number of specific roles academics can play. Providing evidence-based research and longitudinal studies of the effects of regeneration and gentrification is vitally important as a tool to counter the barrage of rhetoric and propaganda which is deployed by developers, lobbyists, and PRs to convince communities that regeneration will have a positive effect. This is not a level playing field, with developers spending large amounts of money on PR, while local communities with no resources have to work around the clock to present an alternative viewpoint. In this context, research and mapping showing the impact of displacement is crucial, and we now have a number of studies which are engaged in this area.

Academics have a key role in creating a critical discourse and raising awareness, both within the academy and more widely. The critical discourse around regeneration and gentrification within academia long preceded the critical debates which are seen as commonplace today.

(3) Your two books *Ground Control* (2012) and *Big Capital* (2017) provide critical intervention in the fields of gentrification and urban studies, what do you think is the impact of your books on resisting displacement caused by gentrification or urban renewal (books as form of resistance)?

My aim has been to take these often complex debates, shrouded in jargon and obscure policy terms, and make them accessible to a mainstream audience. When communities engaged in struggles against displacement read my books, I hope they understand more of the economic and policy context behind what is happening and I hope they feel supported by the stories of others, which illustrate that these forces can be successfully resisted. One of the frustrating aspects of researching this area in the UK is that a lot of the resistance is quite fractured, based around local struggles which lack the resources and infrastructure to link up, but I hope that my work shows the common causation and common themes which run through all of them.

But the impact of my work is limited; my books provide an analysis of the current crisis but I am unable to lay out a blueprint for how to solve it, because that would require a paradigm shift in policy.

(4) What kind of impact do you hope to have from your recent book *Regeneration Songs: Sounds of Investment and Loss from East London*, which brings together a diverse arrays of texts, artistic interventions, and

soundtrack *brings together a diverse arrays of texts, artistic interventions, and soundtrack* (an intersection of popular culture and dissent)?

With *Regeneration Songs*, we show the crucial role played by artists in regeneration. The book had a dual purpose: to show how arts and culture can genuinely empower local communities to resist plans to airbrush them from the area but also to show the extent to which artists can be co-opted by the regeneration process. The project began as a collaboration between myself and artist Albert Duman, after we won a Leverhulme Artist in Residence grant for him to work with me and the MRes Programme I lead at the University of East London. Alberto worked with local musicians in Newham's so-called 'Arc of Opportunity' area to record an album's worth of tracks to accompany a silent promotional film, entitled *Regeneration Supernova*. The film, obtained by Alberto following an FoI request, had been commissioned by Newham Council to show to investors at the Shanghai Expo. The book, which accompanied the film, contains multidisciplinary contributions from academics, artists, writers, and journalists, many of which illustrate Grayson Perry's famous dictum that artists are now 'the shock troops of gentrification'.

(5) What are the emerging agendas around displacement and do you have any potential thoughts regarding the future?

I hope to turn my attention to themes underpinning displacement globally, and while I think the impact of property-based capital will continue to be a dominant force behind displacement and gentrification effects in post-industrial cities, I think that this is going to increasingly intersect with climate change and migration caused both by climate change and war.

As climate change displaces wealthy populations from coastal locations, other parts of cities higher up gentrify, displacing those populations. In countries with a very inadequate or non-existent welfare safety net, the public health impact of displacement can be catastrophic, in turn driving pressures towards migration.

Intervention in property markets and an understanding that housing cannot be provided only through the market is central to mitigating displacement effects. These are the routes taken by some European countries but not at present by more Atlanticist economies.

# 50

'Housing is a Human Right. Here to Stay, Here to Fight': Resisting Housing Displacement Through Gendered, Legal, and Tenured Activism

Mel Nowicki

## Introduction

Recent years have seen housing displacement become increasingly commonplace across the globe, including in the Global North. Particularly since the 1980s, neoliberal ideologies have reconstituted the meaning of housing, from predominately the provision of home and shelter, to a financialised source of profit (Aalbers 2016). This marketisation of housing is reflected in, for example, now-embedded understandings of homeownership as the aspirational tenure and property investment as a sound financial strategy for both individuals and corporations. This reframing of housing's very purpose has subsequently catalysed a range of processes that contribute to housing displacement. These include, but are not limited to, gentrification and 'urban regeneration' projects that displace lower income communities, the mass privatisation of social housing stock, and increasingly punitive welfare systems that financially constrain society's most vulnerable.

It is therefore unsurprising that in this context grassroots activism has developed a multifaceted portfolio of resistance that seeks to reject the increasing normalisation of housing displacement. This chapter draws on a range of methods utilised by communities in various locations around the

---

M. Nowicki (✉)
Oxford Brookes University, Oxford, UK
e-mail: mnowicki@brookes.ac.uk

© The Author(s) 2020
P. Adey et al. (eds.), *The Handbook of Displacement*,
https://doi.org/10.1007/978-3-030-47178-1_50

world to resist and subvert housing displacement. I focus on three methods in particular. First, using examples from Phnom Penh and east London, I explore the ways in which women's presumed natural attachment to the home is utilised as a gendered method of highlighting housing injustice. Second, I concentrate on methods of resistance to housing displacement that focus on interactions with the law. Using examples from a squatter settlement in New Delhi and social media groups in the UK, I explore the ways in which people for whom the law usually acts as a marginalising force repurpose it as a means of protection from displacement. Both sections highlight the ways in which modes that usually exacerbate oppression and vulnerability, such as the control of women through the domestic sphere or the marginalisation of those considered to be illegal subjects, are subverted as means of empowerment and agency in the fight against housing displacement. Third and finally, the chapter examines methods of resistance to housing displacement that seek to fundamentally rework and rethink normative neoliberal constructions of homeownership as the ideal form of tenure. I do so through discussion of squatting and co-housing communities in Copenhagen and Leeds.

This threefold set of examples highlight that, in a time where housing displacement is becoming ever-more normative across a wide range of contexts, resistance has taken increasingly creative forms. Whilst privatisation, speculation, and forced evictions continue to displace people and communities across the world, equally the importance of secure housing continues to be central to activists' fight for a more equitable world.

## Gender Performativity and Housing Activism

Housing and home have long been narrativised as gendered space: a site best placed for women to enact their assumed 'natural' roles as mothers and homemakers (Blunt and Dowling 2006). This 'mutual identification of the woman, the mother and the home' (Morley 2000, p. 63) has historically embedded the oppression of women through the social, cultural, political, and, in some cases, legal denial of their rights to life outside of the domestic sphere. This has been concurrent with the dismissal of domestic work (housework, child-rearing, and so on) as having less socioeconomic worth than paid labour outside of the home (Oakley 1974; Mainardi 1975).

It therefore follows that housing displacement and forced eviction is not a gender-neutral phenomenon. Rather, such processes disproportionately affect women (Brickell et al. 2017; Ryan 2017; Watt 2018). This is due in part to factors relating to expectations of women as primarily responsible for the home and family and mothers being less likely to be in paid employment due to such caring responsibilities (Fernandez Arrigoitia 2017; Watt 2018). However, in the face of increasing housing displacement and forced eviction across the world, women activists have subverted these gendered expectations of their role in the home, utilising traditional assumptions that narrativise women as mothers and homemakers as a performative tool for resisting housing displacement.

For example, Katherine Brickell's (2014, 2020) research with women activists in Cambodia revealed the ways in which women protesting the demolition of their homes in the Boeung Kak Lake region of Phnom Penh deployed gendered strategies of resistance. The Boeung Kak Lake (BKL) region of Cambodia's capital consisted of neighbourhoods reliant on harvesting and fishing industries provided by the lake's resources. In 2007, the lake area was leased for ninety-nine years from the Municipality of Phnom Penh to a Chinese-backed private development company. Over the past decade, the lake has been drained and thousands of locals evicted from their homes to make way for a high-end private development.

The resultant displacement was controversial in its sweeping destruction of both the lake and its surrounding neighbourhoods without public consultation, and the regular use of violence against local residents. In response to the brutal eviction of residents and their subsequent displacement across Phnom Penh, a grassroots activist movement, led by the women of BKL, emerged, whereby wives, mothers, and other female activists fronted a nonviolent social action campaign. Women's central role in resisting displacement in this context was rationalised partly on a practical basis, as a means of reducing violent responses from the municipality and developers and limiting disruption to local men's incomes. However, the campaign was also understood explicitly as a 'women's struggle' due to Cambodian women's traditional responsibilities relating to sustaining a stable and secure home life (Brickell 2014). As Brickell recounts, BKL activists drew upon their traditional connection to home as wives and mothers using a number of strategies, for example, by wearing models of houses, nests, and other imagery associated with the home during protests that alluded to 'the loss of habitat and warmth once provisioned through BKL homes' (Brickell 2014, p. 1265). Other tactics included stripping to their underwear outside the Cambodian parliament as a symbolic expression of being stripped of their housing rights and relatedly their increased exposure and vulnerability brought about by displacement. Although redevelopment and displacement continued in the region, the women activists of BKL nonetheless accrued wide levels of national media attention, bringing to light the injustice of their displacement, a fact which would likely have gone unnoticed were it not for their gendered displays of resistance.

In the UK context, gendered performativity as a method of resisting housing displacement has been utilised by the well-documented London-based activist group Focus E15. The group originally consisted of homeless single mothers housed in a hostel in the east London borough of Newham. When in 2013 the hostel began evicting tenants due to public funding cuts, residents were told that they would be moved as far from London as Birmingham and Hull, miles from their jobs, their children's schools, and family networks. In response, the soon-to-be evicted group set up their campaign and protested regularly outside Newham Council's offices. The group gained high levels of local, and eventually national, media coverage which eventually led to their being rehoused in Newham rather than elsewhere (Watt 2016). In the wake of this victory, Focus E15 became involved in housing activism more broadly, and the group has become ubiquitous in the struggle against gentrification and displacement in London. The group run a weekly street stall, organise protests and public meetings, and their chant 'Housing is a human right. Here to stay, here to fight' has become a well-known call for housing justice in and beyond the capital.

In September 2014, they engaged in their most well-known and widely publicised direct action campaign, occupying a disused block of flats on the Carpenter's Estate in Stratford, east London (close to the site of London's Olympic Park). The block had been earmarked for demolition in order to make way for a new University College London (UCL) campus as part of wider regeneration plans in the area. However, despite the collapse of the UCL contract (in part due to the controversy surrounding the planned evictions), Newham Council continued to decant residents from perfectly functional social housing under the auspices that regeneration of the site would eventually occur (Watt 2013). As a response to this, empty flats on the estate were opened by Focus E15 and used as a social centre, hosting a daily programme of events including workshops, classes, and performances. The two-week occupation had some success, leading to the partial reopening of the flats for people in priority housing need (Watt 2016).

Although their membership now extends beyond that of single mothers, and their current campaigns are far-reaching across different elements of the housing crisis, the Focus E15 collective remain most well-known as a gendered group. Their position within the popular imaginary as young single mothers being evicted and threatened with displacement across the country has proved a large factor in their popularity and influence. Through forefronting their position as young mothers without a home to raise their children,

Focus E15 became symbolic of gendered housing precarity and displacement in London.

The strategies used by the BKL activists and Focus E15 mobilised gender norms, in two disparate cultural contexts, to act as a means of shaming those responsible for housing displacement. They did so by making public the ways in which such displacement actively strips away women's access to domestic life. Such methods directed responsibility towards government and corporate actors whose decision-making has put at risk groups deemed to be particularly in need of housing and home: women and most notably mothers and homemakers. Both the BKL activists and Focus E15 are therefore clear examples of the ways in which modes of oppression themselves, in this case gendered oppression and assumptions relating to women's place in the home, can be subverted as a tool of resistance to housing displacement.

## Reconstructing the Law as a Tool of Resistance

Legal frameworks are often utilised by the state and other bodies of governance as a means of enacting housing displacement and dispossession. This is due in part to legal knowledge and professional support in countering housing displacement regularly being out of reach, financially and socially, for those most vulnerable to eviction and dispossession. This is exacerbated by national legal systems and loopholes that often protect the most powerful stakeholders in relation to property, via for, example, tax breaks for landlords, and wider legal cultures that prioritise the rights of property owners above rights to home or shelter (Forrest and Hirayama 2014). The law, then, tends to be understood as a barrier, rather than an aid, to resisting housing displacement. As Hubbard and Lees note in relation to legal geographies scholarship, 'there is a view that there is a fundamental disconnect between progressive publics and the state, and that the law tends to support extractive, exclusionary and coercive state policies' (Hubbard and Lees 2018, p. 9).

And yet, just as the law is utilised to instil urban spatial injustice and facilitate housing displacement, it can also provide important entry points for activists to reclaim their rights to housing and home. Similarly to the ways in which women activists utilised traditional assumptions relating to gender and the home in their activism strategies, the law, too, is reworked and repurposed by activists resisting housing displacement. This chapter draws out two examples, of a New Delhi squatter settlement and an online community in the UK, to demonstrate the varied ways in which resistance is enacted through the reworking of legal frameworks by those at the sharp end of housing law.

Ayona Datta's research with residents of a squatter settlement on the periphery of New Delhi highlights the ways in which people living in illegal settlements are exceptionalised through law: rendered illegal subjects in need of control and containment through a force of law (Datta 2012; Derrida 1992). For them, the law is not an abstract notion but a material and cultural violence that they must confront on a daily basis. This everyday force of law occurs, for example, through the continuous threat of demolition of residents' homes and their status as illegal subjects limiting their access to legal housing (Datta 2012).

In order to negotiate this everyday violence of law, settlement residents have developed ways of utilising legal frameworks to protect themselves as best they can from displacement and dismissal. Like the women activists of BKL and Newham, the Delhi settlers sought to rework the very structures that made them vulnerable to displacement in the first place. For example, Datta traces the ways in which squatters use their legal exceptionalism to recontextualise the law for their benefit. One method places emphasis on laws and policies pertaining to positive discrimination within the Indian Constitution. For example, residents used their lower caste, low-income, and/or tribal status to legitimise, through legal frameworks, their need to be resettled in formal housing. This was enacted through referencing elements of the Indian Constitution that promote positive discrimination for marginalised groups. As Datta notes, 'the legal subjecthood embodied in the constitution as fundamental rights, as low-castes, or as women has now become the only political resource towards making claims to shelter' (Datta 2012, p. 26). Another iteration of this strategy was residents' connections to and work with NGOs. Links to such organisations were in part a means of legitimising residents as political actors, a performance of legal knowledge and collective action that contributes to residents' moral claims to a legal home if and when large-scale resettlement finally occurs (Datta 2012). The actions of the Delhi squatters are therefore an example of when those vulnerable to displacement through their position as illegal and marginalised subjects attempt to utilise these same vulnerabilities as a legal argument for their protection.

Whilst the squatters documented in Datta's research repurposed the law to frame themselves as having a right to a legal home when resettled, in different contexts, legal spheres have been used to demand a 'right to remain' in order to resist housing displacement altogether. In the UK, for example, this can be seen through legal activism relating to the 'bedroom tax.'

The bedroom tax was introduced in England and Wales in 2013 by the Conservative/Liberal Democrat coalition government (2010–15) in 2012 as an element of the Welfare Reform Act, a series of legislative and policy measures that overhauled the country's welfare system. Borne out of the aftermath of financial recession and the emergence of a social, political, and economic discourse centred on austerity rhetoric, the Act introduced a suite of policies that penalised people in receipt of benefits, further entrenching the stigmatisation of those most likely to be in need of financial support, such as disabled and low-income people (Nowicki 2017, 2018).

The housing component of the Welfare Act, officially termed the 'removal of the spare room subsidy,' but known colloquially as the bedroom tax, applies to social (council or housing association) tenants in receipt of housing benefit in the UK (but excluding Scotland). Since April 2013 (2017 in Northern Ireland), if a social tenant is deemed to have one or more 'spare' bedrooms in their home (according to government guidelines), then the amount of housing benefit they are entitled to is reduced (by 14 per cent for one 'spare' bedroom or 25 per cent for two or more). Ostensibly a solution to issues of overcrowding and under-occupation in the social housing sector, and a method of reducing the national welfare deficit, the bedroom tax has disproportionately affected people living with mental and physical disabilities, who often need extra room to store mobility or medical equipment, or need to sleep in a separate bedroom from their partner due to their condition (Moffatt et al. 2016). Displacement, and the threat of displacement, occurs through eviction as a consequence of rising rent arrears where tenants are no longer able to keep up with rent payments due to cuts to their housing benefit (Nowicki 2017).

Resisting housing displacement in this context is inhibited by the bedroom tax both affecting people disproportionately likely to have mobility issues and the widespread impact the policy has had on geographically disparate communities across the UK. This has in turn impacted the ways in which those opposed to the policy have shaped their resistive strategies. Social media in particular has provided an invaluable means of legal knowledge exchange and resistance among tenants affected by the bedroom tax. For example, Facebook groups have been established as a way of not only sharing grievances regarding the policy's impact but to actively encourage resistance through the exchange and cultivation of legal knowledge. This occurs through people posting details of their specific circumstances on the group pages and asking other members for advice regarding how they might appeal their local authority's decision to implement the bedroom tax. For example, many people often post queries relating to the size and shapes of their rooms, looking for advice on the eligibility of launching an appeal on the basis that what their local authority has deemed a 'spare bedroom' is legally too small to be classed as such. Other group members respond by posting previous disputes that claimants have won on this basis in order to help members build their own case. Using amalgamated knowledge of tribunal decisions around bedroom tax appeal cases, group members encourage one another to take legal action and appeal local authority decisions. Here, an everyday method of communication is utilised to access legal knowledge that many people might otherwise find difficult to access. Social media therefore acts as a site through which resistance to the bedroom tax can be sought and rights to remain (Hubbard and Lees 2018) in their homes re-established.

In both of the examples outlined above, the legal structures so often used to evict and displace vulnerable people are reworked to serve the purposes of those at threat of displacement. Although occurring in different social, cultural, and legal contexts, both instances highlight the importance of understanding the nuanced potential of legal spheres as a means of protecting vulnerable communities from housing displacement.

## Resistance for the Long Term? Challenging Normative Constructions of Tenure

The third and final strategy of resistance to housing displacement discussed in this chapter focuses on challenges to normative neoliberal constructions of tenure. The section examines the ways in which activists seek to break away from presiding narratives of housing as a financial asset above and beyond its function as home (Aalbers 2016). Unlike the previous two sets of examples, the methods utilised here are less a subversion of oppressive practices and more a direct challenge to normative tenure conditions that contribute to increased housing displacement. Specifically, I highlight two, connected, methods of resisting displacement through alternative homemaking that defy the financialisation of housing: squatting and co-operative housing.

Particularly in the Global North context, housing displacement has been underpinned by neoliberal logics that over the past several decades have promoted an agenda of the hyper-privatisation of housing (Jarvis 2013). This has in part led to the mass privatisation of social housing stock in many countries and the displacement of communities from their neighbourhoods through processes such as gentrification and 'urban regeneration' policies that capitalise on the monetary value of inner-city neighbourhoods at the expense of their usually working-class residents (see for example Lees 2000; Hamnett 2003; Minton 2017). Housing and household formations that do not comply with the framing of housing as a profitable entity, rather than an invaluable element of human well-being, are subsequently dismissed and delegitimised (Minton 2017). Political, social, and cultural narratives promote the derision of non-normative tenures, through, for example, now long-entrenched conceptions of social tenants and those in receipt of welfare provision more widely as 'benefits scroungers' and 'welfare queens' (Tyler 2013; Hancock 2004). And yet, there is much in the way of housing activism that seeks to resist displacement through challenging these very notions of housing as financialised product. Activist methods such as squatting in part call for a return to understanding housing as a site of potential collectivisation and care rather than individual profiteering.

Squatting movements and collectives have, throughout history and across a range of geographical contexts, been at the forefront of this non-compliance with housing as a financial good. Particularly in the European context, squatters regularly posit themselves as challenging housing precarity, property speculation, and the increasingly displacing effects of urban regeneration schemes (Vasudevan 2017; Reeve 2009). As Vasudevan notes, 'the history of urban squatting has always been closely connected to housing insecurity and the efforts of ordinary people to secure their own right to housing and the basic fundamentals of survival' (2017, p. 239).

Resisting housing displacement through squatting varies widely in scale and scope. Squatting movements operate on a range of scales, from the occupation of single buildings, to the establishment of entire autonomous squatted neighbourhoods. Copenhagen's Freetown Christiania is perhaps the most well-known example of large-scale squatting in the Western context. A former military barracks, Christiania was squatted in the early 1970s as a response to a lack of affordable housing. The neighbourhood's approximately 1000 residents have over the past four decades developed their own set of rules, separate from Danish law, including an absence of privately owned property (Jarvis 2013). Although no longer technically squatted, as a deal with the Danish state in 2012 enabled residents to collectively purchase the neighbourhood, squatting's anti-capitalist principles continue to be central to Christiania's ethos. This continued commitment to providing an alternative to housing tenure that centres on ownership and profitability was cemented by the inclusion of a clause in the 2012 deal whereby residents have a right to occupy, but not buy or sell, property in the neighbourhood. In this way, the residents of Christiania have continued to enact policies of 'degrowth,' focusing on affordability and community, and 'human relations over market relations' (Jarvis 2017, 2018), placing it in direct opposition to normative constructions of what constitutes a successful housing system.

The legitimisation of the area through the 2012 deal enabled the community to both directly protect themselves from housing displacement, and more broadly challenge normative understandings of housing markets (Jarvis 2013). Christiania, then, is not only a successful example of resistance to housing displacement because of its eventual legitimisation but perhaps more importantly it is evidence that normative models of housing tenure and markets centred on profitability can be overthrown. In short, that it is possible to live in secure housing free from the threats of displacement and eviction that the financialisation of housing has so deeply entrenched and normalised (Aalbers 2016).

Another important, and related, method of resisting housing displacement through the establishment of secure and equitable tenure is that of co-housing. Co-housing schemes seek to work outside of top-down, elite-led models of change, producing grassroots, post-capitalist forms of housing (Chatterton 2016; Jarvis et al. 2016). These models are designed in part to disintegrate the now-normative threat of housing displacement by ensuring residents themselves are the lead decision makers regarding their housing and communities. Unlike neoliberal market-oriented forms of housing tenure, co-housing principles are reoriented towards an 'urban commons' (Kornberger and Borch 2015; Chatterton 2016; Bunce 2016). For example, LILAC, a co-operative development in Leeds in the north of England, operates through what is referred to as a 'mutual homeownership scheme' whereby each member has a lease providing them with a democratic stake in the scheme. Residents pay an equity share in the development based on their income, rather than on any valuation of their property (LILAC 2018). This enables the scheme to remain affordable for lower-income residents and removes the threat of housing displacement. Residents are embedded in their housing and neighbourhoods, rather than disposable subjects under threat of displacement should they lose their jobs or be evicted from their homes by developers or landlords seeking higher rental incomes.

Relatedly, in recent years, there has also been a growing international interest in another form of co-housing, community land trusts (CLT). In recent decades, CLTs have been developed in a range of countries, including the US, Kenya, and the UK (Moore and McKee 2012). CLTs focus on empowering and granting agency to local communities regarding their housing needs through the democratic management of assets. CLTs' aims include suppressing property and land values by retaining a portion of equity from any sales and keeping housing affordable no matter how many times it is sold, regardless of local housing markets (Moore and McKee 2012; Bunce 2016). The rapid growth of CLTs (e.g., half of the CLTs in existence in the US in 2012 were founded after 2000) (Moore and McKee 2012) has in large part been catalysed by growing housing unaffordability, rising insecurity of tenure, and increases in displacement-inducing processes such as gentrification. For example, in the UK, the East London Community Land Trust, London's first CLT, is developing community-led regeneration plans on the site of a former hospital in Tower Hamlets, an area that would ordinarily be classed as an area of high financial value (Bunce 2016).

Although not a new concept, the growing interest in and commitment to co-housing models in part highlights a heightened sense of urgency and the grassroots desire to rethink and rework normative housing systems. Whilst squatting is regularly hampered by its increasing illegality in many countries, including the UK and the Netherlands (Vasudevan 2017; Reeve 2015), co-housing efforts seek to circumvent this by carving out space for secure tenures and the maintenance of communities under threat of displacement through legally and politically legitimate means. Whilst there are certainly issues with co-housing models, most notably in relation to their relatively small scale and the large time and financial commitments required to establish such developments, they nonetheless pose an opportunity for the provision of long-term, sustainable solutions to housing displacement.

## Conclusions

This chapter has sought to highlight the multifaceted ways in which resistance to housing displacement is enacted and performed by an equally wide-ranging group of people. From Cambodia to Copenhagen, Newham to New Delhi, vulnerable communities are developing an array of creative methods in order to protect their homes.

The aim of this chapter, then, has not been to provide a comprehensive overview of all housing activism but rather to highlight the multifaceted nature of resistance to and subversion of housing injustice. In the case of the resistance methods that utilised gendered and legal frameworks, the very structures often used to oppress, displace, and erase people from their homes and neighbourhoods were reformulated as ways in which to resist these violent acts. And in relation to the examples of squatting and co-housing movements in the final section of the chapter, one of the root causes of increased housing displacement, the financialisation of housing, is itself being challenged and alternatives eked out.

However, despite the continued and wide-ranging efforts of activists and communities, the threat of housing displacement remains a normative, everyday experience for many. It is only through the continued dedication and creativity of those seeking to resist such displacement that affordable, sustainable, and equitable forms of housing will infiltrate mainstream housing markets in the long term.

## References

Aalbers, M. (2016). *The financialization of housing: A political economy approach*. London: Routledge.

Blunt, A., & Dowling, R. (2006). *Home*. Abingdon, UK: Routledge.

Brickell, K. (2014). 'The whole world is watching': Intimate geopolitics of forced eviction and women's activism in Cambodia. *Annals of the Association of American Geographers, 104*(6), 1256–1272.

Brickell, K. (2020). *Home SOS: Gender, Violence and Survival in Crisis Ordinary Cambodia*. Hoboken, Chichester: Wiley-Blackwell.

Brickell, K., Fernandez Arrigoitia, M., & Vasudevan, A. (Eds.). (2017). *Geographies of forced eviction: Dispossession, violence, resistance*. London: Palgrave Macmillan.

Bunce, S. (2016). Pursuing urban commons: Politics and alliances in community land trust activism in east London. *Antipode, 48*(1), 134–150.

Chatterton, P. (2016). Building transitions to post-capitalist urban commons. *Transactions of the Institute of British Geographers, 41*(4), 403–415.

Datta, A. (2012). *The illegal city: Space, law and gender in a Delhi squatter settlement*. Farnham, UK: Ashgate.

Derrida, J. (1992). Force of law: The 'mystical foundation of authority'. In D. Cornell, M. Rosenfeld, & D. Carlson (Eds.), *Deconstruction and the possibility of justice* (pp. 3–67). London: Routledge.

Fernandez Arrigoitia, M. (2017). Unsettling resettlements: Community, belonging and livelihood in Rio de Janeiro's *Minha Casa Minha Vida*. In K. Brickell, M. Fernandez Arrigoitia, & A. Vasudevan (Eds.), *Geographies of forced eviction: Dispsossession, violence, resistance*. London: Palgrave Macmillan.

Forrest, R., & Hirayama, Y. (2014). The financialisation of the social project: Embedded liberalism, neoliberalism, and homeownership. *Urban Studies, 52*(2), 233–244.

Hamnett, C. (2003). Gentrification and the middle-class eemaking of inner London, 1961–2001. *Urban Studies, 40*(12), 2401–2426.

Hancock, A. (2004). *The politics of disgust: The public identity of the welfare queen*. New York, NY: New York University Press.

Hubbard, P., & Lees, L. (2018). The right to community? Legal geographies of resistance on London's gentrification frontiers. *City, 22*(1), 8–25.

Jarvis, H. (2013). Against the 'tyranny' of single-family dwelling: Insights from Christiania at 40. *Gender, Place & Culture, 20*(8), 939–959.

Jarvis, H. (2017). Sharing, togetherness and intentional degrowth *Progress in Human Geography*, Online first. https://doi.org/10.1177/0309132517746519.

Jarvis, H. (2018, 3 July). I lived in a commune in Denmark to see how people could make their lives better and longer. *The Independent*. Retrieved from https://www.independent.co.uk/voices/denmark-commune-freetown-christiania-copenhagen-squatters-social-experiment-a8428471.html.

Jarvis, H., Scanlon, K., Fernandez Arrigoitia, M., Chatterton, P., Kear, A., O'Reilly, D. G., Sargisson, L., & Stevenson, F. (2016). *Cohousing: Shared futures*. Economic and Social Research Council.

Kornberger, M., & Borch, C. (2015). Introduction: Urban commons. In C. Borch & M. Kornberger (Eds.), *Urban commons: Rethinking the city* (pp. 1–21). Abingdon, UK: Routledge.

Lees, L. (2000). A reappraisal of gentrification: Towards a 'geography of gentrification'. *Progress in Human Geography, 24*(3), 389–408.

LILAC (Low Impact Living Affordable Community). (2018). Affordable. *LILAC*. Retrieved from http://www.lilac.coop/affordable/.

Mainardi, P. (1975). *The politics of housework*. Boston, MA: New England Free Press.

Minton, A. (2017). *Big capital: Who is London for?* London: Penguin.

Moffatt, S., Lawson, S., Patterson, R., Holding, E., Dennison, A., Sowden, S., & Brown, J. (2016). A qualitative study of the impact of the UK 'bedroom tax'. *Journal of Public Health, 38*(2), 197–205.

Moore, T., & McKee, K. (2012). Empowering local communities? An international review of community land trusts. *Housing Studies, 27*(2), 280–290.

Morley, D. (2000). *Home territories: Media, mobility and identity*. London: Routledge.

Nowicki, M. (2017). Domicide and the coalition: Austerity, citizenship and moralities of forced eviction in inner London. In K. Brickell, M. Fernandez Arrigoitia, & A. Vasudevan (Eds.), *Geographies of forced eviction: Dispossession, violence, resistance* (pp. 121–144). London: Palgrave Macmillan.

Nowicki, M. (2018). A Britain that everyone is proud to call home? The bedroom tax, political rhetoric and home unmaking in UK housing policy. *Social & Cultural Geography, 19*(5), 647–667.

Oakley, A. (1974). *The sociology of housework*. London: Robertson.

Reeve, K. (2009). The UK squatters movement 1968–1980. In L. Van Hoogenhuijze (Ed.), *Kritiek 2009: Jaarboek voor Socialistische Discussie en Analyse* (pp. 137–159). Amsterdam, Netherlands: Aksant.

Reeve, K. (2015). Criminalising the poor: Squatting, homelessness, and social welfare. In L. Fox O'Mahony, D. O'Mahoney, & R. Hickey (Eds.), *Moral rhetoric and the criminalisation of squatting: Vulnerable demons?* (pp. 133–154). London and Abingdon, UK: Routledge.

Ryan, C. (2017). Gendering Palestinian dispossession: Evaluating land loss in the West Bank. *Antipode, 49*(2), 477–498.

Tyler, I. (2013). *Revolting subjects: Social abjection and resistance in neoliberal Britain*. London and New York, NY: Zed Books.

Vasudevan, A. (2017). *The autonomous city*. London: Verso.

Watt, P. (2013). 'It's not for us': Regeneration, the 2012 Olympics and the gentrification of east London. *City, 17*(1), 99–118.

Watt, P. (2016). A nomadic war machine in the metropolis: En/countering London's 21st century housing crisis with Focus E15. *City, 20*(2), 297–320.

Watt, P. (2018). Gendering the right to housing in the city: Homeless female lone parents in post-Olympics, austerity east London. *Cities, 76*, 43–51.

# 51

# Contesting Displacement Through Radical Emplacement and Occupations in Austerity Europe

Mara Ferreri

## Introduction

In cities across the Northern hemisphere, displacement from a place of home has long existed and affected low-income communities, migrants, and ethnic minorities. In different forms, urban policies and models rooted in financialisation, deregulation, limited or dwindling social housing provision, and a pro-development approach to planning have left urban residents vulnerable to predatory transnational investment, privatisation, and new waves of gentrification. The financial crisis of 2007–8 has further exacerbated, and in some cases drastically precipitated, an already unjust situation, and 'brought home to us our vulnerability to displacement and dispossession' (O'Mahony and Sweeney 2011, p. 2). The threat of displacement and evictions has become commonplace for increasingly larger swathes of the population (Brickell et al. 2017; Rolnik 2013). The generalisation of the threat of displacement has been accompanied by a rise in organised resistance which has reignited existing social movement and encouraged the emergence of new ones. From campaigns to 'stay put' in threatened social housing, to protest camps, to solidarity occupations with migrants and refugees, a politics of radical emplacement has come to the fore in the transnational political arena.

M. Ferreri (✉)
Northumbria University, Newcastle upon Tyne, UK
e-mail: mara.ferreri@northumbria.ac.uk

© The Author(s) 2020
P. Adey et al. (eds.), *The Handbook of Displacement*,
https://doi.org/10.1007/978-3-030-47178-1_51

In both northern and southern European cities, resistance to displacement has been articulated through a shared repertoire of action, from anti-eviction resistance to the occupation of public spaces and vacant buildings (European Action Coalition 2016). Anti-eviction work has targeted dispossession through displacement by physically reclaiming place, increasingly through forms of direct action. Similarly to the large-scale mobilisation against home foreclosure and towards the reoccupation of vacant housing in the United States, the 'strategic illegality of "home liberations"' (Roy 2017), practices of occupation have accompanied wider organising to offer spaces to meet, find shelter, and make 'home' under inhospitable conditions (Lancione 2019). In doing so, occupations have become important tactical ways to make visible material and political causes of displacement and for connecting specific and localised instances to wider social justice: citywide, nationally, and transnationally. In this chapter, I present key issues in displacement resistance across northern and southern European cities in response to post-2008 austerity policies and to new processes of gentrification and speculative real estate development. Without attempting an exhaustive overview, in what follows I bring into dialogue interdisciplinary literature on organised practices of displacement resistance that deploy the occupation of spaces. I broaden the gaze from strictly academic contributions to include 'grey literature' produced by displaced people and allies, to give visibility to alternative sites of knowledge production and to understand practices that are often tactical and ephemeral, and produced through complex local and transnational meaning-making processes that may not be neatly captured by academic categories. From the standpoint of organising, 'displacement' is a signifier imbued with multiple meanings and forms of organising, which in widespread austerity-led precarisation build on but exceed familiar repertoires, strategies, and subject positions. Challenged by embodied encounters and practical solidarity, customary notions of displacement are questioned and stretched to encompass wider and multidimensional dynamics, with resistance articulating in response to and against different processes of dispersal or removal of unwanted bodies from cities.

## An Expanded Definition of Displacement

The concept of displacement has long been central to critical scholarship concerned with the disruptive effects of urbanisation dynamics on the lived geographies of individuals and communities, particularly the urban poor and minorities. Classical understandings of gentrification have long placed displacement at the core of the issue, not simply in terms of its direct effects on individuals and households but also in relation to the wider logics of urban change (Lees et al. 2013; Marcuse 1985; Slater 2009). In a more expanded sense, displacement has been defined as the result of processes of unmaking 'place' through stigmatisation, the threat of eviction, and the foreclosure of the possibility of being emplaced (Nowicki 2017; Tyler and Slater 2018; Slater 2013). From claiming a right to 'stay put' (Newman and Wyly 2006) to claiming a right to 'dwell'—understood as 'the right to exert a reasonable level of power over one's basic living conditions, with all the physical and mental benefits that entails' (Baeten et al. 2017, p. 2)—the debate on definitions of displacement and resistance to it are increasingly becoming more nuanced. The need for an expanded understanding of displacement and displacement resistance has been made more urgent by the 2008 global financial crisis, with calls made for a 'reinvigorated critical gentrification studies' to 'explore these new forms of austerity-fuelled dispossession and document ways to resist them' (Annunziata and Lees 2016).

'Displacement' can name urban phenomena that differ greatly and that affect various subject positions in terms of tenure, power relations, scale, and socioeconomic characteristics. The polyvalence of the term has implications for understanding displacement resistance. As noted by Bernt and Holm on defining the urban processes affecting Prenzlauerberg in Berlin, the issue at stake here is not how to define displacement but rather how the term is mobilised and politicised: it is the 'ambiguity in the use of the term displacement that forms the point of entry for politics and makes the definition of displacement a politically controversial issue' (Bernt and Holm 2009, p. 314). Political controversy is furthermore caused by processes of transnational adaption from Anglo-American debates and cities to geographically, socially, and culturally specific contexts, where univocal translation of 'gentrification' and 'displacement' can prove elusive. In Spanish-speaking contexts, for instance, activists have used humour to popularise the former, such as through the art-activist workshops 'Gentrificación no es un nombre de señora' (Gentrification is not a lady's name) run alongside a transnational digital 'Museum of the Displaced' (Left Hand Rotation n.d.). Expanded definitions of 'displacement' and its adaptation to different contexts appear symptomatic of a lived theorising that seeks out material causes to multiple and varied symptoms (see Ferreri 2017), as well as strategies and tactics of organising for urban social justice. In the following, I examine different forms of displacement resistance through spatial occupations, maintaining a broad definition of 'displacement' as grounded in specific conditions and struggles. From responses to displacement from low-income housing, to the rise of anti-eviction platforms and the shift from occupying squares to 'recuperating' vacant housing, to forms of organised and coordinated mass squatting, an expanded understanding of displacement resistance opens the possibility for a broader politics of emplacement and organising for staking claims to the city.

## Resisting Displacement from Low-Income Housing

The relationship between limited or dwindling provision of secure low-income housing and displacement is well-known in the current housing crisis across Europe. From the perspective of activists on the ground, the 'privatisation of social housing is a significant factor contributing to evictions' in eastern Europe, where nationalised housing was dismantled following the collapse of the communist bloc, as much as in western Europe, where demolition and privatisation, often in the name of 'urban regeneration,' is leading to greater tenure insecurity (European Action Coalition 2016, p. 5). The threat of displacement from low-income rental housing has recently been growing also in part as a result of insufficient maintenance and neglect of homes built between the 1950s and the 1970s. Baeten et al. have proposed the term 'renoviction' to define the specificities of the threat of eviction that emerges from rent increases caused by market-led refurbishment of low-income rental housing (Baeten et al. 2017), leading to forms of organised tenants-led resistance, even in Swedish housing, which is generally thought of as a highly regulated sector (Thörn et al. 2016). In Berlin, tenants of high-rise 1970s buildings in Kottbusser Tor in Kreuzberg became internationally known in May 2012 for their Protest Camp Kotti & Co against rent increases. Many affected tenants and members of the campaign were of Turkish origin; they occupied the local public square by building a small wood pavilion, or Gecekondu (Turkish word for houses built overnight), to organise meetings, talks, and public events (Kotti und Ko n.d.).

In Ireland and the United Kingdom, the threat of displacement from public (council) housing estates has become the centre of intensified and networked campaigning and organising (Lees and Ferreri 2016; Norris and Hearne 2016), which often use forms of direct action to raise the profile of estates. Campaigning and symbolic occupations on estates under threat of demolition often start from acts of reclaiming council housing as an object of positive attachment through a recognition of lived experiences, as in the Radical Housing Network's 2016 campaign 'We love Council Housing', which countered the stigmatisation of council housing and the naturalisation of its demolition in order to generate cross-tenure solidarity (Wills 2016). In

London, between 2014 and 2015, a range of high-profile occupations of vacant flats and houses within council estates, such as the Aylesbury and West Hendon, have been used to draw attention to empty flats at times of unprecedented demand (see also Watt and Minton 2016). The slogan 'These homes need people, these people need homes' was deployed in cases such as the high-profile political occupation of vacant flats on the Carpenters Estate, in Stratford, east London, during summer 2014. The occupation was carried out by Focus E15, a group of young homeless mothers that came together in 2013 to resist displacement from the sheltered housing association homes where they lived, after the local authority, Newham Council, reduced its funding for homelessness programmes due to the central government austerity budget (Gillespie et al. 2018). The Carpenters Estate occupation forced the local government to refurbish thirty vacant homes on the estate to rehouse families on the local housing waiting list; the longer-term future of the estate, however, remains uncertain; Focus E15 and allies thus continue campaigning for 'social housing, not social cleansing.'

Resistance to displacement from social rented and other low-income housing threatened with demolition and rebuilding is not a phenomenon circumscribed to central and western European cities. Increasingly, literature is emerging on resistance to privatisation of public housing estates in countries where such provision has always been marginal. In Spain, where public housing accounts for less than 2 per cent of all housing stock, neoliberal frameworks for the management of public resources led to the privatisation of some of its limited social rented housing and have encountered resistance. In Madrid, for example, the 2012 campaign 'Yo no me voy' (I will not go) has been resisting the privatisation of housing belonging to the city's municipal housing company. The campaign prevented the sale of 130 flats and enabled tenants to maintain the existing, more secure rental agreements (Annunziata and Lees 2016). In Milan, Italy, neighbourhood-based cross-sector organising has been able to stop the demolition of municipal housing for rental and successfully demand public funding for refurbishment (Turolla 2017). In Rome, the occupation of publicly owned social rented housing has been a tactical tool of citywide movements for the right to housing since the 1960s, but has seen a resurgence since the 2000s, in an effort to make visible the need for mobilising existing vacant public housing (Mudu 2014; Davoli 2018).

## From Anti-Eviction Platforms and Occupying Squares to 'Recuperating' Housing

Since 2008, the effects of the global financial crisis and ensuing political response have been felt by the rise in evictions and increased insecurity of tenure for low-income groups, migrants, women, and the elderly. In response, a number of anti-eviction groups have emerged across major European cities, constituting what social movement scholars see as a 'new wave' of urban mobilisation. Groups such as 'Abitare nella crisi' in Italy, the 'Plataforma de Afectados por la Hipoteca' (PAH) in Spain, and the 'Droit au Logement' in France have all been active in resisting evictions by picketing homes to prevent displacement and by making visible the increase in eviction proceedings and the greater risk of homelessness among low-income urban residents. In Berlin, the 'Bündnis Zwangsräumung Verhindern' (Platform to stop evictions) was created in 2012 in response to a steep increase in the number of evictions, with official estimates pointing to a 19.5 per cent increase between 2013 and 2014 (Ordóñez et al. 2015; Westenberg and Coers 2014). In the post-2010 cycle of networked social movements, continued occupations of squares and streets through protest camps have become prominent (McCurdy et al. 2016) and have intersected with anti-eviction organising and the spatialities of occupations of residential buildings for living and communal use.

Examples of convergences between 'occupation-based campaigning' (Gonick 2016) and anti-displacement movements have become particularly established in Spain, the country in Europe that has experienced the largest and most intense wave of housing evictions as well as the most organised waves of protest (Colau and Alemany 2014). Due to a combination of economic recession, rising unemployment, and mortgage arrears, according to official statistics, between 2008 and 2013 there were nearly 320,000 evictions (Barbero 2015), a number that has since more than doubled. In this context, the 15M movement which started in March 2011 played a particularly important role in the development and legitimation of urban spatial occupations as part of a wider repertoire of social movements. The camps and occupations organised and promoted by the 15M have transformed a tool of protest into a shared model for self-organisation and direct democracy, in a historical 'virtuous convergence' between the 'movement of occupation' and the 'movement of political squatting' (Martínez and García 2015). Drawing on research in Madrid, Martínez and García have analysed both personal and infrastructural exchanges between activists from the political squatting scene and the new political subject emerging from the crisis of political representation in the wake of the global financial crisis and beginning of the austerity regime in the country. The structural homology between the occupation of squares and of buildings, as well as the novelty of a largely horizontal form of political organising, sustained the conditions for 'chains of accumulation of activist exchanges' (Martínez and García 2015, p. 164) that filtered from public squares—notably, Puerta del Sol in the capital—to neighbourhood and local assemblies that enlarged and transformed their former functions, giving them new meanings as forums for public debate and being in common. Concretely, between May and December 2011, a high number of occupations of vacant buildings across many Spanish cities offered spatial as well as social resources to guarantee continuity in the movement 'after the square.'

This filtering fed into the beginning of a nationwide anti-eviction campaign ('Stop desahucios'), which spread to cities across the country and which by 2014 had successfully stopped over 1100 evictions (de Andrés et al. 2015). The campaign strengthened the gathering of political experiences and collectives around the issue of housing and increased the legitimacy of occupation-based protests, as well as of occupations for living. The PAH played a pivotal role, with the establishment of the Obra Social de la PAH, which organised the occupation of empty housing owned by financial institutions, whose vacancy derived from processes of mortgage foreclosure and eviction. Martinez and García understand this shift as one of 'mutual hybridation' between traditionally political occupation-based protest groups and forms of social occupation by groups that did not have previous experiences of activism (Martínez and García 2015, p. 175). By 2016, the PAH's Obra Social Campaign had rehoused 3500 people throughout the country and maintained forty-nine occupied blocks of flats.

The widening of practical solidarity accompanied a greater legitimacy to forms of reclaiming—*reapropriación o recuperación*—of vacant buildings for housing and 'the collective recuperation of vacant housing that is owned by financial institutions has converted into a distinguishing mark of the contemporary housing struggles' (Pavón and Janoschka 2016, p. 113). Resistance and recuperation of vacant homes have been analysed as transformative and empowering, both on an individual and on a collective level (García-Lamarca 2017). For this dimension of empowerment, as well as for its role in generating municipal citizens' platforms which gained control of key Spanish municipalities during the administrative elections of 2015, the Spanish movement continues to be a point of reference for activists throughout European cities. It must be noted, however, that despite the mass mobilisations, evictions and eviction threats continue to be an everyday reality for many, demonstrating a shift in the housing crisis towards greater precarisation: around 2011, nearly the 90 per cent of all evictions concerned situations of home repossession due to mortgage arrears, in 2015 more than half (53 per cent) derived from rent arrears in the private rented sector (Gutiérrez and Delclòs 2017), leading many to prefer squatting to facing street homelessness (Coordinadora de Vivienda de la Comunidad de Madrid 2017; Obra Social Barcelona 2018).

## A Resurgence of Squatting Between Protest and Deprivation

A significant, if little known, consequence of the 2008 financial crisis across many European cities has been the rise of squatting—the unlawful occupation of land or buildings for shelter and protest—from a marginal and undetected practice to a mass and organised phenomenon. As with other forms of precarious housing, it is a 'product of recurring cycles of creative destruction and accumulation by dispossession, which have repeatedly condemned significant numbers of people to misery and prompted many to seek informal forms of housing and shelter' (Vasudevan 2015, p. 29). In social movement scholarship, squatting that originates in and responds primarily to housing needs has been described as 'deprivation-based squatting' (Prujit 2013) and has generally been distinguished from other forms of 'political' occupation of spaces. While deprivation-based squatting is understood as 'an action that interrupts phases of homelessness, or of living under degraded housing conditions' (Mudu 2014, p. 137), its increasing politicisation has been an important characteristic of the movements of convergence mentioned above. In the Ile-de-France region (Paris), some estimates argue that by the early 2010s there were approximately 2000 'invisible squats,' a percentage of which had been mapped by official institutions and whose causes could be directly linked to the housing crisis (Aguilera 2013). In this context, the French Droit au Logement movement has supported 'deprivation-based squatting' as a tool to negotiate secure housing. In Rome, where coordinated mass squatting has seen an important increase since the early 2000s, the peak of the economic, housing, and migration crises post-2008 led to an intense period of coordinated occupations known as the 'Tsunami Tour'. In 2016, an estimated 6000 households were living in occupations in the capital as part of intersecting organised political networks for the right to housing (Davoli 2018). In Catalonia, a recent questionnaire of 600 households living in occupied properties recorded that in over 75 per cent of cases the occupation was a result of the lack of affordable housing relative to income, and in at least 29 per cent of the cases, the occupation was the direct result of an eviction (Obra Social Barcelona 2018). Organised occupations are increasingly reclaimed by neighbourhood housing collectives as a legitimate response to any form of real estate speculation (Fig. 51.1).

Migrant and ethnic minorities are generally more susceptible to living in insecure or precarious housing conditions; at times of so-called 'migrant crises' (Mountz and Loyd 2014), the intersection of new migratory waves and housing issues has created the conditions for alliances and new solidarities within anti-displacement movements. In Madrid, the 15M square occupations in 2011 have been noted as key inflection points for a widening of intersectional solidarities between political protest movements, and more informal 'occupations for housing,' where shelter and mutual aid among migrants were dominant (Martínez López 2017). In Berlin, since 2012, high-profile political occupations of buildings and public spaces have been deployed to provide self-organised spaces for refugees and supporters to meet, share knowledge and resources, and organise against policies of dispersal which were experienced as isolating. The Refugee Strike House, which occupied a

**Fig. 51.1** Graffiti: 'Occupation against speculation,' Barcelona, March 2019. (Source: Mara Ferreri)

vacant school in Ohlauer Straße 12, and which included a self-organised women-only floor, and the Refugee Protest Camp, a tent city which occupied the public square Oranienplatz in Kreuzberg, from October 2012 to April 2014, were particularly high-profile examples. Refugees and anti-racist organisers deployed the occupations to challenge two specific form of displacement: the first was the threat of deportation; the second, the legal requirement (*Residenzpflicht*, or mandatory residence) for applicants of refugee status to reside within boundaries defined by immigration authorities, which many migrant and refugee advocacy organisations oppose as a violation of fundamental human rights (azozomox 2014). Research in southern European cities such as Rome and Athens has found an exacerbation of discriminatory discourses against ethnic minorities, such as Roma communities (Annunziata and Lees 2016). Despite greater marginalisation, however, Roma people living in informal settlements under threat of eviction participated in building alliances with organised squatting practices and social movements, providing grounds for the emergence of new solidarities. As noted in the case of Rome, activist movements and some NGOs attempted to frame the exclusion of Roma communities in terms of 'socio-economic status rather than through their ethnic identity' (Maestri 2014, p. 819). This reframing has generated convergences between mobilisations that come from evicted informal settlements and wider social movements deploying occupations to struggle for the right to housing.

## Conclusions

Displacement from a place of home has sadly become a common phenomenon across both northern and southern European countries and has been felt with particular intensity in urban centres. Although socially and economically vulnerable individuals and social groups have been most affected, the number of those in insecure housing has swelled to include private tenants and mortgage holders, making eviction and the threat of eviction a reality across a wide set of subject positions in contemporary European cities. The conditions of people at risk of displacement are socially and geographically varied, as have been the responses by civil society and activist groups. The intersecting of multiple 'crises'—financial, economic, housing, and refugee—has made 'displacement' an overarching organising issue related to a diversity of causes, from forced eviction to gentrification, the financialisation of housing, and austerity policies. In this chapter, I have examined the multiple forms of displacement at play in austerity Europe as seen through different strategies and practices of resistance that have emerged since 2008. I have drawn on scholarly as well as activist accounts to blur the line between sanctioned and unsanctioned knowledge production and to highlight the important work of activist and militant-investigative writing.

Displacement resistance occupies streets, squares, and buildings to enact an embodied interruption of processes of home dispossession and homelessness, towards a collective reclaiming of place and emplacement. Attentiveness to geographical and historical specificities shows that displacement has been and continues to be resisted through occupations, undertaken for different aims and through varying strategies, from short-term emergency responses and established tools for negotiation to the establishment of self-organised alternatives. In contrast to individualisation, marginalisation, and stigmatisation, 'taking place' through forms of occupation has become an important tactic for empowerment and emancipation and for generating and sustaining intersecting solidarities. From anti-eviction platforms to anti-gentrification neighbourhood committees, from self-organised homeless women to anti-racism and refugee support, displacement resistance at the local and global scales is reimagined in the intersectional practices of a multiplicity of collectives and groups. Resistance to displacement through occupations is thus both a response to material, immediate issues, and a prefigurative reclaiming that exhibits the possibility of a different coexistence. Rethinking displacement from practices of resistance means to foreground the significance of collectively staking a claim to place, be it a building, a neighbourhood, or a city, often across different subject positions. In austerity Europe, the practical and political convergence of anti-displacement practices is beginning to intersect with solidarity organising, towards a place-based and intersectional politics of radical emplacement.

**Grant Acknowledgment** The writing of this chapter has been made possible by the H2020 MSCA programme (grant No 665919) for the 'Commoning Housing' project.

# References

Aguilera, T. (2013). Configurations of squats in Paris and the Ile-de-France Region. In Squatting Europe Kollective (Ed.), *Squatting in Europe. Radical spaces, urban struggles* (pp. 209–230). Wivenhoe, UK: Minor Composition.

de Andrés, E. Á., Campos, M. J. Z., & Zapata, P. (2015). Stop the evictions! The diffusion of networked social movements and the emergence of a hybrid space: The case of the Spanish Mortgage Victims Group. *Habitat International, 46*, 252–259.

Annunziata, S., & Lees, L. (2016). Resisting 'austerity gentrification' and displacement in southern Europe. *Sociological Research Online, 21*(3).

azozomox. (2014). *Social diversity, precarity and migration within the squatting movements in Berlin*. Retrieved from https://sqek.squat.net.

Baeten, G., Westin, S., Pull, E., & Molina, I. (2017). Pressure and violence: Housing renovation and displacement in Sweden. *Environment and Planning A: Economy and Space, 49*(3), 631–651.

Barbero, I. (2015). When rights need to be (re)claimed: Austerity measures, neoliberal housing policies and anti-eviction activism in Spain. *Critical Social Policy, 35*(2), 270–280.

Bernt, M., & Holm, A. (2009). Is it, or is it not? The conceptualisation of gentrification and displacement and its political implications in the case of Berlin-Prenzlauer Berg. *City, 13*(2–3), 312–324.

Brickell, K., Arrigoitia, M. F., & Vasudevan, A. (2017). Geographies of forced eviction: Dispossession, violence, resistance. In K. Brickell, M. F. Arrigoitia, & A. Vasudevan (Eds.), *Geographies of forced eviction: Dispossession, violence, resistance* (pp. 1–23). London: Palgrave Macmillan.

Colau, A., & Alemany, A. (2014). Mortgaged lives: From the housing bubble to the right to housing [Trans. M. Teran]. *Journal of Aesthetics & Protest Press*. Retrieved from https://www.joaap.org/press/mortgagedlives.html.

Coordinadora de Vivienda de la Comunidad de Madrid. (2017). *La vivienda no es delito. Recuperando un derecho: quién y por qué se okupa en Madrid*. Madrid, Spain: El Viejo Topo. Retrieved from https://www.elviejotopo.com/libro/la-vivienda-no-es-delito/.

Davoli, C. (2018). Le occupazioni abitative a Roma: Una pratica dei movimenti per il diritto all'abitare. In A. Coppola & G. Punziano (Eds.), *Roma in Transizione. Governo, Strategie, Metabolismi e Quadri Di Vita Di Una Metropoli* (pp. 303–316). Roma-Milano, Italy: Planum Publisher.

European Action Coalition. (2016). *Evictions across Europe*. European Action Coalition for the Right to Housing and to the City. Retrieved from https://housingnotprofit.org/files/EvictionsAcrossEurope.pdf.

Ferreri, M. (2017). Beyond 'staying put': Reflections on discursive strategies in recent anti-gentrification movements. *Urbanistica 3, 5*(13), 89–93.

García-Lamarca, M. (2017). Creating political subjects: Collective knowledge and action to enact housing rights in Spain. *Community Development Journal, 52*(3), 421–435.

Gillespie, T., Hardy, K., & Watt, P. (2018). Austerity urbanism and Olympic counter-legacies: Gendering, defending and expanding the urban commons in East London. *Environment and Planning D: Society and Space, 36*(5), 812–830.

Gonick, S. (2016). From occupation to recuperation: Property, politics and provincialization in contemporary Madrid. *International Journal of Urban and Regional Research, 40*(4), 833–848.

Gutiérrez, A., & Delclòs, X. (2017). Geografía de la crisis inmobiliaria en Cataluña: Una lectura a partir de los desahucios por ejecución hipotecaria. *Scripta Nova. Revista Electrónica de Geografía y Ciencias Sociales, 21*, 553–559.

Kotti und Ko. (n.d.). Retrieved from http://kottiundco.net/english/.

Lancione, M. (2019). Weird exoskeletons: Propositional politics and the making of home in underground Bucharest. *International Journal of Urban and Regional Research, 43*(3), 535–550.

Lees, L., & Ferreri, M. (2016). Resisting gentrification on its final frontiers: Learning from the Heygate Estate in London (1974–2013). *Cities, 57*, 14–24.

Lees, L., Slater, T., & Wyly, E. (2013). *Gentrification*. London and New York, NY: Routledge.

Left Hand Rotation. (n.d.). *Museo de los desplazados*. Retrieved from http://www.museodelosdesplazados.com/.

Maestri, G. (2014). The economic crisis as opportunity: How austerity generates new strategies and solidarities for negotiating Roma access to housing in Rome. *City, 18*(6), 808–823. https://doi.org/10.1080/13604813.2014.962895.

Marcuse, P. (1985). Gentrification, abandonment and displacement: Connections, causes and policy responses in New York City. *Journal of Urban and Contemporary Law, 28*, 195–240.

Martínez, M. Á., & García, Á. (2015). Ocupar las plazas, liberar edificios. *ACME: An International Journal for Critical Geographies, 14*(1), 157–184.

Martínez López, M. Á. (2017). Squatters and migrants in Madrid: Interactions, contexts and cycles. *Urban Studies, 54*(11), 2472–2489.

McCurdy, P., Feigenbaum, A., & Frenzel, F. (2016). Protest camps and repertoires of contention. *Social Movement Studies, 15*(1), 97–104.

Mountz, A., & Loyd, J. M. (2014). Constructing the Mediterranean region: Obscuring violence in the bordering of Europe's migration 'crises'. *ACME: An International E-Journal for Critical Geographies, 13*(2), 173–195.

Mudu, P. (2014). Ogni Sfratto Sarà Una Barricata: Squatting for housing and social conflict in Rome. In C. Cattaneo & M. A. Martínez (Eds.), *The squatters' movement in Europe: Commons and autonomy as alternatives to capitalism* (pp. 136–163). London: Pluto Press.

Newman, K., & Wyly, E. K. (2006). The right to stay put, revisited: Gentrification and resistance to displacement in New York City. *Urban Studies, 43*(1), 23–57.

Norris, M., & Hearne, R. (2016). Privatizing public housing redevelopment: Grassroots resistance, co-operation and devastation in three Dublin neighbourhoods. *Cities, 57*, 40–46.

Nowicki, M. (2017). Domicide and the coalition: Austerity, citizenship and moralities of forced eviction in inner London. In K. Brickell, M. Fernandez-Arrigoitia, & A. Vasudevan (Eds.), *Geographies of forced eviction: Dispossession, violence, resistance* (pp. 121–143). London: Palgrave Macmillan.

O'Mahony, L. F., & Sweeney, J. A. (2011). The idea of home in law: Displacement and dispossession. In L. F. O'Mahony & J. A. Sweeney (Eds.), *The idea of home in law: Displacement and dispossession* (pp. 1–11). Farnham, UK: Ashgate.

Obra Social Barcelona. (2018). *¡La vivienda para quien la habita! Informe sobre ocupación de vivienda vacía en Catalunya*. Barcelona. Retrieved from https://obrasocialbcn.net/informe-ockupacio/.

Ordóñez, V., Feenstra, R. A., & Tormey, S. (2015). Citizens against austerity: A comparative reflection on Plataforma de Afectados por la Hipoteca (PAH) and Bündnis Zwangsräumung Verhindern (BZV). *Araucaria, 17*(34), 133–154.

Pavón, I. G. C., & Janoschka, M. (2016). Viviendas en disputa – ¿espacios de emancipación? Un análisis de las luchas por la vivienda en Madrid. *Historia Actual Online, 40*, 113–127.

Prujit, H. (2013). Squatting in Europe. In Squatting Europe Kollective (Ed.), *Squatting in Europe. Radical spaces, urban struggles* (pp. 17–60). Wivenhoe, UK: Minor Composition.

Rolnik, R. (2013). Late neoliberalism: The financialization of homeownership and housing rights. *International Journal of Urban and Regional Research, 37*(3), 1058–1066.

Roy, A. (2017). Dis/possessive collectivism: Property and personhood at city's end. *Geoforum, 80*, A1–A11.

Slater, T. (2009). Missing Marcuse: On gentrification and displacement. *City, 13*(2–3), 292–311.

Slater, T. (2013). Expulsions from public housing: The hidden context of concentrated affluence. *Cities, 35*, 384–390.

Thörn, C., Krusell, M., & Widehammar, M. (2016). *Rätt att bo kvar. En handbok i organisering mot hyreshöjningar och gentrifiering* [The right to stay. A handbook on how to organise against rent increases and gentrification]. Retrieved from http://www.rattattbokvar.se/.

Turolla, T. (2017). 'Abitare attivista' in un quartiere popolare milanese. L'esperienza del comitato Drago e del Giambellino-Lorenteggio. *Antropologia, 4*(3). https://doi.org/10.14672/ada20171346%25p.

Tyler, I., & Slater, T. (2018). Rethinking the sociology of stigma. *The Sociological Review, 66*(4), 721–743.

Vasudevan, A. (2015). *Metropolitan preoccupations: The spatial politics of squatting in Berlin*. Oxford, UK: Wiley-Blackwell.

Watt, P., & Minton, A. (2016). London's housing crisis and its activisms. *City, 20*(2), 204–221.

Westenberg, G. S., & Coers, M. (Directors). (2014). *Mietrebellen* [Documentary film]. Retrieved from http://mietrebellen.de/.

Wills, J. (2016). Building urban power from housing crisis. *City, 20*(2), 292–296.

# 52

# Legal Geographies of Resistance to Gentrification and Displacement: Lessons from the Aylesbury Estate in London

## Loretta Lees and Phil Hubbard

## Introduction

Council estates in London have been called the 'final gentrification frontier' (Lees 2014a), with the last vestiges of truly affordable housing (state-subsidised council estates) being slowly destroyed. This is a process that involves the local state effectively handing estates over to private developers, who demolish the existing housing and replace it with a more dense mix of 'affordable' housing which is cross-subsidised by the market rate homes which make up the majority of properties. For this to happen, the state has to first orchestrate a process euphemistically termed 'decanting': tenants bid for properties elsewhere in the borough, or are moved against their will; private renters in leasehold properties are evicted; whilst leaseholders are bought out, often at unfavourable rates. The displacements this process sets in motion are then both direct and indirect (see Marcuse 1985), and entail phenomenological (Davidson and Lees 2010), as well as physical, dislocation (see Elliot-Cooper et al. 2019).

---

L. Lees (✉)
University of Leicester, Leicester, UK
e-mail: loretta.lees@le.ac.uk

P. Hubbard
King's College London, London, UK
e-mail: philip.hubbard@kcl.ac.uk

The literature on gentrification-induced displacement has undergone a renaissance more recently with the importance of the state in fuelling these displacements coming to the fore (see Lees et al. 2016; Zhang and He 2018). Like displacement, resistance has long been a recurrent theme in gentrification studies, and research on resistance has also undergone a renaissance in recent years (see Annunziata and Rivas-Alonso 2018; plus the special issues in *City* 2016 and *Cities* 2016). However, detailed exploration of the 'fight to stay put' in the face of displacement remains limited. Critical urban scholars and the media, certainly in London, have tended to prefer the stories of those who have taken to the streets (e.g., the publicity given to the E15 mothers protest, Russell Brand's interventions on the New Era Estate, and the 'Cereal Killer Café' anti-gentrification riots on Brick Lane). This overshadows the other important legal battles being fought by ordinary people and their supporters: these rarely get a mention even when there have been significant wins (Hubbard and Lees 2018). Given Chester Hartman's (1974, 1984; see also Hartman et al. 1982) now-infamous writings on how a community organised itself through the courts to resist gentrification—effectively exercising their 'right to stay put'—this is a significant omission in the gentrification literature.

While many legal challenges have been largely unsuccessful, costly, and time-consuming, there have been some glimmers of hope in recent legal adjudications, especially in London. Here, there are emerging signs that the law might be able to align state and institutional power with what Delaney (2016, p. 269) terms 'vectors of justice' and offer a means by which displacement might be legally resisted (see also Bryant and McGee 1983). In this chapter, we zoom in on the legal challenges brought over the gentrification-induced displacements wrought on the Aylesbury Estate in Southwark, London: two public inquiries were held in 2015 and 2018, both of which had wins for those fighting displacement. In so doing, we emphasise the importance of law in resisting displacement and we evaluate the wider implications of the outcomes of both inquiries for fighting gentrification and displacement.

## The Legal Tool of Displacement: Compulsory Purchase Orders

Compulsory purchase (eminent domain) is the power held by the state to acquire title to property without the consent of an owner, it is a key driver of urban change in European cities (Gray and Porter 2015, p. 380). It was, ironically, the lynchpin of the British modernist council housing programme which followed in the wake of the Town and Country Planning Act of 1947 (Christophers 2010, p. 869). Compulsory Purchase Orders (CPOs) were used in post-war slum clearances to make way for new council estates, and they are being used again for twenty-first-century 'new' urban renewal on those same estates to pave the way for private development. The UK has seen numerous well-known CPOs in relation to urban regeneration, for example, London Docklands (Batley 1989); Cardiff's docklands (Imrie and Thomas 1997); the Housing Market Renewal Pathfinder schemes in northern cities (Allen 2008); the London 2012 Olympics (Davis and Thornley 2010); the Glasgow 2014 Commonwealth Games (Gray and Porter 2015); but in recent years, council estate renewal in London seems to have made up a significant proportion of all CPOs served in England and Wales (Lees and Ferreri 2016; Hubbard and Lees 2018). In this chapter, we ask: 'what are CPO inquiries useful for? Delaying development, raising public awareness, bolstering local campaigns, bringing people together on an estate, wresting information from the council or developers, stopping a scheme and saving an estate (possibly!)'.[1]

Council estate renewal can be seen as part of a process of 'accumulation by dispossession' (Harvey 2003), given it involves the release of common and state assets to the market that requires the direct displacement of some, or all, of those who dwell on affected estates. Irrespective that redeveloped estates might retain significant amounts of social (but rarely council) housing, the redevelopment itself involves displacement so that existing housing can be demolished, refurbished, or densified, with the payoff for the developers being (profit) the opportunity to develop speculative private housing aimed at upper-middle-class consumers (see Lees et al. 2008; Watt 2009; Lees 2014b, on this as state-led gentrification). For some, this signifies not just the dismantling of low-income or working-class communities, but the end of council estates as we know them. Here, the CPO plays a key role, in that it allows local councils to 'purchase back' from leaseholders the very properties they originally sold them to (Rendell 2018). As Layard (2018) says, these different legal devices of ownership—freehold and leasehold—provide extraordinary security for one landowner (the freeholder) and possible vulnerability for another (the leaseholder).

Ironically, despite this vulnerability, the greatest potential for resistance lies with leaseholders on council estates who have either exercised their right to buy their property or have bought them from a previous council tenant who exercised their right to buy. For these residents, human rights (notably Article 8, the right to family life and Article 1, Protocol 1 of the 1950 European Convention of Human Rights) can be invoked to demand either procedural or substantive changes by landowners. CPOs were famously used on the

Heygate Estate (see Lees 2014a; Lees et al. 2013; London Tenants Federation et al. 2014), adjacent to the Aylesbury Estate in London, to remove leaseholders pre-demolition:

> The Executive is advised that the Council has a power to compulsorily acquire land and property interests under Section 226(1)(a) of the *Town and Country Planning Act 1990* (as amended by Section 99 of the *Planning and Compulsory Purchase Act 2004*) ('the 1990 Act'). 22. Section 226(1)(a) gives the Council power to acquire compulsorily any land in their area if the Council think that the acquisition will 'facilitate the carrying out of development/re-development, or improvement on, or in relation to, the land'. … In exercising this power the Council must have regard to Section 226(1)(b) of the 1990 Act and must not exercise the power unless it thinks that the development, re-development or improvement is likely to contribute to…the promotion of improvement of the economic well-being of the area; the promotion or improvement of the social well-being of the area; the promotion or improvement of the environmental well-being of the area.[2]

A public inquiry was triggered when remaining leaseholders on the Heygate Estate objected to the order which would see them dispossessed of their homes. The objectors were led by the Heygate Leaseholders Group, supported by expert witnesses, including academics. The CPO-ed leaseholders got compensation offers of, on average, £95,480 for a one-bedroom flat, £107,230 for a two-bedroom flat, £156,833 for a three-bedroom maisonette, and £177,421 for a four-bedroom maisonette (Freedom of Information data collected by campaign groups, see http://heygatewashome.org/displacement.html). A studio flat on the redeveloped Heygate Estate, now renamed Elephant Park, started at £330,000.

The Heygate Estate Public Inquiry—although not resulting in the refusal of the CPO—did act as a legal learning curve for those involved (see Lees and Ferreri 2016), and critically it brought the gentrification debates surrounding council estates out into the open (Layard 2018). Indeed, the outing of the expulsion of 3000 plus residents from the Heygate Estate marked it in history as an infamous example of state-led gentrification. A number of those involved in the Heygate Estate Public Inquiry were subsequently involved in the Aylesbury Estate CPO public inquiries, to which we now turn.

## The First Aylesbury Estate Public Inquiry

The Aylesbury Estate in Southwark is one of the largest public housing estates in Britain, built between 1967 and 1977; the 2700 dwellings were designed to house a population of roughly 10,000 residents. In 1997, the Aylesbury was given 'New Deal for Communities' (NDC) status and studies began on how the stigmatised (as crime-ridden) estate could be regenerated. The NDC was given £56.2m over ten years in order to lever in a further £400m as part of the estate's proposed stock transfer from council to housing association tenure. But the residents voted against the stock transfer of the Aylesbury from Southwark Council in 2001, 73 per cent voting to keep the whole estate in council ownership (with 76 per cent of the estate turning out to vote). Nevertheless, in 2005, the Liberal Democrat-led Southwark Council stated that the estate was too expensive to refurbish and that demolition was the most cost-effective solution. The estate was to be redeveloped as a new, 'mixed community' (see Lees 2014b). The physical regeneration of the estate was to displace approximately 20 per cent of the existing households, including the existing leaseholders who had bought under 'right to buy.' It was intended that those that got to move back on to the original footprint of the estate would have to fit themselves into a new community almost twice the density and in which the majority of the inhabitants would be middle-class residents renting at market rates or owning their own home. On seeing what had happened on the Heygate Estate next door, Aylesbury residents were understandably worried, and a number of leaseholders came together, not simply to get better compensation but to fight to stay put.

The first Aylesbury Public Inquiry was prompted by the Aylesbury Leaseholder's Action Group (ALAG), eleven leaseholders in eight properties earmarked for demolition across different blocks on the estate.[3] The planning inspector, Lesley Coffey, oversaw the inquiry which took place on various dates between April and October 2015. The inquiry was located in 'Arry's Bar (see Fig. 52.1) at the Millwall football ground in south-east London (see Rendell 2017). This unusual location was chosen due to fear of protestors after the occupation of the Aylesbury some time earlier (see Lees and Ferreri 2016); the bar was located inside the football ground gates and the inquiry could be locked down if it was felt necessary. Qualifying objections and one non-qualifying objection to the CPO were received prior to the commencement of the inquiry, with several additional objections made at the inquiry. Although planning permission had already been granted for the demolition and redevelopment of the Aylesbury in 2015, for it to go ahead, the CPO had

**Fig. 52.1** 'Arry's bar: Site of first Aylesbury Estate Public Inquiry with heavy security. (Photo courtesy of 35% campaign)

to be confirmed by the government. As stated earlier in relation to the Heygate CPO, Southwark had to satisfy the following tests under section 226(1A) of the Town and Country Planning Act 1990, with the regeneration for the CPO land having to fulfil one or more of the following: (1) the promotion or improvement of the economic well-being of the area; (2) the promotion or improvement of the social well-being of the area; (3) the promotion or improvement of the environmental well-being of the area.

The main grounds of ALAG's objection related to the failure of the scheme to ensure that social rented housing would be provided on the new development; the viability and deliverability of the scheme; the option of refurbishment not properly being considered; the scheme not promoting the social well-being of the area; the failure of the Acquiring Authority to carry out an Equality Impact Assessment in relation to the leaseholders; and the suggestion that the CPO breached the human rights of the leaseholders.[4]

The arguments made by the objectors that the planning inspector and the Secretary of State took on board are summarised here:

1) The CPO is a breach of human rights: the rights of the Objectors under the European Convention on Human Rights (ECHR) in respect of Articles 1 (right to quiet enjoyment of property) and 8 (right to respect for private and family life); 2) the Council has not taken reasonable steps to acquire the land interests by agreement; 3) Council valuations are too low (sometimes as low as 40% of market price) and the council is not allowing independent valuations, only those done in-house; 4) the CPO's confirmation would deprive leaseholders of their homes, their savings, and displace them from the area; 5) various parts of the scheme do not comply with the council's sunlight and daylight standards, the principles of the AAP and section 7 of the NPPF; 6) the proposed development will have considerable economic and social dis-benefits in terms of consequences for those leaseholders remaining on the Order Land. The council did not undertake an equalities impact assessment as per their public sector equality duty as required by the Equalities Act 2010. The leaseholders are mainly BME. Depriving a BME homeowner of his/her home requires an assessment of whether that homeowner would be more adversely affected than one from a non-predominantly BME estate. Leaseholders from the BME community on the Estate derive cultural advantages from living in the area. They face forced separation from their communities, which in many cases may result in difficulty in retaining contact with a particular culture.[5]

As discussed in Hubbard and Lees (2018), following the inquiry, the inspector recommended that the CPO should not be confirmed because overall there would be too many negative impacts meaning that 'a compelling case in the public interest [had] not been proved.'[6] The Secretary of State for Communities and Local Government's 2016 (initial) decision (withdrawn in 2017) to confirm this recommendation hence represented a significant and surprising victory for the leaseholders and others involved:

> The decision raises some real issues for the CPO industry. It paints an uncomfortable picture of CPO being a tool of gentrification, driving residents and small businesses out of their communities on account of rising land values and rents; the polar opposite of what a CPO is intended to achieve, which should be to improve and restore vitality to a local area. (Vas 2017, n.p.)

The key reasons given by the Secretary of State for his decision were that there had been insufficient negotiation with remaining leaseholders; that Southwark Council had not taken reasonable steps to acquire land interests by agreement; that there would be considerable economic, social, and environmental disbenefits for the leaseholders who would remain on the land; that interference with the human rights of those with an interest in the relevant land was not sufficiently justified; and overall, that the test for a 'compelling case in the public interest' had not been met (as required by CPO policy guidance). This decision stressed the importance of addressing human rights when individuals are affected by a CPO (i.e., Article 8 of the ECHR right to respect 'private and family life') and also highlights the increasing importance of public sector equality duty given the ruling that children, the elderly, and black and ethnic minority residents would be 'disproportionately affected' by the CPO, and that it would have a negative impact on their ability to retain their cultural ties.[7] Issues such as the 'dislocation from family life' and the potential to harm the education of affected children were identified in the decision letter, indicating a much wider approach to assessing the impacts of a CPO than had been the case previously. In the Secretary of State's summation:

> The options for most leaseholders are either to leave the area, or to invest the majority of their savings in a new property. Article 8(1) ... is therefore clearly engaged. In relation to Article 8(2) (which permits interference which is proportionate when balanced against the protection of the rights and freedoms of others), the Secretary of State finds that the interference with residents' (in particular leaseholders') Article 8 rights is not demonstrably necessary or proportionate, taking into account the likelihood that if the scheme is approved, it will probably force many of those concerned to move from this area ... The likelihood that leaseholders will have to move away from the area will result in consequential impacts to family life and, for example, the dislocation from local family, the education of affected children and, potentially, dislocation from their cultural heritage for some residents.[8]

The letter went on to note 'the lack of clear evidence regarding the ethnic and/or age make-up of those who now remain resident at the Estate and who are therefore actually affected by any decision to reject or confirm the Order' but argued that given that '67% of the population living on the Estate were of BME origin' it would be highly likely that there would be a disproportionate impact of the CPO on the elderly and children from these groups. Hence, it was adjudged that it would be those from ethnic minorities who would be most likely to dominate the profile of those remaining on the estate and it is this population who would have to move out of the area if the order was confirmed. In noting this, the Secretary of State stated that 'white British culture is more widely-established across the UK, including at housing sites to which residents may be moved, whereas minority cultural centres are often less widespread, which is likely to make cultural integration harder for those of BME origin who are forced to move than those of a white British origin.' Its implications will clearly be a significant factor in future CPO decisions in London, not least where estate renewal threatens communities where BME residents are present in significant numbers (White and Morton 2016). Indeed, Leary-Owhin (2018) has proclaimed: 'That decision set some precedents which threatened seriously the future of estate regeneration in England!'

Indeed, the initial ruling of the first Aylesbury public inquiry boosted the confidence of others seeking to object to redevelopment proposals on the grounds of failure by an authority to properly comply with its equality duty. For example, it was raised in objections to proposals by the London Borough of Haringey to promote estate regeneration through the Haringey Development Vehicle (HDV).[9] Indeed, the Secretary of State's ruling on the Aylesbury Estate regarding equality duty and human rights under the ECHR 'was a game-changer, for now many authorities in the early stages of preparing CPOs are making greater demands of developers in terms of their proposed relocation and re-housing strategies to avoid similar criticisms of their own schemes' (Thomas 2017). Yet, the importance of robust evidence must not be underestimated, for the 2018 High Court ruling that the Haringey Development Vehicle was lawful, charged that the claimant's complaints regarding public sector equality duty under s149 of the Equality Act 2010 were entirely speculative, with the judge cautioning 'how remote from reality equalities arguments can become forensically.'[10]

## The Revised Aylesbury Estate Public Inquiry

Perhaps unsurprisingly, Southwark Council challenged the 2015 Aylesbury CPO public inquiry decision in the High Court:

> The Secretary of State for Communities and Local Government … notified Southwark Council that he would consent to judgment and ask the court to quash his decision not to confirm the Compulsory Purchase Order for the remaining properties in Phase 1 of the regeneration of the Aylesbury Estate. A Consent Order has been agreed with DCLG and has been sent to the Interested Parties for their agreement. If the Court decides to quash the decision, then in accordance with the terms of the Order the Secretary of State will arrange a new public inquiry to decide the merits of the Compulsory Purchase Order, to be held as soon as practicable. (Pereira and Murphy 2017)

The decision made at the first Aylesbury public inquiry was then overturned on a technicality: the Secretary of State's decision not to confirm the

CPO was quashed as a result of his failure to adequately explain why a change to Southwark Council's £16,000 policy did not alter his decision not to confirm the CPO (given between the close of the CPO public inquiry and the publication of the Secretary of State's decision, Southwark had scrapped the requirement for affected leaseholders to commit all but £16,000 of their savings towards the purchase of new homes).

The second, revised public inquiry sat at various dates in January and April 2018: 'this mixed-communities-led estate regeneration CPO public inquiry (was) probably the biggest and one of the most important ever in the UK' (Leary-Owhin 2018, p. 3). This inquiry was held at Southwark Council's offices in Tooley Street, with a new planning inspector, Martin Whitehead. Like in the first inquiry, there was also heightened security, with security guards on the doors and in the room. Of the four remaining leasehold interests on the order land, two of these were fighting as ALAG and one represented herself. ALAG (Beverley Robinson and Agnes Kabuto) and its supporters this time had proper legal representation from Chris Jacobs, the same barrister from Landmark Chambers, who summed up in the first inquiry (he was paid through fundraising undertaken by ALAG and its supporters). Southwark called nine witnesses, ALAG called twenty-seven witnesses,[11] and Judi Bos, the leaseholder representing herself, called three witnesses.

Understandably concerned by the first-hand (ALAG) and academic evidence that was given to the first inquiry on displacement and community impacts, Southwark Council this time hired their own academic to produce a report on the likely impacts of the 'regeneration.' Southwark used this report to counter the previous and any new evidence on displacement. The report by Owhin Leary spoke to a number of points: first, it was clear that Southwark and their barrister were concerned about the fact that the academic research on displacement had been significant in the first win, as such they set out to counter it with other academic research that would tell a different story (in the inquiry, the intent was clear when Southwark's hired academic said: 'Lees has not presented credible evidence of why people have moved' and that 'only credible, statistically valid data can prove displacement' (verbatim from the taped inquiry)); second, the council sought to present its own research as unbiased while the objectors' research was presented as biased and political (Southwark's barrister tried to make out that Lees was a political activist and had come to the inquiry with a 'posse' of like-minded academic comrades). When Southwark's hired academic stated that displaced Aylesbury residents could stay in touch via social media and that 'social relations can operate over great distances,' those Aylesbury residents in the audience erupted and had to be quietened by the inspector. He also stated that folk should give up their home for 'the greater public benefit,' that the objectors had misused the term 'community' and that 'it's not gentrification, we need a different name for it' (verbatim from the taped inquiry).

But ultimately, the two Aylesbury leaseholders fighting as ALAG reached a confidential agreement with Southwark Council in the middle of the inquiry, after it published a new rehousing policy—a rehousing assistance scheme for homeowners affected by regeneration—that went beyond their statutory rehousing duty under the 1973 Land Compensation Act (amended). That the council made this change in policy[12] is in no small part due to the first and the revised Aylesbury public inquiries.

The new rehousing assistance scheme is discretionary and assists homeowners to find a suitable housing route to avoid having to CPO them. Southwark did this in recognition of 'the trauma and inconvenience caused to displaced homeowners affected by regeneration' (note to ALAG objectors from Southwark 2018). The result was a significant policy change: leaseholders now benefit from improved terms, including a new equity loan scheme to help them buy a replacement home that their families can now inherit. Where Southwark had initially offered displaced leaseholders the opportunity to buy a new council built home with a minimum 50 per cent equity share, this was brought down to 25 per cent. Homeowners no longer have to invest their home-loss payment into the acquisition of a replacement home. Inheritance clauses in the shared equity and equity loan leases were amended to allow inheritance, to be able to pass the property on to a partner or children. Pre-emption clauses (if a leaseholder wanted to sell, they had to offer the property to the council first before putting it on the open market) were removed from shared equity and equity loan schemes. Southwark also committed to covering any additional stamp duty costs due to them opting for the new equity loan model. As she settled, Beverley Robinson resigned as chair of ALAG, but ALAG vowed to support the remaining objector, Judi Bos.

The outcome for the remaining objectors was announced late in 2018: they lost their case. The decision letter laid out the basis on which this was determined. First, it stated that:

> In terms of environmental wellbeing … the Inspector considers that circumstances have changed since the CPO was considered previously by the Secretary of State in 2016, so that most of the buildings on the site are now vacant, in the process of being demolished or have been demolished, and most of the open space is inaccessible to the public due to it being used as part of the demolition site …The Inspector says comparisons with the scheme as proposed through the granted planning application are therefore difficult. However, he concluded that

based on the evidence the scheme would contribute to the improvement of the environmental well-being of the area.[13]

Second, the judgement proceeded to consider the social disbenefits of the CPO and stated that these would be 'limited to just one household' and would have 'a very limited impact on the social well-being of the area.' It went on to argue 'the Order would enable the construction of a new development of housing and community facilities' and 'it would represent a good contribution to the social wellbeing of the area.' Third, the inspector found the direct and indirect economic benefits to be 'far greater than any economic dis-benefit that would occur as a result of the Order.'[14]

On this basis, and having considered 'the Inspector's separate analysis of the environmental, social and economic aspects of wellbeing,' the Secretary of State was satisfied that the public would clearly benefit from the confirmation of the CPO and completion of this phase of the estate's regeneration. This time, in the ruling, the emphasis put on human rights was back-pedalled:

> The Inspector accepted that there would be an interference with the human rights of the remaining objectors. He was, however, also satisfied (IR211) that any interference with their rights under Article 1 of the First Protocol and Article 8 of the ECHR is in accordance with the law, pursuant of a legitimate aim, and proportionate given the scale of the public benefits.[15]

On public sector equality duty (PSED), whilst recognising that BME and older residents were disproportionately affected, he said:

> On balance the mitigation measures have demonstrated that the PSED negative impacts have been adequately addressed, where possible, and would amount to reasonable steps to meet protected groups' needs and mitigate residual disadvantage suffered, advancing equality of opportunity and minimising discriminatory impact.[16]

This was a significant rollback from the ruling after the first inquiry, suggesting that as the community was dispersed and decanted, the strength of the opposition case lost critical mass.

Nevertheless, whilst the CPO has been confirmed, there have been significant gains: an increase in social rented housing in the phase of development in question (albeit with no overall gain across the scheme) and a beneficial change for leaseholders in regeneration estates across the borough, giving them a better chance of remaining in the area they have chosen to live in.

Notting Hill Housing Trust, the housing association redeveloping the estate, also applied for a variation to the S106,[17] which would entail a new shared equity loan scheme option for leaseholders, similar to that introduced by Southwark as part of the ALAG settlement, and this might open the door to resident leaseholders (not yet CPO-ed) remaining on the footprint of the Aylesbury Estate, noting that these two public inquiries only related to the 'First Development Site,' a small part of the 60-acre Aylesbury Estate. There are still hundreds of residents on the rest of the Aylesbury Estate whose homes remain under threat. The evidence withdrawn due to ALAG's settlement can be used again.

## Conclusion

One purpose of legal work in gentrification studies is to investigate whether there are soft spots where we can challenge the developers and state institutions that push gentrification into new neighbourhoods (Layard 2018). This is exactly what those involved in the Aylesbury public inquiries did, yet it is no easy task.[18] But such minor victories in the court room help shift the policy debate in decisive ways. Notably, recent UK policy missives now acknowledge the need to engage with, and protect, existing council estate residents, responding in part to concerns about the impact of estate redevelopment on existing communities. For example, the Department of Communities and Local Government's (2016) *Estate Regeneration National Strategy Resident Engagement and Protection* demands 'more than legal' protections for council estate residents:

> It is a legal requirement for leaseholders to be compensated if their home is demolished. However, we expect that schemes will go further and offer leaseholders a package that enables them to stay on the estate or close by. We also expect leaseholders to be offered the option of an independent valuation of their property. (DCLG 2016, p. 5)

The GLA (2016) also published its *Draft Good Practice Guide to Estate Regeneration*, which notes the potentially disruptive effects of regeneration on existing communities and neighbourhoods, setting forth principles for resident-led regeneration:

> The Mayor believes that for estate regeneration to be a success, there must be resident support for proposals, based on full and transparent consultation. These

proposals should offer full rights to return for displaced tenants and a fair deal for leaseholders, and demolition should only be followed where it does not result in a loss of social housing, or where all other options have been exhausted. (GLA 2016, p. 4)

This shifting ground is testimony to the hard work of those who have fought gentrification through legal process and in so doing highlighted the injustices of displacement. In this chapter, we have given this resistance the attention it deserves. In London, the fight to stay put in the face of state-led gentrification and displacement through the legal system is at an all-time high, and it is having real impact on policy and practice, and also in educating the public more widely about the injustices these displacements enact.

## Notes

1. http://bailey.persona-pi.com/Public-Inquiries/aylesbury-estate/Presented%20Council%20docs/app_0_12.pdf.
2. http://moderngov.southwark.gov.uk/documents/s7807/Heygate%20Estate-%20Compulsory%20Purchase%20orders%20report.pdf.
3. ALAG was supported by a handful of academics (the geographer Loretta Lees, architectural historians Ben Campkin and Jane Rendell, and engineering scientist Kate Crawford all at UCL), housing activists (including the 35% Campaign group), an ex-Conservative Southwark councilor Toby Eckersley, and eventually, on the last day, a pro bono lawyer, Chris Jacobs from Landmark Chambers.
4. In relation to the latter, the Court accepted that a CPO should not be confirmed unless the case in the public interest fairly reflects the necessary element of balance required in the application of article 8 and Article 1 of the First Protocol to the ECHR (*London Borough of Bexley and Sainsbury's v SoSE* [2001] EWHC Admin 323, paragraphs 33–48 and *Pascoe* [2006] EWHC paragraph 66).
5. http://bailey.persona-pi.com/Public-Inquiries/aylesbury-estate/Presented%20Council%20docs/app_0_12.pdf.
6. CPO Report to the Secretary of State for Communities and Local Government 29 January 2016 by Leslie Coffey on Application for the Confirmation of the London Borough of Southwark (Aylesbury Estate Site 1B-1C) Compulsory Purchase Order 2014 NPCU/CPO/A5840/74092.
7. This was also a decisive factor in *R (Harris) v London Borough of Haringey* (Court of Appeal, 5 May 2010) where the court held that the council, when granting planning permission, failed to discharge its duties under section 71(1) of the Race Relations Act 1976 (now replaced by the Equality Act 2010

Public Sector Equality Duty) in terms of considering the potential effects of the scheme on Latin American traders or loss of housing by ethnic minorities. In this instance, permission was quashed on the basis that due regard was not given to the loss of housing by ethnic minority groups (see Ricketts 2016).

8. Letter to Karen Jones, Southwark Council, from Dave Jones, Senior Planner, on behalf of the Secretary of State for Communities and Local Government, 16 September 2016, Ref: NPCU/CPO/A5840/74092.
9. https://www.minutes.haringey.gov.uk/documents/s94484/2e%20Evidence%20from%20Loretta%20Lees.pdf.
10. https://cornerstonebarristers.com/news/haringey-hdv-ruled-lawful/.
11. Including a number of academics: those who were cross-examined (the others were not, as ALAG settled) included Loretta Lees, Richard Baxter, and Ben Campkin (but their evidence was subsequently withdrawn due to the settlement).
12. See http://moderngov.southwark.gov.uk/ieDecisionDetails.aspx?ID=6501.
13. http://bailey.persona-pi.com/Public-Inquiries/aylesbury-estate/general/decision_letter_14-11-18.pdf.
14. http://bailey.persona-pi.com/Public-Inquiries/aylesbury-estate/general/decision_letter_14-11-18.pdf.
15. http://bailey.persona-pi.com/Public-Inquiries/aylesbury-estate/general/decision_letter_14-11-18.pdf.
16. http://bailey.persona-pi.com/Public-Inquiries/aylesbury-estate/general/decision_letter_14-11-18.pdf.
17. On S106 see https://www.local.gov.uk/pas/pas-topics/infrastructure/s106-obligations-overview.
18. See the advice here: http://bailey.persona-pi.com/Public-Inquiries/aylesbury-estate/Presented%20Council%20docs/app_0_12.pdf.

# References

Allen, C. (2008). *Housing market renewal and social class*. London: Routledge.

Annunziata, S., & Rivas-Alonso, C. (2018). Resisting gentrification. In L. Lees with M. Phillips (Eds.), *Handbook of gentrification studies* (pp. 393–412). Cheltenham, UK: Edward Elgar.

Batley, R. (1989). London Docklands: An analysis of power relations between UDCs and local government. *Public Administration, 67*(2), 167–187.

Bryant, D. C., Jr., & McGee, H. W., Jr. (1983). Gentrification and the law: Combatting urban displacement. *Wash. U. Journal of Urban & Contemporary Law, 25*, 43–144.

Christophers, B. (2010). Geographical knowledges and neoliberal tensions: Compulsory land purchase in the context of contemporary urban redevelopment. *Environment and Planning A, 42*(4), 856–873.

Cities. (2016). Resistance to social housing transformation [Special issue]. *Cities, 57,* 1–62.

City. (2016). London's housing crisis and its activisms [Special issue]. *City, 20* (2).

Davidson, M., & Lees, L. (2010). New-build gentrification: Its histories, trajectories, and critical geographies. *Population, Space and Place, 16*(5), 395–411.

Davis, J., & Thornley, A. (2010). Urban regeneration for the London 2010 Olympics: Issues of land acquisition and legacy. *City, Culture and Society, 1*(2), 89–98.

DCLG (Department of Communities and Local Government). (2016). *Estate regeneration national strategy resident engagement and protection.* Retrieved from https://www.gov.uk/government/uploads/system/uploads/attachment_data/file/575578/Estate_Regeneration_National_Strategy_-_Resident_Engagement_and_Protection.pdf.

Delaney, D. (2016). Legal geography II: Discerning injustice. *Progress in Human Geography, 40*(2), 267–274.

Elliot-Cooper, A., Hubbard, P., & Lees, L. (2019). Moving beyond Marcuse: Gentrification, displacement, and the violence of un-homing. *Progress in Human Geography.* https://doi.org/10.1177/0309132519830511.

GLA. (2016). *Draft Good Practice Guide to Estate Regeneration.* Retrieved from https://www.london.gov.uk/sites/default/files/09.draftgoodpracticeestateregenerationguidedec16.pdf.

Gray, N., & Porter, L. (2015). By any means necessary: Urban regeneration and the 'state of exception' in Glasgow's Commonwealth Games 2014. *Antipode, 47*(2), 380–400.

Hartman, C. (1974). *Yerba Buena: Land grab and community resistance in San Francisco.* San Francisco, CA: Glide Publications.

Hartman, C. (1984). The right to stay put. In C. Geisler & F. Popper (Eds.), *Land reform, American style* (pp. 302–318). Totowa, NJ: Rowman and Allanheld.

Hartman, C., Keating, D., & LeGates, R. (1982). *Displacement: How to fight it.* Washington, DC: National Housing Law Project.

Harvey, D. (2003). *The new imperialism.* New York, NY: Oxford University Press.

Hubbard, P., & Lees, L. (2018). The right to community: Legal geographies of resistance on London's final gentrification frontier. *City, 22*(1), 8–25.

Imrie, R., & Thomas, H. (1997). Law, legal struggles, and urban regeneration: Rethinking the relationships. *Urban Studies, 34*(8), 1401–1418.

Layard, A. (2018). Property and planning law in England: Facilitating and countering gentrification. In L. Lees with M. Phillips (Eds.), *Handbook of gentrification studies* (pp. 444–466). Cheltenham, UK: Edward Elgar.

Leary-Owhin, M. (2018). *Public spaces of hope: Reflections on the gentrification glass ceiling.* Paper presented to AESOP 2018 Annual Congress,

Gothenburg, Sweden. Retrieved from http://www.academia.edu/36723125/Public_Spaces_of_Hope_Reflections_on_the_Gentrification_Glass_Ceiling.

Lees, L. (2014a). The death of sustainable communities in London? In R. Imrie & L. Lees (Eds.), *Sustainable London? The future of a global city* (pp. 149–172). Bristol, UK: Policy Press.

Lees, L. (2014b). The urban injustices of New Labour's 'new urban renewal': The case of the Aylesbury Estate in London. *Antipode, 46*(4), 921–947.

Lees, L., & Ferreri, M. (2016). Resisting gentrification on its final frontiers: Lessons from the Heygate Estate in London (1974–2013). *Cities, 57*, 14–24.

Lees, L., Just Space, The London Tenants Federation, & Southwark Notes Archive Group (SNAG). (2013). Challenging 'the new urban renewal': The social cleansing of council estates in London. In B. Campkin, D., Roberts, & R. Ross (Eds.), *Urban Pamphleteer #2 – London: Regeneration realities* (pp. 6–10). London: Urban Lab, UCL.

Lees, L., Slater, T., & Wyly, E. (2008). *Gentrification*. New York, NY: Routledge.

Lees, L., Shin, H., & Lopez-Morales, E. (2016). *Planetary Gentrification*. Cambridge, UK: Polity Press.

Marcuse, P. (1985). Gentrification, abandonment, and displacement: Connections, causes, and policy responses in New York City. *Urban Law Annual: Journal of Urban and Contemporary Law, 28*, 195–240.

Pereira, J., & Murphy, M. (2017, 27 April). *Aylesbury Estate CPO decision: The Secretary of State consents to judgment*. Retrieved from http://ftbchambers.co.uk/news/aylesbury-estate-cpo-decision-secretary-state-consents-judgment.

Rendell, J. (2017). 'Arry's Bar: Condensing and displacing on the Aylesbury Estate. *The Journal of Architecture, 22*(3), 532–554.

Rendell, J. (2018). Figuring speech: Before and after writing. In J. Charley (Ed.), *The Routledge companion on architecture, literature and the city* (pp. 385–410). Abingdon, UK and New York, NY: Routledge.

Ricketts, S. (2016, 29 September). Regeneration X: Failed CPOs. *Local Government Lawyer*. Retrieved from https://www.localgovernmentlawyer.co.uk/projects-and-regeneration/317-projects-features/31801-regeneration-x-failed-cpos.

The London Tenants Federation, Lees, L., Just Space, & SNAG. (2014). *Staying put: An anti-gentrification handbook for council estates in London*. London. Retrieved from https://southwarknotes.files.wordpress.com/2014/06/staying-put-web-version-low.pdf.

Thomas, L. (2017, 17 November). Compulsory purchase: Life after Aylesbury. *Journal of Planning and Environment Law*. Retrieved from https://www.ashurst.com/en/news-and-insights/legal-updates/compulsory-purchase-life-after-aylesbury/.

Vas, M. (2017). CPO – gentrification or regeneration? [Web log post]. *UK Planning Law Blog*. Retrieved from http://www.planninglawblog.com/cpo-gentrification-or-regeneration.

Watt, P. (2009). Housing stock transfers, regeneration and state-led gentrification in London. *Urban Policy and Research, 27*(3), 229–242.

White, M., & Morton, H. (2016). *A new 'right to a community'? Decision by the Secretary of State not to confirm the CPO for Aylesbury Estate.* Retrieved from http://hsfnotes.com/realestatedevelopment/2016/09/28/a-new-right-to-a-community-decision-by-the-secretary-of-state-not-to-confirm-the-cpo-for-aylesbury-estate/#page=1.

Zhang, Z., & He, S. (2018). Gentrification-induced displacement. In L. Lees with P. Phillips (Eds.), *Handbook of gentrification studies* (pp. 134–154). Cheltenham, UK: Edward Elgar.

# 53

# Local Faith Communities and Responses to Displacement

Susanna Trotta and Olivia Wilkinson

## Introduction

Academic interest in local faith communities (LFCs) and refugee response is relatively recent with a special issue of the *Journal of Refugee Studies* in 2011 (Fiddian-Qasmiyeh 2011a) marking a turning point. The focus on LFCs allows scholars to look at the contextual specificity of faith actors in their locations, in interrelation with the other actors in that context and recognising the embeddedness of LFCs in the culture, politics, economics, and society of their place. A new era of ethnography (e.g., Ngo 2018), as well as cross-country comparison (Greatrick et al. 2018), has allowed for deeper understandings of the contextualised actions of LFCs for refugee response.

In international humanitarian action, policy discourse has shifted attention to power imbalances between local and international actors and the need to 'localise' humanitarian response. Localisation was enshrined in the Grand Bargain commitments launched at the World Humanitarian Summit in 2016 and encapsulated in the phrase 'as local as possible, as international as necessary' (UNSG 2016). Yet there is a fundamental gap in localisation as local

---

S. Trotta (✉)
Humboldt University, Berlin, Germany

Joint Learning Initiative on Faith and Local Communities, Washington, DC, USA

O. Wilkinson
Joint Learning Initiative on Faith and Local Communities, Washington, DC, USA
e-mail: olivia@jliflc.com

© The Author(s) 2020
P. Adey et al. (eds.), *The Handbook of Displacement*,
https://doi.org/10.1007/978-3-030-47178-1_53

actors must still abide by the standards and practices of international humanitarianism. As Fiddian-Qasmiyeh puts it, 'There is of course a major paradox inherent in the localisation of aid agenda: it aims to "support" local responses precisely by institutionalising them within the broader paradigm and parameters established by the "international system"' (Fiddian-Qasmiyeh 2018). This paradox has brought to light the need to highlight southern-led humanitarianisms that lie outside the hegemony of the international humanitarian system (Fiddian-Qasmiyeh and Pacitto 2019).

Part of that hegemony is a pervasive and 'functional secularism' (Ager and Ager 2011) that frames international humanitarian discourse. The effects lead to a privatisation, marginalisation, and instrumentalisation of religion in the humanitarian system (Ager and Ager 2015), in which religious belief and practice are consciously and unconsciously ignored by humanitarian decision makers, only coming into view when they can serve a purpose, such as mediating between conflicting parties or offering financial and human resources. The combination of the paradox of localisation and the functional secularism of the humanitarian system means that LFCs that do not fit into the international frameworks of humanitarianism are looked over for partnership. There are many reasons for hesitation from international humanitarian actors to partner with LFCs, not least fears of politicisation around religious groups and partiality in aid distribution (Carpi 2018; Wilkinson 2018). The roles played by local faith actors are complex and multifaceted and more nuanced analyses are needed (Kidwai and Fiddian-Qasmiyeh 2017). Research points towards the lack of understanding from the international humanitarian system about the ways in which LFCs implement humanitarian principles of impartiality and neutrality and provide more effective and appropriate assistance than the humanitarian system itself, in spite of its standards and procedures (El Nakib and Ager 2015; Wilkinson 2017).

To continue to add to the nuancing of our understanding of the ways in which LFCs are responding for refugees and forced migrants, we have been collating and reviewing the literature over the past two years (2016–2018), with a growing database of 550 articles and resources of interest of which approximately 100 feed into our knowledge and background for this chapter. We have noted a general bias in the literature towards studies of Christian communities in the Global North (Wilkinson and Ager 2017, p. 10), partly due to the nature of our searches which were in English, French, Italian, German, and Spanish only. Nevertheless, the bias underlines that researchers in the Global North also have much to do to highlight context and cases outside their own backyard. We include some of the examples from the Global North in the rest of this chapter, but we have also purposely prioritised research with examples from the Global South to correct some of this imbalance in a small way.

International actors are interested in LFCs in principle and several efforts have been made to create conversations on the topic of LFCs in refugee response. Building on the work of the Joint Learning Initiative on Faith and Local Communities' Refugee and Forced Migration Learning Hub, this chapter uses a structure of stages and spaces of displacement (while affirming that the stages are rarely, if ever, linear) to highlight examples of LFCs and encounters with refugees and forced migrants. The stages and spaces are broadly characterised as journeying, arrival and reception, and long-term solutions to displacement.

## Journeying

LFCs are present at various stages of a journey but also travelling with and as part of displaced people's journeys, facilitating travel with transnational religious networks, at borders, in camps, in urban centres, in detention centres, and in non-material places such as states of statelessness and immobility. We have chosen just a few spaces of interest, recognising that religious beliefs and practices travel with a displaced person and community, dynamically evolving with changing contexts and experiences.

Faith-driven initiatives along migration routes often constitute a very concrete source of support to displaced people in the form of shelters, food or clothes distribution, and other immediate services. 'La 72' is a safe space along the migration route from Central America to Mexico or the United States. Situated in the state of Tabasco (Mexico), close to the border with Guatemala, Franciscan friars founded this shelter. It hosts women and minors in dedicated spaces and is a relatively rare example of faith communities providing LGBTIQ programmes for those forced to flee because of their gender identity. In addition to operating as a shelter, as Olayo-Méndez (2017) highlights, La 72 is also engaged in advocating for migrant rights and in fostering dialogue between and among migrants, local communities, and the authorities.

LFCs are not homogenous in their responses, with groups even from the same community or congregation holding very different opinions and ways of working for refugees. According to Rexhepi (2018), the response of Muslim LFCs to the transit of thousands of refugees through the Balkan countries in the summer of 2015 was heterogeneous. While the state-sanctioned IRCM (Islamic Religious Community of Macedonia) proved to be very reluctant in showing and enacting solidarity with the refugees, several local Muslim communities and individuals did organise food and clothes distributions, as well as shelter facilities, sometimes in religious buildings, as in the case of Kumanovo's mosque. This difference exposes existing tensions within Macedonia and in other Balkan states, where the idea of a 'moderate' or 'European' Islam has been mobilised in the interest of the EU's securitising migration policies, as opposed to the threat of 'radical' or 'Arab' Islam. In Rexhepi's words, 'IRCM's primary concern was not organising help for the refugees, but how to distance itself from them to protect local Muslims from the threat of political Islamist ideology' (Rexhepi 2018, p. 13). A nuanced understanding of the position of politically affiliated actors and the immediate support from local communities demonstrate not only the relationships between faith community members and faith leaders but also the tensions behind and beyond faith communities in carefully negotiating their places in society.

The presence of forcibly displaced people at borders in the Global North has caused vast political and media attention in recent years. In Europe, the so-called refugee 'crisis' of 2015 exposed external and internal borders and the inhumane consequences for refugees. At the south-western European border, two Catholic organisations work in support of refugees. One, called ELIN, was founded by two nuns and is based in Ceuta, a Spanish enclave in Moroccan territory. The other one, Asociación pro Derechos Humanos de Andalucía (ADPHA), was founded by a priest and is one of the most important solidarity organisations in the region. As in the case of Muslim communities in the Balkans, these grassroots organisations work independently from institutional faith leaders. As Alcalde and Portos (2018) claim, their initiatives, while not confrontational with the authorities, do not mirror the local bishop's conservative approach towards refugee-related issues. These initiatives are indiscriminate, providing for all regardless of religious affiliation (Alcalde and Portos 2018, p. 175)

LFCs are engaged in education programmes for displaced people in many countries. McCarthy notes that in Turkey, where nearly one million displaced Syrian children of education age currently live, 'education provision has become an essential element of faith-based humanitarianism' (McCarthy 2017, p. 2). Some of the Muslim LFC members he interviewed revealed how the provision of education is also seen as 'a fight against exploitation and assimilation by Western values' (McCarthy 2017, p. 7). McCarthy also argues that, in this context, LFCs providing education to Syrian refugees within state programmes essentially act as an expression of the government's political agenda.

Camps, like borders, are sometimes spaces of transit offering short-term shelter. However, they often become spaces of long-term permanence and of immobility/'stuckness' for displaced people. There is substantial evidence of programmes run in refugee or IDP camps by international faith-based organisations and by their local partners (Fiddian-Qasmiyeh and Pacitto 2019) but less on how religious faiths of displaced people are shaped during displacement. As Zink (2017) explains in his study on Sudan's 'lost boys,' many of the young unaccompanied Dinka, fleeing war and living in refugee camps, were eager to learn and separate from older generations. They joined their camp's Christian community as, in Zink's words, 'it was well established that education led to conversion' (Zink 2017, p. 346). On one hand, studies underline the complexity of conversion for people on the move, such as conversion as a migration strategy to allow for asylum status and conversion among refugees resettled by Christian communities (Connor 2014; Nawyn 2006). On the other hand, research demonstrates the agency of displaced people to resist conversion while also regularly engaging with people of another faith (Kraft 2017) and their agency to perform religion for international actors around them to maximise support from faith-based and secular organisations (Fiddian-Qasmiyeh 2011b).

Stages and spaces can exist where displaced people are neither here nor there, no longer in their country of origin, not yet formally recognised as a refugee, or waiting to hear the results of an asylum claim, or as stateless individuals, with no recognition of where legal procedures can posit their 'belonging.' These are different states, but some will overlap, such as involuntary immobility based on the wait for decisions about their refugee status. Some are imposed by external forces, such as detention and removal centres. McGregor (2012) makes the case that removal centres are places of religious revival in the United Kingdom. In all but one centre, chapels, mosques, and multifaith rooms are provided. Detainees organised their own Pentecostal prayer services, from which they derived a sense of 'shared humanity, a desire for justice, and to assert their right to move freely and to be in Britain' (McGregor 2012, p. 241). In the United States, there are also visitation programmes for direct pastoral initiatives with detainees, in which volunteers visit centres to 'offer friendship and a listening presence' (Snyder 2015, p. 171). This is a significant commitment with statistics from 2013 indicating that Jesuit Refugee Services alone had 'offered 1,071 hours of direct services and teaching, in which 50,566 people participated, as well as 2,741 hours of visits with detainees' (Snyder 2015, pp. 171–172). Overall, volunteers were involved in six categories of work: '(1) visitation programmes, (2) hospitality houses, (3) chaplaincy, (4) post-detention support, (5) vigils, and (6) campaigns and awareness raising' (Snyder et al. 2015, p. 168).

Zoma (2014) reports that faith communities are particularly effective in using their networks to attract volunteers for visits to detention centres but that this can also cause discomfort for detainees. In one example, a volunteer 'was wearing a cross when speaking with an immigrant detainee in a detention centre who identified as Muslim; the detainee was made uncomfortable by the cross, stating that he was Muslim and did not want to be converted' (Zoma 2014, p. 46). This example shows that, while there are advantages in terms of motivation (both for volunteers and detainees) and networks from faith communities can maintain high standards of neutrality, practices while inside detention centres and with detainees must be monitored to ensure that they do not create more harm. It is equally important to make sure that outsourced religious care in immigrant family detention facilities is not taken advantage of by the authorities. In a study on child–parent detention centres in the United States, Cuéllar (2017) describes how pastoral counselling with contracted chaplains can be used as a way to deter the detainees' hopes for freedom, as the chaplains are strictly prevented from providing any advice that concerns their case, and attempts by counsellors to facilitate access to legal help are sanctioned. On the other hand, his account of the Holy Week hunger strike carried out by eighty women at Karned detention centre (Cuéllar 2017, p. 15) shows the disruptive potential of religion-inspired practices.

## Arrival and Reception

LFCs engage in a variety of areas connected to the stages of displacement after arrival. These might concern involvement in welcoming or hostile discourses and practices, playing a role in Refugee Status Determination and in reception arrangements, including the revived tradition of sanctuary, as well as in education and peacebuilding activities, such as interfaith initiatives.

Regarding the legal procedures around asylum and refugee status, Mugnes et al. (2014) provide an example of pastoral counselling for asylum seekers in Registration and Procedure Centres (RPCs) in Switzerland. Following an agreement signed in 1995, all RPCs allow pastoral counselling. Pastoral counsellors give information to asylum seekers on the asylum system and legal counsel available. In Cairo, St. Andrew's Refugee Services provide legal support for displaced people (Wilkinson and Ager 2017, p. 32). Lyck-Bowen and Owen (2019) have analysed some interfaith initiatives in support of refugees in Germany, Sweden, the United Kingdom, and Poland. In particular, the

Swedish project called Good Neighbours is run jointly by a local church and a local mosque and provides a range of services including legal advice and language classes for refugees.

Religious identity itself can play a pivotal role during Refugee Status Determination interviews with government and UNHCR officials. In this regard, Madziva (2017) describes the case of Christian asylum seekers from a Muslim majority country (Pakistan) in the United Kingdom and the mutual dynamics of attribution of religious identity according to stereotyped bodily features between asylum seekers and officials. If Pakistani asylum seekers are identified with Islam (and their English names as suspicious) by the interviewers, the latter are, in turn, thought to be Muslim by the former, especially when the claim's outcome is negative (Madziva 2017, pp. 14–15).

Displaced people have taken sanctuary in religious buildings for centuries, and church asylum is still practiced. For instance, during interviews held in 2016, numerous European pastors and ministers claimed that:

> if an 'acceptable solution' to the refugee crisis was not found in EU policy and law then it was the responsibility of Christians and the churches to intervene through the provision of church asylum. Moreover, the intervention would be made regardless of its legal status (i.e. there was an expressed willingness to act against the law), if ministers, administrators, and parishioners felt that it was their moral duty to do so. (Mitchell 2017, p. 280)

Thus, Mitchell encapsulates sanctuary as a 'spatial politics of the sacred' (Mitchell 2017, p. 282) that favours faith-related moral norms over state legislation. Most recently, the ninety-six-day sanctuary created in a church in the Netherlands to shelter a family of refugees from Armenia received international media attention (Kingsley 2019).

Snyder (2017) has documented how different faith groups in the United Kingdom mobilised to offer spiritual and practical support to refugees, and that 200 members of Christian, Jewish, Muslim, Sikh, Buddhist, and Hindu communities signed a letter in 2016 urging the prime minister to adopt more humane policies towards asylum seekers. Khallouk (2018), in turn, while describing Muslim communities' engagements with refugees in Germany, has highlighted that the cooperation between Muslim and non-Muslim actors in refugee relief services 'attracted public and political attention to Islamic associations,' contributing to a growing awareness of their role in society, and favouring interfaith encounters. In Malaysia, Hoffstaedter relays how UNHCR is relating the need to protect refugees to Muslim concepts of solidarity, in order to offer an alternative narrative of welcome for refugees (Hoffstaedter 2017).

LFCs are not necessarily welcoming, of course, and religion in the host countries can be used to fuel mechanisms of 'othering' of the newly arrived. As Ralston points out, 'the depictions of refugees as a deluge and as threat to Christian and Western values often deploy long-standing Christian tropes against Islam, Muhammad, and Shari'a to describe Muslims as inherently violent and other' (Ralston 2017, p. 33). Religious belief and practice is embedded in other cultural, social, political (Zic 2017), and economic aspects of life, as suggested by Baker's comparison between the arrival of Bosnians in the 1990s and of Muslim refugees today. In her words, 'ideas of race, nationality and religion have intersected to imply that integrating Middle Eastern, North African and central Asian Muslims should inherently be more difficult' (Baker 2017). For instance, she observes that the former were mostly represented as families with children, while the latter are depicted as a mass of adult male individuals, thus adding a gendered and generational dimension to the issue.

## Long-Term Solutions to Displacement

Displaced people's experiences of return to and reintegration in their country of origin, resettlement to a third country, and access to alternative long-term solutions such as private sponsorship programmes are often connected to activities of LFCs to provide, and almost always interlinked with displaced people's personal and community religious beliefs and practices. In research on Machazians in Mozambique after the civil war, Lubkemann acknowledges that there was 'a whole range of socio-spiritual dilemmas motivat[ing] returnees' (2002, pp. 199–200). Returnees went back to the same districts but not exactly the same localities. Returnees were worried about loose *nfukwas* which followed from the deaths in war of soldiers and a subsequent lack of burial. Local communities and churches organised ceremonies to 'send home' these sprits. The gender dynamics of return were also notable in this example. Men had moved to South Africa for extended periods of time and women, some of whom had had children out of wedlock, sought alternative spiritual and social support networks, outside the Machazian sphere. The women in Machaze joined Protestant churches that 'allowed them to interact with other Machazian women like themselves without the threat of spiritual sanctions, by re-casting these women in the role of "church sisters" rather than "Machazian women"' (Lubkemann 2002, p. 205). Commitment to a socio-spiritual realm of relationships shifted for these women, influenced by their experiences in the civil war. It was not a loss of religious relationship following the trauma of war but a relocating of socio-spiritual meaning.

Religious networks have played a role in the establishment of safe and legal routes to refuge. As Moreira (2017) points out, Catholic organisations in Brazil have been major contributors to the growth of the country's commitment to refugee resettlement from the region and beyond. In the United States, different religious groups have advocated for the inclusion of their community members within resettlement programmes:

> FBOs and communities like the Church World Service (CWS), the Hebrew Immigrant Aid Society (HIAS), Lutheran Immigration Refugee Services (LIRS), the United States Conference of Catholic Bishops and non-sectarian organisations like the International Rescue Committee mediated between the government and refugees. These organisations' advocacy has been central to refugee admission and resettlement and has been crucial in defining refugees' identity. (Ray 2018, p. 777)

The case of the humanitarian corridors programme in Italy (Collyer et al. 2017; Trotta 2017) demonstrates how FBOs and their networks can actively promote the opening of *alternative* safe and legal routes to refuge in countries such as Italy, France, and Belgium. These programmes are complementary to the government-led resettlement schemes, in that they use different criteria of access to protection. Notably, in the case of humanitarian corridors refugee status is not a prerequisite for obtaining a visa under the programme. Rather, the faith communities' partner organisations are in charge of selection processes in transit countries such as Lebanon and Libya. This is similar to Canada's private sponsorship programmes, where the majority of Sponsorship Agreement Holders (SAHs) are faith-based organisations (Tito and Cochand 2017).

## Conclusion

LFCs, motivated by faith, are often engaged in the most forward-looking and even 'radical' initiatives and practices in support of displaced people, opposing xenophobia and discrimination towards them. LFC efforts include offering sanctuary, creating and expanding safe and legal routes, and using safe spaces where displaced people can not only receive support but also express themselves and carry out their migratory plans regardless of their legal status, age, gender, or sexual orientation. LFCs help mourn the dead and bring to light the reality of pain and suffering for those displaced in their advocacy work. Their social work to support displaced people has allowed for interfaith cooperation to emerge and bridge-building between communities so that social integration can take place for displaced people.

Yet it is also clear that the root of some xenophobic and discriminatory practices can be traced back to statements by religious leaders and that the perpetuation of discrimination is tied to religio-political roles of faith communities in society. Faith leaders' and institutional faith actors' positions can be less supportive (distanced, but not outwardly discriminatory) through to open hostility to refugees. Some initiatives can become counterproductive and dangerous to displaced people if faith-based actors operate in their own interest or in connection with smuggling networks and/or governments pursuing their own immigration agendas. The evidence has shown how there is diversity even within otherwise closely knit faith communities and that the question of providing protection for displaced people can be a divisive issue. Likewise, conversion and proselytisation are deeply complex issues that displaced people must navigate.

Ultimately, there is no definitive picture of 'good' LFCs and 'bad' LFCs that should be consequently welcomed into or pushed out of the international humanitarian system for refugee response. Instead, this review has revealed the urgent need to engage with the complexity and contextualisation of local faith-based response for refugees. What we can demonstrate from the collation of research from around the world is that LFCs are involved in responses for displacement and that an ignorance or avoidance of LFCs will blind international responses to crucial local actors and potential partners.

# References

Ager, A., & Ager, J. (2011). Faith and the discourse of secular humanitarianism. *Journal of Refugee Studies, 24*(3), 456–472.

Ager, A., & Ager, J. (2015). *Faith, secularism, and humanitarian engagement: Finding the place of religion in the support of displaced communities*. New York, NY: Palgrave Macmillan.

Alcalde, J., & Portos, M. (2018). Refugee solidarity in a multilevel political opportunity structure: The case of Spain. In D. d. Porta (Ed.), *Solidarity mobilizations in the 'refugee crisis'* (Palgrave Studies in European Political Sociology) (pp. 155–182). Cham, Switzerland: Palgrave Macmillan. https://doi.org/10.1007/978-3-319-71752-4_6.

Baker, C. (2017, July 3). Why were Bosniaks treated more favourably than today's Muslim refugees? On differing narratives of identity, religion and security [Web log post]. Retrieved from http://blogs.lse.ac.uk/europpblog/

Carpi, E. (2018). Does faith-based aid provision always localise aid? *Refugee Hosts*. Retrieved from https://refugeehosts.org/2018/01/22/does-faith-based-aid-provision-always-localise-aid/

Collyer, M., Mancinelli, M., & Petito, F. (2017, October 1). *Humanitarian corridors: Safe and legal pathways to Europe* [Policy briefing—Policy@Sussex]. Retrieved March 18, 2018, from https://www.sussex.ac.uk/webteam/gateway/file.php?name=3927-hc-policy-briefing-july2017-web.pdf&site=11

Connor, P. (2014). *Immigrant faith: Patterns of immigrant religion in the United States, Canada, and Western Europe*. New York, NY: NYU Press.

Cuéllar, G. L. (2017). Deportation as a sacrament of the state: The religious instruction of contracted chaplains in U.S. detention facilities. *Journal of Ethnic and Migration Studies, 45*, 1–20. https://doi.org/10.1080/1369183X.2017.1404260.

El Nakib, S., & Ager, A. (2015). *Local faith community and civil society engagement in humanitarian response with Syrian refugees in Irbid, Jordan* (Report to the Henry Luce Foundation). New York, NY: Columbia University, Mailman School of Public Health. Retrieved from https://jliflc.com/wp/wp-content/uploads/2015/06/El-Nakib-Ager-Local-faith-communities-and-humanitarian-response-in-Irbid-.pdf.

Fiddian-Qasmiyeh, E. (2011a). Introduction: Faith-based humanitarianism in contexts of forced displacement. *Journal of Refugee Studies, 24*(3), 429–439. https://doi.org/10.1093/jrs/fer033.

Fiddian-Qasmiyeh, E. (2011b). The pragmatics of performance: Putting 'faith' in aid in the Sahrawi refugee camps. *Journal of Refugee Studies, 24*(3), 533–547.

Fiddian-Qasmiyeh, E. (2018, 16 July). *Histories and spaces of Southern-led responses to displacement*. Southern Responses to Displacement. Retrieved February 25, 2019, from https://southernresponses.org/2018/07/16/histories-and-spaces-of-southern-led-responses-to-displacement/

Fiddian-Qasmiyeh, E., & Pacitto, J. (2019). Southern-led faith-based responses to refugees. In B. Schewel & E. K. Wilson (Eds.), *Religion and European society: A primer*. Newark, NJ: John Wiley & Sons.

Greatrick, A., Fiddian-Qasmiyeh, E., Rowlands, A., Ager, A., & Stonebridge, L. (2018). Local faith community responses to displacement in Lebanon, Jordan and Turkey: Emerging evidence and new approaches. *Refugee Hosts*. Retrieved from https://refugeehosts.files.wordpress.com/2018/06/rh-local-faith-report.pdf

Hoffstaedter, G. (2017). Refugees, Islam, and the state: The role of religion in providing sanctuary in Malaysia. *Journal of Immigrant & Refugee Studies, 15*(3), 287–304. https://doi.org/10.1080/15562948.2017.1302033.

Khallouk, M. (2018). Confronting the current refugee crisis: The importance of Islamic citizens' initiatives in Germany. In U. Schmiedel & G. Smith (Eds.),

*Religion in the European refugee crisis* (pp. 87–103). Cham, Switzerland: Palgrave Macmillan.

Kidwai, S., & Fiddian-Qasmiyeh, E. (2017, June 3). Seeking evidence to provide protection: How can local faith communities support refugees? *Refugee Hosts*. Retrieved April 21, 2019, from https://refugeehosts.org/2017/06/03/seeking-evidence-to-provide-protection-how-can-local-faith-communities-support-refugees/

Kingsley, P. (2019, January 31). 96 days later, nonstop church service to protect refugees finally ends. *The New York Times*. Retrieved from https://www.nytimes.com/2019/01/30/world/europe/netherlands-church-vigil-refugees.html

Kraft, K. (2017). Religious exploration and conversion in forced displacement: A case study of Syrian Muslim refugees in Lebanon receiving assistance from Evangelical Christians. *Journal of Contemporary Religion, 32*(2), 221–235. https://doi.org/10.1080/13537903.2017.1298904.

Lubkemann, S. (2002). Where to be an ancestor? Reconstituting socio-spiritual worlds among displaced Mozambicans. *Journal of Refugee Studies, 15*(2), 189–212. https://doi.org/10.1093/jrs/15.2.189.

Lyck-Bowen, M., & Owen, M. (2019). A multi-religious response to the migrant crisis in Europe: A preliminary examination of potential benefits of multi-religious cooperation on the integration of migrants. *Journal of Ethnic and Migration Studies, 45*(1), 21–41. https://doi.org/10.1080/1369183X.2018.1437344.

Madziva, R. (2017). 'Your name does not tick the box': The intertwining of names, bodies, religion and nationality in the construction of identity within the UK asylum system. *Ethnic and Racial Studies, 41*(5), 938–957. https://doi.org/10.1080/01419870.2017.1318215.

McCarthy, A. T. (2017). Non-state actors and education as a humanitarian response: Role of faith-based organizations in education for Syrian refugees in Turkey. *Journal of International Humanitarian Action, 2*, 13. https://doi.org/10.1186/s41018-017-0028-x.

McGregor, J. (2012). Rethinking detention and deportability: Removal centres as spaces of religious revival. *Political Geography, 31*(4), 236–246. https://doi.org/10.1016/j.polgeo.2012.03.003.

Mitchell, K. (2017). Freedom, faith, and humanitarian governance: The spatial politics of church asylum in Europe. *Space and Polity, 21*(3), 269–288. https://doi.org/10.1080/13562576.2017.1380883.

Moreira, J. B. (2017). Refugee policy in Brazil (1995–2010): Achievements and challenges. *Refugee Survey Quarterly, 36*(4), 25–44. https://doi.org/10.1093/rsq/hdx014.

Mugnes, S., Prosperio, F., & Deponti, L. (2014). An ecumenical organisation for asylum seekers in Switzerland. *Forced Migration Review, 48*, 61. Retrieved from http://www.fmreview.org/faith/mugnes-proserpio-deponti.html.

Nawyn, S. J. (2006). Faith, ethnicity, and culture in refugee resettlement. *American Behavioral Scientist, 49*(11), 1509–1527. https://doi.org/10.1177/0002764206288462.

Ngo, M. (2018). *Between humanitarianism and evangelism in faith-based organisations: A case from the African migration route*. Abingdon, UK and New York, NY: Routledge.

Olayo-Méndez, A. (2017). La 72: An oasis along the migration routes in Mexico. *Forced Migration Review, 56*, 10–11. Retrieved from https://www.fmreview.org/sites/fmr/files/FMRdownloads/en/latinamerica-caribbean/olayomendez.pdf.

Ralston, J. (2017). Bearing witness: Reframing Christian–Muslim encounter in light of the refugee crisis. *Theology Today, 74*(1), 22–35. https://doi.org/10.1177/0040573616689837.

Ray, M. (2018). Crossing borders: Family migration strategies and routes from Burma to the US. *Journal of Ethnic and Migration Studies, 44*(5), 773–791. https://doi.org/10.1080/1369183X.2017.1314815.

Rexhepi, P. (2018). Arab others at European borders: Racializing religion and refugees along the Balkan Route. *Ethnic and Racial Studies, 41*(12), 2215–2234. https://doi.org/10.1080/01419870.2017.1415455.

Snyder, S. (2015). Looking through the bars: Immigration detention and the ethics of mysticism. *Journal of the Society of Christian Ethics, 35*(1), 167–187. https://doi.org/10.1353/sce.2015.0002.

Snyder, S. (2017, February 8). How faith communities in the UK are responding to the refugee crisis [Web log post]. *Religion and the Public Sphere*. Retrieved from http://blogs.lse.ac.uk/religionpublicsphere/

Snyder, S., Bell, H., & Busch-Armendariz, N. (2015). Immigration detention and faith-based organizations. *Social Work, 60*(2), 165–173. https://doi.org/10.1093/sw/swv004.

Tito, S., & Cochand, S. (2017). The story of a small Canadian congregation sponsoring a refugee family. *Forced Migration Review, 54*, 60–61. Retrieved from https://www.fmreview.org/sites/fmr/files/FMRdownloads/en/resettlement/tito-cochand.pdf.

Trotta, S. (2017). *Safe and legal passages to Europe: A case study of faith-based humanitarian corridors to Italy*. University College London Migration Research Unit. Retrieved from http://www.geog.ucl.ac.uk/research/research-centres/migration-research-unit/working-papers/MRU%20WP%20Susanna%20Trotta%202017%205.pdf/view.

UNSG (UN Secretary-General). (2016). *One humanity: Shared responsibility | Report of the UN Secretary-General for the World Humanitarian Summit*. United Nations. Retrieved from http://www.alnap.org/resource/21845

Wilkinson, O. (2017). 'Faith can come in, but not religion': Secularity and its effects on the disaster response to Typhoon Haiyan. *Disasters, 42*(3), 459–474. https://doi.org/10.1111/disa.12258.

Wilkinson, O. (2018, 7 February). When local faith actors meet localisation. *Refugee Hosts*. Retrieved April 15, 2018, from https://refugeehosts.org/2018/02/07/when-local-faith-actors-meet-localisation/

Wilkinson, O., & Ager, J. (2017). *Scoping study on local faith communities in urban displacement: Evidence on localisation and urbanisation*. Washington, DC: Joint Learning Initiative on Faith and Local Communities. Retrieved from https://jliflc.com/resources/jli-refugee-scoping-lfc-urban/.

Zic, B. (2017). The political impact of displacement: Wartime IDPs, religiosity, and post-war politics in Bosnia. *Politics and Religion, 10*(4), 862–886. https://doi.org/10.1017/S1755048317000335.

Zink, J. (2017). Lost boys, found church: Dinka refugees and religious change in Sudan's second civil war. *The Journal of Ecclesiastical History, 68*(2), 340–360. https://doi.org/10.1017/S0022046916000683.

Zoma, M. (2014). Respecting faiths, avoiding harm: Psychosocial assistance in Jordan and the United States. *Forced Migration Review, 48*, 45–47. Retrieved from https://www.fmreview.org/faith/zoma.

# 54

# Hosting the Displaced: From Sanctuary Cities to Hospitable Homes

Jonathan Darling

In December 2015, twenty-four Syrian families arrived on the Scottish island of Bute. With a population of 6500 people, the Isle of Bute is a small island situated in the Forth of Clyde. It was not a location, or a community, experienced in supporting the displaced. Nevertheless, the local authority of Argyll and Bute was one of the first to agree to resettle Syrian refugees as part of the United Kingdom's Syrian Vulnerable Person's Resettlement Scheme (SVPRS). Through the SVPRS, families are resettled directly from refugee camps on the borders of Syria to local authorities across the United Kingdom (Brooks 2015). Centred on a commitment to resettle 20,000 Syrian refugees by 2020, the SVPRS received widespread public and media support when announced in the days following the global circulation of an image of the body of toddler Alan Kurdi, discovered on a Turkish beach. Through the SVPRS, refugees are provided with housing through local authorities, language training, integration workshops, and basic furnishings. Beyond this, local residents on Bute collected donations of clothing, furniture, and homeware to support the new arrivals in settling on the island (Brooks 2015). The Bute case is one among many local authorities who have supported Syrian refugees resettled to the United Kingdom, yet it has also been a case that has captured public attention. I open with it here to point to the ways in which hosting displaced

J. Darling (✉)
Durham University, Durham, UK
e-mail: jonathan.m.darling@durham.ac.uk

populations can at times confound expectations and make significant demands on communities. Often those demands are met, such that the story of displacement is not only one of exclusionary policies and border walls but also one of new beginnings and hospitable openings. In this chapter, I outline some of these hospitable dimensions of displacement in order to highlight the political and ethical tensions that run through efforts to accommodate asylum seekers and refugees.

The chapter develops as follows. I begin by outlining a number of formalised processes of resettlement through which nation-states have sought to meet their obligations to refugees. Such resettlement routes offer a highly selective means of managing and supporting the displaced. With the limits of these in mind, I then discuss the work of sanctuary movements that seek to position cities as sites for supporting the displaced. In examining sanctuary movements, I refer to a range of advocacy and activist groups across Europe and North America, which have sought to protect asylum seekers, refugees, and undocumented migrants from immigration enforcement. In some cases, this activism involves directly sheltering vulnerable individuals, and, in others, it has focused on changing attitudes towards displaced people through intercultural work. Finally, the chapter turns to a range of more localised examples of accommodation, from solidarity squatting in Athens to a network of community hosting in the United Kingdom. In the former instance, solidarity squatting, an activist framing of solidarity between citizens and non-citizens, is at the heart of providing shared spaces of informal accommodation. These squats, created through the appropriation of unused and often abandoned buildings such as former hotels, offer spaces of temporary protection for the displaced and sites of connection between displaced people and a range of other disadvantaged residents. In doing so, these examples of welcoming often test the limits of nation-state obligations and assert in their place a moral responsibility to support the displaced. I begin though, with the formal accommodations of the nation-state.

## Refugee Resettlement and Dispersal

In the majority of countries across the Global North, accommodation for asylum seekers and refugees takes two predominant forms. In the first instance, refugee resettlement has been argued to present one of the few 'durable solutions' for refugee groups who would otherwise be confined to indefinite periods in refugee camps (UNHCR 2007). Through resettlement processes, preselected groups of refugees are assisted in moving to a new country by charities, NGOs, state agencies, and UNHCR. The logistics of this process differ internationally, with some cases focusing on an annual quota of refugees and being coordinated by governments working directly with UNHCR, whilst others draw upon community sponsorship of refugees to encourage resettlement. In the United States, for example, resettlement is based on a public–private partnership model where local community networks work with US government agencies to resettle refugees. In this context, faith-based organisations have been at the forefront of resettlement efforts, both in providing assistance and in advocating for resettlement rights (Eby et al. 2011). Resettlement has therefore often been framed as a humanitarian operation, centred on securing the safety of those deemed 'most vulnerable,' either by UNHCR or by prospective 'host' governments.

Often significant conditions are attached to resettlement, such as the pre-screening of potential refugee groups to assess their needs, their ability to integrate, and the 'urgency' of their resettlement requirement. Thus, whilst resettlement may offer a more secure future for those displaced, it remains rare in practice. UNHCR estimate that they are only able to put forward for resettlement around 1 per cent of the refugees who fall within their remit, and not all of these are selected by 'host' governments (UNHCR 2018). Crucially, the selectivity of resettlement is only set to increase as the growth of populist political parties, anti-immigrant sentiment, and nativist bordering have placed resettlement programmes under political threat. At present, despite agreements to resettle 160,000 refugees hosted in camps in Greece and Italy across the EU, there has been very little progress towards resettlement, leading Crawley (2016) to decry a European 'crisis of solidarity' in addressing refugee mobility at the borders of Europe. At the same time, resistance towards resettlement reflects a concern over the costs of supporting refugees in times of austerity. Just as resettlement involves a high level of scrutiny, resettlement often also involves a relatively high level of support for those arriving. Those refugees resettled through the United Kingdom's Gateway Protection Programme, for instance, are provided with housing support, a cultural orientation program, language training, and support to move into education or employment.

The significance of resettlement politically is that it is often presented as an idealised form of state-managed response to refugees, with resettled refugees portrayed as safe, secure, and pre-screened individuals who arrive in regulated numbers. Yet through such rhetoric, those refugees and asylum seekers who travel through other migratory routes are all the more easily presented as less 'worthy' of a hospitality that is strictly limited. At the same time, the mobilisation of a category of 'vulnerability' to distinguish and select refugees most 'in need' serves to reiterate an image of the 'genuine' refugee as a vulnerable victim awaiting humanitarian intervention. Such an image not only undermines the political agency of refugees (Nyers 2006) but also simplifies the complexity of migratory journeys and motivations into a singular category of vulnerability. Doing so, as Squire (2018) argues, enables humanitarian interventions to appear morally virtuous, whilst not grappling with the political causes of displacement in the first place.

These tensions are brought further to the fore if we consider resettlement alongside the systems of asylum dispersal that operate in many European countries. Dispersal programmes are in place in Denmark, Germany, the Netherlands, Sweden, and the United Kingdom and are argued to represent attempts to 'spread the burden' of asylum seeker accommodation (Arnoldus et al. 2003; Huizinga and van Hoven 2018; Robinson 2003; van Liempt 2011; Wren 2003). Beyond this economic argument, there are a series of historical connections between the emergence of dispersal as a policy and other forms of migration management. For instance, in Germany, the dispersal of asylum seekers developed in the wake of dispersal policies to regulate the mobility of guest workers across the country (Boswell 2001). Whilst in the United Kingdom, asylum seeker dispersal began in 2000 but was informed by the experience of smaller scale 'emergency' responses to refugee groups from Vietnam in the 1970s and Kosovo in the 1990s (Robinson 1998). In both cases, efforts to manage the mobility of the displaced were informed by histories of control over the movement of non-citizens.

It is this regulatory function that is most often highlighted in considerations of asylum dispersal. For example, Schuster (2005) argues that dispersal needs to be viewed as part of a larger governmental formation that produces the marginality of asylum seekers through interconnected mechanisms of dispersal, detention, and deportation. At the same time, part of the policy rationale behind dispersal in many European countries has been to promote social integration and cohesion, with Huizinga and van Hoven (2018, p. 313) noting that in the Netherlands:

> By dispersing rather than concentrating refugees, the Dutch government aims to develop more effective social integration trajectories based on the assumption that refugees will be encouraged to initiate social interaction with people from the host society.

The evidence of such an effect is extremely mixed however, with Huizinga and van Hoven (2018) finding some examples of prosaic social interactions but also noting the limits of these forms of engagement and the concurrent isolation felt by many of those dispersed. Similarly, Hynes's (2009) study of the experiences of dispersed asylum seekers in the United Kingdom suggests that the compulsory mobility of dispersal creates a 'policy-imposed liminality,' as individuals struggle to establish social networks within alien, and at times hostile, new surroundings. Experiences of dispersal are argued in this context to vary considerably between cities, as a result of both the importance of place identity and localised cultures of welcome (Platts-Fowler and Robinson 2015; Darling 2018) and the uneven and fragmented nature of support and legal services across the United Kingdom (Burridge and Gill 2017; Darling 2016).

This is not, however, to suggest that resettled refugees have a necessarily comfortable experience. Indeed, the challenges of integration and finding accommodation for resettled refugees are often stark, especially in contexts where accommodation is not provided by the state and refugees must enter into competition for affordable housing with other socially marginalised groups (Carter and Osborne 2009; Forrest et al. 2013; Logan and Murdie 2016). Rather, by considering the relation between resettlement and dispersal, we might see the emergence of a two-track system of refugee protection and support in much of the Global North. For those refugees classified as the 'most vulnerable,' refugee resettlement may offer a rare opportunity to gain protection and some sense of security in a new country. By contrast, those still seeking asylum are accommodated in poorer conditions, with restrictions on mobility, the right to work, and on opportunities for education and language learning. This interaction illustrates that whilst international obligations may be seen to be met in principle, in practice, such rights are only afforded to an increasingly select few of the globally displaced (see Coddington 2018). At the same time, the meeting of such obligations for those deemed 'most vulnerable' becomes a means to deny the rights and entitlements of others. With this in mind, I turn to two examples that seek to challenge the limits of refugee resettlement and asylum dispersal. The first being the growth of sanctuary movements in Europe and North America, and the second being the emergence of a series of more informal practices of hosting and occupation.

## Sanctuary Cities

A growing focus on political activism and advocacy under the banner of sanctuary can be seen to varying degrees across a range of international contexts, including the work of the New Sanctuary Movement in the United States and Canada, the Cities of Refuge initiative across Europe, and the United Kingdom's City of Sanctuary movement. Whilst distinct in their practices and approaches, each of these movements advocate for the rights of asylum seekers and refugees, often through a concern with urban presence and through a language of hospitality (Bauder 2017; Darling 2017). For example, the UK City of Sanctuary movement explicitly seeks to inculcate a 'culture of hospitality' whereby refugees and asylum seekers are welcomed in towns and cities (Darling 2010; Squire 2011). This grassroots movement focuses on intercultural events, awareness raising, and providing volunteering and training opportunities for refugee and asylum groups. In doing so, they practice a model of hospitality based on opportunities for refugees and asylum seekers to interact with the cities in which they are accommodated through dispersal. City of Sanctuary is one example of a wider trend towards sanctuary, with a key distinction being between the legal protections of 'sanctuary cities' in North America and the cultural effects of sanctuary movements in Europe (Bauder and Gonzalez 2018).

In the first instance, the New Sanctuary Movement in North America, a network of social movements with roots in faith-based communities that focus on offering protection from deportation to individual refugees and undocumented migrants (Bauder 2017), coexists alongside a range of sanctuary cities that have attempted to legislate for access to municipal services regardless of status and to protect residents through non-cooperation with immigration authorities (Cunningham 2013; Mancina 2013). Of these, San Francisco was an early exemplar, but the sanctuary city designation has grown considerably and has taken on an increasingly contentious political position since the Trump administration's threat to defund such municipalities. Examinations of sanctuary cities have highlighted the opportunities to establish rights to remain through restricting the reach of federal immigration enforcement and offering services based on residency rather than status as essential elements in protecting refugees (Nyers 2010; Ridgley 2008). From access to local healthcare services and social support to the provision of locally bound drivers' licences for undocumented residents, the legislation on which a 'sanctuary' designation is based alters between urban jurisdictions (Bauder 2017). Yet, what remains constant is a focus on protecting those present in the city from arrest and deportation, and in doing so, foregrounding the needs of refugees and undocumented migrants as urban issues that should concern all citizens. In this sense, as Ridgley (2008, p. 56) argues, the sanctuary city 'is not only a space of protection from an increasingly anti-immigrant national security agenda, but also a potential line of flight out of which alternative futures can be materialized,' as it serves to embed sanctuary into the very legal and social fabric of a city.

By contrast, in Europe, the growth in sanctuary-orientated movements does not carry the same legislative weight as in the North American context (Bauder 2017; Pyykkönen 2009). Often emerging from charitable and religious organisations, this European strand of sanctuary tends to emphasise the role of asylum seekers and refugees as contributing to the social and cultural life of their 'host' communities (Goodall 2010). For example, in Sweden, Lundberg and Strange (2017) illustrate how a range of initiatives to promote values of hospitality are taken by different cities. In Stockholm, groups wanting to align themselves with Refuge Stockholm were given a list of requirements to meet before being able to be part of the movement. These included providing free or discounted services to undocumented people and not demanding social security numbers for services in an echo of some of the requirements of North American sanctuary cities. This model of gaining support from local organisations relied upon spreading the word about sanctuary events, initiatives, and opportunities through everyday spaces, such as cafes, gyms, cultural associations, and universities. To be part of Refuge Stockholm, organisations were required to offer opportunities for undocumented migrants and refugees to be involved in their activities (Lundberg and Strange 2017), mirroring some of the practices of City of Sanctuary where volunteering was seen as a valuable resource to feel part of urban life (Darling and Squire 2013). Similarly, in Malmo, city officials worked with sanctuary groups to gain access to the library and ensure that undocumented individuals could borrow books despite lacking formal status (Lundberg and Strange 2017, p. 357).

In these cases, the work of sanctuary movements can be vitally important in trying to promote values of hospitality within cities and in offering often piecemeal yet important concessions to those seeking protection, from the right to borrow books to the ability to seek health care without risking deportation. However, such work has also been critically challenged. First, it has been argued that sanctuary may represent a means of governing through the assertion of humanitarian intentions (Darling 2017). In legislative terms, Chavin and Garcés-Mascareñas (2012, p. 244) argue that 'local incorporation practices' reflect 'regulatory imperatives and worries over public safety.' Through enabling undocumented migrants to access services and support, cities can be seen to 'manage' an undocumented population for the wider 'good' of the city, thereby allaying concerns over public health and public order as Mancina (2013) argues in the case of San Francisco. Seen through this critical lens, the language of the sanctuary city becomes a means to govern the presence of irregular migrants and refugees. As a result, the question of who 'deserves' the support of the sanctuary city comes to the fore in debating the limits of urban hospitality (Bagelman 2013; Marrow 2012; Yukich 2013).

This question of 'deservingness' also speaks to the second basis on which sanctuary has been critiqued—as an exemplar of a humanitarian logic that positions asylum seekers as passive recipients of support. Thus, whilst sanctuary movements can be significant in supporting those seeking refuge, they can also risk denying agency to asylum seekers and refugees in a similar manner to the categorising role of 'vulnerability' in refugee resettlement (Nordling et al. 2017).

These varied examples of sanctuary each reflect different political contexts and opportunity structures. Yet what they hold in common is a focus on considering how a city might offer protection to those seeking refuge. In practice, such forms of welcome may assuage the effects of repressive immigration controls, but they also risk the reiteration of categorical assumptions over who is 'deserving' of welcome. Distinctions such as these are central to the categorising processes that shape ideals of hospitality (Derrida 2001) and illustrate how progressive imaginaries of the city may be enfolded into state-centric logics of citizenship. Building on these discussions, I want to consider a final set of localised efforts to support the displaced, centred on the political potentials of hosting, and orientated around examples from the United Kingdom and, first, from Athens.

## Hosting, Solidarity, and Support

Athens offers an interesting case because it is a city that has faced multiple forms of social, political, and financial crisis in its recent history, whilst also being located at the heart of the so-called 'refugee crisis' in Europe. As routes into northern Europe were closed to refugees throughout 2015 and 2016, many found themselves stranded in the towns and cities of Greece and Italy, unable to move onwards through the continent (Trilling 2017). As a result, cities like Athens saw the growth of informal settlements of refugees on the streets, leading the Greek government to adopt a policy of housing for all refugees in camps on the outskirts of cities (Koptyaeva 2017). As Tsavdaroglou (2018, p. 382) notes, in the Athenian context, these state-run refugee camps included 'overcrowded dilapidated factories, old military bases and an abandoned airport.' Yet at the same time, as a result of the socio-spatial crises to have affected Athens, a range of public and private buildings, including schools, hospitals, and hotels, were abandoned and able to be occupied by diverse groups of activists (Kaika 2012; Tsavdaroglou 2018), who employed occupation as a means of political resistance to the exclusions of austerity policies. From late 2015 onwards, by transforming buildings into squats, these activists were able to house refugees and offer an informal alternative to the state's policy of housing refugees in camps (Koptyaeva 2017). As a result, Athens has seen the emergence of squats that combine solidarity activism between migrant groups and other marginalised populations, with a desire to reclaim the visibility of refugees as present in the city (for similar activist connections in Madrid, see Lopez 2016). One effect of this collective form of urban occupation has been a cycle of displacement and reoccupation, as police evictions seek to encourage refugees to move to the camps, but instead see the formation of new squats in response (Tsavdaroglou 2018, p. 392). To draw out the political significance of such squats further, I focus on one of the most high-profile of these, City Plaza.

City Plaza is a disused hotel that has been occupied since April 2016. By May 2016, 380 refugees were living in the hotel, supported by a range of local and visiting activists from across Europe (Squire 2018, p. 120). By contrast to the forms of humanitarian accommodation seen in refugee resettlement processes, City Plaza does not include people on the basis of vulnerability but 'rather in terms of the diversity that they bring to the collective,' with reasons for migration not considered but 'attention is paid to ensuring a mix of nationalities, a gender balance, and a combination of religious beliefs' (Squire 2018, p. 121). Whilst inclusion is not based on categorises of vulnerability, an expectation of involvement in the collective life of the squat is present. Thus, residents are assigned activities, including cleaning public spaces, shifts in the communal kitchen, and the upkeep of the squat more generally. As Koptyaeva (2017, p. 38) emphasises, for families in particular, these 'activities may be seen as part of the attempt to create the feeling of shared space or, in other words, the understanding of "being at home," a home that should be kept clean and comfortable.' The significance of such activities is partly in offering some means of occupying the time spent waiting for a decision from the state on one's asylum application but also in creating the sense of a shared space of collective investment.

The significance of City Plaza has thus been argued to lie in the forms of solidarity that its approach to accommodating the displaced develops. In this vein, Squire (2018) argues that City Plaza is a space that allows refugees to reject the status of the vulnerable or the victim often conferred upon them through more formal processes of resettlement or encampment. In part, this is because, unlike other refugee squats in the city, City Plaza has a range of residents, thus, 'as more than simply a refugee squat it [City Plaza] is occupied by a collective of refugee, student, and solidarity activists, and involves a relatively sophisticated approach to communal living and community decision-making' (Squire 2018, p. 120). The upshot of which is a shared sense of collective life not confined to assignments of immigration status, vulnerability, or migratory history. The precariousness of City Plaza as a squat always at risk of being dismantled by the state is, in these terms, a precarity that cuts across identifications and is experienced through a form of communal living that rejects the choice between a refugee camp and a humanitarian model of assistance (Squire 2018). In this sense, City Plaza, and the other refugee squats precariously maintained throughout Athens, speak to not just survival in the city but also an assertion of agency through the rejection of a choice between enclosure or victimhood. This assertiveness draws on the longer history of squatting as a practice that mobilises the act of occupation as a critique of existing social and political relations (Vasudevan 2017).

The second example I want to discuss is less radical in its orientation but no less important for the lives of the refugees it accommodates. The No Accommodation Network (NACCOM) is a network across the United Kingdom, aiming to prevent destitution among asylum seekers and refugees (NACCOM 2017). NACCOM brings together a range of voluntary organisations that all provide shelter and accommodation to asylum seekers and refugees. The reasons for such destitution are manifold, with delays in gaining support upon receiving refugee status a common issue (Doyle 2014) and the government's policy of removing support to asylum seekers who have had negative decisions on their claims leading to a significant destitute population reliant on charities for survival (Lewis 2007; Darling 2009). In this context, NACCOM brings together the work of small organisations across the country who run night shelters for destitute asylum seekers, host refugees, and asylum seekers in spare bedrooms, and in some cases have even been able to purchase property to enable asylum seekers and refugees to have a more secure future. At the same time, NACCOM targets policy change through campaigns to end destitution and raises issues of migrant homelessness with local authorities, MPs, and government ministers. Whilst less radical in orientation than the squatting of Athens, the work of advocates and activists to house asylum seekers and refugees through NACCOM displays a similar critique of the exclusions of accommodation policies at national levels. It illustrates how forms of hosting and supporting the displaced have extended beyond the boundaries of official accounts of resettlement or dispersal, to form alternative networks that stand in opposition to the exclusions of filtering, selection, and categorisation that these systems enforce in the United Kingdom. Thus, as with the activists of Athens and the organisers of sanctuary movements across Europe and North America, we witness here how minor political actions of accommodating, hosting, and giving space can not only support asylum seekers and refugees but can also focus attention on the need for support in the first place.

## Conclusion

In this chapter, I have sought to examine some of the ways in which asylum seekers and refugees have been accommodated and supported in countries in the Global North. From state-led resettlement programmes to community-focused initiatives to house the otherwise destitute, a series of tensions run through the practice of hosting the displaced. Challenges of integration and adaptation to new environments are common in all such cases, but the question of who is accommodated, and of who performs the welcome of accommodating, is a critical ethical and political concern in these examples. For whilst efforts at hospitality at the level of the nation-state may be limited and highly selective, similar tensions over inclusion exist within more radical forms of refugee squatting and movements for sanctuary. The political and moral decisions taken in such contexts are not new, as Derrida (2001) reminds us in his deconstruction of hospitality, and they resist any complete or coherent resolution. Rather, they point towards the need to recognise that hospitality is a negotiated relation and, as such, cannot be predicted or regulated in advance. Just as the Syrian refugees making a new life on the Isle of Bute are, in many small ways, remaking that island and its communities through their presence, so efforts to accommodate, host, and support the displaced create a range of unpredictable effects, for newcomers, established communities, and political activists alike.

## References

Arnoldus, M., Dukes, T., & Musterd, S. (2003). Dispersal policies in the Netherlands. In V. Robinson, R. Andersson, & S. Musterd (Eds.), *Spreading the 'burden'? A review of policies to disperse asylum seekers and refugees* (pp. 25–64). Bristol, UK: Policy Press.

Bagelman, J. (2013). Sanctuary: A politics of ease? *Alternatives: Global, Local, Political, 38*(1), 49–62.

Bauder, H. (2017). Sanctuary cities: Policies and practices in international perspective. *International Migration, 55*(2), 174–187.

Bauder, H., & Gonzalez, D. A. (2018). Municipal responses to 'illegality': Urban sanctuary across national contexts. *Social Inclusion, 6*(1), 124–134.

Boswell, C. (2001). *Spreading the costs of asylum seekers: A critical assessment of dispersal policies in Germany and the UK*. London: Anglo-German Foundation for the Study of Industrial Society.

Brooks, L. (2015, November 14). 'People want to help': Scottish town prepares to host Syrian refugees. *The Guardian*. Retrieved November 15, 2018, from https://www.theguardian.com/uk-news/2015/nov/14/scotland-rothesay-bute-refugees-syria-welcome

Burridge, A., & Gill, N. (2017). Conveyor-belt justice: Precarity, access to justice, and uneven geographies of legal aid in UK asylum appeals. *Antipode, 49*(1), 23–42.

Carter, T. S., & Osborne, J. (2009). Housing and neighbourhood challenges of refugee resettlement in declining inner city neighbourhoods: A Winnipeg case study. *Journal of Immigrant & Refugee Studies, 7*(3), 308–327.

Chavin, S., & Garcés-Mascareñas, B. (2012). Beyond informal citizenship: The new moral economy of migrant illegality. *International Political Sociology, 6*(3), 241–259.

Coddington, K. (2018). Landscapes of refugee protection. *Transactions of the Institute of British Geographers, 43*(3), 326–340.

Crawley, H. (2016). Managing the unmanageable? Understanding Europe's response to the migration 'crisis'. *Human Geography, 9*(2), 13–23.

Cunningham, H. (2013). The emergence of the Ontario sanctuary coalition. In R. K. Lippert & S. Rehaag (Eds.), *Sanctuary practices in international perspectives: Migration, citizenship and social movements* (pp. 162–174). Abingdon, UK: Routledge.

Darling, J. (2009). Becoming bare life: Asylum, hospitality and the politics of encampment. *Environment and Planning D: Society and Space, 27*(4), 649–665.

Darling, J. (2010). A City of Sanctuary: The relational re-imagining of Sheffield's asylum politics. *Transactions of the Institute of British Geographers, 35*(1), 125–140.

Darling, J. (2016). Asylum in austere times: Instability, privatization and experimentation within the UK asylum dispersal system. *Journal of Refugee Studies, 29*(4), 483–505.

Darling, J. (2017). Forced migration and the city: Irregularity, informality, and the politics of presence. *Progress in Human Geography, 41*(2), 178–198.

Darling, J. (2018). The fragility of welcome—commentary to Gill. *Fennia, 196*(2), 220–224.

Darling, J., & Squire, V. (2013). Everyday enactments of sanctuary: The UK City of Sanctuary movement. In R. K. Lippert & S. Rehaag (Eds.), *Sanctuary practices in international perspectives: Migration, citizenship and social movements* (pp. 191–204). Abingdon, UK: Routledge.

Derrida, J. (2001). *On cosmopolitanism and forgiveness* (M. Dooley & M. Hughes, Trans.). London: Routledge.

Doyle, L. (2014). *28 days later: Experiences of new refugees in the UK*. London: Refugee Council.

Eby, J., Iverson, E., Smyers, J., & Kekic, E. (2011). The faith community's role in refugee resettlement in the United States. *Journal of Refugee Studies, 24*(3), 586–605.

Forrest, J., Hermes, K., Johnston, R., & Poulsen, M. (2013). The housing resettlement experience of refugee immigrants to Australia. *Journal of Refugee Studies, 26*(2), 187–206.

Goodall, C. (2010). *The coming of the stranger: Asylum seekers, trust and hospitality in a British city* (New Issues in Refugee Research—Research Paper 195). UN High Commissioner for Refugees, Geneva, Switzerland.

Huizinga, R. P., & van Hoven, B. (2018). Everyday geographies of belonging: Syrian refugee experiences in the northern Netherlands. *Geoforum, 96*, 309–317.

Hynes, P. (2009). Contemporary compulsory dispersal and the absence of space for the restoration of trust. *Journal of Refugee Studies, 22*(1), 97–121.

Kaika, M. (2012). The economic crisis seen from the everyday. *City, 16*(4), 422–430.

Koptyaeva, A. (2017). Collective homemaking in transit. *Forced Migration Review, 55*, 37–38.

Lewis, H. (2007). *Destitution in Leeds*. York, UK: Joseph Rowntree Foundation.

Logan, J., & Murdie, R. (2016). Home in Canada? The settlement experiences of Tibetans in Parkdale, Toronto. *Journal of International Migration and Integration, 17*(1), 95–113.

Lopez, M. A. M. (2016). Squatters and migrants in Madrid: Interactions, contexts and cycles. *Urban Studies, 54*(11), 2472–2489.

Lundberg, A., & Strange, M. (2017). Who provides the conditions for human life? Sanctuary movements in Sweden as both contesting and working with state agencies. *Politics, 37*(3), 347–362.

Mancina, P. (2013). The birth of a sanctuary city: A history of governmental sanctuary in San Francisco. In R. K. Lippert & S. Rehaag (Eds.), *Sanctuary practices in international perspectives: Migration, citizenship and social movements* (pp. 205–218). Abingdon, UK: Routledge.

Marrow, H. B. (2012). Deserving to a point: Unauthorized immigrants in San Francisco's universal access healthcare model. *Social Science & Medicine, 74*(6), 846–854.

NACCOM. (2017). NACCOM: The no accommodation network. *NACCOM*. Retrieved November 24, 2018, from https://naccom.org.uk/about/.

Nordling, V., Sager, M., & Söderman, E. (2017). From citizenship to mobile commons: Reflections on the local struggles of undocumented migrants in the city of Malmö, Sweden. *Citizenship Studies, 21*(6), 710–726.

Nyers, P. (2006). *Rethinking refugees: Beyond state of emergency*. London: Routledge.

Nyers, P. (2010). No one is illegal between city and nation. *Studies in Social Justice, 4*(2), 127–143.

Platts-Fowler, D., & Robinson, D. (2015). A place for integration: Refugee experiences in two English cities. *Population, Space and Place, 21*(5), 476–491.

Pyykkönen, M. (2009). Deportation vs. sanctuary: The rationalities, technologies, and subjects of Finnish sanctuary practices. *Refuge, 26*(1), 20–32.

Ridgley, J. (2008). Cities of refuge: Immigration enforcement, police, and the insurgent genealogies of citizenship in U.S. Sanctuary Cities. *Urban Geography, 29*(1), 53–77.

Robinson, V. (1998). The importance of information in the resettlement of refugees in the UK. *Journal of Refugee Studies, 11*(2), 146–160.

Robinson, V. (2003). Dispersal policies in the UK. In V. Robinson, R. Andersson, & S. Musterd (Eds.), *Spreading the 'burden'? A review of policies to disperse asylum seekers and refugees* (pp. 103–148). Bristol, UK: Policy Press.

Schuster, L. (2005). A sledgehammer to crack a nut: Deportation, detention and dispersal in Europe. *Social Policy & Administration, 39*(6), 606–621.

Squire, V. (2011). From community cohesion to mobile solidarities: The City of Sanctuary network and the Strangers into Citizens campaign. *Political Studies, 59*(2), 290–307.

Squire, V. (2018). Mobile solidarities and precariousness at City Plaza: Beyond vulnerable and disposable lives. *Studies in Social Justice, 12*(1), 111–132.

Trilling, D. (2017). *Lights in the distance: Exile and refuge at the borders of Europe.* London: Picador.

Tsavdaroglou, C. (2018). The newcomers' right to the common space: The case of Athens during the refugee crisis. *ACME, 17*(2), 376–401.

UNHCR. (2007). *The 10-point plan.* Geneva, Switzerland: UNHCR.

UNHCR. (2018). *UNHCR resettlement handbook.* Geneva, Switzerland: UNHCR.

van Liempt, I. (2011). From Dutch dispersal to ethnic enclaves in the UK: The relationship between segregation and integration examined through the eyes of Somalis. *Urban Studies, 4*(16), 3385–3398.

Vasudevan, A. (2017). *The autonomous city: A history of urban squatting.* London: Verso.

Wren, K. (2003). Refugee dispersal in Denmark: From macro- to micro-scale analysis. *International Journal of Population Geography, 9*(1), 57–75.

Yukich, G. (2013). Constructing the model immigrant: Movement strategy and immigrant deservingness in the New Sanctuary Movement. *Social Problems, 60*(3), 302–320.

# 55

# Food and the Politics of Refuge: The Transformative Power of Asylum Seeker and Refugee Food Initiatives

Fiona Murphy

## Introduction

This chapter examines the important role the politics of food plays in awakening knowledge and solidarity with the experiences of asylum seekers and refugees. Food, as a shared universal, has the potency to engender empathy, solidarity, and indeed, action and activism on behalf of and with asylum seekers and refugees. In recent years, the issue of forced displacement has figured largely in public debate and a broad range of food initiatives have emerged in response to the broader displacement of people globally, which the UNHCR now indicates is at 68.5 million. To the backdrop of heightened xenophobia, racism, and the growth of right-wing populism (Vieten 2018), politically mobilising food movements which centre on issues of displacement are now burgeoning in Europe, Australia, North America, and beyond; such movements should feature as a critical part of the analysis of the broader dynamics of forced displacement.

From Eat Offbeat and Foodhini in the United States to the Refugee Kitchen in Sydney and Food Justice Truck in Melbourne, Australia,[1] to refugee 'solidarity' dinners and pop-up restaurants right across the world and the international Refugee Food Festival, many initiatives are seeking to transform political and public narratives about asylum seekers and refugees through food. 'Foodie culture' coupled with the rise of social media, in particular

F. Murphy (✉)
Queen's University Belfast, Belfast, UK

platforms such as Instagram, and the notion of a 'pop-up' food experience, has expanded food horizons that have opened up space for discussion and debate in its championing of particular kinds of rights. Additionally, in places of waiting for the forcibly displaced, in particular border sites such as Calais, the provision of food aid by small volunteer groups in an attempt to deliver food justice to refugees has become a key feature of everyday survival for refugees in these regions. While a dense literature exists on the relationship between food and migration (see, for example, Caplan 1997; Charon Cardona 2004; Diner 2001; Douglas 1984; Harbottle 2000; Kershen 2002; Searles 2002; Thomas 2004) more broadly, fewer analyses exist of the way in which the politics of food is being instrumentalised in the context of forced displacement (see McKay et al. 2018). In drawing on food as a universal, but also as a way into other worlds, asylum seeker and refugee food initiatives are opening up new spaces of encounter and solidarity thereby creating new conversations about what forced displacement means in reality.

Within the social sciences more generally, there is a long history of scholarly analyses of the role of food in maintaining ethnic, regional, or national identities in spite of migration (Abrahams 1984; Counihan 1999). However, much of this literature is not focused on issues of forced displacement in the context of asylum and refuge but rather broader migration journeys. This is, however, starting to change with the rapid growth of interest in the many aspects of everyday life experience for asylum seekers and refugees. Additionally, the ways in which 'foodways' (defined as the cultural and historical trajectories of food communities) (Abrahams 1984, p. 20) impact diasporic identities and communities in their maintenance of collective identity and forging of a symbolic, historical, and cultural connection to their 'homeland' has featured large (Calvo 1982). Foodways in migration contexts exist as a deeply affective form of social connection (Bell and Valentine 1997), as well as a deeply emotive collective memory which, as Charon Cardona (2004, p. 51) argues, is a 'construction and re-creation of their remembered and reconstituted' ethnic identity. Such reflections on the role of food in the maintenance of culture and identity figure quite large in my research on the politics of food in the context of asylum seekers and refugees in Ireland. In mobilising their food projects and broader movements, many of my research participants utilise such notions, coupled with a call for empathy and solidarity, in order to underpin the broader goals of their initiatives.

In addition to this approach, the issue of food security figures largely in work examining food and displacement. In particular, asylum seekers and refugees living in detention and camps are vulnerable to food insecurity. This food insecurity figures in different ways, in terms of the lack of availability of nutritious food but also in terms of an individual's agency over its provision in spaces of waiting and detention. Ultimately, this absence of agency diminishes personal and bodily autonomy in eroding and startling ways. There is a vast interdisciplinary literature on this topic, which goes beyond the scope of this chapter but nonetheless is key to any discussion on the relationship between food and forced displacement (Cuny and Hill 1999; Eade and Vaux 2007; Martin 2004). Concerns about food security are growing globally amidst the intersecting crises of climate change, conflict, displacement, and austerity, and feature prominently in humanitarian and development contexts. Furthermore, there is a body of literature that reflects on the relationship between asylum seekers, refugees, host communities, and food security (Ruiz and Vargas Silva 2013; Verwimp and Muñoz-Mora 2013). Urban refugees are also vulnerable to food poverty, particularly in an Ireland where homelessness and poverty have increased; many of the food projects discussed in this chapter have emerged in response to some of these challenges.

Asylum seeker and refugee food projects also carve out a pathway towards new kinds of food entrepreneurship (see also Basu 2002) and creativity. In the Irish context, many of these food projects have also arisen as a direct response to living in an asylum seeker accommodation system called direct provision. As a system of institutionalised living, many of these accommodation centres do not have kitchen facilities; hence residents have to eat canteen food and do not have individual autonomy around their food choices (however, this is starting to change in some centres). Food insecurity then figures large in the everyday lives of asylum seekers and refugees in Ireland. Asylum seekers in Ireland gained the right to work in 2018, but the system of direct provision remains intact, in spite of growing protest amongst mainstream Irish society. This chapter examines a number of the innovative responses to the challenge of food insecurity and weak labour opportunities for asylum seekers and refugees in Ireland by examining a number of projects such as the Refugee Food Festival, Our Table, Cooking for Freedom, and the Sligo Global Kitchen (to name only a few in Ireland).[2] These projects are a mix of asylum seeker and refugee led or collaborative (with NGOs or activist groups). These particular examples, whilst anchored in an Irish context, open the space for comparative analysis on the wider global issues attached to forced displacement. The politics of food is an excellent frame through which to understand the everyday life challenges of asylum seekers and refugees.

# The Irish System of Asylum Management: Direct Provision

Since 2010, Ellie Kisyombe, an asylum seeker in Ireland originally from Malawi, and the founder of the Our Table food initiative, has been living in direct provision, a system of detention and waiting which forms part of the Irish system of 'migration management.' Individuals and families live in direct provision, often for lengthy periods, as they await decisions on their asylum application. Overseen by the Reception and Integration Agency (RIA), direct provision provides accommodation and board, as well as other essential services such as medical care. There are thirty-eight direct provision centres in Ireland (at the time of writing). The centres are fashioned out of old hotel premises, guesthouses, a convent, a holiday camp, and a caravan site (to name a few), all sites unsuitable to house individuals and families for long periods of time.

When introduced in 2000, direct provision was only ever meant to be a temporary measure. But, in 2018, it continues to exist amongst a global web of immigration detention. The increased privatisation of the system means inconsistent and often substandard living conditions for asylum seekers while private companies make a profit. This connects to a broader global culture of the commercialisation of the asylum system which sees widespread profiteering through private contracts whilst conditions for asylum seekers deteriorate. In 2015, the McMahon report highlighted just how detrimental to individuals' and families' well-being the Irish system of direct provision is, through its 175 recommendations on how to improve the asylum process (McMahon 2015). A 2017 report from the Ombudsman[3] shows that direct provision is not fit for purpose. It also highlights high incidences of food poverty and insecurity, issues which have been long highlighted by both grassroots organisations and NGOs but remain poorly addressed by the Irish state:

> Up until 2017, residents of direct provision could not cook their own food but, with the introduction of the Reception and Integration Agency's so-called independent living scheme, a number of direct provision centres have had kitchens installed. However, this limited intervention still means that many residents of direct provision do not have individual autonomy over their food. Residents of direct provision continually report large amounts of unhealthy foods (such as chips) being served almost on a daily basis, thus many of my interviewees have spoken about weight gain and a general deterioration of their health due to poor diet. Individuals and families with specific dietary requirements often struggle with the food they are served and would much prefer to be in a situation where

they could buy and cook their own food. Over the course of my research, I have spoken to a range of people living in direct provision who struggle with the lack of autonomy that this engenders (in particular parents of ill children and mothers who would like to impart culture through food) and, to date, very little has been done to remedy this complex, often dehumanising aspect of direct provision.

Ellie has lived in a number of these centres in her time in the asylum system. She details how direct provision has become a system of warehousing human beings seeking refuge in Ireland and describes her experience of direct provision:

> You have to find ways to cope with the system of direct provision, particularly for your mental health, so I started to try and think of ways to help other asylum seekers—activities such as gardening, sewing (…) to help them. Not being able to cook or taste your own food is terrible. We are living a life of poverty in direct provision.

The overarching sentiment from critics of direct provision is that this is a system that strips people of their agency, autonomy, and, in many cases, dignity. This is what has motivated Ellie to become one of the more vocal campaigners against direct provision in Ireland, alongside her colleagues in MASI (Movement of Asylum Seekers in Ireland). Her passion for cooking coupled with her despair at direct provision has called forth an activism articulated through food, as a very direct response to the often pernicious, eroding experiences of life in direct provision. As a system of asylum management, direct provision does not provide the security and safety that people seeking refuge from forced displacement require. Not unique to Ireland, such systems of institutionalised living or detention further inflict harm on individuals and families fleeing persecution in their home countries.

## Founding Our Table

Ellie's food initiative, called Our Table, is one that endeavours to show the power of food entrepreneurship as an alternative pathway to labour integration in Ireland for asylum seekers and refugees, but it is also a project that seeks to cultivate awareness of the challenging, often dehumanising, conditions asylum seekers inhabit in the Irish system of 'asylum management'—direct provision. For Ellie, her food project is a way of changing preconceived narratives on displacement, a space where she can take back agency amidst a cacophony of misrepresentations and misunderstandings around the experience of seeking asylum. It is also a space where she can reassume her agency as a strong feminist activist, a business woman, and a chef.

Ellie used to dream of fresh vegetables. She imagined a lush garden punctuated with clusters of vegetables and fruits from her home country. She could see herself running her fingers along the jagged edged of a pineapple breathing in the elixir of a healthier life. She dreamed of cooking. She could see herself kneading, mixing, folding her way into nourishment. Crackling with courage and hope from the fresh taste of her dreams, she embarked on her protests against direct provision, a place where people cannot choose or cook their own food. A place that assumes people seeking refuge have only vulnerabilities and not capabilities including cultural ways of being and identities to sustain.

Together with an Irish businesswoman Michelle Darmody, Ellie created the idea of Our Table. An initiative based on food and cooking, it served as a form of solidarity and protest, a creative response to direct provision that highlights a failure of political listening to asylum seekers and refugees in contemporary Ireland. Her carving out of a protest brought her to work alongside the Irish Refugee Council. As part of her internship, she attempted to turn her dreams of gardens and cooking into a reality for asylum seekers stifled by life in direct provision. A fortuitous encounter with Michelle, who had established Dublin's Cake Café, brought a like-minded venture into being, of cooking as a form of activism, education, and solidarity building.

For a number of months, together with other women from direct provision, Ellie and Michelle cooked at a number of community events, including a pop-up event at one of Dublin's main art galleries (IMMA). After years in direct provision, Ellie found the process of cooking traditional food from Malawi both liberating and emotional. Cooking for her became a lavish gift of agency shot through with the ability to protest and resist the hardships of life as an asylum seeker. It became a way of living a second life beyond the confines of direct provision.

As word spread, the pair were encouraged to reimagine their venture on a larger scale and this initially became a two-day pop-up restaurant in Dublin's Project Arts Centre in April 2016. Ellie, Michelle, and a team of volunteers created a menu which reflected the diverse food traditions of people in direct provision. It was a form of world-building and resistance through food, an event shared with the public as well as politicians, journalists, and members of the civil sector society. Such was the success of this two-day event that the project then ran for a three-month period, thereby highlighting the power of such food initiatives to create bridges of solidarity. Furthermore, playing a key role is in pushing against barriers of isolation and separation created by spaces of warehousing such as direct provision.

Ellie recalls the date of the launch of Our Table by well-known Northern Ireland actor Stephen Rea as the day the world heard about the election of Donald Trump as American president. For Ellie, this made the goals of Our Table, a celebration of diversity and resistance, even more important in the face of global challenges ahead, and it gave her and the other women from direct provision back their power: 'Our Table helped me gain confidence … there are times in direct provision where one feels as if they could give up, my training in Ballymaloe also gave me power and energy.' She believes she and the other volunteers have become a guild of sorts, beckoning towards a space of resistance.

Recalling Stephen Rea's words on how Irish people, in particular, should understand the significance of forced migration and food poverty because of their own history of famine and forced displacement, Ellie points towards the ignorance that exists regarding the experiences of asylum seekers and refugees in Ireland. Direct provision enables this kind of ignorance through the distance it creates between asylum seekers and citizens. In some respects, the very system itself attempts to narrow citizen responses to its existence. But the success of Our Table comes through its ability to radicalise the range of responses through something we can all identify with, namely food.

Ellie realised her dream of becoming a trained chef in Darina Allen's cookery school in Ballymaloe. She has run an Our Table food stall at Christchurch, and, now, alongside her team, is in the process of developing food products for retail distribution. In a seismic rejigging of her every day, her dreams about cooking and food have become a reality, but she continues to live in direct provision.

## Cooking for Freedom

Similar stories and initiatives are unfolding elsewhere in Ireland and across the globe. Through different, imaginative food initiatives, asylum seekers and refugees are instrumentalising a politics of food to build awareness about food insecurity in forced displacement contexts and its attendant issues of health and identity/cultural politics. In Cork and Galway, Syrian chefs have come together to cook, fundraise, and create awareness of the refugee experience. Bay Leaf, a Syrian restaurant in Bandon, has conducted a number of fundraising and awareness-building nights. In Sligo, an initiative called the Sligo

Global Kitchen holds public food events once a month through the local arts centre, the Model, to create awareness about direct provision.

One of the members of this group, Mabel Chah, explained that 'there are a lot of barriers when you live in direct provision' and these food events provide a connection to the local community in Sligo which facilitate friendship and solidarity. Awarded the Irish Food Writers Guild community award in March 2018, Sligo Global Kitchen is now looking to further their venture in a more sustainable manner. Similar initiatives have emerged in Limerick, Athlone, Millstreet, and Killarney—all working towards remedying the challenge of food poverty in direct provision.

In Dublin, a group of women from a direct provision centre have created an initiative called Cooking for Freedom, which sees the women coming together to use an external kitchen to cook their own food and feed their families the well-being and culture denied to them.

One of the organisers of Cooking for Freedom, Mavis, explains that:

> The idea came because we can't cook for ourselves in direct provision. With the help of MASI and Refugee and Migrant Solidarity Ireland (RAMSI), we found a space to come together to cook. We are not losing our skills, we get a lot of mental health benefits from it too, it brings us together, helps us to discuss the challenges we have living in direct provision. Above all else, we get great pleasure from eating food from our own native countries that we have eaten ourselves. Even buying the ingredients brings such pleasure.

Mavis describes how Cooking for Freedom has alleviated some of the challenges experienced by families of children with food allergies. It has also brought women and families living in direct provision together to form a network of support and solidarity in a space where everyday life presents constant challenges and insecurity. Unfortunately, all of these laudable projects struggle for funding in their attempt to transform the wanton disregard of asylum seekers' well-being in direct provision.

One of the key strengths of these projects is in their ability to create sanctuary spaces for asylum seekers and refugees who participate therein, specifically for women. Both Mavis and Ellie describe their ventures as places that create inclusive spaces that are at once movements for political resistance and a safe space for asylum seekers and refugees to come together, to cook, chat, share, and be. In the midst of what might be considered a new era of solidarity, at least in the Irish context, food entrepreneurship projects such as Our Table, Cooking for Freedom, and the Sligo Global Kitchen fashion a solidarity that focuses on shared commonalities, thereby nourishing both the personal and political, in a way that evolves alternative socialities and networks that move to different forms of action.

## Refugee Food Festival

In the summer of 2018, Ireland held its first 'Refugee Food Festival' with much media focus and praise from the participants. The festival, an idea initially borne in France with UNHCR support, was based on a concept which saw refugee chefs coming into famous French restaurants to cook for a day. One of the key aims of the festival was to challenge negative perceptions of asylum seekers and refugees by showcasing talent and diverse food. The criteria for participating in the festival are stringent, calling for excellent culinary skills and restaurant management. A broader aim of the festival was the professional labour integration of participating chefs. Since the first Paris-based festival in 2016, the event has now evolved to include sixteen different countries. A key part of the festival is cooperation and collaboration with NGOS, as well as UNHCR support, which is significant in promoting the festival as a rights-oriented one.[4]

As part of the research for a podcast my colleague Aylisha Hogan and I documented the first Refugee Food Festival in Ireland, interviewing participating members including restaurants and different direct provision centres. Feedback on the festival was overwhelmingly positive with asylum seekers and refugees pointing to the importance of the festival in accelerating their training in the food and service industry and also on the role of the festival in giving visibility to the challenges that asylum seekers and refugees face. Events were held in restaurants in different cities and towns in Ireland, including Dublin, Cork, Galway, Waterford, Mayo, Sligo, and Louth, in restaurants, cafes, and a direct provision centre. One of the organisers, Jess Murphy, of Kai Galway, articulates the objectives of the festival as way of bringing attention to direct provision and also as a way of raising funds towards providing training, education, and scholarships to asylum seekers and refugees in Ireland:

> I got all the top restaurants that I could together, and everyone said yes. Having really high end restaurants showcase refugee's talents is great. Like people know they are here, but they don't really understand much. With the participation of the high end restaurants the festival will be kept in the press which is great (…) There are lots of people involved, volunteering and working for free, supported by the UNHCR. (…) 'Doing something' from this is the main thing, like we

can send people for training or to University with scholarships from this as well as building awareness about direct provision.

Alongside her co-organiser, Lisa Regan, Jess worked hard to create new networks and a form of solidarity that encompassed the very practical outcome of providing training, as well as a broader aim of recasting preconceived narratives about asylum seekers and refugees. In our documenting of the event, we also spoke to attendees asking them what motivated them to attend; answers varied from a passing interest to wanting to learn more about the experience of asylum seekers and refugees in Ireland to those more directly involved as activists. For the organisers and the participants, new kinds of collaboration and solidarities emerged, and many of the participants still living in direct provision stated that they hoped the experience would pave the pathway to finding employment within the sector. Many of our asylum seeker and refugee interviewees saw in the festival an expression of solidarity but also a crystallisation of solidarity between different kinds of organisations and spaces which allowed for a reframing of their personal agency in spite of living life in direct provision.

While the refugee food festival is not asylum seeker or refugee led in the same way that other initiatives are and therefore produces solidarity in a different fashion (in that its focus is on *giving voice* or *making visible* refugee talents and skills), it nonetheless highlights the value of the *process* of producing solidarities across time and through events like this. While a cynical take on these kinds of festivals might be that it is merely a performance of solidarity or an exercise in PR, our research documented the social and political value of events such as the Refugee Food Festival to its diverse range of participants from asylum seekers to restaurateurs. The Refugee Food Festival was generative of mutually transformative solidarities and relationships (Featherstone 2012) that spanned social, class, and even spatial divisions (between direct provision and the host localities) which re-stitched harmful narratives about asylum seekers and refugees. The first refugee food festival happened in Ireland at a key moment where the debate about right to work permits for asylum seekers had made significant progress. More broadly, as a movement across Europe, it also points to the benefits of a more transnational solidarity with respect to the experience of asylum seekers and refugees in what is becoming an increasingly hostile climate contaminated by populist right-wing politics.

## Conclusion

While many of these food initiatives have emerged in response to the concrete and pressing material needs of asylum seekers in direct provision, they point to the possibility of new ways of political participation, self-organisation, and of labour inclusion. Food, as a universal, transcends difference, as do these food projects, which while all very specifically anchored in the experience of life in direct provision, point far beyond their situatedness and locality. Ultimately, notions of solidarity and action and their boundedness become reformed by these very important ventures. They also point to a global system of asylum system with serious failings. What these projects make clear in the Irish instance is that an alternative to direct provision must be urgently found. To do so is the only dignified response to one of the more urgent crises of our time. More broadly, these projects speak to the importance of delivering on food justice for the forcibly displaced in a world where food insecurity, lack of autonomy and agency with respect to food provision, and health and well-being issues feature large.

## Notes

1. See the following links to access these food initiatives—eatoffbeat https://eat-offbeat.com/ (Accessed on 30 January 2019), Foodhini https://foodhini.com/ (Accessed on 30 January 2019), Refugee Kitchen Sydney http://www.thetradingcircle.com.au/fourbravewomen (Accessed on 30 January 2019), Food Justice Truck https://www.asrc.org.au/foodjustice/ (Accessed on 30 January 2019).
2. A version of the research data here has been published by RTÉ Brainstorm (https://www.rte.ie/brainstorm/). I reproduce it here with the full permission of RTÉ Brainstorm.
3. The Ombudsman is an official who is charged with representing and dealing with the concerns of the public or anyone who feels they have been treated unfairly by an organisation dealing in public services.
4. See the Refugee festival kit for further information on how the festival is organised and run https://www.unhcr.org/be/wp-content/uploads/sites/46/2017/09/kit-refugee-food-festival-en.pdf (Accessed on 25 November 2018).

# References

Abrahams, R. (1984). Equal opportunity eating: A structural excursus on things of the mouth. In L. K. Brown & K. Mussell (Eds.), *Ethnic and regional foodways in the United States* (pp. 19–36). Knoxville, TN: The University of Tennessee Press.

Basu, A. (2002). Immigrant entrepreneurs in the food sector: Breaking the mould. In A. J. Kershen (Ed.), *Food in the migrant experience*. Aldershot, UK: Ashgate Publishing Ltd.

Bell, D., & Valentine, G. (1997). *Consuming geographies: We are where we eat*. London: Routledge.

Calvo, M. (1982). 'Migration et alimentation', Information sur les sciences sociales. *Social Science Information, 21*(3), 383–446.

Caplan, P. (Ed.). (1997). *Food, health and identity*. London and New York, NY: Routledge.

Charon Cardona, E. T. (2004). Re-encountering Cuban tastes in Australia. *The Australian Journal of Anthropology, 15*(1), 40–53.

Counihan, C. M. (1999). *The anthropology of food and body: Gender, meaning, and power*. New York, NY and London: Routledge.

Cuny, F. C., & Hill, R. B. (1999). *Famine, conflict, and response: A basic guide*. West Hartford, CT: Kumarian Press.

Diner, H. R. (2001). *Hungering for America: Italian, Irish and Jewish foodways in the age of migration*. Cambridge, MA and London: Harvard University Press.

Douglas, M. (Ed.). (1984). *Food in the social order: Studies of food and festivities in three American communities*. New York, NY: Russel Sage Foundation.

Eade, D., & Vaux, T. (2007). *Development and humanitarianism: Practical issues*. Bloomfield, CT: Kumarian Press.

Featherstone, D. (2012). *Solidarity: Hidden histories and geographies of internationalism*. London: Zed Books.

Harbottle, L. (2000). *Food for health, food for wealth: Ethnic and gender identities in British Iranian communities*. New York, NY: Berghahn Books.

Kershen, A. J. (Ed.). (2002). *Food in the migrant experience*. Aldershot, UK: Ashgate Publishing Ltd.

Martin, S. F. (2004). *Refugee women*. Lanham, MD: Lexington Books.

Mckay, F., Lippi, K., Dunn, M., Haines, B. C., & Lindberg, R. (2018). Food-based social enterprises and asylum seekers: The food justice truck. *Nutrients, 10*(6), 756.

McMahon, B. (2015). *Working group to report to government on improvements to the protection process, including direct provision and supports to asylum seekers: Final report*. Dublin, Ireland: Department of Justice.

Ruiz, I., & Vargas Silva, C. (2013). The economics of forced migration. *Journal of Development Studies, 49*(6), 772–784.

Searles, E. (2002). Food and the making of modern Inuit identities. *Food and Foodways, 10*(1–2), 55–78.

Thomas, M. (2004). Transitions in taste in Vietnam and the diaspora. *The Australian Journal of Anthropology, 15*(1), 54–67.

Verwimp, P., & Muñoz-Mora, J. C. (2013). *Returning home after civil war: Food security, nutrition and poverty among Burundian households* (Working Paper 123). Households in Conflict Network, Brighton, UK.

Vieten, U. (2018). Ambivalences of cosmopolitianisms, elites and far right populisms in twenty first century Europe. In G. Fitzi, J. Mackert, & B. S. Turner (Eds.), *Populism and the crisis of democracy* (pp. 101–108). London: Routledge.

# Index[1]

A

Abolition, 23, 48, 274, 494, 502–503
Aboriginal/Indigenous, 20, 384–387, 390, 403, 607, 696
Academics, 3, 7, 8, 11–13, 19, 23, 30, 89, 101, 132, 134, 179, 199, 200, 207, 294, 297, 306n1, 313, 354, 365, 416, 419, 422, 423, 433, 434, 436, 445–447, 464, 465, 497, 502, 647, 656, 658, 660, 661, 667, 690, 702, 705, 709, 714, 722, 723, 740, 756, 762, 766n3, 767n11, 771
Activism, 8, 28, 30, 105, 390, 503, 543, 590, 591, 653–658, 674, 707, 725–736, 745, 786, 789, 793, 799, 803, 804
Adaptation, 56, 59, 200, 203, 205, 207, 385, 386, 392, 676, 741, 795
Administrative boundaries, 435, 442, 443
Administrative data, 21, 431–447
Advocacy, 9, 29, 60, 61, 164, 200, 480, 662, 683, 701–714, 748, 779, 780, 786, 789
Aesthetics, 14, 15, 27, 28, 42, 59, 173–176, 179, 190, 257, 384, 385, 392, 394, 417, 419, 423, 453–457, 661, 662, 664, 674, 678, 689, 702, 705
Affect, 4, 11, 18, 28, 48, 49, 73, 81, 91, 99–105, 144, 148, 149, 174, 204, 225, 247, 262, 316, 334, 337–339, 366, 416, 478, 485, 498, 499, 503, 547, 586, 589, 608, 674, 726, 741
Affordable housing, 30, 258, 464, 721, 733, 746, 753, 789
Agency, 3, 7, 17, 21, 23–25, 28, 31, 47, 70, 72, 74, 85, 91, 93, 94, 105, 115–118, 132, 165, 182, 228, 237, 240–244, 257, 262, 272, 275, 298, 305, 313, 322, 337, 349, 352, 363, 366, 367, 419, 421, 424, 454, 465, 480, 494, 496, 497, 499–503, 507, 508, 511, 512, 515, 516, 550–558, 574, 576, 585, 586, 592, 600, 601, 606, 608, 632, 633, 635, 637, 641, 642, 645–647, 661, 662, 674, 675, 679, 681, 683, 711, 726, 734, 775, 787, 788, 792, 794, 801, 803, 804, 808, 809

---

[1] Note: Page numbers followed by 'n' refer to notes.

© The Author(s) 2020
P. Adey et al. (eds.), *The Handbook of Displacement*,
https://doi.org/10.1007/978-3-030-47178-1

# 814  Index

Animal agency, 551–558
Anti-eviction, 281, 740, 741, 744–746, 749
Archaeology, 555
Architecture, 12, 16, 20, 26, 187, 190, 248, 383–394, 445, 569, 629, 635
Arctic, 90, 92, 93
Arendt, Hannah, 100, 101, 248, 692
Art, 12, 28, 42, 73, 179–185, 187, 188, 192, 193, 193n1, 257, 294, 562, 600, 665, 677, 687–699, 702, 704, 706, 707, 710, 711, 714, 723, 804, 806
Art-based and creative methods, 705
Artists, 11, 28, 42, 174, 179–182, 184–187, 192, 193n1, 387, 417, 561–563, 660, 666, 667, 669, 681, 687–690, 692–699, 710, 712, 723
Arts-based research, 701–714
Artwashing, 15, 179–194
Assemblage, 7, 11, 43, 89–95, 406, 495, 557, 571
Atrocity, 4, 24, 521–533, 589, 664, 665, 675
Austerity, 22, 47, 181, 184, 278, 434, 441, 459, 464, 466, 467, 470, 721, 731, 739–749, 787, 792, 801
Authenticity, 161, 662, 674

## B

Bailiffs, 7, 493–503
Bedouin, 103, 556
Bedroom Tax, 437–441, 730–732
Benefits, 21, 28, 50, 70, 150, 193, 208, 248, 366, 408, 434–441, 445, 446, 447n1, 462, 469, 478, 562, 631, 632, 639, 674, 721, 730, 731, 741, 763, 764, 806, 808
Biennales, 192, 384, 688–690, 698

*Big Capital*, 721, 722
Biometric scans, 161
Biopolitics, 14, 55, 161, 507, 508, 511, 512, 514, 515, 546
Border regime, 19–21, 237, 238, 349, 413, 416, 420, 423, 424, 591
Bordering, 44–47, 221, 238, 302, 453, 787
Borders, 1, 41, 68, 111, 123, 150, 159, 200, 218–221, 223, 237, 238, 298, 315, 335, 336, 347, 364, 383, 413, 425n1, 431, 454, 463, 495, 507, 530, 571, 586, 600, 614, 669, 674, 702, 773, 785, 800
Brazil, 13, 146, 163, 271, 276, 278, 779
Bucharest, 29, 653–658

## C

Calais, 315, 318–320, 322, 323, 561, 665, 800
Cambodia, 13, 25, 206, 393, 526, 533n3, 541–547, 727, 735
Camps, 1, 2, 4, 25, 80, 84, 102, 121, 122, 163, 220, 225, 290, 292, 294, 298, 300, 302, 305, 314, 367, 370–373, 386, 388, 391–393, 395n5, 395n7, 462, 464, 465, 467, 485, 524, 526, 528, 533n6, 551–558, 564n12, 570–572, 574–580, 600, 603, 604, 609n2, 640, 643, 655, 665, 688, 690, 692–696, 698, 728, 739, 744, 773, 775, 787, 792, 793, 800
Cash transfers, 161, 467
Circulation, 8, 222, 333, 415, 417, 575, 691, 693, 702, 785
Cities, 5, 7, 16, 30, 46–48, 100, 102, 122–124, 135, 163, 181, 183, 191, 208, 255, 257–262, 264,

266, 272–275, 277–281, 291, 298, 317, 319, 330, 332, 333, 369, 384, 387, 415, 441, 443, 466, 468, 485, 510, 512, 514, 525, 526, 530, 533n3, 542, 631, 634, 653, 656, 678, 689, 697, 721, 723, 739–746, 748, 749, 754, 755, 785–795, 807

Climate change, 2, 15, 24, 56, 58, 59, 81, 83, 89–95, 199–205, 207–210, 365, 408, 533, 559–561, 630, 647, 687, 691, 723, 801

Climate refugees, 7, 89, 200–204, 463, 466, 533

Coastal, 83, 638, 723

Co-housing, 734, 735

Commoning, 44, 49–50

Community activism, 279

Community/ies, 2, 59, 70, 92, 103, 109, 123, 132, 160, 179, 199, 229, 240, 255, 261–266, 272, 294, 333, 334, 364, 384, 390, 416, 460, 475, 523, 543, 571, 586, 600, 629, 653, 675, 703, 722, 725, 739, 754, 771–780, 785, 800

Connected refugees, 584

Conservative-led coalition, 721

Containment, 46, 136, 137, 222, 261, 315, 321–323, 352, 383, 385, 387, 389, 391, 730

Contemporary art, 28, 687–699

Convicts, 20, 330–332, 386, 388–390, 394n2

Co-production, 24, 294, 606

Corporeality, 248, 664, 674

Critical discourse, 722

Cultural citizenship, 705–706, 714

Culture, 42, 144, 145, 150, 182, 183, 247, 331, 333, 384–386, 389, 394, 394n3, 400, 401, 406, 408, 494, 526, 531, 542, 545, 569, 570, 574, 575, 577, 607, 616, 618, 632, 675, 688, 689, 701–703, 706, 723, 729, 759, 760, 771, 789, 800, 802, 803, 806

D

Data access, 447n3

Data archiving, 444

Data processing, 442

Debates, 3, 5, 10, 11, 14, 29–32, 61, 89, 90, 186, 188, 200, 202, 218, 220, 221, 246, 297, 306n1, 315, 348, 367, 459, 465, 531, 592, 656, 658, 678, 701, 722, 741, 745, 756, 765, 799, 800, 808

De-identified data, 439, 442

Deportation, 10, 16, 25, 43, 47–50, 136, 223, 226, 237–249, 332, 414–417, 425n3, 476, 478, 480–483, 485, 486n1, 507–509, 511–513, 515, 516, 522, 531, 541, 542, 544–547, 591, 748, 788, 790, 791

Detention, 15, 16, 23, 42, 43, 45, 47–50, 79, 102, 121, 122, 136, 217–219, 222–224, 226, 228, 229, 240, 242, 245–247, 304, 316, 319, 322–324, 325n5, 330, 332, 334, 337, 383, 386, 393, 425n3, 463, 465, 476, 480–485, 507–509, 511–513, 515, 544, 545, 590, 690, 693–695, 698, 773, 775, 776, 788, 800–803

De-territorialisation, 41, 92, 94

Developers, 31, 182–185, 190, 193, 257, 258, 682, 722, 727, 734, 753, 755, 761, 765

Digital informatics, 161

Disasters, 4, 5, 19, 33, 56–58, 85, 89, 113, 122, 125, 126, 131, 134, 136, 158, 160, 162, 164, 174, 175, 200, 203, 206, 208, 209, 314, 315, 323, 324n3, 364, 369, 381, 432, 433, 464, 510, 511, 561, 600, 629–647, 691

Disorientation, 17, 18, 247, 336, 368, 678, 683

Displaced energy, 569–580

Displaced migrants, 2, 15, 227, 583–585, 587–589, 591

Displacement atrocity, 24, 527–529

Displacement economies, 67–74, 75n5

Dispossession, 15, 23, 30, 32, 56, 58, 59, 68, 71, 79, 83, 86, 179–184, 186, 188–190, 224, 256, 258, 273, 279, 281, 314–316, 322, 324, 387, 389, 403, 405–407, 493, 494, 497, 498, 668, 670, 691, 696, 729, 739–741, 746, 749, 755

Documenta, 688, 690, 693, 697

Documentary, 1, 28, 655, 656, 665, 682, 698

Domestic violence, 7, 21, 80, 135–137, 432, 434–436, 438, 441–445

Duman, Albert, 723

E

East London, 15, 180, 187–189, 191, 193, 726, 728, 743

Ecological justice, 55

Economic displacement, 478, 482

Elephant & Castle, 721

Elephant Park, 721, 756

Embodiment, 44, 55, 100, 101, 105, 220, 258

Emergence, 32, 41, 42, 102, 221, 319, 330, 352, 558, 731, 739, 748, 788, 789, 793

Emotions, 61, 100, 245, 404, 498, 585, 587–589, 593, 601, 606–608, 615, 617, 621, 623, 664, 666, 707

Empathy, 28, 423, 664, 674, 676, 677, 682, 683, 799, 800

Empower, 161, 723

Enforcement, 23, 69, 115, 221, 228, 237, 239, 240, 243, 298, 303, 304, 318, 319, 350, 351, 421, 475, 479–483, 485, 493–503, 507–516, 641, 663, 786, 790

Entertainment, 28, 674

Estate regeneration, 721, 761, 762, 765

Ethnic cleansing, 24, 180, 523, 530–532

Ethnicity, 11, 22, 41, 132, 136, 143–151, 246, 431, 439

Ethnography, 245, 246, 418, 419, 501, 585, 616, 708, 771

EU borders, 349–351

Europe, 1, 13, 14, 17, 18, 28, 30, 45, 68, 144, 145, 149, 158, 159, 227, 230, 230n1, 246, 248, 273, 274, 299, 304, 306n3, 313–324, 324n4, 348, 349, 371, 384, 391, 392, 394, 415, 417, 468, 496, 500, 507, 508, 510–511, 513, 515, 516n1, 551, 561, 573, 584–586, 588, 601, 606, 641, 656–659, 662, 665, 669, 676, 702, 739–749, 774, 786, 787, 789–794, 799, 808

Eviction, 1, 4, 23, 33, 43, 48–50, 58, 71, 72, 79, 82, 132, 174, 271–281, 319, 432–434, 438, 441, 466, 493–500, 502, 503, 653, 655–658, 727–729, 731, 734, 739, 741, 742, 744–748, 793

Eviction enforcement, 493–503

Evidence, 15, 73, 132, 144–146, 150, 204, 206, 207, 237, 263, 302, 333, 336, 371, 414, 420, 433, 437, 457, 601–603, 607, 679, 710, 734, 760, 762, 764, 765, 767n11, 775, 780, 788

Experience, 2, 42, 67, 82, 92, 99, 122, 143, 157, 174, 205, 224, 245, 258, 267, 294, 299, 313, 330, 338, 339, 346, 364, 392, 399, 413, 434, 453, 480, 498, 550, 576, 600, 613–625, 656, 673–683, 696, 701, 711–713, 736, 773, 778, 788, 799

Extractivism, 59

F

Family photographs, 599–608, 609n1

Film, 28, 180, 182, 192, 609n2, 656, 657, 660, 663, 665–667, 669, 673, 682, 695, 696, 698, 710, 723

Finance, 22, 277, 467, 498

Force, 2, 7, 10, 17, 18, 30, 45, 79, 85, 89, 91, 93, 94, 112, 123–125, 164, 165, 204, 238, 242, 243, 267, 276, 291, 298, 317, 322, 333, 336, 346, 347, 353, 384, 385, 401, 460, 477, 478, 494, 496, 497, 499, 500, 513, 524, 527–529, 531, 532, 542, 552, 556, 574, 586, 600, 601, 608, 636, 640, 647, 654, 675, 679, 693, 722, 723, 726, 730, 760, 775

Forced displacement, 23, 24, 68, 109, 112, 135, 143–151, 217, 220, 222, 224, 226, 229, 294, 305, 324, 388, 389, 393, 460, 463, 521–533, 569, 574, 654, 688, 696, 702, 704, 712, 799–801, 803, 805

Forced displacement and terrorism, 294

Forced eviction, 25, 114, 432, 493, 494, 499, 501, 541–543, 547, 655, 658, 726, 727, 748

Forced migration, 42, 43, 102, 136, 157, 164, 165, 174, 297, 303, 306n1, 315, 316, 350, 355, 399, 403, 404, 431, 441–444, 559, 584, 586, 614, 805

Foreign investors/investment, 721

Forensic/s, 419, 679

G

Gambling, 304

Gamification, 679

Gaming, 682

Gender, 5, 11, 22, 69, 72, 73, 86, 111, 121, 123, 125–127, 131–137, 146, 149, 175, 206, 246, 262, 266, 301, 302, 355, 363, 459, 465, 475, 479, 513, 583, 586, 615, 617–620, 624, 625, 632, 662, 697, 726–729, 778, 779, 793

Gender-Based Violence and Displacement, 294

Gender identity, 121, 124, 773

Genocide, 24, 385, 386, 389, 406, 521–523, 525, 526, 528, 529, 531, 532, 533n1, 544, 546, 604

Gentrification, 4, 5, 15, 30, 31, 44, 49, 131, 179–194, 257–259, 262, 263, 265, 277, 278, 281, 432, 435, 493, 722, 723, 725, 728, 732, 735, 739–741, 748, 753–766

Geographic displacement, 404, 480, 485, 588, 689

Global capital, 721

Globalisation, 28, 41, 46, 219, 238, 302, 498, 687, 688, 691–693, 697, 699

Global South, 13, 30, 31, 74, 83, 144, 146, 150, 160, 241, 248, 299, 303, 349, 432, 433, 453, 462, 464–466, 469, 561, 773
Governance, 3, 4, 22–24, 32, 41, 44, 47, 55, 56, 58–60, 90, 221, 244, 247, 281, 365, 403, 459–464, 466, 467, 470, 508, 729
Government, 2, 21, 33, 48, 60, 72, 84, 113, 114, 117, 121, 122, 126, 127, 132, 158–160, 162, 164, 185, 187, 193, 199, 203, 218, 222, 241, 247, 249, 262, 275, 279–281, 316–318, 321, 332, 350, 364, 366, 367, 386, 392, 393, 394n4, 403, 421, 433–435, 437–442, 445, 446, 461, 462, 477, 480, 495, 501, 507–509, 511, 515, 557, 558, 587, 603, 631, 636, 638, 640, 641, 645, 654, 660, 694, 695, 729–731, 743, 758, 774, 777, 779, 780, 787, 788, 792, 794
Governmental mobility, 218, 220–227, 229, 230, 316
*Ground Control*, 722

H

Haptic, 28, 674, 677, 681, 683
Health, 18, 100, 113, 123, 124, 126, 132, 246, 262, 363–374, 393, 405, 408, 409, 435, 436, 445, 465, 512, 561, 586, 589, 630–632, 634, 640–642, 723, 791, 802, 803, 805, 806, 809
Health security, 374
Heygate Estate, 182, 721, 756, 757
Hidden displacement, 433, 441
Home (geographies of), 400, 401, 619, 621
Homeless, 10, 49, 80, 122, 124, 220, 258, 332, 436, 438, 441, 654, 721, 728, 743, 749
Homes, 5, 42, 73, 79, 91, 103, 112, 123, 134, 159, 180, 244, 262, 271, 290, 331, 335, 367, 387, 400–403, 415, 432, 462, 494, 509, 523, 542, 552, 571, 587, 599, 613–624, 633, 653, 677, 721, 725, 739, 753, 785–795, 803
Hospitality, 31, 571, 660, 775, 787, 790–792, 795
Hosting, 31, 191, 278, 367, 681, 728, 785–795
Housing, 4, 15, 22, 26, 30, 48, 81, 82, 91, 92, 114, 136, 162, 180, 181, 185–193, 193n1, 224, 228, 249, 258, 272, 275–278, 280, 281, 363, 372, 373, 393, 402, 405, 408, 432–441, 445, 447n1, 460, 462, 464–466, 469, 470, 479, 494–498, 507, 510, 513, 523, 553, 561, 616, 620, 629–647, 653–658, 721, 723, 725–736, 740–748, 753–755, 757, 758, 760, 763–765, 766n3, 767n7, 785, 787, 789, 792, 793
Housing benefit, 21, 434, 437–441, 445, 446, 447n1, 721, 731
Human–animal relations, 561
Humanitarian action, 161–164, 771
Humanitarian design, 383
Humanitarianism, 351, 352, 354, 681, 772, 774
Humanitarian law, 109, 110
Humanitarian standards of practice, 161–162
Humanitarian technology, 161–162

Human rights, 3, 4, 7, 47, 48, 82, 111–115, 121, 158, 160, 162–164, 203, 237, 238, 281, 335, 336, 338, 363, 393, 404, 431, 432, 479, 494, 521, 532, 584, 590, 591, 702, 725–736, 748, 755, 758–761, 764

I

*Immersion*, 673–683
Immigrants, 23, 124, 136, 145, 147, 149, 150, 334, 371, 384–386, 388, 390, 392, 394n1, 475–485, 547n3, 590, 673, 689, 693, 776
Immigration, 23, 136, 144, 146–150, 237, 239–241, 244, 245, 247, 303, 330, 332, 334, 349, 393, 394n1, 404, 475–478, 480–485, 512, 545, 663, 748, 780, 786, 790, 792, 794, 802
Immigration bureaucracy, 23, 475–483, 485
Immigration law/immigration legislation, 47, 149, 463, 475, 476, 478, 483, 678
Immobilities, 2, 4, 8, 9, 42, 49, 69, 222, 228, 303, 315, 316, 332, 511, 710, 773, 775
Indigenous, 21, 25, 26, 43, 49, 60, 61, 91, 92, 136, 146, 147, 271, 273, 384–386, 389–391, 399–409, 432, 526, 560, 696
Indigenous knowledge, 403, 406, 580
Infrastructure, 9, 11, 17, 25, 26, 41, 42, 47, 56, 58, 59, 90, 91, 161, 238–240, 248, 271, 272, 314, 320, 323, 325n5, 331, 335, 337, 364, 367, 369, 370, 467, 495, 498, 508, 527, 575, 583–593, 629, 631, 632, 634, 635, 637, 639, 640, 642, 643, 646, 647, 669, 722

Interaction, 18, 25, 89, 90, 93, 102, 210, 248, 306n6, 588, 601, 602, 607, 608, 616, 674, 677, 681–683, 726, 788, 789
Internal frontiers, 17, 313–324
Internally displaced persons, in Iraq, 294
Internally displaced persons/people, 7, 600
Internal migration, 433, 435, 443, 444
Internment, 386, 388, 391, 395n5, 600, 604
Intervention, 2, 7, 11–14, 19, 22–25, 29, 56, 68, 73, 105, 158–161, 165, 173–176, 181, 186, 274, 280, 290–295, 347, 351, 352, 381, 402, 405, 439, 453–457, 494, 502, 541–547, 575, 631–634, 653–658, 681, 721–723, 754, 777, 788, 802
Intimate partner violence, 81, 84
Involuntary migration, 399, 400, 403–405, 408
Israel, 299, 302, 670

J

Journeys, 8, 16–20, 32, 123, 136, 174–176, 205, 209, 224, 292, 297–306, 317, 318, 320–323, 329–339, 345, 346, 348, 350, 355, 363, 364, 366, 368, 369, 371–374, 406, 413–418, 420, 423, 424, 441–445, 549, 550, 558, 570, 579, 585, 586, 601, 606, 621, 670, 673, 676, 697, 702, 773, 788, 800
Judicial displacement, 476, 483–485
Justice system, 333, 445

K

Knowledge co-production, 294

## L

Labels, 21, 183, 210, 242, 303–305, 368, 459, 461–465, 470, 647
Land, 2, 16, 18, 21, 43, 45, 46, 50, 56–60, 71, 72, 82, 83, 91, 102, 163, 204, 220, 258, 276, 346, 389–391, 402, 403, 405–409, 432, 434, 494–496, 498, 499, 503, 527, 531, 561, 638–642, 696, 697, 759, 760, 762
Languages of valuation, 60
Latino/a, 23, 475, 480–482, 485
Law, 6, 7, 16, 31, 43, 50, 102, 109–112, 114, 115, 117, 148–150, 163, 228, 238, 241, 256, 257, 260, 263, 267, 277, 315, 348, 394n3, 403, 415, 419, 421, 436, 438, 461, 463, 465, 475, 476, 478, 479, 484, 485, 494, 509, 510, 513, 524, 531, 654, 656, 693, 703, 726, 729–732, 754, 764, 777
Leeds City Council, 437, 439
Legal frameworks, 24, 110, 117, 267, 347, 365, 495, 729, 730, 735
Legal geographies, 729, 753–766
Leverhulme, 723
Lines of flight, 90, 92, 94–95
Livelihoods, 25, 30, 56, 59, 69, 70, 72, 92–94, 109, 201, 208, 209, 467, 469, 470, 552, 554, 555, 557, 560, 574, 630, 632–634, 637, 639, 642, 646, 659
Lobbyists, 722
Local authorities, 183–185, 220, 367, 433, 435–445, 447, 654, 731, 732, 743, 785, 794
Local faith communities (LFCs), 31, 771–780
Localisation, 771, 772
Local struggles, 722
London, 15, 31, 181, 191, 193, 436, 443, 602, 695, 721
Longitudinal studies, 722
Luxury apartment blocks, 721
Luxury development, 721

## M

Making, 2, 21, 29, 46, 75n2, 95, 99, 122, 246, 275, 319, 336, 371, 373, 385, 414, 416, 421, 477, 482, 493, 502, 503, 511, 551, 555, 557, 560, 574, 580, 603, 605, 608, 614, 615, 619, 621, 639, 655, 656, 677, 695, 704, 729, 730, 744, 748
Mapping, 21, 161, 184, 239–243, 259–260, 354, 413–424, 437–441, 502, 512, 516, 585, 722
Maritime displacement, 18, 345–355
Market rent, 721
Massumi, B., 99, 104, 105
Materiality, 24, 91, 93, 223, 266, 347, 405, 569, 580, 614–625
Mediterranean migration, 348–350, 355
Mega-events, 272, 278–281
Migrants, 1, 2, 4, 10, 14, 15, 17, 19, 21–24, 31, 33, 46–49, 68, 89, 124, 132, 145, 146, 148–150, 157, 165, 175, 176, 199–211, 217–230, 237, 242, 245, 247, 249, 298, 300, 302–306, 306n6, 313–324, 331, 350, 354, 355, 365, 368, 371, 374, 387, 395n7, 395n8, 404, 409, 413–419, 421, 423, 424, 453–457, 457n1, 461, 463, 469, 476, 480, 507–516, 551, 569–571, 583–589, 591, 593, 600–602, 605, 606, 608, 614–625, 674, 693, 702, 703, 710, 739, 744, 747, 748, 772, 773, 786, 790, 791, 793, 794

Migration, 4, 42, 58, 102, 122, 132, 146, 157, 199, 217, 237, 271, 297, 314, 332, 345, 384, 399, 415, 431, 453, 467, 476, 507–516, 558, 569, 584, 604, 614, 702, 773, 788

Migration policy, 15, 202, 217, 218, 223, 225, 229

Mobility, 2, 42, 72, 90, 122, 135, 162, 174, 199, 217, 238, 279, 298, 314, 330–332, 336, 349, 372, 433, 508, 571, 584, 608, 731, 788

Mobility justice, 9, 44, 49–50

Movement, 2, 8–10, 18, 21, 23, 28, 31–33, 41–45, 47–50, 55, 67, 70, 92, 102, 103, 114, 121–124, 136, 149, 160, 164, 173, 174, 201, 205, 206, 209, 219, 222, 224, 227–229, 239, 240, 244, 248, 249, 258, 272, 273, 280, 281, 298–300, 303–305, 313, 315, 316, 318, 320–324, 324n4, 329–332, 350, 352, 354, 369, 389, 399, 407, 413–415, 418, 419, 424, 431–435, 439, 453, 460, 462, 467, 494, 500, 501, 503, 508, 522, 528, 529, 552, 558, 571, 586, 588, 592, 600, 608, 618, 619, 641, 669, 677, 687, 688, 693, 696, 697, 702, 727, 733, 735, 739, 743–748, 786, 788–792, 794, 795

Museums, 15, 179–181, 186–193, 278, 419

N

*Naqab*, 103

Navigation, 261, 298, 303–305, 306n7, 585

Necropolitics, 255, 258, 259, 514, 515

Neoliberalism, 22, 46, 184, 255, 459, 460, 467, 470, 654

Newham, 188, 191, 193, 723, 728, 730, 735

O

Occupation, 30, 49, 50, 61, 102, 147, 160, 220, 280, 319, 554, 556, 660, 693, 728, 733, 739–749, 757, 792–794

P

Palestine, 102, 103, 105, 670

Paradoxes of displacement, 73

Participation, 50, 193n1, 272, 633, 678, 679, 681, 701, 705, 707, 711, 807, 809

Perception, 29, 347, 385, 404, 406, 501, 661, 677, 680, 691, 704, 807

Performance, 134, 223, 546, 592, 601, 665, 667, 680, 690, 694, 704, 728, 730, 808

Perry, Grayson, 723

Photography, 26, 600, 602–608, 665, 675, 690, 704, 707, 710, 712

Physical displacement, 72, 400, 404, 478, 677

Placement, 41, 43–47, 49, 50, 391

Play, 4, 9, 10, 25, 99, 137, 144, 147, 158, 160, 162, 165, 208, 220, 241, 290, 292, 300, 314, 318, 333, 346, 348, 352, 353, 465, 550, 551, 555, 556, 560, 571, 583, 617, 624, 631, 661, 664, 665, 680, 688, 703, 722, 748, 777

Policies of exclusion, 259, 260, 265

Policy, 7, 15, 27, 68, 71, 75n3, 90, 115, 121, 122, 125, 132, 147, 150, 182, 183, 199, 201–205, 210, 217, 219, 223, 228, 229, 239, 243, 275, 277, 297, 299, 305, 306n1, 313, 315–317, 323, 384, 390, 393, 406, 416, 417, 433, 437, 438, 440, 441, 445, 446, 459–461, 467, 468, 470, 484, 493, 496, 502, 514, 551, 584, 632, 638, 660, 681, 694, 696, 721, 722, 730, 731, 760, 762, 763, 765, 766, 771, 777, 788, 789, 792–794

Political ecology, 11, 55–61, 133

Political violence, 24, 71, 72, 521, 524, 532

Postcolonial displacement, 157–165

Post-industrial cities, 723

Praxis, 8, 407, 707

Primitive accumulation, 273

Prisoners, 329, 331, 333–339, 386, 391, 512, 600, 665, 688, 692, 693

Procedural rhetoric, 682

Propaganda, 247, 367, 687, 722

Property markets, 23, 723

Property prices, 721

Protection, 7, 20, 46–48, 68, 75n1, 104, 122, 149, 157–165, 202, 203, 227, 237, 279, 280, 302, 304–306, 306n1, 315, 348, 364, 366, 403, 409, 432, 461–463, 465, 466, 476, 478, 484, 485, 494, 512, 515, 533n5, 586, 631, 632, 641, 703, 726, 730, 760, 765, 779, 780, 786, 789–792

Public health, 113, 368, 791

Public relations, 179

Q

Quantitative analysis, 592

Quantitative Easing, 721

Queer, 11, 43, 44, 49, 121–127, 136

R

Race, 11, 16, 41, 73, 86, 110, 124, 127, 132, 143–151, 159, 175, 239, 257, 261, 355, 386–388, 431, 459, 461, 465, 475, 476, 479, 512, 583, 619, 625, 666, 667, 688, 697, 778

Racialisation, 145, 149, 347

Raising awareness, 665, 722

Rearrangement, 90, 91, 93–95

Re-enactment, 680

Refugee camps, 17, 22, 71, 79, 290–295, 303, 383, 392, 404, 460, 462, 464–469, 552, 553, 555, 558, 570, 575–577, 621, 665, 679, 680, 683, 693, 695, 698, 775, 785, 786, 792, 794

Refugee law, 109–111, 114, 115, 150, 348, 703

Refugee response, 771, 773, 780

Refugees, 1, 47, 70, 79, 89, 99, 110, 122, 131, 148, 158, 175, 200, 217, 237, 290–295, 298, 315, 364, 383, 404, 432, 459, 533, 549, 570, 583, 600, 621, 631, 659–670, 673, 687, 702, 747, 771, 785–789, 795, 799–809

Refugees, in Kurdistan Region, 290–295

Regeneration, 30, 179, 181, 184, 185, 188, 189, 193n1, 432, 434, 435, 646, 721–723, 728, 733, 735, 755, 757, 758, 761–765

*Regeneration Supernova*, 723

Relationality, 42, 43, 49
Religion, 22, 110, 144, 145, 159, 239, 324n3, 355, 403, 404, 431, 461, 543, 583, 591, 772, 775, 776, 778
Rents, 30, 147, 432, 438–441, 465, 498, 502, 653, 654, 721, 731, 742, 746, 759
Repair, 570, 572, 639, 643, 644
Representation, 13–15, 25, 28, 32, 41, 133, 163, 174, 175, 188, 202, 245, 338, 417, 423, 445, 481, 526, 550–558, 561, 601, 603, 609n1, 623, 659–670, 674, 675, 680–683, 687, 689, 691, 702–704, 706–707, 709, 713, 744, 762
Repression, 113, 276, 386, 388, 500, 694
Research methods, 422
Residential displacement, 21, 30, 431–447
Resistance, 23, 30, 32, 49, 55, 56, 60–61, 69, 80, 85, 105, 137, 246, 247, 249, 276, 322, 331, 335–337, 339, 494, 498–502, 513, 516, 521, 524, 531, 533n2, 591, 655, 656, 658, 701, 722, 725–735, 739–743, 745, 749, 753–766, 787, 792, 804–806
Resisting/resistance, 25, 29–32, 134, 180, 242, 500, 542, 714, 722, 725–736, 742–744, 754
Resources, 22, 24, 29, 56, 57, 59, 60, 69, 70, 74, 83, 94, 113, 114, 144, 206–208, 226, 256, 259, 261, 264, 301, 304, 367, 368, 373, 435, 447, 462, 464, 466, 527, 528, 560, 575, 586, 588, 631, 722, 727, 730, 743, 745, 747, 772, 791
Re-territorialisation, 90–92, 95
Rhythm, 103, 404, 416, 501

Robin Hood Gardens, 15, 180–182, 184–187, 189, 191, 192, 194
Romania, 228, 499, 653–658
Roma people, 653, 655, 658, 748

S

Safety net, 723
Sanctuary, 31, 48, 101, 245, 776, 777, 779, 785–795, 806
Scholar-activism, 22
Second World War, 45, 110, 160, 384, 391, 603, 629, 692
Securitisation, 149, 202, 203, 300, 349–351, 414–416, 493, 497, 508
Service monitoring, 436, 443, 444, 446
Settler colonialism, 21, 384, 399, 403, 406, 407, 494
Sexuality, 16, 123, 125, 132, 135, 266, 355
Shanghai Expo, 723
Smartphones, 25, 26, 421, 583–593, 601, 623
Smart refugees, 584, 586
Social anthropology, 570
Social displacement, 478, 482
Social housing, 30, 181, 185, 186, 189, 192, 435–440, 445, 468, 654, 725, 728, 731, 732, 739, 742, 743, 766
Social network, 206, 276, 298, 301, 330, 421, 424, 432, 789
Social policy, 21, 406, 445
Social rent, 440, 721
Socio-natures, 58
Solidarity, 30–33, 145, 249, 320, 337, 347, 354, 547, 562, 584, 592, 655, 661, 739, 740, 742, 745, 747–749, 773, 774, 778, 786, 787, 792–795, 799, 800, 804–806, 808, 809

Soundscapes, 674, 681–683
Sovereignty, 47, 71, 110, 145, 163, 176, 238, 243, 258, 390, 515
Space, 8, 41, 56, 73, 80, 91, 99, 123, 133, 161, 173, 182, 206, 220, 238, 255–267, 278, 302, 314, 329, 330, 332, 339, 351, 373, 384, 400, 417, 435, 464, 485, 493, 511, 521, 550, 569, 585, 602, 613, 635, 656, 665, 673, 689, 701–714, 726, 763, 773, 786, 800
Spatiality, 79, 80, 84, 239, 262, 384, 402, 601–604, 606–608, 744
Spectatorship, 28, 660, 664, 675, 677
Spiritual homelessness, 400–402, 405
Squatting, 43, 50, 497, 502, 726, 732, 733, 735, 742, 744, 746–748, 786, 794, 795
Stories, 10, 13, 17, 19, 24, 93, 105, 184, 200, 259, 294–295, 298, 313, 355, 390, 416–418, 423, 551, 556, 557, 590, 604, 655, 659–670, 675–677, 679, 681, 695, 697, 702, 704, 705, 707, 708, 712, 713, 722, 754, 762, 786, 805
Street-level bureaucracy, 509
Support services, 126, 442
Surveillance, 11, 18, 21, 26, 161–162, 222, 223, 225, 230, 349, 351, 353, 354, 414, 420–422, 424, 454, 455, 457, 480, 481, 485, 552, 585 587, 593, 695, 696

T

Tagging, 413, 415, 420, 424
Technology, 6, 9, 11, 14–16, 18, 23, 25, 26, 28, 29, 32, 45, 161–162, 218, 221–225, 237–249, 255–267, 272, 274, 314, 316, 318, 319, 321–324, 351, 354, 383, 384, 407, 414, 420, 421, 423, 424, 455, 467, 480, 495, 497, 498, 500, 508, 569, 571, 574–576, 580, 584–586, 591, 593, 621, 634, 673, 675, 676, 678–680, 682, 695
Temporality, 8, 19, 79–85, 100, 105, 133, 135–137, 239, 245, 247, 248, 255, 262, 314, 321, 413–416, 425n3, 601, 604–606, 608, 624
Territory, 14, 46, 58, 91, 92, 94, 106n1, 112, 145, 149, 159, 204, 221, 227, 228, 238, 239, 241, 243, 244, 246, 276, 280, 332, 387, 388, 401, 406, 417, 436, 453, 508–511, 513, 514, 522, 523, 528, 606, 614, 674, 692, 774
Theatre, 73, 391, 676, 680
Tracing, 19, 315, 413–416, 434–436, 441–444, 453–457, 513
Tracking, 11, 21, 132, 158, 413, 414, 419–421, 424, 439
Trajectories, 10, 16, 18, 21, 158, 298–301, 303–305, 320, 413–424, 455, 498, 573, 580, 585, 661, 668, 707, 712, 788, 800
Transgression, 125, 238, 297–306, 709
Trauma, 2, 8, 17, 23, 26, 57, 82, 84–85, 100, 105, 132, 133, 245, 301, 368, 393, 404–406, 498, 516, 524, 544, 607, 679, 691, 763, 779

U

Undocumented, 31, 46, 48, 82, 243, 248, 299, 368, 414, 415, 417, 425n2, 477, 478, 480, 482, 483, 509, 515, 546, 547n3, 590, 786, 790, 791

United Nations (UN), 59, 121, 135, 160, 162, 201, 203, 313, 347, 365, 366, 432–434, 462, 522, 523, 531, 532, 633, 679
United States (US), 13, 23, 27, 33, 45–47, 68, 84, 101, 111, 124, 132, 136, 145, 150, 158, 191, 192, 220, 224, 226, 227, 240, 259, 260, 264, 273–275, 299, 345, 369, 384, 391, 392, 415, 419–422, 425n2, 432, 465, 475–485, 495, 497, 503, 541–547, 592, 618, 641, 642, 673, 734, 740, 773, 775, 776, 779, 787, 789, 799
Uprooting, 42–44, 50, 298, 304, 524, 696
Urban displacement, 255, 256, 273, 274, 280, 560
Urban renewal, 15, 256, 274–277, 281, 722, 755
Urban studies, 4, 68, 584, 722

V

V&A, 15, 180, 181, 185–193
Ventimiglia, 227, 317–320, 323
Video games, 682
Violence, 2, 4, 6–8, 11, 19, 21, 23, 24, 33, 43, 44, 46, 48, 69, 71, 72, 79–86, 111, 122, 124, 126, 133, 135–137, 144, 160, 164, 223, 243, 257, 258, 267, 273, 294, 300–303, 331, 383, 384, 386, 394, 405, 408, 409, 424, 431, 453–457, 480, 483–485, 494, 495, 497, 499, 501, 503, 513, 515, 521–525, 529–533, 546, 600, 613, 617, 618, 665, 666, 683, 693–697, 727, 730

Virtual reality (VR), 11, 28, 673–683
Visual culture, 607, 701, 702
Vulnerability, 22, 32, 67, 69, 83, 84, 113, 115, 134, 135, 149, 199, 200, 302, 345, 350, 364, 508, 512, 516, 550, 584, 634, 640, 642, 726, 727, 730, 739, 755, 787, 788, 792–794, 804

W

Waiting, 8, 79–85, 100, 134–136, 230, 243, 298, 416, 438, 510, 511, 575, 588, 663, 668, 693, 743, 775, 793, 800–802
War, 4, 7, 19, 25, 28, 67, 71, 82, 103, 110, 112, 125, 145, 160, 202, 290, 294, 324n3, 386, 391–393, 405, 454, 464, 482, 510, 511, 521, 523, 524, 530, 531, 533n1, 541, 549, 551, 559, 589, 664, 690, 693, 695, 723, 775, 778, 779
Welfare, 184, 240, 432–434, 437–441, 445, 468, 478, 640, 702, 723, 725, 731, 733
Will, 608
Women, 30, 43, 48, 70, 72, 80, 82, 84, 85, 132–137, 263, 290–295, 301, 302, 371, 373, 388, 432, 434, 435, 438, 441–445, 481, 483, 513, 530, 531, 541–547, 550, 554, 556, 578, 602, 617–620, 622, 624, 654, 662, 726, 727, 729, 730, 744, 749, 773, 776, 778, 804–806

Y

Yazidi women, 289, 293